Digital Image Processing

Third Edition

Rafael C. Gonzalez
University of Tennessee

Richard E. Woods
MedData Interactive

PEARSON
Prentice
Hall

Upper Saddle River, NJ 07458

Library of Congress Cataloging-in-Publication Data on File

Vice President and Editorial Director, ECS: *Marcia J. Horton*
Executive Editor: *Michael McDonald*
Associate Editor: *Alice Dworkin*
Editorial Assistant: *William Opaluch*
Managing Editor: *Scott Disanno*
Production Editor: *Rose Kernan*
Director of Creative Services: *Paul Belfanti*
Creative Director: *Juan Lopez*
Art Director: *Heather Scott*
Art Editors: *Gregory Dulles* and *Thomas Benfatti*
Manufacturing Manager: *Alexis Heydt-Long*
Manufacturing Buyer: *Lisa McDowell*
Senior Marketing Manager: *Tim Galligan*

© 2008 by Pearson Education, Inc.
Pearson Prentice Hall
Pearson Education, Inc.
Upper Saddle River, New Jersey 07458

The authors and publisher of this book have used their best efforts in preparing this book. These efforts include the development, research, and testing of the theories and programs to determine their effectiveness. The authors and publisher make no warranty of any kind, expressed or implied, with regard to these programs or the documentation contained in this book. The authors and publisher shall not be liable in any event for incidental or consequential damages with, or arising out of, the furnishing, performance, or use of these programs.

Printed in the United States of America.
10 9 8 7 6 5 4 3

ISBN 0-13-168728-X
        978-0-13-168728-8

Pearson Education Ltd., *London*
Pearson Education Australia Pty. Ltd., *Sydney*
Pearson Education Singapore, Pte., Ltd.
Pearson Education North Asia Ltd., *Hong Kong*
Pearson Education Canada, Inc., *Toronto*
Pearson Educación de Mexico, S.A. de C.V.
Pearson Education—Japan, *Tokyo*
Pearson Education Malaysia, Pte. Ltd.
Pearson Education, Inc., *Upper Saddle River, New Jersey*

To Samantha
and
To Janice, David, and Jonathan

Contents

3 *Intensity Transformations and Spatial Filtering 104*

4 *Filtering in the Frequency Domain*   199

5 *Image Restoration and Reconstruction* 311

6 *Color Image Processing   394*

7 *Wavelets and Multiresolution Processing*   461

8 *Image Compression*   525

9 *Morphological Image Processing   627*

10 Image Segmentation 689

11 Representation and Description 795

12 *Object Recognition   861*

Preface

When something can be read without effort,
great effort has gone into its writing.

Enrique Jardiel Poncela

This edition of *Digital Image Processing* is a major revision of the book. As in the 1977 and 1987 editions by Gonzalez and Wintz, and the 1992 and 2002 editions by Gonzalez and Woods, this fifth-generation edition was prepared with students and instructors in mind. The principal objectives of the book continue to be to provide an introduction to basic concepts and methodologies for digital image processing, and to develop a foundation that can be used as the basis for further study and research in this field. To achieve these objectives, we focused again on material that we believe is fundamental and whose scope of application is not limited to the solution of specialized problems. The mathematical complexity of the book remains at a level well within the grasp of college seniors and first-year graduate students who have introductory preparation in mathematical analysis, vectors, matrices, probability, statistics, linear systems, and computer programming. The book Web site provides tutorials to support readers needing a review of this background material.

One of the principal reasons this book has been the world leader in its field for more than 30 years is the level of attention we pay to the changing educational needs of our readers. The present edition is based on the most extensive survey we have ever conducted. The survey involved faculty, students, and independent readers of the book in 134 institutions from 32 countries. The major findings of the survey indicated a need for:

- A more comprehensive introduction early in the book to the mathematical tools used in image processing.
- An expanded explanation of histogram processing techniques.
- Stating complex algorithms in step-by-step summaries.
- An expanded explanation of spatial correlation and convolution.
- An introduction to fuzzy set theory and its application to image processing.
- A revision of the material dealing with the frequency domain, starting with basic principles and showing how the discrete Fourier transform follows from data sampling.
- Coverage of computed tomography (CT).
- Clarification of basic concepts in the wavelets chapter.
- A revision of the data compression chapter to include more video compression techniques, updated standards, and watermarking.
- Expansion of the chapter on morphology to include morphological reconstruction and a revision of gray-scale morphology.

- Expansion of the coverage on image segmentation to include more advanced edge detection techniques such as Canny's algorithm, and a more comprehensive treatment of image thresholding.
- An update of the chapter dealing with image representation and description.
- Streamlining the material dealing with structural object recognition.

The new and reorganized material that resulted in the present edition is our attempt at providing a reasonable degree of balance between rigor, clarity of presentation, and the findings of the market survey, while at the same time keeping the length of the book at a manageable level. The major changes in this edition of the book are as follows.

Chapter 1: A few figures were updated and part of the text was rewritten to correspond to changes in later chapters.

Chapter 2: Approximately 50% of this chapter was revised to include new images and clearer explanations. Major revisions include a new section on image interpolation and a comprehensive new section summarizing the principal mathematical tools used in the book. Instead of presenting "dry" mathematical concepts one after the other, however, we took this opportunity to bring into Chapter 2 a number of image processing applications that were scattered throughout the book. For example, image averaging and image subtraction were moved to this chapter to illustrate arithmetic operations. This follows a trend we began in the second edition of the book to move as many applications as possible early in the discussion not only as illustrations, but also as motivation for students. After finishing the newly organized Chapter 2, a reader will have a basic understanding of how digital images are manipulated and processed. This is a solid platform upon which the rest of the book is built.

Chapter 3: Major revisions of this chapter include a detailed discussion of spatial correlation and convolution, and their application to image filtering using spatial masks. We also found a consistent theme in the market survey asking for numerical examples to illustrate histogram equalization and specification, so we added several such examples to illustrate the mechanics of these processing tools. Coverage of fuzzy sets and their application to image processing was also requested frequently in the survey. We included in this chapter a new section on the foundation of fuzzy set theory, and its application to intensity transformations and spatial filtering, two of the principal uses of this theory in image processing.

Chapter 4: The topic we heard most about in comments and suggestions during the past four years dealt with the changes we made in Chapter 4 from the first to the second edition. Our objective in making those changes was to simplify the presentation of the Fourier transform and the frequency domain. Evidently, we went too far, and numerous users of the book complained that the new material was too superficial. We corrected that problem in the present edition. The material now begins with the Fourier transform of one continuous variable and proceeds to derive the discrete Fourier transform starting with basic concepts of sampling and convolution. A byproduct of the flow of this material is an intuitive derivation of the sampling theorem and its implications. The 1-D material is then extended to 2-D, where we give a number of examples to illustrate the effects of sampling on digital images, including aliasing and moiré patterns. The 2-D discrete Fourier transform is then illustrated and a number of important properties are derived and summarized. These concepts are then used as the basis for filtering in the frequency domain. Finally, we discuss implementation issues such as transform decomposition and the derivation of a fast Fourier transform algorithm. At the end of this chapter, the reader will have progressed from sampling of 1-D functions through a clear derivation of the foundation of the discrete Fourier transform and some of its most important uses in digital image processing.

Chapter 5: The major revision in this chapter was the addition of a section dealing with image reconstruction from projections, with a focus on computed tomography (CT). Coverage of CT starts with an intuitive example of the underlying principles of image reconstruction from projections and the various imaging modalities used in practice. We then derive the Radon transform and the Fourier slice theorem and use them as the basis for formulating the concept of filtered backprojections. Both parallel- and fan-beam reconstruction are discussed and illustrated using several examples. Inclusion of this material was long overdue and represents an important addition to the book.

Chapter 6: Revisions to this chapter were limited to clarifications and a few corrections in notation. No new concepts were added.

Chapter 7: We received numerous comments regarding the fact that the transition from previous chapters into wavelets was proving difficult for beginners. Several of the foundation sections were rewritten in an effort to make the material clearer.

Chapter 8: This chapter was rewritten completely to bring it up to date. New coding techniques, expanded coverage of video, a revision of the section on standards, and an introduction to image watermarking are among the major changes. The new organization will make it easier for beginning students to follow the material.

Chapter 9: The major changes in this chapter are the inclusion of a new section on morphological reconstruction and a complete revision of the section on gray-scale morphology. The inclusion of morphological reconstruction for both binary and gray-scale images made it possible to develop more complex and useful morphological algorithms than before.

Chapter 10: This chapter also underwent a major revision. The organization is as before, but the new material includes greater emphasis on basic principles as well as discussion of more advanced segmentation techniques. Edge models are discussed and illustrated in more detail, as are properties of the gradient. The Marr-Hildreth and Canny edge detectors are included to illustrate more advanced edge detection techniques. The section on thresholding was rewritten also to include Otsu's method, an optimum thresholding technique whose popularity has increased significantly over the past few years. We introduced this approach in favor of optimum thresholding based on the Bayes classification rule, not only because it is easier to understand and implement, but also because it is used considerably more in practice. The Bayes approach was moved to Chapter 12, where the Bayes decision rule is discussed in more detail. We also added a discussion on how to use edge information to improve thresholding and several new adaptive thresholding examples. Except for minor clarifications, the sections on morphological watersheds and the use of motion for segmentation are as in the previous edition.

Chapter 11: The principal changes in this chapter are the inclusion of a boundary-following algorithm, a detailed derivation of an algorithm to fit a minimum-perimeter polygon to a digital boundary, and a new section on co-occurrence matrices for texture description. Numerous examples in Sections 11.2 and 11.3 are new, as are all the examples in Section 11.4.

Chapter 12: Changes in this chapter include a new section on matching by correlation and a new example on using the Bayes classifier to recognize regions of interest in multispectral images. The section on structural classification now limits discussion only to string matching.

All the revisions just mentioned resulted in over 400 new images, over 200 new line drawings and tables, and more than 80 new homework problems. Where appropriate, complex processing procedures were summarized in the form of step-by-step algorithm formats. The references at the end of all chapters were updated also.

The book Web site, established during the launch of the second edition, has been a success, attracting more than 20,000 visitors each month. The site was redesigned and upgraded to correspond to the launch of this edition. For more details on features and content, see *The Book Web Site*, following the *Acknowledgments*.

This edition of *Digital Image Processing* is a reflection of how the educational needs of our readers have changed since 2002. As is usual in a project such as this, progress in the field continues after work on the manuscript stops. One of the reasons why this book has been so well accepted since it first appeared in 1977 is its continued emphasis on fundamental concepts—an approach that, among other things, attempts to provide a measure of stability in a rapidly-evolving body of knowledge. We have tried to follow the same principle in preparing this edition of the book.

R. C. G.
R. E. W.

Acknowledgments

We are indebted to a number of individuals in academic circles as well as in industry and government who have contributed to this edition of the book. Their contributions have been important in so many different ways that we find it difficult to acknowledge them in any other way but alphabetically. In particular, we wish to extend our appreciation to our colleagues Mongi A. Abidi, Steven L. Eddins, Yongmin Kim, Bryan Morse, Andrew Oldroyd, Ali M. Reza, Edgardo Felipe Riveron, Jose Ruiz Shulcloper, and Cameron H. G. Wright for their many suggestions on how to improve the presentation and/or the scope of coverage in the book.

Numerous individuals and organizations provided us with valuable assistance during the writing of this edition. Again, we list them alphabetically. We are particularly indebted to Courtney Esposito and Naomi Fernandes at The Mathworks for providing us with MATLAB software and support that were important in our ability to create or clarify many of the examples and experimental results included in this edition of the book. A significant percentage of the new images used in this edition (and in some cases their history and interpretation) were obtained through the efforts of individuals whose contributions are sincerely appreciated. In particular, we wish to acknowledge the efforts of Serge Beucher, Melissa D. Binde, James Blankenship, Uwe Boos, Ernesto Bribiesca, Michael E. Casey, Michael W. Davidson, Susan L. Forsburg, Thomas R. Gest, Lalit Gupta, Daniel A. Hammer, Zhong He, Roger Heady, Juan A. Herrera, John M. Hudak, Michael Hurwitz, Chris J. Johannsen, Rhonda Knighton, Don P. Mitchell, Ashley Mohamed, A. Morris, Curtis C. Ober, Joseph E. Pascente, David. R. Pickens, Michael Robinson, Barrett A. Schaefer, Michael Shaffer, Pete Sites, Sally Stowe, Craig Watson, David K. Wehe, and Robert A. West. We also wish to acknowledge other individuals and organizations cited in the captions of numerous figures throughout the book for their permission to use that material.

Special thanks go to Vince O'Brien, Rose Kernan, Scott Disanno, Michael McDonald, Joe Ruddick, Heather Scott, and Alice Dworkin, at Prentice Hall. Their creativity, assistance, and patience during the production of this book are truly appreciated.

R.C.G.
R.E.W.

The Book Web Site

www.prenhall.com/gonzalezwoods
or its mirror site,
www.imageprocessingplace.com

Digital Image Processing is a completely self-contained book. However, the companion Web site offers additional support in a number of important areas.

For the Student or Independent Reader the site contains

- Reviews in areas such as probability, statistics, vectors, and matrices.
- Complete solutions to selected problems.
- Computer projects.
- A Tutorials section containing dozens of tutorials on most of the topics discussed in the book.
- A database containing all the images in the book.

For the Instructor the site contains

- An *Instructor's Manual* with complete solutions to all the problems in the book, as well as course and laboratory teaching guidelines. The manual is available free of charge to instructors who have adopted the book for classroom use.
- Classroom presentation materials in PowerPoint format.
- Material removed from previous editions, downloadable in convenient PDF format.
- Numerous links to other educational resources.

For the Practitioner the site contains additional specialized topics such as

- Links to commercial sites.
- Selected new references.
- Links to commercial image databases.

The Web site is an ideal tool for keeping the book current between editions by including new topics, digital images, and other relevant material that has appeared after the book was published. Although considerable care was taken in the production of the book, the Web site is also a convenient repository for any errors that may be discovered between printings. References to the book Web site are designated in the book by the following icon:

About the Authors

Rafael C. Gonzalez

R. C. Gonzalez received the B.S.E.E. degree from the University of Miami in 1965 and the M.E. and Ph.D. degrees in electrical engineering from the University of Florida, Gainesville, in 1967 and 1970, respectively. He joined the Electrical and Computer Engineering Department at the University of Tennessee, Knoxville (UTK) in 1970, where he became Associate Professor in 1973, Professor in 1978, and Distinguished Service Professor in 1984. He served as Chairman of the department from 1994 through 1997. He is currently a Professor Emeritus at UTK.

Gonzalez is the founder of the Image & Pattern Analysis Laboratory and the Robotics & Computer Vision Laboratory at the University of Tennessee. He also founded Perceptics Corporation in 1982 and was its president until 1992. The last three years of this period were spent under a full-time employment contract with Westinghouse Corporation, who acquired the company in 1989.

Under his direction, Perceptics became highly successful in image processing, computer vision, and laser disk storage technology. In its initial ten years, Perceptics introduced a series of innovative products, including: The world's first commercially-available computer vision system for automatically reading license plates on moving vehicles; a series of large-scale image processing and archiving systems used by the U.S. Navy at six different manufacturing sites throughout the country to inspect the rocket motors of missiles in the Trident II Submarine Program; the market-leading family of imaging boards for advanced Macintosh computers; and a line of trillion-byte laser disk products.

He is a frequent consultant to industry and government in the areas of pattern recognition, image processing, and machine learning. His academic honors for work in these fields include the 1977 UTK College of Engineering Faculty Achievement Award; the 1978 UTK Chancellor's Research Scholar Award; the 1980 Magnavox Engineering Professor Award; and the 1980 M.E. Brooks Distinguished Professor Award. In 1981 he became an IBM Professor at the University of Tennessee and in 1984 he was named a Distinguished Service Professor there. He was awarded a Distinguished Alumnus Award by the University of Miami in 1985, the Phi Kappa Phi Scholar Award in 1986, and the University of Tennessee's Nathan W. Dougherty Award for Excellence in Engineering in 1992.

Honors for industrial accomplishment include the 1987 IEEE Outstanding Engineer Award for Commercial Development in Tennessee; the 1988 Albert Rose Nat'l Award for Excellence in Commercial Image Processing; the 1989 B. Otto Wheeley Award for Excellence in Technology Transfer; the 1989 Coopers and Lybrand Entrepreneur of the Year Award; the 1992 IEEE Region 3 Outstanding Engineer Award; and the 1993 Automated Imaging Association National Award for Technology Development.

Gonzalez is author or co-author of over 100 technical articles, two edited books, and four textbooks in the fields of pattern recognition, image processing, and robotics. His books are used in over 1000 universities and research institutions throughout the world. He is listed in the prestigious Marquis *Who's Who in America*, Marquis *Who's Who in Engineering*, Marquis *Who's Who in the World*, and in 10 other national and international biographical citations. He is the co-holder of two U.S. Patents, and has been an associate editor of the *IEEE Transactions on Systems, Man and Cybernetics*, and the *International Journal of Computer and Information Sciences*. He is a member of numerous professional and honorary societies, including Tau Beta Pi, Phi Kappa Phi, Eta Kappa Nu, and Sigma Xi. He is a Fellow of the IEEE.

Richard E. Woods

Richard E. Woods earned his B.S., M.S., and Ph.D. degrees in Electrical Engineering from the University of Tennessee, Knoxville. His professional experiences range from entrepreneurial to the more traditional academic, consulting, governmental, and industrial pursuits. Most recently, he founded MedData Interactive, a high technology company specializing in the development of handheld computer systems for medical applications. He was also a founder and Vice President of Perceptics Corporation, where he was responsible for the development of many of the company's quantitative image analysis and autonomous decision-making products.

Prior to Perceptics and MedData, Dr. Woods was an Assistant Professor of Electrical Engineering and Computer Science at the University of Tennessee and prior to that, a computer applications engineer at Union Carbide Corporation. As a consultant, he has been involved in the development of a number of special-purpose digital processors for a variety of space and military agencies, including NASA, the Ballistic Missile Systems Command, and the Oak Ridge National Laboratory.

Dr. Woods has published numerous articles related to digital signal processing and is a member of several professional societies, including Tau Beta Pi, Phi Kappa Phi, and the IEEE. In 1986, he was recognized as a Distinguished Engineering Alumnus of the University of Tennessee.

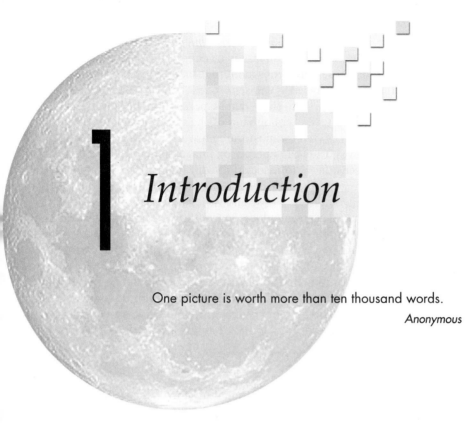

1 Introduction

One picture is worth more than ten thousand words.

Anonymous

Preview

Interest in digital image processing methods stems from two principal application areas: improvement of pictorial information for human interpretation; and processing of image data for storage, transmission, and representation for autonomous machine perception. This chapter has several objectives: (1) to define the scope of the field that we call image processing; (2) to give a historical perspective of the origins of this field; (3) to give you an idea of the state of the art in image processing by examining some of the principal areas in which it is applied; (4) to discuss briefly the principal approaches used in digital image processing; (5) to give an overview of the components contained in a typical, general-purpose image processing system; and (6) to provide direction to the books and other literature where image processing work normally is reported.

1.1 What Is Digital Image Processing?

An image may be defined as a two-dimensional function, $f(x, y)$, where x and y are *spatial* (plane) coordinates, and the amplitude of f at any pair of coordinates (x, y) is called the *intensity* or *gray level* of the image at that point. When x, y, and the intensity values of f are all finite, discrete quantities, we call the image a *digital image*. The field of *digital image processing* refers to processing digital images by means of a digital computer. Note that a digital image is composed of a finite number of elements, each of which has a particular location and value. These elements are called *picture elements*, *image elements*, *pels*, and *pixels*. *Pixel* is the term used most widely to denote the elements of a digital image. We consider these definitions in more formal terms in Chapter 2.

Vision is the most advanced of our senses, so it is not surprising that images play the single most important role in human perception. However, unlike humans, who are limited to the visual band of the electromagnetic (EM) spectrum, imaging machines cover almost the entire EM spectrum, ranging from gamma to radio waves. They can operate on images generated by sources that humans are not accustomed to associating with images. These include ultrasound, electron microscopy, and computer-generated images. Thus, digital image processing encompasses a wide and varied field of applications.

There is no general agreement among authors regarding where image processing stops and other related areas, such as image analysis and computer vision, start. Sometimes a distinction is made by defining image processing as a discipline in which both the input and output of a process are images. We believe this to be a limiting and somewhat artificial boundary. For example, under this definition, even the trivial task of computing the average intensity of an image (which yields a single number) would not be considered an image processing operation. On the other hand, there are fields such as computer vision whose ultimate goal is to use computers to emulate human vision, including learning and being able to make inferences and take actions based on visual inputs. This area itself is a branch of artificial intelligence (AI) whose objective is to emulate human intelligence. The field of AI is in its earliest stages of infancy in terms of development, with progress having been much slower than originally anticipated. The area of image analysis (also called image understanding) is in between image processing and computer vision.

There are no clear-cut boundaries in the continuum from image processing at one end to computer vision at the other. However, one useful paradigm is to consider three types of computerized processes in this continuum: low-, mid-, and high-level processes. Low-level processes involve primitive operations such as image preprocessing to reduce noise, contrast enhancement, and image sharpening. A low-level process is characterized by the fact that both its inputs and outputs are images. Mid-level processing on images involves tasks such as segmentation (partitioning an image into regions or objects), description of those objects to reduce them to a form suitable for computer processing, and classification (recognition) of individual objects. A mid-level process is characterized by the fact that its inputs generally are images, but its outputs are attributes extracted from those images (e.g., edges, contours, and the identity of individual objects). Finally, higher-level processing involves "making sense" of an ensemble of recognized objects, as in image analysis, and, at the far end of the continuum, performing the cognitive functions normally associated with vision.

Based on the preceding comments, we see that a logical place of overlap between image processing and image analysis is the area of recognition of individual regions or objects in an image. Thus, what we call in this book *digital image processing* encompasses processes whose inputs and outputs are images and, in addition, encompasses processes that extract attributes from images, up to and including the recognition of individual objects. As an illustration to clarify these concepts, consider the area of automated analysis of text. The processes of acquiring an image of the area containing the text, preprocessing that image, extracting (segmenting) the individual characters, describing the characters in a form suitable for computer processing, and recognizing those individual characters are in the scope of what we call digital image processing in this book. Making sense of the content of the page may be viewed as being in the domain of image analysis and even computer vision, depending on the level of complexity implied by the statement "making sense." As will become evident shortly, digital image processing, as we have defined it, is used successfully in a broad range of areas of exceptional social and economic value. The concepts developed in the following chapters are the foundation for the methods used in those application areas.

1.2   The Origins of Digital Image Processing

One of the first applications of digital images was in the newspaper industry, when pictures were first sent by submarine cable between London and New York. Introduction of the Bartlane cable picture transmission system in the early 1920s reduced the time required to transport a picture across the Atlantic from more than a week to less than three hours. Specialized printing equipment coded pictures for cable transmission and then reconstructed them at the receiving end. Figure 1.1 was transmitted in this way and reproduced on a telegraph printer fitted with typefaces simulating a halftone pattern.

Some of the initial problems in improving the visual quality of these early digital pictures were related to the selection of printing procedures and the distribution of intensity levels. The printing method used to obtain Fig. 1.1 was abandoned toward the end of 1921 in favor of a technique based on photographic reproduction made from tapes perforated at the telegraph receiving terminal. Figure 1.2 shows an image obtained using this method. The improvements over Fig. 1.1 are evident, both in tonal quality and in resolution.

FIGURE 1.1 A digital picture produced in 1921 from a coded tape by a telegraph printer with special type faces. (McFarlane.[†])

[†]References in the Bibliography at the end of the book are listed in alphabetical order by authors' last names.

FIGURE 1.2 A digital picture made in 1922 from a tape punched after the signals had crossed the Atlantic twice. (McFarlane.)

The early Bartlane systems were capable of coding images in five distinct levels of gray. This capability was increased to 15 levels in 1929. Figure 1.3 is typical of the type of images that could be obtained using the 15-tone equipment. During this period, introduction of a system for developing a film plate via light beams that were modulated by the coded picture tape improved the reproduction process considerably.

Although the examples just cited involve digital images, they are not considered digital image processing results in the context of our definition because computers were not involved in their creation. Thus, the history of digital image processing is intimately tied to the development of the digital computer. In fact, digital images require so much storage and computational power that progress in the field of digital image processing has been dependent on the development of digital computers and of supporting technologies that include data storage, display, and transmission.

The idea of a computer goes back to the invention of the abacus in Asia Minor, more than 5000 years ago. More recently, there were developments in the past two centuries that are the foundation of what we call a computer today. However, the basis for what we call a *modern* digital computer dates back to only the 1940s with the introduction by John von Neumann of two key concepts: (1) a memory to hold a stored program and data, and (2) conditional branching. These two ideas are the foundation of a central processing unit (CPU), which is at the heart of computers today. Starting with von Neumann, there were a series of key advances that led to computers powerful enough to

FIGURE 1.3 Unretouched cable picture of Generals Pershing and Foch, transmitted in 1929 from London to New York by 15-tone equipment. (McFarlane.)

be used for digital image processing. Briefly, these advances may be summarized as follows: (1) the invention of the transistor at Bell Laboratories in 1948; (2) the development in the 1950s and 1960s of the high-level programming languages COBOL (Common Business-Oriented Language) and FORTRAN (Formula Translator); (3) the invention of the integrated circuit (IC) at Texas Instruments in 1958; (4) the development of operating systems in the early 1960s; (5) the development of the microprocessor (a single chip consisting of the central processing unit, memory, and input and output controls) by Intel in the early 1970s; (6) introduction by IBM of the personal computer in 1981; and (7) progressive miniaturization of components, starting with large scale integration (LI) in the late 1970s, then very large scale integration (VLSI) in the 1980s, to the present use of ultra large scale integration (ULSI). Concurrent with these advances were developments in the areas of mass storage and display systems, both of which are fundamental requirements for digital image processing.

The first computers powerful enough to carry out meaningful image processing tasks appeared in the early 1960s. The birth of what we call digital image processing today can be traced to the availability of those machines and to the onset of the space program during that period. It took the combination of those two developments to bring into focus the potential of digital image processing concepts. Work on using computer techniques for improving images from a space probe began at the Jet Propulsion Laboratory (Pasadena, California) in 1964 when pictures of the moon transmitted by *Ranger 7* were processed by a computer to correct various types of image distortion inherent in the on-board television camera. Figure 1.4 shows the first image of the moon taken by *Ranger 7* on July 31, 1964 at 9:09 A.M. Eastern Daylight Time (EDT), about 17 minutes before impacting the lunar surface (the markers, called *reseau* marks, are used for geometric corrections, as discussed in Chapter 2). This also is the first image of the moon taken by a U.S. spacecraft. The imaging lessons learned with *Ranger 7* served as the basis for improved methods used to enhance and restore images from the Surveyor missions to the moon, the Mariner series of flyby missions to Mars, the Apollo manned flights to the moon, and others.

FIGURE 1.4 The first picture of the moon by a U.S. spacecraft. *Ranger 7* took this image on July 31, 1964 at 9:09 A.M. EDT, about 17 minutes before impacting the lunar surface. (Courtesy of NASA.)

In parallel with space applications, digital image processing techniques began in the late 1960s and early 1970s to be used in medical imaging, remote Earth resources observations, and astronomy. The invention in the early 1970s of computerized axial tomography (CAT), also called computerized tomography (CT) for short, is one of the most important events in the application of image processing in medical diagnosis. Computerized axial tomography is a process in which a ring of detectors encircles an object (or patient) and an X-ray source, concentric with the detector ring, rotates about the object. The X-rays pass through the object and are collected at the opposite end by the corresponding detectors in the ring. As the source rotates, this procedure is repeated. Tomography consists of algorithms that use the sensed data to construct an image that represents a "slice" through the object. Motion of the object in a direction perpendicular to the ring of detectors produces a set of such slices, which constitute a three-dimensional (3-D) rendition of the inside of the object. Tomography was invented independently by Sir Godfrey N. Hounsfield and Professor Allan M. Cormack, who shared the 1979 Nobel Prize in Medicine for their invention. It is interesting to note that X-rays were discovered in 1895 by Wilhelm Conrad Roentgen, for which he received the 1901 Nobel Prize for Physics. These two inventions, nearly 100 years apart, led to some of the most important applications of image processing today.

From the 1960s until the present, the field of image processing has grown vigorously. In addition to applications in medicine and the space program, digital image processing techniques now are used in a broad range of applications. Computer procedures are used to enhance the contrast or code the intensity levels into color for easier interpretation of X-rays and other images used in industry, medicine, and the biological sciences. Geographers use the same or similar techniques to study pollution patterns from aerial and satellite imagery. Image enhancement and restoration procedures are used to process degraded images of unrecoverable objects or experimental results too expensive to duplicate. In archeology, image processing methods have successfully restored blurred pictures that were the only available records of rare artifacts lost or damaged after being photographed. In physics and related fields, computer techniques routinely enhance images of experiments in areas such as high-energy plasmas and electron microscopy. Similarly successful applications of image processing concepts can be found in astronomy, biology, nuclear medicine, law enforcement, defense, and industry.

These examples illustrate processing results intended for human interpretation. The second major area of application of digital image processing techniques mentioned at the beginning of this chapter is in solving problems dealing with machine perception. In this case, interest is on procedures for extracting from an image information in a form suitable for computer processing. Often, this information bears little resemblance to visual features that humans use in interpreting the content of an image. Examples of the type of information used in machine perception are statistical moments, Fourier transform coefficients, and multidimensional distance measures. Typical problems in machine perception that routinely utilize image processing techniques are automatic character recognition, industrial machine vision for product assembly and inspection, military recognizance, automatic processing of fingerprints, screening of X-rays and blood samples, and machine processing of aerial and satellite imagery for weather prediction and environmental assessment. The continuing decline in the ratio of computer price to performance and the expansion of networking and communication bandwidth via the World Wide Web and the Internet have created unprecedented opportunities for continued growth of digital image processing. Some of these application areas are illustrated in the following section.

1.3 Examples of Fields that Use Digital Image Processing

Today, there is almost no area of technical endeavor that is not impacted in some way by digital image processing. We can cover only a few of these applications in the context and space of the current discussion. However, limited as it is, the material presented in this section will leave no doubt in your mind regarding the breadth and importance of digital image processing. We show in this section numerous areas of application, each of which routinely utilizes the digital image processing techniques developed in the following chapters. Many of the images shown in this section are used later in one or more of the examples given in the book. All images shown are digital.

The areas of application of digital image processing are so varied that some form of organization is desirable in attempting to capture the breadth of this field. One of the simplest ways to develop a basic understanding of the extent of image processing applications is to categorize images according to their source (e.g., visual, X-ray, and so on). The principal energy source for images in use today is the electromagnetic energy spectrum. Other important sources of energy include acoustic, ultrasonic, and electronic (in the form of electron beams used in electron microscopy). Synthetic images, used for modeling and visualization, are generated by computer. In this section we discuss briefly how images are generated in these various categories and the areas in which they are applied. Methods for converting images into digital form are discussed in the next chapter.

Images based on radiation from the EM spectrum are the most familiar, especially images in the X-ray and visual bands of the spectrum. Electromagnetic waves can be conceptualized as propagating sinusoidal waves of varying wavelengths, or they can be thought of as a stream of massless particles, each traveling in a wavelike pattern and moving at the speed of light. Each massless particle contains a certain amount (or bundle) of energy. Each bundle of energy is called a *photon*. If spectral bands are grouped according to energy per photon, we obtain the spectrum shown in Fig. 1.5, ranging from gamma rays (highest energy) at one end to radio waves (lowest energy) at the other.

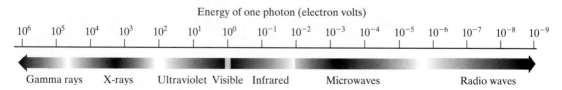

FIGURE 1.5 The electromagnetic spectrum arranged according to energy per photon.

The bands are shown shaded to convey the fact that bands of the EM spectrum are not distinct but rather transition smoothly from one to the other.

1.3.1 Gamma-Ray Imaging

Major uses of imaging based on gamma rays include nuclear medicine and astronomical observations. In nuclear medicine, the approach is to inject a patient with a radioactive isotope that emits gamma rays as it decays. Images are produced from the emissions collected by gamma ray detectors. Figure 1.6(a) shows an image of a complete bone scan obtained by using gamma-ray imaging. Images of this sort are used to locate sites of bone pathology, such as infections

a b
c d

FIGURE 1.6
Examples of gamma-ray imaging. (a) Bone scan. (b) PET image. (c) Cygnus Loop. (d) Gamma radiation (bright spot) from a reactor valve. (Images courtesy of (a) G.E. Medical Systems, (b) Dr. Michael E. Casey, CTI PET Systems, (c) NASA, (d) Professors Zhong He and David K. Wehe, University of Michigan.)

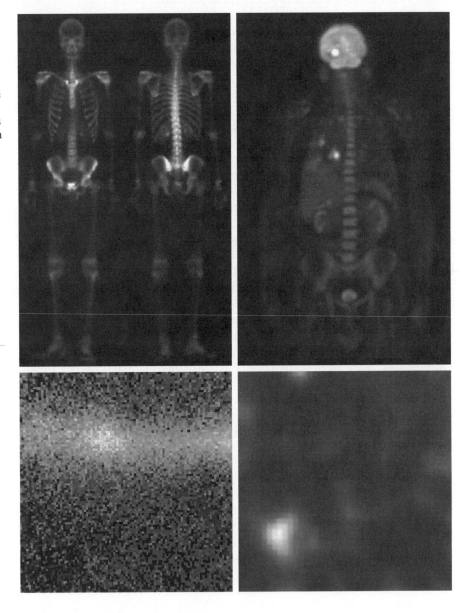

or tumors. Figure 1.6(b) shows another major modality of nuclear imaging called positron emission tomography (PET). The principle is the same as with X-ray tomography, mentioned briefly in Section 1.2. However, instead of using an external source of X-ray energy, the patient is given a radioactive isotope that emits positrons as it decays. When a positron meets an electron, both are annihilated and two gamma rays are given off. These are detected and a tomographic image is created using the basic principles of tomography. The image shown in Fig. 1.6(b) is one sample of a sequence that constitutes a 3-D rendition of the patient. This image shows a tumor in the brain and one in the lung, easily visible as small white masses.

A star in the constellation of Cygnus exploded about 15,000 years ago, generating a superheated stationary gas cloud (known as the Cygnus Loop) that glows in a spectacular array of colors. Figure 1.6(c) shows an image of the Cygnus Loop in the gamma-ray band. Unlike the two examples in Figs. 1.6(a) and (b), this image was obtained using the natural radiation of the object being imaged. Finally, Fig. 1.6(d) shows an image of gamma radiation from a valve in a nuclear reactor. An area of strong radiation is seen in the lower left side of the image.

1.3.2 X-Ray Imaging

X-rays are among the oldest sources of EM radiation used for imaging. The best known use of X-rays is medical diagnostics, but they also are used extensively in industry and other areas, like astronomy. X-rays for medical and industrial imaging are generated using an X-ray tube, which is a vacuum tube with a cathode and anode. The cathode is heated, causing free electrons to be released. These electrons flow at high speed to the positively charged anode. When the electrons strike a nucleus, energy is released in the form of X-ray radiation. The energy (penetrating power) of X-rays is controlled by a voltage applied across the anode, and by a current applied to the filament in the cathode. Figure 1.7(a) shows a familiar chest X-ray generated simply by placing the patient between an X-ray source and a film sensitive to X-ray energy. The intensity of the X-rays is modified by absorption as they pass through the patient, and the resulting energy falling on the film develops it, much in the same way that light develops photographic film. In digital radiography, digital images are obtained by one of two methods: (1) by digitizing X-ray films; or (2) by having the X-rays that pass through the patient fall directly onto devices (such as a phosphor screen) that convert X-rays to light. The light signal in turn is captured by a light-sensitive digitizing system. We discuss digitization in more detail in Chapters 2 and 4.

Angiography is another major application in an area called contrast-enhancement radiography. This procedure is used to obtain images (called *angiograms*) of blood vessels. A catheter (a small, flexible, hollow tube) is inserted, for example, into an artery or vein in the groin. The catheter is threaded into the blood vessel and guided to the area to be studied. When the catheter reaches the site under investigation, an X-ray contrast medium is injected through the tube. This enhances contrast of the blood vessels and enables the radiologist to see any irregularities or blockages. Figure 1.7(b) shows an example of an aortic angiogram. The catheter can be seen being inserted into the

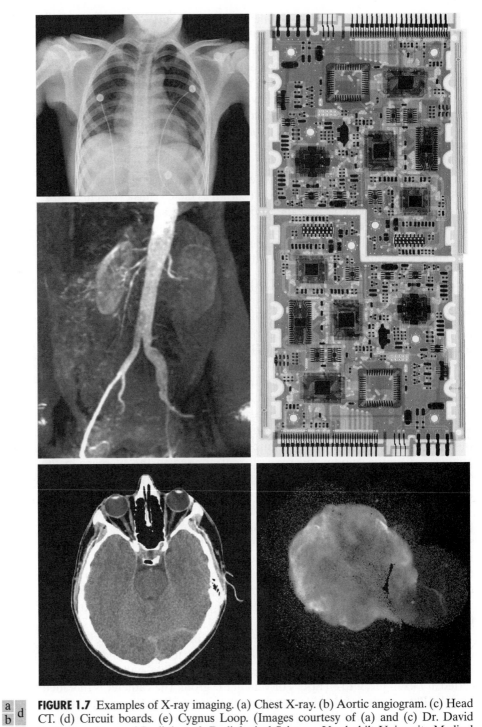

FIGURE 1.7 Examples of X-ray imaging. (a) Chest X-ray. (b) Aortic angiogram. (c) Head CT. (d) Circuit boards. (e) Cygnus Loop. (Images courtesy of (a) and (c) Dr. David R. Pickens, Dept. of Radiology & Radiological Sciences, Vanderbilt University Medical Center; (b) Dr. Thomas R. Gest, Division of Anatomical Sciences, University of Michigan Medical School; (d) Mr. Joseph E. Pascente, Lixi, Inc.; and (e) NASA.)

large blood vessel on the lower left of the picture. Note the high contrast of the large vessel as the contrast medium flows up in the direction of the kidneys, which are also visible in the image. As discussed in Chapter 2, angiography is a major area of digital image processing, where image subtraction is used to enhance further the blood vessels being studied.

Another important use of X-rays in medical imaging is computerized axial tomography (CAT). Due to their resolution and 3-D capabilities, CAT scans revolutionized medicine from the moment they first became available in the early 1970s. As noted in Section 1.2, each CAT image is a "slice" taken perpendicularly through the patient. Numerous slices are generated as the patient is moved in a longitudinal direction. The ensemble of such images constitutes a 3-D rendition of the inside of the body, with the longitudinal resolution being proportional to the number of slice images taken. Figure 1.7(c) shows a typical head CAT slice image.

Techniques similar to the ones just discussed, but generally involving higher-energy X-rays, are applicable in industrial processes. Figure 1.7(d) shows an X-ray image of an electronic circuit board. Such images, representative of literally hundreds of industrial applications of X-rays, are used to examine circuit boards for flaws in manufacturing, such as missing components or broken traces. Industrial CAT scans are useful when the parts can be penetrated by X-rays, such as in plastic assemblies, and even large bodies, like solid-propellant rocket motors. Figure 1.7(e) shows an example of X-ray imaging in astronomy. This image is the Cygnus Loop of Fig. 1.6(c), but imaged this time in the X-ray band.

1.3.3 Imaging in the Ultraviolet Band

Applications of ultraviolet "light" are varied. They include lithography, industrial inspection, microscopy, lasers, biological imaging, and astronomical observations. We illustrate imaging in this band with examples from microscopy and astronomy.

Ultraviolet light is used in fluorescence microscopy, one of the fastest growing areas of microscopy. Fluorescence is a phenomenon discovered in the middle of the nineteenth century, when it was first observed that the mineral fluorspar fluoresces when ultraviolet light is directed upon it. The ultraviolet light itself is not visible, but when a photon of ultraviolet radiation collides with an electron in an atom of a fluorescent material, it elevates the electron to a higher energy level. Subsequently, the excited electron relaxes to a lower level and emits light in the form of a lower-energy photon in the visible (red) light region. The basic task of the fluorescence microscope is to use an excitation light to irradiate a prepared specimen and then to separate the much weaker radiating fluorescent light from the brighter excitation light. Thus, only the emission light reaches the eye or other detector. The resulting fluorescing areas shine against a dark background with sufficient contrast to permit detection. The darker the background of the nonfluorescing material, the more efficient the instrument.

Fluorescence microscopy is an excellent method for studying materials that can be made to fluoresce, either in their natural form (primary fluorescence) or when treated with chemicals capable of fluorescing (secondary fluorescence). Figures 1.8(a) and (b) show results typical of the capability of fluorescence microscopy. Figure 1.8(a) shows a fluorescence microscope image of normal corn, and Fig. 1.8(b) shows corn infected by "smut," a disease of cereals, corn,

FIGURE 1.8
Examples of
ultraviolet
imaging.
(a) Normal corn.
(b) Smut corn.
(c) Cygnus Loop.
(Images courtesy
of (a) and
(b) Dr. Michael
W. Davidson,
Florida State
University,
(c) NASA.)

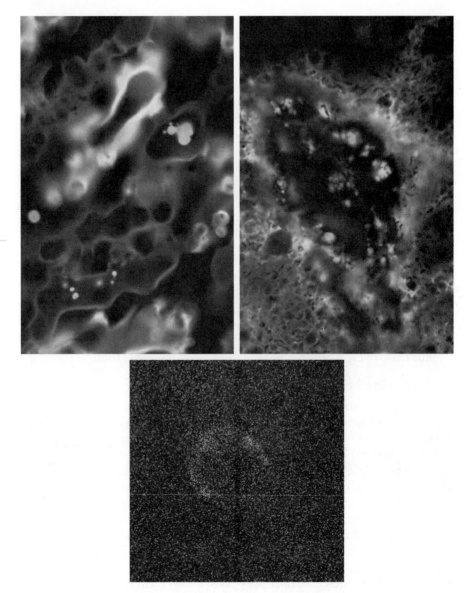

grasses, onions, and sorghum that can be caused by any of more than 700 species of parasitic fungi. Corn smut is particularly harmful because corn is one of the principal food sources in the world. As another illustration, Fig. 1.8(c) shows the Cygnus Loop imaged in the high-energy region of the ultraviolet band.

1.3.4 Imaging in the Visible and Infrared Bands

Considering that the visual band of the electromagnetic spectrum is the most familiar in all our activities, it is not surprising that imaging in this band out-weighs by far all the others in terms of breadth of application. The infrared band often is used in conjunction with visual imaging, so we have grouped the visible and infrared bands in this section for the purpose of illustration. We consider in the following discussion applications in light microscopy, astronomy, remote sensing, industry, and law enforcement.

Figure 1.9 shows several examples of images obtained with a light microscope. The examples range from pharmaceuticals and microinspection to materials characterization. Even in microscopy alone, the application areas are too numerous to detail here. It is not difficult to conceptualize the types of processes one might apply to these images, ranging from enhancement to measurements.

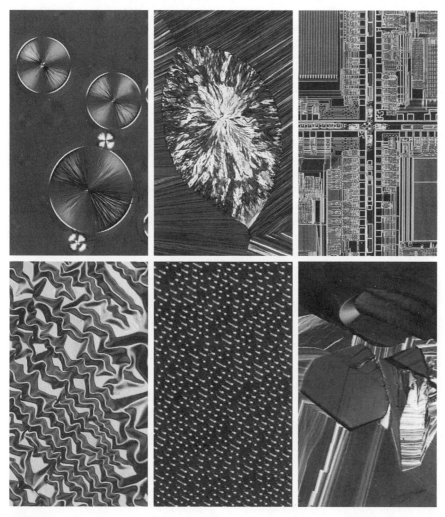

a b c
d e f

FIGURE 1.9 Examples of light microscopy images. (a) Taxol (anticancer agent), magnified 250×. (b) Cholesterol—40×. (c) Microprocessor—60×. (d) Nickel oxide thin film—600×. (e) Surface of audio CD—1750×. (f) Organic superconductor—450×. (Images courtesy of Dr. Michael W. Davidson, Florida State University.)

TABLE 1.1
Thematic bands
in NASA's
LANDSAT
satellite.

Band No.	Name	Wavelength (μm)	Characteristics and Uses
1	Visible blue	0.45–0.52	Maximum water penetration
2	Visible green	0.52–0.60	Good for measuring plant vigor
3	Visible red	0.63–0.69	Vegetation discrimination
4	Near infrared	0.76–0.90	Biomass and shoreline mapping
5	Middle infrared	1.55–1.75	Moisture content of soil and vegetation
6	Thermal infrared	10.4–12.5	Soil moisture; thermal mapping
7	Middle infrared	2.08–2.35	Mineral mapping

Another major area of visual processing is remote sensing, which usually includes several bands in the visual and infrared regions of the spectrum. Table 1.1 shows the so-called *thematic bands* in NASA's LANDSAT satellite. The primary function of LANDSAT is to obtain and transmit images of the Earth from space for purposes of monitoring environmental conditions on the planet. The bands are expressed in terms of wavelength, with 1 μm being equal to 10^{-6} m (we discuss the wavelength regions of the electromagnetic spectrum in more detail in Chapter 2). Note the characteristics and uses of each band in Table 1.1.

In order to develop a basic appreciation for the power of this type of *multispectral* imaging, consider Fig. 1.10, which shows one image for each of

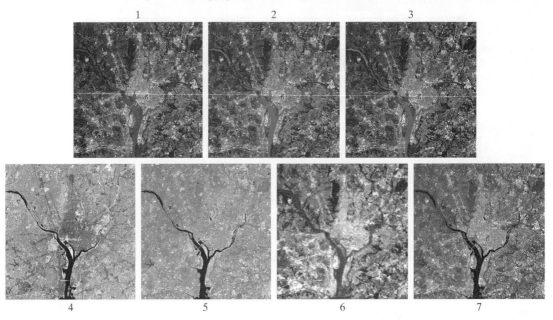

FIGURE 1.10 LANDSAT satellite images of the Washington, D.C. area. The numbers refer to the thematic bands in Table 1.1. (Images courtesy of NASA.)

FIGURE 1.11
Satellite image
of Hurricane
Katrina taken on
August 29, 2005.
(Courtesy of
NOAA.)

the spectral bands in Table 1.1. The area imaged is Washington D.C., which in-cludes features such as buildings, roads, vegetation, and a major river (the Po-tomac) going though the city. Images of population centers are used routinely (over time) to assess population growth and shift patterns, pollution, and other factors harmful to the environment. The differences between visual and in-frared image features are quite noticeable in these images. Observe, for exam-ple, how well defined the river is from its surroundings in Bands 4 and 5.

Weather observation and prediction also are major applications of multi-spectral imaging from satellites. For example, Fig. 1.11 is an image of Hurricane Katrina one of the most devastating storms in recent memory in the Western Hemisphere. This image was taken by a National Oceanographic and Atmos-pheric Administration (NOAA) satellite using sensors in the visible and in-frared bands. The eye of the hurricane is clearly visible in this image.

Figures 1.12 and 1.13 show an application of infrared imaging. These images are part of the *Nighttime Lights of the World* data set, which provides a global inventory of human settlements. The images were generated by the infrared imaging system mounted on a NOAA DMSP (Defense Meteorological Satel-lite Program) satellite. The infrared imaging system operates in the band 10.0 to 13.4 μm, and has the unique capability to observe faint sources of visible-near infrared emissions present on the Earth's surface, including cities, towns, villages, gas flares, and fires. Even without formal training in image processing, it is not difficult to imagine writing a computer program that would use these im-ages to estimate the percent of total electrical energy used by various regions of the world.

A major area of imaging in the visual spectrum is in automated visual in-spection of manufactured goods. Figure 1.14 shows some examples. Figure 1.14(a) is a controller board for a CD-ROM drive. A typical image processing task with products like this is to inspect them for missing parts (the black square on the top, right quadrant of the image is an example of a missing component).

FIGURE 1.12
Infrared satellite
images of the
Americas. The
small gray map is
provided for
reference.
(Courtesy of
NOAA.)

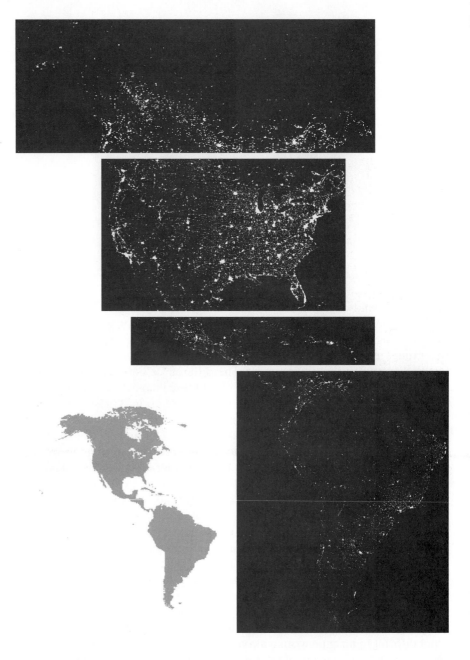

Figure 1.14(b) is an imaged pill container. The objective here is to have a machine look for missing pills. Figure 1.14(c) shows an application in which image processing is used to look for bottles that are not filled up to an acceptable level. Figure 1.14(d) shows a clear-plastic part with an unacceptable number of air pockets in it. Detecting anomalies like these is a major theme of industrial inspection that includes other products such as wood and cloth. Figure 1.14(e)

FIGURE 1.13
Infrared satellite
images of the
remaining
populated part of
the world. The
small gray map is
provided for
reference.
(Courtesy of
NOAA.)

shows a batch of cereal during inspection for color and the presence of anomalies such as burned flakes. Finally, Fig. 1.14(f) shows an image of an intraocular
implant (replacement lens for the human eye). A "structured light" illumination technique was used to highlight for easier detection flat lens deformations
toward the center of the lens. The markings at 1 o'clock and 5 o'clock are
tweezer damage. Most of the other small speckle detail is debris. The objective
in this type of inspection is to find damaged or incorrectly manufactured im-
plants automatically, prior to packaging.

As a final illustration of image processing in the visual spectrum, consider
Fig. 1.15. Figure 1.15(a) shows a thumb print. Images of fingerprints are routinely processed by computer, either to enhance them or to find features that
aid in the automated search of a database for potential matches. Figure 1.15(b)
shows an image of paper currency. Applications of digital image processing in
this area include automated counting and, in law enforcement, the reading of
the serial number for the purpose of tracking and identifying bills. The two vehicle images shown in Figs. 1.15 (c) and (d) are examples of automated license
plate reading. The light rectangles indicate the area in which the imaging system

FIGURE 1.14
Some examples
of manufactured
goods often
checked using
digital image
processing.
(a) A circuit
board controller.
(b) Packaged pills.
(c) Bottles.
(d) Air bubbles
in a clear-plastic
product.
(e) Cereal.
(f) Image of
intraocular
implant.
(Fig. (f) courtesy
of Mr. Pete Sites,
Perceptics
Corporation.)

detected the plate. The black rectangles show the results of automated reading of the plate content by the system. License plate and other applications of character recognition are used extensively for traffic monitoring and surveillance.

1.3.5 Imaging in the Microwave Band

The dominant application of imaging in the microwave band is radar. The unique feature of imaging radar is its ability to collect data over virtually any region at any time, regardless of weather or ambient lighting conditions. Some

a b
c
d

FIGURE 1.15
Some additional examples of imaging in the visual spectrum. (a) Thumb print. (b) Paper currency. (c) and (d) Automated license plate reading. (Figure (a) courtesy of the National Institute of Standards and Technology. Figures (c) and (d) courtesy of Dr. Juan Herrera, Perceptics Corporation.)

radar waves can penetrate clouds, and under certain conditions can also see through vegetation, ice, and dry sand. In many cases, radar is the only way to explore inaccessible regions of the Earth's surface. An imaging radar works like a flash camera in that it provides its own illumination (microwave pulses) to illuminate an area on the ground and take a snapshot image. Instead of a camera lens, a radar uses an antenna and digital computer processing to record its images. In a radar image, one can see only the microwave energy that was reflected back toward the radar antenna.

Figure 1.16 shows a spaceborne radar image covering a rugged mountainous area of southeast Tibet, about 90 km east of the city of Lhasa. In the lower right corner is a wide valley of the Lhasa River, which is populated by Tibetan farmers and yak herders and includes the village of Menba. Mountains in this area reach about 5800 m (19,000 ft) above sea level, while the valley floors lie about 4300 m (14,000 ft) above sea level. Note the clarity and detail of the image, unencumbered by clouds or other atmospheric conditions that normally interfere with images in the visual band.

FIGURE 1.16
Spaceborne radar
image of
mountains in
southeast Tibet.
(Courtesy of
NASA.)

1.3.6 Imaging in the Radio Band

As in the case of imaging at the other end of the spectrum (gamma rays), the major applications of imaging in the radio band are in medicine and astronomy. In medicine, radio waves are used in magnetic resonance imaging (MRI). This technique places a patient in a powerful magnet and passes radio waves through his or her body in short pulses. Each pulse causes a responding pulse of radio waves to be emitted by the patient's tissues. The location from which these signals originate and their strength are determined by a computer, which produces a two-dimensional picture of a section of the patient. MRI can produce pictures in any plane. Figure 1.17 shows MRI images of a human knee and spine.

The last image to the right in Fig. 1.18 shows an image of the Crab Pulsar in the radio band. Also shown for an interesting comparison are images of the same region but taken in most of the bands discussed earlier. Note that each image gives a totally different "view" of the Pulsar.

1.3.7 Examples in which Other Imaging Modalities Are Used

Although imaging in the electromagnetic spectrum is dominant by far, there are a number of other imaging modalities that also are important. Specifically, we discuss in this section acoustic imaging, electron microscopy, and synthetic (computer-generated) imaging.

Imaging using "sound" finds application in geological exploration, industry, and medicine. Geological applications use sound in the low end of the sound spectrum (hundreds of Hz) while imaging in other areas use ultrasound (millions of Hz). The most important commercial applications of image processing in geology are in mineral and oil exploration. For image acquisition over land, one of the main approaches is to use a large truck and a large flat steel plate. The plate is pressed on the ground by the truck, and the truck is vibrated through a frequency spectrum up to 100 Hz. The strength and speed of the

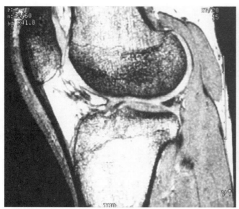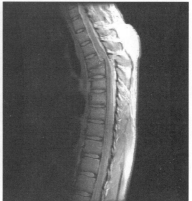

a b

FIGURE 1.17 MRI images of a human (a) knee, and (b) spine. (Image (a) courtesy of Dr. Thomas R. Gest, Division of Anatomical Sciences, University of Michigan Medical School, and (b) courtesy of Dr. David R. Pickens, Department of Radiology and Radiological Sciences, Vanderbilt University Medical Center.)

returning sound waves are determined by the composition of the Earth below the surface. These are analyzed by computer, and images are generated from the resulting analysis.

For marine acquisition, the energy source consists usually of two air guns towed behind a ship. Returning sound waves are detected by hydrophones placed in cables that are either towed behind the ship, laid on the bottom of the ocean, or hung from buoys (vertical cables). The two air guns are alternately pressurized to ~2000 psi and then set off. The constant motion of the ship provides a transversal direction of motion that, together with the returning sound waves, is used to generate a 3-D map of the composition of the Earth below the bottom of the ocean.

Figure 1.19 shows a cross-sectional image of a well-known 3-D model against which the performance of seismic imaging algorithms is tested. The arrow points to a hydrocarbon (oil and/or gas) trap. This target is brighter than the surrounding layers because the change in density in the target region is

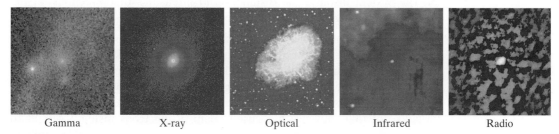

| Gamma | X-ray | Optical | Infrared | Radio |

FIGURE 1.18 Images of the Crab Pulsar (in the center of each image) covering the electromagnetic spectrum. (Courtesy of NASA.)

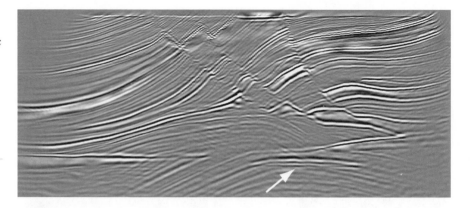

larger. Seismic interpreters look for these "bright spots" to find oil and gas. The layers above also are bright, but their brightness does not vary as strongly across the layers. Many seismic reconstruction algorithms have difficulty imaging this target because of the faults above it.

Although ultrasound imaging is used routinely in manufacturing, the best known applications of this technique are in medicine, especially in obstetrics, where unborn babies are imaged to determine the health of their development. A byproduct of this examination is determining the sex of the baby. Ultrasound images are generated using the following basic procedure:

1. The ultrasound system (a computer, ultrasound probe consisting of a source and receiver, and a display) transmits high-frequency (1 to 5 MHz) sound pulses into the body.
2. The sound waves travel into the body and hit a boundary between tissues (e.g., between fluid and soft tissue, soft tissue and bone). Some of the sound waves are reflected back to the probe, while some travel on further until they reach another boundary and get reflected.
3. The reflected waves are picked up by the probe and relayed to the computer.
4. The machine calculates the distance from the probe to the tissue or organ boundaries using the speed of sound in tissue (1540 m/s) and the time of each echo's return.
5. The system displays the distances and intensities of the echoes on the screen, forming a two-dimensional image.

In a typical ultrasound image, millions of pulses and echoes are sent and received each second. The probe can be moved along the surface of the body and angled to obtain various views. Figure 1.20 shows several examples.

We continue the discussion on imaging modalities with some examples of electron microscopy. Electron microscopes function as their optical counterparts, except that they use a focused beam of electrons instead of light to image a specimen. The operation of electron microscopes involves the following basic steps: A stream of electrons is produced by an electron source and accelerated toward the specimen using a positive electrical potential. This stream

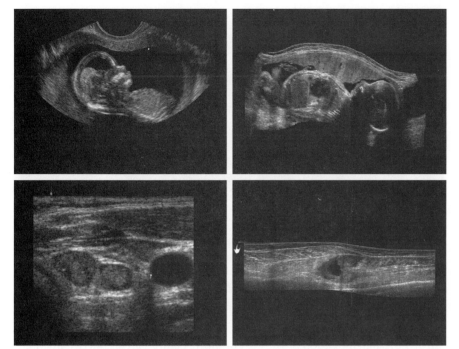

FIGURE 1.20
Examples of
ultrasound
imaging. (a) Baby.
(b) Another
view of baby.
(c) Thyroids.
(d) Muscle layers
showing lesion.
(Courtesy of
Siemens Medical
Systems, Inc.,
Ultrasound
Group.)

is confined and focused using metal apertures and magnetic lenses into a thin, monochromatic beam. This beam is focused onto the sample using a magnetic lens. Interactions occur inside the irradiated sample, affecting the electron beam. These interactions and effects are detected and transformed into an image, much in the same way that light is reflected from, or absorbed by, objects in a scene. These basic steps are carried out in all electron microscopes.

A *transmission electron microscope* (TEM) works much like a slide projector. A projector shines (transmits) a beam of light through a slide; as the light passes through the slide, it is modulated by the contents of the slide. This transmitted beam is then projected onto the viewing screen, forming an enlarged image of the slide. TEMs work the same way, except that they shine a beam of electrons through a specimen (analogous to the slide). The fraction of the beam transmitted through the specimen is projected onto a phosphor screen. The interaction of the electrons with the phosphor produces light and, therefore, a viewable image. A *scanning electron microscope* (SEM), on the other hand, actually scans the electron beam and records the interaction of beam and sample at each location. This produces one dot on a phosphor screen. A complete image is formed by a raster scan of the beam through the sample, much like a TV camera. The electrons interact with a phosphor screen and produce light. SEMs are suitable for "bulky" samples, while TEMs require very thin samples.

Electron microscopes are capable of very high magnification. While light microscopy is limited to magnifications on the order 1000×, electron microscopes

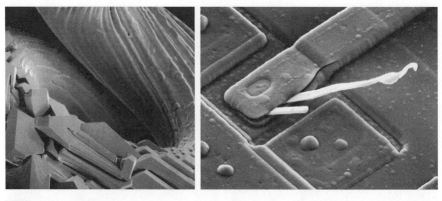

a b

FIGURE 1.21 (a) 250× SEM image of a tungsten filament following thermal failure (note the shattered pieces on the lower left). (b) 2500× SEM image of damaged integrated circuit. The white fibers are oxides resulting from thermal destruction. (Figure (a) courtesy of Mr. Michael Shaffer, Department of Geological Sciences, University of Oregon, Eugene; (b) courtesy of Dr. J. M. Hudak, McMaster University, Hamilton, Ontario, Canada.)

can achieve magnification of 10,000× or more. Figure 1.21 shows two SEM images of specimen failures due to thermal overload.

We conclude the discussion of imaging modalities by looking briefly at images that are not obtained from physical objects. Instead, they are generated by computer. *Fractals* are striking examples of computer-generated images (Lu [1997]). Basically, a fractal is nothing more than an iterative reproduction of a basic pattern according to some mathematical rules. For instance, *tiling* is one of the simplest ways to generate a fractal image. A square can be subdivided into four square subregions, each of which can be further subdivided into four smaller square regions, and so on. Depending on the complexity of the rules for filling each subsquare, some beautiful tile images can be generated using this method. Of course, the geometry can be arbitrary. For instance, the fractal image could be grown radially out of a center point. Figure 1.22(a) shows a fractal grown in this way. Figure 1.22(b) shows another fractal (a "moonscape") that provides an interesting analogy to the images of space used as illustrations in some of the preceding sections.

Fractal images tend toward artistic, mathematical formulations of "growth" of subimage elements according to a set of rules. They are useful sometimes as random textures. A more structured approach to image generation by computer lies in 3-D modeling. This is an area that provides an important intersection between image processing and computer graphics and is the basis for many 3-D visualization systems (e.g., flight simulators). Figures 1.22(c) and (d) show examples of computer-generated images. Since the original object is created in 3-D, images can be generated in any perspective from plane projections of the 3-D volume. Images of this type can be used for medical training and for a host of other applications, such as criminal forensics and special effects.

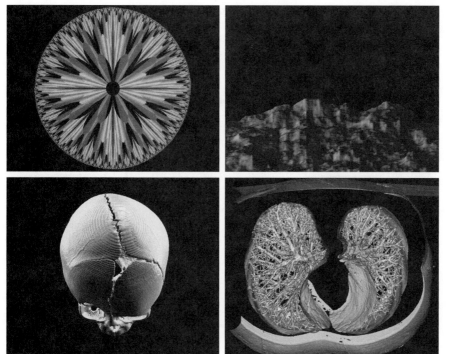

FIGURE 1.22
(a) and (b) Fractal images. (c) and (d) Images generated from 3-D computer models of the objects shown. (Figures (a) and (b) courtesy of Ms. Melissa D. Binde, Swarthmore College; (c) and (d) courtesy of NASA.)

1.4   Fundamental Steps in Digital Image Processing

It is helpful to divide the material covered in the following chapters into the two broad categories defined in Section 1.1: methods whose input and output are images, and methods whose inputs may be images but whose outputs are attributes extracted from those images. This organization is summarized in Fig. 1.23. The diagram does not imply that every process is applied to an image. Rather, the intention is to convey an idea of all the methodologies that can be applied to images for different purposes and possibly with different objectives. The discussion in this section may be viewed as a brief overview of the material in the remainder of the book.

Image acquisition is the first process in Fig. 1.23. The discussion in Section 1.3 gave some hints regarding the origin of digital images. This topic is considered in much more detail in Chapter 2, where we also introduce a number of basic digital image concepts that are used throughout the book. Note that acquisition could be as simple as being given an image that is already in digital form. Generally, the image acquisition stage involves preprocessing, such as scaling.

Image enhancement is the process of manipulating an image so that the result is more suitable than the original for a specific application. The word *specific* is important here, because it establishes at the outset that enhancement techniques are problem oriented. Thus, for example, a method that is quite useful for enhancing X-ray images may not be the best approach for enhancing satellite images taken in the infrared band of the electromagnetic spectrum.

FIGURE 1.23
Fundamental
steps in digital
image processing.
The chapter(s)
indicated in the
boxes is where the
material
described in the
box is discussed.

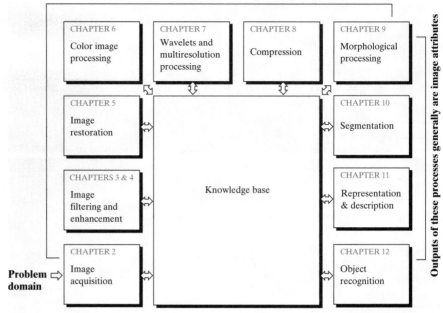

    There is no general "theory" of image enhancement. When an image is
processed for visual interpretation, the viewer is the ultimate judge of how
well a particular method works. Enhancement techniques are so varied, and
use so many different image processing approaches, that it is difficult to assemble a meaningful body of techniques suitable for enhancement in one
chapter without extensive background development. For this reason, and also
because beginners in the field of image processing generally find enhancement applications visually appealing, interesting, and relatively simple to understand, we use image enhancement as examples when introducing new
concepts in parts of Chapter 2 and in Chapters 3 and 4. The material in the
latter two chapters span many of the methods used traditionally for image enhancement. Therefore, using examples from image enhancement to introduce
new image processing methods developed in these early chapters not only
saves having an extra chapter in the book dealing with image enhancement
but, more importantly, is an effective approach for introducing newcomers to
the details of processing techniques early in the book. However, as you will
see in progressing through the rest of the book, the material developed in
these chapters is applicable to a much broader class of problems than just
image enhancement.

    Image restoration is an area that also deals with improving the appearance
of an image. However, unlike enhancement, which is subjective, image restoration is objective, in the sense that restoration techniques tend to be based on
mathematical or probabilistic models of image degradation. Enhancement, on
the other hand, is based on human subjective preferences regarding what constitutes a "good" enhancement result.

Color image processing is an area that has been gaining in importance because of the significant increase in the use of digital images over the Internet. Chapter 6 covers a number of fundamental concepts in color models and basic color processing in a digital domain. Color is used also in later chapters as the basis for extracting features of interest in an image.

Wavelets are the foundation for representing images in various degrees of resolution. In particular, this material is used in this book for image data compression and for pyramidal representation, in which images are subdivided successively into smaller regions.

Compression, as the name implies, deals with techniques for reducing the storage required to save an image, or the bandwidth required to transmit it. Although storage technology has improved significantly over the past decade, the same cannot be said for transmission capacity. This is true particularly in uses of the Internet, which are characterized by significant pictorial content. Image compression is familiar (perhaps inadvertently) to most users of computers in the form of image file extensions, such as the jpg file extension used in the JPEG (Joint Photographic Experts Group) image compression standard.

Morphological processing deals with tools for extracting image components that are useful in the representation and description of shape. The material in this chapter begins a transition from processes that output images to processes that output image attributes, as indicated in Section 1.1.

Segmentation procedures partition an image into its constituent parts or objects. In general, autonomous segmentation is one of the most difficult tasks in digital image processing. A rugged segmentation procedure brings the process a long way toward successful solution of imaging problems that require objects to be identified individually. On the other hand, weak or erratic segmentation algorithms almost always guarantee eventual failure. In general, the more accurate the segmentation, the more likely recognition is to succeed.

Representation and description almost always follow the output of a segmentation stage, which usually is raw pixel data, constituting either the boundary of a region (i.e., the set of pixels separating one image region from another) or all the points in the region itself. In either case, converting the data to a form suitable for computer processing is necessary. The first decision that must be made is whether the data should be represented as a boundary or as a complete region. Boundary representation is appropriate when the focus is on external shape characteristics, such as corners and inflections. Regional representation is appropriate when the focus is on internal properties, such as texture or skeletal shape. In some applications, these representations complement each other. Choosing a representation is only part of the solution for transforming raw data into a form suitable for subsequent computer processing. A method must also be specified for describing the data so that features of interest are highlighted. *Description*, also called *feature selection*, deals with extracting attributes that result in some quantitative information of interest or are basic for differentiating one class of objects from another.

Recognition is the process that assigns a label (e.g., "vehicle") to an object based on its descriptors. As detailed in Section 1.1, we conclude our coverage of digital image processing with the development of methods for recognition of individual objects.

So far we have said nothing about the need for prior knowledge or about the interaction between the *knowledge base* and the processing modules in Fig. 1.23. Knowledge about a problem domain is coded into an image processing system in the form of a knowledge database. This knowledge may be as simple as detailing regions of an image where the information of interest is known to be located, thus limiting the search that has to be conducted in seeking that information. The knowledge base also can be quite complex, such as an interrelated list of all major possible defects in a materials inspection problem or an image database containing high-resolution satellite images of a region in connection with change-detection applications. In addition to guiding the operation of each processing module, the knowledge base also controls the interaction between modules. This distinction is made in Fig. 1.23 by the use of double-headed arrows between the processing modules and the knowledge base, as opposed to single-headed arrows linking the processing modules.

Although we do not discuss image display explicitly at this point, it is important to keep in mind that viewing the results of image processing can take place at the output of any stage in Fig. 1.23. We also note that not all image processing applications require the complexity of interactions implied by Fig. 1.23. In fact, not even all those modules are needed in many cases. For example, image enhancement for human visual interpretation seldom requires use of any of the other stages in Fig. 1.23. In general, however, as the complexity of an image processing task increases, so does the number of processes required to solve the problem.

1.5 Components of an Image Processing System

As recently as the mid-1980s, numerous models of image processing systems being sold throughout the world were rather substantial peripheral devices that attached to equally substantial host computers. Late in the 1980s and early in the 1990s, the market shifted to image processing hardware in the form of single boards designed to be compatible with industry standard buses and to fit into engineering workstation cabinets and personal computers. In addition to lowering costs, this market shift also served as a catalyst for a significant number of new companies specializing in the development of software written specifically for image processing.

Although large-scale image processing systems still are being sold for massive imaging applications, such as processing of satellite images, the trend continues toward miniaturizing and blending of general-purpose small computers with specialized image processing hardware. Figure 1.24 shows the basic components comprising a typical *general-purpose* system used for digital image processing. The function of each component is discussed in the following paragraphs, starting with image sensing.

With reference to *sensing*, two elements are required to acquire digital images. The first is a physical device that is sensitive to the energy radiated by the object we wish to image. The second, called a *digitizer*, is a device for converting

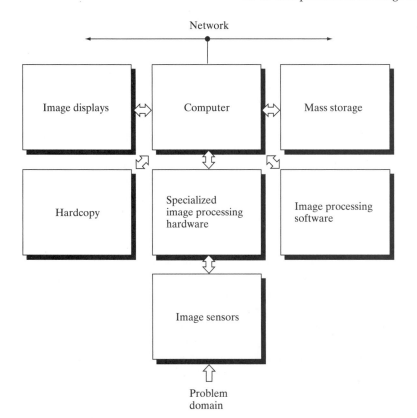

FIGURE 1.24
Components of a
general-purpose
image processing
system.

the output of the physical sensing device into digital form. For instance, in a digital video camera, the sensors produce an electrical output proportional to light intensity. The digitizer converts these outputs to digital data. These topics are covered in Chapter 2.

Specialized image processing hardware usually consists of the digitizer just mentioned, plus hardware that performs other primitive operations, such as an arithmetic logic unit (ALU), that performs arithmetic and logical operations in parallel on entire images. One example of how an ALU is used is in averaging images as quickly as they are digitized, for the purpose of noise reduction. This type of hardware sometimes is called a *front-end subsystem*, and its most distinguishing characteristic is speed. In other words, this unit performs functions that require fast data throughputs (e.g., digitizing and averaging video images at 30 frames/s) that the typical main computer cannot handle.

The *computer* in an image processing system is a general-purpose computer and can range from a PC to a supercomputer. In dedicated applications, sometimes custom computers are used to achieve a required level of performance, but our interest here is on general-purpose image processing systems. In these systems, almost any well-equipped PC-type machine is suitable for off-line image processing tasks.

Software for image processing consists of specialized modules that perform specific tasks. A well-designed package also includes the capability for the user to write code that, as a minimum, utilizes the specialized modules. More sophisticated software packages allow the integration of those modules and general-purpose software commands from at least one computer language.

Mass storage capability is a must in image processing applications. An image of size 1024 × 1024 pixels, in which the intensity of each pixel is an 8-bit quantity, requires one megabyte of storage space if the image is not compressed. When dealing with thousands, or even millions, of images, providing adequate storage in an image processing system can be a challenge. Digital storage for image processing applications falls into three principal categories: (1) short-term storage for use during processing, (2) on-line storage for relatively fast recall, and (3) archival storage, characterized by infrequent access. Storage is measured in bytes (eight bits), Kbytes (one thousand bytes), Mbytes (one million bytes), Gbytes (meaning giga, or one billion, bytes), and Tbytes (meaning tera, or one trillion, bytes).

One method of providing short-term storage is computer memory. Another is by specialized boards, called *frame buffers*, that store one or more images and can be accessed rapidly, usually at video rates (e.g., at 30 complete images per second). The latter method allows virtually instantaneous image *zoom*, as well as *scroll* (vertical shifts) and *pan* (horizontal shifts). Frame buffers usually are housed in the specialized image processing hardware unit in Fig. 1.24. On-line storage generally takes the form of magnetic disks or optical-media storage. The key factor characterizing on-line storage is frequent access to the stored data. Finally, archival storage is characterized by massive storage requirements but infrequent need for access. Magnetic tapes and optical disks housed in "jukeboxes" are the usual media for archival applications.

Image displays in use today are mainly color (preferably flat screen) TV monitors. Monitors are driven by the outputs of image and graphics display cards that are an integral part of the computer system. Seldom are there requirements for image display applications that cannot be met by display cards available commercially as part of the computer system. In some cases, it is necessary to have stereo displays, and these are implemented in the form of headgear containing two small displays embedded in goggles worn by the user.

Hardcopy devices for recording images include laser printers, film cameras, heat-sensitive devices, inkjet units, and digital units, such as optical and CD-ROM disks. Film provides the highest possible resolution, but paper is the obvious medium of choice for written material. For presentations, images are displayed on film transparencies or in a digital medium if image projection equipment is used. The latter approach is gaining acceptance as the standard for image presentations.

Networking is almost a default function in any computer system in use today. Because of the large amount of data inherent in image processing applications, the key consideration in image transmission is bandwidth. In dedicated networks, this typically is not a problem, but communications with remote sites via the Internet are not always as efficient. Fortunately, this situation is improving quickly as a result of optical fiber and other broadband technologies.

Summary

The main purpose of the material presented in this chapter is to provide a sense of perspective about the origins of digital image processing and, more important, about current and future areas of application of this technology. Although the coverage of these topics in this chapter was necessarily incomplete due to space limitations, it should have left you with a clear impression of the breadth and practical scope of digital image processing. As we proceed in the following chapters with the development of image processing theory and applications, numerous examples are provided to keep a clear focus on the utility and promise of these techniques. Upon concluding the study of the final chapter, a reader of this book will have arrived at a level of understanding that is the foundation for most of the work currently underway in this field.

References and Further Reading

References at the end of later chapters address specific topics discussed in those chapters, and are keyed to the Bibliography at the end of the book. However, in this chapter we follow a different format in order to summarize in one place a body of journals that publish material on image processing and related topics. We also provide a list of books from which the reader can readily develop a historical and current perspective of activities in this field. Thus, the reference material cited in this chapter is intended as a general-purpose, easily accessible guide to the published literature on image processing.

Major refereed journals that publish articles on image processing and related topics include: *IEEE Transactions on Image Processing; IEEE Transactions on Pattern Analysis and Machine Intelligence; Computer Vision, Graphics, and Image Processing* (prior to 1991); *Computer Vision and Image Understanding; IEEE Transactions on Systems, Man and Cybernetics; Artificial Intelligence; Pattern Recognition; Pattern Recognition Letters; Journal of the Optical Society of America* (prior to 1984); *Journal of the Optical Society of America—A: Optics, Image Science and Vision; Optical Engineering; Applied Optics—Information Processing; IEEE Transactions on Medical Imaging; Journal of Electronic Imaging; IEEE Transactions on Information Theory; IEEE Transactions on Communications; IEEE Transactions on Acoustics, Speech and Signal Processing; Proceedings of the IEEE;* and issues of the *IEEE Transactions on Computers* prior to 1980. Publications of the International Society for Optical Engineering (SPIE) also are of interest.

The following books, listed in reverse chronological order (with the number of books being biased toward more recent publications), contain material that complements our treatment of digital image processing. These books represent an easily accessible overview of the area for the past 30-plus years and were selected to provide a variety of treatments. They range from textbooks, which cover foundation material; to handbooks, which give an overview of techniques; and finally to edited books, which contain material representative of current research in the field.

Prince, J. L. and Links, J. M. [2006]. *Medical Imaging, Signals, and Systems*, Prentice Hall, Upper Saddle River, NJ.

Bezdek, J. C. et al. [2005]. *Fuzzy Models and Algorithms for Pattern Recognition and Image Processing*, Springer, New York.

Davies, E. R. [2005]. *Machine Vision: Theory, Algorithms, Practicalities*, Morgan Kaufmann, San Francisco, CA.

Rangayyan, R. M. [2005]. *Biomedical Image Analysis*, CRC Press, Boca Raton, FL.

Umbaugh, S. E. [2005]. *Computer Imaging: Digital Image Analysis and Processing*, CRC Press, Boca Raton, FL.

Gonzalez, R. C., Woods, R. E., and Eddins, S. L. [2004]. *Digital Image Processing Using MATLAB*, Prentice Hall, Upper Saddle River, NJ.

Snyder, W. E. and Qi, Hairong [2004]. *Machine Vision*, Cambridge University Press, New York.

Klette, R. and Rosenfeld, A. [2004]. *Digital Geometry—Geometric Methods for Digital Picture Analysis*, Morgan Kaufmann, San Francisco, CA.

Won, C. S. and Gray, R. M. [2004]. *Stochastic Image Processing*, Kluwer Academic/Plenum Publishers, New York.

Soille, P. [2003]. *Morphological Image Analysis: Principles and Applications*, 2nd ed., Springer-Verlag, New York.

Dougherty, E. R. and Lotufo, R. A. [2003]. *Hands-on Morphological Image Processing*, SPIE—The International Society for Optical Engineering, Bellingham, WA.

Gonzalez, R. C. and Woods, R. E. [2002]. *Digital Image Processing*, 2nd ed., Prentice Hall, Upper Saddle River, NJ.

Forsyth, D. F. and Ponce, J. [2002]. *Computer Vision—A Modern Approach*, Prentice Hall, Upper Saddle River, NJ.

Duda, R. O., Hart, P. E., and Stork, D. G. [2001]. *Pattern Classification*, 2nd ed., John Wiley & Sons, New York.

Pratt, W. K. [2001]. *Digital Image Processing*, 3rd ed., John Wiley & Sons, New York.

Ritter, G. X. and Wilson, J. N. [2001]. *Handbook of Computer Vision Algorithms in Image Algebra*, CRC Press, Boca Raton, FL.

Shapiro, L. G. and Stockman, G. C. [2001]. *Computer Vision*, Prentice Hall, Upper Saddle River, NJ.

Dougherty, E. R. (ed.) [2000]. *Random Processes for Image and Signal Processing*, IEEE Press, New York.

Etienne, E. K. and Nachtegael, M. (eds.). [2000]. *Fuzzy Techniques in Image Processing*, Springer-Verlag, New York.

Goutsias, J., Vincent, L., and Bloomberg, D. S. (eds.). [2000]. *Mathematical Morphology and Its Applications to Image and Signal Processing*, Kluwer Academic Publishers, Boston, MA.

Mallot, A. H. [2000]. *Computational Vision*, The MIT Press, Cambridge, MA.

Marchand-Maillet, S. and Sharaiha, Y. M. [2000]. *Binary Digital Image Processing: A Discrete Approach*, Academic Press, New York.

Mitra, S. K. and Sicuranza, G. L. (eds.) [2000]. *Nonlinear Image Processing*, Academic Press, New York.

Edelman, S. [1999]. *Representation and Recognition in Vision*, The MIT Press, Cambridge, MA.

Lillesand, T. M. and Kiefer, R. W. [1999]. *Remote Sensing and Image Interpretation*, John Wiley & Sons, New York.

Mather, P. M. [1999]. *Computer Processing of Remotely Sensed Images: An Introduction*, John Wiley & Sons, New York.

Petrou, M. and Bosdogianni, P. [1999]. *Image Processing: The Fundamentals*, John Wiley & Sons, UK.

Russ, J. C. [1999]. *The Image Processing Handbook*, 3rd ed., CRC Press, Boca Raton, FL.

Smirnov, A. [1999]. *Processing of Multidimensional Signals*, Springer-Verlag, New York.

Sonka, M., Hlavac, V., and Boyle, R. [1999]. *Image Processing, Analysis, and Computer Vision*, PWS Publishing, New York.

Haskell, B. G. and Netravali, A. N. [1997]. *Digital Pictures: Representation, Compression, and Standards*, Perseus Publishing, New York.

Jahne, B. [1997]. *Digital Image Processing: Concepts, Algorithms, and Scientific Applications*, Springer-Verlag, New York.

Castleman, K. R. [1996]. *Digital Image Processing*, 2nd ed., Prentice Hall, Upper Saddle River, NJ.

Geladi, P. and Grahn, H. [1996]. *Multivariate Image Analysis*, John Wiley & Sons, New York.

Bracewell, R. N. [1995]. *Two-Dimensional Imaging*, Prentice Hall, Upper Saddle River, NJ.

Sid-Ahmed, M. A. [1995]. *Image Processing: Theory, Algorithms, and Architectures*, McGraw-Hill, New York.

Jain, R., Rangachar, K., and Schunk, B. [1995]. *Computer Vision*, McGraw-Hill, New York.

Mitiche, A. [1994]. *Computational Analysis of Visual Motion*, Perseus Publishing, New York.

Baxes, G. A. [1994]. *Digital Image Processing: Principles and Applications*, John Wiley & Sons, New York.

Gonzalez, R. C. and Woods, R. E. [1992]. *Digital Image Processing*, Addison-Wesley, Reading, MA.

Haralick, R. M. and Shapiro, L. G. [1992]. *Computer and Robot Vision*, vols. 1 & 2, Addison-Wesley, Reading, MA.

Pratt, W. K. [1991] *Digital Image Processing*, 2nd ed., Wiley-Interscience, New York.

Lim, J. S. [1990]. *Two-Dimensional Signal and Image Processing*, Prentice Hall, Upper Saddle River, NJ.

Jain, A. K. [1989]. *Fundamentals of Digital Image Processing*, Prentice Hall, Upper Saddle River, NJ.

Schalkoff, R. J. [1989]. *Digital Image Processing and Computer Vision*, John Wiley & Sons, New York.

Giardina, C. R. and Dougherty, E. R. [1988]. *Morphological Methods in Image and Signal Processing*, Prentice Hall, Upper Saddle River, NJ.

Levine, M. D. [1985]. *Vision in Man and Machine*, McGraw-Hill, New York.

Serra, J. [1982]. *Image Analysis and Mathematical Morphology*, Academic Press, New York.

Ballard, D. H. and Brown, C. M. [1982]. *Computer Vision*, Prentice Hall, Upper Saddle River, NJ.

Fu, K. S. [1982]. *Syntactic Pattern Recognition and Applications*, Prentice Hall, Upper Saddle River, NJ.

Nevatia, R. [1982]. *Machine Perception*, Prentice Hall, Upper Saddle River, NJ.

Pavlidis, T. [1982]. *Algorithms for Graphics and Image Processing*, Computer Science Press, Rockville, MD.

Rosenfeld, A. and Kak, A. C. [1982]. *Digital Picture Processing*, 2nd ed., vols. 1 & 2, Academic Press, New York.

Hall, E. L. [1979]. *Computer Image Processing and Recognition*, Academic Press, New York.

Gonzalez, R. C. and Thomason, M. G. [1978]. *Syntactic Pattern Recognition: An Introduction*, Addison-Wesley, Reading, MA.

Andrews, H. C. and Hunt, B. R. [1977]. *Digital Image Restoration*, Prentice Hall, Upper Saddle River, NJ.

Pavlidis, T. [1977]. *Structural Pattern Recognition*, Springer-Verlag, New York.

Tou, J. T. and Gonzalez, R. C. [1974]. *Pattern Recognition Principles*, Addison-Wesley, Reading, MA.

Andrews, H. C. [1970]. *Computer Techniques in Image Processing*, Academic Press, New York.

2 Digital Image Fundamentals

Those who wish to succeed must ask the right preliminary questions.

Aristotle

Preview

The purpose of this chapter is to introduce you to a number of basic concepts in digital image processing that are used throughout the book. Section 2.1 summarizes the mechanics of the human visual system, including image formation in the eye and its capabilities for brightness adaptation and discrimination. Section 2.2 discusses light, other components of the electromagnetic spectrum, and their imaging characteristics. Section 2.3 discusses imaging sensors and how they are used to generate digital images. Section 2.4 introduces the concepts of uniform image sampling and intensity quantization. Additional topics discussed in that section include digital image representation, the effects of varying the number of samples and intensity levels in an image, the concepts of spatial and intensity resolution, and the principles of image interpolation. Section 2.5 deals with a variety of basic relationships between pixels. Finally, Section 2.6 is an introduction to the principal mathematical tools we use throughout the book. A second objective of that section is to help you begin developing a "feel" for how these tools are used in a variety of basic image processing tasks. The scope of these tools and their application are expanded as needed in the remainder of the book.

2.1   Elements of Visual Perception

Although the field of digital image processing is built on a foundation of mathematical and probabilistic formulations, human intuition and analysis play a central role in the choice of one technique versus another, and this choice often is made based on subjective, visual judgments. Hence, developing a basic understanding of human visual perception as a first step in our journey through this book is appropriate. Given the complexity and breadth of this topic, we can only aspire to cover the most rudimentary aspects of human vision. In particular, our interest is in the mechanics and parameters related to how images are formed and perceived by humans. We are interested in learning the physical limitations of human vision in terms of factors that also are used in our work with digital images. Thus, factors such as how human and electronic imaging devices compare in terms of resolution and ability to adapt to changes in illumination are not only interesting, they also are important from a practical point of view.

2.1.1 Structure of the Human Eye

Figure 2.1 shows a simplified horizontal cross section of the human eye. The eye is nearly a sphere, with an average diameter of approximately 20 mm. Three membranes enclose the eye: the *cornea* and *sclera* outer cover; the *choroid*; and the *retina*. The cornea is a tough, transparent tissue that covers

FIGURE 2.1
Simplified
diagram of a cross
section of the
human eye.

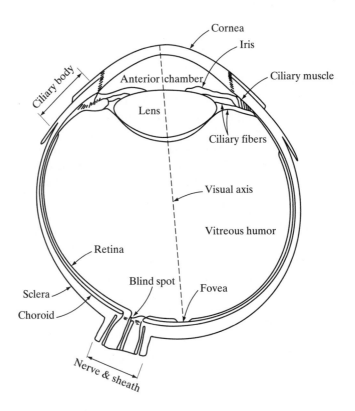

the anterior surface of the eye. Continuous with the cornea, the sclera is an opaque membrane that encloses the remainder of the optic globe.

The choroid lies directly below the sclera. This membrane contains a network of blood vessels that serve as the major source of nutrition to the eye. Even superficial injury to the choroid, often not deemed serious, can lead to severe eye damage as a result of inflammation that restricts blood flow. The choroid coat is heavily pigmented and hence helps to reduce the amount of extraneous light entering the eye and the backscatter within the optic globe. At its anterior extreme, the choroid is divided into the *ciliary body* and the *iris*. The latter contracts or expands to control the amount of light that enters the eye. The central opening of the iris (the pupil) varies in diameter from approximately 2 to 8 mm. The front of the iris contains the visible pigment of the eye, whereas the back contains a black pigment.

The *lens* is made up of concentric layers of fibrous cells and is suspended by fibers that attach to the ciliary body. It contains 60 to 70% water, about 6% fat, and more protein than any other tissue in the eye. The lens is colored by a slightly yellow pigmentation that increases with age. In extreme cases, excessive clouding of the lens, caused by the affliction commonly referred to as *cataracts*, can lead to poor color discrimination and loss of clear vision. The lens absorbs approximately 8% of the visible light spectrum, with relatively higher absorption at shorter wavelengths. Both infrared and ultraviolet light are absorbed appreciably by proteins within the lens structure and, in excessive amounts, can damage the eye.

The innermost membrane of the eye is the *retina*, which lines the inside of the wall's entire posterior portion. When the eye is properly focused, light from an object outside the eye is imaged on the retina. Pattern vision is afforded by the distribution of discrete light receptors over the surface of the retina. There are two classes of receptors: *cones* and *rods*. The cones in each eye number between 6 and 7 million. They are located primarily in the central portion of the retina, called the *fovea*, and are highly sensitive to color. Humans can resolve fine details with these cones largely because each one is connected to its own nerve end. Muscles controlling the eye rotate the eyeball until the image of an object of interest falls on the fovea. Cone vision is called *photopic* or bright-light vision.

The number of rods is much larger: Some 75 to 150 million are distributed over the retinal surface. The larger area of distribution and the fact that several rods are connected to a single nerve end reduce the amount of detail discernible by these receptors. Rods serve to give a general, overall picture of the field of view. They are not involved in color vision and are sensitive to low levels of illumination. For example, objects that appear brightly colored in daylight when seen by moonlight appear as colorless forms because only the rods are stimulated. This phenomenon is known as *scotopic* or dim-light vision.

Figure 2.2 shows the density of rods and cones for a cross section of the right eye passing through the region of emergence of the optic nerve from the eye. The absence of receptors in this area results in the so-called *blind spot* (see Fig. 2.1). Except for this region, the distribution of receptors is radially symmetric about the fovea. Receptor density is measured in degrees from the

FIGURE 2.2
Distribution of
rods and cones in
the retina.

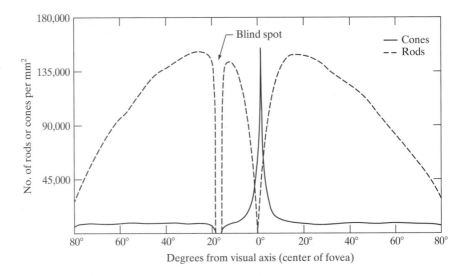

fovea (that is, in degrees off axis, as measured by the angle formed by the visual axis and a line passing through the center of the lens and intersecting the retina). Note in Fig. 2.2 that cones are most dense in the center of the retina (in the center area of the fovea). Note also that rods increase in density from the center out to approximately 20° off axis and then decrease in density out to the extreme periphery of the retina.

The fovea itself is a circular indentation in the retina of about 1.5 mm in diameter. However, in terms of future discussions, talking about square or rectangular arrays of sensing elements is more useful. Thus, by taking some liberty in interpretation, we can view the fovea as a square sensor array of size 1.5 mm $\times$ 1.5 mm. The density of cones in that area of the retina is approximately 150,000 elements per mm^2. Based on these approximations, the number of cones in the region of highest acuity in the eye is about 337,000 elements. Just in terms of raw resolving power, a charge-coupled device (CCD) imaging chip of medium resolution can have this number of elements in a receptor array no larger than 5 mm $\times$ 5 mm. While the ability of humans to integrate intelligence and experience with vision makes these types of number comparisons somewhat superficial, keep in mind for future discussions that the basic ability of the eye to resolve detail certainly is comparable to current electronic imaging sensors.

2.1.2 Image Formation in the Eye

In an ordinary photographic camera, the lens has a fixed focal length, and focusing at various distances is achieved by varying the distance between the lens and the imaging plane, where the film (or imaging chip in the case of a digital camera) is located. In the human eye, the converse is true; the distance between the lens and the imaging region (the retina) is fixed, and the focal length needed to achieve proper focus is obtained by varying the shape of the lens. The fibers in the ciliary body accomplish this, flattening or thickening the

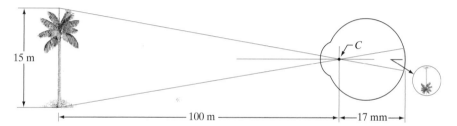

FIGURE 2.3
Graphical
representation of
the eye looking at
a palm tree. Point
C is the optical
center of the lens.

lens for distant or near objects, respectively. The distance between the center of the lens and the retina along the visual axis is approximately 17 mm. The range of focal lengths is approximately 14 mm to 17 mm, the latter taking place when the eye is relaxed and focused at distances greater than about 3 m.

The geometry in Fig. 2.3 illustrates how to obtain the dimensions of an image formed on the retina. For example, suppose that a person is looking at a tree 15 m high at a distance of 100 m. Letting h denote the height of that object in the retinal image, the geometry of Fig. 2.3 yields $15/100 = h/17$ or $h = 2.55$ mm. As indicated in Section 2.1.1, the retinal image is focused primarily on the region of the fovea. Perception then takes place by the relative excitation of light receptors, which transform radiant energy into electrical impulses that ultimately are decoded by the brain.

2.1.3 Brightness Adaptation and Discrimination

Because digital images are displayed as a discrete set of intensities, the eye's ability to discriminate between different intensity levels is an important consideration in presenting image processing results. The range of light intensity levels to which the human visual system can adapt is enormous—on the order of 10^{10}—from the scotopic threshold to the glare limit. Experimental evidence indicates that *subjective brightness* (intensity as *perceived* by the human visual system) is a logarithmic function of the light intensity incident on the eye. Figure 2.4, a plot

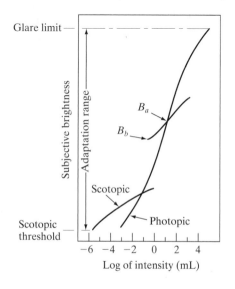

FIGURE 2.4
Range of
subjective
brightness
sensations
showing a
particular
adaptation level.

of light intensity versus subjective brightness, illustrates this characteristic. The long solid curve represents the range of intensities to which the visual system can adapt. In photopic vision alone, the range is about 10^6. The transition from scotopic to photopic vision is gradual over the approximate range from 0.001 to 0.1 millilambert (-3 to -1 mL in the log scale), as the double branches of the adaptation curve in this range show.

The essential point in interpreting the impressive dynamic range depicted in Fig. 2.4 is that the visual system cannot operate over such a range *simultaneously.* Rather, it accomplishes this large variation by changing its overall sensitivity, a phenomenon known as *brightness adaptation.* The total range of distinct intensity levels the eye can discriminate simultaneously is rather small when compared with the total adaptation range. For any given set of conditions, the current sensitivity level of the visual system is called the *brightness adaptation level,* which may correspond, for example, to brightness B_a in Fig. 2.4. The short intersecting curve represents the range of subjective brightness that the eye can perceive when adapted to this level. This range is rather restricted, having a level B_b at and below which all stimuli are perceived as indistinguishable blacks. The upper portion of the curve is not actually restricted but, if extended too far, loses its meaning because much higher intensities would simply raise the adaptation level higher than B_a.

The ability of the eye to discriminate between *changes* in light intensity at any specific adaptation level is also of considerable interest. A classic experiment used to determine the capability of the human visual system for brightness discrimination consists of having a subject look at a flat, uniformly illuminated area large enough to occupy the entire field of view. This area typically is a diffuser, such as opaque glass, that is illuminated from behind by a light source whose intensity, I, can be varied. To this field is added an increment of illumination, ΔI, in the form of a short-duration flash that appears as a circle in the center of the uniformly illuminated field, as Fig. 2.5 shows.

If ΔI is not bright enough, the subject says "no," indicating no perceivable change. As ΔI gets stronger, the subject may give a positive response of "yes," indicating a perceived change. Finally, when ΔI is strong enough, the subject will give a response of "yes" all the time. The quantity $\Delta I_c/I$, where ΔI_c is the increment of illumination discriminable 50% of the time with background illumination I, is called the *Weber ratio.* A small value of $\Delta I_c/I$ means that a small percentage change in intensity is discriminable. This represents "good" brightness discrimination. Conversely, a large value of $\Delta I_c/I$ means that a large percentage change in intensity is required. This represents "poor" brightness discrimination.

FIGURE 2.5 Basic experimental setup used to characterize brightness discrimination.

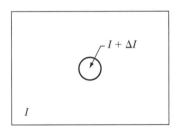

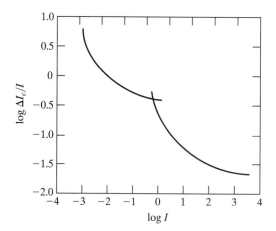

FIGURE 2.6
Typical Weber
ratio as a function
of intensity.

A plot of $\log \Delta I_c/I$ as a function of $\log I$ has the general shape shown in Fig. 2.6. This curve shows that brightness discrimination is poor (the Weber ratio is large) at low levels of illumination, and it improves significantly (the Weber ratio decreases) as background illumination increases. The two branches in the curve reflect the fact that at low levels of illumination vision is carried out by the rods, whereas at high levels (showing better discrimination) vision is the function of cones.

If the background illumination is held constant and the intensity of the other source, instead of flashing, is now allowed to vary incrementally from never being perceived to always being perceived, the typical observer can discern a total of one to two dozen different intensity changes. Roughly, this result is related to the number of different intensities a person can see at any one point in a monochrome image. This result does not mean that an image can be represented by such a small number of intensity values because, as the eye roams about the image, the average background changes, thus allowing a *different* set of incremental changes to be detected at each new adaptation level. The net consequence is that the eye is capable of a much broader range of *overall* intensity discrimination. In fact, we show in Section 2.4.3 that the eye is capable of detecting objectionable contouring effects in monochrome images whose overall intensity is represented by fewer than approximately two dozen levels.

Two phenomena clearly demonstrate that perceived brightness is not a simple function of intensity. The first is based on the fact that the visual system tends to undershoot or overshoot around the boundary of regions of different intensities. Figure 2.7(a) shows a striking example of this phenomenon. Although the intensity of the stripes is constant, we actually perceive a brightness pattern that is strongly scalloped near the boundaries [Fig. 2.7(c)]. These seemingly scalloped bands are called *Mach bands* after Ernst Mach, who first described the phenomenon in 1865.

The second phenomenon, called *simultaneous contrast*, is related to the fact that a region's perceived brightness does not depend simply on its intensity, as Fig. 2.8 demonstrates. All the center squares have exactly the same intensity.

a
b
c

FIGURE 2.7
Illustration of the
Mach band effect.
Perceived
intensity is not a
simple function of
actual intensity.

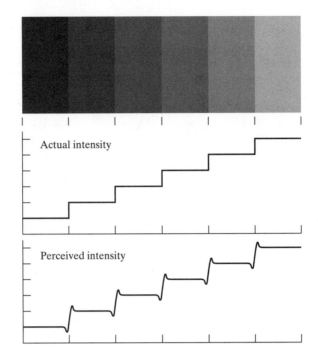

However, they appear to the eye to become darker as the background gets lighter. A more familiar example is a piece of paper that seems white when lying on a desk, but can appear totally black when used to shield the eyes while looking directly at a bright sky.

Other examples of human perception phenomena are optical illusions, in which the eye fills in nonexisting information or wrongly perceives geometrical properties of objects. Figure 2.9 shows some examples. In Fig. 2.9(a), the outline of a square is seen clearly, despite the fact that no lines defining such a figure are part of the image. The same effect, this time with a circle, can be seen in Fig. 2.9(b); note how just a few lines are sufficient to give the illusion of a

a b c

FIGURE 2.8 Examples of simultaneous contrast. All the inner squares have the same intensity, but they appear progressively darker as the background becomes lighter.

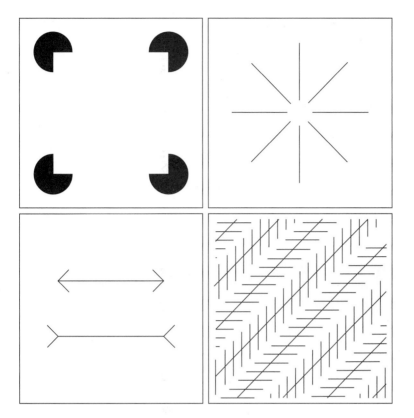

FIGURE 2.9 Some well-known optical illusions.

complete circle. The two horizontal line segments in Fig. 2.9(c) are of the same length, but one appears shorter than the other. Finally, all lines in Fig. 2.9(d) that are oriented at 45° are equidistant and parallel. Yet the crosshatching creates the illusion that those lines are far from being parallel. Optical illusions are a characteristic of the human visual system that is not fully understood.

2.2 Light and the Electromagnetic Spectrum

The electromagnetic spectrum was introduced in Section 1.3. We now consider this topic in more detail. In 1666, Sir Isaac Newton discovered that when a beam of sunlight is passed through a glass prism, the emerging beam of light is not white but consists instead of a continuous spectrum of colors ranging from violet at one end to red at the other. As Fig. 2.10 shows, the range of colors we perceive in visible light represents a very small portion of the electromagnetic spectrum. On one end of the spectrum are radio waves with wavelengths billions of times longer than those of visible light. On the other end of the spectrum are gamma rays with wavelengths millions of times smaller than those of visible light. The electromagnetic spectrum can be expressed in terms of wavelength, frequency, or energy. Wavelength (λ) and frequency (ν) are related by the expression

$$\lambda = \frac{c}{\nu} \qquad (2.2\text{-}1)$$

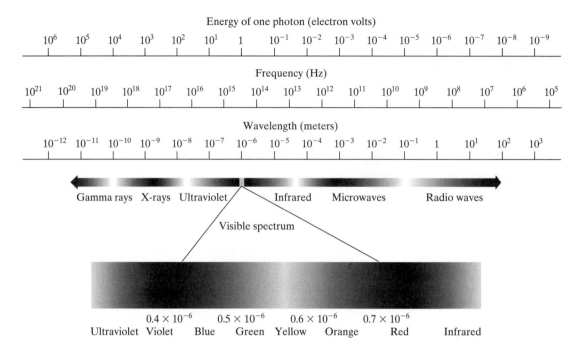

Energy of one photon (electron volts)

| 10^6 | 10^5 | 10^4 | 10^3 | 10^2 | 10^1 | 1 | 10^{-1} | 10^{-2} | 10^{-3} | 10^{-4} | 10^{-5} | 10^{-6} | 10^{-7} | 10^{-8} | 10^{-9} |

Frequency (Hz)

| 10^{21} | 10^{20} | 10^{19} | 10^{18} | 10^{17} | 10^{16} | 10^{15} | 10^{14} | 10^{13} | 10^{12} | 10^{11} | 10^{10} | 10^9 | 10^8 | 10^7 | 10^6 | 10^5 |

Wavelength (meters)

| 10^{-12} | 10^{-11} | 10^{-10} | 10^{-9} | 10^{-8} | 10^{-7} | 10^{-6} | 10^{-5} | 10^{-4} | 10^{-3} | 10^{-2} | 10^{-1} | 1 | 10^1 | 10^2 | 10^3 |

Gamma rays   X-rays   Ultraviolet   Infrared   Microwaves   Radio waves

Visible spectrum

| 0.4 × 10^{-6} | | 0.5 × 10^{-6} | | 0.6 × 10^{-6} | | 0.7 × 10^{-6} | |
Ultraviolet   Violet   Blue   Green   Yellow   Orange   Red   Infrared

FIGURE 2.10 The electromagnetic spectrum. The visible spectrum is shown zoomed to facilitate explanation, but note that the visible spectrum is a rather narrow portion of the EM spectrum.

where c is the speed of light (2.998×10^8 m/s). The energy of the various components of the electromagnetic spectrum is given by the expression

$$E = h\nu \qquad (2.2\text{-}2)$$

where h is Planck's constant. The units of wavelength are meters, with the terms *microns* (denoted μm and equal to 10^{-6} m) and *nanometers* (denoted nm and equal to 10^{-9} m) being used just as frequently. Frequency is measured in Hertz (Hz), with one Hertz being equal to one cycle of a sinusoidal wave per second. A commonly used unit of energy is the electron-volt.

Electromagnetic waves can be visualized as propagating sinusoidal waves with wavelength λ (Fig. 2.11), or they can be thought of as a stream of massless particles, each traveling in a wavelike pattern and moving at the speed of light. Each massless particle contains a certain amount (or bundle) of energy. Each

FIGURE 2.11
Graphical
representation of
one wavelength.

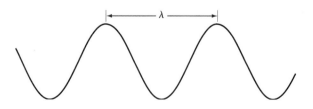

bundle of energy is called a *photon*. We see from Eq. (2.2-2) that energy is proportional to frequency, so the higher-frequency (shorter wavelength) electromagnetic phenomena carry more energy per photon. Thus, radio waves have photons with low energies, microwaves have more energy than radio waves, infrared still more, then visible, ultraviolet, X-rays, and finally gamma rays, the most energetic of all. This is the reason why gamma rays are so dangerous to living organisms.

Light is a particular type of electromagnetic radiation that can be sensed by the human eye. The visible (color) spectrum is shown expanded in Fig. 2.10 for the purpose of discussion (we consider color in much more detail in Chapter 6). The visible band of the electromagnetic spectrum spans the range from approximately 0.43 μm (violet) to about 0.79 μm (red). For convenience, the color spectrum is divided into six broad regions: violet, blue, green, yellow, orange, and red. No color (or other component of the electromagnetic spectrum) ends abruptly, but rather each range blends smoothly into the next, as shown in Fig. 2.10.

The colors that humans perceive in an object are determined by the nature of the light *reflected* from the object. A body that reflects light relatively balanced in all visible wavelengths appears white to the observer. However, a body that favors reflectance in a limited range of the visible spectrum exhibits some shades of color. For example, green objects reflect light with wavelengths primarily in the 500 to 570 nm range while absorbing most of the energy at other wavelengths.

Light that is void of color is called *monochromatic* (or *achromatic*) *light*. The only attribute of monochromatic light is its *intensity* or amount. Because the intensity of monochromatic light is perceived to vary from black to grays and finally to white, the term *gray level* is used commonly to denote monochromatic intensity. We use the terms *intensity* and *gray level* interchangeably in subsequent discussions. The range of measured values of monochromatic light from black to white is usually called the *gray scale*, and monochromatic images are frequently referred to as *gray-scale images*.

Chromatic (*color*) *light* spans the electromagnetic energy spectrum from approximately 0.43 to 0.79 μm, as noted previously. In addition to frequency, three basic quantities are used to describe the quality of a chromatic light source: radiance, luminance, and brightness. *Radiance* is the total amount of energy that flows from the light source, and it is usually measured in watts (W). *Luminance*, measured in lumens (lm), gives a measure of the amount of energy an observer *perceives* from a light source. For example, light emitted from a source operating in the far infrared region of the spectrum could have significant energy (radiance), but an observer would hardly perceive it; its luminance would be almost zero. Finally, as discussed in Section 2.1, *brightness* is a subjective descriptor of light perception that is practically impossible to measure. It embodies the achromatic notion of intensity and is one of the key factors in describing color sensation.

Continuing with the discussion of Fig. 2.10, we note that at the short-wavelength end of the electromagnetic spectrum, we have gamma rays and X-rays. As discussed in Section 1.3.1, gamma radiation is important for medical and astronomical imaging, and for imaging radiation in nuclear environments.

Hard (high-energy) X-rays are used in industrial applications. Chest and dental X-rays are in the lower energy (soft) end of the X-ray band. The soft X-ray band transitions into the far ultraviolet light region, which in turn blends with the visible spectrum at longer wavelengths. Moving still higher in wavelength, we encounter the infrared band, which radiates heat, a fact that makes it useful in imaging applications that rely on "heat signatures." The part of the infrared band close to the visible spectrum is called the *near-infrared* region. The opposite end of this band is called the *far-infrared* region. This latter region blends with the microwave band. This band is well known as the source of energy in microwave ovens, but it has many other uses, including communication and radar. Finally, the radio wave band encompasses television as well as AM and FM radio. In the higher energies, radio signals emanating from certain stellar bodies are useful in astronomical observations. Examples of images in most of the bands just discussed are given in Section 1.3.

In principle, if a sensor can be developed that is capable of detecting energy radiated by a band of the electromagnetic spectrum, we can image events of interest in that band. It is important to note, however, that the wavelength of an electromagnetic wave required to "see" an object must be of the same size as or smaller than the object. For example, a water molecule has a diameter on the order of 10^{-10} m. Thus, to study molecules, we would need a source capable of emitting in the far ultraviolet or soft X-ray region. This limitation, along with the physical properties of the sensor material, establishes the fundamental limits on the capability of imaging sensors, such as visible, infrared, and other sensors in use today.

Although imaging is based predominantly on energy radiated by electromagnetic waves, this is not the only method for image generation. For example, as discussed in Section 1.3.7, sound reflected from objects can be used to form ultrasonic images. Other major sources of digital images are electron beams for electron microscopy and synthetic images used in graphics and visualization.

2.3 Image Sensing and Acquisition

Most of the images in which we are interested are generated by the combination of an "illumination" source and the reflection or absorption of energy from that source by the elements of the "scene" being imaged. We enclose *illumination* and *scene* in quotes to emphasize the fact that they are considerably more general than the familiar situation in which a visible light source illuminates a common everyday 3-D (three-dimensional) scene. For example, the illumination may originate from a source of electromagnetic energy such as radar, infrared, or X-ray system. But, as noted earlier, it could originate from less traditional sources, such as ultrasound or even a computer-generated illumination pattern. Similarly, the scene elements could be familiar objects, but they can just as easily be molecules, buried rock formations, or a human brain. Depending on the nature of the source, illumination energy is reflected from, or transmitted through, objects. An example in the first category is light reflected from a planar surface. An example in the second category is when X-rays pass through a patient's body for the purpose of generating a diagnostic X-ray film. In some applications, the reflected or transmitted energy is focused onto a photoconverter (e.g., a phosphor screen), which converts the energy into visible light. Electron microscopy and some applications of gamma imaging use this approach.

Figure 2.12 shows the three principal sensor arrangements used to transform illumination energy into digital images. The idea is simple: Incoming energy is transformed into a voltage by the combination of input electrical power and sensor material that is responsive to the particular type of energy being detected. The output voltage waveform is the response of the sensor(s), and a digital quantity is obtained from each sensor by digitizing its response. In this section, we look at the principal modalities for image sensing and generation. Image digitizing is discussed in Section 2.4.

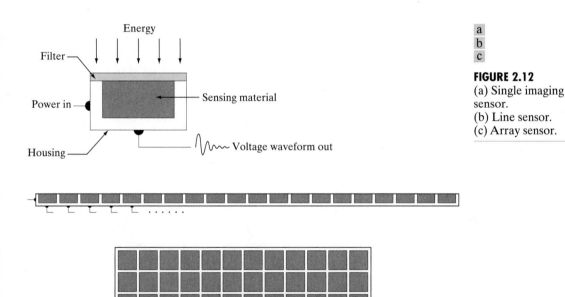

a
b
c

FIGURE 2.12
(a) Single imaging sensor.
(b) Line sensor.
(c) Array sensor.

FIGURE 2.13
Combining a
single sensor with
motion to
generate a 2-D
image.

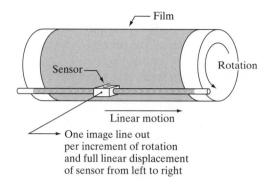

Film

Rotation

Sensor

Linear motion

One image line out
per increment of rotation
and full linear displacement
of sensor from left to right

2.3.1 Image Acquisition Using a Single Sensor

Figure 2.12(a) shows the components of a single sensor. Perhaps the most familiar sensor of this type is the photodiode, which is constructed of silicon materials and whose output voltage waveform is proportional to light. The use of a filter in front of a sensor improves selectivity. For example, a green (pass) filter in front of a light sensor favors light in the green band of the color spectrum. As a consequence, the sensor output will be stronger for green light than for other components in the visible spectrum.

In order to generate a 2-D image using a single sensor, there has to be relative displacements in both the *x*- and *y*-directions between the sensor and the area to be imaged. Figure 2.13 shows an arrangement used in high-precision scanning, where a film negative is mounted onto a drum whose mechanical rotation provides displacement in one dimension. The single sensor is mounted on a lead screw that provides motion in the perpendicular direction. Because mechanical motion can be controlled with high precision, this method is an inexpensive (but slow) way to obtain high-resolution images. Other similar mechanical arrangements use a flat bed, with the sensor moving in two linear directions. These types of mechanical digitizers sometimes are referred to as *microdensitometers*.

Another example of imaging with a single sensor places a laser source coincident with the sensor. Moving mirrors are used to control the outgoing beam in a scanning pattern and to direct the reflected laser signal onto the sensor. This arrangement can be used also to acquire images using strip and array sensors, which are discussed in the following two sections.

2.3.2 Image Acquisition Using Sensor Strips

A geometry that is used much more frequently than single sensors consists of an in-line arrangement of sensors in the form of a sensor strip, as Fig. 2.12(b) shows. The strip provides imaging elements in one direction. Motion perpendicular to the strip provides imaging in the other direction, as shown in Fig. 2.14(a). This is the type of arrangement used in most flat bed scanners. Sensing devices with 4000 or more in-line sensors are possible. In-line sensors are used routinely in airborne imaging applications, in which the imaging system is mounted on an aircraft that

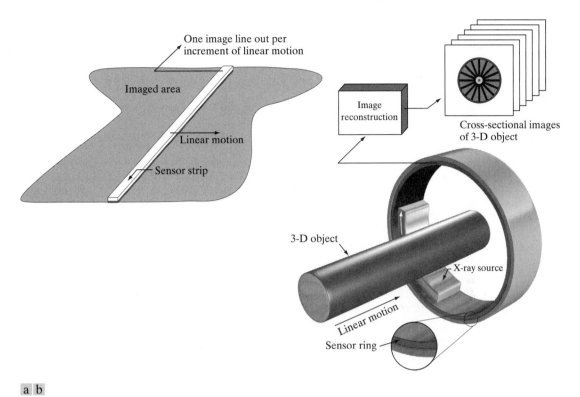

FIGURE 2.14 (a) Image acquisition using a linear sensor strip. (b) Image acquisition using a circular sensor strip.

flies at a constant altitude and speed over the geographical area to be imaged. One-dimensional imaging sensor strips that respond to various bands of the electromagnetic spectrum are mounted perpendicular to the direction of flight. The imaging strip gives one line of an image at a time, and the motion of the strip completes the other dimension of a two-dimensional image. Lenses or other focusing schemes are used to project the area to be scanned onto the sensors.

Sensor strips mounted in a ring configuration are used in medical and industrial imaging to obtain cross-sectional ("slice") images of 3-D objects, as Fig. 2.14(b) shows. A rotating X-ray source provides illumination and the sensors opposite the source collect the X-ray energy that passes through the object (the sensors obviously have to be sensitive to X-ray energy). This is the basis for medical and industrial computerized axial tomography (CAT) imaging as indicated in Sections 1.2 and 1.3.2. It is important to note that the output of the sensors must be processed by reconstruction algorithms whose objective is to transform the sensed data into meaningful cross-sectional images (see Section 5.11). In other words, images are not obtained directly from the sensors by motion alone; they require extensive processing. A 3-D digital volume consisting of stacked images is generated as the object is moved in a direction perpendicular to the sensor ring. Other modalities of imaging based on the CAT principle include magnetic resonance imaging (MRI) and positron emission tomography (PET). The illumination sources, sensors, and types of images are different, but conceptually they are very similar to the basic imaging approach shown in Fig. 2.14(b).

2.3.3 Image Acquisition Using Sensor Arrays

Figure 2.12(c) shows individual sensors arranged in the form of a 2-D array. Numerous electromagnetic and some ultrasonic sensing devices frequently are arranged in an array format. This is also the predominant arrangement found in digital cameras. A typical sensor for these cameras is a CCD array, which can be manufactured with a broad range of sensing properties and can be packaged in rugged arrays of 4000×4000 elements or more. CCD sensors are used widely in digital cameras and other light sensing instruments. The response of each sensor is proportional to the integral of the light energy projected onto the surface of the sensor, a property that is used in astronomical and other applications requiring low noise images. Noise reduction is achieved by letting the sensor integrate the input light signal over minutes or even hours. Because the sensor array in Fig. 2.12(c) is two-dimensional, its key advantage is that a complete image can be obtained by focusing the energy pattern onto the surface of the array. Motion obviously is not necessary, as is the case with the sensor arrangements discussed in the preceding two sections.

The principal manner in which array sensors are used is shown in Fig. 2.15. This figure shows the energy from an illumination source being reflected from a scene element (as mentioned at the beginning of this section, the energy also could be transmitted through the scene elements). The first function performed by the imaging system in Fig. 2.15(c) is to collect the incoming energy and focus it onto an image plane. If the illumination is light, the front end of the imaging system is an optical lens that projects the viewed scene onto the lens focal plane, as Fig. 2.15(d) shows. The sensor array, which is coincident with the focal plane, produces outputs proportional to the integral of the light received at each sensor. Digital and analog circuitry sweep these outputs and convert them to an analog signal, which is then digitized by another section of the imaging system. The output is a digital image, as shown diagrammatically in Fig. 2.15(e). Conversion of an image into digital form is the topic of Section 2.4.

In some cases, we image the source directly, as in obtaining images of the sun.

Image intensities can become negative during processing or as a result of interpretation. For example, in radar images objects moving toward a radar system often are interpreted as having negative velocities while objects moving away are interpreted as having positive velocities. Thus, a velocity image might be coded as having both positive and negative values. When storing and displaying images, we normally scale the intensities so that the smallest negative value becomes 0 (see Section 2.6.3 regarding intensity scaling).

2.3.4 A Simple Image Formation Model

As introduced in Section 1.1, we denote images by two-dimensional functions of the form $f(x, y)$. The value or amplitude of f at spatial coordinates (x, y) is a positive scalar quantity whose physical meaning is determined by the source of the image. When an image is generated from a physical process, its intensity values are proportional to energy radiated by a physical source (e.g., electromagnetic waves). As a consequence, $f(x, y)$ must be nonzero

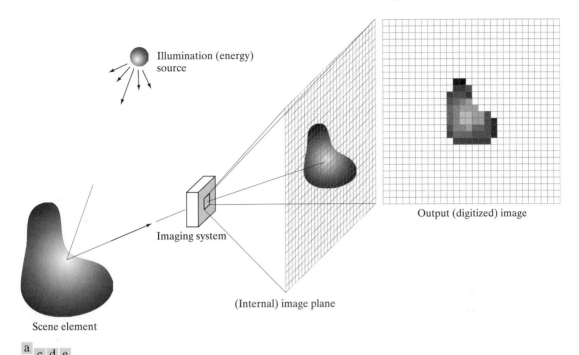

a
b c d e

FIGURE 2.15 An example of the digital image acquisition process. (a) Energy ("illumination") source. (b) An element of a scene. (c) Imaging system. (d) Projection of the scene onto the image plane. (e) Digitized image.

and finite; that is,

$$0 < f(x, y) < \infty \qquad (2.3\text{-}1)$$

   The function $f(x, y)$ may be characterized by two components: (1) the amount of source illumination incident on the scene being viewed, and (2) the amount of illumination reflected by the objects in the scene. Appropriately, these are called the *illumination* and *reflectance* components and are denoted by $i(x, y)$ and $r(x, y)$, respectively. The two functions combine as a product to form $f(x, y)$:

$$f(x, y) = i(x, y)r(x, y) \qquad (2.3\text{-}2)$$

where

$$0 < i(x, y) < \infty \qquad (2.3\text{-}3)$$

and

$$0 < r(x, y) < 1 \qquad (2.3\text{-}4)$$

Equation (2.3-4) indicates that reflectance is bounded by 0 (total absorption) and 1 (total reflectance). The nature of $i(x, y)$ is determined by the illumination source, and $r(x, y)$ is determined by the characteristics of the imaged objects. It is noted that these expressions also are applicable to images formed via transmission of the illumination through a medium, such as a chest X-ray.

In this case, we would deal with a *transmissivity* instead of a *reflectivity* function, but the limits would be the same as in Eq. (2.3-4), and the image function formed would be modeled as the product in Eq. (2.3-2).

EXAMPLE 2.1:
Some typical
values of
illumination and
reflectance.

■ The values given in Eqs. (2.3-3) and (2.3-4) are theoretical bounds. The following *average* numerical figures illustrate some typical ranges of $i(x, y)$ for visible light. On a clear day, the sun may produce in excess of 90,000 lm/m^2 of illumination on the surface of the Earth. This figure decreases to less than 10,000 lm/m^2 on a cloudy day. On a clear evening, a full moon yields about 0.1 lm/m^2 of illumination. The typical illumination level in a commercial office is about 1000 lm/m^2. Similarly, the following are typical values of $r(x, y)$: 0.01 for black velvet, 0.65 for stainless steel, 0.80 for flat-white wall paint, 0.90 for silver-plated metal, and 0.93 for snow. ■

Let the intensity (gray level) of a monochrome image at any coordinates (x_0, y_0) be denoted by

$$\ell = f(x_0, y_0) \tag{2.3-5}$$

From Eqs. (2.3-2) through (2.3-4), it is evident that ℓ lies in the range

$$L_{min} \le \ell \le L_{max} \tag{2.3-6}$$

In theory, the only requirement on L_{min} is that it be positive, and on L_{max} that it be finite. In practice, $L_{min} = i_{min} r_{min}$ and $L_{max} = i_{max} r_{max}$. Using the preceding average office illumination and range of reflectance values as guidelines, we may expect $L_{min} \approx 10$ and $L_{max} \approx 1000$ to be typical limits for indoor values in the absence of additional illumination.

The interval $[L_{min}, L_{max}]$ is called the *gray* (or *intensity*) *scale*. Common practice is to shift this interval numerically to the interval $[0, L - 1]$, where $\ell = 0$ is considered black and $\ell = L - 1$ is considered white on the gray scale. All intermediate values are shades of gray varying from black to white.

2.4 Image Sampling and Quantization

The discussion of
sampling in this section is
of an intuitive nature. We
consider this topic in
depth in Chapter 4.

From the discussion in the preceding section, we see that there are numerous ways to acquire images, but our objective in all is the same: to generate digital images from sensed data. The output of most sensors is a continuous voltage waveform whose amplitude and spatial behavior are related to the physical phenomenon being sensed. To create a digital image, we need to convert the continuous sensed data into digital form. This involves two processes: *sampling* and *quantization*.

2.4.1 Basic Concepts in Sampling and Quantization

The basic idea behind sampling and quantization is illustrated in Fig. 2.16. Figure 2.16(a) shows a continuous image f that we want to convert to digital form. An image may be continuous with respect to the x- and y-coordinates, and also in amplitude. To convert it to digital form, we have to sample the

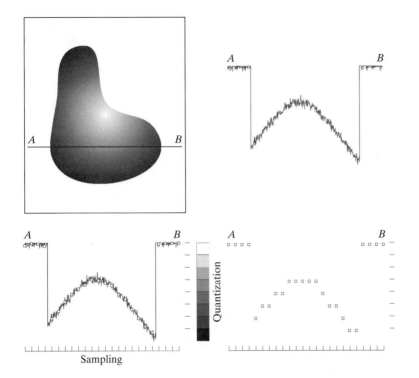

a b
c d

FIGURE 2.16
Generating a
digital image.
(a) Continuous
image. (b) A scan
line from A to B
in the continuous
image, used to
illustrate the
concepts of
sampling and
quantization.
(c) Sampling and
quantization.
(d) Digital
scan line.

function in both coordinates and in amplitude. Digitizing the coordinate values
is called *sampling*. Digitizing the amplitude values is called *quantization*.

   The one-dimensional function in Fig. 2.16(b) is a plot of amplitude (intensity
level) values of the continuous image along the line segment AB in Fig. 2.16(a).
The random variations are due to image noise. To sample this function, we take
equally spaced samples along line AB, as shown in Fig. 2.16(c). The spatial location of each sample is indicated by a vertical tick mark in the bottom part of the
figure. The samples are shown as small white squares superimposed on the function. The set of these discrete locations gives the sampled function. However, the
values of the samples still span (vertically) a continuous range of intensity values. In order to form a digital function, the intensity values also must be converted (*quantized*) into discrete quantities. The right side of Fig. 2.16(c) shows
the intensity scale divided into eight discrete intervals, ranging from black to
white. The vertical tick marks indicate the specific value assigned to each of the
eight intensity intervals. The continuous intensity levels are quantized by assigning one of the eight values to each sample. The assignment is made depending on
the vertical proximity of a sample to a vertical tick mark. The digital samples
resulting from both sampling and quantization are shown in Fig. 2.16(d). Starting at the top of the image and carrying out this procedure line by line produces
a two-dimensional digital image. It is implied in Fig. 2.16 that, in addition to the
number of discrete levels used, the accuracy achieved in quantization is highly
dependent on the noise content of the sampled signal.

   Sampling in the manner just described assumes that we have a continuous
image in both coordinate directions as well as in amplitude. In practice, the

method of sampling is determined by the sensor arrangement used to generate the image. When an image is generated by a single sensing element combined with mechanical motion, as in Fig. 2.13, the output of the sensor is quantized in the manner described above. However, spatial sampling is accomplished by selecting the number of individual mechanical increments at which we activate the sensor to collect data. Mechanical motion can be made very exact so, in principle, there is almost no limit as to how fine we can sample an image using this approach. In practice, limits on sampling accuracy are determined by other factors, such as the quality of the optical components of the system.

When a sensing strip is used for image acquisition, the number of sensors in the strip establishes the sampling limitations in one image direction. Mechanical motion in the other direction can be controlled more accurately, but it makes little sense to try to achieve sampling density in one direction that exceeds the sampling limits established by the number of sensors in the other. Quantization of the sensor outputs completes the process of generating a digital image.

When a sensing array is used for image acquisition, there is no motion and the number of sensors in the array establishes the limits of sampling in both directions. Quantization of the sensor outputs is as before. Figure 2.17 illustrates this concept. Figure 2.17(a) shows a continuous image projected onto the plane of an array sensor. Figure 2.17(b) shows the image after sampling and quantization. Clearly, the quality of a digital image is determined to a large degree by the number of samples and discrete intensity levels used in sampling and quantization. However, as we show in Section 2.4.3, image content is also an important consideration in choosing these parameters.

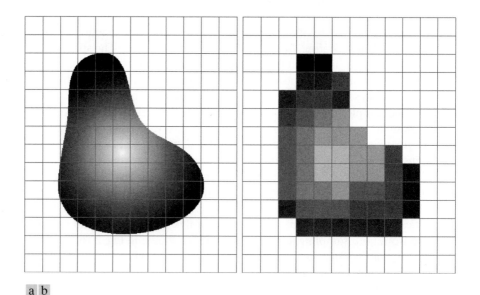

a b

FIGURE 2.17 (a) Continuous image projected onto a sensor array. (b) Result of image sampling and quantization.

2.4.2 Representing Digital Images

Let $f(s, t)$ represent a continuous image function of two continuous variables, s and t. We convert this function into a *digital image* by sampling and quantization, as explained in the previous section. Suppose that we sample the continuous image into a 2-D array, $f(x, y)$, containing M rows and N columns, where (x, y) are discrete coordinates. For notational clarity and convenience, we use integer values for these discrete coordinates: $x = 0, 1, 2, \ldots, M - 1$ and $y = 0, 1, 2, \ldots, N - 1$. Thus, for example, the value of the digital image at the origin is $f(0, 0)$, and the next coordinate value along the first row is $f(0, 1)$. Here, the notation $(0, 1)$ is used to signify the second sample along the first row. It *does not* mean that these are the values of the physical coordinates when the image was sampled. In general, the value of the image at any coordinates (x, y) is denoted $f(x, y)$, where x and y are integers. The section of the real plane spanned by the coordinates of an image is called the *spatial domain*, with x and y being referred to as *spatial variables* or *spatial coordinates*.

As Fig. 2.18 shows, there are three basic ways to represent $f(x, y)$. Figure 2.18(a) is a plot of the function, with two axes determining spatial location

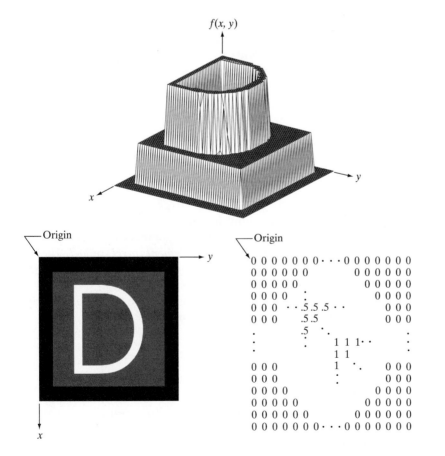

a
b c

FIGURE 2.18
(a) Image plotted as a surface.
(b) Image displayed as a visual intensity array.
(c) Image shown as a 2-D numerical array (0, .5, and 1 represent black, gray, and white, respectively).

and the third axis being the values of f (intensities) as a function of the two spatial variables x and y. Although we can infer the structure of the image in this example by looking at the plot, complex images generally are too detailed and difficult to interpret from such plots. This representation is useful when working with gray-scale sets whose elements are expressed as triplets of the form (x, y, z), where x and y are spatial coordinates and z is the value of f at coordinates (x, y). We work with this representation in Section 2.6.4.

The representation in Fig. 2.18(b) is much more common. It shows $f(x, y)$ as it would appear on a monitor or photograph. Here, the intensity of each point is proportional to the value of f at that point. In this figure, there are only three equally spaced intensity values. If the intensity is normalized to the interval $[0, 1]$, then each point in the image has the value 0, 0.5, or 1. A monitor or printer simply converts these three values to black, gray, or white, respectively, as Fig. 2.18(b) shows. The third representation is simply to display the numerical values of $f(x, y)$ as an array (matrix). In this example, f is of size 600×600 elements, or 360,000 numbers. Clearly, printing the complete array would be cumbersome and convey little information. When developing algorithms, however, this representation is quite useful when only parts of the image are printed and analyzed as numerical values. Figure 2.18(c) conveys this concept graphically.

We conclude from the previous paragraph that the representations in Figs. 2.18(b) and (c) are the most useful. Image displays allow us to view results at a glance. Numerical arrays are used for processing and algorithm development. In equation form, we write the representation of an $M \times N$ numerical array as

$$f(x, y) = \begin{bmatrix} f(0, 0) & f(0, 1) & \cdots & f(0, N - 1) \\ f(1, 0) & f(1, 1) & \cdots & f(1, N - 1) \\ \vdots & \vdots & & \vdots \\ f(M - 1, 0) & f(M - 1, 1) & \cdots & f(M - 1, N - 1) \end{bmatrix} \qquad (2.4\text{-}1)$$

Both sides of this equation are equivalent ways of expressing a digital image quantitatively. The right side is a matrix of real numbers. Each element of this matrix is called an *image element, picture element, pixel,* or *pel*. The terms *image* and *pixel* are used throughout the book to denote a digital image and its elements.

In some discussions it is advantageous to use a more traditional matrix notation to denote a digital image and its elements:

$$\mathbf{A} = \begin{bmatrix} a_{0, 0} & a_{0, 1} & \cdots & a_{0, N-1} \\ a_{1, 0} & a_{1, 1} & \cdots & a_{1, N-1} \\ \vdots & \vdots & & \vdots \\ a_{M-1, 0} & a_{M-1, 1} & \cdots & a_{M-1, N-1} \end{bmatrix} \qquad (2.4\text{-}2)$$

Clearly, $a_{ij} = f(x = i, y = j) = f(i, j)$, so Eqs. (2.4-1) and (2.4-2) are identical matrices. We can even represent an image as a vector, $\mathbf{v}$. For example, a column vector of size $MN \times 1$ is formed by letting the first M elements of $\mathbf{v}$ be the first column of $\mathbf{A}$, the next M elements be the second column, and so on. Alternatively, we can use the rows instead of the columns of $\mathbf{A}$ to form such a vector. Either representation is valid, as long as we are consistent.

Returning briefly to Fig. 2.18, note that the origin of a digital image is at the top left, with the positive x-axis extending downward and the positive y-axis extending to the right. This is a conventional representation based on the fact that many image displays (e.g., TV monitors) sweep an image starting at the top left and moving to the right one row at a time. More important is the fact that the first element of a matrix is by convention at the top left of the array, so choosing the origin of $f(x, y)$ at that point makes sense mathematically. Keep in mind that this representation is the standard right-handed Cartesian coordinate system with which you are familiar.[†] We simply show the axes pointing downward and to the right, instead of to the right and up.

Expressing sampling and quantization in more formal mathematical terms can be useful at times. Let Z and R denote the set of integers and the set of real numbers, respectively. The sampling process may be viewed as partitioning the xy-plane into a grid, with the coordinates of the center of each cell in the grid being a pair of elements from the Cartesian product Z^2, which is the set of all ordered pairs of elements (z_i, z_j), with z_i and z_j being integers from Z. Hence, $f(x, y)$ is a digital image if (x, y) are integers from Z^2 and f is a function that assigns an intensity value (that is, a real number from the set of real numbers, R) to each distinct pair of coordinates (x, y). This functional assignment is the quantization process described earlier. If the intensity levels also are integers (as usually is the case in this and subsequent chapters), Z replaces R, and a digital image then becomes a 2-D function whose coordinates and amplitude values are integers.

This digitization process requires that decisions be made regarding the values for M, N, and for the number, L, of discrete intensity levels. There are no restrictions placed on M and N, other than they have to be positive integers. However, due to storage and quantizing hardware considerations, the number of intensity levels typically is an integer power of 2:

$$L = 2^k \qquad (2.4-3)$$

We assume that the discrete levels are equally spaced and that they are integers in the interval $[0, L - 1]$. Sometimes, the range of values spanned by the gray scale is referred to informally as the dynamic range. This is a term used in different ways in different fields. Here, we define the *dynamic range* of an imaging system to be the ratio of the maximum measurable intensity to the minimum

Often, it is useful for computation or for algorithm development purposes to scale the L intensity values to the range $[0, 1]$, in which case they cease to be integers. However, in most cases these values are scaled back to the integer range $[0, L - 1]$ for image storage and display.

[†]Recall that a right-handed coordinate system is such that, when the index of the right hand points in the direction of the positive x-axis and the middle finger points in the (perpendicular) direction of the positive y-axis, the thumb points up. As Fig. 2.18(a) shows, this indeed is the case in our image coordinate system.

FIGURE 2.19 An image exhibiting saturation and noise. Saturation is the highest value beyond which all intensity levels are clipped (note how the entire saturated area has a high, *constant* intensity level). Noise in this case appears as a grainy texture pattern. Noise, especially in the darker regions of an image (e.g., the stem of the rose) masks the lowest detectable true intensity level.

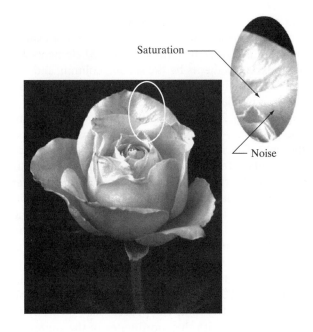

detectable intensity level in the system. As a rule, the upper limit is determined by *saturation* and the lower limit by *noise* (see Fig. 2.19). Basically, dynamic range establishes the lowest and highest intensity levels that a system can represent and, consequently, that an image can have. Closely associated with this concept is image *contrast*, which we define as the difference in intensity between the highest and lowest intensity levels in an image. When an appreciable number of pixels in an image have a high dynamic range, we can expect the image to have high contrast. Conversely, an image with low dynamic range typically has a dull, washed-out gray look. We discuss these concepts in more detail in Chapter 3.

The number, b, of bits required to store a digitized image is

$$b = M \times N \times k \qquad (2.4\text{-}4)$$

When $M = N$, this equation becomes

$$b = N^2 k \qquad (2.4\text{-}5)$$

Table 2.1 shows the number of bits required to store square images with various values of N and k. The number of intensity levels corresponding to each value of k is shown in parentheses. When an image can have 2^k intensity levels, it is common practice to refer to the image as a "k-bit image." For example, an image with 256 possible discrete intensity values is called an 8-bit image. Note that storage requirements for 8-bit images of size 1024×1024 and higher are not insignificant.

TABLE 2.1
Number of storage bits for various values of N and k. L is the number of intensity levels.

N/k	1 ($L = 2$)	2 ($L = 4$)	3 ($L = 8$)	4 ($L = 16$)	5 ($L = 32$)	6 ($L = 64$)	7 ($L = 128$)	8 ($L = 256$)
32	1,024	2,048	3,072	4,096	5,120	6,144	7,168	8,192
64	4,096	8,192	12,288	16,384	20,480	24,576	28,672	32,768
128	16,384	32,768	49,152	65,536	81,920	98,304	114,688	131,072
256	65,536	131,072	196,608	262,144	327,680	393,216	458,752	524,288
512	262,144	524,288	786,432	1,048,576	1,310,720	1,572,864	1,835,008	2,097,152
1024	1,048,576	2,097,152	3,145,728	4,194,304	5,242,880	6,291,456	7,340,032	8,388,608
2048	4,194,304	8,388,608	12,582,912	16,777,216	20,971,520	25,165,824	29,369,128	33,554,432
4096	16,777,216	33,554,432	50,331,648	67,108,864	83,886,080	100,663,296	117,440,512	134,217,728
8192	67,108,864	134,217,728	201,326,592	268,435,456	335,544,320	402,653,184	469,762,048	536,870,912

2.4.3 Spatial and Intensity Resolution

Intuitively, spatial resolution is a measure of the smallest discernible detail in an image. Quantitatively, *spatial resolution* can be stated in a number of ways, with *line pairs per unit distance*, and *dots (pixels) per unit distance* being among the most common measures. Suppose that we construct a chart with alternating black and white vertical lines, each of width W units (W can be less than 1). The width of a *line pair* is thus $2W$, and there are $1/2W$ line pairs per unit distance. For example, if the width of a line is 0.1 mm, there are 5 line pairs per unit distance (mm). A widely used definition of image resolution is the largest number of *discernible* line pairs per unit distance (e.g., 100 line pairs per mm). Dots per unit distance is a measure of image resolution used commonly in the printing and publishing industry. In the U.S., this measure usually is expressed as *dots per inch* (dpi). To give you an idea of quality, newspapers are printed with a resolution of 75 dpi, magazines at 133 dpi, glossy brochures at 175 dpi, and the book page at which you are presently looking is printed at 2400 dpi.

The key point in the preceding paragraph is that, to be meaningful, measures of spatial resolution must be stated with respect to spatial units. Image size by itself does not tell the complete story. To say that an image has, say, a resolution 1024 × 1024 pixels is not a meaningful statement without stating the spatial dimensions encompassed by the image. Size by itself is helpful only in making comparisons between imaging capabilities. For example, a digital camera with a 20-megapixel CCD imaging chip can be expected to have a higher capability to resolve detail than an 8-megapixel camera, assuming that both cameras are equipped with comparable lenses and the comparison images are taken at the same distance.

Intensity resolution similarly refers to the smallest discernible change in intensity level. We have considerable discretion regarding the number of samples used to generate a digital image, but this is not true regarding the number of intensity levels. Based on hardware considerations, the number of intensity levels usually is an integer power of two, as mentioned in the previous section. The most common number is 8 bits, with 16 bits being used in some applications in which enhancement of specific intensity ranges is necessary. Intensity quantization using 32 bits is rare. Sometimes one finds systems that can digitize the intensity levels of an image using 10 or 12 bits, but these are the exception, rather than the rule. Unlike spatial resolution, which must be based on a per unit of distance basis to be meaningful, it is common practice to refer to the number of bits used to quantize intensity as the *intensity resolution*. For example, it is common to say that an image whose intensity is quantized into 256 levels has 8 bits of intensity resolution. Because true discernible changes in intensity are influenced not only by noise and saturation values but also by the capabilities of human perception (see Section 2.1), saying than an image has 8 bits of intensity resolution is nothing more than a statement regarding the ability of an 8-bit system to quantize intensity in fixed increments of 1/256 units of intensity amplitude.

The following two examples illustrate individually the comparative effects of image size and intensity resolution on discernable detail. Later in this section, we discuss how these two parameters interact in determining perceived image quality.

EXAMPLE 2.2:
Illustration of the effects of reducing image spatial resolution.

■ Figure 2.20 shows the effects of reducing spatial resolution in an image. The images in Figs. 2.20(a) through (d) are shown in 1250, 300, 150, and 72 dpi, respectively. Naturally, the lower resolution images are smaller than the original. For example, the original image is of size 3692×2812 pixels, but the 72 dpi image is an array of size 213×162. In order to facilitate comparisons, all the smaller images were zoomed back to the original size (the method used for zooming is discussed in Section 2.4.4). This is somewhat equivalent to "getting closer" to the smaller images so that we can make comparable statements about visible details.

There are some small visual differences between Figs. 2.20(a) and (b), the most notable being a slight distortion in the large black needle. For the most part, however, Fig. 2.20(b) is quite acceptable. In fact, 300 dpi is the typical minimum image spatial resolution used for book publishing, so one would not expect to see much difference here. Figure 2.20(c) begins to show visible degradation (see, for example, the round edges of the chronometer and the small needle pointing to 60 on the right side). Figure 2.20(d) shows degradation that is visible in most features of the image. As we discuss in Section 4.5.4, when printing at such low resolutions, the printing and publishing industry uses a number of "tricks" (such as locally varying the pixel size) to produce much better results than those in Fig. 2.20(d). Also, as we show in Section 2.4.4, it is possible to improve on the results of Fig. 2.20 by the choice of interpolation method used.   ■

a b
c d

FIGURE 2.20 Typical effects of reducing spatial resolution. Images shown at: (a) 1250 dpi, (b) 300 dpi, (c) 150 dpi, and (d) 72 dpi. The thin black borders were added for clarity. They are not part of the data.

EXAMPLE 2.3:
Typical effects of varying the number of intensity levels in a digital image.

■ In this example, we keep the number of samples constant and reduce the number of intensity levels from 256 to 2, in integer powers of 2. Figure 2.21(a) is a 452 × 374 CT projection image, displayed with $k = 8$ (256 intensity levels). Images such as this are obtained by fixing the X-ray source in one position, thus producing a 2-D image in any desired direction. Projection images are used as guides to set up the parameters for a CT scanner, including tilt, number of slices, and range.

Figures 2.21(b) through (h) were obtained by reducing the number of bits from $k = 7$ to $k = 1$ while keeping the image size constant at 452 × 374 pixels. The 256-, 128-, and 64-level images are visually identical for all practical purposes. The 32-level image in Fig. 2.21(d), however, has an imperceptible set of

a b
c d

FIGURE 2.21
(a) 452 × 374, 256-level image. (b)–(d) Image displayed in 128, 64, and 32 intensity levels, while keeping the image size constant.

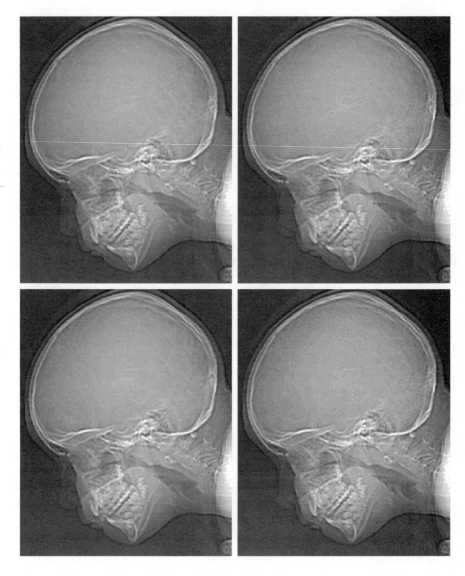

very fine ridge-like structures in areas of constant or nearly constant intensity (particularly in the skull). This effect, caused by the use of an insufficient number of intensity levels in smooth areas of a digital image, is called *false contouring*, so called because the ridges resemble topographic contours in a map. False contouring generally is quite visible in images displayed using 16 or less uniformly spaced intensity levels, as the images in Figs. 2.21(e) through (h) show.

As a very rough rule of thumb, and assuming integer powers of 2 for convenience, images of size 256×256 pixels with 64 intensity levels and printed on a size format on the order of 5×5 cm are about the lowest spatial and intensity resolution images that can be expected to be reasonably free of objectionable sampling checkerboards and false contouring.   ■

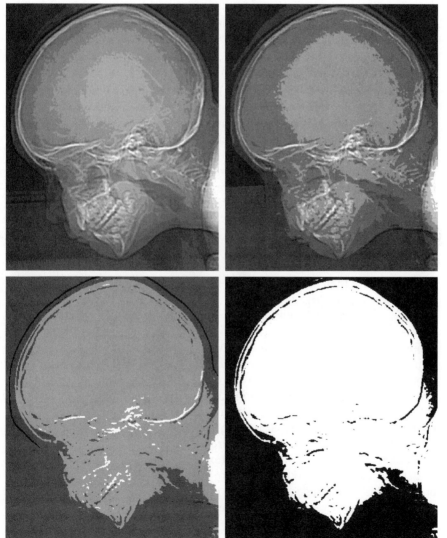

e f
g h

FIGURE 2.21
(*Continued*)
(e)–(h) Image displayed in 16, 8, 4, and 2 intensity levels. (Original courtesy of Dr. David R. Pickens, Department of Radiology & Radiological Sciences, Vanderbilt University Medical Center.)

a b c

FIGURE 2.22 (a) Image with a low level of detail. (b) Image with a medium level of detail. (c) Image with a relatively large amount of detail. (Image (b) courtesy of the Massachusetts Institute of Technology.)

The results in Examples 2.2 and 2.3 illustrate the effects produced on image quality by varying N and k independently. However, these results only partially answer the question of how varying N and k affects images because we have not considered yet any relationships that might exist between these two parameters. An early study by Huang [1965] attempted to quantify experimentally the effects on image quality produced by varying N and k simultaneously. The experiment consisted of a set of subjective tests. Images similar to those shown in Fig. 2.22 were used. The woman's face is representative of an image with relatively little detail; the picture of the cameraman contains an intermediate amount of detail; and the crowd picture contains, by comparison, a large amount of detail.

Sets of these three types of images were generated by varying N and k, and observers were then asked to rank them according to their subjective quality. Results were summarized in the form of so-called *isopreference curves* in the Nk-plane. (Figure 2.23 shows average isopreference curves representative of curves corresponding to the images in Fig. 2.22.) Each point in the Nk-plane represents an image having values of N and k equal to the coordinates of that point. Points lying on an isopreference curve correspond to images of equal subjective quality. It was found in the course of the experiments that the isopreference curves tended to shift right and upward, but their shapes in each of the three image categories were similar to those in Fig. 2.23. This is not unexpected, because a shift up and right in the curves simply means larger values for N and k, which implies better picture quality.

The key point of interest in the context of the present discussion is that isopreference curves tend to become more vertical as the detail in the image increases. This result suggests that for images with a large amount of detail only a few intensity levels may be needed. For example, the isopreference curve in Fig. 2.23 corresponding to the crowd is nearly vertical. This indicates that, for a fixed value of N, the perceived quality for this type of image is

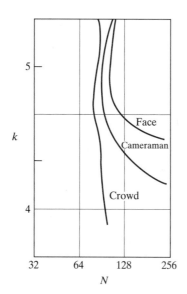

FIGURE 2.23
Typical
isopreference
curves for the
three types of
images in
Fig. 2.22.

nearly independent of the number of intensity levels used (for the range of intensity levels shown in Fig. 2.23). It is of interest also to note that perceived quality in the other two image categories remained the same in some intervals in which the number of samples was increased, but the number of intensity levels actually decreased. The most likely reason for this result is that a decrease in k tends to increase the apparent contrast, a visual effect that humans often perceive as improved quality in an image.

2.4.4 Image Interpolation

Interpolation is a basic tool used extensively in tasks such as zooming, shrinking, rotating, and geometric corrections. Our principal objective in this section is to introduce interpolation and apply it to image resizing (shrinking and zooming), which are basically image *resampling* methods. Uses of interpolation in applications such as rotation and geometric corrections are discussed in Section 2.6.5. We also return to this topic in Chapter 4, where we discuss image resampling in more detail.

Fundamentally, *interpolation* is the process of using known data to estimate values at unknown locations. We begin the discussion of this topic with a simple example. Suppose that an image of size 500×500 pixels has to be enlarged 1.5 times to 750×750 pixels. A simple way to visualize zooming is to create an imaginary 750×750 grid with the same pixel spacing as the original, and then shrink it so that it fits exactly over the original image. Obviously, the pixel spacing in the shrunken 750×750 grid will be less than the pixel spacing in the original image. To perform intensity-level assignment for any point in the overlay, we look for its closest pixel in the original image and assign the intensity of that pixel to the new pixel in the 750×750 grid. When we are finished assigning intensities to all the points in the overlay grid, we expand it to the original specified size to obtain the zoomed image.

The method just discussed is called *nearest neighbor interpolation* because it assigns to each new location the intensity of its nearest neighbor in the original image (pixel neighborhoods are discussed formally in Section 2.5). This approach is simple but, as we show later in this section, it has the tendency to produce undesirable artifacts, such as severe distortion of straight edges. For this reason, it is used infrequently in practice. A more suitable approach is *bilinear interpolation*, in which we use the four nearest neighbors to estimate the intensity at a given location. Let (x, y) denote the coordinates of the location to which we want to assign an intensity value (think of it as a point of the grid described previously), and let $v(x, y)$ denote that intensity value. For bilinear interpolation, the assigned value is obtained using the equation

$$v(x, y) = ax + by + cxy + d \qquad (2.4\text{-}6)$$

Contrary to what the name suggests, note that bilinear interpolation is not *linear because of the* xy *term.*

where the four coefficients are determined from the four equations in four unknowns that can be written using the four nearest neighbors of point (x, y). As you will see shortly, bilinear interpolation gives much better results than nearest neighbor interpolation, with a modest increase in computational burden.

The next level of complexity is *bicubic interpolation*, which involves the sixteen nearest neighbors of a point. The intensity value assigned to point (x, y) is obtained using the equation

$$v(x, y) = \sum_{i=0}^{3} \sum_{j=0}^{3} a_{ij} x^i y^j \qquad (2.4\text{-}7)$$

where the sixteen coefficients are determined from the sixteen equations in sixteen unknowns that can be written using the sixteen nearest neighbors of point (x, y). Observe that Eq. (2.4-7) reduces in form to Eq. (2.4-6) if the limits of both summations in the former equation are 0 to 1. Generally, bicubic interpolation does a better job of preserving fine detail than its bilinear counterpart. Bicubic interpolation is the standard used in commercial image editing programs, such as Adobe Photoshop and Corel Photopaint.

EXAMPLE 2.4:
Comparison of interpolation approaches for image shrinking and zooming.

■ Figure 2.24(a) is the same image as Fig. 2.20(d), which was obtained by reducing the resolution of the 1250 dpi image in Fig. 2.20(a) to 72 dpi (the size shrank from the original size of 3692 × 2812 to 213 × 162 pixels) and then zooming the reduced image back to its original size. To generate Fig. 2.20(d) we used nearest neighbor interpolation both to shrink and zoom the image. As we commented before, the result in Fig. 2.24(a) is rather poor. Figures 2.24(b) and (c) are the results of repeating the same procedure but using, respectively, bilinear and bicubic interpolation for both shrinking and zooming. The result obtained by using bilinear interpolation is a significant improvement over nearest neighbor interpolation. The bicubic result is slightly sharper than the bilinear image. Figure 2.24(d) is the same as Fig. 2.20(c), which was obtained using nearest neighbor interpolation for both shrinking and zooming. We commented in discussing that figure that reducing the resolution to 150 dpi began showing degradation in the image. Figures 2.24(e) and (f) show the results of using

a b c
d e f

FIGURE 2.24 (a) Image reduced to 72 dpi and zoomed back to its original size (3692 × 2812 pixels) using nearest neighbor interpolation. This figure is the same as Fig. 2.20(d). (b) Image shrunk and zoomed using bilinear interpolation. (c) Same as (b) but using bicubic interpolation. (d)–(f) Same sequence, but shrinking down to 150 dpi instead of 72 dpi [Fig. 2.24(d) is the same as Fig. 2.20(c)]. Compare Figs. 2.24(e) and (f), especially the latter, with the original image in Fig. 2.20(a).

bilinear and bicubic interpolation, respectively, to shrink and zoom the image. In spite of a reduction in resolution from 1250 to 150, these last two images compare reasonably favorably with the original, showing once again the power of these two interpolation methods. As before, bicubic interpolation yielded slightly sharper results. ■

It is possible to use more neighbors in interpolation, and there are more complex techniques, such as using splines and wavelets, that in some instances can yield better results than the methods just discussed. While preserving fine detail is an exceptionally important consideration in image generation for 3-D graphics (Watt [1993], Shirley [2002]) and in medical image processing (Lehmann et al. [1999]), the extra computational burden seldom is justifiable for general-purpose digital image processing, where bilinear or bicubic interpolation typically are the methods of choice.

2.5 Some Basic Relationships between Pixels

In this section, we consider several important relationships between pixels in a digital image. As mentioned before, an image is denoted by $f(x, y)$. When referring in this section to a particular pixel, we use lowercase letters, such as p and q.

2.5.1 Neighbors of a Pixel

A pixel p at coordinates (x, y) has four *horizontal* and *vertical* neighbors whose coordinates are given by

$$(x + 1, y), (x - 1, y), (x, y + 1), (x, y - 1)$$

This set of pixels, called the 4-*neighbors* of p, is denoted by $N_4(p)$. Each pixel is a unit distance from (x, y), and some of the neighbor locations of p lie outside the digital image if (x, y) is on the border of the image. We deal with this issue in Chapter 3.

The four *diagonal* neighbors of p have coordinates

$$(x + 1, y + 1), (x + 1, y - 1), (x - 1, y + 1), (x - 1, y - 1)$$

and are denoted by $N_D(p)$. These points, together with the 4-neighbors, are called the 8-*neighbors* of p, denoted by $N_8(p)$. As before, some of the neighbor locations in $N_D(p)$ and $N_8(p)$ fall outside the image if (x, y) is on the border of the image.

2.5.2 Adjacency, Connectivity, Regions, and Boundaries

Let V be the set of intensity values used to define adjacency. In a binary image, $V = \{1\}$ if we are referring to adjacency of pixels with value 1. In a gray-scale image, the idea is the same, but set V typically contains more elements. For example, in the adjacency of pixels with a range of possible intensity values 0 to 255, set V could be any subset of these 256 values. We consider three types of adjacency:

We use the symbols $\cap$ and $\cup$ to denote set intersection and union, respectively. Given sets A and B, recall that their *intersection* is the set of elements that are members of both A and B. The *union* of these two sets is the set of elements that are members of A, of B, or of both. We discuss sets in more detail in Section 2.6.4.

(a) 4-*adjacency*. Two pixels p and q with values from V are 4-adjacent if q is in the set $N_4(p)$.

(b) 8-*adjacency*. Two pixels p and q with values from V are 8-adjacent if q is in the set $N_8(p)$.

(c) *m-adjacency* (mixed adjacency). Two pixels p and q with values from V are *m*-adjacent if

   (i) q is in $N_4(p)$, *or*

   (ii) q is in $N_D(p)$ *and* the set $N_4(p) \cap N_4(q)$ has no pixels whose values are from V.

Mixed adjacency is a modification of 8-adjacency. It is introduced to eliminate the ambiguities that often arise when 8-adjacency is used. For example, consider the pixel arrangement shown in Fig. 2.25(a) for $V = \{1\}$. The three pixels at the top of Fig. 2.25(b) show multiple (ambiguous) 8-adjacency, as indicated by the dashed lines. This ambiguity is removed by using m-adjacency, as shown in Fig. 2.25(c).

A *(digital) path* (or *curve*) from pixel p with coordinates (x, y) to pixel q with coordinates (s, t) is a sequence of distinct pixels with coordinates

$$(x_0, y_0), (x_1, y_1), \ldots, (x_n, y_n)$$

where $(x_0, y_0) = (x, y)$, $(x_n, y_n) = (s, t)$, and pixels (x_i, y_i) and (x_{i-1}, y_{i-1}) are adjacent for $1 \le i \le n$. In this case, n is the *length* of the path. If $(x_0, y_0) = (x_n, y_n)$, the path is a *closed* path. We can define 4-, 8-, or m-paths depending on the type of adjacency specified. For example, the paths shown in Fig. 2.25(b) between the top right and bottom right points are 8-paths, and the path in Fig. 2.25(c) is an m-path.

Let S represent a subset of pixels in an image. Two pixels p and q are said to be *connected* in S if there exists a path between them consisting entirely of pixels in S. For any pixel p in S, the *set* of pixels that are connected to it in S is called a *connected component* of S. If it only has one connected component, then set S is called a *connected set*.

Let R be a subset of pixels in an image. We call R a *region* of the image if R is a connected set. Two regions, R_i and R_j are said to be *adjacent* if their union forms a connected set. Regions that are not adjacent are said to be *disjoint*. We consider 4- and 8-adjacency when referring to regions. For our definition to make sense, the type of adjacency used must be specified. For example, the two regions (of 1s) in Fig. 2.25(d) are adjacent only if 8-adjacency is used (according to the definition in the previous paragraph, a 4-path between the two regions does not exist, so their union is not a connected set).

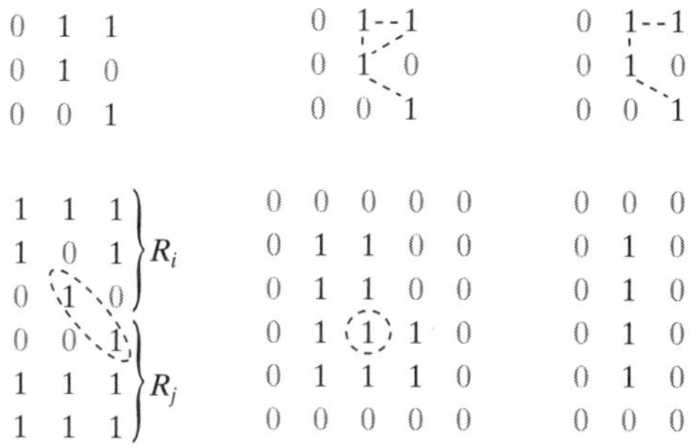

a b c
d e f

FIGURE 2.25 (a) An arrangement of pixels. (b) Pixels that are 8-adjacent (adjacency is shown by dashed lines; note the ambiguity). (c) m-adjacency. (d) Two regions (of 1s) that are adjacent if 8-adjecency is used. (e) The circled point is part of the boundary of the 1-valued pixels only if 8-adjacency between the region and background is used. (f) The inner boundary of the 1-valued region does not form a closed path, but its outer boundary does.

Suppose that an image contains K disjoint regions, $R_k, k = 1, 2, \ldots, K$, none of which touches the image border.[†] Let R_u denote the union of all the K regions, and let $(R_u)^c$ denote its complement (recall that the *complement* of a set S is the set of points that are not in S). We call all the points in R_u the *foreground*, and all the points in $(R_u)^c$ the *background* of the image.

The *boundary* (also called the *border* or *contour*) of a region R is the set of points that are adjacent to points in the complement of R. Said another way, the border of a region is the set of pixels in the region that have at least one background neighbor. Here again, we must specify the connectivity being used to define adjacency. For example, the point circled in Fig. 2.25(e) is not a member of the border of the 1-valued region if 4-connectivity is used between the region and its background. As a rule, adjacency between points in a region and its background is defined in terms of 8-connectivity to handle situations like this.

The preceding definition sometimes is referred to as the *inner border* of the region to distinguish it from its *outer border*, which is the corresponding border in the background. This distinction is important in the development of border-following algorithms. Such algorithms usually are formulated to follow the outer boundary in order to guarantee that the result will form a closed path. For instance, the inner border of the 1-valued region in Fig. 2.25(f) is the region itself. This border does not satisfy the definition of a closed path given earlier. On the other hand, the outer border of the region does form a closed path around the region.

If R happens to be an entire image (which we recall is a rectangular set of pixels), then its boundary is defined as the set of pixels in the first and last rows and columns of the image. This extra definition is required because an image has no neighbors beyond its border. Normally, when we refer to a region, we are referring to a subset of an image, and any pixels in the boundary of the region that happen to coincide with the border of the image are included implicitly as part of the region boundary.

The concept of an *edge* is found frequently in discussions dealing with regions and boundaries. There is a key difference between these concepts, however. The boundary of a finite region forms a closed path and is thus a "global" concept. As discussed in detail in Chapter 10, edges are formed from pixels with derivative values that exceed a preset threshold. Thus, the idea of an edge is a "local" concept that is based on a measure of intensity-level discontinuity at a point. It is possible to link edge points into edge segments, and sometimes these segments are linked in such a way that they correspond to boundaries, but this is not always the case. The one exception in which edges and boundaries correspond is in binary images. Depending on the type of connectivity and edge operators used (we discuss these in Chapter 10), the edge extracted from a binary region will be the same as the region boundary.

[†]We make this assumption to avoid having to deal with special cases. This is done without loss of generality because if one or more regions touch the border of an image, we can simply pad the image with a 1-pixel-wide border of background values.

This is intuitive. Conceptually, until we arrive at Chapter 10, it is helpful to think of edges as intensity discontinuities and boundaries as closed paths.

2.5.3 Distance Measures

For pixels p, q, and z, with coordinates (x, y), (s, t), and (v, w), respectively, D is a *distance function* or *metric* if

(a) $D(p, q) \geq 0$ $(D(p, q) = 0$ iff $p = q)$,
(b) $D(p, q) = D(q, p)$, and
(c) $D(p, z) \leq D(p, q) + D(q, z)$.

The *Euclidean distance* between p and q is defined as

$$D_e(p, q) = \left[(x - s)^2 + (y - t)^2\right]^{\frac{1}{2}} \tag{2.5-1}$$

For this distance measure, the pixels having a distance less than or equal to some value r from (x, y) are the points contained in a disk of radius r centered at (x, y).

The D_4 *distance* (called the *city-block distance*) between p and q is defined as

$$D_4(p, q) = |x - s| + |y - t| \tag{2.5-2}$$

In this case, the pixels having a D_4 distance from (x, y) less than or equal to some value r form a diamond centered at (x, y). For example, the pixels with D_4 distance ≤ 2 from (x, y) (the center point) form the following contours of constant distance:

```
            2
         2  1  2
      2  1  0  1  2
         2  1  2
            2
```

The pixels with $D_4 = 1$ are the 4-neighbors of (x, y).

The D_8 *distance* (called the *chessboard distance*) between p and q is defined as

$$D_8(p, q) = \max(|x - s|, |y - t|) \tag{2.5-3}$$

In this case, the pixels with D_8 distance from (x, y) less than or equal to some value r form a square centered at (x, y). For example, the pixels with D_8 distance ≤ 2 from (x, y) (the center point) form the following contours of constant distance:

```
      2  2  2  2  2
      2  1  1  1  2
      2  1  0  1  2
      2  1  1  1  2
      2  2  2  2  2
```

The pixels with $D_8 = 1$ are the 8-neighbors of (x, y).

Note that the D_4 and D_8 distances between p and q are independent of any paths that might exist between the points because these distances involve only the coordinates of the points. If we elect to consider m-adjacency, however, the D_m distance between two points is defined as the shortest m-path between the points. In this case, the distance between two pixels will depend on the values of the pixels along the path, as well as the values of their neighbors. For instance, consider the following arrangement of pixels and assume that p, p_2, and p_4 have value 1 and that p_1 and p_3 can have a value of 0 or 1:

$$
\begin{array}{cc}
p_3 & p_4 \\
p_1 & p_2 \\
p &
\end{array}
$$

Suppose that we consider adjacency of pixels valued 1 (i.e., $V = \{1\}$). If p_1 and p_3 are 0, the length of the shortest m-path (the D_m distance) between p and p_4 is 2. If p_1 is 1, then p_2 and p will no longer be m-adjacent (see the definition of m-adjacency) and the length of the shortest m-path becomes 3 (the path goes through the points $pp_1p_2p_4$). Similar comments apply if p_3 is 1 (and p_1 is 0); in this case, the length of the shortest m-path also is 3. Finally, if both p_1 and p_3 are 1, the length of the shortest m-path between p and p_4 is 4. In this case, the path goes through the sequence of points $pp_1p_2p_3p_4$.

2.6 An Introduction to the Mathematical Tools Used in Digital Image Processing

Before proceeding, you may find it helpful to download and study the review material available in the Tutorials section of the book Web site. The review covers introductory material on matrices and vectors, linear systems, set theory, and probability.

This section has two principal objectives: (1) to introduce you to the various mathematical tools we use throughout the book; and (2) to help you begin developing a "feel" for how these tools are used by applying them to a variety of basic image-processing tasks, some of which will be used numerous times in subsequent discussions. We expand the scope of the tools and their application as necessary in the following chapters.

2.6.1 Array versus Matrix Operations

An *array* operation involving one or more images is carried out on a *pixel-by-pixel* basis. We mentioned earlier in this chapter that images can be viewed equivalently as matrices. In fact, there are many situations in which operations between images are carried out using matrix theory (see Section 2.6.6). It is for this reason that a clear distinction must be made between array and matrix operations. For example, consider the following 2×2 images:

$$
\begin{bmatrix} a_{11} & a_{12} \\ a_{21} & a_{22} \end{bmatrix} \quad \text{and} \quad \begin{bmatrix} b_{11} & b_{12} \\ b_{21} & b_{22} \end{bmatrix}
$$

The *array product* of these two images is

$$
\begin{bmatrix} a_{11} & a_{12} \\ a_{21} & a_{22} \end{bmatrix} \begin{bmatrix} b_{11} & b_{12} \\ b_{21} & b_{22} \end{bmatrix} = \begin{bmatrix} a_{11}b_{11} & a_{12}b_{12} \\ a_{21}b_{21} & a_{22}b_{22} \end{bmatrix}
$$

On the other hand, the *matrix product* is given by

$$\begin{bmatrix} a_{11} & a_{12} \\ a_{21} & a_{22} \end{bmatrix} \begin{bmatrix} b_{11} & b_{12} \\ b_{21} & b_{22} \end{bmatrix} = \begin{bmatrix} a_{11}b_{11} + a_{12}b_{21} & a_{11}b_{12} + a_{12}b_{22} \\ a_{21}b_{11} + a_{22}b_{21} & a_{21}b_{12} + a_{22}b_{22} \end{bmatrix}$$

We assume array operations throughout the book, unless stated otherwise. For example, when we refer to raising an image to a power, we mean that each individual pixel is raised to that power; when we refer to dividing an image by another, we mean that the division is between corresponding pixel pairs, and so on.

2.6.2 Linear versus Nonlinear Operations

One of the most important classifications of an image-processing method is whether it is linear or nonlinear. Consider a general operator, H, that produces an output image, $g(x, y)$, for a given input image, $f(x, y)$:

$$H[f(x, y)] = g(x, y) \tag{2.6-1}$$

H is said to be a *linear operator* if

$$H[a_i f_i(x, y) + a_j f_j(x, y)] = a_i H[f_i(x, y)] + a_j H[f_j(x, y)]$$
$$= a_i g_i(x, y) + a_j g_j(x, y) \tag{2.6-2}$$

where a_i, a_j, $f_i(x, y)$, and $f_j(x, y)$ are arbitrary constants and images (of the same size), respectively. Equation (2.6-2) indicates that the output of a linear operation due to the sum of two inputs is the same as performing the operation on the inputs individually and then summing the results. In addition, the output of a linear operation to a constant times an input is the same as the output of the operation due to the original input multiplied by that constant. The first property is called the property of *additivity* and the second is called the property of *homogeneity*.

As a simple example, suppose that H is the sum operator, Σ; that is, the function of this operator is simply to sum its inputs. To test for linearity, we start with the left side of Eq. (2.6-2) and attempt to prove that it is equal to the right side:

$$\sum [a_i f_i(x, y) + a_j f_j(x, y)] = \sum a_i f_i(x, y) + \sum a_j f_j(x, y)$$

$$= a_i \sum f_i(x, y) + a_j \sum f_j(x, y)$$

$$= a_i g_i(x, y) + a_j g_j(x, y)$$

These are array summations, not the sums of all the elements of the images. As such, the sum of a single image is the image itself.

where the first step follows from the fact that summation is distributive. So, an expansion of the left side is equal to the right side of Eq. (2.6-2), and we conclude that the sum operator is linear.

On the other hand, consider the max operation, whose function is to find the maximum value of the pixels in an image. For our purposes here, the simplest way to prove that this operator is nonlinear, is to find an example that fails the test in Eq. (2.6-2). Consider the following two images

$$f_1 = \begin{bmatrix} 0 & 2 \\ 2 & 3 \end{bmatrix} \quad \text{and} \quad f_2 = \begin{bmatrix} 6 & 5 \\ 4 & 7 \end{bmatrix}$$

and suppose that we let $a_1 = 1$ and $a_2 = -1$. To test for linearity, we again start with the left side of Eq. (2.6-2):

$$\max\left\{(1)\begin{bmatrix} 0 & 2 \\ 2 & 3 \end{bmatrix} + (-1)\begin{bmatrix} 6 & 5 \\ 4 & 7 \end{bmatrix}\right\} = \max\left\{\begin{bmatrix} -6 & -3 \\ -2 & -4 \end{bmatrix}\right\}$$

$$= -2$$

Working next with the right side, we obtain

$$(1)\max\left\{\begin{bmatrix} 0 & 2 \\ 2 & 3 \end{bmatrix}\right\} + (-1)\max\left\{\begin{bmatrix} 6 & 5 \\ 4 & 7 \end{bmatrix}\right\} = 3 + (-1)7$$

$$= -4$$

The left and right sides of Eq. (2.6-2) are not equal in this case, so we have proved that in general the max operator is nonlinear.

As you will see in the next three chapters, especially in Chapters 4 and 5, linear operations are exceptionally important because they are based on a large body of theoretical and practical results that are applicable to image processing. Nonlinear systems are not nearly as well understood, so their scope of application is more limited. However, you will encounter in the following chapters several nonlinear image processing operations whose performance far exceeds what is achievable by their linear counterparts.

2.6.3 Arithmetic Operations

Arithmetic operations between images are array operations which, as discussed in Section 2.6.1, means that arithmetic operations are carried out between corresponding pixel pairs. The four arithmetic operations are denoted as

$$s(x, y) = f(x, y) + g(x, y)$$
$$d(x, y) = f(x, y) - g(x, y)$$
$$p(x, y) = f(x, y) \times g(x, y) \tag{2.6-3}$$
$$v(x, y) = f(x, y) \div g(x, y)$$

It is understood that the operations are performed between corresponding pixel pairs in f and g for $x = 0, 1, 2, \ldots, M - 1$ and $y = 0, 1, 2, \ldots, N - 1$

where, as usual, M and N are the row and column sizes of the images. Clearly, s, d, p, and v are images of size $M \times N$ also. Note that image arithmetic in the manner just defined involves images of the same size. The following examples are indicative of the important role played by arithmetic operations in digital image processing.

◼ Let $g(x, y)$ denote a corrupted image formed by the addition of noise, $\eta(x, y)$, to a noiseless image $f(x, y)$; that is,

$$g(x, y) = f(x, y) + \eta(x, y) \tag{2.6-4}$$

EXAMPLE 2.5:
Addition
(averaging) of
noisy images for
noise reduction.

where the assumption is that at every pair of coordinates (x, y) the noise is uncorrelated[†] and has zero average value. The objective of the following procedure is to reduce the noise content by adding a set of noisy images, $\{g_i(x, y)\}$. This is a technique used frequently for image enhancement.

If the noise satisfies the constraints just stated, it can be shown (Problem 2.20) that if an image $\bar{g}(x, y)$ is formed by averaging K different noisy images,

$$\bar{g}(x, y) = \frac{1}{K} \sum_{i=1}^{K} g_i(x, y) \tag{2.6-5}$$

then it follows that

$$E\{\bar{g}(x, y)\} = f(x, y) \tag{2.6-6}$$

and

$$\sigma^2_{\bar{g}(x,y)} = \frac{1}{K} \sigma^2_{\eta(x,y)} \tag{2.6-7}$$

where $E\{\bar{g}(x, y)\}$ is the expected value of $\bar{g}$, and $\sigma^2_{\bar{g}(x,y)}$ and $\sigma^2_{\eta(x,y)}$ are the variances of $\bar{g}$ and η, respectively, all at coordinates (x, y). The standard deviation (square root of the variance) at any point in the average image is

$$\sigma_{\bar{g}(x,y)} = \frac{1}{\sqrt{K}} \sigma_{\eta(x,y)} \tag{2.6-8}$$

As K increases, Eqs. (2.6-7) and (2.6-8) indicate that the variability (as measured by the variance or the standard deviation) of the pixel values at each location (x, y) decreases. Because $E\{\bar{g}(x, y)\} = f(x, y)$, this means that $\bar{g}(x, y)$ approaches $f(x, y)$ as the number of noisy images used in the averaging process increases. In practice, the images $g_i(x, y)$ must be *registered* (aligned) in order to avoid the introduction of blurring and other artifacts in the output image.

[†]Recall that the variance of a random variable z with mean m is defined as $E[(z - m)^2]$, where $E\{\cdot\}$ is the expected value of the argument. The covariance of two random variables z_i and z_j is defined as $E[(z_i - m_i)(z_j - m_j)]$. If the variables are *uncorrelated*, their covariance is 0.

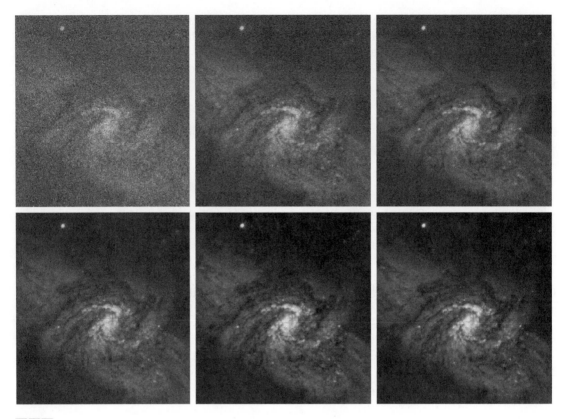

a b c
d e f

FIGURE 2.26 (a) Image of Galaxy Pair NGC 3314 corrupted by additive Gaussian noise. (b)–(f) Results of averaging 5, 10, 20, 50, and 100 noisy images, respectively. (Original image courtesy of NASA.)

The images shown in this example are from a galaxy pair called NGC 3314, taken by NASA's Hubble Space Telescope. NGC 3314 lies about 140 million light-years from Earth, in the direction of the southern-hemisphere constellation Hydra. The bright stars forming a pinwheel shape near the center of the front galaxy were formed from interstellar gas and dust.

An important application of image averaging is in the field of astronomy, where imaging under very low light levels frequently causes sensor noise to render single images virtually useless for analysis. Figure 2.26(a) shows an 8-bit image in which corruption was simulated by adding to it Gaussian noise with zero mean and a standard deviation of 64 intensity levels. This image, typical of noisy images taken under low light conditions, is useless for all practical purposes. Figures 2.26(b) through (f) show the results of averaging 5, 10, 20, 50, and 100 images, respectively. We see that the result in Fig. 2.26(e), obtained with $K = 50$, is reasonably clean. The image Fig. 2.26(f), resulting from averaging 100 noisy images, is only a slight improvement over the image in Fig. 2.26(e).

Addition is a discrete version of continuous integration. In astronomical observations, a process equivalent to the method just described is to use the integrating capabilities of CCD (see Section 2.3.3) or similar sensors for noise reduction by observing the same scene over long periods of time. Cooling also is used to reduce sensor noise. The net effect, however, is analogous to averaging a set of noisy digital images.

■ A frequent application of image subtraction is in the enhancement of *differences* between images. For example, the image in Fig. 2.27(b) was obtained by setting to zero the least-significant bit of every pixel in Fig. 2.27(a). Visually, these images are indistinguishable. However, as Fig. 2.27(c) shows, subtracting one image from the other clearly shows their differences. Black (0) values in this difference image indicate locations where there is no difference between the images in Figs. 2.27(a) and (b).

EXAMPLE 2.6:
Image subtraction
for enhancing
differences.

As another illustration, we discuss briefly an area of medical imaging called *mask mode radiography*, a commercially successful and highly beneficial use of image subtraction. Consider image differences of the form

$$g(x, y) = f(x, y) - h(x, y) \tag{2.6-9}$$

In this case $h(x, y)$, the *mask*, is an X-ray image of a region of a patient's body captured by an intensified TV camera (instead of traditional X-ray film) located opposite an X-ray source. The procedure consists of injecting an X-ray contrast medium into the patient's bloodstream, taking a series of images called *live images* [samples of which are denoted as $f(x, y)$] of the same anatomical region as $h(x, y)$, and subtracting the mask from the series of incoming live images after injection of the contrast medium. The net effect of subtracting the mask from each sample live image is that the areas that are different between $f(x, y)$ and $h(x, y)$ appear in the output image, $g(x, y)$, as enhanced detail. Because images can be captured at TV rates, this procedure in essence gives a movie showing how the contrast medium propagates through the various arteries in the area being observed.

Change detection via image subtraction is used also in image segmentation, which is the topic of Chapter 10.

Figure 2.28(a) shows a mask X-ray image of the top of a patient's head prior to injection of an iodine medium into the bloodstream, and Fig. 2.28(b) is a sample of a live image taken after the medium was injected. Figure 2.28(c) is

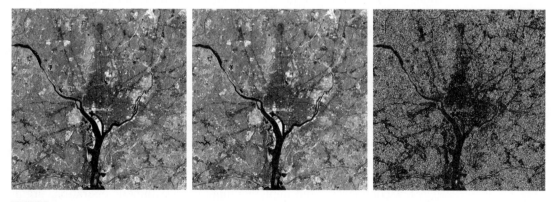

a b c

FIGURE 2.27 (a) Infrared image of the Washington, D.C. area. (b) Image obtained by setting to zero the least significant bit of every pixel in (a). (c) Difference of the two images, scaled to the range $[0, 255]$ for clarity.

a b
c d

FIGURE 2.28
Digital
subtraction
angiography.
(a) Mask image.
(b) A live image.
(c) Difference
between (a) and
(b). (d) Enhanced
difference image.
(Figures (a) and
(b) courtesy of
The Image
Sciences Institute,
University
Medical Center,
Utrecht, The
Netherlands.)

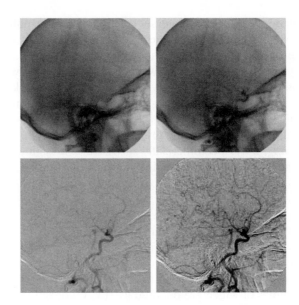

the difference between (a) and (b). Some fine blood vessel structures are visible in this image. The difference is clear in Fig. 2.28(d), which was obtained by enhancing the contrast in (c) (we discuss contrast enhancement in the next chapter). Figure 2.28(d) is a clear "map" of how the medium is propagating through the blood vessels in the subject's brain. ■

EXAMPLE 2.7:
Using image
multiplication and
division for
shading
correction.

■ An important application of image multiplication (and division) is *shading correction*. Suppose that an imaging sensor produces images that can be modeled as the product of a "perfect image," denoted by $f(x, y)$, times a shading function, $h(x, y)$; that is, $g(x, y) = f(x, y)h(x, y)$. If $h(x, y)$ is known, we can obtain $f(x, y)$ by multiplying the sensed image by the inverse of $h(x, y)$ (i.e., dividing g by h). If $h(x, y)$ is not known, but access to the imaging system is possible, we can obtain an approximation to the shading function by imaging a target of constant intensity. When the sensor is not available, we often can estimate the shading pattern directly from the image, as we discuss in Section 9.6. Figure 2.29 shows an example of shading correction.

Another common use of image multiplication is in *masking*, also called *region of interest* (ROI), operations. The process, illustrated in Fig. 2.30, consists simply of multiplying a given image by a mask image that has 1s in the ROI and 0s elsewhere. There can be more than one ROI in the mask image, and the shape of the ROI can be arbitrary, although rectangular shapes are used frequently for ease of implementation. ■

A few comments about implementing image arithmetic operations are in order before we leave this section. In practice, most images are displayed using 8 bits (even 24-bit color images consist of three separate 8-bit channels). Thus, we expect image values to be in the range from 0 to 255. When images

a b c

FIGURE 2.29 Shading correction. (a) Shaded SEM image of a tungsten filament and support, magnified approximately 130 times. (b) The shading pattern. (c) Product of (a) by the reciprocal of (b). (Original image courtesy of Michael Shaffer, Department of Geological Sciences, University of Oregon, Eugene.)

are saved in a standard format, such as TIFF or JPEG, conversion to this range is automatic. However, the approach used for the conversion depends on the system used. For example, the values in the difference of two 8-bit images can range from a minimum of -255 to a maximum of 255, and the values of a sum image can range from 0 to 510. Many software packages simply set all negative values to 0 and set to 255 all values that exceed this limit when converting images to 8 bits. Given an image f, an approach that guarantees that the full range of an arithmetic operation between images is "captured" into a fixed number of bits is as follows. First, we perform the operation

$$f_m = f - \min(f) \tag{2.6-10}$$

a b c

FIGURE 2.30 (a) Digital dental X-ray image. (b) ROI mask for isolating teeth with fillings (white corresponds to 1 and black corresponds to 0). (c) Product of (a) and (b).

which creates an image whose minimum value is 0. Then, we perform the operation

$$f_s = K\left[f_m/\max(f_m) \right] \tag{2.6-11}$$

which creates a scaled image, f_s, whose values are in the range $[0, K]$. When working with 8-bit images, setting $K = 255$ gives us a scaled image whose intensities span the full 8-bit scale from 0 to 255. Similar comments apply to 16-bit images or higher. This approach can be used for all arithmetic operations. When performing division, we have the extra requirement that a small number should be added to the pixels of the divisor image to avoid division by 0.

2.6.4 Set and Logical Operations

In this section, we introduce briefly some important set and logical operations. We also introduce the concept of a fuzzy set.

Basic set operations

Let A be a set composed of *ordered pairs* of real numbers. If $a = (a_1, a_2)$ is an *element* of A, then we write

$$a \in A \tag{2.6-12}$$

Similarly, if a is not an element of A, we write

$$a \notin A \tag{2.6-13}$$

The set with no elements is called the *null* or *empty set* and is denoted by the symbol $\varnothing$.

A set is specified by the contents of two braces: $\{ \cdot \}$. For example, when we write an expression of the form $C = \{w | w = -d, d \in D\}$, we mean that set C is the set of elements, w, such that w is formed by multiplying each of the elements of set D by -1. One way in which sets are used in image processing is to let the elements of sets be the *coordinates* of pixels (ordered pairs of integers) representing regions (objects) in an image.

If every element of a set A is also an element of a set B, then A is said to be a *subset* of B, denoted as

$$A \subseteq B \tag{2.6-14}$$

The *union* of two sets A and B, denoted by

$$C = A \cup B \tag{2.6-15}$$

is the set of elements belonging to either A, B, *or* both. Similarly, the *intersection* of two sets A and B, denoted by

$$D = A \cap B \tag{2.6-16}$$

is the set of elements belonging to both A *and* B. Two sets A and B are said to be *disjoint* or *mutually exclusive* if they have no common elements, in which case,

$$A \cap B = \varnothing \tag{2.6-17}$$

The *set universe, U*, is the set of all elements in a given application. By definition, all set elements in a given application are members of the universe defined for that application. For example, if you are working with the set of real numbers, then the set universe is the real line, which contains all the real numbers. In image processing, we typically define the universe to be the rectangle containing all the pixels in an image.

The *complement* of a set A is the set of elements that are not in A:

$$A^c = \{w | w \notin A\} \tag{2.6-18}$$

The *difference* of two sets A and B, denoted $A - B$, is defined as

$$A - B = \{w | w \in A, w \notin B\} = A \cap B^c \tag{2.6-19}$$

We see that this is the set of elements that belong to A, but not to B. We could, for example, define A^c in terms of U and the set difference operation: $A^c = U - A$.

Figure 2.31 illustrates the preceding concepts, where the universe is the set of coordinates contained within the rectangle shown, and sets A and B are the sets of coordinates contained within the boundaries shown. The result of the set operation indicated in each figure is shown in gray.[†]

In the preceding discussion, set membership is based on position (coordinates). An implicit assumption when working with images is that the intensity of all pixels in the sets is the same, as we have not defined set operations involving intensity values (e.g., we have not specified what the intensities in the intersection of two sets is). The only way that the operations illustrated in Fig. 2.31 can make sense is if the images containing the sets are binary, in which case we can talk about set membership based on coordinates, the assumption being that all member of the sets have the same intensity. We discuss this in more detail in the following subsection.

When dealing with gray-scale images, the preceding concepts are not applicable, because we have to specify the intensities of all the pixels resulting from a set operation. In fact, as you will see in Sections 3.8 and 9.6, the union and intersection operations for gray-scale values usually are defined as the max and min of corresponding pixel pairs, respectively, while the complement is defined as the pairwise differences between a constant and the intensity of every pixel in an image. The fact that we deal with corresponding pixel pairs tells us that gray-scale set operations are array operations, as defined in Section 2.6.1. The following example is a brief illustration of set operations involving gray-scale images. We discuss these concepts further in the two sections mentioned above.

[†]The operations in Eqs. (2.6-12)–(2.6-19) are the basis for the algebra of sets, which starts with properties such as the *commutative laws*: $A \cup B = B \cup A$ and $A \cap B = B \cap A$, and from these develops a broad theory based on set operations. A treatment of the algebra of sets is beyond the scope of the present discussion, but you should be aware of its existence.

a b c
d e

FIGURE 2.31
(a) Two sets of coordinates, A and B, in 2-D space. (b) The union of A and B. (c) The intersection of A and B. (d) The complement of A. (e) The difference between A and B. In (b)–(e) the shaded areas represent the members of the set operation indicated.

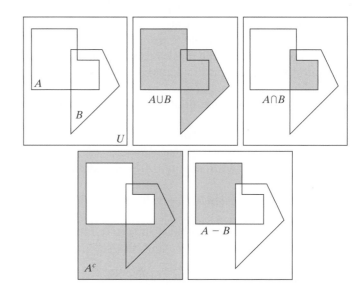

EXAMPLE 2.8:
Set operations involving image intensities.

■ Let the elements of a gray-scale image be represented by a set A whose elements are triplets of the form (x, y, z), where x and y are spatial coordinates and z denotes intensity, as mentioned in Section 2.4.2. We can define the *complement* of A as the set $A^c = \{(x, y, K - z) | (x, y, z) \in A\}$, which simply denotes the set of pixels of A whose intensities have been subtracted from a constant K. This constant is equal to $2^k - 1$, where k is the number of intensity bits used to represent z. Let A denote the 8-bit gray-scale image in Fig. 2.32(a), and suppose that we want to form the negative of A using set

a b c

FIGURE 2.32 Set operations involving gray-scale images. (a) Original image. (b) Image negative obtained using set complementation. (c) The union of (a) and a constant image. (Original image courtesy of G.E. Medical Systems.)

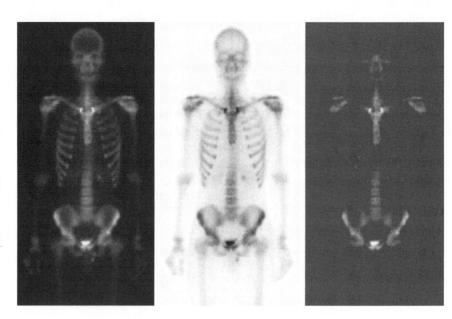

operations. We simply form the set $A_n = A^c = \{(x, y, 255 - z)|(x, y, z) \in A\}$. Note that the coordinates are carried over, so A_n is an image of the same size as A. Figure 2.32(b) shows the result.

The union of two gray-scale sets A and B may be defined as the set

$$A \cup B = \left\{ \max_{z}(a, b)|a \in A, b \in B \right\}$$

That is, the union of two gray-scale sets (images) is an array formed from the maximum intensity between pairs of spatially corresponding elements. Again, note that coordinates carry over, so the union of A and B is an image of the same size as these two images. As an illustration, suppose that A again represents the image in Fig. 2.32(a), and let B denote a rectangular array of the same size as A, but in which all values of z are equal to 3 times the mean intensity, m, of the elements of A. Figure 2.32(c) shows the result of performing the set union, in which all values exceeding $3m$ appear as values from A and all other pixels have value $3m$, which is a mid-gray value. ■

Logical operations

When dealing with binary images, we can think of *foreground* (1-valued) and *background* (0-valued) sets of pixels. Then, if we define regions (objects) as being composed of foreground pixels, the set operations illustrated in Fig. 2.31 become operations between the coordinates of objects in a binary image. When dealing with binary images, it is common practice to refer to union, intersection, and complement as the OR, AND, and NOT *logical* operations, where "logical" arises from logic theory in which 1 and 0 denote true and false, respectively.

Consider two regions (sets) A and B composed of foreground pixels. The OR of these two sets is the set of elements (coordinates) belonging either to A or B or to both. The AND operation is the set of elements that are common to A and B. The NOT operation of a set A is the set of elements not in A. Because we are dealing with images, if A is a given set of foreground pixels, NOT(A) is the set of all pixels in the image that are not in A, these pixels being background pixels and possibly other foreground pixels. We can think of this operation as turning all elements in A to 0 (black) and all the elements not in A to 1 (white). Figure 2.33 illustrates these operations. Note in the fourth row that the result of the operation shown is the set of foreground pixels that belong to A but not to B, which is the definition of set difference in Eq. (2.6-19). The last row in the figure is the XOR (exclusive OR) operation, which is the set of foreground pixels belonging to A or B, but not both. Observe that the preceding operations are between regions, which clearly can be irregular and of different sizes. This is as opposed to the gray-scale operations discussed earlier, which are array operations and thus require sets whose spatial dimensions are the same. That is, gray-scale set operations involve complete images, as opposed to regions of images.

We need be concerned in theory only with the cability to implement the AND, OR, and NOT logic operators because these three operators are *functionally*

FIGURE 2.33
Illustration of
logical operations
involving
foreground
(white) pixels.
Black represents
binary 0s and
white binary 1s.
The dashed lines
are shown for
reference only.
They are not part
of the result.

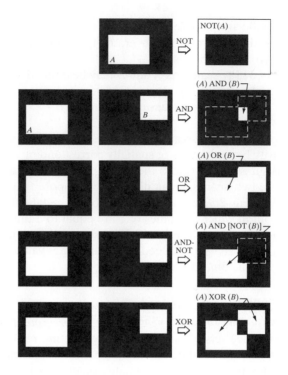

complete. In other words, any other logic operator can be implemented by using only these three basic functions, as in the fourth row of Fig. 2.33, where we implemented the set difference operation using AND and NOT. Logic operations are used extensively in image morphology, the topic of Chapter 9.

Fuzzy sets

The preceding set and logical results are *crisp* concepts, in the sense that elements either are or are not members of a set. This presents a serious limitation in some applications. Consider a simple example. Suppose that we wish to categorize all people in the world as being young or not young. Using crisp sets, let U denote the set of all people and let A be a subset of U, which we call the *set of young people*. In order to form set A, we need a *membership function* that assigns a value of 1 or 0 to every element (person) in U. If the value assigned to an element of U is 1, then that element is a member of A; otherwise it is not. Because we are dealing with a bi-valued logic, the membership function simply defines a threshold at or below which a person is considered young, and above which a person is considered not young. Suppose that we define as young any person of age 20 or younger. We see an immediate difficulty. A person whose age is 20 years and 1 sec would not be a member of the set of young people. This limitation arises regardless of the age threshold we use to classify a person as being young. What we need is more flexibility in what we mean by "young," that is, we need a *gradual* transition from young to not young. The theory of *fuzzy sets* implements this concept by utilizing membership functions that are gradual between the limit values of 1 (definitely young) to 0 (definite-ly not young). Using fuzzy sets, we can make a statement such as a person being 50% young (in the middle of the transition between young and not young). In other words, age is an imprecise concept, and fuzzy logic provides the tools to deal with such concepts. We explore fuzzy sets in detail in Section 3.8.

2.6.5 Spatial Operations

Spatial operations are performed directly on the pixels of a given image. We classify spatial operations into three broad categories: (1) single-pixel opera-tions, (2) neighborhood operations, and (3) geometric spatial transformations.

Single-pixel operations

The simplest operation we perform on a digital image is to alter the values of its individual pixels based on their intensity. This type of process may be ex-pressed as a transformation function, T, of the form:

$$s = T(z) \tag{2.6-20}$$

where z is the intensity of a pixel in the original image and s is the (mapped) intensity of the corresponding pixel in the processed image. For example, Fig. 2.34 shows the transformation used to obtain the negative of an 8-bit image, such as the image in Fig. 2.32(b), which we obtained using set operations. We discuss in Chapter 3 a number of techniques for specifying intensity trans-formation functions.

Neighborhood operations

Let S_{xy} denote the set of coordinates of a neighborhood centered on an arbi-trary point (x, y) in an image, f. Neighborhood processing generates a corres-ponding pixel at the same coordinates in an output (processed) image, g, such that the value of that pixel is determined by a specified operation involving the pixels in the input image with coordinates in S_{xy}. For example, suppose that the specified operation is to compute the average value of the pixels in a rec-tangular neighborhood of size $m \times n$ centered on (x, y). The locations of pixels

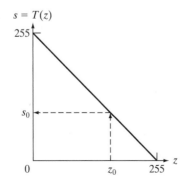

FIGURE 2.34 Intensity transformation function used to obtain the negative of an 8-bit image. The dashed arrows show transformation of an arbitrary input intensity value z_0 into its corresponding output value s_0.

a b
c d

FIGURE 2.35
Local averaging
using
neighborhood
processing. The
procedure is
illustrated in
(a) and (b) for a
rectangular
neighborhood.
(c) The aortic
angiogram
discussed in
Section 1.3.2.
(d) The result of
using Eq. (2.6-21)
with $m = n = 41$.
The images are of
size 790×686
pixels.

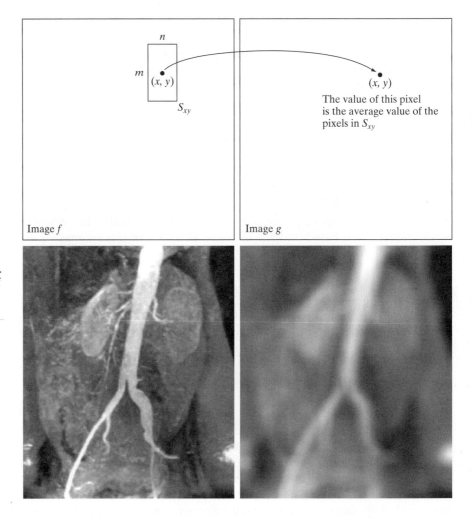

in this region constitute the set S_{xy}. Figures 2.35(a) and (b) illustrate the process. We can express this operation in equation form as

$$g(x, y) = \frac{1}{mn} \sum_{(r,c) \in S_{xy}} f(r, c) \qquad (2.6\text{-}21)$$

where r and c are the row and column coordinates of the pixels whose coordinates are members of the set S_{xy}. Image g is created by varying the coordinates (x, y) so that the center of the neighborhood moves from pixel to pixel in image f, and repeating the neighborhood operation at each new location. For instance, the image in Fig. 2.35(d) was created in this manner using a neighborhood of size 41×41. The net effect is to perform local blurring in the original image. This type of process is used, for example, to eliminate small details and thus render "blobs" corresponding to the largest regions of an image. We discuss neighborhood processing in Chapters 3 and 5, and in several other places in the book.

Geometric spatial transformations and image registration

Geometric transformations modify the spatial relationship between pixels in an image. These transformations often are called *rubber-sheet* transformations because they may be viewed as analogous to "printing" an image on a sheet of rubber and then stretching the sheet according to a predefined set of rules. In terms of digital image processing, a geometric transformation consists of two basic operations: (1) a spatial transformation of coordinates and (2) intensity interpolation that assigns intensity values to the spatially transformed pixels.

The transformation of coordinates may be expressed as

$$(x, y) = T\{(v, w)\} \tag{2.6-22}$$

where (v, w) are pixel coordinates in the original image and (x, y) are the corresponding pixel coordinates in the transformed image. For example, the transformation $(x, y) = T\{(v, w)\} = (v/2, w/2)$ shrinks the original image to half its size in both spatial directions. One of the most commonly used spatial coordinate transformations is the *affine transform* (Wolberg [1990]), which has the general form

$$[x \ y \ 1] = [v \ w \ 1] \, \mathbf{T} = [v \ w \ 1] \begin{bmatrix} t_{11} & t_{12} & 0 \\ t_{21} & t_{22} & 0 \\ t_{31} & t_{32} & 1 \end{bmatrix} \tag{2.6-23}$$

This transformation can scale, rotate, translate, or sheer a set of coordinate points, depending on the value chosen for the elements of matrix $\mathbf{T}$. Table 2.2 illustrates the matrix values used to implement these transformations. The real power of the matrix representation in Eq. (2.6-23) is that it provides the framework for concatenating together a sequence of operations. For example, if we want to resize an image, rotate it, and move the result to some location, we simply form a 3×3 matrix equal to the product of the scaling, rotation, and translation matrices from Table 2.2.

The preceding transformations relocate pixels on an image to new locations. To complete the process, we have to assign intensity values to those locations. This task is accomplished using intensity interpolation. We already discussed this topic in Section 2.4.4. We began that section with an example of zooming an image and discussed the issue of intensity assignment to new pixel locations. Zooming is simply scaling, as detailed in the second row of Table 2.2, and an analysis similar to the one we developed for zooming is applicable to the problem of assigning intensity values to the relocated pixels resulting from the other transformations in Table 2.2. As in Section 2.4.4, we consider nearest neighbor, bilinear, and bicubic interpolation techniques when working with these transformations.

In practice, we can use Eq. (2.6-23) in two basic ways. The first, called a *forward mapping*, consists of scanning the pixels of the input image and, at

TABLE 2.2
Affine transformations based on Eq. (2.6-23).

Transformation Name	Affine Matrix, T	Coordinate Equations	Example
Identity	$\begin{bmatrix} 1 & 0 & 0 \\ 0 & 1 & 0 \\ 0 & 0 & 1 \end{bmatrix}$	$x = v$ $y = w$	
Scaling	$\begin{bmatrix} c_x & 0 & 0 \\ 0 & c_y & 0 \\ 0 & 0 & 1 \end{bmatrix}$	$x = c_x v$ $y = c_y w$	
Rotation	$\begin{bmatrix} \cos\theta & \sin\theta & 0 \\ -\sin\theta & \cos\theta & 0 \\ 0 & 0 & 1 \end{bmatrix}$	$x = v\cos\theta - w\sin\theta$ $y = v\sin\theta + w\cos\theta$	
Translation	$\begin{bmatrix} 1 & 0 & 0 \\ 0 & 1 & 0 \\ t_x & t_y & 1 \end{bmatrix}$	$x = v + t_x$ $y = w + t_y$	
Shear (vertical)	$\begin{bmatrix} 1 & 0 & 0 \\ s_v & 1 & 0 \\ 0 & 0 & 1 \end{bmatrix}$	$x = v + s_v w$ $y = w$	
Shear (horizontal)	$\begin{bmatrix} 1 & s_h & 0 \\ 0 & 1 & 0 \\ 0 & 0 & 1 \end{bmatrix}$	$x = v$ $y = s_h v + w$	

each location, (v, w), computing the spatial location, (x, y), of the corresponding pixel in the output image using Eq. (2.6-23) directly. A problem with the forward mapping approach is that two or more pixels in the input image can be transformed to the same location in the output image, raising the question of how to combine multiple output values into a single output pixel. In addition, it is possible that some output locations may not be assigned a pixel at all. The second approach, called *inverse mapping*, scans the output pixel locations and, at each location, (x, y), computes the corresponding location in the input image using $(v, w) = T^{-1}(x, y)$. It then interpolates (using one of the techniques discussed in Section 2.4.4) among the nearest input pixels to determine the intensity of the output pixel value. Inverse mappings are more efficient to implement than forward mappings and are used in numerous commercial implementations of spatial transformations (for example, MATLAB uses this approach).

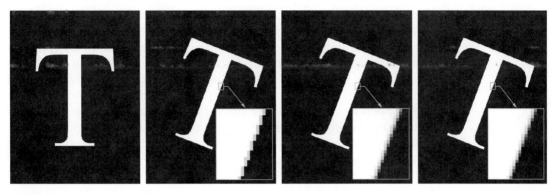

a b c d

FIGURE 2.36 (a) A 300 dpi image of the letter T. (b) Image rotated 21° using nearest neighbor interpolation to assign intensity values to the spatially transformed pixels. (c) Image rotated 21° using bilinear interpolation. (d) Image rotated 21° using bicubic interpolation. The enlarged sections show edge detail for the three interpolation approaches.

■ The objective of this example is to illustrate image rotation using an affine transform. Figure 2.36(a) shows a 300 dpi image and Figs. 2.36(b)–(d) are the results of rotating the original image by 21°, using nearest neighbor, bilinear, and bicubic interpolation, respectively. Rotation is one of the most demanding geometric transformations in terms of preserving straight-line features. As we see in the figure, nearest neighbor interpolation produced the most jagged edges and, as in Section 2.4.4, bilinear interpolation yielded significantly improved results. As before, using bicubic interpolation produced slightly sharper results. In fact, if you compare the enlarged detail in Figs. 2.36(c) and (d), you will notice in the middle of the subimages that the number of vertical gray "blocks" that provide the intensity transition from light to dark in Fig. 2.36(c) is larger than the corresponding number of blocks in (d), indicting that the latter is a sharper edge. Similar results would be obtained with the other spatial transformations in Table 2.2 that require interpolation (the identity transformation does not, and neither does the translation transformation if the increments are an integer number of pixels). This example was implemented using the inverse mapping approach discussed in the preceding paragraph. ■

EXAMPLE 2.9: Image rotation and intensity interpolation.

Image registration is an important application of digital image processing used to align two or more images of the same scene. In the preceding discussion, the form of the transformation function required to achieve a desired geometric transformation was known. In image registration, we have available the input and output images, but the specific transformation that produced the output image from the input generally is unknown. The problem, then, is to estimate the transformation function and then use it to register the two images. To clarify terminology, the input image is the image that we wish to transform, and what we call the *reference* image is the image against which we want to register the input.

For example, it may be of interest to align (register) two or more images taken at approximately the same time, but using different imaging systems, such as an MRI (magnetic resonance imaging) scanner and a PET (positron emission tomography) scanner. Or, perhaps the images were taken at different times using the same instrument, such as satellite images of a given location taken several days, months, or even years apart. In either case, combining the images or performing quantitative analysis and comparisons between them requires compensating for geometric distortions caused by differences in viewing angle, distance, and orientation; sensor resolution; shift in object positions; and other factors.

One of the principal approaches for solving the problem just discussed is to use *tie points* (also called *control points*), which are corresponding points whose locations are known precisely in the input and reference images. There are numerous ways to select tie points, ranging from interactively selecting them to applying algorithms that attempt to detect these points automatically. In some applications, imaging systems have physical artifacts (such as small metallic objects) embedded in the imaging sensors. These produce a set of *known* points (called *reseau marks*) directly on all images captured by the system, which can be used as guides for establishing tie points.

The problem of estimating the transformation function is one of modeling. For example, suppose that we have a set of four tie points each in an input and a reference image. A simple model based on a bilinear approximation is given by

$$x = c_1 v + c_2 w + c_3 vw + c_4 \tag{2.6-24}$$

and

$$y = c_5 v + c_6 w + c_7 vw + c_8 \tag{2.6-25}$$

where, during the estimation phase, (v, w) and (x, y) are the coordinates of tie points in the input and reference images, respectively. If we have four pairs of corresponding tie points in both images, we can write eight equations using Eqs. (2.6-24) and (2.6-25) and use them to solve for the eight unknown coefficients, $c_1, c_2, \ldots, c_8$. These coefficients constitute the model that transforms the pixels of one image into the locations of the pixels of the other to achieve registration.

Once we have the coefficients, Eqs. (2.6-24) and (2.6-25) become our vehicle for transforming all the pixels in the input image to generate the desired new image, which, if the tie points were selected correctly, should be registered with the reference image. In situations where four tie points are insufficient to obtain satisfactory registration, an approach used frequently is to select a larger number of tie points and then treat the quadrilaterals formed by groups of four tie points as subimages. The subimages are processed as above, with all the pixels within a quadrilateral being transformed using the coefficients determined from those tie points. Then we move to another set of four tie points and repeat the procedure until all quadrilateral regions have been processed. Of course, it is possible to use regions that are more complex than quadrilaterals and employ more complex models, such as polynomials fitted by least squares algorithms. In general, the number of control points and sophistication of the model required to solve a problem is dependent on the severity of the geometric distortion. Finally, keep in mind that the transformation defined by Eqs. (2.6-24) and (2.6-25), or any other model for that matter, simply maps the spatial coordinates of the pixels in the input image. We still need to perform intensity interpolation using any of the methods discussed previously to assign intensity values to those pixels.

■ Figure 2.37(a) shows a reference image and Fig. 2.37(b) shows the same image, but distorted geometrically by vertical and horizontal shear. Our objective is to use the reference image to obtain tie points and then use the tie points to register the images. The tie points we selected (manually) are shown as small white squares near the corners of the images (we needed only four tie

EXAMPLE 2.10:
Image
registration.

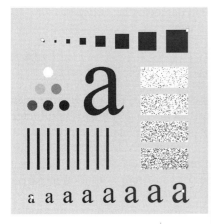

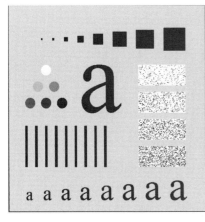

a b
c d

FIGURE 2.37
Image
registration.
(a) Reference
image. (b) Input
(geometrically
distorted image).
Corresponding tie
points are shown
as small white
squares near the
corners.
(c) Registered
image (note the
errors in the
border).
(d) Difference
between (a) and
(c), showing more
registration
errors.

points because the distortion is linear shear in both directions). Figure 2.37(c) shows the result of using these tie points in the procedure discussed in the preceding paragraphs to achieve registration. We note that registration was not perfect, as is evident by the black edges in Fig. 2.37(c). The difference image in Fig. 2.37(d) shows more clearly the slight lack of registration between the reference and corrected images. The reason for the discrepancies is error in the manual selection of the tie points. It is difficult to achieve perfect matches for tie points when distortion is so severe.   ■

2.6.6 Vector and Matrix Operations

Consult the Tutorials section in the book Web site for a brief tutorial on vectors and matrices.

Multispectral image processing is a typical area in which vector and matrix operations are used routinely. For example, you will learn in Chapter 6 that color images are formed in RGB color space by using red, green, and blue component images, as Fig. 2.38 illustrates. Here we see that *each* pixel of an RGB image has three components, which can be organized in the form of a *column vector*

$$\mathbf{z} = \begin{bmatrix} z_1 \\ z_2 \\ z_3 \end{bmatrix} \qquad (2.6\text{-}26)$$

where z_1 is the intensity of the pixel in the red image, and the other two elements are the corresponding pixel intensities in the green and blue images, respectively. Thus an RGB color image of size $M \times N$ can be represented by three component images of this size, or by a total of MN 3-D vectors. A general multispectral case involving n component images (e.g., see Fig. 1.10) will result in n-dimensional vectors. We use this type of vector representation in parts of Chapters 6, 10, 11, and 12.

Once pixels have been represented as vectors we have at our disposal the tools of vector-matrix theory. For example, the *Euclidean distance, D,* between a pixel vector $\mathbf{z}$ and an arbitrary point $\mathbf{a}$ in n-dimensional space is defined as the vector product

$$D(\mathbf{z}, \mathbf{a}) = \left[(\mathbf{z} - \mathbf{a})^T (\mathbf{z} - \mathbf{a}) \right]^{\frac{1}{2}}$$
$$= \left[(z_1 - a_1)^2 + (z_2 - a_2)^2 + \cdots + (z_n - a_n)^2 \right]^{\frac{1}{2}} \qquad (2.6\text{-}27)$$

FIGURE 2.38
Formation of a vector from corresponding pixel values in three RGB component images.

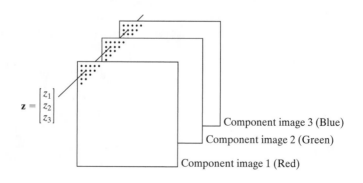

$$\mathbf{z} = \begin{bmatrix} z_1 \\ z_2 \\ z_3 \end{bmatrix}$$

Component image 3 (Blue)

Component image 2 (Green)

Component image 1 (Red)

We see that this is a generalization of the 2-D Euclidean distance defined in Eq. (2.5-1). Equation (2.6-27) sometimes is referred to as a *vector norm*, denoted by $\|\mathbf{z} - \mathbf{a}\|$. We will use distance computations numerous times in later chapters.

Another important advantage of pixel vectors is in linear transformations, represented as

$$\mathbf{w} = \mathbf{A}(\mathbf{z} - \mathbf{a}) \tag{2.6-28}$$

where $\mathbf{A}$ is a matrix of size $m \times n$ and $\mathbf{z}$ and $\mathbf{a}$ are column vectors of size $n \times 1$. As you will learn later, transformations of this type have a number of useful applications in image processing.

As noted in Eq. (2.4-2), entire images can be treated as matrices (or, equivalently, as vectors), a fact that has important implication in the solution of numerous image processing problems. For example, we can express an image of size $M \times N$ as a vector of dimension $MN \times 1$ by letting the first row of the image be the first N elements of the vector, the second row the next N elements, and so on. With images formed in this manner, we can express a broad range of linear processes applied to an image by using the notation

$$\mathbf{g} = \mathbf{Hf} + \mathbf{n} \tag{2.6-29}$$

where $\mathbf{f}$ is an $MN \times 1$ vector representing an input image, $\mathbf{n}$ is an $MN \times 1$ vector representing an $M \times N$ noise pattern, $\mathbf{g}$ is an $MN \times 1$ vector representing a processed image, and $\mathbf{H}$ is an $MN \times MN$ matrix representing a linear process applied to the input image (see Section 2.6.2 regarding linear processes). It is possible, for example, to develop an entire body of generalized techniques for image restoration starting with Eq. (2.6-29), as we discuss in Section 5.9. We touch on the topic of using matrices again in the following section, and show other uses of matrices for image processing in Chapters 5, 8, 11, and 12.

2.6.7 Image Transforms

All the image processing approaches discussed thus far operate directly on the pixels of the input image; that is, they work directly in the *spatial domain*. In some cases, image processing tasks are best formulated by transforming the input images, carrying the specified task in a *transform domain*, and applying the inverse transform to return to the spatial domain. You will encounter a number of different transforms as you proceed through the book. A particularly important class of 2-D linear transforms, denoted $T(u, v)$, can be expressed in the general form

$$T(u, v) = \sum_{x=0}^{M-1} \sum_{y=0}^{N-1} f(x, y) r(x, y, u, v) \tag{2.6-30}$$

where $f(x, y)$ is the input image, $r(x, y, u, v)$ is called the *forward transformation kernel*, and Eq. (2.6-30) is evaluated for $u = 0, 1, 2, \ldots, M - 1$ and $v = 0, 1, 2, \ldots, N - 1$. As before, x and y are spatial variables, while M and N

FIGURE 2.39
General approach for operating in the linear transform domain.

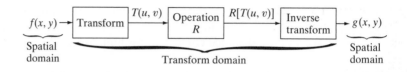

are the row and column dimensions of f. Variables u and v are called the *transform variables*. $T(u, v)$ is called the *forward transform* of $f(x, y)$. Given $T(u, v)$, we can recover $f(x, y)$ using the *inverse transform* of $T(u, v)$,

$$f(x, y) = \sum_{u=0}^{M-1} \sum_{v=0}^{N-1} T(u, v)s(x, y, u, v) \qquad (2.6\text{-}31)$$

for $x = 0, 1, 2, \ldots, M - 1$ and $y = 0, 1, 2, \ldots, N - 1$, where $s(x, y, u, v)$ is called the *inverse transformation kernel*. Together, Eqs. (2.6-30) and (2.6-31) are called a *transform pair*.

Figure 2.39 shows the basic steps for performing image processing in the linear transform domain. First, the input image is transformed, the transform is then modified by a predefined operation, and, finally, the output image is obtained by computing the inverse of the modified transform. Thus, we see that the process goes from the spatial domain to the transform domain and then back to the spatial domain.

EXAMPLE 2.11:
Image processing in the transform domain.

■ Figure 2.40 shows an example of the steps in Fig. 2.39. In this case the transform used was the Fourier transform, which we mention briefly later in this section and discuss in detail in Chapter 4. Figure 2.40(a) is an image corrupted

a b
c d

FIGURE 2.40
(a) Image corrupted by sinusoidal interference. (b) Magnitude of the Fourier transform showing the bursts of energy responsible for the interference. (c) Mask used to eliminate the energy bursts. (d) Result of computing the inverse of the modified Fourier transform. (Original image courtesy of NASA.)

by sinusoidal interference, and Fig. 2.40(b) is the magnitude of its Fourier transform, which is the output of the first stage in Fig. 2.39. As you will learn in Chapter 4, sinusoidal interference in the spatial domain appears as bright bursts of intensity in the transform domain. In this case, the bursts are in a circular pattern that can be seen in Fig. 2.40(b). Figure 2.40(c) shows a mask image (called a *filter*) with white and black representing 1 and 0, respectively. For this example, the operation in the second box of Fig. 2.39 is to multiply the mask by the transform, thus eliminating the bursts responsible for the interference. Figure 2.40(d) shows the final result, obtained by computing the inverse of the modified transform. The interference is no longer visible, and important detail is quite clear. In fact, you can even see the *fiducial marks* (faint crosses) that are used for image alignment.   ■

The forward transformation kernel is said to be *separable* if

$$r(x, y, u, v) = r_1(x, u)r_2(y, v) \tag{2.6-32}$$

In addition, the kernel is said to be *symmetric* if $r_1(x, y)$ is functionally equal to $r_2(x, y)$, so that

$$r(x, y, u, v) = r_1(x, u)r_1(y, v) \tag{2.6-33}$$

Identical comments apply to the inverse kernel by replacing r with s in the preceding equations.

The 2-D Fourier transform discussed in Example 2.11 has the following forward and inverse kernels:

$$r(x, y, u, v) = e^{-j2\pi(ux/M+vy/N)} \tag{2.6-34}$$

and

$$s(x, y, u, v) = \frac{1}{MN}e^{j2\pi(ux/M+vy/N)} \tag{2.6-35}$$

respectively, where $j = \sqrt{-1}$, so these kernels are complex. Substituting these kernels into the general transform formulations in Eqs. (2.6-30) and (2.6-31) gives us the *discrete Fourier transform pair*:

$$T(u, v) = \sum_{x=0}^{M-1} \sum_{y=0}^{N-1} f(x, y)e^{-j2\pi(ux/M+vy/N)} \tag{2.6-36}$$

and

$$f(x, y) = \frac{1}{MN} \sum_{u=0}^{M-1} \sum_{v=0}^{N-1} T(u, v)e^{j2\pi(ux/M+vy/N)} \tag{2.6-37}$$

These equations are of fundamental importance in digital image processing, and we devote most of Chapter 4 to deriving them starting from basic principles and then using them in a broad range of applications.

It is not difficult to show that the Fourier kernels are separable and symmetric (Problem 2.25), and that separable and symmetric kernels allow 2-D transforms to be computed using 1-D transforms (Problem 2.26). When the forward and inverse kernels of a transform pair satisfy these two conditions, and $f(x, y)$ is a square image of size $M \times M$, Eqs. (2.6-30) and (2.6-31) can be expressed in matrix form:

$$\mathbf{T} = \mathbf{AFA} \qquad (2.6\text{-}38)$$

where $\mathbf{F}$ is an $M \times M$ matrix containing the elements of $f(x, y)$ [see Eq. (2.4-2)], $\mathbf{A}$ is an $M \times M$ matrix with elements $a_{ij} = r_1(i, j)$, and $\mathbf{T}$ is the resulting $M \times M$ transform, with values $T(u, v)$ for $u, v = 0, 1, 2, \ldots, M - 1$.

To obtain the inverse transform, we pre- and post-multiply Eq. (2.6-38) by an inverse transformation matrix $\mathbf{B}$:

$$\mathbf{BTB} = \mathbf{BAFAB} \qquad (2.6\text{-}39)$$

If $\mathbf{B} = \mathbf{A}^{-1}$,

$$\mathbf{F} = \mathbf{BTB} \qquad (2.6\text{-}40)$$

indicating that $\mathbf{F}$ [whose elements are equal to image $f(x, y)$] can be recovered completely from its forward transform. If $\mathbf{B}$ is not equal to $\mathbf{A}^{-1}$, then use of Eq. (2.6-40) yields an approximation:

$$\hat{\mathbf{F}} = \mathbf{BAFAB} \qquad (2.6\text{-}41)$$

In addition to the Fourier transform, a number of important transforms, including the Walsh, Hadamard, discrete cosine, Haar, and slant transforms, can be expressed in the form of Eqs. (2.6-30) and (2.6-31) or, equivalently, in the form of Eqs. (2.6-38) and (2.6-40). We discuss several of these and some other types of image transforms in later chapters.

2.6.8 Probabilistic Methods

Consult the Tutorials section in the book Web site for a brief overview of probability theory.

Probability finds its way into image processing work in a number of ways. The simplest is when we treat intensity values as random quantities. For example, let $z_i, i = 0, 1, 2, \ldots, L - 1$, denote the values of all possible intensities in an $M \times N$ digital image. The probability, $p(z_k)$, of intensity level z_k occurring in a given image is estimated as

$$p(z_k) = \frac{n_k}{MN} \qquad (2.6\text{-}42)$$

where n_k is the number of times that intensity z_k occurs in the image and MN is the total number of pixels. Clearly,

$$\sum_{k=0}^{L-1} p(z_k) = 1 \qquad (2.6\text{-}43)$$

Once we have $p(z_k)$, we can determine a number of important image characteristics. For example, the mean (average) intensity is given by

$$m = \sum_{k=0}^{L-1} z_k \, p(z_k) \qquad (2.6\text{-}44)$$

Similarly, the variance of the intensities is

$$\sigma^2 = \sum_{k=0}^{L-1} (z_k - m)^2 p(z_k) \qquad (2.6\text{-}45)$$

The variance is a measure of the spread of the values of z about the mean, so it is a useful measure of image contrast. In general, the nth moment of random variable z about the mean is defined as

$$\mu_n(z) = \sum_{k=0}^{L-1} (z_k - m)^n p(z_k) \qquad (2.6\text{-}46)$$

We see that $\mu_0(z) = 1$, $\mu_1(z) = 0$, and $\mu_2(z) = \sigma^2$. Whereas the mean and variance have an immediately obvious relationship to visual properties of an image, higher-order moments are more subtle. For example, a positive third moment indicates that the intensities are biased to values higher than the mean, a negative third moment would indicate the opposite condition, and a zero third moment would tell us that the intensities are distributed approximately equally on both sides of the mean. These features are useful for computational purposes, but they do not tell us much about the appearance of an image in general.

The units of the variance are in intensity values squared. When comparing contrast values, we usually use the standard deviation, σ (square root of the variance), instead because its dimensions are directly in terms of intensity values.

■ Figure 2.41 shows three 8-bit images exhibiting low, medium, and high contrast, respectively. The standard deviations of the pixel intensities in the three images are 14.3, 31.6, and 49.2 intensity levels, respectively. The corresponding variance values are 204.3, 997.8, and 2424.9, respectively. Both sets of values tell the same story but, given that the range of possible intensity values in these images is $[0, 255]$, the standard deviation values relate to this range much more intuitively than the variance.   ■

EXAMPLE 2.12:
Comparison of standard deviation values as measures of image intensity contrast.

As you will see in progressing through the book, concepts from probability play a central role in the development of image processing algorithms. For example, in Chapter 3 we use the probability measure in Eq. (2.6-42) to derive intensity transformation algorithms. In Chaper 5, we use probability and matrix formulations to develop image restoration algorithms. In Chapter 10, probability is used for image segmentation, and in Chapter 11 we use it for texture description. In Chapter 12, we derive optimum object recognition techniques based on a probabilistic formulation.

  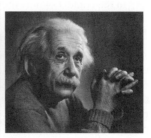

a b c

FIGURE 2.41
Images exhibiting
(a) low contrast,
(b) medium
contrast, and
(c) high contrast.

Thus far, we have addressed the issue of applying probability to a single random variable (intensity) over a single 2-D image. If we consider sequences of images, we may interpret the third variable as time. The tools needed to handle this added complexity are *stochastic* image processing techniques (the word *stochastic* is derived from a Greek word meaning roughly "to aim at a target," implying randomness in the outcome of the process). We can go a step further and consider an *entire* image (as opposed to a point) to be a spatial random event. The tools needed to handle formulations based on this concept are techniques from *random fields*. We give one example in Section 5.8 of how to treat entire images as random events, but further discussion of stochastic processes and random fields is beyond the scope of this book. The references at the end of this chapter provide a starting point for reading about these topics.

Summary

The material in this chapter is primarily background for subsequent discussions. Our treatment of the human visual system, although brief, provides a basic idea of the capabilities of the eye in perceiving pictorial information. The discussion on light and the electromagnetic spectrum is fundamental in understanding the origin of the many images we use in this book. Similarly, the image model developed in Section 2.3.4 is used in the Chapter 4 as the basis for an image enhancement technique called *homomorphic filtering*.

The sampling and interpolation ideas introduced in Section 2.4 are the foundation for many of the digitizing phenomena you are likely to encounter in practice. We will return to the issue of sampling and many of its ramifications in Chapter 4, after you have mastered the Fourier transform and the frequency domain.

The concepts introduced in Section 2.5 are the basic building blocks for processing techniques based on pixel neighborhoods. For example, as we show in the following chapter, and in Chapter 5, neighborhood processing methods are at the core of many image enhancement and restoration procedures. In Chapter 9, we use neighborhood operations for image morphology; in Chapter 10, we use them for image segmentation; and in Chapter 11 for image description. When applicable, neighborhood processing is favored in commercial applications of image processing because of their operational speed and simplicity of implementation in hardware and/or firmware.

The material in Section 2.6 will serve you well in your journey through the book. Although the level of the discussion was strictly introductory, you are now in a position to conceptualize what it means to process a digital image. As we mentioned in that section, the tools introduced there are expanded as necessary in the following chapters. Rather than dedicate an entire chapter or appendix to develop a comprehensive treatment of mathematical concepts in one place, you will find it considerably more meaningful to learn the necessary extensions of the mathematical tools from Section 2.6 in later chapters, in the context of how they are applied to solve problems in image processing.

References and Further Reading

Additional reading for the material in Section 2.1 regarding the structure of the human eye may be found in Atchison and Smith [2000] and Oyster [1999]. For additional reading on visual perception, see Regan [2000] and Gordon [1997]. The book by Hubel [1988] and the classic book by Cornsweet [1970] also are of interest. Born and Wolf [1999] is a basic reference that discusses light in terms of electromagnetic theory. Electromagnetic energy propagation is covered in some detail by Felsen and Marcuvitz [1994].

The area of image sensing is quite broad and very fast moving. An excellent source of information on optical and other imaging sensors is the Society for Optical Engineering (SPIE). The following are representative publications by the SPIE in this area: Blouke et al. [2001], Hoover and Doty [1996], and Freeman [1987].

The image model presented in Section 2.3.4 is from Oppenheim, Schafer, and Stockham [1968]. A reference for the illumination and reflectance values used in that section is the *IESNA Lighting Handbook* [2000]. For additional reading on image sampling and some of its effects, such as aliasing, see Bracewell [1995]. We discuss this topic in more detail in Chapter 4. The early experiments mentioned in Section 2.4.3 on perceived image quality as a function of sampling and quatization were reported by Huang [1965]. The issue of reducing the number of samples and intensity levels in an image while minimizing the ensuing degradation is still of current interest, as exemplified by Papamarkos and Atsalakis [2000]. For further reading on image shrinking and zooming, see Sid-Ahmed [1995], Unser et al. [1995], Umbaugh [2005], and Lehmann et al. [1999]. For further reading on the topics covered in Section 2.5, see Rosenfeld and Kak [1982], Marchand-Maillet and Sharaiha [2000], and Ritter and Wilson [2001].

Additional reading on linear systems in the context of image processing (Section 2.6.2) may be found in Castleman [1996]. The method of noise reduction by image averaging (Section 2.6.3) was first proposed by Kohler and Howell [1963]. See Peebles [1993] regarding the expected value of the mean and variance of a sum of random variables. Image subtraction (Section 2.6.3) is a generic image processing tool used widely for change detection. For image subtraction to make sense, it is necessary that the images being subtracted be registered or, alternatively, that any artifacts due to motion be identified. Two papers by Meijering et al. [1999, 2001] are illustrative of the types of techniques used to achieve these objectives.

A basic reference for the material in Section 2.6.4 is Cameron [2005]. For more advanced reading on this topic, see Tourlakis [2003]. For an introduction to fuzzy sets, see Section 3.8 and the corresponding references in Chapter 3. For further details on single-point and neighborhood processing (Section 2.6.5), see Sections 3.2 through 3.4 and the references on these topics in Chapter 3. For geometric spatial transformations, see Wolberg [1990].

Noble and Daniel [1988] is a basic reference for matrix and vector operations (Section 2.6.6). See Chapter 4 for a detailed discussion on the Fourier transform (Section 2.6.7), and Chapters 7, 8, and 11 for examples of other types of transforms used in digital image processing. Peebles [1993] is a basic introduction to probability and random variables (Section 2.6.8) and Papoulis [1991] is a more advanced treatment of this topic. For foundation material on the use of stochastic and random fields for image processing, see Rosenfeld and Kak [1982], Jähne [2002], and Won and Gray [2004].

For details of software implementation of many of the techniques illustrated in this chapter, see Gonzalez, Woods, and Eddins [2004].

Problems

★**2.1**   Using the background information provided in Section 2.1, and thinking purely in geometric terms, estimate the diameter of the smallest printed dot that the eye can discern if the page on which the dot is printed is 0.2 m away from the eyes. Assume for simplicity that the visual system ceases to detect the dot when the image of the dot on the fovea becomes smaller than the diameter of one receptor (cone) in that area of the retina. Assume further that the fovea can be

Detailed solutions to the problems marked with a star can be found in the book Web site. The site also contains suggested projects based on the material in this chapter.

modeled as a square array of dimensions 1.5 mm $\times$ 1.5 mm, and that the cones and spaces between the cones are distributed uniformly throughout this array.

2.2   When you enter a dark theater on a bright day, it takes an appreciable interval of time before you can see well enough to find an empty seat. Which of the visual processes explained in Section 2.1 is at play in this situation?

★**2.3**   Although it is not shown in Fig. 2.10, alternating current certainly is part of the electromagnetic spectrum. Commercial alternating current in the United States has a frequency of 60 Hz. What is the wavelength in kilometers of this component of the spectrum?

2.4   You are hired to design the front end of an imaging system for studying the boundary shapes of cells, bacteria, viruses, and protein. The front end consists, in this case, of the illumination source(s) and corresponding imaging camera(s). The diameters of circles required to enclose individual specimens in each of these categories are $50, 1, 0.1$, and 0.01 μm, respectively.

   (a) Can you solve the imaging aspects of this problem with a single sensor and camera? If your answer is yes, specify the illumination wavelength band and the type of camera needed. By "type," we mean the band of the electromagnetic spectrum to which the camera is most sensitive (e.g., infrared).

   (b) If your answer in (a) is no, what type of illumination sources and corresponding imaging sensors would you recommend? Specify the light sources and cameras as requested in part (a). Use the *minimum* number of illumination sources and cameras needed to solve the problem.

   By "solving the problem," we mean being able to detect circular details of diameter $50, 1, 0.1$, and 0.01 μm, respectively.

2.5   A CCD camera chip of dimensions 7×7 mm, and having 1024×1024 elements, is focused on a square, flat area, located 0.5 m away. How many line pairs per mm will this camera be able to resolve? The camera is equipped with a 35-mm lens. (*Hint:* Model the imaging process as in Fig. 2.3, with the focal length of the camera lens substituting for the focal length of the eye.)

★**2.6**   An automobile manufacturer is automating the placement of certain components on the bumpers of a limited-edition line of sports cars. The components are color coordinated, so the robots need to know the color of each car in order to select the appropriate bumper component. Models come in only four colors: blue, green, red, and white. You are hired to propose a solution based on imaging. How would you solve the problem of automatically determining the color of each car, keeping in mind that *cost* is the most important consideration in your choice of components?

2.7   Suppose that a flat area with center at (x_0, y_0) is illuminated by a light source with intensity distribution

$$i(x, y) = Ke^{-[(x-x_0)^2+(y-y_0)^2]}$$

Assume for simplicity that the reflectance of the area is constant and equal to 1.0, and let $K = 255$. If the resulting image is digitized with k bits of intensity resolution, and the eye can detect an abrupt change of eight shades of intensity between adjacent pixels, what value of k will cause visible false contouring?

2.8   Sketch the image in Problem 2.7 for $k = 2$.

★**2.9**   A common measure of transmission for digital data is the *baud rate*, defined as the number of bits transmitted per second. Generally, transmission is accomplished in packets consisting of a start bit, a byte (8 bits) of information, and a stop bit. Using these facts, answer the following:

(a) How many minutes would it take to transmit a 1024×1024 image with 256 intensity levels using a 56K baud modem?

(b) What would the time be at 3000K baud, a representative medium speed of a phone DSL (Digital Subscriber Line) connection?

2.10 High-definition television (HDTV) generates images with 1125 horizontal TV lines interlaced (where every other line is painted on the tube face in each of two fields, each field being 1/60th of a second in duration). The width-to-height aspect ratio of the images is 16:9. The fact that the number of horizontal lines is fixed determines the vertical resolution of the images. A company has designed an image capture system that generates digital images from HDTV images. The resolution of each TV (horizontal) line in their system is in proportion to vertical resolution, with the proportion being the width-to-height ratio of the images. Each pixel in the color image has 24 bits of intensity resolution, 8 bits each for a red, a green, and a blue image. These three "primary" images form a color image. How many bits would it take to store a 2-hour HDTV movie?

★2.11 Consider the two image subsets, S_1 and S_2, shown in the following figure. For $V = \{1\}$, determine whether these two subsets are (a) 4-adjacent, (b) 8-adjacent, or (c) *m*-adjacent.

```
          S₁                    S₂
0 ┌ 0   0   0   0 ┐ 0   0   1   1 ┐ 0
1 │ 0   0   1   0 │ 0   1   0   0 │ 1
1 │ 0   0   1   0 │ 1   1   0   0 │ 0
0 └ 0   1   1   1 ┘ 0   0   0   0 ┘ 0
0   0   1   1   1   0   0   1   1   1
```

★2.12 Develop an algorithm for converting a one-pixel-thick 8-path to a 4-path.

2.13 Develop an algorithm for converting a one-pixel-thick *m*-path to a 4-path.

2.14 Refer to the discussion at the end of Section 2.5.2, where we defined the background as $(R_u)^c$, the complement of the union of all the regions in an image. In some applications, it is advantageous to define the background as the subset of pixels $(R_u)^c$ that are not region hole pixels (informally, think of holes as sets of background pixels surrounded by region pixels). How would you modify the definition to exclude hole pixels from $(R_u)^c$? An answer such as "the background is the subset of pixels of $(R_u)^c$ that are not hole pixels" is not acceptable. (*Hint:* Use the concept of connectivity.)

2.15 Consider the image segment shown.

★(a) Let $V = \{0, 1\}$ and compute the lengths of the shortest 4-, 8-, and *m*-path between p and q. If a particular path does not exist between these two points, explain why.

(b) Repeat for $V = \{1, 2\}$.

```
      3   1   2   1 (q)
      2   2   0   2
      1   2   1   1
  (p) 1   0   1   2
```

2.16 ★**(a)** Give the condition(s) under which the D_4 distance between two points p and q is equal to the shortest 4-path between these points.

(b) Is this path unique?

2.17 Repeat Problem 2.16 for the D_8 distance.

★**2.18** In the next chapter, we will deal with operators whose function is to compute the sum of pixel values in a small subimage area, S. Show that these are linear operators.

2.19 The median, ζ, of a set of numbers is such that half the values in the set are below ζ and the other half are above it. For example, the median of the set of values $\{2, 3, 8, 20, 21, 25, 31\}$ is 20. Show that an operator that computes the median of a subimage area, S, is nonlinear.

★**2.20** Prove the validity of Eqs. (2.6-6) and (2.6-7). [*Hint:* Start with Eq. (2.6-4) and use the fact that the expected value of a sum is the sum of the expected values.]

2.21 Consider two 8-bit images whose intensity levels span the full range from 0 to 255.

(a) Discuss the limiting effect of repeatedly subtracting image (2) from image (1). Assume that the result is represented also in eight bits.

(b) Would reversing the order of the images yield a different result?

★**2.22** Image subtraction is used often in industrial applications for detecting missing components in product assembly. The approach is to store a "golden" image that corresponds to a correct assembly; this image is then subtracted from incoming images of the same product. Ideally, the differences would be zero if the new products are assembled correctly. Difference images for products with missing components would be nonzero in the area where they differ from the golden image. What conditions do you think have to be met in practice for this method to work?

2.23 ★**(a)** With reference to Fig. 2.31, sketch the set $(A \cap B) \cup (A \cup B)^c$.

(b) Give expressions for the sets shown shaded in the following figure in terms of sets A, B, and C. The shaded areas in each figure constitute one set, so give one expression for each of the three figures.

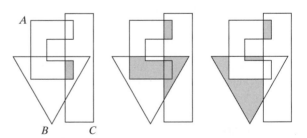

2.24 What would be the equations analogous to Eqs. (2.6-24) and (2.6-25) that would result from using triangular instead of quadrilateral regions?

2.25 Prove that the Fourier kernels in Eqs. (2.6-34) and (2.6-35) are separable and symmetric.

★**2.26** Show that 2-D transforms with separable, symmetric kernels can be computed by (1) computing 1-D transforms along the individual rows (columns) of the input, followed by (2) computing 1-D transforms along the columns (rows) of the result from step (1).

2.27   A plant produces a line of translucent miniature polymer squares. Stringent quality requirements dictate 100% visual inspection, and the plant manager finds the
use of human inspectors increasingly expensive. Inspection is semiautomated. At
each inspection station, a robotic mechanism places each polymer square over a
light located under an optical system that produces a magnified image of the
square. The image completely fills a viewing screen measuring 80 × 80 mm. Defects appear as dark circular blobs, and the inspector's job is to look at the screen
and reject any sample that has one or more such dark blobs with a diameter of
0.8 mm or larger, as measured on the scale of the screen. The manager believes
that if she can find a way to automate the process completely, she will increase
profits by 50%. She also believes that success in this project will aid her climb up
the corporate ladder. After much investigation, the manager decides that the way
to solve the problem is to view each inspection screen with a CCD TV camera
and feed the output of the camera into an image processing system capable of detecting the blobs, measuring their diameter, and activating the accept/reject but-
tons previously operated by an inspector. She is able to find a system that can do
the job, as long as the smallest defect occupies an area of at least 2 × 2 pixels in
the digital image. The manager hires you to help her specify the camera and lens
system, but requires that you use off-the-shelf components. For the lenses, assume that this constraint means any integer multiple of 25 mm or 35 mm, up to
200 mm. For the cameras, it means resolutions of 512 × 512, 1024 × 1024, or
2048 × 2048 pixels. The *individual* imaging elements in these cameras are
squares measuring 8 × 8 μm, and the spaces between imaging elements are
2 μm. For this application, the cameras cost much more than the lenses, so the
problem should be solved with the lowest-resolution camera possible, based on
the choice of lenses. As a consultant, you are to provide a written recommenda-
tion, showing in reasonable detail the analysis that led to your conclusion. Use
the same imaging geometry suggested in Problem 2.5.

3 Intensity Transformations and Spatial Filtering

It makes all the difference whether one sees darkness through the light or brightness through the shadows.

David Lindsay

Preview

The term *spatial domain* refers to the image plane itself, and image processing methods in this category are based on direct manipulation of pixels in an image. This is in contrast to image processing in a *transform domain* which, as introduced in Section 2.6.7 and discussed in more detail in Chapter 4, involves first transforming an image into the transform domain, doing the processing there, and obtaining the inverse transform to bring the results back into the spatial domain. Two principal categories of spatial processing are intensity transformations and spatial filtering. As you will learn in this chapter, intensity transformations operate on single pixels of an image, principally for the purpose of contrast manipulation and image thresholding. Spatial filtering deals with performing operations, such as image sharpening, by working in a neighborhood of every pixel in an image. In the sections that follow, we discuss a number of "classical" techniques for intensity transformations and spatial filtering. We also discuss in some detail fuzzy techniques that allow us to incorporate imprecise, knowledge-based information in the formulation of intensity transformations and spatial filtering algorithms.

3.1 Background

3.1.1 The Basics of Intensity Transformations and Spatial Filtering

All the image processing techniques discussed in this section are implemented in the *spatial domain*, which we know from the discussion in Section 2.4.2 is simply the plane containing the pixels of an image. As noted in Section 2.6.7, spatial domain techniques operate directly on the pixels of an image as opposed, for example, to the frequency domain (the topic of Chapter 4) in which operations are performed on the Fourier transform of an image, rather than on the image itself. As you will learn in progressing through the book, some image processing tasks are easier or more meaningful to implement in the spatial domain while others are best suited for other approaches. Generally, spatial domain techniques are more efficient computationally and require less processing resources to implement.

The spatial domain processes we discuss in this chapter can be denoted by the expression

$$g(x, y) = T[f(x, y)] \qquad (3.1\text{-}1)$$

where $f(x, y)$ is the input image, $g(x, y)$ is the output image, and T is an operator on f defined over a neighborhood of point (x, y). The operator can apply to a single image (our principal focus in this chapter) or to a set of images, such as performing the pixel-by-pixel sum of a sequence of images for noise reduction, as discussed in Section 2.6.3. Figure 3.1 shows the basic implementation of Eq. (3.1-1) on a single image. The point (x, y) shown is an arbitrary location in the image, and the small region shown containing the point is a neighborhood of (x, y), as explained in Section 2.6.5. Typically, the neighborhood is rectangular, centered on (x, y), and much smaller in size than the image.

Other neighborhood shapes, such as digital approximations to circles, are used sometimes, but rectangular shapes are by far the most prevalent because they are much easier to implement computationally.

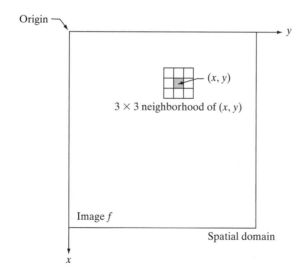

Origin

y

(x, y)

3×3 neighborhood of (x, y)

Image f

Spatial domain

x

FIGURE 3.1
A 3×3 neighborhood about a point (x, y) in an image in the spatial domain. The neighborhood is moved from pixel to pixel in the image to generate an output image.

The process that Fig. 3.1 illustrates consists of moving the origin of the neighborhood from pixel to pixel and applying the operator T to the pixels in the neighborhood to yield the output at that location. Thus, for any specific location (x, y), the value of the output image g at those coordinates is equal to the result of applying T to the neighborhood with origin at (x, y) in f. For example, suppose that the neighborhood is a square of size 3×3, and that operator T is defined as "compute the average intensity of the neighborhood." Consider an arbitrary location in an image, say $(100, 150)$. Assuming that the origin of the neighborhood is at its center, the result, $g(100, 150)$, at that location is computed as the sum of $f(100, 150)$ and its 8-neighbors, divided by 9 (i.e., the average intensity of the pixels encompassed by the neighborhood). The origin of the neighborhood is then moved to the next location and the procedure is repeated to generate the next value of the output image g. Typically, the process starts at the top left of the input image and proceeds pixel by pixel in a horizontal scan, one row at a time. When the origin of the neighborhood is at the border of the image, part of the neighborhood will reside outside the image. The procedure is either to ignore the outside neighbors in the computations specified by T, or to pad the image with a border of 0s or some other specified intensity values. The thickness of the padded border depends on the size of the neighborhood. We will return to this issue in Section 3.4.1.

As we discuss in detail in Section 3.4, the procedure just described is called *spatial filtering*, in which the neighborhood, along with a predefined operation, is called a *spatial filter* (also referred to as a *spatial mask, kernel, template*, or *window*). The type of operation performed in the neighborhood determines the nature of the filtering process.

The smallest possible neighborhood is of size 1×1. In this case, g depends only on the value of f at a single point (x, y) and T in Eq. (3.1-1) becomes an *intensity* (also called *gray-level* or *mapping*) *transformation function* of the form

$$s = T(r) \qquad (3.1\text{-}2)$$

where, for simplicity in notation, s and r are variables denoting, respectively, the intensity of g and f at any point (x, y). For example, if $T(r)$ has the form in Fig. 3.2(a), the effect of applying the transformation to every pixel of f to generate the corresponding pixels in g would be to produce an image of

a b

FIGURE 3.2
Intensity
transformation
functions.
(a) Contrast-
stretching
function.
(b) Thresholding
function.

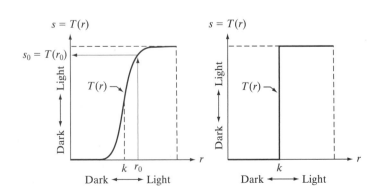

higher contrast than the original by darkening the intensity levels below k and brightening the levels above k. In this technique, sometimes called *contrast stretching* (see Section 3.2.4), values of r lower than k are compressed by the transformation function into a narrow range of s, toward black. The opposite is true for values of r higher than k. Observe how an intensity value r_0 is mapped to obtain the corresponding value s_0. In the limiting case shown in Fig. 3.2(b), $T(r)$ produces a two-level (binary) image. A mapping of this form is called a *thresholding* function. Some fairly simple, yet powerful, processing approaches can be formulated with intensity transformation functions. In this chapter, we use intensity transformations principally for image enhancement. In Chapter 10, we use them for image segmentation. Approaches whose results depend only on the intensity at a point sometimes are called *point processing* techniques, as opposed to the *neighborhood processing* techniques discussed earlier in this section.

3.1.2 About the Examples in This Chapter

Although intensity transformations and spatial filtering span a broad range of applications, most of the examples in this chapter are applications to image enhancement. *Enhancement* is the process of manipulating an image so that the result is more suitable than the original for a specific application. The word *specific* is important here because it establishes at the outset that enhancement techniques are problem oriented. Thus, for example, a method that is quite useful for enhancing X-ray images may not be the best approach for enhancing satellite images taken in the infrared band of the electromagnetic spectrum. There is no general "theory" of image enhancement. When an image is processed for visual interpretation, the viewer is the ultimate judge of how well a particular method works. When dealing with machine perception, a given technique is easier to quantify. For example, in an automated character-recognition system, the most appropriate enhancement method is the one that results in the best recognition rate, leaving aside other considerations such as computational requirements of one method over another.

Regardless of the application or method used, however, image enhancement is one of the most visually appealing areas of image processing. By its very nature, beginners in image processing generally find enhancement applications interesting and relatively simple to understand. Therefore, using examples from image enhancement to illustrate the spatial processing methods developed in this chapter not only saves having an extra chapter in the book dealing with image enhancement but, more importantly, is an effective approach for introducing newcomers to the details of processing techniques in the spatial domain. As you will see as you progress through the book, the basic material developed in this chapter is applicable to a much broader scope than just image enhancement.

3.2 Some Basic Intensity Transformation Functions

Intensity transformations are among the simplest of all image processing techniques. The values of pixels, before and after processing, will be denoted by r and s, respectively. As indicated in the previous section, these values are related by an expression of the form $s = T(r)$, where T is a transformation that maps a pixel value r into a pixel value s. Because we are dealing with digital quantities, values of a transformation function typically are stored in a one-dimensional array and the mappings from r to s are implemented via table lookups. For an 8-bit environment, a lookup table containing the values of T will have 256 entries.

As an introduction to intensity transformations, consider Fig. 3.3, which shows three basic types of functions used frequently for image enhancement: linear (negative and identity transformations), logarithmic (log and inverse-log transformations), and power-law (*n*th power and *n*th root transformations). The identity function is the trivial case in which output intensities are identical to input intensities. It is included in the graph only for completeness.

3.2.1 Image Negatives

The negative of an image with intensity levels in the range $[0, L - 1]$ is obtained by using the negative transformation shown in Fig. 3.3, which is given by the expression

$$s = L - 1 - r \tag{3.2-1}$$

Reversing the intensity levels of an image in this manner produces the equivalent of a photographic negative. This type of processing is particularly suited for enhancing white or gray detail embedded in dark regions of an

FIGURE 3.3 Some basic intensity transformation functions. All curves were scaled to fit in the range shown.

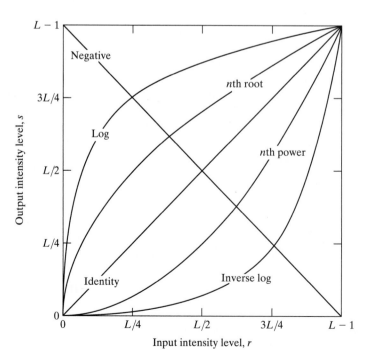

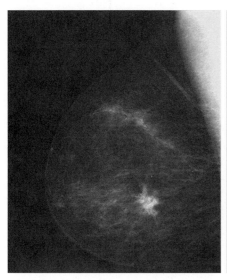 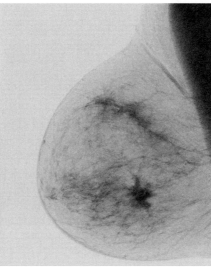

FIGURE 3.4
(a) Original digital mammogram.
(b) Negative image obtained using the negative transformation in Eq. (3.2-1). (Courtesy of G.E. Medical Systems.)

image, especially when the black areas are dominant in size. Figure 3.4 shows an example. The original image is a digital mammogram showing a small lesion. In spite of the fact that the visual content is the same in both images, note how much easier it is to analyze the breast tissue in the negative image in this particular case.

3.2.2 Log Transformations

The general form of the log transformation in Fig. 3.3 is

$$s = c \log(1 + r) \tag{3.2-2}$$

where c is a constant, and it is assumed that $r \geq 0$. The shape of the log curve in Fig. 3.3 shows that this transformation maps a narrow range of low intensity values in the input into a wider range of output levels. The opposite is true of higher values of input levels. We use a transformation of this type to expand the values of dark pixels in an image while compressing the higher-level values. The opposite is true of the inverse log transformation.

Any curve having the general shape of the log functions shown in Fig. 3.3 would accomplish this spreading/compressing of intensity levels in an image, but the power-law transformations discussed in the next section are much more versatile for this purpose. The log function has the important characteristic that it compresses the dynamic range of images with large variations in pixel values. A classic illustration of an application in which pixel values have a large dynamic range is the Fourier spectrum, which will be discussed in Chapter 4. At the moment, we are concerned only with the image characteristics of spectra. It is not unusual to encounter spectrum values that range from 0 to 10^6 or higher. While processing numbers such as these presents no problems for a computer, image display systems generally will not be able to reproduce

a b

FIGURE 3.5
(a) Fourier
spectrum.
(b) Result of
applying the log
transformation in
Eq. (3.2-2) with
$c = 1$.

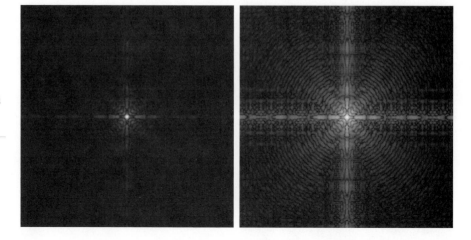

faithfully such a wide range of intensity values. The net effect is that a significant degree of intensity detail can be lost in the display of a typical Fourier
spectrum.

As an illustration of log transformations, Fig. 3.5(a) shows a Fourier spectrum with values in the range 0 to 1.5×10^6. When these values are scaled linearly for display in an 8-bit system, the brightest pixels will dominate the
display, at the expense of lower (and just as important) values of the spectrum. The effect of this dominance is illustrated vividly by the relatively small
area of the image in Fig. 3.5(a) that is not perceived as black. If, instead of displaying the values in this manner, we first apply Eq. (3.2-2) (with $c = 1$ in this
case) to the spectrum values, then the range of values of the result becomes 0
to 6.2, which is more manageable. Figure 3.5(b) shows the result of scaling this
new range linearly and displaying the spectrum in the same 8-bit display. The
wealth of detail visible in this image as compared to an unmodified display of
the spectrum is evident from these pictures. Most of the Fourier spectra seen
in image processing publications have been scaled in just this manner.

3.2.3 Power-Law (Gamma) Transformations

Power-law transformations have the basic form

$$s = c r^{\gamma} \tag{3.2-3}$$

where c and γ are positive constants. Sometimes Eq. (3.2-3) is written as
$s = c(r + \varepsilon)^{\gamma}$ to account for an offset (that is, a measurable output when the
input is zero). However, offsets typically are an issue of display calibration
and as a result they are normally ignored in Eq. (3.2-3). Plots of s versus r for
various values of γ are shown in Fig. 3.6. As in the case of the log transformation, power-law curves with fractional values of γ map a narrow range of dark
input values into a wider range of output values, with the opposite being true
for higher values of input levels. Unlike the log function, however, we notice

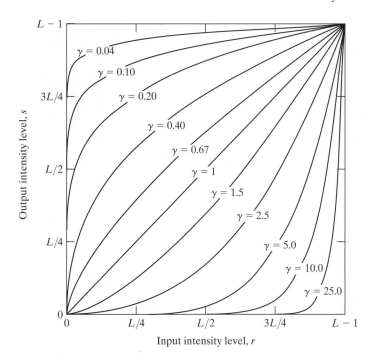

FIGURE 3.6 Plots of the equation $s = cr^\gamma$ for various values of γ ($c = 1$ in all cases). All curves were scaled to fit in the range shown.

here a family of possible transformation curves obtained simply by varying γ. As expected, we see in Fig. 3.6 that curves generated with values of $\gamma > 1$ have exactly the opposite effect as those generated with values of $\gamma < 1$. Finally, we note that Eq. (3.2-3) reduces to the identity transformation when $c = \gamma = 1$.

A variety of devices used for image capture, printing, and display respond according to a power law. By convention, the exponent in the power-law equation is referred to as *gamma* [hence our use of this symbol in Eq. (3.2-3)]. The process used to correct these power-law response phenomena is called *gamma correction*. For example, cathode ray tube (CRT) devices have an intensity-to-voltage response that is a power function, with exponents varying from approximately 1.8 to 2.5. With reference to the curve for $\gamma = 2.5$ in Fig. 3.6, we see that such display systems would tend to produce images that are darker than intended. This effect is illustrated in Fig. 3.7. Figure 3.7(a) shows a simple intensity-ramp image input into a monitor. As expected, the output of the monitor appears darker than the input, as Fig. 3.7(b) shows. Gamma correction in this case is straightforward. All we need to do is preprocess the input image before inputting it into the monitor by performing the transformation $s = r^{1/2.5} = r^{0.4}$. The result is shown in Fig. 3.7(c). When input into the same monitor, this gamma-corrected input produces an output that is close in appearance to the original image, as Fig. 3.7(d) shows. A similar analysis would apply to other imaging devices such as scanners and printers. The only difference would be the device-dependent value of gamma (Poynton [1996]).

FIGURE 3.7
(a) Intensity ramp image. (b) Image as viewed on a simulated monitor with a gamma of 2.5. (c) Gamma-corrected image. (d) Corrected image as viewed on the same monitor. Compare (d) and (a).

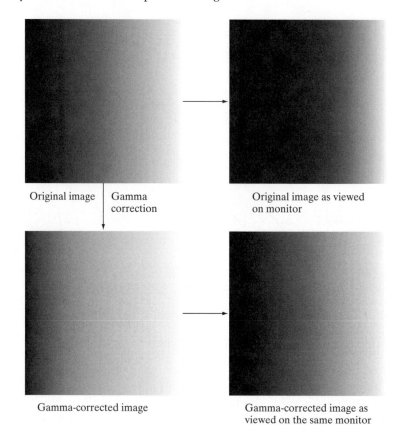

Original image | Gamma correction

Original image as viewed on monitor

Gamma-corrected image

Gamma-corrected image as viewed on the same monitor

Gamma correction is important if displaying an image accurately on a computer screen is of concern. Images that are not corrected properly can look either bleached out, or, what is more likely, too dark. Trying to reproduce colors accurately also requires some knowledge of gamma correction because varying the value of gamma changes not only the intensity, but also the ratios of red to green to blue in a color image. Gamma correction has become increasingly important in the past few years, as the use of digital images for commercial purposes over the Internet has increased. It is not unusual that images created for a popular Web site will be viewed by millions of people, the majority of whom will have different monitors and/or monitor settings. Some computer systems even have partial gamma correction built in. Also, current image standards do not contain the value of gamma with which an image was created, thus complicating the issue further. Given these constraints, a reasonable approach when storing images in a Web site is to preprocess the images with a gamma that represents an "average" of the types of monitors and computer systems that one expects in the open market at any given point in time.

■ In addition to gamma correction, power-law transformations are useful for general-purpose contrast manipulation. Figure 3.8(a) shows a magnetic resonance image (MRI) of an upper thoracic human spine with a fracture dislocation and spinal cord impingement. The fracture is visible near the vertical center of the spine, approximately one-fourth of the way down from the top of the picture. Because the given image is predominantly dark, an expansion of intensity levels is desirable. This can be accomplished with a power-law transformation with a fractional exponent. The other images shown in the figure were obtained by processing Fig. 3.8(a) with the power-law transformation

EXAMPLE 3.1:
Contrast enhancement using power-law transformations.

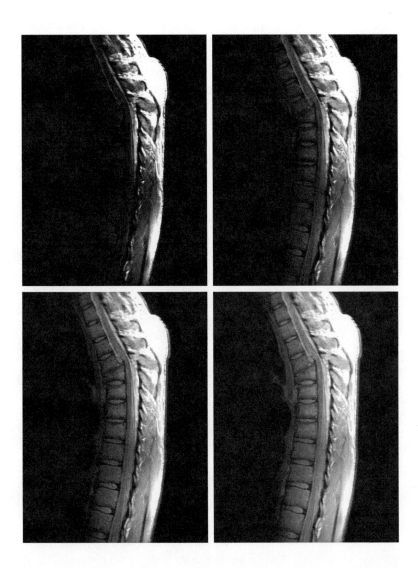

a b
c d

FIGURE 3.8
(a) Magnetic resonance image (MRI) of a fractured human spine.
(b)–(d) Results of applying the transformation in Eq. (3.2-3) with $c = 1$ and $\gamma = 0.6, 0.4$, and 0.3, respectively. (Original image courtesy of Dr. David R. Pickens, Department of Radiology and Radiological Sciences, Vanderbilt University Medical Center.)

function of Eq. (3.2-3). The values of gamma corresponding to images (b) through (d) are 0.6, 0.4, and 0.3, respectively (the value of c was 1 in all cases). We note that, as gamma decreased from 0.6 to 0.4, more detail became visible. A further decrease of gamma to 0.3 enhanced a little more detail in the background, but began to reduce contrast to the point where the image started to have a very slight "washed-out" appearance, especially in the background. By comparing all results, we see that the best enhancement in terms of contrast and discernable detail was obtained with $\gamma = 0.4$. A value of $\gamma = 0.3$ is an approximate limit below which contrast in this particular image would be reduced to an unacceptable level. ■

EXAMPLE 3.2:
Another illustration of power-law transformations.

■ Figure 3.9(a) shows the opposite problem of Fig. 3.8(a). The image to be processed now has a washed-out appearance, indicating that a compression of intensity levels is desirable. This can be accomplished with Eq. (3.2-3) using values of γ greater than 1. The results of processing Fig. 3.9(a) with $\gamma = 3.0$, 4.0, and 5.0 are shown in Figs. 3.9(b) through (d). Suitable results were obtained with gamma values of 3.0 and 4.0, the latter having a slightly

a b
c d

FIGURE 3.9
(a) Aerial image.
(b)–(d) Results of applying the transformation in Eq. (3.2-3) with $c = 1$ and $\gamma = 3.0$, 4.0, and 5.0, respectively. (Original image for this example courtesy of NASA.)

more appealing appearance because it has higher contrast. The result obtained with $\gamma = 5.0$ has areas that are too dark, in which some detail is lost. The dark region to the left of the main road in the upper left quadrant is an example of such an area.   ■

3.2.4 Piecewise-Linear Transformation Functions

A complementary approach to the methods discussed in the previous three sections is to use piecewise linear functions. The principal advantage of piecewise linear functions over the types of functions we have discussed thus far is that the form of piecewise functions can be arbitrarily complex. In fact, as you will see shortly, a practical implementation of some important transformations can be formulated only as piecewise functions. The principal disadvantage of piecewise functions is that their specification requires considerably more user input.

Contrast stretching

One of the simplest piecewise linear functions is a contrast-stretching transformation. Low-contrast images can result from poor illumination, lack of dynamic range in the imaging sensor, or even the wrong setting of a lens aperture during image acquisition. *Contrast stretching* is a process that expands the range of intensity levels in an image so that it spans the full intensity range of the recording medium or display device.

Figure 3.10(a) shows a typical transformation used for contrast stretching. The locations of points (r_1, s_1) and (r_2, s_2) control the shape of the transformation function. If $r_1 = s_1$ and $r_2 = s_2$, the transformation is a linear function that produces no changes in intensity levels. If $r_1 = r_2$, $s_1 = 0$ and $s_2 = L - 1$, the transformation becomes a *thresholding function* that creates a binary image, as illustrated in Fig. 3.2(b). Intermediate values of (r_1, s_1) and (r_2, s_2) produce various degrees of spread in the intensity levels of the output image, thus affecting its contrast. In general, $r_1 \leq r_2$ and $s_1 \leq s_2$ is assumed so that the function is single valued and monotonically increasing. This condition preserves the order of intensity levels, thus preventing the creation of intensity artifacts in the processed image.

Figure 3.10(b) shows an 8-bit image with low contrast. Figure 3.10(c) shows the result of contrast stretching, obtained by setting $(r_1, s_1) = (r_{min}, 0)$ and $(r_2, s_2) = (r_{max}, L - 1)$, where r_{min} and r_{max} denote the minimum and maximum intensity levels in the image, respectively. Thus, the transformation function stretched the levels linearly from their original range to the full range $[0, L - 1]$. Finally, Fig. 3.10(d) shows the result of using the thresholding function defined previously, with $(r_1, s_1) = (m, 0)$ and $(r_2, s_2) = (m, L - 1)$, where m is the mean intensity level in the image. The original image on which these results are based is a scanning electron microscope image of pollen, magnified approximately 700 times.

Intensity-level slicing

Highlighting a specific range of intensities in an image often is of interest. Applications include enhancing features such as masses of water in satellite imagery and enhancing flaws in X-ray images. The process, often called *intensity-level*

a b
c d

FIGURE 3.10
Contrast stretching.
(a) Form of
transformation
function. (b) A
low-contrast image.
(c) Result of
contrast stretching.
(d) Result of
thresholding.
(Original image
courtesy of Dr.
Roger Heady,
Research School of
Biological Sciences,
Australian National
University,
Canberra,
Australia.)

slicing, can be implemented in several ways, but most are variations of two basic
themes. One approach is to display in one value (say, white) all the values in the
range of interest and in another (say, black) all other intensities. This transformation, shown in Fig. 3.11(a), produces a binary image. The second approach,
based on the transformation in Fig. 3.11(b), brightens (or darkens) the desired
range of intensities but leaves all other intensity levels in the image unchanged.

a b

FIGURE 3.11 (a) This
transformation
highlights intensity
range $[A, B]$ and
reduces all other
intensities to a lower
level. (b) This
transformation
highlights range
$[A, B]$ and preserves
all other intensity
levels.

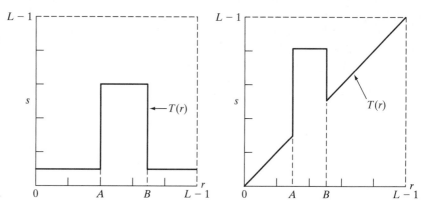

■ Figure 3.12(a) is an aortic angiogram near the kidney area (see Section 1.3.2 for a more detailed explanation of this image). The objective of this example is to use intensity-level slicing to highlight the major blood vessels that appear brighter as a result of an injected contrast medium. Figure 3.12(b) shows the result of using a transformation of the form in Fig. 3.11(a), with the selected band near the top of the scale, because the range of interest is brighter than the background. The net result of this transformation is that the blood vessel and parts of the kidneys appear white, while all other intensities are black. This type of enhancement produces a binary image and is useful for studying the *shape* of the flow of the contrast medium (to detect blockages, for example).

If, on the other hand, interest lies in the actual intensity values of the region of interest, we can use the transformation in Fig. 3.11(b). Figure 3.12(c) shows the result of using such a transformation in which a band of intensities in the mid-gray region around the mean intensity was set to black, while all other intensities were left unchanged. Here, we see that the gray-level tonality of the major blood vessels and part of the kidney area were left intact. Such a result might be useful when interest lies in measuring the actual flow of the contrast medium as a function of time in a series of images.   ■

EXAMPLE 3.3:
Intensity-level
slicing.

Bit-plane slicing

Pixels are digital numbers composed of bits. For example, the intensity of each pixel in a 256-level gray-scale image is composed of 8 bits (i.e., one byte). Instead of highlighting intensity-level ranges, we could highlight the contribution

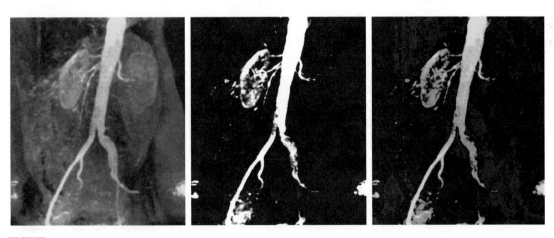

a b c

FIGURE 3.12 (a) Aortic angiogram. (b) Result of using a slicing transformation of the type illustrated in Fig. 3.11(a), with the range of intensities of interest selected in the upper end of the gray scale. (c) Result of using the transformation in Fig. 3.11(b), with the selected area set to black, so that grays in the area of the blood vessels and kidneys were preserved. (Original image courtesy of Dr. Thomas R. Gest, University of Michigan Medical School.)

3.3.1 Histogram Equalization

Consider for a moment continuous intensity values and let the variable r denote the intensities of an image to be processed. As usual, we assume that r is in the range $[0, L - 1]$, with $r = 0$ representing black and $r = L - 1$ representing white. For r satisfying these conditions, we focus attention on transformations (intensity mappings) of the form

$$s = T(r) \quad 0 \leq r \leq L - 1 \tag{3.3-1}$$

that produce an output intensity level s for every pixel in the input image having intensity r. We assume that:

(a) $T(r)$ is a monotonically[†] increasing function in the interval $0 \leq r \leq L - 1$; and

(b) $0 \leq T(r) \leq L - 1$ for $0 \leq r \leq L - 1$.

In some formulations to be discussed later, we use the inverse

$$r = T^{-1}(s) \quad 0 \leq s \leq L - 1 \tag{3.3-2}$$

in which case we change condition (a) to

(a′) $T(r)$ is a strictly monotonically increasing function in the interval $0 \leq r \leq L - 1$.

The requirement in condition (a) that $T(r)$ be monotonically increasing guarantees that output intensity values will never be less than corresponding input values, thus preventing artifacts created by reversals of intensity. Condition (b) guarantees that the range of output intensities is the same as the input. Finally, condition (a′) guarantees that the mappings from s back to r will be one-to-one, thus preventing ambiguities. Figure 3.17(a) shows a function

a b

FIGURE 3.17
(a) Monotonically increasing function, showing how multiple values can map to a single value.
(b) Strictly monotonically increasing function. This is a one-to-one mapping, both ways.

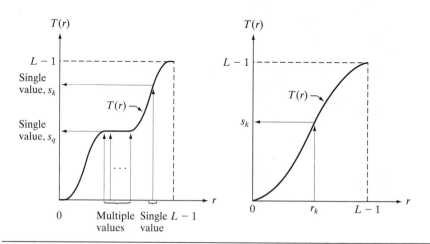

[†] Recall that a function $T(r)$ is *monotonically increasing* if $T(r_2) \geq T(r_1)$ for $r_2 > r_1$. $T(r)$ is a *strictly monotonically increasing* function if $T(r_2) > T(r_1)$ for $r_2 > r_1$. Similar definitions apply to monotonically decreasing functions.

that satisfies conditions (a) and (b). Here, we see that it is possible for multiple values to map to a single value and still satisfy these two conditions. That is, a monotonic transformation function performs a one-to-one or many-to-one mapping. This is perfectly fine when mapping from r to s. However, Fig. 3.17(a) presents a problem if we wanted to recover the values of r uniquely from the mapped values (inverse mapping can be visualized by reversing the direction of the arrows). This would be possible for the inverse mapping of s_k in Fig. 3.17(a), but the inverse mapping of s_q is a *range* of values, which, of course, prevents us in general from recovering the original value of r that resulted in s_q. As Fig. 3.17(b) shows, requiring that $T(r)$ be strictly monotonic guarantees that the inverse mappings will be *single valued* (i.e., the mapping is one-to-one in both directions). This is a theoretical requirement that allows us to derive some important histogram processing techniques later in this chapter. Because in practice we deal with integer intensity values, we are forced to round all results to their nearest integer values. Therefore, when strict monotonicity is not satisfied, we address the problem of a nonunique inverse transformation by looking for the closest integer matches. Example 3.8 gives an illustration of this.

The intensity levels in an image may be viewed as random variables in the interval $[0, L - 1]$. A fundamental descriptor of a random variable is its probability density function (PDF). Let $p_r(r)$ and $p_s(s)$ denote the PDFs of r and s, respectively, where the subscripts on p are used to indicate that p_r and p_s are different functions in general. A fundamental result from basic probability theory is that if $p_r(r)$ and $T(r)$ are known, and $T(r)$ is continuous and differentiable over the range of values of interest, then the PDF of the transformed (mapped) variable s can be obtained using the simple formula

$$p_s(s) = p_r(r) \left| \frac{dr}{ds} \right| \tag{3.3-3}$$

Thus, we see that the PDF of the output intensity variable, s, is determined by the PDF of the input intensities and the transformation function used [recall that r and s are related by $T(r)$].

A transformation function of particular importance in image processing has the form

$$s = T(r) = (L - 1) \int_0^r p_r(w) \, dw \tag{3.3-4}$$

where w is a dummy variable of integration. The right side of this equation is recognized as the cumulative distribution function (CDF) of random variable r. Because PDFs always are positive, and recalling that the integral of a function is the area under the function, it follows that the transformation function of Eq. (3.3-4) satisfies condition (a) because the area under the function cannot decrease as r increases. When the upper limit in this equation is $r = (L - 1)$, the integral evaluates to 1 (the area under a PDF curve always is 1), so the maximum value of s is $(L - 1)$ and condition (b) is satisfied also.

To find the $p_s(s)$ corresponding to the transformation just discussed, we use Eq. (3.3-3). We know from Leibniz's rule in basic calculus that the derivative of a definite integral with respect to its upper limit is the integrand evaluated at the limit. That is,

$$\frac{ds}{dr} = \frac{dT(r)}{dr}$$

$$= (L - 1)\frac{d}{dr}\left[\int_0^r p_r(w)\, dw\right] \qquad (3.3\text{-}5)$$

$$= (L - 1)p_r(r)$$

Substituting this result for dr/ds in Eq. (3.3-3), and keeping in mind that all probability values are positive, yields

$$p_s(s) = p_r(r)\left|\frac{dr}{ds}\right|$$

$$= p_r(r)\left|\frac{1}{(L - 1)p_r(r)}\right| \qquad (3.3\text{-}6)$$

$$= \frac{1}{L - 1} \qquad 0 \le s \le L - 1$$

We recognize the form of $p_s(s)$ in the last line of this equation as a *uniform* probability density function. Simply stated, we have demonstrated that performing the intensity transformation in Eq. (3.3-4) yields a random variable, s, characterized by a uniform PDF. It is important to note from this equation that $T(r)$ depends on $p_r(r)$ but, as Eq. (3.3-6) shows, the resulting $p_s(s)$ *always* is uniform, *independently* of the form of $p_r(r)$. Figure 3.18 illustrates these concepts.

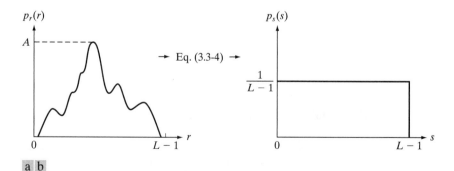

a b

FIGURE 3.18 (a) An arbitrary PDF. (b) Result of applying the transformation in Eq. (3.3-4) to all intensity levels, r. The resulting intensities, s, have a uniform PDF, independently of the form of the PDF of the r's.

■ To fix ideas, consider the following simple example. Suppose that the (continuous) intensity values in an image have the PDF

$$p_r(r) = \begin{cases} \dfrac{2r}{(L-1)^2} & \text{for } 0 \le r \le L-1 \\ 0 & \text{otherwise} \end{cases}$$

From Eq. (3.3-4),

$$s = T(r) = (L-1) \int_0^r p_r(w)\, dw = \frac{2}{L-1} \int_0^r w\, dw = \frac{r^2}{L-1}$$

Suppose next that we form a new image with intensities, s, obtained using this transformation; that is, the s values are formed by squaring the corresponding intensity values of the input image and dividing them by $(L-1)$. For example, consider an image in which $L = 10$, and suppose that a pixel in an arbitrary location (x, y) in the input image has intensity $r = 3$. Then the pixel in that location in the new image is $s = T(r) = r^2/9 = 1$. We can verify that the PDF of the intensities in the new image is uniform simply by substituting $p_r(r)$ into Eq. (3.3-6) and using the fact that $s = r^2/(L-1)$; that is,

$$p_s(s) = p_r(r) \left| \frac{dr}{ds} \right| = \frac{2r}{(L-1)^2} \left| \left[\frac{ds}{dr} \right]^{-1} \right|$$

$$= \frac{2r}{(L-1)^2} \left| \left[\frac{d}{dr} \frac{r^2}{L-1} \right]^{-1} \right|$$

$$= \frac{2r}{(L-1)^2} \left| \frac{(L-1)}{2r} \right| = \frac{1}{L-1}$$

where the last step follows from the fact that r is nonnegative and we assume that $L > 1$. As expected, the result is a uniform PDF. ■

For discrete values, we deal with probabilities (histogram values) and summations instead of probability density functions and integrals.[†] As mentioned earlier, the probability of occurrence of intensity level r_k in a digital image is approximated by

$$p_r(r_k) = \frac{n_k}{MN} \qquad k = 0, 1, 2, \ldots, L-1 \tag{3.3-7}$$

where MN is the total number of pixels in the image, n_k is the number of pixels that have intensity r_k, and L is the number of possible intensity levels in the image (e.g., 256 for an 8-bit image). As noted in the beginning of this section, a plot of $p_r(r_k)$ versus r_k is commonly referred to as a *histogram*.

[†]The conditions of monotonicity stated earlier apply also in the discrete case. We simply restrict the values of the variables to be discrete.

The discrete form of the transformation in Eq. (3.3-4) is

$$s_k = T(r_k) = (L - 1) \sum_{j=0}^{k} p_r(r_j)$$

$$= \frac{(L - 1)}{MN} \sum_{j=0}^{k} n_j \qquad k = 0, 1, 2, \dots, L - 1 \tag{3.3-8}$$

Thus, a processed (output) image is obtained by mapping each pixel in the input image with intensity r_k into a corresponding pixel with level s_k in the output image, using Eq. (3.3-8). The transformation (mapping) $T(r_k)$ in this equation is called a *histogram equalization* or *histogram linearization* transformation. It is not difficult to show (Problem 3.10) that this transformation satisfies conditions (a) and (b) stated previously in this section.

EXAMPLE 3.5:
A simple illustration of histogram equalization.

▪ Before continuing, it will be helpful to work through a simple example. Suppose that a 3-bit image ($L = 8$) of size 64×64 pixels ($MN = 4096$) has the intensity distribution shown in Table 3.1, where the intensity levels are integers in the range $[0, L - 1] = [0, 7]$.

The histogram of our hypothetical image is sketched in Fig. 3.19(a). Values of the histogram equalization transformation function are obtained using Eq. (3.3-8). For instance,

$$s_0 = T(r_0) = 7 \sum_{j=0}^{0} p_r(r_j) = 7p_r(r_0) = 1.33$$

Similarly,

$$s_1 = T(r_1) = 7 \sum_{j=0}^{1} p_r(r_j) = 7p_r(r_0) + 7p_r(r_1) = 3.08$$

and $s_2 = 4.55$, $s_3 = 5.67$, $s_4 = 6.23$, $s_5 = 6.65$, $s_6 = 6.86$, $s_7 = 7.00$. This transformation function has the staircase shape shown in Fig. 3.19(b).

TABLE 3.1
Intensity distribution and histogram values for a 3-bit, 64×64 digital image.

r_k	n_k	$p_r(r_k) = n_k/MN$
$r_0 = 0$	790	0.19
$r_1 = 1$	1023	0.25
$r_2 = 2$	850	0.21
$r_3 = 3$	656	0.16
$r_4 = 4$	329	0.08
$r_5 = 5$	245	0.06
$r_6 = 6$	122	0.03
$r_7 = 7$	81	0.02

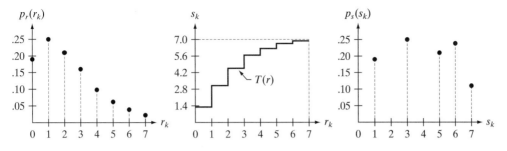

a b c

FIGURE 3.19 Illustration of histogram equalization of a 3-bit (8 intensity levels) image. (a) Original histogram. (b) Transformation function. (c) Equalized histogram.

At this point, the s values still have fractions because they were generated by summing probability values, so we round them to the nearest integer:

$$s_0 = 1.33 \rightarrow 1 \qquad s_4 = 6.23 \rightarrow 6$$

$$s_1 = 3.08 \rightarrow 3 \qquad s_5 = 6.65 \rightarrow 7$$

$$s_2 = 4.55 \rightarrow 5 \qquad s_6 = 6.86 \rightarrow 7$$

$$s_3 = 5.67 \rightarrow 6 \qquad s_7 = 7.00 \rightarrow 7$$

These are the values of the equalized histogram. Observe that there are only five distinct intensity levels. Because $r_0 = 0$ was mapped to $s_0 = 1$, there are 790 pixels in the histogram equalized image with this value (see Table 3.1). Also, there are in this image 1023 pixels with a value of $s_1 = 3$ and 850 pixels with a value of $s_2 = 5$. However both r_3 and r_4 were mapped to the same value, 6, so there are $(656 + 329) = 985$ pixels in the equalized image with this value. Similarly, there are $(245 + 122 + 81) = 448$ pixels with a value of 7 in the histogram equalized image. Dividing these numbers by $MN = 4096$ yielded the equalized histogram in Fig. 3.19(c).

Because a histogram is an approximation to a PDF, and no new allowed intensity levels are created in the process, perfectly flat histograms are rare in practical applications of histogram equalization. Thus, unlike its continuous counterpart, it cannot be proved (in general) that discrete histogram equalization results in a uniform histogram. However, as you will see shortly, using Eq. (3.3-8) has the general tendency to spread the histogram of the input image so that the intensity levels of the equalized image span a wider range of the intensity scale. The net result is contrast enhancement. ■

We discussed earlier in this section the many advantages of having intensity values that cover the entire gray scale. In addition to producing intensities that have this tendency, the method just derived has the additional advantage that it is fully "automatic." In other words, given an image, the process of histogram equalization consists simply of implementing Eq. (3.3-8), which is based on information that can be extracted directly from the given image, without the need for further parameter specifications. We note also the simplicity of the computations required to implement the technique.

The *inverse transformation* from s back to r is denoted by

$$r_k = T^{-1}(s_k) \qquad k = 0, 1, 2, \ldots, L - 1 \tag{3.3-9}$$

It can be shown (Problem 3.10) that this inverse transformation satisfies conditions (a') and (b) only if none of the levels, $r_k, k = 0, 1, 2, \ldots, L - 1$, are missing from the input image, which in turn means that none of the components of the image histogram are zero. Although the inverse transformation is not used in histogram equalization, it plays a central role in the histogram-matching scheme developed in the next section.

EXAMPLE 3.6:
Histogram
equalization.

■ The left column in Fig. 3.20 shows the four images from Fig. 3.16, and the center column shows the result of performing histogram equalization on each of these images. The first three results from top to bottom show significant improvement. As expected, histogram equalization did not have much effect on the fourth image because the intensities of this image already span the full intensity scale. Figure 3.21 shows the transformation functions used to generate the equalized images in Fig. 3.20. These functions were generated using Eq. (3.3-8). Observe that transformation (4) has a nearly linear shape, indicating that the inputs were mapped to nearly equal outputs.

The third column in Fig. 3.20 shows the histograms of the equalized images. It is of interest to note that, while all these histograms are different, the histogram-equalized images themselves are visually very similar. This is not unexpected because the basic difference between the images on the left column is one of contrast, not content. In other words, because the images have the same content, the increase in contrast resulting from histogram equalization was enough to render any intensity differences in the equalized images visually indistinguishable. Given the significant contrast differences between the original images, this example illustrates the power of histogram equalization as an adaptive contrast enhancement tool.   ■

3.3.2 Histogram Matching (Specification)

As indicated in the preceding discussion, histogram equalization automatically determines a transformation function that seeks to produce an output image that has a uniform histogram. When automatic enhancement is desired, this is a good approach because the results from this technique are predictable and the method is simple to implement. We show in this section that there are applications in which attempting to base enhancement on a uniform histogram is not the best approach. In particular, it is useful sometimes to be able to specify the shape of the histogram that we wish the processed image to have. The method used to generate a processed image that has a specified histogram is called *histogram matching* or *histogram specification*.

FIGURE 3.20 Left column: images from Fig. 3.16. Center column: corresponding histogram-equalized images. Right column: histograms of the images in the center column.

FIGURE 3.21
Transformation
functions for
histogram
equalization.
Transformations
(1) through (4)
were obtained from
the histograms of
the images (from
top to bottom) in
the left column of
Fig. 3.20 using
Eq. (3.3-8).

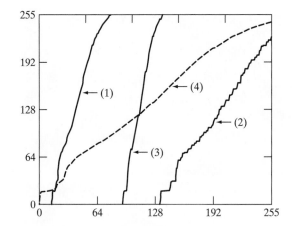

Let us return for a moment to continuous intensities r and z (considered continuous random variables), and let $p_r(r)$ and $p_z(z)$ denote their corresponding continuous probability density functions. In this notation, r and z denote the intensity levels of the input and output (processed) images, respectively. We can estimate $p_r(r)$ from the given input image, while $p_z(z)$ is the *specified* probability density function that we wish the output image to have.

Let s be a random variable with the property

$$s = T(r) = (L - 1) \int_0^r p_r(w)\, dw \tag{3.3-10}$$

where, as before, w is a dummy variable of integration. We recognize this expression as the continuous version of histogram equalization given in Eq. (3.3-4).

Suppose next that we define a random variable z with the property

$$G(z) = (L - 1) \int_0^z p_z(t)\, dt = s \tag{3.3-11}$$

where t is a dummy variable of integration. It then follows from these two equations that $G(z) = T(r)$ and, therefore, that z must satisfy the condition

$$z = G^{-1}[T(r)] = G^{-1}(s) \tag{3.3-12}$$

The transformation $T(r)$ can be obtained from Eq. (3.3-10) once $p_r(r)$ has been estimated from the input image. Similarly, the transformation function $G(z)$ can be obtained using Eq. (3.3-11) because $p_z(z)$ is given.

Equations (3.3-10) through (3.3-12) show that an image whose intensity levels have a specified probability density function can be obtained from a given image by using the following procedure:

1. Obtain $p_r(r)$ from the input image and use Eq. (3.3-10) to obtain the values of s.
2. Use the specified PDF in Eq. (3.3-11) to obtain the transformation function $G(z)$.

3. Obtain the inverse transformation $z = G^{-1}(s)$; because z is obtained from s, this process is a *mapping* from s to z, the latter being the desired values.

4. Obtain the output image by first equalizing the input image using Eq. (3.3-10); the pixel values in this image are the s values. For each pixel with value s in the equalized image, perform the inverse mapping $z = G^{-1}(s)$ to obtain the corresponding pixel in the output image. When all pixels have been thus processed, the PDF of the output image will be equal to the specified PDF.

■ Assuming continuous intensity values, suppose that an image has the intensity PDF $p_r(r) = 2r/(L - 1)^2$ for $0 \le r \le (L - 1)$ and $p_r(r) = 0$ for other values of r. Find the transformation function that will produce an image whose intensity PDF is $p_z(z) = 3z^2/(L - 1)^3$ for $0 \le z \le (L - 1)$ and $p_z(z) = 0$ for other values of z.

EXAMPLE 3.7:
Histogram specification.

First, we find the histogram equalization transformation for the interval $[0, L - 1]$:

$$s = T(r) = (L - 1) \int_0^r p_r(w) \, dw = \frac{2}{(L - 1)} \int_0^r w \, dw = \frac{r^2}{(L - 1)}$$

By definition, this transformation is 0 for values outside the range $[0, L - 1]$. Squaring the values of the input intensities and dividing them by $(L - 1)^2$ will produce an image whose intensities, s, have a uniform PDF because this is a histogram-equalization transformation, as discussed earlier.

We are interested in an image with a specified histogram, so we find next

$$G(z) = (L - 1) \int_0^z p_z(w) \, dw = \frac{3}{(L - 1)^2} \int_0^z w^2 \, dw = \frac{z^3}{(L - 1)^2}$$

over the interval $[0, L - 1]$; this function is 0 elsewhere by definition. Finally, we require that $G(z) = s$, but $G(z) = z^3/(L - 1)^2$; so $z^3/(L - 1)^2 = s$, and we have

$$z = \left[(L - 1)^2 s\right]^{1/3}$$

So, if we multiply every histogram equalized pixel by $(L - 1)^2$ and raise the product to the power $1/3$, the result will be an image whose intensities, z, have the PDF $p_z(z) = 3z^2/(L - 1)^3$ in the interval $[0, L - 1]$, as desired.

Because $s = r^2/(L - 1)$ we can generate the z's directly from the intensities, r, of the input image:

$$z = \left[(L - 1)^2 s\right]^{1/3} = \left[(L - 1)^2 \frac{r^2}{(L - 1)}\right]^{1/3} = \left[(L - 1)r^2\right]^{1/3}$$

Thus, squaring the value of each pixel in the original image, multiplying the result by $(L - 1)$, and raising the product to the power $1/3$ will yield an image whose intensity levels, z, have the specified PDF. We see that the intermediate step of equalizing the input image can be skipped; all we need is to obtain the transformation function $T(r)$ that maps r to s. Then, the two steps can be combined into a single transformation from r to z.   ■

As the preceding example shows, histogram specification is straightforward in principle. In practice, a common difficulty is finding meaningful analytical expressions for $T(r)$ and G^{-1}. Fortunately, the problem is simplified significantly when dealing with discrete quantities. The price paid is the same as for histogram equalization, where only an approximation to the desired histogram is achievable. In spite of this, however, some very useful results can be obtained, even with crude approximations.

The discrete formulation of Eq. (3.3-10) is the histogram equalization transformation in Eq. (3.3-8), which we repeat here for convenience:

$$s_k = T(r_k) = (L-1)\sum_{j=0}^{k} p_r(r_j)$$

(3.3-13)

$$= \frac{(L-1)}{MN} \sum_{j=0}^{k} n_j \quad k = 0, 1, 2, \ldots, L-1$$

where, as before, MN is the total number of pixels in the image, n_j is the number of pixels that have intensity value r_j, and L is the total number of possible intensity levels in the image. Similarly, given a specific value of s_k, the discrete formulation of Eq. (3.3-11) involves computing the transformation function

$$G(z_q) = (L-1)\sum_{i=0}^{q} p_z(z_i)$$

(3.3-14)

for a value of q, so that

$$G(z_q) = s_k$$

(3.3-15)

where $p_z(z_i)$, is the ith value of the specified histogram. As before, we find the desired value z_q by obtaining the inverse transformation:

$$z_q = G^{-1}(s_k)$$

(3.3-16)

In other words, this operation gives a value of z for each value of s; thus, it performs a *mapping* from s to z.

In practice, we do not need to compute the inverse of G. Because we deal with intensity levels that are integers (e.g., 0 to 255 for an 8-bit image), it is a simple matter to compute all the possible values of G using Eq. (3.3-14) for $q = 0, 1, 2, \ldots, L-1$. These values are scaled and rounded to their nearest integer values spanning the range $[0, L-1]$. The values are stored in a table. Then, given a particular value of s_k, we look for the closest match in the values stored in the table. If, for example, the 64th entry in the table is the closest to s_k, then $q = 63$ (recall that we start counting at 0) and z_{63} is the best solution to Eq. (3.3-15). Thus, the given value s_k would be associated with z_{63} (i.e., that specific value of s_k would *map* to z_{63}). Because the zs are intensities used as the basis for specifying the histogram $p_z(z)$, it follows that $z_0 = 0$, $z_1 = 1, \ldots, z_{L-1} = L - 1$, so z_{63} would have the intensity value 63. By repeating this procedure, we would find the mapping of each value of s_k to the value of z_q that is the closest solution to Eq. (3.3-15). These mappings are the solution to the histogram-specification problem.

Recalling that the s_ks are the values of the histogram-equalized image, we may summarize the histogram-specification procedure as follows:

1. Compute the histogram $p_r(r)$ of the given image, and use it to find the histogram equalization transformation in Eq. (3.3-13). Round the resulting values, s_k, to the integer range $[0, L - 1]$.
2. Compute all values of the transformation function G using the Eq. (3.3-14) for $q = 0, 1, 2, \ldots, L - 1$, where $p_z(z_i)$ are the values of the specified histogram. Round the values of G to integers in the range $[0, L - 1]$. Store the values of G in a table.
3. For every value of s_k, $k = 0, 1, 2, \ldots, L - 1$, use the stored values of G from step 2 to find the corresponding value of z_q so that $G(z_q)$ is closest to s_k and store these mappings from s to z. When more than one value of z_q satisfies the given s_k (i.e., the mapping is not unique), choose the smallest value by convention.
4. Form the histogram-specified image by first histogram-equalizing the input image and then mapping every equalized pixel value, s_k, of this image to the corresponding value z_q in the histogram-specified image using the mappings found in step 3. As in the continuous case, the intermediate step of equalizing the input image is conceptual. It can be skipped by combining the two transformation functions, T and G^{-1}, as Example 3.8 shows.

As mentioned earlier, for G^{-1} to satisfy conditions (a') and (b), G has to be strictly monotonic, which, according to Eq. (3.3-14), means that none of the values $p_z(z_i)$ of the specified histogram can be zero (Problem 3.10). When working with discrete quantities, the fact that this condition may not be satisfied is not a serious implementation issue, as step 3 above indicates. The following example illustrates this numerically.

■ Consider again the 64×64 hypothetical image from Example 3.5, whose histogram is repeated in Fig. 3.22(a). It is desired to transform this histogram so that it will have the values specified in the second column of Table 3.2. Figure 3.22(b) shows a sketch of this histogram.

EXAMPLE 3.8:
A simple example of histogram specification.

The first step in the procedure is to obtain the scaled histogram-equalized values, which we did in Example 3.5:

$$s_0 = 1 \quad s_2 = 5 \quad s_4 = 7 \quad s_6 = 7$$

$$s_1 = 3 \quad s_3 = 6 \quad s_5 = 7 \quad s_7 = 7$$

FIGURE 3.22
(a) Histogram of a
3-bit image. (b)
Specified
histogram.
(c) Transformation
function obtained
from the specified
histogram.
(d) Result of
performing
histogram
specification.
Compare
(b) and (d).

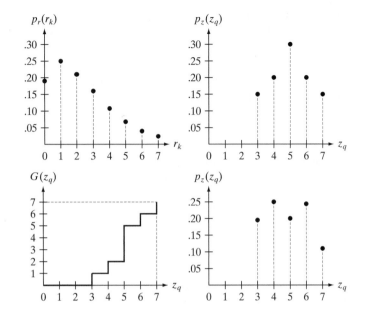

In the next step, we compute all the values of the transformation function, G, using Eq. (3.3-14):

$$G(z_0) = 7\sum_{j=0}^{0} p_z(z_j) = 0.00$$

Similarly,

$$G(z_1) = 7\sum_{j=0}^{1} p_z(z_j) = 7\left[p(z_0) + p(z_1)\right] = 0.00$$

and

$$G(z_2) = 0.00 \quad G(z_4) = 2.45 \quad G(z_6) = 5.95$$

$$G(z_3) = 1.05 \quad G(z_5) = 4.55 \quad G(z_7) = 7.00$$

TABLE 3.2
Specified and
actual histograms
(the values in the
third column are
from the
computations
performed in the
body of Example
3.8).

z_q	Specified $p_z(z_q)$	Actual $p_z(z_k)$
$z_0 = 0$	0.00	0.00
$z_1 = 1$	0.00	0.00
$z_2 = 2$	0.00	0.00
$z_3 = 3$	0.15	0.19
$z_4 = 4$	0.20	0.25
$z_5 = 5$	0.30	0.21
$z_6 = 6$	0.20	0.24
$z_7 = 7$	0.15	0.11

As in Example 3.5, these fractional values are converted to integers in our valid range, $[0, 7]$. The results are:

$$G(z_0) = 0.00 \rightarrow 0 \qquad\qquad G(z_4) = 2.45 \rightarrow 2$$
$$G(z_1) = 0.00 \rightarrow 0 \qquad\qquad G(z_5) = 4.55 \rightarrow 5$$
$$G(z_2) = 0.00 \rightarrow 0 \qquad\qquad G(z_6) = 5.95 \rightarrow 6$$
$$G(z_3) = 1.05 \rightarrow 1 \qquad\qquad G(z_7) = 7.00 \rightarrow 7$$

These results are summarized in Table 3.3, and the transformation function is sketched in Fig. 3.22(c). Observe that G is not strictly monotonic, so condition (a′) is violated. Therefore, we make use of the approach outlined in step 3 of the algorithm to handle this situation.

In the third step of the procedure, we find the smallest value of z_q so that the value $G(z_q)$ is the closest to s_k. We do this for every value of s_k to create the required mappings from s to z. For example, $s_0 = 1$, and we see that $G(z_3) = 1$, which is a perfect match in this case, so we have the correspondence $s_0 \rightarrow z_3$. That is, every pixel whose value is 1 in the histogram equalized image would map to a pixel valued 3 (in the corresponding location) in the histogram-specified image. Continuing in this manner, we arrive at the mappings in Table 3.4.

In the final step of the procedure, we use the mappings in Table 3.4 to map every pixel in the histogram equalized image into a corresponding pixel in the newly created histogram-specified image. The values of the resulting histogram are listed in the third column of Table 3.2, and the histogram is sketched in Fig. 3.22(d). The values of $p_z(z_q)$ were obtained using the same procedure as in Example 3.5. For instance, we see in Table 3.4 that $s = 1$ maps to $z = 3$, and there are 790 pixels in the histogram-equalized image with a value of 1. Therefore, $p_z(z_3) = 790/4096 = 0.19$.

Although the final result shown in Fig. 3.22(d) does not match the specified histogram exactly, the general trend of moving the intensities toward the high end of the intensity scale definitely was achieved. As mentioned earlier, obtaining the histogram-equalized image as an intermediate step is useful for explaining the procedure, but this is not necessary. Instead, we could list the mappings from the rs to the ss and from the ss to the zs in a three-column

z_q	$G(z_q)$
$z_0 = 0$	0
$z_1 = 1$	0
$z_2 = 2$	0
$z_3 = 3$	1
$z_4 = 4$	2
$z_5 = 5$	5
$z_6 = 6$	6
$z_7 = 7$	7

TABLE 3.3
All possible values of the transformation function G scaled, rounded, and ordered with respect to z.

TABLE 3.4
Mappings of all
the values of s_k
into corresponding
values of z_q.

s_k	$\rightarrow$	z_q
1	$\rightarrow$	3
3	$\rightarrow$	4
5	$\rightarrow$	5
6	$\rightarrow$	6
7	$\rightarrow$	7

table. Then, we would use those mappings to map the original pixels directly into the pixels of the histogram-specified image.   ■

EXAMPLE 3.9:
Comparison
between
histogram
equalization and
histogram
matching.

■ Figure 3.23(a) shows an image of the Mars moon, Phobos, taken by NASA's *Mars Global Surveyor*. Figure 3.23(b) shows the histogram of Fig. 3.23(a). The image is dominated by large, dark areas, resulting in a histogram characterized by a large concentration of pixels in the dark end of the gray scale. At first glance, one might conclude that histogram equalization would be a good approach to enhance this image, so that details in the dark areas become more visible. It is demonstrated in the following discussion that this is not so.

Figure 3.24(a) shows the histogram equalization transformation [Eq. (3.3-8) or (3.3-13)] obtained from the histogram in Fig. 3.23(b). The most relevant characteristic of this transformation function is how fast it rises from intensity level 0 to a level near 190. This is caused by the large concentration of pixels in the input histogram having levels near 0. When this transformation is applied to the levels of the input image to obtain a histogram-equalized result, the net effect is to map a very narrow interval of dark pixels into the upper end of the gray scale of the output image. Because numerous pixels in the input image have levels precisely in this interval, we would expect the result to be an image with a light, washed-out appearance. As Fig. 3.24(b) shows, this is indeed the

a b

FIGURE 3.23
(a) Image of the
Mars moon
Phobos taken by
NASA's *Mars
Global Surveyor.*
(b) Histogram.
(Original image
courtesy of
NASA.)

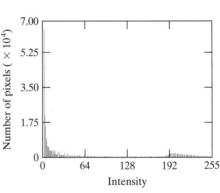

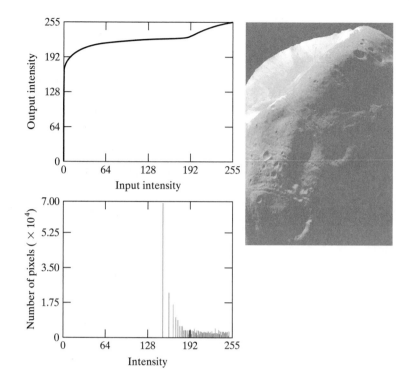

a b
c

FIGURE 3.24
(a) Transformation
function for
histogram
equalization.
(b) Histogram-
equalized image
(note the washed-
out appearance).
(c) Histogram
of (b).

case. The histogram of this image is shown in Fig. 3.24(c). Note how all the intensity levels are biased toward the upper one-half of the gray scale.

Because the problem with the transformation function in Fig. 3.24(a) was
caused by a large concentration of pixels in the original image with levels near
0, a reasonable approach is to modify the histogram of that image so that it
does not have this property. Figure 3.25(a) shows a *manually specified* function
that preserves the general shape of the original histogram, but has a smoother
transition of levels in the dark region of the gray scale. Sampling this function
into 256 equally spaced discrete values produced the desired specified histogram. The transformation function $G(z)$ obtained from this histogram using
Eq. (3.3-14) is labeled transformation (1) in Fig. 3.25(b). Similarly, the inverse
transformation $G^{-1}(s)$ from Eq. (3.3-16) (obtained using the step-by-step procedure discussed earlier) is labeled transformation (2) in Fig. 3.25(b). The enhanced image in Fig. 3.25(c) was obtained by applying transformation (2) to
the pixels of the histogram-equalized image in Fig. 3.24(b). The improvement
of the histogram-specified image over the result obtained by histogram equalization is evident by comparing these two images. It is of interest to note that a
rather modest change in the original histogram was all that was required to
obtain a significant improvement in appearance. Figure 3.25(d) shows the histogram of Fig. 3.25(c). The most distinguishing feature of this histogram is
how its low end has shifted right toward the lighter region of the gray scale
(but not excessively so), as desired.   ■

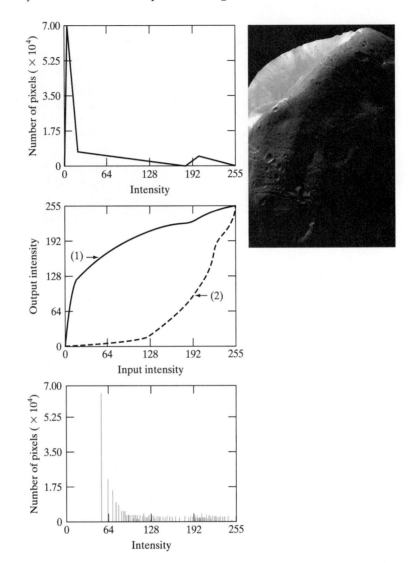

a c
b
d

FIGURE 3.25
(a) Specified
histogram.
(b) Transformations.
(c) Enhanced image
using mappings
from curve (2).
(d) Histogram of (c).

Although it probably is obvious by now, we emphasize before leaving this section that histogram specification is, for the most part, a trial-and-error process. One can use guidelines learned from the problem at hand, just as we did in the preceding example. At times, there may be cases in which it is possible to formulate what an "average" histogram should look like and use that as the specified histogram. In cases such as these, histogram specification becomes a straightforward process. In general, however, there are no rules for specifying histograms, and one must resort to analysis on a case-by-case basis for any given enhancement task.

3.3.3 Local Histogram Processing

The histogram processing methods discussed in the previous two sections are *global*, in the sense that pixels are modified by a transformation function based on the intensity distribution of an entire image. Although this global approach is suitable for overall enhancement, there are cases in which it is necessary to enhance details over small areas in an image. The number of pixels in these areas may have negligible influence on the computation of a global transformation whose shape does not necessarily guarantee the desired local enhancement. The solution is to devise transformation functions based on the intensity distribution in a neighborhood of every pixel in the image.

The histogram processing techniques previously described are easily adapted to local enhancement. The procedure is to define a neighborhood and move its center from pixel to pixel. At each location, the histogram of the points in the neighborhood is computed and either a histogram equalization or histogram specification transformation function is obtained. This function is then used to map the intensity of the pixel centered in the neighborhood. The center of the neighborhood region is then moved to an adjacent pixel location and the procedure is repeated. Because only one row or column of the neighborhood changes during a pixel-to-pixel translation of the neighborhood, updating the histogram obtained in the previous location with the new data introduced at each motion step is possible (Problem 3.12). This approach has obvious advantages over repeatedly computing the histogram of all pixels in the neighborhood region each time the region is moved one pixel location. Another approach used sometimes to reduce computation is to utilize nonoverlapping regions, but this method usually produces an undesirable "blocky" effect.

■ Figure 3.26(a) shows an 8-bit, 512 × 512 image that at first glance appears to contain five black squares on a gray background. The image is slightly noisy, but the noise is imperceptible. Figure 3.26(b) shows the result of global histogram equalization. As often is the case with histogram equalization of smooth, noisy regions, this image shows significant enhancement of the noise. Aside from the noise, however, Fig. 3.26(b) does not reveal any new significant details from the original, other than a very faint hint that the top left and bottom right squares contain an object. Figure 3.26(c) was obtained using local histogram equalization with a neighborhood of size 3 × 3. Here, we see significant detail contained within the dark squares. The intensity values of these objects were too close to the intensity of the large squares, and their sizes were too small, to influence global histogram equalization significantly enough to show this detail. ■

EXAMPLE 3.10:
Local histogram
equalization.

3.3.4 Using Histogram Statistics for Image Enhancement

Statistics obtained directly from an image histogram can be used for image enhancement. Let r denote a discrete random variable representing intensity values in the range $[0, L - 1]$, and let $p(r_i)$ denote the normalized histogram

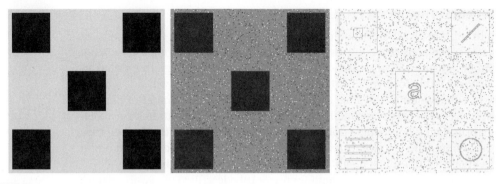

a b c

FIGURE 3.26 (a) Original image. (b) Result of global histogram equalization. (c) Result of local histogram equalization applied to (a), using a neighborhood of size 3 × 3.

component corresponding to value r_i. As indicated previously, we may view $p(r_i)$ as an estimate of the probability that intensity r_i occurs in the image from which the histogram was obtained.

As we discussed in Section 2.6.8, the nth moment of r about its mean is defined as

$$\mu_n(r) = \sum_{i=0}^{L-1} (r_i - m)^n p(r_i) \qquad (3.3\text{-}17)$$

We follow convention in using m for the mean value. Do not confuse it with the same symbol used to denote the number of rows in an $m \times n$ neighborhood, in which we also follow notational convention.

where m is the mean (average intensity) value of r (i.e., the average intensity of the pixels in the image):

$$m = \sum_{i=0}^{L-1} r_i p(r_i) \qquad (3.3\text{-}18)$$

The second moment is particularly important:

$$\mu_2(r) = \sum_{i=0}^{L-1} (r_i - m)^2 p(r_i) \qquad (3.3\text{-}19)$$

We recognize this expression as the intensity variance, normally denoted by σ^2 (recall that the standard deviation is the square root of the variance). Whereas the mean is a measure of average intensity, the variance (or standard deviation) is a measure of contrast in an image. Observe that all moments are computed easily using the preceding expressions once the histogram has been obtained from a given image.

When working with only the mean and variance, it is common practice to estimate them directly from the sample values, without computing the histogram. Appropriately, these estimates are called the *sample mean* and *sample variance*. They are given by the following familiar expressions from basic statistics:

$$m = \frac{1}{MN} \sum_{x=0}^{M-1} \sum_{y=0}^{N-1} f(x, y) \qquad (3.3\text{-}20)$$

and

The denominator in Eq. (3.3-21) is written sometimes as $MN - 1$ instead of MN. This is done to obtain a so-called *unbiased* estimate of the variance. However, we are more interested in Eqs. (3.3-21) and (3.3-19) agreeing when the histogram in the latter equation is computed from the same image used in Eq. (3.3-21). For this we require the MN term. The difference is negligible for any image of practical size.

$$\sigma^2 = \frac{1}{MN} \sum_{x=0}^{M-1} \sum_{y=0}^{N-1} \left[f(x, y) - m \right]^2 \qquad (3.3\text{-}21)$$

for $x = 0, 1, 2, \ldots, M - 1$ and $y = 0, 1, 2, \ldots, N - 1$. In other words, as we know, the mean intensity of an image can be obtained simply by summing the values of all its pixels and dividing the sum by the total number of pixels in the image. A similar interpretation applies to Eq. (3.3-21). As we illustrate in the following example, the results obtained using these two equations are identical to the results obtained using Eqs. (3.3-18) and (3.3-19), provided that the histogram used in these equations is computed from the same image used in Eqs. (3.3-20) and (3.3-21).

■ Before proceeding, it will be useful to work through a simple numerical example to fix ideas. Consider the following 2-bit image of size 5×5:

EXAMPLE 3.11:
Computing histogram statistics.

$$
\begin{array}{ccccc}
0 & 0 & 1 & 1 & 2 \\
1 & 2 & 3 & 0 & 1 \\
3 & 3 & 2 & 2 & 0 \\
2 & 3 & 1 & 0 & 0 \\
1 & 1 & 3 & 2 & 2 \\
\end{array}
$$

The pixels are represented by 2 bits; therefore, $L = 4$ and the intensity levels are in the range $[0, 3]$. The total number of pixels is 25, so the histogram has the components

$$p(r_0) = \frac{6}{25} = 0.24; \quad p(r_1) = \frac{7}{25} = 0.28;$$

$$p(r_2) = \frac{7}{25} = 0.28; \quad p(r_3) = \frac{5}{25} = 0.20$$

where the numerator in $p(r_i)$ is the number of pixels in the image with intensity level r_i. We can compute the average value of the intensities in the image using Eq. (3.3-18):

$$m = \sum_{i=0}^{3} r_i p(r_i)$$

$$= (0)(0.24) + (1)(0.28) + (2)(0.28) + (3)(0.20)$$

$$= 1.44$$

Letting $f(x, y)$ denote the preceding 5×5 array and using Eq. (3.3-20), we obtain

$$m = \frac{1}{25} \sum_{x=0}^{4} \sum_{y=0}^{4} f(x, y)$$

$$= 1.44$$

As expected, the results agree. Similarly, the result for the variance is the same (1.1264) using either Eq. (3.3-19) or (3.3-21).   ■

We consider two uses of the mean and variance for enhancement purposes. The *global* mean and variance are computed over an entire image and are useful for gross adjustments in overall intensity and contrast. A more powerful use of these parameters is in local enhancement, where the *local* mean and variance are used as the basis for making changes that depend on image characteristics in a neighborhood about each pixel in an image.

Let (x, y) denote the coordinates of any pixel in a given image, and let S_{xy} denote a neighborhood (subimage) of specified size, centered on (x, y). The mean value of the pixels in this neighborhood is given by the expression

$$m_{S_{xy}} = \sum_{i=0}^{L-1} r_i \, p_{S_{xy}}(r_i) \tag{3.3-22}$$

where $p_{S_{xy}}$ is the histogram of the pixels in region S_{xy}. This histogram has L components, corresponding to the L possible intensity values in the input image. However, many of the components are 0, depending on the size of S_{xy}. For example, if the neighborhood is of size 3×3 and $L = 256$, only between 1 and 9 of the 256 components of the histogram of the neighborhood will be nonzero. These non-zero values will correspond to the number of different intensities in S_{xy} (the maximum number of possible different intensities in a 3×3 region is 9, and the minimum is 1).

The variance of the pixels in the neighborhood similarly is given by

$$\sigma_{S_{xy}}^2 = \sum_{i=0}^{L-1} (r_i - m_{S_{xy}})^2 \, p_{S_{xy}}(r_i) \tag{3.3-23}$$

As before, the local mean is a measure of average intensity in neighborhood S_{xy}, and the local variance (or standard deviation) is a measure of intensity contrast in that neighborhood. Expressions analogous to (3.3-20) and (3.3-21) can be written for neighborhoods. We simply use the pixel values in the neighborhoods in the summations and the number of pixels in the neighborhood in the denominator.

As the following example illustrates, an important aspect of image processing using the local mean and variance is the flexibility they afford in developing simple, yet powerful enhancement techniques based on statistical measures that have a close, predictable correspondence with image appearance.

EXAMPLE 3.12:
Local enhancement using histogram statistics.

■ Figure 3.27(a) shows an SEM (scanning electron microscope) image of a tungsten filament wrapped around a support. The filament in the center of the image and its support are quite clear and easy to study. There is another filament structure on the right, dark side of the image, but it is almost imperceptible, and its size and other characteristics certainly are not easily discernable. Local enhancement by contrast manipulation is an ideal approach to problems such as this, in which parts of an image may contain hidden features.

a b c

FIGURE 3.27 (a) SEM image of a tungsten filament magnified approximately 130×. (b) Result of global histogram equalization. (c) Image enhanced using local histogram statistics. (Original image courtesy of Mr. Michael Shaffer, Department of Geological Sciences, University of Oregon, Eugene.)

In this particular case, the problem is to enhance dark areas while leaving the light area as unchanged as possible because it does not require enhancement. We can use the concepts presented in this section to formulate an enhancement method that can tell the difference between dark and light and, at the same time, is capable of enhancing only the dark areas. A measure of whether an area is relatively light or dark at a point (x, y) is to compare the average local intensity, $m_{S_{xy}}$, to the average image intensity, called the *global mean* and denoted m_G. This quantity is obtained with Eq. (3.3-18) or (3.3-20) using the entire image. Thus, we have the first element of our enhancement scheme: We will consider the pixel at a point (x, y) as a candidate for processing if $m_{S_{xy}} \leq k_0 m_G$, where k_0 is a positive constant with value less than 1.0.

Because we are interested in enhancing areas that have low contrast, we also need a measure to determine whether the contrast of an area makes it a candidate for enhancement. We consider the pixel at a point (x, y) as a candidate for enhancement if $\sigma_{S_{xy}} \leq k_2 \sigma_G$, where σ_G is the *global standard deviation* obtained using Eqs. (3.3-19) or (3.3-21) and k_2 is a positive constant. The value of this constant will be greater than 1.0 if we are interested in enhancing light areas and less than 1.0 for dark areas.

Finally, we need to restrict the lowest values of contrast we are willing to accept; otherwise the procedure would attempt to enhance constant areas, whose standard deviation is zero. Thus, we also set a lower limit on the local standard deviation by requiring that $k_1 \sigma_G \leq \sigma_{S_{xy}}$, with $k_1 < k_2$. A pixel at (x, y) that meets all the conditions for local enhancement is processed simply by multiplying it by a specified constant, E, to increase (or decrease) the value of its intensity level relative to the rest of the image. Pixels that do not meet the enhancement conditions are not changed.

We summarize the preceding approach as follows. Let $f(x, y)$ represent the value of an image at any image coordinates (x, y), and let $g(x, y)$ represent the corresponding enhanced value at those coordinates. Then,

$$g(x, y) = \begin{cases} E \cdot f(x, y) & \text{if } m_{S_{xy}} \leq k_0 m_G \text{ AND } k_1 \sigma_G \leq \sigma_{S_{xy}} \leq k_2 \sigma_G \\ f(x, y) & \text{otherwise} \end{cases} \qquad (3.3\text{-}24)$$

for $x = 0, 1, 2, \ldots, M - 1$ and $y = 0, 1, 2, \ldots, N - 1$, where, as indicated above, E, k_0, k_1, and k_2 are specified parameters, m_G is the global mean of the input image, and σ_G is its standard deviation. Parameters $m_{S_{xy}}$ and $\sigma_{S_{xy}}$ are the local mean and standard deviation, respectively. As usual, M and N are the row and column image dimensions.

Choosing the parameters in Eq. (3.3-24) generally requires a bit of experimentation to gain familiarity with a given image or class of images. In this case, the following values were selected: $E = 4.0$, $k_0 = 0.4$, $k_1 = 0.02$, and $k_2 = 0.4$. The relatively low value of 4.0 for E was chosen so that, when it was multiplied by the levels in the areas being enhanced (which are dark), the result would still tend toward the dark end of the scale, and thus preserve the general visual balance of the image. The value of k_0 was chosen as less than half the global mean because we can see by looking at the image that the areas that require enhancement definitely are dark enough to be below half the global mean. A similar analysis led to the choice of values for k_1 and k_2. Choosing these constants is not difficult in general, but their choice definitely must be guided by a logical analysis of the enhancement problem at hand. Finally, the size of the local area S_{xy} should be as small as possible in order to preserve detail and keep the computational burden as low as possible. We chose a region of size 3×3.

As a basis for comparison, we enhanced the image using global histogram equalization. Figure 3.27(b) shows the result. The dark area was improved but details still are difficult to discern, and the light areas were changed, something we did not want to do. Figure 3.27(c) shows the result of using the local statistics method explained above. In comparing this image with the original in Fig. 3.27(a) or the histogram equalized result in Fig. 3.27(b), we note the obvious detail that has been brought out on the right side of Fig. 3.27(c). Observe, for example, the clarity of the ridges in the dark filaments. It is noteworthy that the light-intensity areas on the left were left nearly intact, which was one of our initial objectives. ■

3.4 Fundamentals of Spatial Filtering

In this section, we introduce several basic concepts underlying the use of spatial filters for image processing. Spatial filtering is one of the principal tools used in this field for a broad spectrum of applications, so it is highly advisable that you develop a solid understanding of these concepts. As mentioned at the beginning of this chapter, the examples in this section deal mostly with the use of spatial filters for image enhancement. Other applications of spatial filtering are discussed in later chapters.

The name *filter* is borrowed from frequency domain processing, which is the topic of the next chapter, where "filtering" refers to accepting (passing) or rejecting certain frequency components. For example, a filter that passes low frequencies is called a *lowpass* filter. The net effect produced by a lowpass filter is to blur (smooth) an image. We can accomplish a similar smoothing directly on the image itself by using spatial filters (also called spatial *masks*, *kernels*, *templates*, and *windows*). In fact, as we show in Chapter 4, there is a one-to-one correspondence between linear spatial filters and filters in the frequency domain. However, spatial filters offer considerably more versatility because, as you will see later, they can be used also for nonlinear filtering, something we cannot do in the frequency domain.

See Section 2.6.2 regarding linearity.

3.4.1 The Mechanics of Spatial Filtering

In Fig. 3.1, we explained briefly that a spatial filter consists of (1) a *neighborhood*, (typically a small rectangle), and (2) a *predefined operation* that is performed on the image pixels encompassed by the neighborhood. Filtering creates a new pixel with coordinates equal to the coordinates of the center of the neighborhood, and whose value is the result of the filtering operation.[†] A processed (filtered) image is generated as the center of the filter visits each pixel in the input image. If the operation performed on the image pixels is linear, then the filter is called a *linear spatial filter*. Otherwise, the filter is *nonlinear*. We focus attention first on linear filters and then illustrate some simple nonlinear filters. Section 5.3 contains a more comprehensive list of nonlinear filters and their application.

Figure 3.28 illustrates the mechanics of linear spatial filtering using a 3×3 neighborhood. At any point (x, y) in the image, the response, $g(x, y)$, of the filter is the sum of products of the filter coefficients and the image pixels encompassed by the filter:

$$g(x, y) = w(-1, -1)f(x - 1, y - 1) + w(-1, 0)f(x - 1, y) + \ldots$$
$$+ w(0, 0)f(x, y) + \ldots + w(1, 1)f(x + 1, y + 1)$$

Observe that the center coefficient of the filter, $w(0, 0)$, aligns with the pixel at location (x, y). For a mask of size $m \times n$, we assume that $m = 2a + 1$ and $n = 2b + 1$, where a and b are positive integers. This means that our focus in the following discussion is on filters of odd size, with the smallest being of size 3×3. In general, linear spatial filtering of an image of size $M \times N$ with a filter of size $m \times n$ is given by the expression:

$$g(x, y) = \sum_{s=-a}^{a} \sum_{t=-b}^{b} w(s, t) f(x + s, y + t)$$

where x and y are varied so that each pixel in w visits every pixel in f.

It certainly is possible to work with filters of even size or mixed even and odd sizes. However, working with odd sizes simplifies indexing and also is more intuitive because the filters have centers falling on integer values.

[†] The filtered pixel value typically is assigned to a corresponding location in a new image created to hold the results of filtering. It is seldom the case that filtered pixels replace the values of the corresponding location in the original image, as this would change the content of the image while filtering still is being performed.

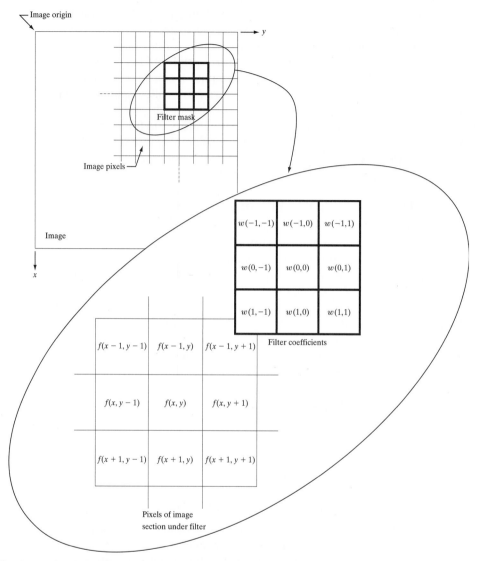

FIGURE 3.28 The mechanics of linear spatial filtering using a 3 × 3 filter mask. The form chosen to denote the coordinates of the filter mask coefficients simplifies writing expressions for linear filtering.

3.4.2 Spatial Correlation and Convolution

There are two closely related concepts that must be understood clearly when performing linear spatial filtering. One is *correlation* and the other is *convolution*. Correlation is the process of moving a filter mask over the image and computing the sum of products at each location, exactly as explained in the previous section. The mechanics of convolution are the same, except that the filter is first rotated by 180°. The best way to explain the differences between the two concepts is by example. We begin with a 1-D illustration.

Figure 3.29(a) shows a 1-D function, f, and a filter, w, and Fig. 3.29(b) shows the starting position to perform correlation. The first thing we note is that there

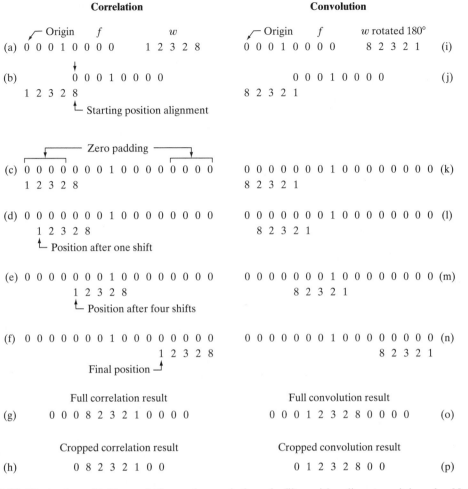

FIGURE 3.29 Illustration of 1-D correlation and convolution of a filter with a discrete unit impulse. Note that correlation and convolution are functions of *displacement*.

are parts of the functions that do not overlap. The solution to this problem is to pad f with enough 0s on either side to allow each pixel in w to visit every pixel in f. If the filter is of size m, we need $m - 1$ 0s on either side of f. Figure 3.29(c) shows a properly padded function. The first value of correlation is the sum of products of f and w for the initial position shown in Fig. 3.29(c) (the sum of products is 0). This corresponds to a displacement $x = 0$. To obtain the second value of correlation, we shift w one pixel location to the right (a displacement of $x = 1$) and compute the sum of products. The result again is 0. In fact, the first nonzero result is when $x = 3$, in which case the 8 in w overlaps the 1 in f and the result of correlation is 8. Proceeding in this manner, we obtain the full correlation result in Fig. 3.29(g). Note that it took 12 values of x (i.e., $x = 0, 1, 2, \ldots, 11$) to fully slide w past f so that each pixel in w visited every pixel in f. Often, we like to work with correlation arrays that are the same size as f, in which case we crop the full correlation to the size of the original function, as Fig. 3.29(h) shows.

Zero padding is not the only option. For example, we could duplicate the value of the first and last element $m - 1$ times on each side of f, or mirror the first and last $m - 1$ elements and use the mirrored values for padding.

There are two important points to note from the discussion in the preceding paragraph. First, correlation is a function of *displacement* of the filter. In other words, the first value of correlation corresponds to zero displacement of the filter, the second corresponds to one unit displacement, and so on. The second thing to notice is that correlating a filter w with a function that contains all 0s and a single 1 yields a result that is a *copy* of w, but *rotated* by 180°. We call a function that contains a single 1 with the rest being 0s a *discrete unit impulse*. So we conclude that correlation of a function with a discrete unit impulse yields a rotated version of the function at the location of the impulse.

The concept of convolution is a cornerstone of linear system theory. As you will learn in Chapter 4, a fundamental property of convolution is that convolving a function with a unit impulse yields a copy of the function at the location of the impulse. We saw in the previous paragraph that correlation yields a copy

Note that rotation by 180° is equivalent to flipping the function horizontally.

of the function also, but rotated by 180°. Therefore, if we *pre-rotate* the filter and perform the same sliding sum of products operation, we should be able to obtain the desired result. As the right column in Fig. 3.29 shows, this indeed is the case. Thus, we see that to perform convolution all we do is rotate one function by 180° and perform the same operations as in correlation. As it turns out, it makes no difference which of the two functions we rotate.

The preceding concepts extend easily to images, as Fig. 3.30 shows. For a filter of size $m \times n$, we pad the image with a minimum of $m - 1$ rows of 0s at the top and bottom and $n - 1$ columns of 0s on the left and right. In this case, m and n are equal to 3, so we pad f with two rows of 0s above and below and two columns of 0s to the left and right, as Fig. 3.30(b) shows. Figure 3.30(c) shows the initial position of the filter mask for performing correlation, and Fig. 3.30(d) shows the full correlation result. Figure 3.30(e) shows the corre-

In 2-D, rotation by 180° is equivalent to flipping the mask along one axis and then the other.

sponding cropped result. Note again that the result is rotated by 180°. For convolution, we pre-rotate the mask as before and repeat the sliding sum of products just explained. Figures 3.30(f) through (h) show the result. You see again that convolution of a function with an impulse copies the function at the location of the impulse. It should be clear that, if the filter mask is symmetric, correlation and convolution yield the same result.

If, instead of containing a single 1, image f in Fig. 3.30 had contained a region identically equal to w, the value of the correlation function (after normalization) would have been maximum when w was centered on that region of f. Thus, as you will see in Chapter 12, correlation can be used also to find *matches* between images.

Summarizing the preceding discussion in equation form, we have that the correlation of a filter $w(x, y)$ of size $m \times n$ with an image $f(x, y)$, denoted as $w(x, y) \star f(x, y)$, is given by the equation listed at the end of the last section, which we repeat here for convenience:

$$w(x, y) \star f(x, y) = \sum_{s=-a}^{a} \sum_{t=-b}^{b} w(s, t)f(x + s, y + t) \qquad (3.4\text{-}1)$$

This equation is evaluated for all values of the displacement variables x and y so that all elements of w visit every pixel in f, where we assume that f has been padded appropriately. As explained earlier, $a = (m - 1)/2, b = (n - 1)/2$, and we assume for notational convenience that m and n are odd integers.

Padded f

FIGURE 3.30
Correlation
(middle row) and
convolution (last
row) of a 2-D
filter with a 2-D
discrete, unit
impulse. The 0s
are shown in gray
to simplify visual
analysis.

```
              Padded f
              0 0 0 0 0 0 0 0 0
              0 0 0 0 0 0 0 0 0
              0 0 0 0 0 0 0 0 0
 Origin  f(x, y)    0 0 0 0 0 0 0 0 0
0 0 0 0 0      0 0 0 0 1 0 0 0 0
0 0 0 0 0  w(x, y)  0 0 0 0 0 0 0 0 0
0 0 1 0 0   1 2 3   0 0 0 0 0 0 0 0 0
0 0 0 0 0   4 5 6   0 0 0 0 0 0 0 0 0
0 0 0 0 0   7 8 9   0 0 0 0 0 0 0 0 0
   (a)              (b)
```

```
Initial position for w    Full correlation result   Cropped correlation result
1 2 3 0 0 0 0 0 0    0 0 0 0 0 0 0 0 0    0 0 0 0 0
4 5 6 0 0 0 0 0 0    0 0 0 0 0 0 0 0 0    0 9 8 7 0
7 8 9 0 0 0 0 0 0    0 0 0 0 0 0 0 0 0    0 6 5 4 0
0 0 0 0 0 0 0 0 0    0 0 0 9 8 7 0 0 0    0 3 2 1 0
0 0 0 0 1 0 0 0 0    0 0 0 6 5 4 0 0 0    0 0 0 0 0
0 0 0 0 0 0 0 0 0    0 0 0 3 2 1 0 0 0
0 0 0 0 0 0 0 0 0    0 0 0 0 0 0 0 0 0
0 0 0 0 0 0 0 0 0    0 0 0 0 0 0 0 0 0
0 0 0 0 0 0 0 0 0    0 0 0 0 0 0 0 0 0
   (c)                  (d)                  (e)
```

```
Rotated w                 Full convolution result   Cropped convolution result
9 8 7 0 0 0 0 0 0    0 0 0 0 0 0 0 0 0    0 0 0 0 0
6 5 4 0 0 0 0 0 0    0 0 0 0 0 0 0 0 0    0 1 2 3 0
3 2 1 0 0 0 0 0 0    0 0 0 0 0 0 0 0 0    0 4 5 6 0
0 0 0 0 0 0 0 0 0    0 0 0 1 2 3 0 0 0    0 7 8 9 0
0 0 0 0 1 0 0 0 0    0 0 0 4 5 6 0 0 0    0 0 0 0 0
0 0 0 0 0 0 0 0 0    0 0 0 7 8 9 0 0 0
0 0 0 0 0 0 0 0 0    0 0 0 0 0 0 0 0 0
0 0 0 0 0 0 0 0 0    0 0 0 0 0 0 0 0 0
0 0 0 0 0 0 0 0 0    0 0 0 0 0 0 0 0 0
   (f)                  (g)                  (h)
```

In a similar manner, the convolution of $w(x, y)$ and $f(x, y)$, denoted by $w(x, y) \star f(x, y)$,[†] is given by the expression

$$w(x, y) \star f(x, y) = \sum_{s=-a}^{a} \sum_{t=-b}^{b} w(s, t)f(x - s, y - t) \qquad (3.4-2)$$

where the minus signs on the right flip f (i.e., rotate it by $180°$). Flipping and shifting f instead of w is done for notational simplicity and also to follow convention. The result is the same. As with correlation, this equation is evaluated for all values of the displacement variables x and y so that every element of w visits every pixel in f, which we assume has been padded appropriately. You should expand Eq. (3.4-2) for a 3×3 mask and convince yourself that the result using this equation is identical to the example in Fig. 3.30. In practice, we frequently work with an algorithm that implements

Often, when the meaning is clear, we denote the result of correlation or convolution by a function $g(x, y)$, instead of writing $w(x, y) ☆ f(x, y)$ or $w(x, y) \star f(x, y)$. For example, see the equation at the end of the previous section, and Eq. (3.5-1).

[†] Because correlation and convolution are commutative, we have that $w(x, y) ☆ f(x, y) = f(x, y) ☆ w(x, y)$ and $w(x, y) \star f(x, y) = f(x, y) \star w(x, y)$.

Eq. (3.4-1). If we want to perform correlation, we input w into the algorithm; for convolution, we input w rotated by 180°. The reverse is true if an algorithm that implements Eq. (3.4-2) is available instead.

As mentioned earlier, convolution is a cornerstone of linear system theory. As you will learn in Chapter 4, the property that the convolution of a function with a unit impulse copies the function at the location of the impulse plays a central role in a number of important derivations. We will revisit convolution in Chapter 4 in the context of the Fourier transform and the convolution theorem. Unlike Eq. (3.4-2), however, we will be dealing with convolution of functions that are of the same size. The form of the equation is the same, but the limits of summation are different.

Using correlation or convolution to perform spatial filtering is a matter of preference. In fact, because either Eq. (3.4-1) or (3.4-2) can be made to perform the function of the other by a simple rotation of the filter, what is important is that the filter mask used in a given filtering task be specified in a way that corresponds to the intended operation. All the linear spatial filtering results in this chapter are based on Eq. (3.4-1).

Finally, we point out that you are likely to encounter the terms, *convolution filter*, *convolution mask* or *convolution kernel* in the image processing literature. As a rule, these terms are used to denote a spatial filter, and not necessarily that the filter will be used for true convolution. Similarly, "convolving a mask with an image" often is used to denote the sliding, sum-of-products process we just explained, and does not necessarily differentiate between correlation and convolution. Rather, it is used generically to denote either of the two operations. This imprecise terminology is a frequent source of confusion.

3.4.3 Vector Representation of Linear Filtering

When interest lies in the characteristic response, R, of a mask either for correlation or convolution, it is convenient sometimes to write the sum of products as

$$R = w_1 z_1 + w_2 z_2 + \ldots + w_{mn} z_{mn}$$

Consult the Tutorials section of the book Web site for a brief review of vectors and matrices.

$$= \sum_{k=1}^{mn} w_k z_k \qquad (3.4\text{-}3)$$

$$= \mathbf{w}^T \mathbf{z}$$

where the ws are the coefficients of an $m \times n$ filter and the zs are the corresponding image intensities encompassed by the filter. If we are interested in using Eq. (3.4-3) for correlation, we use the mask as given. To use the same equation for convolution, we simply rotate the mask by 180°, as explained in the last section. It is implied that Eq. (3.4-3) holds for a particular pair of coordinates (x, y). You will see in the next section why this notation is convenient for explaining the characteristics of a given linear filter.

FIGURE 3.31
Another
representation of
a general 3 × 3
filter mask.

As an example, Fig. 3.31 shows a general 3 × 3 mask with coefficients labeled as above. In this case, Eq. (3.4-3) becomes

$$R = w_1 z_1 + w_2 z_2 + \ldots + w_9 z_9$$

$$= \sum_{k=1}^{9} w_k z_k \qquad\qquad (3.4\text{-}4)$$

$$= \mathbf{w}^T \mathbf{z}$$

where $\mathbf{w}$ and $\mathbf{z}$ are 9-dimensional vectors formed from the coefficients of the mask and the image intensities encompassed by the mask, respectively.

3.4.4 Generating Spatial Filter Masks

Generating an $m \times n$ *linear* spatial filter requires that we specify mn mask coefficients. In turn, these coefficients are selected based on what the filter is supposed to do, keeping in mind that all we can do with linear filtering is to implement a sum of products. For example, suppose that we want to replace the pixels in an image by the average intensity of a 3 × 3 neighborhood centered on those pixels. The average value at any location (x, y) in the image is the sum of the nine intensity values in the 3 × 3 neighborhood centered on (x, y) divided by 9. Letting $z_i, i = 1, 2, \ldots, 9$, denote these intensities, the average is

$$R = \frac{1}{9} \sum_{i=1}^{9} z_i$$

But this is the same as Eq. (3.4-4) with coefficient values $w_i = 1/9$. In other words, a linear filtering operation with a 3 × 3 mask whose coefficients are $1/9$ implements the desired averaging. As we discuss in the next section, this operation results in image smoothing. We discuss in the following sections a number of other filter masks based on this basic approach.

In some applications, we have a continuous function of two variables, and the objective is to obtain a spatial filter mask based on that function. For example, a Gaussian function of two variables has the basic form

$$h(x, y) = e^{-\frac{x^2 + y^2}{2\sigma^2}}$$

where σ is the standard deviation and, as usual, we assume that coordinates x and y are integers. To generate, say, a 3 × 3 filter mask from this function, we sample it about its center. Thus, $w_1 = h(-1, -1)$, $w_2 = h(-1, 0)$, . . . , $w_9 = h(1, 1)$. An $m \times n$ filter mask is generated in a similar manner. Recall that a 2-D Gaussian function has a bell shape, and that the standard deviation controls the "tightness" of the bell.

Generating a *nonlinear* filter requires that we specify the size of a neighborhood and the operation(s) to be performed on the image pixels contained in the neighborhood. For example, recalling that the max operation is nonlinear (see Section 2.6.2), a 5×5 max filter centered at an arbitrary point (x, y) of an image obtains the maximum intensity value of the 25 pixels and assigns that value to location (x, y) in the processed image. Nonlinear filters are quite powerful, and in some applications can perform functions that are beyond the capabilities of linear filters, as we show later in this chapter and in Chapter 5.

3.5 Smoothing Spatial Filters

Smoothing filters are used for blurring and for noise reduction. Blurring is used in preprocessing tasks, such as removal of small details from an image prior to (large) object extraction, and bridging of small gaps in lines or curves. Noise reduction can be accomplished by blurring with a linear filter and also by nonlinear filtering.

3.5.1 Smoothing Linear Filters

The output (response) of a smoothing, linear spatial filter is simply the average of the pixels contained in the neighborhood of the filter mask. These filters sometimes are called *averaging filters*. As mentioned in the previous section, they also are referred to a *lowpass filters*.

The idea behind smoothing filters is straightforward. By replacing the value of every pixel in an image by the average of the intensity levels in the neighborhood defined by the filter mask, this process results in an image with reduced "sharp" transitions in intensities. Because random noise typically consists of sharp transitions in intensity levels, the most obvious application of smoothing is noise reduction. However, edges (which almost always are desirable features of an image) also are characterized by sharp intensity transitions, so averaging filters have the undesirable side effect that they blur edges. Another application of this type of process includes the smoothing of false contours that result from using an insufficient number of intensity levels, as discussed in Section 2.4.3. A major use of averaging filters is in the reduction of "irrelevant" detail in an image. By "irrelevant" we mean pixel regions that are small with respect to the size of the filter mask. This latter application is illustrated later in this section.

Figure 3.32 shows two 3×3 smoothing filters. Use of the first filter yields the standard average of the pixels under the mask. This can best be seen by substituting the coefficients of the mask into Eq. (3.4-4):

$$R = \frac{1}{9} \sum_{i=1}^{9} z_i$$

which is the average of the intensity levels of the pixels in the 3×3 neighborhood defined by the mask, as discussed earlier. Note that, instead of being 1/9,

FIGURE 3.32 Two 3 × 3 smoothing (averaging) filter masks. The constant multiplier in front of each mask is equal to 1 divided by the sum of the values of its coefficients, as is required to compute an average.

the coefficients of the filter are all 1s. The idea here is that it is computationally more efficient to have coefficients valued 1. At the end of the filtering process the entire image is divided by 9. An $m \times n$ mask would have a normalizing constant equal to $1/mn$. A spatial averaging filter in which all coefficients are equal sometimes is called a *box filter*.

The second mask in Fig. 3.32 is a little more interesting. This mask yields a so-called *weighted average*, terminology used to indicate that pixels are multiplied by different coefficients, thus giving more importance (weight) to some pixels at the expense of others. In the mask shown in Fig. 3.32(b) the pixel at the center of the mask is multiplied by a higher value than any other, thus giving this pixel more importance in the calculation of the average. The other pixels are inversely weighted as a function of their distance from the center of the mask. The diagonal terms are further away from the center than the orthogonal neighbors (by a factor of $\sqrt{2}$) and, thus, are weighed less than the immediate neighbors of the center pixel. The basic strategy behind weighing the center point the highest and then reducing the value of the coefficients as a function of increasing distance from the origin is simply an attempt to reduce blurring in the smoothing process. We could have chosen other weights to accomplish the same general objective. However, the sum of all the coefficients in the mask of Fig. 3.32(b) is equal to 16, an attractive feature for computer implementation because it is an integer power of 2. In practice, it is difficult in general to see differences between images smoothed by using either of the masks in Fig. 3.32, or similar arrangements, because the area spanned by these masks at any one location in an image is so small.

With reference to Eq. (3.4-1), the general implementation for filtering an $M \times N$ image with a weighted averaging filter of size $m \times n$ (m and n odd) is given by the expression

$$g(x, y) = \frac{\sum_{s=-a}^{a} \sum_{t=-b}^{b} w(s, t)f(x + s, y + t)}{\sum_{s=-a}^{a} \sum_{t=-b}^{b} w(s, t)} \tag{3.5-1}$$

The parameters in this equation are as defined in Eq. (3.4-1). As before, it is understood that the complete filtered image is obtained by applying Eq. (3.5-1) for $x = 0, 1, 2, \ldots, M - 1$ and $y = 0, 1, 2, \ldots, N - 1$. The denominator in

Eq. (3.5-1) is simply the sum of the mask coefficients and, therefore, it is a constant that needs to be computed only once.

EXAMPLE 3.13:
Image smoothing with masks of various sizes.

■ The effects of smoothing as a function of filter size are illustrated in Fig. 3.33, which shows an original image and the corresponding smoothed results obtained using square averaging filters of sizes $m = 3, 5, 9, 15$, and 35 pixels, respectively. The principal features of these results are as follows: For $m = 3$, we note a general slight blurring throughout the entire image but, as expected, details that are of approximately the same size as the filter mask are affected considerably more. For example, the 3×3 and 5×5 black squares in the image, the small letter "a," and the fine grain noise show significant blurring when compared to the rest of the image. Note that the noise is less pronounced, and the jagged borders of the characters were pleasingly smoothed.

The result for $m = 5$ is somewhat similar, with a slight further increase in blurring. For $m = 9$ we see considerably more blurring, and the 20% black circle is not nearly as distinct from the background as in the previous three images, illustrating the blending effect that blurring has on objects whose intensities are close to that of its neighboring pixels. Note the significant further smoothing of the noisy rectangles. The results for $m = 15$ and 35 are extreme with respect to the sizes of the objects in the image. This type of aggresive blurring generally is used to eliminate small objects from an image. For instance, the three small squares, two of the circles, and most of the noisy rectangle areas have been blended into the background of the image in Fig. 3.33(f). Note also in this figure the pronounced black border. This is a result of padding the border of the original image with 0s (black) and then trimming off the padded area after filtering. Some of the black was blended into all filtered images, but became truly objectionable for the images smoothed with the larger filters. ■

As mentioned earlier, an important application of spatial averaging is to blur an image for the purpose of getting a gross representation of objects of interest, such that the intensity of smaller objects blends with the background and larger objects become "bloblike" and easy to detect. The size of the mask establishes the relative size of the objects that will be blended with the background. As an illustration, consider Fig. 3.34(a), which is an image from the Hubble telescope in orbit around the Earth. Figure 3.34(b) shows the result of applying a 15×15 averaging mask to this image. We see that a number of objects have either blended with the background or their intensity has diminished considerably. It is typical to follow an operation like this with thresholding to eliminate objects based on their intensity. The result of using the thresholding function of Fig. 3.2(b) with a threshold value equal to 25% of the highest intensity in the blurred image is shown in Fig. 3.34(c). Comparing this result with the original image, we see that it is a reasonable representation of what we would consider to be the largest, brightest objects in that image.

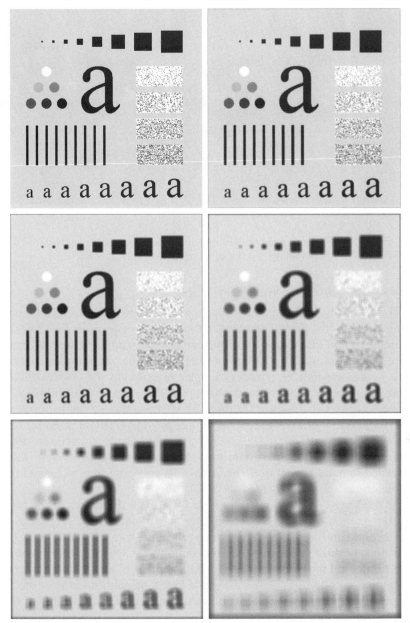

FIGURE 3.33 (a) Original image, of size 500×500 pixels. (b)–(f) Results of smoothing with square averaging filter masks of sizes $m = 3, 5, 9, 15$, and 35, respectively. The black squares at the top are of sizes 3, 5, 9, 15, 25, 35, 45, and 55 pixels, respectively; their borders are 25 pixels apart. The letters at the bottom range in size from 10 to 24 points, in increments of 2 points; the large letter at the top is 60 points. The vertical bars are 5 pixels wide and 100 pixels high; their separation is 20 pixels. The diameter of the circles is 25 pixels, and their borders are 15 pixels apart; their intensity levels range from 0% to 100% black in increments of 20%. The background of the image is 10% black. The noisy rectangles are of size 50×120 pixels.

a	b
c	d
e	f

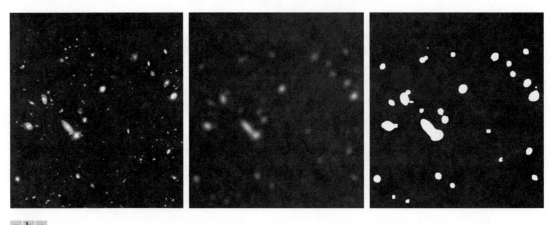

a b c

FIGURE 3.34 (a) Image of size 528 × 485 pixels from the Hubble Space Telescope. (b) Image filtered with a 15 × 15 averaging mask. (c) Result of thresholding (b). (Original image courtesy of NASA.)

3.5.2 Order-Statistic (Nonlinear) Filters

Order-statistic filters are nonlinear spatial filters whose response is based on ordering (ranking) the pixels contained in the image area encompassed by the filter, and then replacing the value of the center pixel with the value determined by the ranking result. The best-known filter in this category is the *median filter*, which, as its name implies, replaces the value of a pixel by the median of the intensity values in the neighborhood of that pixel (the original value of the pixel is included in the computation of the median). Median filters are quite popular because, for certain types of random noise, they provide excellent noise-reduction capabilities, with considerably less blurring than linear smoothing filters of similar size. Median filters are particularly effective in the presence of *impulse noise*, also called *salt-and-pepper noise* because of its appearance as white and black dots superimposed on an image.

The median, ξ, of a set of values is such that half the values in the set are less than or equal to ξ, and half are greater than or equal to ξ. In order to perform median filtering at a point in an image, we first sort the values of the pixel in the neighborhood, determine their median, and assign that value to the corresponding pixel in the filtered image. For example, in a 3×3 neighborhood the median is the 5th largest value, in a 5×5 neighborhood it is the 13th largest value, and so on. When several values in a neighborhood are the same, all equal values are grouped. For example, suppose that a 3×3 neighborhood has values $(10, 20, 20, 20, 15, 20, 20, 25, 100)$. These values are sorted as $(10, 15, 20, 20, 20, 20, 20, 25, 100)$, which results in a median of 20. Thus, the principal function of median filters is to force points with distinct intensity levels to be more like their neighbors. In fact, isolated clusters of pixels that are light or dark with respect to their neighbors, and whose area is less than $m^2/2$ (one-half the filter area), are eliminated by an $m \times m$ median filter. In this case "eliminated" means forced to the median intensity of the neighbors. Larger clusters are affected considerably less.

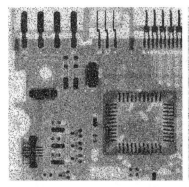 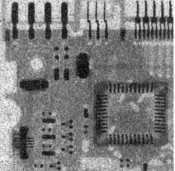 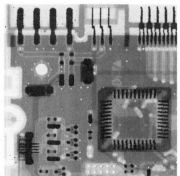

a b c

FIGURE 3.35 (a) X-ray image of circuit board corrupted by salt-and-pepper noise. (b) Noise reduction with a 3 × 3 averaging mask. (c) Noise reduction with a 3 × 3 median filter. (Original image courtesy of Mr. Joseph E. Pascente, Lixi, Inc.)

Although the median filter is by far the most useful order-statistic filter in image processing, it is by no means the only one. The median represents the 50th percentile of a ranked set of numbers, but recall from basic statistics that ranking lends itself to many other possibilities. For example, using the 100th percentile results in the so-called *max filter*, which is useful for finding the brightest points in an image. The response of a 3 × 3 max filter is given by $R = \max\{z_k | k = 1, 2, \ldots, 9\}$. The 0th percentile filter is the *min filter*, used for the opposite purpose. Median, max, min, and several other nonlinear filters are considered in more detail in Section 5.3.

See Section 10.3.5 regarding percentiles.

■ Figure 3.35(a) shows an X-ray image of a circuit board heavily corrupted by salt-and-pepper noise. To illustrate the point about the superiority of median filtering over average filtering in situations such as this, we show in Fig. 3.35(b) the result of processing the noisy image with a 3 × 3 neighborhood averaging mask, and in Fig. 3.35(c) the result of using a 3 × 3 median filter. The averaging filter blurred the image and its noise reduction performance was poor. The superiority in all respects of median over average filtering in this case is quite evident. In general, median filtering is much better suited than averaging for the removal of salt-and-pepper noise. ■

EXAMPLE 3.14: Use of median filtering for noise reduction.

3.6 Sharpening Spatial Filters

The principal objective of sharpening is to highlight transitions in intensity. Uses of image sharpening vary and include applications ranging from electronic printing and medical imaging to industrial inspection and autonomous guidance in military systems. In the last section, we saw that image blurring could be accomplished in the spatial domain by pixel averaging in a neighborhood. Because averaging is analogous to integration, it is logical to conclude that sharpening can be accomplished by spatial differentiation. This, in fact, is the case, and the discussion in this section deals with various ways of defining and implementing operators for sharpening by digital differentiation. Fundamentally, the strength of the response of a derivative operator is proportional to the degree of intensity discontinuity of the image at the point at which the operator is applied. Thus, image differentiation enhances edges and other discontinuities (such as noise) and deemphasizes areas with slowly varying intensities.

3.6.1 Foundation

In the two sections that follow, we consider in some detail sharpening filters that are based on first- and second-order derivatives, respectively. Before proceeding with that discussion, however, we stop to look at some of the fundamental properties of these derivatives in a digital context. To simplify the explanation, we focus attention initially on one-dimensional derivatives. In particular, we are interested in the behavior of these derivatives in areas of constant intensity, at the onset and end of discontinuities (step and ramp discontinuities), and along intensity ramps. As you will see in Chapter 10, these types of discontinuities can be used to model noise points, lines, and edges in an image. The behavior of derivatives during transitions into and out of these image features also is of interest.

The derivatives of a digital function are defined in terms of differences. There are various ways to define these differences. However, we require that any definition we use for a *first derivative* (1) must be zero in areas of constant intensity; (2) must be nonzero at the onset of an intensity step or ramp; and (3) must be nonzero along ramps. Similarly, any definition of a *second derivative* (1) must be zero in constant areas; (2) must be nonzero at the onset *and* end of an intensity step or ramp; and (3) must be zero along ramps of constant slope. Because we are dealing with digital quantities whose values are finite, the maximum possible intensity change also is finite, and the shortest distance over which that change can occur is between adjacent pixels.

A basic definition of the first-order derivative of a one-dimensional function $f(x)$ is the difference

We return to Eq. (3.6-1) in Section 10.2.1 and show how it follows from a Taylor series expansion. For now, we accept it as a definition.

$$\frac{\partial f}{\partial x} = f(x + 1) - f(x) \tag{3.6-1}$$

We used a partial derivative here in order to keep the notation the same as when we consider an image function of two variables, $f(x, y)$, at which time we will be dealing with partial derivatives along the two spatial axes. Use of a partial derivative in the present discussion does not affect in any way the nature of what we are trying to accomplish. Clearly, $\partial f / \partial x = df/dx$ when there is only one variable in the function; the same is true for the second derivative.

We define the second-order derivative of $f(x)$ as the difference

$$\frac{\partial^2 f}{\partial x^2} = f(x + 1) + f(x - 1) - 2f(x) \tag{3.6-2}$$

It is easily verified that these two definitions satisfy the conditions stated above. To illustrate this, and to examine the similarities and differences between

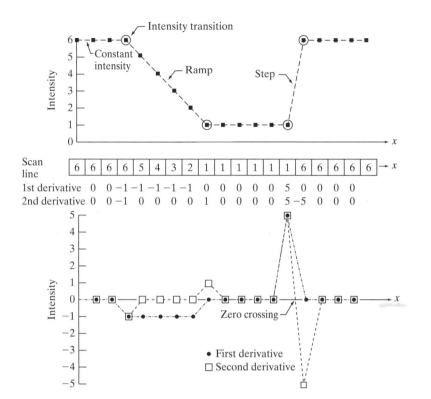

FIGURE3.36
Illustration of the first and second derivatives of a 1-D digital function representing a section of a horizontal intensity profile from an image. In (a) and (c) data points are joined by dashed lines as a visualization aid.

first- and second-order derivatives of a digital function, consider the example in Fig. 3.36.

Figure 3.36(b) (center of the figure) shows a section of a scan line (intensity profile). The values inside the small squares are the intensity values in the scan line, which are plotted as black dots above it in Fig. 3.36(a). The dashed line connecting the dots is included to aid visualization. As the figure shows, the scan line contains an intensity ramp, three sections of constant intensity, and an intensity step. The circles indicate the onset or end of intensity transitions. The first- and second-order derivatives computed using the two preceding definitions are included below the scan line in Fig. 3.36(b), and are plotted in Fig. 3.36(c). When computing the first derivative at a location x, we subtract the value of the function at that location from the next point. So this is a "look-ahead" operation. Similarly, to compute the second derivative at x, we use the previous and the next points in the computation. To avoid a situation in which the previous or next points are outside the range of the scan line, we show derivative computations in Fig. 3.36 from the second through the penultimate points in the sequence.

Let us consider the properties of the first and second derivatives as we traverse the profile from left to right. First, we encounter an area of constant intensity and, as Figs. 3.36(b) and (c) show, both derivatives are zero there, so condition (1) is satisfied for both. Next, we encounter an intensity ramp followed by a step, and we note that the first-order derivative is nonzero at the onset of the ramp and the step; similarly, the second derivative is nonzero at the onset *and* end of both the ramp and the step; therefore, property (2) is satisfied for both derivatives. Finally, we see that property (3) is satisfied also for both derivatives because the first derivative is nonzero and the second is zero along the ramp. Note that the sign of the second derivative changes at the onset and end of a step or ramp. In fact, we see in Fig. 3.36(c) that in a step transition a line joining these two values crosses the horizontal axis midway between the two extremes. This *zero crossing* property is quite useful for locating edges, as you will see in Chapter 10.

Edges in digital images often are ramp-like transitions in intensity, in which case the first derivative of the image would result in thick edges because the derivative is nonzero along a ramp. On the other hand, the second derivative would produce a double edge one pixel thick, separated by zeros. From this, we conclude that the second derivative enhances fine detail much better than the first derivative, a property that is ideally suited for sharpening images. Also, as you will learn later in this section, second derivatives are much easier to implement than first derivates, so we focus our attention initially on second derivatives.

3.6.2 Using the Second Derivative for Image Sharpening—The Laplacian

In this section we consider the implementation of 2-D, second-order derivatives and their use for image sharpening. We return to this derivative in Chapter 10, where we use it extensively for image segmentation. The approach basically consists of defining a discrete formulation of the second-order derivative and then constructing a filter mask based on that formulation. We are interested in *isotropic* filters, whose response is independent of the direction of the discontinuities in the image to which the filter is applied. In other words, isotropic filters are *rotation invariant*, in the sense that rotating the image and then applying the filter gives the same result as applying the filter to the image first and then rotating the result.

It can be shown (Rosenfeld and Kak [1982]) that the simplest isotropic derivative operator is the Laplacian, which, for a function (image) $f(x, y)$ of two variables, is defined as

$$\nabla^2 f = \frac{\partial^2 f}{\partial x^2} + \frac{\partial^2 f}{\partial y^2} \tag{3.6-3}$$

Because derivatives of any order are linear operations, the Laplacian is a linear operator. To express this equation in discrete form, we use the definition in Eq. (3.6-2), keeping in mind that we have to carry a second variable. In the *x*-direction, we have

$$\frac{\partial^2 f}{\partial x^2} = f(x + 1, y) + f(x - 1, y) - 2f(x, y) \tag{3.6-4}$$

and, similarly, in the *y*-direction we have

$$\frac{\partial^2 f}{\partial y^2} = f(x, y + 1) + f(x, y - 1) - 2f(x, y) \tag{3.6-5}$$

Therefore, it follows from the preceding three equations that the discrete Laplacian of two variables is

$$\nabla^2 f(x, y) = f(x + 1, y) + f(x - 1, y) + f(x, y + 1) + f(x, y - 1)$$
$$-4f(x, y) \tag{3.6-6}$$

This equation can be implemented using the filter mask in Fig. 3.37(a), which gives an isotropic result for rotations in increments of 90°. The mechanics of implementation are as in Section 3.5.1 for linear smoothing filters. We simply are using different coefficients here.

The diagonal directions can be incorporated in the definition of the digital Laplacian by adding two more terms to Eq. (3.6-6), one for each of the two diagonal directions. The form of each new term is the same as either Eq. (3.6-4) or (3.6-5), but the coordinates are along the diagonals. Because each diagonal term also contains a $-2f(x, y)$ term, the total subtracted from the difference terms now would be $-8f(x, y)$. Figure 3.37(b) shows the filter mask used to implement this new definition. This mask yields isotropic results in increments of 45°. You are likely to see in practice the Laplacian masks in Figs. 3.37(c) and (d). They are obtained from definitions of the second derivatives that are the negatives of the ones we used in Eqs. (3.6-4) and (3.6-5). As such, they yield equivalent results, but the difference in sign must be kept in mind when combining (by addition or subtraction) a Laplacian-filtered image with another image.

Because the Laplacian is a derivative operator, its use highlights intensity discontinuities in an image and deemphasizes regions with slowly varying intensity levels. This will tend to produce images that have grayish edge lines and other discontinuities, all superimposed on a dark, featureless background. Background features can be "recovered" while still preserving the sharpening

0	1	0
1	−4	1
0	1	0

1	1	1
1	−8	1
1	1	1

0	−1	0
−1	4	−1
0	−1	0

−1	−1	−1
−1	8	−1
−1	−1	−1

a b
c d

FIGURE 3.37
(a) Filter mask used to implement Eq. (3.6-6).
(b) Mask used to implement an extension of this equation that includes the diagonal terms.
(c) and (d) Two other implementations of the Laplacian found frequently in practice.

effect of the Laplacian simply by adding the Laplacian image to the original. As noted in the previous paragraph, it is important to keep in mind which definition of the Laplacian is used. If the definition used has a negative center coefficient, then we *subtract*, rather than add, the Laplacian image to obtain a sharpened result. Thus, the basic way in which we use the Laplacian for image sharpening is

$$g(x, y) = f(x, y) + c\left[\nabla^2 f(x, y)\right] \qquad (3.6-7)$$

where $f(x, y)$ and $g(x, y)$ are the input and sharpened images, respectively. The constant is $c = -1$ if the Laplacian filters in Fig. 3.37(a) or (b) are used, and $c = 1$ if either of the other two filters is used.

EXAMPLE 3.15:
Image sharpening using the Laplacian.

■ Figure 3.38(a) shows a slightly blurred image of the North Pole of the moon. Figure 3.38(b) shows the result of filtering this image with the Laplacian mask in Fig. 3.37(a). Large sections of this image are black because the Laplacian contains both positive and negative values, and all negative values are clipped at 0 by the display.

A typical way to scale a Laplacian image is to add to it its minimum value to bring the new minimum to zero and then scale the result to the full $[0, L - 1]$ intensity range, as explained in Eqs. (2.6-10) and (2.6-11). The image in Fig. 3.38(c) was scaled in this manner. Note that the dominant features of the image are edges and sharp intensity discontinuities. The background, previously black, is now gray due to scaling. This grayish appearance is typical of Laplacian images that have been scaled properly. Figure 3.38(d) shows the result obtained using Eq. (3.6-7) with $c = -1$. The detail in this image is unmistakably clearer and sharper than in the original image. Adding the original image to the Laplacian restored the overall intensity variations in the image, with the Laplacian increasing the contrast at the locations of intensity discontinuities. The net result is an image in which small details were enhanced and the background tonality was reasonably preserved. Finally, Fig. 3.38(e) shows the result of repeating the preceding procedure with the filter in Fig. 3.37(b). Here, we note a significant improvement in sharpness over Fig. 3.38(d). This is not unexpected because using the filter in Fig. 3.37(b) provides additional differentiation (sharpening) in the diagonal directions. Results such as those in Figs. 3.38(d) and (e) have made the Laplacian a tool of choice for sharpening digital images. ■

3.6.3 Unsharp Masking and Highboost Filtering

A process that has been used for many years by the printing and publishing industry to sharpen images consists of subtracting an unsharp (smoothed) version of an image from the original image. This process, called *unsharp masking*, consists of the following steps:

1. Blur the original image.
2. Subtract the blurred image from the original (the resulting difference is called the *mask*.)
3. Add the mask to the original.

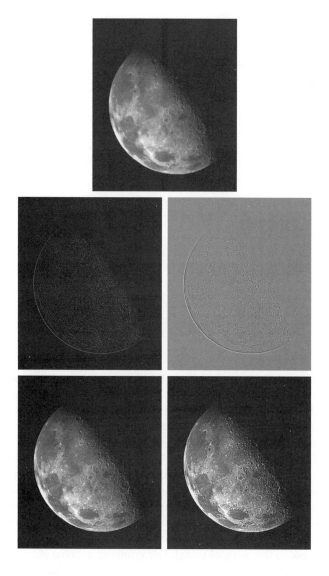

a
b c
d e

FIGURE 3.38
(a) Blurred image of the North Pole of the moon.
(b) Laplacian without scaling.
(c) Laplacian with scaling. (d) Image sharpened using the mask in Fig. 3.37(a). (e) Result of using the mask in Fig. 3.37(b). (Original image courtesy of NASA.)

Letting $\overline{f}(x, y)$ denote the blurred image, unsharp masking is expressed in equation form as follows. First we obtain the mask:

$$g_{\text{mask}}(x, y) = f(x, y) - \overline{f}(x, y) \tag{3.6-8}$$

Then we add a weighted portion of the mask back to the original image:

$$g(x, y) = f(x, y) + k * g_{\text{mask}}(x, y) \tag{3.6-9}$$

where we included a weight, k ($k \geq 0$), for generality. When $k = 1$, we have unsharp masking, as defined above. When $k > 1$, the process is referred to as

a
b
c
d

FIGURE 3.39 1-D illustration of the mechanics of unsharp masking. (a) Original signal. (b) Blurred signal with original shown dashed for reference. (c) Unsharp mask. (d) Sharpened signal, obtained by adding (c) to (a).

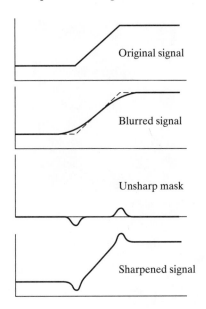

Original signal

Blurred signal

Unsharp mask

Sharpened signal

highboost filtering. Choosing $k < 1$ de-emphasizes the contribution of the unsharp mask.

Figure 3.39 explains how unsharp masking works. The intensity profile in Fig. 3.39(a) can be interpreted as a horizontal scan line through a vertical edge that transitions from a dark to a light region in an image. Figure 3.39(b) shows the result of smoothing, superimposed on the original signal (shown dashed) for reference. Figure 3.39(c) is the unsharp mask, obtained by subtracting the blurred signal from the original. By comparing this result with the section of Fig. 3.36(c) corresponding to the ramp in Fig. 3.36(a), we note that the unsharp mask in Fig. 3.39(c) is very similar to what we would obtain using a second-order derivative. Figure 3.39(d) is the final sharpened result, obtained by adding the mask to the original signal. The points at which a change of slope in the intensity occurs in the signal are now emphasized (sharpened). Observe that negative values were added to the original. Thus, it is possible for the final result to have negative intensities if the original image has any zero values or if the value of k is chosen large enough to emphasize the peaks of the mask to a level larger than the minimum value in the original. Negative values would cause a dark halo around edges, which, if k is large enough, can produce objectionable results.

EXAMPLE 3.16:
Image sharpening using unsharp masking.

■ Figure 3.40(a) shows a slightly blurred image of white text on a dark gray background. Figure 3.40(b) was obtained using a Gaussian smoothing filter (see Section 3.4.4) of size 5×5 with $\sigma = 3$. Figure 3.40(c) is the unsharp mask, obtained using Eq. (3.6-8). Figure 3.40(d) was obtained using unsharp

a
b
c
d
e

FIGURE 3.40
(a) Original
image.
(b) Result of
blurring with a
Gaussian filter.
(c) Unsharp
mask. (d) Result
of using unsharp
masking.
(e) Result of
using highboost
filtering.

masking [Eq. (3.6-9) with $k = 1$]. This image is a slight improvement over the
original, but we can do better. Figure 3.40(e) shows the result of using Eq. (3.6-9)
with $k = 4.5$, the largest possible value we could use and still keep positive all the
values in the final result. The improvement in this image over the original is
significant. ■

3.6.4 Using First-Order Derivatives for (Nonlinear) Image Sharpening—The Gradient

First derivatives in image processing are implemented using the magnitude of
the gradient. For a function $f(x, y)$, the gradient of f at coordinates (x, y) is defined as the two-dimensional column *vector*

$$\nabla f \equiv \text{grad}(f) \equiv \begin{bmatrix} g_x \\ g_y \end{bmatrix} = \begin{bmatrix} \dfrac{\partial f}{\partial x} \\ \dfrac{\partial f}{\partial y} \end{bmatrix} \qquad (3.6\text{-}10)$$

We discuss the gradient
in detail in Section
10.2.5. Here, we are interested only in using the
magnitude of the gradi-
ent for image sharpening.

This vector has the important geometrical property that it points in the direction of the greatest rate of change of f at location (x, y).

The *magnitude* (*length*) of vector ∇f, denoted as $M(x, y)$, where

$$M(x, y) = \text{mag}(\nabla f) = \sqrt{g_x^2 + g_y^2} \qquad (3.6\text{-}11)$$

is the *value* at (x, y) of the rate of change in the direction of the gradient vector. Note that $M(x, y)$ is an image of the same size as the original, created when
x and y are allowed to vary over all pixel locations in f. It is common practice
to refer to this image as the *gradient image* (or simply as the *gradient* when the
meaning is clear).

Because the components of the gradient vector are derivatives, they are linear operators. However, the magnitude of this vector is not because of the squaring and square root operations. On the other hand, the partial derivatives in Eq. (3.6-10) are not rotation invariant (isotropic), but the magnitude of the gradient vector is. In some implementations, it is more suitable computationally to approximate the squares and square root operations by absolute values:

$$M(x, y) \approx |g_x| + |g_y| \tag{3.6-12}$$

This expression still preserves the relative changes in intensity, but the isotropic property is lost in general. However, as in the case of the Laplacian, the isotropic properties of the discrete gradient defined in the following paragraph are preserved only for a limited number of rotational increments that depend on the filter masks used to approximate the derivatives. As it turns out, the most popular masks used to approximate the gradient are isotropic at multiples of 90°. These results are independent of whether we use Eq. (3.6-11) or (3.6-12), so nothing of significance is lost in using the latter equation if we choose to do so.

As in the case of the Laplacian, we now define discrete approximations to the preceding equations and from there formulate the appropriate filter masks. In order to simplify the discussion that follows, we will use the notation in Fig. 3.41(a) to denote the intensities of image points in a 3 × 3 region. For

FIGURE 3.41
A 3 × 3 region of an image (the zs are intensity values).
(b)–(c) Roberts cross gradient operators.
(d)–(e) Sobel operators. All the mask coefficients sum to zero, as expected of a derivative operator.

z_1	z_2	z_3
z_4	z_5	z_6
z_7	z_8	z_9

−1	0
0	1

0	−1
1	0

−1	−2	−1
0	0	0
1	2	1

−1	0	1
−2	0	2
−1	0	1

example, the center point, z_5, denotes $f(x, y)$ at an arbitrary location, (x, y); z_1 denotes $f(x - 1, y - 1)$; and so on, using the notation introduced in Fig. 3.28. As indicated in Section 3.6.1, the simplest approximations to a first-order derivative that satisfy the conditions stated in that section are $g_x = (z_8 - z_5)$ and $g_y = (z_6 - z_5)$. Two other definitions proposed by Roberts [1965] in the early development of digital image processing use cross differences:

$$g_x = (z_9 - z_5) \quad \text{and} \quad g_y = (z_8 - z_6) \tag{3.6-13}$$

If we use Eqs. (3.6-11) and (3.6-13), we compute the gradient image as

$$M(x, y) = \left[(z_9 - z_5)^2 + (z_8 - z_6)^2 \right]^{1/2} \tag{3.6-14}$$

If we use Eqs. (3.6-12) and (3.6-13), then

$$M(x, y) \approx |z_9 - z_5| + |z_8 - z_6| \tag{3.6-15}$$

where it is understood that x and y vary over the dimensions of the image in the manner described earlier. The partial derivative terms needed in equation (3.6-13) can be implemented using the two linear filter masks in Figs. 3.41(b) and (c). These masks are referred to as the *Roberts cross-gradient operators*.

Masks of even sizes are awkward to implement because they do not have a center of symmetry. The smallest filter masks in which we are interested are of size 3 × 3. Approximations to g_x and g_y using a 3 × 3 neighborhood centered on z_5 are as follows:

$$g_x = \frac{\partial f}{\partial x} = (z_7 + 2z_8 + z_9) - (z_1 + 2z_2 + z_3) \tag{3.6-16}$$

and

$$g_y = \frac{\partial f}{\partial y} = (z_3 + 2z_6 + z_9) - (z_1 + 2z_4 + z_7) \tag{3.6-17}$$

These equations can be implemented using the masks in Figs. 3.41(d) and (e). The difference between the third and first rows of the 3 × 3 image region implemented by the mask in Fig. 3.41(d) approximates the partial derivative in the x-direction, and the difference between the third and first columns in the other mask approximates the derivative in the y-direction. After computing the partial derivatives with these masks, we obtain the magnitude of the gradient as before. For example, substituting g_x and g_y into Eq. (3.6-12) yields

$$M(x, y) \approx |(z_7 + 2z_8 + z_9) - (z_1 + 2z_2 + z_3)| \\ + |(z_3 + 2z_6 + z_9) - (z_1 + 2z_4 + z_7)| \tag{3.6-18}$$

The masks in Figs. 3.41(d) and (e) are called the *Sobel operators*. The idea behind using a weight value of 2 in the center coefficient is to achieve some smoothing by giving more importance to the center point (we discuss this in more detail in Chapter 10). Note that the coefficients in all the masks shown in Fig. 3.41 sum to 0, indicating that they would give a response of 0 in an area of constant intensity, as is expected of a derivative operator.

As mentioned earlier, the computations of g_x and g_y are linear operations because they involve derivatives and, therefore, can be implemented as a sum of products using the spatial masks in Fig. 3.41. The nonlinear aspect of sharpening with the gradient is the computation of $M(x, y)$ involving squaring and square roots, or the use of absolute values, all of which are nonlinear operations. These operations are performed *after* the linear process that yields g_x and g_y.

EXAMPLE 3.17:
Use of the
gradient for edge
enhancement.

■ The gradient is used frequently in industrial inspection, either to aid humans in the detection of defects or, what is more common, as a preprocessing step in automated inspection. We will have more to say about this in Chapters 10 and 11. However, it will be instructive at this point to consider a simple example to show how the gradient can be used to enhance defects and eliminate slowly changing background features. In this example, enhancement is used as a preprocessing step for automated inspection, rather than for human analysis.

Figure 3.42(a) shows an optical image of a contact lens, illuminated by a lighting arrangement designed to highlight imperfections, such as the two edge defects in the lens boundary seen at 4 and 5 o'clock. Figure 3.42(b) shows the gradient obtained using Eq. (3.6-12) with the two Sobel masks in Figs. 3.41(d) and (e). The edge defects also are quite visible in this image, but with the added advantage that constant or slowly varying shades of gray have been eliminated, thus simplifying considerably the computational task required for automated inspection. The gradient can be used also to highlight small specs that may not be readily visible in a gray-scale image (specs like these can be foreign matter, air pockets in a supporting solution, or miniscule imperfections in the lens). The ability to enhance small discontinuities in an otherwise flat gray field is another important feature of the gradient.   ■

a b

FIGURE 3.42
(a) Optical image
of contact lens
(note defects on
the boundary at 4
and 5 o'clock).
(b) Sobel
gradient.
(Original image
courtesy of Pete
Sites, Perceptics
Corporation.)

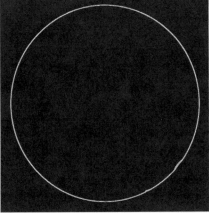

3.7 Combining Spatial Enhancement Methods

With a few exceptions, like combining blurring with thresholding (Fig. 3.34), we have focused attention thus far on individual approaches. Frequently, a given task will require application of several complementary techniques in order to achieve an acceptable result. In this section we illustrate by means of an example how to combine several of the approaches developed thus far in this chapter to address a difficult image enhancement task.

The image in Fig. 3.43(a) is a nuclear whole body bone scan, used to detect diseases such as bone infection and tumors. Our objective is to enhance this image by sharpening it and by bringing out more of the skeletal detail. The narrow dynamic range of the intensity levels and high noise content make this image difficult to enhance. The strategy we will follow is to utilize the Laplacian to highlight fine detail, and the gradient to enhance prominent edges. For reasons that will be explained shortly, a smoothed version of the gradient image will be used to mask the Laplacian image (see Fig. 2.30 regarding masking). Finally, we will attempt to increase the dynamic range of the intensity levels by using an intensity transformation.

Figure 3.43(b) shows the Laplacian of the original image, obtained using the filter in Fig. 3.37(d). This image was scaled (for display only) using the same technique as in Fig. 3.38(c). We can obtain a sharpened image at this point simply by adding Figs. 3.43(a) and (b), according to Eq. (3.6-7). Just by looking at the noise level in Fig. 3.43(b), we would expect a rather noisy sharpened image if we added Figs. 3.43(a) and (b), a fact that is confirmed by the result in Fig. 3.43(c). One way that comes immediately to mind to reduce the noise is to use a median filter. However, median filtering is a nonlinear process capable of removing image features. This is unacceptable in medical image processing.

An alternate approach is to use a mask formed from a smoothed version of the gradient of the original image. The motivation behind this is straightforward and is based on the properties of first- and second-order derivatives explained in Section 3.6.1. The Laplacian, being a second-order derivative operator, has the definite advantage that it is superior in enhancing fine detail. However, this causes it to produce noisier results than the gradient. This noise is most objectionable in smooth areas, where it tends to be more visible. The gradient has a stronger average response in areas of significant intensity transitions (ramps and steps) than does the Laplacian. The response of the gradient to noise and fine detail is lower than the Laplacian's and can be lowered further by smoothing the gradient with an averaging filter. The idea, then, is to smooth the gradient and multiply it by the Laplacian image. In this context, we may view the smoothed gradient as a mask image. The product will preserve details in the strong areas while reducing noise in the relatively flat areas. This process can be interpreted roughly as combining the best features of the Laplacian and the gradient. The result is added to the original to obtain a final sharpened image.

Figure 3.43(d) shows the Sobel gradient of the original image, computed using Eq. (3.6-12). Components g_x and g_y were obtained using the masks in Figs. 3.41(d) and (e), respectively. As expected, edges are much more dominant

FIGURE 3.43
(a) Image of
whole body bone
scan.
(b) Laplacian of
(a). (c) Sharpened
image obtained by
adding (a) and (b).
(d) Sobel gradient
of (a).

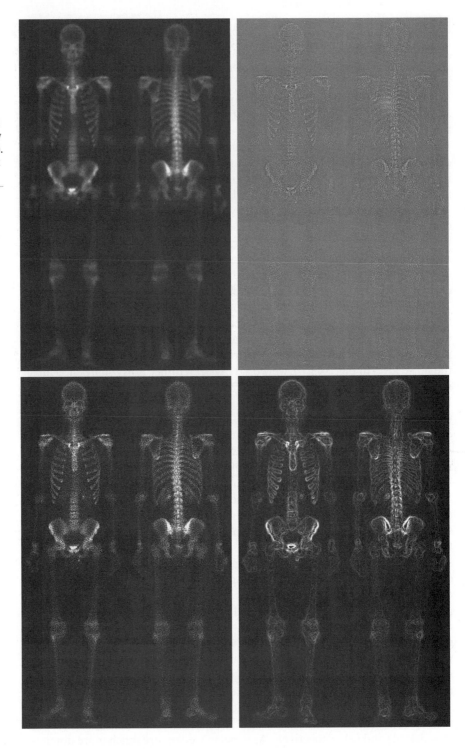

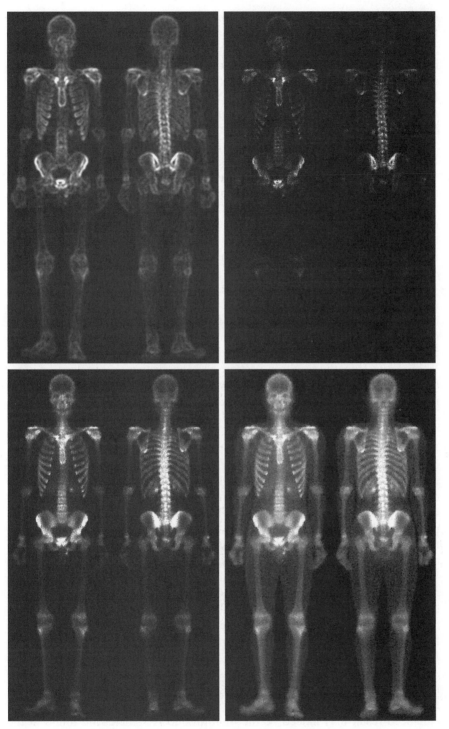

FIGURE 3.43
(Continued)
(e) Sobel image smoothed with a 5 × 5 averaging filter. (f) Mask image formed by the product of (c) and (e).
(g) Sharpened image obtained by the sum of (a) and (f). (h) Final result obtained by applying a power-law transformation to (g). Compare (g) and (h) with (a). (Original image courtesy of G.E. Medical Systems.)

in this image than in the Laplacian image. The smoothed gradient image in Fig. 3.43(e) was obtained by using an averaging filter of size 5 × 5. The two gradient images were scaled for display in the same manner as the Laplacian image. Because the smallest possible value of a gradient image is 0, the background is black in the scaled gradient images, rather than gray as in the scaled Laplacian. The fact that Figs. 3.43(d) and (e) are much brighter than Fig. 3.43(b) is again evidence that the gradient of an image with significant edge content has values that are higher in general than in a Laplacian image.

The product of the Laplacian and smoothed-gradient image is shown in Fig. 3.43(f). Note the dominance of the strong edges and the relative lack of visible noise, which is the key objective behind masking the Laplacian with a smoothed gradient image. Adding the product image to the original resulted in the sharpened image shown in Fig. 3.43(g). The significant increase in sharpness of detail in this image over the original is evident in most parts of the image, including the ribs, spinal cord, pelvis, and skull. This type of improvement would not have been possible by using the Laplacian or the gradient alone.

The sharpening procedure just discussed does not affect in an appreciable way the dynamic range of the intensity levels in an image. Thus, the final step in our enhancement task is to increase the dynamic range of the sharpened image. As we discussed in some detail in Sections 3.2 and 3.3, there are a number of intensity transformation functions that can accomplish this objective. We do know from the results in Section 3.3.2 that histogram equalization is not likely to work well on images that have dark intensity distributions like our images have here. Histogram specification could be a solution, but the dark characteristics of the images with which we are dealing lend themselves much better to a power-law transformation. Since we wish to spread the intensity levels, the value of γ in Eq. (3.2-3) has to be less than 1. After a few trials with this equation, we arrived at the result in Fig. 3.43(h), obtained with $\gamma = 0.5$ and $c = 1$. Comparing this image with Fig. 3.43(g), we see that significant new detail is visible in Fig. 3.43(h). The areas around the wrists, hands, ankles, and feet are good examples of this. The skeletal bone structure also is much more pronounced, including the arm and leg bones. Note also the faint definition of the outline of the body, and of body tissue. Bringing out detail of this nature by expanding the dynamic range of the intensity levels also enhanced noise, but Fig. 3.43(h) represents a significant visual improvement over the original image.

The approach just discussed is representative of the types of processes that can be linked in order to achieve results that are not possible with a single technique. The way in which the results are used depends on the application. The final user of the type of images shown in this example is likely to be a radiologist. For a number of reasons that are beyond the scope of our discussion, physicians are unlikely to rely on enhanced results to arrive at a diagnosis. However, enhanced images are quite useful in highlighting details that can serve as clues for further analysis in the original image or sequence of images. In other areas, the enhanced result may indeed be the final product. Examples are found in the printing industry, in image-based product inspection, in forensics, in microscopy, in surveillance, and in a host of other areas where the principal objective of enhancement is to obtain an image with a higher content of visual detail.

3.8   Using Fuzzy Techniques for Intensity Transformations and Spatial Filtering

We conclude this chapter with an introduction to fuzzy sets and their application to intensity transformations and spatial filtering, which are the main topics of discussion in the preceding sections. As it turns out, these two applications are among the most frequent areas in which fuzzy techniques for image processing are applied. The references at the end of this chapter provide an entry point to the literature on fuzzy sets and to other applications of fuzzy techniques in image processing. As you will see in the following discussion, fuzzy sets provide a framework for incorporating human knowledge in the solution of problems whose formulation is based on imprecise concepts.

3.8.1 Introduction

As noted in Section 2.6.4, a *set* is a collection of objects (elements) and *set theory* is the set of tools that deals with operations on and among sets. Set theory, along with mathematical logic, is one of the axiomatic foundations of classical mathematics. Central to set theory is the notion of set membership. We are used to dealing with so-called "crisp" sets, whose membership only can be true or false in the traditional sense of bi-valued Boolean logic, with 1 typically indicating true and 0 indicating false. For example, let Z denote the set of all people, and suppose that we want to define a subset, A, of Z, called the "set of young people." In order to form this subset, we need to define a *membership function* that assigns a value of 1 or 0 to every element, z, of Z. Because we are dealing with a bi-valued logic, the membership function simply defines a threshold at or below which a person is considered young, and above which a person is considered not young. Figure 3.44(a) summarizes this concept using an age threshold of 20 years and letting $\mu_A(z)$ denote the membership function just discussed.

Membership functions also are called characteristic functions.

We see an immediate difficulty with this formulation: A person 20 years of age is considered young, but a person whose age is 20 years and 1 second is not a member of the set of young people. This is a fundamental problem with crisp sets that limits the use of classical set theory in many practical applications.

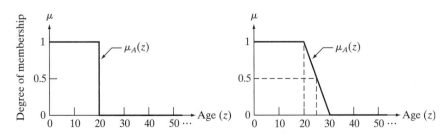

a b

FIGURE 3.44
Membership functions used to generate (a) a crisp set, and (b) a fuzzy set.

What we need is more flexibility in what we mean by "young," that is, a *gradual* transition from young to not young. Figure 3.44(b) shows one possibility. The key feature of this function is that it is infinite valued, thus allowing a continuous transition between young and not young. This makes it possible to have *degrees* of "youngness." We can make statements now such as a person being young (upper flat end of the curve), relatively young (toward the beginning of the ramp), 50% young (in the middle of the ramp), not so young (toward the end of the ramp), and so on (note that decreasing the slope of the curve in Fig. 3.44(b) introduces more vagueness in what we mean by "young.") These types of vague (*fuzzy*) statements are more in line with what humans use when talking imprecisely about age. Thus, we may interpret infinite-valued membership functions as being the foundation of a *fuzzy logic*, and the sets generated using them may be viewed as *fuzzy sets*. These ideas are formalized in the following section.

3.8.2 Principles of Fuzzy Set Theory

Fuzzy set theory was introduced by L. A. Zadeh in a paper more than four decades ago (Zadeh [1965]). As the following discussion shows, fuzzy sets provide a formalism for dealing with imprecise information.

Definitions

We follow conventional fuzzy set notation in using Z, instead of the more traditional set notation U, to denote the set universe in a given application.

Let Z be a set of elements (objects), with a generic element of Z denoted by z; that is, $Z = \{z\}$. This set is called the *universe of discourse*. A *fuzzy set*[†] A in Z is characterized by a *membership function*, $\mu_A(z)$, that associates with each element of Z a real number in the interval $[0, 1]$. The value of $\mu_A(z)$ at z represents the *grade of membership* of z in A. The nearer the value of $\mu_A(z)$ is to unity, the higher the membership grade of z in A, and conversely when the value of $\mu_A(z)$ is closer to zero. The concept of "belongs to," so familiar in ordinary sets, does not have the same meaning in fuzzy set theory. With ordinary sets, we say that an element either belongs or does not belong to a set. With fuzzy sets, we say that all zs for which $\mu_A(z) = 1$ are *full* members of the set, all zs for which $\mu_A(z) = 0$ are *not* members of the set, and all zs for which $\mu_A(z)$ is between 0 and 1 have *partial* membership in the set. Therefore, a fuzzy set is an *ordered pair* consisting of values of z and a corresponding membership function that assigns a grade of membership to each z. That is,

$$A = \{z, \mu_A(z) | z \in Z\} \tag{3.8-1}$$

When the variables are continuous, the set A in this equation can have an infinite number of elements. When the values of z are discrete, we can show the elements of A explicitly. For instance, if age increments in Fig. 3.44 were limited to integer years, then we would have

$$A = \{(1, 1), (2, 1), (3, 1), \ldots, (20, 1), (21, 0.9), (22, 0.8), \ldots, (25, 0.5)(24, 0.4), \ldots, (29, 0.1)\}$$

[†]The term *fuzzy subset* is also used in the literature, indicating that A is as subset of Z. However, *fuzzy set* is used more frequently.

where, for example, the element $(22, 0.8)$ denotes that age 22 has a 0.8 degree of membership in the set. All elements with ages 20 and under are full members of the set and those with ages 30 and higher are not members of the set. Note that a plot of this set would simply be discrete points lying on the curve of Fig. 3.44(b), so $\mu_A(z)$ completely defines A. Viewed another way, we see that a (discrete) fuzzy set is nothing more than the set of points of a function that maps each element of the problem domain (universe of discourse) into a number greater than 0 and less than or equal to 1. Thus, one often sees the terms *fuzzy set* and *membership function* used interchangeably.

When $\mu_A(z)$ can have only two values, say 0 and 1, the membership function reduces to the familiar characteristic function of an ordinary (crisp) set A. Thus, ordinary sets are a special case of fuzzy sets. Next, we consider several definitions involving fuzzy sets that are extensions of the corresponding definitions from ordinary sets.

Empty set: A fuzzy set is *empty* if and only if its membership function is identically zero in Z.

Equality: Two fuzzy sets A and B are *equal*, written $A = B$, if and only if $\mu_A(z) = \mu_B(z)$ for all $z \in Z$.

The notation "for all $z \in Z$" reads: "for all z belonging to Z."

Complement: The *complement* (NOT) of a fuzzy set A, denoted by $\overline{A}$, or NOT(A), is defined as the set whose membership function is

$$\mu_{\overline{A}}(z) = 1 - \mu_A(z) \tag{3.8-2}$$

for all $z \in Z$.

Subset: A fuzzy set A is a *subset* of a fuzzy set B if and only if

$$\mu_A(z) \leq \mu_B(z) \tag{3.8-3}$$

for all $z \in Z$.

Union: The *union* (OR) of two fuzzy sets A and B, denoted $A \cup B$, or A OR B, is a fuzzy set U with membership function

$$\mu_U(z) = \max[\mu_A(z), \mu_B(z)] \tag{3.8-4}$$

for all $z \in Z$.

Intersection: The *intersection* (AND) of two fuzzy sets A and B, denoted $A \cap B$, or A AND B, is a fuzzy set I with membership function

$$\mu_I(z) = \min[\mu_A(z), \mu_B(z)] \tag{3.8-5}$$

for all $z \in Z$.

Note that the familiar terms NOT, OR, and AND are used interchangeably when working with fuzzy sets to denote complementation, union, and intersection, respectively.

a b
c d

FIGURE 3.45
(a) Membership
functions of two
sets, A and B. (b)
Membership
function of the
complement of A.
(c) and (d)
Membership
functions of the
union and
intersection of the
two sets.

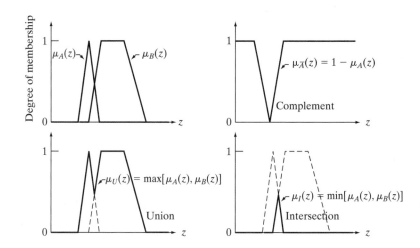

EXAMPLE 3.18:
Illustration of
fuzzy set
definitions.

■ Figure 3.45 illustrates some of the preceding definitions. Figure 3.45(a) shows the membership functions of two sets, A and B, and Fig. 3.45(b) shows the membership function of the complement of A. Figure 3.45(c) shows the membership function of the union of A and B, and Fig. 3.45(d) shows the corresponding result for the intersection of these two sets. Note that these figures are consistent with our familiar notion of complement, union, and intersection of crisp sets.[†]   ■

Although fuzzy logic and probability operate over the same $[0, 1]$ interval, there is a significant distinction to be made between the two. Consider the example from Fig. 3.44. A probabilistic statement might read: "There is a 50% chance that a person is young," while a fuzzy statement would read "A person's degree of membership within the set of young people is 0.5." The difference between these two statements is important. In the first statement, a person is considered to be either in the set of young or the set of not young people; we simply have only a 50% chance of knowing to which set the person belongs. The second statement presupposes that a person is young to some degree, with that degree being in this case 0.5. Another interpretation is to say that this is an "average" young person: not really young, but not too near being not young. In other words, fuzzy logic is not probabilistic at all; it just deals with degrees of membership in a set. In this sense, we see that fuzzy logic concepts find application in situations characterized by vagueness and imprecision, rather than by randomness.

[†]You are likely to encounter examples in the literature in which the area under the curve of the membership function of, say, the intersection of two fuzzy sets, is shaded to indicate the result of the operation. This is a carryover from ordinary set operations and is incorrect. Only the points along the membership function itself are applicable when dealing with fuzzy sets.

Some common membership functions

Types of membership functions used in practice include the following.

Triangular:

$$\mu(z) = \begin{cases} 1 - (a - z)/b & a - b \le z < a \\ 1 - (z - a)/c & a \le z \le a + c \\ 0 & \text{otherwise} \end{cases} \qquad (3.8\text{-}6)$$

Trapezoidal:

$$\mu(z) = \begin{cases} 1 - (a - z)/c & a - c \le z < a \\ 1 & a \le z < b \\ 1 - (z - b)/d & b \le z \le b + d \\ 0 & \text{otherwise} \end{cases} \qquad (3.8\text{-}7)$$

Sigma:

$$\mu(z) = \begin{cases} 1 - (a - z)/b & a - b \le z \le a \\ 1 & z > a \\ 0 & \text{otherwise} \end{cases} \qquad (3.8\text{-}8)$$

S-shape:

$$S(z; a, b, c) = \begin{cases} 0 & z < a \\ 2\left(\dfrac{z - a}{c - a}\right)^2 & a \le z \le b \\ 1 - 2\left(\dfrac{z - c}{c - a}\right)^2 & b < z \le c \\ 1 & z > c \end{cases} \qquad (3.8\text{-}9)$$

Bell-shape:

$$\mu(z) = \begin{cases} S(z; c - b, c - b/2, c) & z \le c \\ 1 - S(z; c, c + b/2, c + b) & z > c \end{cases} \qquad (3.8\text{-}10)$$

The bell-shape function sometimes is referred to as the Π (or π) function.

Truncated Gaussian:

$$\mu(z) = \begin{cases} e^{-\frac{(z - a)^2}{2b^2}} & a - c \le z \le a + c \\ 0 & \text{otherwise} \end{cases} \qquad (3.8\text{-}11)$$

Typically, only the independent variable, z, is included when writing $\mu(z)$ in order to simplify equations. We made an exception in Eq. (3.8-9) in order to use its form in Eq. (3.8-10). Figure 3.46 shows examples of the membership

FIGURE 3.46
Membership functions corresponding to Eqs. (3.8-6)–(3.8-11).

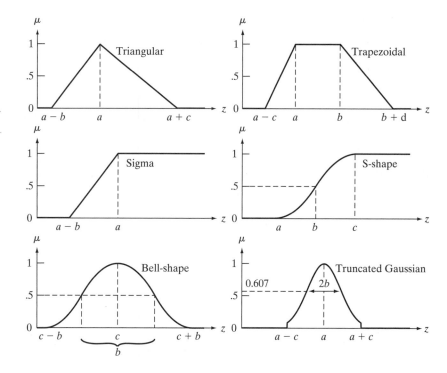

functions just discussed. The first three functions are piecewise linear, the next two functions are smooth, and the last function is a truncated Gaussian function. Equation (3.8-9) describes an important S-shape function that it used frequently when working with fuzzy sets. The value of $z = b$ at which $S = 0.5$ in this equation is called the *crossover point*. As Fig. 3.46(d) shows, this is the point at which the curve changes inflection. It is not difficult to show (Problem 3.31) that $b = (a + c)/2$. In the bell-shape curve of Fig. 3.46(e), the value of b defines the *bandwidth* of the curve.

3.8.3 Using Fuzzy Sets

In this section, we lay the foundation for using fuzzy sets and illustrate the resulting concepts with examples from simple, familiar situations. We then apply the results to image processing in Sections 3.8.4 and 3.8.5. Approaching the presentation in this way makes the material much easier to understand, especially for readers new to this area.

Suppose that we are interested in using color to categorize a given type of fruit into three groups: verdant, half-mature, and mature. Assume that observations of fruit at various stages of maturity have led to the conclusion that verdant fruit is green, half-mature fruit is yellow, and mature fruit is red. The labels *green*, *yellow*, and *red* are vague descriptions of color sensation. As a starting point, these labels have to be expressed in a fuzzy format. That is, they have to be *fuzzified*. This is achieved by defining membership as a function of

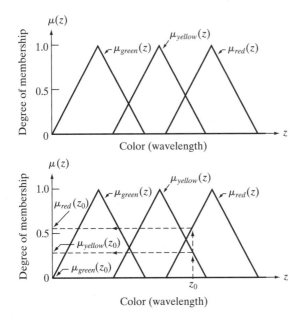

a
b

FIGURE 3.47
(a) Membership functions used to fuzzify color. (b) Fuzzifying a specific color z_0. (Curves describing color sensation are bell shaped; see Section 6.1 for an example. However, using triangular shapes as an approximation is common practice when working with fuzzy sets.)

color (wavelength of light), as Fig. 3.47(a) shows. In this context, *color* is a *linguistic variable*, and a particular color (e.g., *red* at a fixed wavelength) is a *linguistic value*. A linguistic value, z_0, is fuzzified by using a membership functions to map it to the interval $[0, 1]$, as Fig. 3.47(b) shows.

The problem-specific *knowledge* just explained can be formalized in the form of the following *fuzzy IF-THEN rules*:

R_1: IF the color is *green*, THEN the fruit is *verdant*.

OR

R_2: IF the color is *yellow*, THEN the fruit is *half-mature*.

OR

R_3: IF the color is *red*, THEN the fruit is *mature*.

The part of an IF-THEN rule to the left of THEN often is referred to as the *antecedent* (or *premise*). The part to the right is called the *consequent* (or *conclusion*.)

These rules represent the sum total of our knowledge about this problem; they are really nothing more than a formalism for a thought process.

The next step of the procedure is to find a way to use inputs (color) and the knowledge base represented by the IF-THEN rules to create the output of the fuzzy system. This process is known as *implication* or *inference*. However, before implication can be applied, the antecedent of each rule has to be processed to yield a *single* value. As we show at the end of this section, multiple parts of an antecedent are linked by ANDs and ORs. Based on the definitions from Section 3.8.2, this means performing min and max operations. To simplify the explanation, we deal initially with rules whose antecedents contain only one part.

Because we are dealing with fuzzy inputs, the outputs themselves are fuzzy, so membership functions have to be defined for the outputs as well. Figure 3.48

FIGURE 3.48
Membership
functions
characterizing the
outputs *verdant*,
half-mature, and
mature.

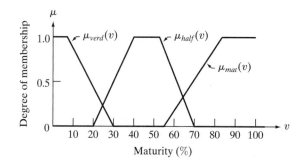

shows the membership functions of the fuzzy outputs we are going to use in this
example. Note that the independent variable of the outputs is *maturity*, which is
different from the independent variable of the inputs.

Figures 3.47 and 3.48, together with the rule base, contain all the information required to relate inputs and outputs. For example, we note that the expression *red* AND *mature* is nothing more than the intersection (AND)
operation defined earlier. In the present case, the independent variables of the
membership functions of inputs and outputs are different, so the result will be
two-dimensional. For instance, Figs. 3.49(a) and (b) show the membership
functions of *red* and *mature*, and Fig. 3.49(c) shows how they relate in two dimensions. To find the result of the AND operation between these two functions, recall from Eq. (3.8-5) that AND is defined as the minimum of the two
membership functions; that is,

$$\mu_3(z, v) = \min\{\mu_{red}(z), \mu_{mat}(v)\} \qquad (3.8\text{-}12)$$

a b
c d

FIGURE 3.49
(a) Shape of the
membership function
associated with the
color red, and
(b) corresponding
output membership
function. These two
functions are
associated by rule R_3.
(c) Combined
representation of the
two functions. The
representation is 2-D
because the
independent
variables in (a) and
(b) are different.
(d) The AND of (a)
and (b), as defined in
Eq. (3.8-5).

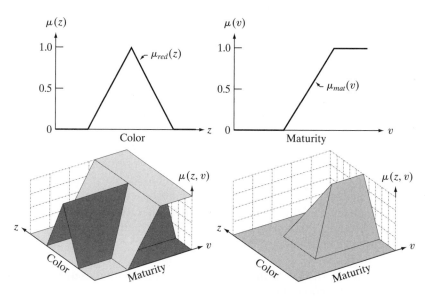

where 3 in the subscript denotes that this is the result of rule R_3 in the knowledge base. Figure 3.49(d) shows the result of the AND operation.[†]

Equation (3.8-12) is a general result involving two membership functions. In practice, we are interested in the output resulting from a *specific* input. Let z_0 denote a specific value of *red*. The degree of membership of the red color component in response to this input is simply a scalar value, $\mu_{red}(z_0)$. We find the output corresponding to rule R_3 and this specific input by performing the AND operation between $\mu_{red}(z_0)$ and the general result, $\mu_3(z, v)$, evaluated also at z_0. As noted before, the AND operation is implemented using the minimum operation:

$$Q_3(v) = \min\{\mu_{red}(z_0), \mu_3(z_0, v)\} \qquad (3.8\text{-}13)$$

where $Q_3(v)$ denotes the fuzzy output due to rule R_3 and a specific input. The only variable in Q_3 is the output variable, v, as expected.

To interpret Eq. (3.8-13) graphically, consider Fig. 3.49(d) again, which shows the general function $\mu_3(z, v)$. Performing the minimum operation of a positive constant, c, and this function would clip all values of $\mu_3(z, v)$ above that constant, as Fig. 3.50(a) shows. However, we are interested only in one value (z_0) along the color axis, so the relevant result is a cross section of the truncated function along the maturity axis, with the cross section placed at z_0, as Fig. 3.50(b) shows [because Fig. 3.50(a) corresponds to rule R_3, it follows that $c = \mu_{red}(z_0)$]. Equation (3.8-13) is the expression for this cross section.

Using the same line of reasoning, we obtain the fuzzy responses due to the other two rules and the specific input z_0, as follows:

$$Q_2(v) = \min\{\mu_{yellow}(z_0), \mu_2(z_0, v)\} \qquad (3.8\text{-}14)$$

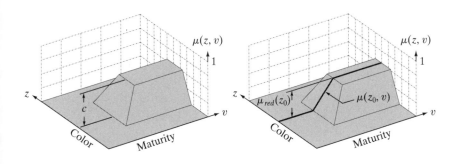

a b

FIGURE 3.50
(a) Result of computing the minimum of an arbitrary constant, c, and function $\mu_3(z, v)$ from Eq. (3.8-12). The minimum is equivalent to an AND operation. (b) Cross section (dark line) at a specific color, z_0.

[†] Note that Eq. (3.8-12) is formed from ordered pairs of values $\{\mu_{red}(z), \mu_{mat}(v)\}$, and recall that a set of ordered pairs is commonly called a *Cartesian product*, denoted by $X \times V$, where X is a set of values $\{\mu_{red}(z_1), \mu_{red}(z_2), \ldots, \mu_{red}(z_n)\}$ generated from $\mu_{red}(z)$ by varying z, and V is a similar set of n values generated from $\mu_{med}(v)$ by varying v. Thus, $X \times V = \{(\mu_{red}(z_1), \mu_{med}(v_1)), \ldots, (\mu_{red}(z_n), \mu_{med}(v_n))\}$, and we see from Fig. 3.49(d) that the AND operation involving two variables can be expresses as a mapping from $X \times V$ to the range $[0, 1]$, denoted as $X \times V \rightarrow [0, 1]$. Although we do not use this notation in the present discussion, we mention it here because you are likely to encounter it in the literature on fuzzy sets.

and

$$Q_1(v) = \min\{\mu_{green}(z_0), \mu_1(z_0, v)\} \qquad (3.8\text{-}15)$$

Each of these equations is the output associated with a particular rule and a specific input. That is, they represent the result of the implication process mentioned a few paragraphs back. Keep in mind that each of these three responses is a fuzzy set, even though the input is a scalar value.

To obtain the overall response, we *aggregate* the individual responses. In the rule base given at the beginning of this section the three rules are associated by the OR operation. Thus, the complete (aggregated) fuzzy output is given by

$$Q = Q_1 \text{ OR } Q_2 \text{ OR } Q_3 \qquad (3.8\text{-}16)$$

and we see that the overall response is the union of three individual fuzzy sets. Because OR is defined as a max operation, we can write this result as

$$Q(v) = \max_r\Big\{\min_s\{\mu_s(z_0), \mu_r(z_0, v)\}\Big\} \qquad (3.8\text{-}17)$$

for $r = \{1, 2, 3\}$ and $s = \{green, yellow, red\}$. Although it was developed in the context of an example, this expression is perfectly general; to extend it to n rules, we simply let $r = \{1, 2, \ldots, n\}$; similarly, we can expand s to include any finite number of membership functions. Equations (3.8-16) and (3.8-17) say the same thing: The response, Q, of our fuzzy system is the union of the individual fuzzy sets resulting from each rule by the implication process.

Figure 3.51 summarizes graphically the discussion up to this point. Figure 3.51(a) shows the three input membership functions evaluated at z_0, and Fig. 3.51(b) shows the outputs in response to input z_0. These fuzzy sets are the clipped cross sections discussed in connection with Fig. 3.50(b). Note that, numerically, Q_1 consists of all 0s because $\mu_{green}(z_0) = 0$; that is, Q_1 is *empty*, as defined in Section 3.8.2. Figure 3.51(c) shows the final result, Q, itself a fuzzy set formed from the union of Q_1, Q_2, and Q_3.

We have successfully obtained the complete output corresponding to a specific input, but we are still dealing with a fuzzy set. The last step is to obtain a crisp output, v_0, from fuzzy set Q using a process appropriately called *defuzzification*. There are a number of ways to defuzzify Q to obtain a crisp output. One of the approaches used most frequently is to compute the center of gravity of this set (the references cited at the end of this chapter discuss others). Thus, if $Q(v)$ from Eq. (3.8-17) can have K possible values, $Q(1), Q(2), \ldots Q(K)$, its center of gravity is given by

$$v_0 = \frac{\sum_{v=1}^{K} v Q(v)}{\sum_{v=1}^{K} Q(v)} \qquad (3.8\text{-}18)$$

Evaluating this equation with the (discrete)[†] values of Q in Fig. 3.51(c) yields $v_0 = 72.3$, indicating that the given color z_0 implies a fruit maturity of approximately 72%.

[†] Fuzzy set Q in Fig. 3.51(c) is shown as a solid curve for clarity, but keep in mind that we are dealing with digital quantities in this book, so Q is a digital function.

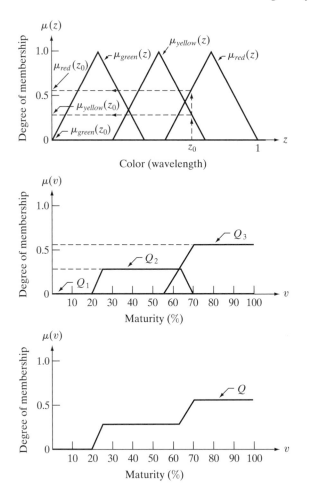

FIGURE 3.51
(a) Membership functions with a specific color, z_0, selected.
(b) Individual fuzzy sets obtained from Eqs. (3.8-13)–(3.8-15). (c) Final fuzzy set obtained by using Eq. (3.8-16) or (3.8-17).

Up to this point, we have considered IF-THEN rules whose antecedents have only one part, such as "IF the color is *red*." Rules containing more than one part must be combined to yield a *single* number that represents the entire antecedent for that rule. For example, suppose that we have the rule: IF the color is *red* OR the consistency is *soft,* THEN the fruit is *mature*. A membership function would have to be defined for the linguistic variable *soft*. Then, to obtain a single number for this rule that takes into account both parts of the antecedent, we first evaluate a given input color value of *red* using the *red* membership function and a given value of *consistency* using the *soft* membership function. Because the two parts are linked by OR, we use the maximum of the two resulting values.[†] This value is then used in the implication process to "clip" the *mature* output membership function, which is the function associated with this rule. The rest of the procedure is as before, as the following summary illustrates.

[†]Antecedents whose parts are connected by ANDs are similarly evaluated using the min operation.

Figure 3.52 shows the fruit example using two inputs: *color* and *consistency*. We can use this figure and the preceding material to summarize the principal steps followed in the application of rule-based fuzzy logic:

1. *Fuzzify the inputs:* For each scalar input, find the corresponding fuzzy values by mapping that input to the interval [0, 1], using the applicable membership functions in each rule, as the first two columns of Fig. 3.52 show.
2. *Perform any required fuzzy logical operations:* The outputs of all parts of an antecedent must be combined to yield a *single* value using the max or min operation, depending on whether the parts are connected by ORs or by ANDs. In Fig. 3.52, all the parts of the antecedents are connected by

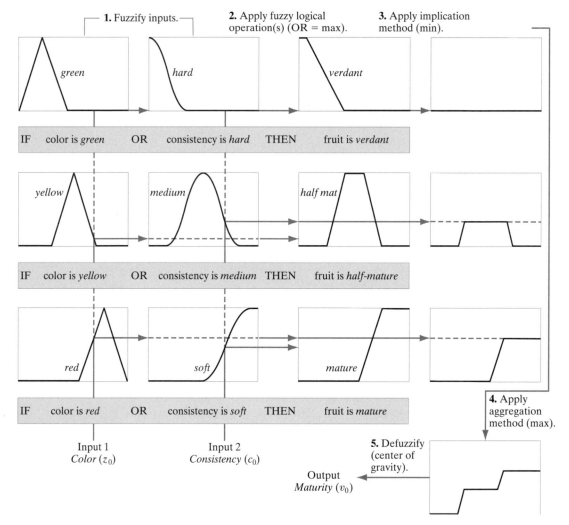

FIGURE 3.52 Example illustrating the five basic steps used typically to implement a fuzzy, rule-based system: (1) fuzzification, (2) logical operations (only OR was used in this example), (3) implication, (4) aggregation, and (5) defuzzification.

ORs, so the max operation is used throughout. The number of parts of an antecedent and the type of logic operator used to connect them can be different from rule to rule.

3. *Apply an implication method:* The single output of the antecedent of each rule is used to provide the output corresponding to that rule. We use AND for implication, which is defined as the min operation. This clips the corresponding output membership function at the value provided by the antecedent, as the third and fourth columns in Fig. 3.52 show.

4. *Apply an aggregation method to the fuzzy sets from step 3:* As the last column in Fig. 3.52 shows, the output of each rule is a fuzzy set. These must be combined to yield a single output fuzzy set. The approach used here is to OR the individual outputs, so the max operation is employed.

5. *Defuzzify the final output fuzzy set:* In this final step, we obtain a crisp, scalar output. This is achieved by computing the center of gravity of the aggregated fuzzy set from step 4.

When the number of variables is large, it is common practice to use the shorthand notation (variable, fuzzy set) to pair a variable with its corresponding membership function. For example, the rule IF the color is *green* THEN the fruit is *verdant* would be written as IF $(z, green)$ THEN $(v, verdant)$ where, as before, variables z and v represent color and degree of maturity, respectively, while *green* and *verdant* are the two fuzzy sets defined by the membership functions $\mu_{green}(z)$ and $\mu_{verd}(v)$, respectively.

In general, when dealing with M IF-THEN rules, N input variables, $z_1, z_2, \ldots z_N$, and one output variable, v, the type of fuzzy rule formulation used most frequently in image processing has the form

IF (z_1, A_{11}) AND (z_2, A_{12}) AND $\ldots$ AND (z_N, A_{1N}) THEN (v, B_1)

IF (z_1, A_{21}) AND (z_2, A_{22}) AND $\ldots$ AND (z_N, A_{2N}) THEN (v, B_2)

$$\cdots\cdots \qquad\qquad\qquad (3.8\text{-}19)$$

IF (z_1, A_{M1}) AND (z_2, A_{M2}) AND $\ldots$ AND (z_N, A_{MN}) THEN (v, B_M)

$$\text{ELSE } (v, B_E)$$

where A_{ij} is the fuzzy set associated with the ith rule and the jth input variable, B_i is the fuzzy set associated with the output of the ith rule, and we have assumed that the components of the rule antecedents are linked by ANDs. Note that we have introduced an ELSE rule, with associated fuzzy set B_E. This rule is executed when none of the preceding rules is completely satisfied; its output is explained below.

As indicated earlier, all the elements of the antecedent of each rule must be evaluated to yield a single scalar value. In Fig. 3.52, we used the max operation because the rules were based on fuzzy ORs. The formulation in Eq. (3.8-19) uses ANDs, so we have to use the min operator. Evaluating the antecedents of the ith rule in Eq. (3.8-19) produces a scalar output, λ_i, given by

$$\lambda_i = \min\{\mu_{A_{ij}}(z_j); \quad j = 1, 2, \ldots, M\} \qquad (3.8\text{-}20)$$

The use of OR or AND in the rule set depends on how the rules are stated, which in turn depends on the problem at hand. We used ORs in Fig. 3.52 and ANDs in Eq. (3.8-19) to give you familiarity with both formulations.

for $i = 1, 2, \ldots, M$, where $\mu_{A_{ij}}(z_j)$ is the membership function of fuzzy set A_{ij} evaluated at the value of the jth input. Often, λ_i is called the *strength level* (or *firing level*) of the ith rule. With reference to the preceding discussion, λ_i is simply the value used to clip the output function of the ith rule.

The ELSE rule is executed when the conditions of the THEN rules are weakly satisfied (we give a detailed example of how ELSE rules are used in Section 3.8.5). Its response should be strong when all the others are weak. In a sense, one can view an ELSE rule as performing a NOT operation on the results of the other rules. We know from Section 3.8.2 that $\mu_{\text{NOT}(A)} = \mu_{\bar{A}}(z) = 1 - \mu_A(z)$. Then, using this idea in combining (ANDing) all the levels of the THEN rules gives the following strength level for the ELSE rule:

$$\lambda_E = \min\left\{1 - \lambda_i; \quad i = 1, 2, \ldots, M\right\} \tag{3.8-21}$$

We see that if all the THEN rules fire at "full strength" (all their responses are 1), then the response of the ELSE rule is 0, as expected. As the responses of the THEN rules weaken, the strength of the ELSE rule increases. This is the fuzzy counterpart of the familiar IF-THEN-ELSE rules used in software programming.

When dealing with ORs in the antecedents, we simply replace the ANDs in Eq. (3.8-19) by ORs and the min in Eq. (3.8-20) by a max; Eq. (3.8-21) does not change. Although one could formulate more complex antecedents and consequents than the ones discussed here, the formulations we have developed using only ANDs or ORs are quite general and are used in a broad spectrum of image processing applications. The references at the end of this chapter contain additional (but less used) definitions of fuzzy logical operators, and discuss other methods for implication (including multiple outputs) and defuzzification. The introduction presented in this section is fundamental and serves as a solid base for more advanced reading on this topic. In the next two sections, we show how to apply fuzzy concepts to image processing.

3.8.4 Using Fuzzy Sets for Intensity Transformations

Consider the general problem of contrast enhancement, one of the principal applications of intensity transformations. We can state the process of enhancing the contrast of a gray-scale image using the following rules:

> IF a pixel is dark, THEN make it *darker*.
> IF a pixel is gray, THEN make it gray.
> IF a pixel is bright, THEN make it brighter.

Keeping in mind that these are fuzzy terms, we can express the concepts of *dark*, *gray*, and *bright* by the membership functions in Fig. 3.53(a).

In terms of the output, we can consider *darker* as being degrees of a dark intensity value (100% black being the limiting shade of dark), *brighter*, as being degrees of a bright shade (100% white being the limiting value), and *gray* as being degrees of an intensity in the middle of the gray scale. What we mean by

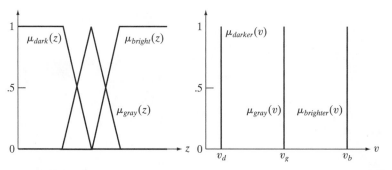

FIGURE 3.53
(a) Input and
(b) output
membership
functions for
fuzzy, rule-based
contrast
enhancement.

"degrees" here is the amount of one specific intensity. For example, 80% black is a very dark gray. When interpreted as *constant* intensities whose strength is modified, the output membership functions are *singletons* (membership functions that are *constant*), as Fig. 3.53(b) shows. The various degrees of an intensity in the range $[0, 1]$ occur when the singletons are clipped by the strength of the response from their corresponding rules, as in the fourth column of Fig. 3.52 (but keep in mind that we are working here with only one input, not two, as in the figure). Because we are dealing with constants in the output membership functions, it follows from Eq. (3.8-18) that the output, v_0, to any input, z_0, is given by

$$v_0 = \frac{\mu_{dark}(z_0) \times v_d + \mu_{gray}(z_0) \times v_g + \mu_{bright}(z_0) \times v_b}{\mu_{dark}(z_0) + \mu_{gray}(z_0) + \mu_{bright}(z_0)} \qquad (3.8-22)$$

The summations in the numerator and denominator in this expressions are simpler than in Eq. (3.8-18) because the output membership functions are constants modified (clipped) by the fuzzified values.

Fuzzy image processing is computationally intensive because the entire process of fuzzification, processing the antecedents of all rules, implication, aggregation, and defuzzification must be applied to *every* pixel in the input image. Thus, using singletons as in Eq. (3.8-22) significantly reduces computational requirements by simplifying implication, aggregation, and defuzzification. These savings can be significant in applications where processing speed is an important requirement.

■ Figure 3.54(a) shows an image whose intensities span a narrow range of the gray scale [see the image histogram in Fig. 3.55(a)], thus giving the image an appearance of low contrast. As a basis for comparison, Fig. 3.54(b) is the result of histogram equalization. As the histogram of this result shows [Fig. 3.55(b)], expanding the entire gray scale does increase contrast, but introduces intensities in the high and low end that give the image an "overexposed" appearance. For example, the details in Professor Einstein's forehead and hair are mostly lost. Figure 3.54(c) shows the result of using the rule-based contrast modification approach discussed in the preceding paragraphs. Figure 3.55(c) shows the input membership functions used, superimposed on the histogram of the original image. The output singletons were selected at $v_d = 0$ (black), $v_g = 127$ (mid gray), and $v_b = 255$ (white).

EXAMPLE 3.19:
Illustration of
image
enhancement
using fuzzy, rule-based contrast
modification.

a b c

FIGURE 3.54 (a) Low-contrast image. (b) Result of histogram equalization. (c) Result of using fuzzy, rule-based contrast enhancement.

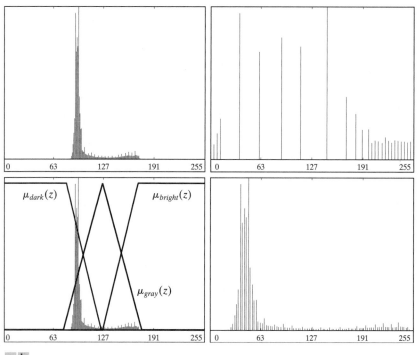

a b
c d

FIGURE 3.55 (a) and (b) Histograms of Figs. 3.54(a) and (b). (c) Input membership functions superimposed on (a). (d) Histogram of Fig. 3.54(c).

Comparing Figs. 3.54(b) and 3.54(c), we see in the latter a considerable improvement in tonality. Note, for example, the level of detail in the forehead and hair, as compared to the same regions in Fig. 3.54(b). The reason for the improvement can be explained easily by studying the histogram of Fig. 3.54(c), shown in Fig. 3.55(d). Unlike the histogram of the equalized image, this histogram has kept the same basic characteristics of the histogram of the original image. However, it is quite evident that the dark levels (talk peaks in the low end of the histogram) were moved left, thus darkening the levels. The opposite was true for bright levels. The mid grays were spread slightly, but much less than in histogram equalization.

The price of this improvement in performance is considerably more processing complexity. A practical approach to follow when processing speed and image throughput are important considerations is to use fuzzy techniques to determine what the histograms of well-balanced images should look like. Then, faster techniques, such as histogram specification, can be used to achieve similar results by mapping the histograms of the input images to one or more of the "ideal" histograms determined using a fuzzy approach.   ■

3.8.5 Using Fuzzy Sets for Spatial Filtering

When applying fuzzy sets to spatial filtering, the basic approach is to define neighborhood properties that "capture" the essence of what the filters are supposed to detect. For example, consider the problem of detecting boundaries between regions in an image. This is important in numerous applications of image processing, such as sharpening, as discussed earlier in this section, and in image segmentation, as discussed in Chapter 10.

We can develop a boundary extraction algorithm based on a simple fuzzy concept: *If a pixel belongs to a uniform region, then make it white; else make it black*, where, *black* and *white* are fuzzy sets. To express the concept of a "uniform region" in fuzzy terms, we can consider the intensity differences between the pixel at the center of a neighborhood and its neighbors. For the 3×3 neighborhood in Fig. 3.56(a), the differences between the center pixel (labeled z_5) and each of the neighbors forms the subimage of size 3×3 in Fig. 3.56(b), where d_i denotes the intensity difference between the ith neighbor and the center point (i.e., $d_i = z_i - z_5$, where the zs are intensity values). A simple set of four IF-THEN rules and one ELSE rule implements the essence of the fuzzy concept mentioned at the beginning of this paragraph:

IF d_2 is *zero* AND d_6 is *zero* THEN z_5 is *white*

IF d_6 is *zero* AND d_8 is *zero* THEN z_5 is *white*

IF d_8 is *zero* AND d_4 is *zero* THEN z_5 is *white*

IF d_4 is *zero* AND d_2 is *zero* THEN z_5 is *white*

ELSE z_5 is *black*

We used only the intensity differences between the 4-neighbors and the center point to simplify the example. Using the 8-neighbors would be a direct extension of the approach shown here.

z_1	z_2	z_3
z_4	z_5	z_6
z_7	z_8	z_9

d_1	d_2	d_3
d_4	0	d_6
d_7	d_8	d_9

Pixel neighborhood   Intensity differences

a b

FIGURE 3.56 (a) A 3 × 3 pixel neighborhood, and (b) corresponding intensity differences between the center pixels and its neighbors. Only $d_2, d_4, d_6,$ and d_8 were used in the present application to simplify the discussion.

where *zero* is a fuzzy set also. The consequent of each rule defines the values to which the intensity of the center pixel (z_5) is mapped. That is, the statement "THEN z_5 is white" means that the intensity of the pixel located at the center of the mask is mapped to white. These rules simply state that the center pixel is considered to be part of a uniform region if the intensity differences just mentioned are zero (in a fuzzy sense); otherwise it is considered a boundary pixel.

Figure 3.57 shows possible membership functions for the fuzzy sets *zero*, *black*, and *white*, respectively, where we used ZE, BL, and WH to simplify notation. Note that the range of the independent variable of the fuzzy set ZE for an image with L possible intensity levels is $[-L + 1, L - 1]$ because intensity differences can range between $-(L - 1)$ and $(L - 1)$. On the other hand, the range of the output intensities is $[0, L - 1]$, as in the original image. Figure 3.58 shows graphically the rules stated above, where the box labeled z_5 indicates that the intensity of the center pixel is mapped to the output value WH or BL.

EXAMPLE 3.20: ■ Figure 3.59(a) shows a 512 × 512 CT scan of a human head, and Fig. 3.59(b) is
Illustration of the result of using the fuzzy spatial filtering approach just discussed. Note the ef-
boundary fectiveness of the method in extracting the boundaries between regions, including
enhancement the contour of the brain (inner gray region). The constant regions in the image ap-
using fuzzy, rule- pear as gray because when the intensity differences discussed earlier are near
based spatial zero, the THEN rules have a strong response. These responses in turn clip function
filtering. WH. The output (the center of gravity of the clipped triangular regions) is a con-
stant between $(L - 1)/2$ and $(L - 1)$, thus producing the grayish tone seen in the image. The contrast of this image can be improved significantly by expanding the

a b

FIGURE 3.57
(a) Membership function of the fuzzy set *zero*.
(b) Membership functions of the fuzzy sets *black* and *white*.

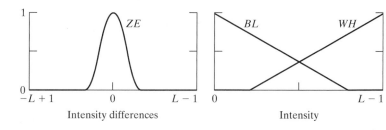

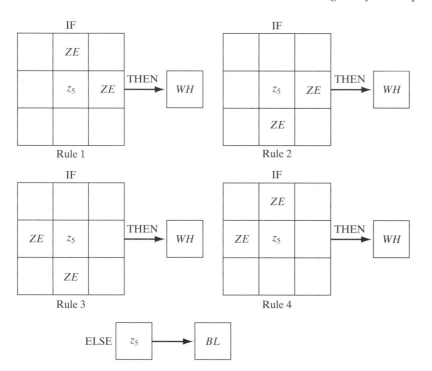

FIGURE 3.58
Fuzzy rules for
boundary
detection.

gray scale. For example, Fig. 3.59(c) was obtained by performing the intensity scaling defined in Eqs. (2.6-10) and (2.6-11), with $K = L - 1$. The net result is that intensity values in Fig. 3.59(c) span the full gray scale from 0 to $(L - 1)$. ■

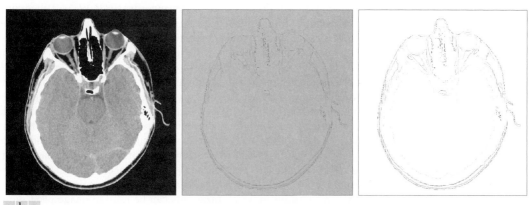

a b c

FIGURE 3.59 (a) CT scan of a human head. (b) Result of fuzzy spatial filtering using the membership functions in Fig. 3.57 and the rules in Fig. 3.58. (c) Result after intensity scaling. The thin black picture borders in (b) and (c) were added for clarity; they are not part of the data. (Original image courtesy of Dr. David R. Pickens, Vanderbilt University.)

Summary

The material you have just learned is representative of current techniques used for intensity transformations and spatial filtering. The topics included in this chapter were selected for their value as fundamental material that would serve as a foundation in an evolving field. Although most of the examples used in this chapter were related to image enhancement, the techniques presented are perfectly general, and you will encounter them again throughout the remaining chapters in contexts totally unrelated to enhancement. In the following chapter, we look again at filtering, but using concepts from the frequency domain. As you will see, there is a one-to-one correspondence between the linear spatial filters studied here and frequency domain filters.

References and Further Reading

The material in Section 3.1 is from Gonzalez [1986]. Additional reading for the material in Section 3.2 may be found in Schowengerdt [1983], Poyton [1996], and Russ [1999]. See also the paper by Tsujii et al. [1998] regarding the optimization of image displays. Early references on histogram processing are Hummel [1974], Gonzalez and Fittes [1977], and Woods and Gonzalez [1981]. Stark [2000] gives some interesting generalizations of histogram equalization for adaptive contrast enhancement. Other approaches for contrast enhancement are exemplified by Centeno and Haertel [1997] and Cheng and Xu [2000]. For further reading on exact histogram specification see Coltuc, Bolon, and Chassery [2006]. For extensions of the local histogram equalization method, see Caselles et al. [1999], and Zhu et al. [1999]. See Narendra and Fitch [1981] on the use and implementation of local statistics for image processing. Kim et al. [1997] present an interesting approach combining the gradient with local statistics for image enhancement.

For additional reading on linear spatial filters and their implementation, see Umbaugh [2005], Jain [1989], and Rosenfeld and Kak [1982]. Rank-order filters are discussed in these references as well. Wilburn [1998] discusses generalizations of rank-order filters. The book by Pitas and Venetsanopoulos [1990] also deals with median and other nonlinear spatial filters. A special issue of the *IEEE Transactions in Image Processing* [1996] is dedicated to the topic of nonlinear image processing. The material on high boost filtering is from Schowengerdt [1983]. We will encounter again many of the spatial filters introduced in this chapter in discussions dealing with image restoration (Chapter 5) and edge detection (Chapter 10).

Fundamental references for Section 3.8 are three papers on fuzzy logic by L. A. Zadeh (Zadeh [1965, 1973, 1976]). These papers are well written and worth reading in detail, as they established the foundation for fuzzy logic and some of its applications. An overview of a broad range of applications of fuzzy logic in image processing can be found in the book by Kerre and Nachtegael [2000]. The example in Section 3.8.4 is based on a similar application described by Tizhoosh [2000]. The example in Section 3.8.5 is basically from Russo and Ramponi [1994]. For additional examples of applications of fuzzy sets to intensity transformations and image filtering, see Patrascu [2004] and Nie and Barner [2006], respectively. The preceding range of references from 1965 through 2006 is a good starting point for more detailed study of the many ways in which fuzzy sets can be used in image processing. Software implementation of most of the methods discussed in this chapter can be found in Gonzalez, Woods, and Eddins [2004].

Problems

★3.1   Give a single intensity transformation function for spreading the intensities of an image so the lowest intensity is 0 and the highest is $L - 1$.

3.2   Exponentials of the form e^{-ar^2}, with a a positive constant, are useful for constructing smooth intensity transformation functions. Start with this basic function and construct transformation functions having the general shapes shown in the following figures. The constants shown are *input* parameters, and your proposed transformations must include them in their specification. (For simplicity in your answers, L_0 is not a required parameter in the third curve.)

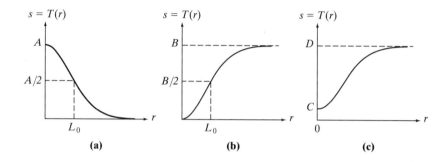

(a)   (b)   (c)

3.3   ★(a) Give a continuous function for implementing the contrast stretching transformation shown in Fig. 3.2(a). In addition to m, your function must include a parameter, E, for controlling the slope of the function as it transitions from low to high intensity values. Your function should be normalized so that its minimum and maximum values are 0 and 1, respectively.

(b) Sketch a family of transformations as a function of parameter E, for a fixed value $m = L/2$, where L is the number of intensity levels in the image.

(c) What is the smallest value of E that will make your function *effectively* perform as the function in Fig. 3.2(b)? In other words, your function does not have to be identical to Fig. 3.2(b). It just has to yield the same result of producing a binary image. Assume that you are working with 8-bit images, and let $m = 128$. Let C denote the smallest positive number representable in the computer you are using.

3.4   Propose a set of intensity-slicing transformations capable of producing all the individual bit planes of an 8-bit monochrome image. (For example, a transformation function with the property $T(r) = 0$ for r in the range $[0, 127]$, and $T(r) = 255$ for r in the range $[128, 255]$ produces an image of the 8th bit plane in an 8-bit image.)

3.5   ★(a) What effect would setting to zero the lower-order bit planes have on the histogram of an image in general?

(b) What would be the effect on the histogram if we set to zero the higher-order bit planes instead?

★3.6   Explain why the discrete histogram equalization technique does not, in general, yield a flat histogram.

3.7　Suppose that a digital image is subjected to histogram equalization. Show that a second pass of histogram equalization (on the histogram-equalized image) will produce exactly the same result as the first pass.

3.8　In some applications it is useful to model the histogram of input images as Gaussian probability density functions of the form

$$p_r(r) = \frac{1}{\sqrt{2\pi}\sigma} e^{-\frac{(r-m)^2}{2\sigma^2}}$$

where m and σ are the mean and standard deviation of the Gaussian PDF. The approach is to let m and σ be measures of average intensity and contrast of a given image. What is the transformation function you would use for histogram equalization?

★**3.9**　Assuming continuous values, show by example that it is possible to have a case in which the transformation function given in Eq. (3.3-4) satisfies conditions (a) and (b) in Section 3.3.1, but its inverse may fail condition (a′).

3.10　(a) Show that the discrete transformation function given in Eq. (3.3-8) for histogram equalization satisfies conditions (a) and (b) in Section 3.3.1.

　　★(b) Show that the inverse discrete transformation in Eq. (3.3-9) satisfies conditions (a′) and (b) in Section 3.3.1 only if none of the intensity levels $r_k, k = 0, 1, \ldots, L - 1$, are missing.

3.11　An image with intensities in the range [0, 1] has the PDF $p_r(r)$ shown in the following diagram. It is desired to transform the intensity levels of this image so that they will have the specified $p_z(z)$ shown. Assume continuous quantities and find the transformation (in terms of r and z) that will accomplish this.

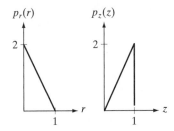

★**3.12**　Propose a method for updating the local histogram for use in the local enhancement technique discussed in Section 3.3.3.

3.13　Two images, $f(x, y)$ and $g(x, y)$, have histograms h_f and h_g. Give the conditions under which you can determine the histograms of

　　★(a) $f(x, y) + g(x, y)$

　　(b) $f(x, y) - g(x, y)$

　　(c) $f(x, y) \times g(x, y)$

　　(d) $f(x, y) \div g(x, y)$

in terms of h_f and h_g. Explain how to obtain the histogram in each case.

3.14　The images shown on the next page are quite different, but their histograms are the same. Suppose that each image is blurred with a 3×3 averaging mask.

　　(a) Would the histograms of the blurred images still be equal? Explain.

　　(b) If your answer is no, sketch the two histograms.

3.15 The implementation of linear spatial filters requires moving the center of a mask throughout an image and, at each location, computing the sum of products of the mask coefficients with the corresponding pixels at that location (see Section 3.4). A lowpass filter can be implemented by setting all coefficients to 1, allowing use of a so-called *box-filter* or *moving-average* algorithm, which consists of updating only the part of the computation that changes from one location to the next.

★ **(a)** Formulate such an algorithm for an $n \times n$ filter, showing the nature of the computations involved and the scanning sequence used for moving the mask around the image.

(b) The ratio of the number of computations performed by a brute-force implementation to the number of computations performed by the box-filter algorithm is called the *computational advantage*. Obtain the computational advantage in this case and plot it as a function of n for $n > 1$. The $1/n^2$ scaling factor is common to both approaches, so you need not consider it in obtaining the computational advantage. Assume that the image has an outer border of zeros that is wide enough to allow you to ignore border effects in your analysis.

3.16 ★ **(a)** Suppose that you filter an image, $f(x, y)$, with a spatial filter mask, $w(x, y)$, using convolution, as defined in Eq. (3.4-2), where the mask is smaller than the image in both spatial directions. Show the important property that, if the coefficients of the mask sum to zero, then the sum of all the elements in the resulting convolution array (filtered image) will be zero also (you may ignore computational inaccuracies). Also, you may assume that the border of the image has been padded with the appropriate number of zeros.

(b) Would the result to (a) be the same if the filtering is implemented using correlation, as defined in Eq. (3.4-1)?

3.17 Discuss the limiting effect of repeatedly applying a 3×3 lowpass spatial filter to a digital image. You may ignore border effects.

3.18 ★ **(a)** It was stated in Section 3.5.2 that isolated clusters of dark or light (with respect to the background) pixels whose area is less than one-half the area of a median filter are eliminated (forced to the median value of the neighbors) by the filter. Assume a filter of size $n \times n$, with n odd, and explain why this is so.

(b) Consider an image having various sets of pixel clusters. Assume that all points in a cluster are lighter or darker than the background (but not both simultaneously in the same cluster), and that the area of each cluster is less than or equal to $n^2/2$. In terms of n, under what condition would one or more of these clusters cease to be isolated in the sense described in part (a)?

★**3.19**   (a) Develop a procedure for computing the median of an $n \times n$ neighborhood.

   (b) Propose a technique for updating the median as the center of the neighborhood is moved from pixel to pixel.

3.20   (a) In a character recognition application, text pages are reduced to binary form using a thresholding transformation function of the form shown in Fig. 3.2(b). This is followed by a procedure that thins the characters until they become strings of binary 1s on a background of 0s. Due to noise, the binarization and thinning processes result in broken strings of characters with gaps ranging from 1 to 3 pixels. One way to "repair" the gaps is to run an averaging mask over the binary image to blur it, and thus create bridges of nonzero pixels between gaps. Give the (odd) size of the smallest averaging mask capable of performing this task.

   (b) After bridging the gaps, it is desired to threshold the image in order to convert it back to binary form. For your answer in (a), what is the minimum value of the threshold required to accomplish this, without causing the segments to break up again?

★**3.21**   The three images shown were blurred using square averaging masks of sizes $n = 23$, 25, and 45, respectively. The vertical bars on the left lower part of (a) and (c) are blurred, but a clear separation exists between them. However, the bars have merged in image (b), in spite of the fact that the mask that produced this image is significantly smaller than the mask that produced image (c). Explain the reason for this.

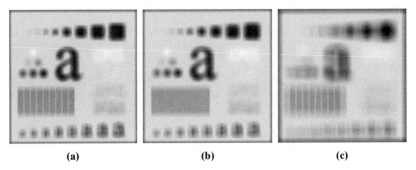

   (a)   (b)   (c)

3.22   Consider an application such as the one shown in Fig. 3.34, in which it is desired to eliminate objects smaller than those enclosed by a square of size $q \times q$ pixels. Suppose that we want to reduce the average intensity of those objects to one-tenth of their original average value. In this way, those objects will be closer to the intensity of the background and they can then be eliminated by thresholding. Give the (odd) size of the smallest averaging mask that will accomplish the desired reduction in average intensity in only one pass of the mask over the image.

3.23   In a given application an averaging mask is applied to input images to reduce noise, and then a Laplacian mask is applied to enhance small details. Would the result be the same if the order of these operations were reversed?

★ **3.24** Show that the Laplacian defined in Eq. (3.6-3) is isotropic (invariant to rotation). You will need the following equations relating coordinates for axis rotation by an angle θ:

$$x = x' \cos \theta - y' \sin \theta$$
$$y = x' \sin \theta + y' \cos \theta$$

where (x, y) are the unrotated and (x', y') are the rotated coordinates.

★ **3.25** You saw in Fig. 3.38 that the Laplacian with a -8 in the center yields sharper results than the one with a -4 in the center. Explain the reason in detail.

3.26 With reference to Problem 3.25,

(a) Would using a larger "Laplacian-like" mask, say, of size 5×5 with a -24 in the center, yield an even sharper result? Explain in detail.

(b) How does this type of filtering behave as a function of mask size?

3.27 Give a 3×3 mask for performing unsharp masking in a single pass through an image. Assume that the average image is obtained using the filter in Fig. 3.32(a).

★ **3.28** Show that subtracting the Laplacian from an image is proportional to unsharp masking. Use the definition for the Laplacian given in Eq. (3.6-6).

3.29 (a) Show that the magnitude of the gradient given in Eq. (3.6-11) is an isotropic operation. (See Problem 3.24.)

(b) Show that the isotropic property is lost in general if the gradient is computed using Eq. (3.6-12).

3.30 A CCD TV camera is used to perform a long-term study by observing the same area 24 hours a day, for 30 days. Digital images are captured and transmitted to a central location every 5 minutes. The illumination of the scene changes from natural daylight to artificial lighting. At no time is the scene without illumination, so it is always possible to obtain an image. Because the range of illumination is such that it is always in the linear operating range of the camera, it is decided not to employ any compensating mechanisms on the camera itself. Rather, it is decided to use image processing techniques to post-process, and thus normalize, the images to the equivalent of constant illumination. Propose a method to do this. You are at liberty to use any method you wish, but state clearly all the assumptions you made in arriving at your design.

3.31 Show that the crossover point in Fig. 3.46(d) is given by $b = (a + c)/2$.

3.32 Use the fuzzy set definitions in Section 3.8.2 and the basic membership functions in Fig. 3.46 to form the membership functions shown below.

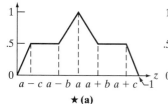

★ (a)

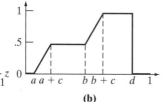

(b)

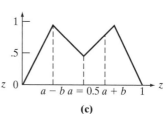

(c)

★**3.33**   What would be the effect of increasing the neighborhood size in the fuzzy filtering approach discussed in Section 3.8.5? Explain the reasoning for your answer (you may use an example to support your answer).

3.34   Design a fuzzy, rule-based system for reducing the effects of impulse noise on a noisy image with intensity values in the interval $[0, L - 1]$. As in Section 3.8.5, use only the differences $d_2, d_4, d_6,$ and d_8 in a 3×3 neighborhood in order to simplify the problem. Let z_5 denote the intensity at the center of the neighborhood, anywhere in the image. The corresponding output intensity values should be $z'_5 = z_5 + v$, where v is the output of your fuzzy system. That is, the output of your fuzzy system is a correction factor used to *reduce* the effect of a noise spike that may be present at the center of the 3×3 neighborhood. Assume that the noise spikes occur sufficiently apart so that you need not be concerned with multiple noise spikes being present in the same neighborhood. The spikes can be dark or light. Use triangular membership functions throughout.

★**(a)**   Give a fuzzy statement for this problem.

★**(b)**   Specify the IF-THEN and ELSE rules.

(c)   Specify the membership functions graphically, as in Fig. 3.57.

(d)   Show a graphical representation of the rule set, as in Fig. 3.58.

(e)   Give a summary diagram of your fuzzy system similar to the one in Fig. 3.52.

4 Filtering in the Frequency Domain

Filter: A device or material for suppressing or minimizing waves or oscillations of certain frequencies.

Frequency: The number of times that a periodic function repeats the same sequence of values during a unit variation of the independent variable.

Webster's New Collegiate Dictionary

Preview

Although significant effort was devoted in the previous chapter to spatial filtering, a thorough understanding of this area is impossible without having at least a working knowledge of how the Fourier transform and the frequency domain can be used for image filtering. You can develop a solid understanding of this topic without having to become a signal processing expert. The key lies in focusing on the fundamentals and their relevance to digital image processing. The notation, usually a source of trouble for beginners, is clarified significantly in this chapter by emphasizing the connection between image characteristics and the mathematical tools used to represent them. This chapter is concerned primarily with establishing a foundation for the Fourier transform and how it is used in basic image filtering. Later, in Chapters 5, 8, 10, and 11, we discuss other applications of the Fourier transform. We begin the discussion with a brief outline of the origins of the Fourier transform and its impact on countless branches of mathematics, science, and engineering. Next, we start from basic principles of function sampling and proceed step-by-step to derive the one- and two-dimensional discrete Fourier transforms, the basic staples of frequency domain processing. During this development, we also touch upon several important aspects of sampling, such as aliasing, whose treatment requires an understanding of the frequency domain and thus are best covered in this chapter. This material is followed by a formulation of filtering in the frequency domain and the development of sections that parallel the spatial smoothing and sharpening filtering techniques discussed in Chapter 3. We conclude the chapter with a discussion of issues related to implementing the Fourier transform in the context of image processing. Because the material in Sections 4.2 through 4.4 is basic background, readers familiar with the concepts of 1-D signal processing, including the Fourier transform, sampling, aliasing, and the convolution theorem, can proceed to Section 4.5, where we begin a discussion of the 2-D Fourier transform and its application to digital image processing.

4.1   Background

4.1.1 A Brief History of the Fourier Series and Transform

The French mathematician Jean Baptiste Joseph Fourier was born in 1768 in the town of Auxerre, about midway between Paris and Dijon. The contribution for which he is most remembered was outlined in a memoir in 1807 and published in 1822 in his book, *La Théorie Analitique de la Chaleur* (*The Analytic Theory of Heat*). This book was translated into English 55 years later by Freeman (see Freeman [1878]). Basically, Fourier's contribution in this field states that any periodic function can be expressed as the sum of sines and/or cosines of different frequencies, each multiplied by a different coefficient (we now call this sum a *Fourier series*). It does not matter how complicated the function is; if it is periodic and satisfies some mild mathematical conditions, it can be represented by such a sum. This is now taken for granted but, at the time it first appeared, the concept that complicated functions could be represented as a sum of simple sines and cosines was not at all intuitive (Fig. 4.1), so it is not surprising that Fourier's ideas were met initially with skepticism.

Even functions that are not periodic (but whose area under the curve is finite) can be expressed as the integral of sines and/or cosines multiplied by a weighing function. The formulation in this case is the *Fourier transform*, and its utility is even greater than the Fourier series in many theoretical and applied disciplines. Both representations share the important characteristic that a function, expressed in either a Fourier series or transform, can be reconstructed (recovered) completely via an inverse process, with no loss of information. This is one of the most important characteristics of these representations because it allows us to work in the "Fourier domain" and then return to the original domain of the function without losing any information. Ultimately, it was the utility of the Fourier series and transform in solving practical problems that made them widely studied and used as fundamental tools.

The initial application of Fourier's ideas was in the field of heat diffusion, where they allowed the formulation of differential equations representing heat flow in such a way that solutions could be obtained for the first time. During the past century, and especially in the past 50 years, entire industries and academic disciplines have flourished as a result of Fourier's ideas. The advent of digital computers and the "discovery" of a fast Fourier transform (FFT) algorithm in the early 1960s (more about this later) revolutionized the field of signal processing. These two core technologies allowed for the first time practical processing of

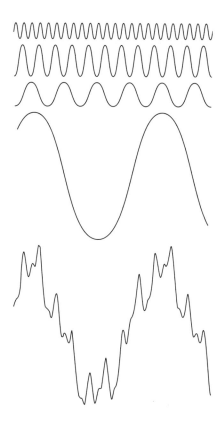

FIGURE 4.1 The function at the bottom is the sum of the four functions above it. Fourier's idea in 1807 that periodic functions could be represented as a weighted sum of sines and cosines was met with skepticism.

a host of signals of exceptional importance, ranging from medical monitors and scanners to modern electronic communications.

We will be dealing only with functions (images) of finite duration, so the Fourier transform is the tool in which we are interested. The material in the following section introduces the Fourier transform and the frequency domain. It is shown that Fourier techniques provide a meaningful and practical way to study and implement a host of image processing approaches. In some cases, these approaches are similar to the ones we developed in Chapter 3.

4.1.2 About the Examples in this Chapter

As in Chapter 3, most of the image filtering examples in this chapter deal with image enhancement. For example, smoothing and sharpening are traditionally associated with image enhancement, as are techniques for contrast manipulation. By its very nature, beginners in digital image processing find enhancement to be interesting and relatively simple to understand. Therefore, using examples from image enhancement in this chapter not only saves having an extra chapter in the book but, more importantly, is an effective tool for introducing newcomers to filtering techniques in the frequency domain. We use frequency domain processing methods for other applications in Chapters 5, 8, 10, and 11.

4.2 Preliminary Concepts

In order to simplify the progression of ideas presented in this chapter, we pause briefly to introduce several of the basic concepts that underlie the material that follows in later sections.

4.2.1 Complex Numbers

A complex number, C, is defined as

$$C = R + jI \tag{4.2-1}$$

where R and I are real numbers, and j is an imaginary number equal to the square of -1; that is, $j = \sqrt{-1}$. Here, R denotes the *real part* of the complex number and I its *imaginary part*. Real numbers are a subset of complex numbers in which $I = 0$. The *conjugate* of a complex number C, denoted C^*, is defined as

$$C^* = R - jI \tag{4.2-2}$$

Complex numbers can be viewed geometrically as points in a plane (called the *complex plane*) whose abscissa is the *real axis* (values of R) and whose ordinate is the *imaginary axis* (values of I). That is, the complex number $R + jI$ is point (R, I) in the rectangular coordinate system of the complex plane.

Sometimes, it is useful to represent complex numbers in polar coordinates,

$$C = |C|(\cos\theta + j\sin\theta) \tag{4.2-3}$$

where $|C| = \sqrt{R^2 + I^2}$ is the length of the vector extending from the origin of the complex plane to point (R, I), and θ is the angle between the vector and the real axis. Drawing a simple diagram of the real and complex axes with the vector in the first quadrant will reveal that $\tan\theta = (I/R)$ or $\theta = \arctan(I/R)$. The arctan function returns angles in the range $[-\pi/2, \pi/2]$. However, because I and R can be positive and negative independently, we need to be able to obtain angles in the full range $[-\pi, \pi]$. This is accomplished simply by keeping track of the sign of I and R when computing θ. Many programming languages do this automatically via so called *four-quadrant arctangent* functions. For example, MATLAB provides the function atan2(Imag, Real) for this purpose.

Using Euler's formula,

$$e^{j\theta} = \cos\theta + j\sin\theta \tag{4.2-4}$$

where $e = 2.71828\ldots$, gives the following familiar representation of complex numbers in polar coordinates,

$$C = |C|e^{j\theta} \tag{4.2-5}$$

where $|C|$ and θ are as defined above. For example, the polar representation of the complex number $1 + j2$ is $\sqrt{3}e^{j\theta}$, where $\theta = 64.4°$ or 1.1 radians. The preceding equations are applicable also to complex functions. For example, a complex function, $F(u)$, of a variable u, can be expressed as the sum $F(u) = R(u) + jI(u)$, where $R(u)$ and $I(u)$ are the real and imaginary component functions. As previously noted, the complex conjugate is $F^*(u) = R(u) - jI(u)$, the magnitude is $|F(u)| = \sqrt{R(u)^2 + I(u)^2}$, and the angle is $\theta(u) = \arctan[I(u)/R(u)]$. We return to complex functions several times in the course of this and the next chapter.

4.2.2 Fourier Series

As indicated in Section 4.1.1, a function $f(t)$ of a continuous variable t that is periodic with period, T, can be expressed as the sum of sines and cosines multiplied by appropriate coefficients. This sum, known as a *Fourier series,* has the form

$$f(t) = \sum_{n=-\infty}^{\infty} c_n e^{j\frac{2\pi n}{T}t} \tag{4.2-6}$$

where

$$c_n = \frac{1}{T}\int_{-T/2}^{T/2} f(t) e^{-j\frac{2\pi n}{T}t}\, dt \qquad \text{for } n = 0, \pm 1, \pm 2, \dots \tag{4.2-7}$$

are the coefficients. The fact that Eq. (4.2-6) is an expansion of sines and cosines follows from Euler's formula, Eq. (4.2-4). We will return to the Fourier series later in this section.

4.2.3 Impulses and Their Sifting Property

Central to the study of linear systems and the Fourier transform is the concept of an impulse and its sifting property. A *unit impulse* of a continuous variable t located at $t = 0$, denoted $\delta(t)$, is *defined* as

An impulse is not a function in the usual sense. A more accurate name is a *distribution* or *generalized function.* However, one often finds in the literature the names *impulse function*, *delta function*, and *Dirac delta function*, despite the misnomer.

$$\delta(t) = \begin{cases} \infty & \text{if } t = 0 \\ 0 & \text{if } t \neq 0 \end{cases} \tag{4.2-8a}$$

and is constrained also to satisfy the identity

$$\int_{-\infty}^{\infty} \delta(t)\, dt = 1 \tag{4.2-8b}$$

Physically, if we interpret t as time, an impulse may be viewed as a spike of infinity amplitude and zero duration, having unit area. An impulse has the so-called *sifting property* with respect to integration,

To *sift* means literally to separate, or to separate out by putting through a sieve.

$$\int_{-\infty}^{\infty} f(t)\delta(t)\, dt = f(0) \tag{4.2-9}$$

provided that $f(t)$ is continuous at $t = 0$, a condition typically satisfied in practice. Sifting simply yields the *value* of the function $f(t)$ at the *location* of the impulse (i.e., the origin, $t = 0$, in the previous equation). A more general statement of the sifting property involves an impulse located at an arbitrary point t_0, denoted by $\delta(t - t_0)$. In this case, the sifting property becomes

$$\int_{-\infty}^{\infty} f(t)\delta(t - t_0)\, dt = f(t_0) \tag{4.2-10}$$

which yields the value of the function at the impulse location, t_0. For instance, if $f(t) = \cos(t)$, using the impulse $\delta(t - \pi)$ in Eq. (4.2-10) yields the result $f(\pi) = \cos(\pi) = -1$. The power of the sifting concept will become quite evident shortly.

Let x represent a *discrete* variable. The *unit discrete impulse*, $\delta(x)$, serves the same purposes in the context of discrete systems as the impulse $\delta(t)$ does when working with continuous variables. It is defined as

$$\delta(x) = \begin{cases} 1 & x = 0 \\ 0 & x \neq 0 \end{cases} \tag{4.2-11a}$$

Clearly, this definition also satisfies the discrete equivalent of Eq. (4.2-8b):

$$\sum_{x=-\infty}^{\infty} \delta(x) = 1 \tag{4.2-11b}$$

The sifting property for discrete variables has the form

$$\sum_{x=-\infty}^{\infty} f(x)\delta(x) = f(0) \tag{4.2-12}$$

or, more generally using a discrete impulse located at $x = x_0$,

$$\sum_{x=-\infty}^{\infty} f(x)\delta(x - x_0) = f(x_0) \tag{4.2-13}$$

As before, we see that the sifting property simply yields the value of the function at the location of the impulse. Figure 4.2 shows the unit discrete impulse diagrammatically. Unlike its continuous counterpart, the discrete impulse is an ordinary function.

Of particular interest later in this section is an *impulse train*, $s_{\Delta T}(t)$, defined as the sum of infinitely many *periodic* impulses ΔT units apart:

$$s_{\Delta T}(t) = \sum_{n=-\infty}^{\infty} \delta(t - n\Delta T) \tag{4.2-14}$$

FIGURE 4.2
A unit discrete
impulse located at
$x = x_0$. Variable x
is discrete, and δ
is 0 everywhere
except at $x = x_0$.

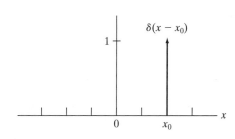

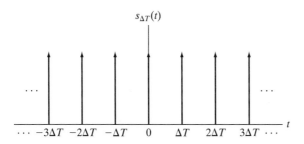

$s_{\Delta T}(t)$

$\cdots -3\Delta T \quad -2\Delta T \quad -\Delta T \quad 0 \quad \Delta T \quad 2\Delta T \quad 3\Delta T \cdots$

FIGURE 4.3 An
impulse train.

Figure 4.3 shows an impulse train. The impulses can be continuous or discrete.

4.2.4 The Fourier Transform of Functions of One Continuous Variable

The *Fourier transform* of a continuous function $f(t)$ of a continuous variable, t, denoted $\Im\{f(t)\}$, is *defined* by the equation[†]

$$\Im\{f(t)\} = \int_{-\infty}^{\infty} f(t) e^{-j2\pi\mu t}\, dt \tag{4.2-15}$$

where μ is also a continuous variable. Because t is integrated out, $\Im\{f(t)\}$ is a function only of μ. We denote this fact explicitly by writing the Fourier transform as $\Im\{f(t)\} = F(\mu)$; that is, the Fourier transform of $f(t)$ may be written for convenience as

$$F(\mu) = \int_{-\infty}^{\infty} f(t) e^{-j2\pi\mu t}\, dt \tag{4.2-16}$$

Conversely, given $F(\mu)$, we can obtain $f(t)$ back using the *inverse Fourier transform*, $f(t) = \Im^{-1}\{F(\mu)\}$, written as

$$f(t) = \int_{-\infty}^{\infty} F(\mu) e^{j2\pi\mu t}\, d\mu \tag{4.2-17}$$

where we made use of the fact that variable μ is integrated out in the inverse transform and wrote simple $f(t)$, rather than the more cumbersome notation $f(t) = \Im^{-1}\{F(\mu)\}$. Equations (4.2-16) and (4.2-17) comprise the so-called *Fourier transform pair*. They indicate the important fact mentioned in Section 4.1 that a function can be recovered from its transform.

Using Euler's formula we can express Eq. (4.2-16) as

$$F(\mu) = \int_{-\infty}^{\infty} f(t)\big[\cos(2\pi\mu t) - j\sin(2\pi\mu t)\big]\, dt \tag{4.2-18}$$

[†]Conditions for the existence of the Fourier transform are complicated to state in general (Champeney [1987]), but a sufficient condition for its existence is that the integral of the absolute value of $f(t)$, or the integral of the square of $f(t)$, be finite. Existence is seldom an issue in practice, except for idealized signals, such as sinusoids that extend forever. These are handled using generalized impulse functions. Our primary interest is in the discrete Fourier transform pair which, as you will see shortly, is guaranteed to exist for all finite functions.

If $f(t)$ is real, we see that its transform in general is complex. Note that the Fourier transform is an expansion of $f(t)$ multiplied by sinusoidal terms whose frequencies are determined by the values of μ (variable t is integrated out, as mentioned earlier). Because the only variable left after integration is frequency, we say that the domain of the Fourier transform is the *frequency domain*. We discuss the frequency domain and its properties in more detail later in this chapter. In our discussion, t can represent any continuous variable, and the units of the frequency variable μ depend on the units of t. For example, if t represents time in seconds, the units of μ are cycles/sec or Hertz (Hz). If t represents distance in meters, then the units of μ are cycles/meter, and so on. In other words, the units of the frequency domain are cycles per unit of the independent variable of the input function.

For consistency in terminology used in the previous two chapters, and to be used later in this chapter in connection with images, we refer to the domain of variable t in general as the *spatial domain*.

EXAMPLE 4.1:
Obtaining the Fourier transform of a simple function.

▨ The Fourier transform of the function in Fig. 4.4(a) follows from Eq. (4.2-16):

$$F(\mu) = \int_{-\infty}^{\infty} f(t) e^{-j2\pi\mu t}\, dt = \int_{-W/2}^{W/2} A e^{-j2\pi\mu t}\, dt$$

$$= \frac{-A}{j2\pi\mu} \left[e^{-j2\pi\mu t} \right]_{-W/2}^{W/2} = \frac{-A}{j2\pi\mu} \left[e^{-j\pi\mu W} - e^{j\pi\mu W} \right]$$

$$= \frac{A}{j2\pi\mu} \left[e^{j\pi\mu W} - e^{-j\pi\mu W} \right]$$

$$= AW \frac{\sin(\pi\mu W)}{(\pi\mu W)}$$

where we used the trigonometric identity $\sin\theta = (e^{j\theta} - e^{-j\theta})/2j$. In this case the complex terms of the Fourier transform combined nicely into a real sine

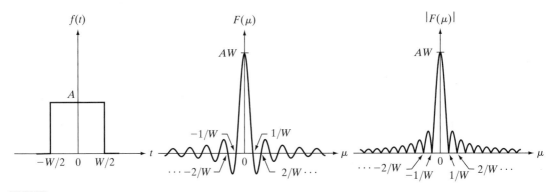

a b c

FIGURE 4.4 (a) A simple function; (b) its Fourier transform; and (c) the spectrum. All functions extend to infinity in both directions.

function. The result in the last step of the preceding expression is known as the *sinc* function:

$$\text{sinc}(m) = \frac{\sin(\pi m)}{(\pi m)} \qquad (4.2\text{-}19)$$

where $\text{sinc}(0) = 1$, and $\text{sinc}(m) = 0$ for all other *integer* values of m. Figure 4.4(b) shows a plot of $F(\mu)$.

In general, the Fourier transform contains complex terms, and it is customary for display purposes to work with the magnitude of the transform (a real quantity), which is called the *Fourier spectrum* or the *frequency spectrum*:

$$|F(\mu)| = AT \left| \frac{\sin(\pi \mu W)}{(\pi \mu W)} \right|$$

Figure 4.4(c) shows a plot of $|F(\mu)|$ as a function of frequency. The key properties to note are that the locations of the zeros of both $F(\mu)$ and $|F(\mu)|$ are *inversely* proportional to the width, W, of the "box" function, that the height of the lobes decreases as a function of distance from the origin, and that the function extends to infinity for both positive and negative values of μ. As you will see later, these properties are quite helpful in interpreting the spectra of two-dimensional Fourier transforms of images.   ■

■ The Fourier transform of a unit impulse located at the origin follows from Eq. (4.2-16):

EXAMPLE 4.2:
Fourier transform of an impulse and of an impulse train.

$$F(\mu) = \int_{-\infty}^{\infty} \delta(t) e^{-j2\pi \mu t} dt$$

$$= \int_{-\infty}^{\infty} e^{-j2\pi \mu t} \delta(t) dt$$

$$= e^{-j2\pi \mu 0} = e^{0}$$

$$= 1$$

where the third step follows from the sifting property in Eq. (4.2-9). Thus, we see that the Fourier transform of an impulse located at the origin of the spatial domain is a constant in the frequency domain. Similarly, the Fourier transform of an impulse located at $t = t_0$ is

$$F(\mu) = \int_{-\infty}^{\infty} \delta(t - t_0) e^{-j2\pi \mu t} dt$$

$$= \int_{-\infty}^{\infty} e^{-j2\pi \mu t} \delta(t - t_0) dt$$

$$= e^{-j2\pi \mu t_0}$$

$$= \cos(2\pi \mu t_0) - j \sin(2\pi \mu t_0)$$

where the third line follows from the sifting property in Eq. (4.2-10) and the last line follows from Euler's formula. These last two lines are equivalent representations of a unit circle centered on the origin of the complex plane.

In Section 4.3, we make use of the Fourier transform of a periodic impulse train. Obtaining this transform is not as straightforward as we just showed for individual impulses. However, understanding how to derive the transform of an impulse train is quite important, so we take the time to derive it in detail here. We start by noting that the only difference in the *form* of Eqs. (4.2-16) and (4.2-17) is the sign of the exponential. Thus, if a function $f(t)$ has the Fourier transform $F(\mu)$, then the latter function evaluated at t, that is, $F(t)$, must have the transform $f(-\mu)$. Using this *symmetry* property and given, as we showed above, that the Fourier transform of an impulse $\delta(t - t_0)$ is $e^{-j2\pi\mu t_0}$, it follows that the function $e^{-j2\pi t_0 t}$ has the transform $\delta(-\mu - t_0)$. By letting $-t_0 = a$, it follows that the transform of $e^{j2\pi a t}$ is $\delta(-\mu + a) = \delta(\mu - a)$, where the last step is true because δ is not zero only when $\mu = a$, which is the same result for either $\delta(-\mu + a)$ or $\delta(\mu - a)$, so the two forms are equivalent.

The impulse train $s_{\Delta T}(t)$ in Eq. (4.2-14) is periodic with period ΔT, so we know from Section 4.2.2 that it can be expressed as a Fourier series:

$$s_{\Delta T}(t) = \sum_{n=-\infty}^{\infty} c_n e^{j\frac{2\pi n}{\Delta T}t}$$

where

$$c_n = \frac{1}{\Delta T} \int_{-\Delta T/2}^{\Delta T/2} s_{\Delta T}(t) e^{-j\frac{2\pi n}{\Delta T}t} dt$$

With reference to Fig. 4.3, we see that the integral in the interval $[-\Delta T/2, \Delta T/2]$ encompasses only the impulse of $s_{\Delta T}(t)$ that is located at the origin. Therefore, the preceding equation becomes

$$c_n = \frac{1}{\Delta T} \int_{-\Delta T/2}^{\Delta T/2} \delta(t) e^{-j\frac{2\pi n}{\Delta T}t} dt$$

$$= \frac{1}{\Delta T} e^0$$

$$= \frac{1}{\Delta T}$$

The Fourier series expansion then becomes

$$s_{\Delta T}(t) = \frac{1}{\Delta T} \sum_{n=-\infty}^{\infty} e^{j\frac{2\pi n}{\Delta T}t}$$

Our objective is to obtain the Fourier transform of this expression. Because summation is a linear process, obtaining the Fourier transform of a sum is the same as obtaining the sum of the transforms of the individual components. These components are exponentials, and we established earlier in this example that

$$\Im\left\{e^{j\frac{2\pi n}{\Delta T}t}\right\} = \delta\left(\mu - \frac{n}{\Delta T}\right)$$

So, $S(\mu)$, the Fourier transform of the periodic impulse train $s_{\Delta T}(t)$, is

$$S(\mu) = \Im\left\{s_{\Delta T}(t)\right\}$$

$$= \Im\left\{\frac{1}{\Delta T}\sum_{n=-\infty}^{\infty}e^{j\frac{2\pi n}{\Delta T}t}\right\}$$

$$= \frac{1}{\Delta T}\Im\left\{\sum_{n=-\infty}^{\infty}e^{j\frac{2\pi n}{\Delta T}t}\right\}$$

$$= \frac{1}{\Delta T}\sum_{n=-\infty}^{\infty}\delta\left(\mu - \frac{n}{\Delta T}\right)$$

This fundamental result tells us that the Fourier transform of an impulse train with period ΔT *is also an impulse train*, whose period is $1/\Delta T$. This inverse proportionality between the periods of $s_{\Delta T}(t)$ and $S(\mu)$ is analogous to what we found in Fig. 4.4 in connection with a box function and its transform. This property plays a fundamental role in the remainder of this chapter.   ■

4.2.5 Convolution

We need one more building block before proceeding. We introduced the idea of convolution in Section 3.4.2. You learned in that section that convolution of two functions involves flipping (rotating by 180°) one function about its origin and sliding it past the other. At each displacement in the sliding process, we perform a computation, which in the case of Chapter 3 was a sum of products. In the present discussion, we are interested in the convolution of two continuous functions, $f(t)$ and $h(t)$, of one *continuous* variable, t, so we have to use integration instead of a summation. The convolution of these two functions, denoted as before by the operator ★, is *defined* as

$$f(t) \star h(t) = \int_{-\infty}^{\infty} f(\tau)h(t-\tau)\,d\tau \qquad (4.2\text{-}20)$$

where the minus sign accounts for the flipping just mentioned, t is the *displacement* needed to slide one function past the other, and τ is a dummy variable that is integrated out. We assume for now that the functions extend from $-\infty$ to ∞.

We illustrated the basic mechanics of convolution in Section 3.4.2, and we will do so again later in this chapter and in Chapter 5. At the moment, we are

interested in finding the Fourier transform of Eq. (4.2-20). We start with
Eq. (4.2-15):

$$\Im\{f(t) \star h(t)\} = \int_{-\infty}^{\infty}\left[\int_{-\infty}^{\infty} f(\tau)h(t-\tau)\,d\tau\right]e^{-j2\pi\mu t}\,dt$$

$$= \int_{-\infty}^{\infty} f(\tau)\left[\int_{-\infty}^{\infty} h(t-\tau)e^{-j2\pi\mu t}\,dt\right]d\tau$$

The term inside the brackets is the Fourier transform of $h(t-\tau)$. We show
later in this chapter that $\Im\{h(t-\tau)\} = H(\mu)e^{-j2\pi\mu\tau}$, where $H(\mu)$ is the
Fourier transform of $h(t)$. Using this fact in the preceding equation gives us

The same result would
be obtained if the order
of $f(t)$ and $h(t)$ were
reversed, so convolution
is commutative.

$$\Im\{f(t) \star h(t)\} = \int_{-\infty}^{\infty} f(\tau)\left[H(\mu)e^{-j2\pi\mu\tau}\right]d\tau$$

$$= H(\mu)\int_{-\infty}^{\infty} f(\tau)e^{-j2\pi\mu\tau}\,d\tau$$

$$= H(\mu)F(\mu)$$

Recalling from Section 4.2.4 that we refer to the domain of t as the spatial domain, and the domain of μ as the frequency domain, the preceding equation
tells us that the Fourier transform of the convolution of two functions in the
spatial domain is equal to the product in the frequency domain of the Fourier
transforms of the two functions. Conversely, if we have the product of the two
transforms, we can obtain the convolution in the spatial domain by computing
the inverse Fourier transform. In other words, $f(t) \star h(t)$ and $H(u)F(u)$ are a
Fourier transform pair. This result is one-half of the *convolution theorem* and
is written as

$$f(t) \star h(t) \Leftrightarrow H(\mu)F(\mu) \tag{4.2-21}$$

The double arrow is used to indicate that the expression on the right is obtained by taking the Fourier transform of the expression on the left, while the
expression on the left is obtained by taking the *inverse* Fourier transform of
the expression on the right.

Following a similar development would result in the other half of the convolution theorem:

$$f(t)h(t) \Leftrightarrow H(\mu) \star F(\mu) \tag{4.2-22}$$

which states that convolution in the frequency domain is analogous to multiplication in the spatial domain, the two being related by the forward and inverse Fourier transforms, respectively. As you will see later in this chapter, the
convolution theorem is the foundation for filtering in the frequency domain.

4.3 Sampling and the Fourier Transform of Sampled Functions

In this section, we use the concepts from Section 4.2 to formulate a basis for expressing sampling mathematically. This will lead us, starting from basic principles, to the Fourier transform of sampled functions.

4.3.1 Sampling

Continuous functions have to be converted into a sequence of discrete values before they can be processed in a computer. This is accomplished by using sampling and quantization, as introduced in Section 2.4. In the following discussion, we examine sampling in more detail.

With reference to Fig. 4.5, consider a continuous function, $f(t)$, that we wish to sample at uniform intervals (ΔT) of the independent variable t. We

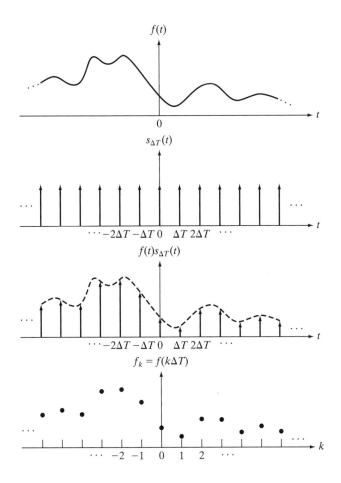

a
b
c
d

FIGURE 4.5
(a) A continuous function. (b) Train of impulses used to model the sampling process. (c) Sampled function formed as the product of (a) and (b). (d) Sample values obtained by integration and using the sifting property of the impulse. (The dashed line in (c) is shown for reference. It is not part of the data.)

assume that the function extends from $-\infty$ to ∞ with respect to t. One way to model sampling is to multiply $f(t)$ by a *sampling function* equal to a train of impulses ΔT units apart, as discussed in Section 4.2.3. That is,

$$\tilde{f}(t) = f(t)s_{\Delta T}(t) = \sum_{n=-\infty}^{\infty} f(t)\delta(t - n\Delta T) \qquad (4.3\text{-}1)$$

where $\tilde{f}(t)$ denotes the sampled function. Each component of this summation is an impulse weighted by the value of $f(t)$ at the location of the impulse, as Fig. 4.5(c) shows. The *value* of each sample is then given by the "strength" of the weighted impulse, which we obtain by integration. That is, the value, f_k, of an arbitrary sample in the sequence is given by

$$f_k = \int_{-\infty}^{\infty} f(t)\delta(t - k\Delta T)\,dt$$

$$= f(k\Delta T) \qquad (4.3\text{-}2)$$

where we used the sifting property of δ in Eq. (4.2-10). Equation (4.3-2) holds for any integer value $k = \ldots, -2, -1, 0, 1, 2, \ldots$. Figure 4.5(d) shows the result, which consists of equally-spaced samples of the original function.

4.3.2 The Fourier Transform of Sampled Functions

Let $F(\mu)$ denote the Fourier transform of a continuous function $f(t)$. As discussed in the previous section, the corresponding sampled function, $\tilde{f}(t)$, is the product of $f(t)$ and an impulse train. We know from the convolution theorem in Section 4.2.5 that the Fourier transform of the product of two functions in the spatial domain is the convolution of the transforms of the two functions in the frequency domain. Thus, the Fourier transform, $\tilde{F}(\mu)$, of the sampled function $\tilde{f}(t)$ is:

$$\tilde{F}(\mu) = \Im\{\tilde{f}(t)\}$$

$$= \Im\{f(t)s_{\Delta T}(t)\} \qquad (4.3\text{-}3)$$

$$= F(\mu) \star S(\mu)$$

where, from Example 4.2,

$$S(\mu) = \frac{1}{\Delta T}\sum_{n=-\infty}^{\infty} \delta\left(\mu - \frac{n}{\Delta T}\right) \qquad (4.3\text{-}4)$$

is the Fourier transform of the impulse train $s_{\Delta T}(t)$. We obtain the convolution of $F(\mu)$ and $S(\mu)$ directly from the definition in Eq. (4.2-20):

$$\tilde{F}(\mu) = F(\mu) \star S(\mu)$$

$$= \int_{-\infty}^{\infty} F(\tau) S(\mu - \tau)\, d\tau$$

$$= \frac{1}{\Delta T} \int_{-\infty}^{\infty} F(\tau) \sum_{n=-\infty}^{\infty} \delta\left(\mu - \tau - \frac{n}{\Delta T}\right) d\tau \qquad (4.3\text{-}5)$$

$$= \frac{1}{\Delta T} \sum_{n=-\infty}^{\infty} \int_{-\infty}^{\infty} F(\tau) \delta\left(\mu - \tau - \frac{n}{\Delta T}\right) d\tau$$

$$= \frac{1}{\Delta T} \sum_{n=-\infty}^{\infty} F\left(\mu - \frac{n}{\Delta T}\right)$$

where the final step follows from the sifting property of the impulse, as given in Eq. (4.2-10).

The summation in the last line of Eq. (4.3-5) shows that the Fourier transform $\tilde{F}(\mu)$ of the sampled function $\tilde{f}(t)$ is an *infinite*, *periodic* sequence of *copies* of $F(\mu)$, the transform of the original, continuous function. The separation between copies is determined by the value of $1/\Delta T$. Observe that although $\tilde{f}(t)$ is a sampled function, its transform $\tilde{F}(\mu)$ is *continuous* because it consists of copies of $F(\mu)$ which is a continuous function.

Figure 4.6 is a graphical summary of the preceding results.[†] Figure 4.6(a) is a sketch of the Fourier transform, $F(\mu)$, of a function $f(t)$, and Fig. 4.6(b) shows the transform, $\tilde{F}(\mu)$, of the sampled function. As mentioned in the previous section, the quantity $1/\Delta T$ is the sampling rate used to generate the sampled function. So, in Fig. 4.6(b) the sampling rate was high enough to provide sufficient separation between the periods and thus preserve the integrity of $F(\mu)$. In Fig. 4.6(c), the sampling rate was just enough to preserve $F(\mu)$, but in Fig. 4.6(d), the sampling rate was below the minimum required to maintain distinct copies of $F(\mu)$ and thus failed to preserve the original transform. Figure 4.6(b) is the result of an *over-sampled* signal, while Figs. 4.6(c) and (d) are the results of *critically-sampling* and *under-sampling* the signal, respectively. These concepts are the basis for the material in the following section.

4.3.3 The Sampling Theorem

We introduced the idea of sampling intuitively in Section 2.4. Now we consider the sampling process formally and establish the conditions under which a continuous function can be *recovered uniquely* from a set of its samples.

[†]For the sake of clarity in illustrations, sketches of Fourier transforms in Fig. 4.6, and other similar figures in this chapter, ignore the fact that transforms typically are complex functions.

FIGURE 4.6
(a) Fourier
transform of a
band-limited
function.
(b)–(d)
Transforms of the
corresponding
sampled function
under the
conditions of
over-sampling,
critically-
sampling, and
under-sampling,
respectively.

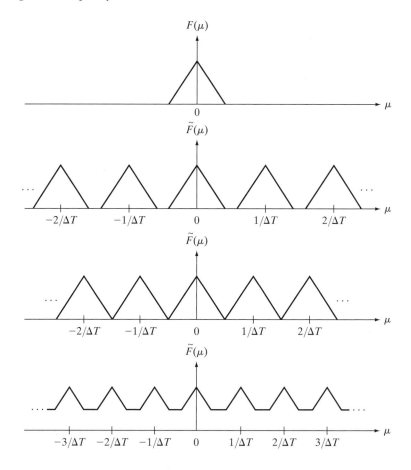

A function $f(t)$ whose Fourier transform is zero for values of frequencies outside a finite interval (band) $[-\mu_{max}, \mu_{max}]$ about the origin is called a *band-limited* function. Figure 4.7(a), which is a magnified section of Fig. 4.6(a), is such a function. Similarly, Fig. 4.7(b) is a more detailed view of the transform of a critically-sampled function shown in Fig. 4.6(c). A lower value of $1/\Delta T$ would cause the periods in $\tilde{F}(\mu)$ to merge; a higher value would provide a clean separation between the periods.

We can recover $f(t)$ from its sampled version if we can isolate a copy of $F(\mu)$ from the periodic sequence of copies of this function contained in $\tilde{F}(\mu)$, the transform of the sampled function $\tilde{f}(t)$. Recall from the discussion in the previous section that $\tilde{F}(\mu)$ is a *continuous*, *periodic* function with period $1/\Delta T$. Therefore, all we need is one complete period to characterize the entire transform. This implies that we can recover $f(t)$ from that single period by using the inverse Fourier transform.

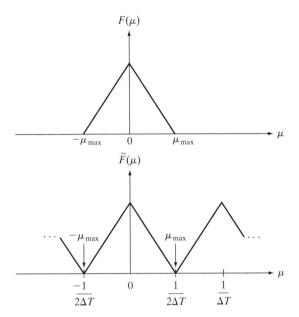

$F(\mu)$

$-\mu_{\max}$ 0 $\mu_{\max}$ μ

$\tilde{F}(\mu)$

$\cdots$ $-\mu_{\max}$ $\mu_{\max}$ $\cdots$ μ

$\dfrac{-1}{2\Delta T}$ 0 $\dfrac{1}{2\Delta T}$ $\dfrac{1}{\Delta T}$

a
b

FIGURE 4.7
(a) Transform of a band-limited function.
(b) Transform resulting from critically sampling the same function.

Extracting from $\tilde{F}(\mu)$ a single period that is equal to $F(\mu)$ is possible if the separation between copies is sufficient (see Fig. 4.6). In terms of Fig. 4.7(b), sufficient separation is guaranteed if $1/2\Delta T > \mu_{\max}$ or

$$\frac{1}{\Delta T} > 2\mu_{\max} \tag{4.3-6}$$

This equation indicates that a continuous, band-limited function can be re-covered completely from a set of its samples if the samples are acquired at a rate exceeding twice the highest frequency content of the function. This result is known as the *sampling theorem*.[†] We can say based on this result that no in-formation is lost if a continuous, band-limited function is represented by sam-ples acquired at a rate greater than twice the highest frequency content of the function. Conversely, we can say that the *maximum* frequency that can be "captured" by sampling a signal at a rate $1/\Delta T$ is $\mu_{\max} = 1/2\Delta T$. Sampling at the Nyquist rate sometimes is sufficient for perfect function recovery, but there are cases in which this leads to difficulties, as we illustrate later in Example 4.3. Thus, the sampling theorem specifies that sampling must exceed the Nyquist rate.

A sampling rate equal to *exactly* twice the highest frequency is called the *Nyquist rate*.

[†]The sampling theorem is a cornerstone of digital signal processing theory. It was first formulated in 1928 by Harry Nyquist, a Bell Laboratories scientist and engineer. Claude E. Shannon, also from Bell Labs, proved the theorem formally in 1949. The renewed interest in the sampling theorem in the late 1940s was motivated by the emergence of early digital computing systems and modern communications, which created a need for methods dealing with digital (sampled) data.

FIGURE 4.8
Extracting one
period of the
transform of a
band-limited
function using an
ideal lowpass
filter.

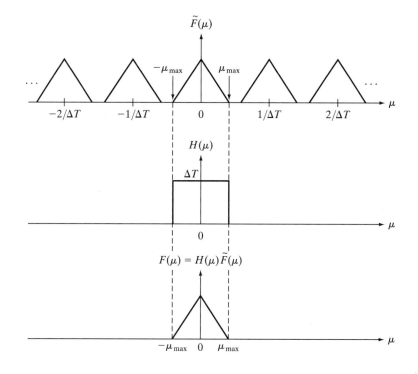

To see how the recovery of $F(\mu)$ from $\tilde{F}(\mu)$ is possible in principle, consider
Fig. 4.8, which shows the Fourier transform of a function sampled at a rate slightly
higher than the Nyquist rate. The function in Fig. 4.8(b) is defined by the equation

The ΔT in Eq. (4.3-7)
cancels out the $1/\Delta T$ in
Eq. (4.3-5).

$$H(\mu) = \begin{cases} \Delta T & -\mu_{max} \le \mu \le \mu_{max} \\ 0 & \text{otherwise} \end{cases} \tag{4.3-7}$$

When multiplied by the periodic sequence in Fig. 4.8(a), this function isolates
the period centered on the origin. Then, as Fig. 4.8(c) shows, we obtain $F(\mu)$ by
multiplying $\tilde{F}(\mu)$ by $H(\mu)$:

$$F(\mu) = H(\mu)\tilde{F}(\mu) \tag{4.3-8}$$

Once we have $F(\mu)$ we can recover $f(t)$ by using the inverse Fourier transform:

$$f(t) = \int_{-\infty}^{\infty} F(\mu)e^{j2\pi\mu t}d\mu \tag{4.3-9}$$

Equations (4.3-7) through (4.3-9) prove that, theoretically, it is possible to
recover a band-limited function from samples of the function obtained at a
rate exceeding twice the highest frequency content of the function. As we
discuss in the following section, the requirement that $f(t)$ must be band-
limited implies in general that $f(t)$ must extend from $-\infty$ to ∞, a condition

that cannot be met in practice. As you will see shortly, having to limit the duration of a function prevents perfect recovery of the function, except in some special cases.

Function $H(\mu)$ is called a *lowpass filter* because it passes frequencies at the low end of the frequency range but it eliminates (filters out) all higher frequencies. It is called also an *ideal* lowpass filter because of its infinitely rapid transitions in amplitude (between 0 and ΔT at location $-\mu_{\max}$ and the reverse at $\mu_{\max}$), a characteristic that cannot be achieved with physical electronic components. We can simulate ideal filters in software, but even then there are limitations, as we explain in Section 4.7.2. We will have much more to say about filtering later in this chapter. Because they are instrumental in recovering (reconstructing) the original function from its samples, filters used for the purpose just discussed are called *reconstruction filters*.

4.3.4 Aliasing

A logical question at this point is: What happens if a band-limited function is sampled at a rate that is less than twice its highest frequency? This corresponds to the under-sampled case discussed in the previous section. Figure 4.9(a) is the same as Fig. 4.6(d), which illustrates this condition. The net effect of lowering the sampling rate below the Nyquist rate is that the periods now overlap, and it becomes impossible to isolate a single period of the transform, regardless of the filter used. For instance, using the ideal lowpass filter in Fig. 4.9(b) would result in a transform that is corrupted by frequencies from adjacent periods, as Fig. 4.9(c) shows. The inverse transform would then yield a corrupted function of t. This effect, caused by under-sampling a function, is known as *frequency aliasing* or simply as *aliasing*. In words, aliasing is a process in which high frequency components of a continuous function "masquerade" as lower frequencies in the sampled function. This is consistent with the common use of the term *alias*, which means "a false identity."

Unfortunately, except for some special cases mentioned below, aliasing is always present in sampled signals because, even if the original sampled function is band-limited, infinite frequency components are introduced the moment we limit the duration of the function, which we always have to do in practice. For example, suppose that we want to limit the duration of a band-limited function $f(t)$ to an interval, say $[0, T]$. We can do this by multiplying $f(t)$ by the function

$$h(t) = \begin{cases} 1 & 0 \le t \le T \\ 0 & \text{otherwise} \end{cases} \tag{4.3-10}$$

This function has the same basic shape as Fig. 4.4(a) whose transform, $H(\mu)$, has frequency components extending to infinity, as Fig. 4.4(b) shows. From the convolution theorem we know that the transform of the product of $h(t)f(t)$ is the convolution of the transforms of the functions. Even if the transform of $f(t)$ is band-limited, convolving it with $H(\mu)$, which involves sliding one function across the other, will yield a result with frequency

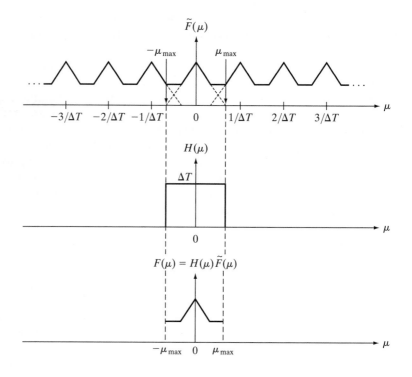

FIGURE 4.9 (a) Fourier transform of an under-sampled, band-limited function. (Interference from adjacent periods is shown dashed in this figure). (b) The same ideal lowpass filter used in Fig. 4.8(b). (c) The product of (a) and (b). The interference from adjacent periods results in aliasing that prevents perfect recovery of $F(\mu)$ and, therefore, of the original, band-limited continuous function. Compare with Fig. 4.8.

components extending to infinity. Therefore, no function of finite duration can be band-limited. Conversely, a function that is band-limited must extend from $-\infty$ to ∞.[†]

We conclude that aliasing is an inevitable fact of working with sampled records of finite length for the reasons stated in the previous paragraph. In practice, the effects of aliasing can be *reduced* by smoothing the input function to attenuate its higher frequencies (e.g., by defocusing in the case of an image). This process, called *anti-aliasing*, has to be done *before* the function is sampled because aliasing is a sampling issue that cannot be "undone after the fact" using computational techniques.

[†]An important special case is when a function that extends from $-\infty$ to ∞ is band-limited *and* periodic. In this case, the function can be truncated and still be band-limited, *provided* that the truncation encompasses *exactly* an integral number of periods. A single truncated period (and thus the function) can be represented by a set of discrete samples satisfying the sampling theorem, taken over the truncated interval.

■ Figure 4.10 shows a classic illustration of aliasing. A pure sine wave extending infinitely in both directions has a single frequency so, obviously, it is band-limited. Suppose that the sine wave in the figure (ignore the large dots for now) has the equation $\sin(\pi t)$, and that the horizontal axis corresponds to time, t, in seconds. The function crosses the axis at $t = \ldots -1, 0, 1, 2, 3 \ldots$.

EXAMPLE 4.3:
Aliasing.

The period, P, of $\sin(\pi t)$ is 2 s, and its frequency is $1/P$, or 1/2 cycles/s. According to the sampling theorem, we can recover this signal from a set of its samples if the sampling rate, $1/\Delta T$, exceeds twice the highest frequency of the signal. This means that a sampling rate greater than 1 sample/s $[2 \times (1/2) = 1]$, or $\Delta T < 1$ s, is required to recover the signal. Observe that sampling this signal at *exactly* twice the frequency (1 sample/s), with samples taken at $t = \ldots -1, 0, 1, 2, 3 \ldots$, results in $\ldots \sin(-\pi)$, $\sin(0)$, $\sin(\pi)$, $\sin(2\pi), \ldots$, which are all 0. This illustrates the reason why the sampling theorem requires a sampling rate that exceeds twice the highest frequency, as mentioned earlier.

Recall that 1 cycle/s is defined as 1 Hz.

The large dots in Fig. 4.10 are samples taken uniformly at a rate of less than 1 sample/s (in fact, the separation between samples exceeds 2 s, which gives a sampling rate lower than 1/2 samples/s). The sampled signal *looks* like a sine wave, but its frequency is about *one-tenth* the frequency of the original. This sampled signal, having a frequency well below anything present in the original continuous function is an example of aliasing. Given just the samples in Fig. 4.10, the seriousness of aliasing in a case such as this is that we would have no way of knowing that these samples are not a true representation of the original function. As you will see in later in this chapter, aliasing in images can produce similarly misleading results.   ■

4.3.5 Function Reconstruction (Recovery) from Sampled Data

In this section, we show that reconstruction of a function from a set of its samples reduces in practice to interpolating between the samples. Even the simple act of displaying an image requires reconstruction of the image from its samples

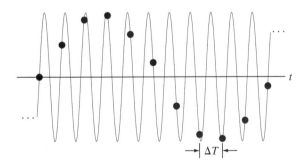

FIGURE 4.10 Illustration of aliasing. The under-sampled function (black dots) looks like a sine wave having a frequency much lower than the frequency of the continuous signal. The period of the sine wave is 2 s, so the zero crossings of the horizontal axis occur every second. ΔT is the separation between samples.

by the display medium. Therefore, it is important to understand the fundamentals of sampled data reconstruction. Convolution is central to developing this understanding, showing again the importance of this concept.

The discussion of Fig. 4.8 and Eq. (4.3-8) outlines the procedure for perfect recovery of a band-limited function from its samples using frequency domain methods. Using the convolution theorem, we can obtain the equivalent result in the spatial domain. From Eq. (4.3-8), $F(\mu) = H(\mu)\tilde{F}(\mu)$, so it follows that

$$
\begin{aligned}
f(t) &= \Im^{-1}\{F(\mu)\} \\
&= \Im^{-1}\{H(\mu)\tilde{F}(\mu)\} \\
&= h(t) \star \tilde{f}(t)
\end{aligned}
\tag{4.3-11}
$$

where the last step follows from the convolution theorem, Eq. (4.2-21). It can be shown (Problem 4.6) that substituting Eq. (4.3-1) for $\tilde{f}(t)$ into Eq. (4.3-11) and then using Eq. (4.2-20) leads to the following *spatial domain* expression for $f(t)$:

$$
f(t) = \sum_{n=-\infty}^{\infty} f(n\,\Delta T)\, \mathrm{sinc}\big[(t - n\,\Delta T)/n\,\Delta T\big]
\tag{4.3-12}
$$

where the sinc function is defined in Eq. (4.2-19). This result is not unexpected because the inverse Fourier transform of the box filter, $H(\mu)$, is a sinc function (see Example 4.1). Equation (4.3-12) shows that the perfectly reconstructed function is an infinite sum of sinc functions weighted by the sample values, and has the important property that the reconstructed function is identically equal to the sample values at multiple integer increments of ΔT. That is, for any $t = k\,\Delta T$, where k is an integer, $f(t)$ is equal to the kth sample $f(k\Delta T)$. This follows from Eq. (4.3-12) because $\mathrm{sinc}(0) = 1$ and $\mathrm{sinc}(m) = 0$ for any other integer value of m. Between sample points, values of $f(t)$ are *interpolations* formed by the sum of the sinc functions.

Equation (4.3-12) requires an infinite number of terms for the interpolations between samples. In practice, this implies that we have to look for approximations that are finite interpolations between samples. As we discussed in Section 2.4.4, the principal interpolation approaches used in image processing are nearest-neighbor, bilinear, and bicubic interpolation. We discuss the effects of interpolation on images in Section 4.5.4.

4.4   The Discrete Fourier Transform (DFT) of One Variable

One of the key goals of this chapter is the derivation of the *discrete Fourier transform* (DFT) starting from basic principles. The material up to this point may be viewed as the foundation of those basic principles, so now we have in place the necessary tools to derive the DFT.

4.4.1 Obtaining the DFT from the Continuous Transform of a Sampled Function

As discussed in Section 4.3.2, the Fourier transform of a sampled, band-limited function extending from $-\infty$ to ∞ is a *continuous, periodic* function that also extends from $-\infty$ to ∞. In practice, we work with a finite number of samples, and the objective of this section is to derive the DFT corresponding to such sample sets.

Equation (4.3-5) gives the transform, $\tilde{F}(\mu)$ of sampled data in terms of the transform of the original function, but it does not give us an expression for $\tilde{F}(\mu)$ in terms of the sampled function $\tilde{f}(t)$ itself. We find such an expression directly from the definition of the Fourier transform in Eq. (4.2-16):

$$\tilde{F}(\mu) = \int_{-\infty}^{\infty} \tilde{f}(t) e^{-j2\pi\mu t}\, dt \tag{4.4-1}$$

By substituting Eq. (4.3-1) for $\tilde{f}(t)$, we obtain

$$\begin{aligned}
\tilde{F}(\mu) &= \int_{-\infty}^{\infty} \tilde{f}(t) e^{-j2\pi\mu t}\, dt \\[2mm]
&= \int_{-\infty}^{\infty} \sum_{n=-\infty}^{\infty} f(t)\delta(t - n\Delta T) e^{-j2\pi\mu t}\, dt \\[2mm]
&= \sum_{n=-\infty}^{\infty} \int_{-\infty}^{\infty} f(t)\delta(t - n\Delta T) e^{-j2\pi\mu t}\, dt \\[2mm]
&= \sum_{n=-\infty}^{\infty} f_n e^{-j2\pi\mu n\Delta T}
\end{aligned} \tag{4.4-2}$$

where the last step follows from Eq. (4.3-2). Although f_n is a discrete function, its Fourier $\tilde{F}(\mu)$ is continuous and infinitely periodic with period $1/\Delta T$, as we know from Eq. (4.3-5). Therefore, all we need to characterize $\tilde{F}(\mu)$ is one period, and sampling one period is the basis for the DFT.

Suppose that we want to obtain M equally spaced samples of $\tilde{F}(\mu)$ taken over the period $\mu = 0$ to $\mu = 1/\Delta T$. This is accomplished by taking the samples at the following frequencies:

$$\mu = \frac{m}{M\Delta T} \qquad m = 0, 1, 2, \ldots, M-1 \tag{4.4-3}$$

Substituting this result for μ into Eq. (4.4-2) and letting F_m denote the result yields

$$F_m = \sum_{n=0}^{M-1} f_n e^{-j2\pi mn/M} \qquad m = 0, 1, 2, \ldots, M-1 \tag{4.4-4}$$

This expression is the discrete Fourier transform we are seeking.[†] Given a set $\{f_n\}$ consisting of M samples of $f(t)$, Eq. (4.4-4) yields a sample set $\{F_m\}$ of M complex discrete values corresponding to the discrete Fourier transform of the input sample set. Conversely, given $\{F_m\}$, we can recover the sample set $\{f_n\}$ by using the *inverse discrete Fourier transform* (IDFT)

$$f_n = \frac{1}{M}\sum_{m=0}^{M-1} F_m e^{j2\pi mn/M} \qquad n = 0, 1, 2, \ldots, M-1 \qquad (4.4\text{-}5)$$

It is not difficult to show (Problem 4.8) that substituting Eq. (4.4-5) for f_n into Eq. (4.4-4) gives the identity $F_m \equiv F_m$. Similarly, substituting Eq. (4.4-4) into Eq. (4.4-5) for F_m yields $f_n \equiv f_n$. This implies that Eqs. (4.4-4) and (4.4-5) constitute a *discrete Fourier transform pair*. Furthermore, these identities indicate that the forward and inverse Fourier transforms exist for any set of samples whose values are finite. Note that neither expression depends explicitly on the sampling interval ΔT nor on the frequency intervals of Eq. (4.4-3). Therefore, the DFT pair is applicable to *any* finite set of discrete samples taken uniformly.

We used m and n in the preceding development to denote discrete variables because it is typical to do so for derivations. However, it is more intuitive, especially in two dimensions, to use the notation x and y for image coordinate variables and u and v for frequency variables, where these are understood to be *integers*.[‡] Then, Eqs. (4.4-4) and (4.4-5) become

$$F(u) = \sum_{x=0}^{M-1} f(x) e^{-j2\pi ux/M} \qquad u = 0, 1, 2, \ldots, M-1 \qquad (4.4\text{-}6)$$

and

$$f(x) = \frac{1}{M}\sum_{u=0}^{M-1} F(u) e^{j2\pi ux/M} \qquad x = 0, 1, 2, \ldots, M-1 \qquad (4.4\text{-}7)$$

where we used functional notation instead of subscripts for simplicity. Clearly, $F(u) \equiv F_m$ and $f(x) \equiv f_n$. From this point on, we use Eqs. (4.4-6) and (4.4-7) to denote the 1-D DFT pair. Some authors include the $1/M$ term in Eq. (4.4-6) instead of the way we show it in Eq. (4.4-7). That does not affect the proof that the two equations form a Fourier transform pair.

[†]Note from Fig. 4.6(b) that the interval $[0, 1/\Delta T]$ covers two back-to-back *half* periods of the transform. This means that the data in F_m requires re-ordering to obtain samples that are ordered from the lowest the highest frequency of a period. This is the price paid for the notational convenience of taking the samples at $m = 0, 1, \ldots, M-1$, instead of using samples on either side of the origin, which would require the use of negative notation. The procedure to order the transform data is discussed in Section 4.6.3.

[‡]We have been careful in using t for *continuous* spatial variables and μ for the corresponding *continuous* frequency variable. From this point on, we will use x and u to denote one-dimensional *discrete* spatial and frequency variables, respectively. When dealing with two-dimensional functions, we will use (t, z) and (μ, ν) to denote *continuous* spatial and frequency domain variables, respectively. Similarly, we will use (x, y) and (u, v) to denote their *discrete* counterparts.

It can be shown (Problem 4.9) that both the forward and inverse discrete transforms are infinitely periodic, with period M. That is,

$$F(u) = F(u + kM) \tag{4.4-8}$$

and

$$f(x) = f(x + kM) \tag{4.4-9}$$

where k is an integer.

The discrete equivalent of the convolution in Eq. (4.2-20) is

$$f(x) \star h(x) = \sum_{m=0}^{M-1} f(m)h(x - m) \tag{4.4-10}$$

for $x = 0, 1, 2, \ldots, M - 1$. Because in the preceding formulations the functions are periodic, their convolution also is periodic. Equation (4.4-10) gives one period of the periodic convolution. For this reason, the process inherent in this equation often is referred to as *circular convolution*, and is a direct result of the periodicity of the DFT and its inverse. This is in contrast with the convolution you studied in Section 3.4.2, in which values of the displacement, x, were determined by the requirement of sliding one function completely past the other, and were not fixed to the range $[0, M - 1]$ as in circular convolution. We discuss this difference and its significance in Section 4.6.3 and in Fig. 4.28.

Finally, we point out that the convolution theorem given in Eqs. (4.2-21) and (4.2-22) is applicable also to discrete variables (Problem 4.10).

It is not obvious why the discrete function $f(x)$ should be periodic, considering that the continuous function from which it was sampled may not be. One informal way to reason this out is to keep in mind that sampling results in a periodic DFT. It is logical that $f(x)$, which is the inverse DFT, has to be periodic also for the DFT pair to exist.

4.4.2 Relationship Between the Sampling and Frequency Intervals

If $f(x)$ consists of M samples of a function $f(t)$ taken ΔT units apart, the duration of the record comprising the set $\{f(x)\}$, $x = 0, 1, 2, \ldots, M - 1$, is

$$T = M\Delta T \tag{4.4-11}$$

The corresponding spacing, Δu, in the discrete frequency domain follows from Eq. (4.4-3):

$$\Delta u = \frac{1}{M\Delta T} = \frac{1}{T} \tag{4.4-12}$$

The entire frequency range spanned by the M components of the DFT is

$$\Omega = M\Delta u = \frac{1}{\Delta T} \tag{4.4-13}$$

Thus, we see from Eqs. (4.4-12) and (4.4-13) that that the resolution in frequency, Δu, of the DFT depends on the duration T over which the continuous function, $f(t)$, is sampled, and the range of frequencies spanned by the DFT depends on the sampling interval ΔT. Observe that both expressions exhibit *inverse* relationships with respect to T and ΔT.

EXAMPLE 4.4:
The mechanics of computing the DFT.

■ Figure 4.11(a) shows four samples of a continuous function, $f(t)$, taken ΔT units apart. Figure 4.11(b) shows the sampled values in the x-domain. Note that the values of x are 0, 1, 2, and 3, indicating that we could be referring to any four samples of $f(t)$.

From Eq. (4.4-6),

$$F(0) = \sum_{x=0}^{3} f(x) = \left[f(0) + f(1) + f(2) + f(3) \right]$$

$$= 1 + 2 + 4 + 4 = 11$$

The next value of $F(u)$ is

$$F(1) = \sum_{x=0}^{3} f(x) e^{-j2\pi(1)x/4}$$

$$= 1e^0 + 2e^{-j\pi/2} + 4e^{-j\pi} + 4e^{-j3\pi/2} = -3 + 2j$$

Similarly, $F(2) = -(1 + 0j)$ and $F(3) = -(3 + 2j)$. Observe that *all* values of $f(x)$ are used in computing *each* term of $F(u)$.

If instead we were given $F(u)$ and were asked to compute its inverse, we would proceed in the same manner, but using the inverse transform. For instance,

$$f(0) = \frac{1}{4} \sum_{u=0}^{3} F(u) e^{j2\pi u(0)}$$

$$= \frac{1}{4} \sum_{u=0}^{3} F(u)$$

$$= \frac{1}{4} [11 - 3 + 2j - 1 - 3 - 2j]$$

$$= \frac{1}{4} [4] = 1$$

which agrees with Fig. 4.11(b). The other values of $f(x)$ are obtained in a similar manner.   ■

a b

FIGURE 4.11
(a) A function, and (b) samples in the x-domain. In (a), t is a continuous variable; in (b), x represents integer values.

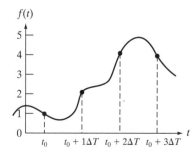
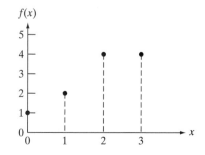

4.5 Extension to Functions of Two Variables

In this section, we extend to two variables the concepts introduced in Sections 4.2 through 4.4.

4.5.1 The 2-D Impulse and Its Sifting Property

The impulse, $\delta(t, z)$, of two continuous variables, t and z, is defined as in Eq. (4.2-8):

$$\delta(t, z) = \begin{cases} \infty & \text{if } t = z = 0 \\ 0 & \text{otherwise} \end{cases} \tag{4.5-1a}$$

and

$$\int_{-\infty}^{\infty} \int_{-\infty}^{\infty} \delta(t, z) \, dt \, dz = 1 \tag{4.5-1b}$$

As in the 1-D case, the 2-D impulse exhibits the *sifting property* under integration,

$$\int_{-\infty}^{\infty} \int_{-\infty}^{\infty} f(t, z) \delta(t, z) \, dt \, dz = f(0, 0) \tag{4.5-2}$$

or, more generally for an impulse located at coordinates (t_0, z_0),

$$\int_{-\infty}^{\infty} \int_{-\infty}^{\infty} f(t, z) \delta(t - t_0, z - z_0) \, dt \, dz = f(t_0, z_0) \tag{4.5-3}$$

As before, we see that the sifting property yields the value of the function $f(t, z)$ at the location of the impulse.

For discrete variables x and y, the 2-D discrete impulse is defined as

$$\delta(x, y) = \begin{cases} 1 & \text{if } x = y = 0 \\ 0 & \text{otherwise} \end{cases} \tag{4.5-4}$$

and its sifting property is

$$\sum_{x=-\infty}^{\infty} \sum_{y=-\infty}^{\infty} f(x, y) \delta(x, y) = f(0, 0) \tag{4.5-5}$$

where $f(x, y)$ is a function of discrete variables x and y. For an impulse located at coordinates (x_0, y_0) (see Fig. 4.12) the sifting property is

$$\sum_{x=-\infty}^{\infty} \sum_{y=-\infty}^{\infty} f(x, y) \delta(x - x_0, y - y_0) = f(x_0, y_0) \tag{4.5-6}$$

As before, the sifting property of a discrete impulse yields the value of the discrete function $f(x, y)$ at the location of the impulse.

FIGURE 4.12
Two-dimensional
unit discrete
impulse. Variables
x and y are
discrete, and δ is
zero everywhere
except at
coordinates
(x_0, y_0).

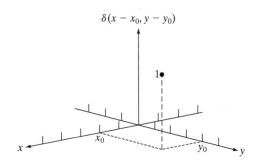

4.5.2 The 2-D Continuous Fourier Transform Pair

Let $f(t, z)$ be a continuous function of two continuous variables, t and z. The two-dimensional, continuous Fourier transform pair is given by the expressions

$$F(\mu, \nu) = \int_{-\infty}^{\infty} \int_{-\infty}^{\infty} f(t, z) e^{-j2\pi(\mu t + \nu z)} \, dt \, dz \qquad (4.5\text{-}7)$$

and

$$f(t, z) = \int_{-\infty}^{\infty} \int_{-\infty}^{\infty} F(\mu, \nu) e^{j2\pi(\mu t + \nu z)} \, d\mu \, d\nu \qquad (4.5\text{-}8)$$

where μ and ν are the frequency variables. When referring to images, t and z are interpreted to be continuous *spatial* variables. As in the 1-D case, the domain of the variables μ and ν defines the *continuous frequency domain*.

EXAMPLE 4.5:
Obtaining the 2-D
Fourier transform
of a simple
function.

■ Figure 4.13(a) shows a 2-D function analogous to the 1-D case in Example 4.1. Following a procedure similar to the one used in that example gives the result

$$F(\mu, \nu) = \int_{-\infty}^{\infty} \int_{-\infty}^{\infty} f(t, z) e^{-j2\pi(\mu t + \nu z)} \, dt \, dz$$

$$= \int_{-T/2}^{T/2} \int_{-Z/2}^{Z/2} A e^{-j2\pi(\mu t + \nu z)} \, dt \, dz$$

$$= ATZ \left[\frac{\sin(\pi \mu T)}{(\pi \mu T)} \right] \left[\frac{\sin(\pi \nu Z)}{(\pi \nu Z)} \right]$$

The magnitude (spectrum) is given by the expression

$$|F(\mu, \nu)| = ATZ \left| \frac{\sin(\pi \mu T)}{(\pi \mu T)} \right| \left| \frac{\sin(\pi \nu Z)}{(\pi \nu Z)} \right|$$

Figure 4.13(b) shows a portion of the spectrum about the origin. As in the 1-D case, the locations of the zeros in the spectrum are inversely proportional to

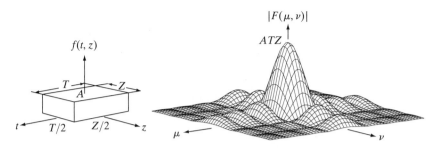

a b

FIGURE 4.13 (a) A 2-D function, and (b) a section of its spectrum (not to scale). The block is longer along the *t*-axis, so the spectrum is more "contracted" along the μ-axis. Compare with Fig. 4.4.

the values of T and Z. Thus, the larger T and Z are, the more "contracted" the spectrum will become, and vice versa.   ■

4.5.3 Two-Dimensional Sampling and the 2-D Sampling Theorem

In a manner similar to the 1-D case, sampling in two dimensions can be modeled using the sampling function (2-D impulse train):

$$s_{\Delta T \Delta Z}(t, z) = \sum_{m=-\infty}^{\infty} \sum_{n=-\infty}^{\infty} \delta(t - m\Delta T, z - n\Delta Z) \qquad (4.5\text{-}9)$$

where ΔT and ΔZ are the separations between samples along the *t*- and *z*-axis of the continuous function $f(t, z)$. Equation (4.5-9) describes a set of periodic impulses extending infinitely along the two axes (Fig. 4.14). As in the 1-D case illustrated in Fig. 4.5, multiplying $f(t, z)$ by $s_{\Delta T \Delta Z}(t, z)$ yields the sampled function.

Function $f(t, z)$ is said to be *band-limited* if its Fourier transform is 0 outside a rectangle established by the intervals $[-\mu_{max}, \mu_{max}]$ and $[-\nu_{max}, \nu_{max}]$; that is,

$$F(\mu, \nu) = 0 \quad \text{for } |\mu| \geq \mu_{max} \text{ and } |\nu| \geq \nu_{max} \qquad (4.5\text{-}10)$$

The *two-dimensional sampling theorem* states that a continuous, band-limited function $f(t, z)$ can be recovered with no error from a set of its samples if the sampling intervals are

$$\Delta T < \frac{1}{2\mu_{max}} \qquad (4.5\text{-}11)$$

and

$$\Delta Z < \frac{1}{2\nu_{max}} \qquad (4.5\text{-}12)$$

or, expressed in terms of the sampling rate, if

FIGURE 4.14
Two-dimensional
impulse train.

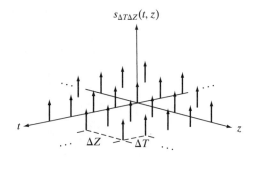

$$\frac{1}{\Delta T} > 2\mu_{\max} \tag{4.5-13}$$

and

$$\frac{1}{\Delta Z} > 2\nu_{\max} \tag{4.5-14}$$

Stated another way, we say that no information is lost if a 2-D, band-limited, continuous function is represented by samples acquired at rates greater than twice the highest frequency content of the function in both the μ- and ν-directions.

Figure 4.15 shows the 2-D equivalents of Figs. 4.6(b) and (d). A 2-D ideal box filter has the form illustrated in Fig. 4.13(a). The dashed portion of Fig. 4.15(a) shows the location of the filter to achieve the necessary isolation of a single period of the transform for reconstruction of a band-limited function from its samples, as in Section 4.3.3. From Section 4.3.4, we know that if the function is under-sampled the periods overlap, and it becomes impossible to isolate a single period, as Fig. 4.15(b) shows. Aliasing would result under such conditions.

4.5.4 Aliasing in Images

In this section, we extend the concept of aliasing to images and discuss several aspects related to image sampling and resampling.

a b

FIGURE 4.15
Two-dimensional
Fourier transforms
of (a) an over-
sampled, and
(b) under-sampled
band-limited
function.

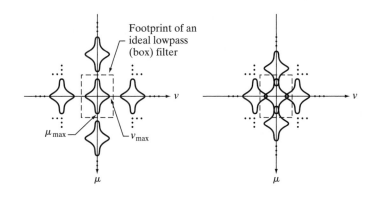

Extension from 1-D aliasing

As in the 1-D case, a continuous function $f(t, z)$ of two continuous variables, t and z, can be band-limited in general only if it extends infinitely in both coordinate directions. The very act of limiting the duration of the function introduces corrupting frequency components extending to infinity in the frequency domain, as explained in Section 4.3.4. Because we cannot sample a function infinitely, aliasing is always present in digital images, just as it is present in sampled 1-D functions. There are two principal manifestations of aliasing in images: spatial aliasing and temporal aliasing. *Spatial aliasing* is due to under-sampling, as discussed in Section 4.3.4. *Temporal aliasing* is related to time intervals between images in a sequence of images. One of the most common examples of temporal aliasing is the "wagon wheel" effect, in which wheels with spokes in a sequence of images (for example, in a movie) appear to be rotating backwards. This is caused by the frame rate being too low with respect to the speed of wheel rotation in the sequence.

Our focus in this chapter is on spatial aliasing. The key concerns with spatial aliasing in images are the introduction of artifacts such as jaggedness in line features, spurious highlights, and the appearance of frequency patterns not present in the original image. The following example illustrates aliasing in images.

■ Suppose that we have an imaging system that is perfect, in the sense that it is noiseless and produces an exact digital image of what it sees, but the number of samples it can take is fixed at 96×96 pixels. If we use this system to digitize checkerboard patterns, it will be able to resolve patterns that are up to 96×96 squares, in which the size of each square is 1×1 pixels. In this limiting case, each pixel in the resulting image will correspond to one square in the pattern. We are interested in examining what happens when the detail (the size of the checkerboard squares) is less than one camera pixel; that is, when the imaging system is asked to digitize checkerboard patterns that have more than 96×96 squares in the field of view.

Figures 4.16(a) and (b) show the result of sampling checkerboards whose squares are of size 16 and 6 pixels on the side, respectively. These results are as expected. However, when the size of the squares is reduced to slightly less than one camera pixel a severely aliased image results, as Fig. 4.16(c) shows. Finally, reducing the size of the squares to slightly less than 0.5 pixels on the side yielded the image in Fig. 4.16(d). In this case, the aliased result looks like a normal checkerboard pattern. In fact, this image would result from sampling a checkerboard image whose squares were 12 pixels on the side. This last image is a good reminder that aliasing can create results that may be quite misleading. ■

EXAMPLE 4.6:
Aliasing in images.

This example should not be construed as being unrealistic. Sampling a "perfect" scene under noiseless, distortion-free conditions is common when converting computer-generated models and vector drawings to digital images.

The effects of aliasing can be *reduced* by slightly defocusing the scene to be digitized so that high frequencies are attenuated. As explained in Section 4.3.4, anti-aliasing filtering has to be done at the "front-end," *before* the image is sampled. There are no such things as after-the-fact software anti-aliasing filters that can be used to reduce the effects of aliasing caused by violations of the sampling theorem. Most commercial digital image manipulation packages do have a feature called "anti-aliasing." However, as illustrated in Examples 4.7

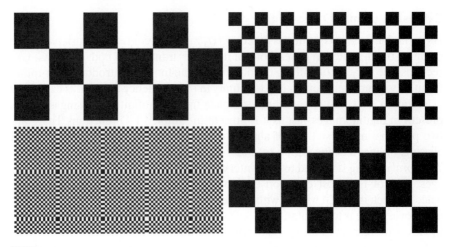

a b
c d

FIGURE 4.16 Aliasing in images. In (a) and (b), the lengths of the sides of the squares are 16 and 6 pixels, respectively, and aliasing is visually negligible. In (c) and (d), the sides of the squares are 0.9174 and 0.4798 pixels, respectively, and the results show significant aliasing. Note that (d) masquerades as a "normal" image.

and 4.8, this term is related to blurring a *digital* image to reduce additional aliasing artifacts caused by resampling. The term does not apply to reducing aliasing in the original sampled image. A significant number of commercial digital cameras have true anti-aliasing filtering built in, either in the lens or on the surface of the sensor itself. For this reason, it is difficult to illustrate aliasing using images obtained with such cameras.

Image interpolation and resampling

As in the 1-D case, perfect reconstruction of a band-limited image function from a set of its samples requires 2-D convolution in the spatial domain with a sinc function. As explained in Section 4.3.5, this theoretically perfect reconstruction requires interpolation using infinite summations which, in practice, forces us to look for approximations. One of the most common applications of 2-D interpolation in image processing is in image resizing (zooming and shrinking). Zooming may be viewed as over-sampling, while shrinking may be viewed as under-sampling. The key difference between these two operations and the sampling concepts discussed in previous sections is that zooming and shrinking are applied to *digital* images.

Interpolation was explained in Section 2.4.4. Our interest there was to illustrate the performance of nearest neighbor, bilinear, and bicubic interpolation. In this section, we give some additional examples with a focus on sampling and anti-aliasing issues. A special case of nearest neighbor interpolation that ties in nicely with over-sampling is zooming by *pixel replication*, which is applicable when we want to increase the size of an image an integer number of times. For instance, to double the size of an image, we duplicate each column. This doubles the image size in the horizontal direction. Then, we duplicate each row of the enlarged image to double the size in the vertical direction. The same procedure is used to enlarge the image any integer number of times. The intensity-level assignment of each pixel is predetermined by the fact that new locations are exact duplicates of old locations.

Image shrinking is done in a manner similar to zooming. Under-sampling is achieved by row-column deletion (e.g., to shrink an image by one-half, we delete every other row and column). We can use the zooming grid analogy in Section 2.4.4 to visualize the concept of shrinking by a non-integer factor, except that we now expand the grid to fit over the original image, do intensity-level interpolation, and then shrink the grid back to its specified size. To reduce aliasing, it is a good idea to blur an image slightly before shrinking it (we discuss frequency domain blurring in Section 4.8). An alternate technique is to *super-sample* the original scene and then reduce (resample) its size by row and column deletion. This can yield sharper results than with smoothing, but it clearly requires access to the original scene. Clearly, if we have no access to the original scene (as typically is the case in practice) super-sampling is not an option.

The process of resampling an image without using band-limiting blurring is called *decimation*.

■ The effects of aliasing generally are worsened when the size of a digital image is reduced. Figure 4.17(a) is an image purposely created to illustrate the effects of aliasing (note the thinly-spaced parallel lines in all garments worn by the subject). There are no objectionable artifacts in Fig. 4.17(a), indicating that

EXAMPLE 4.7:
Illustration of aliasing in resampled images.

a b c

FIGURE 4.17 Illustration of aliasing on resampled images. (a) A digital image with negligible visual aliasing. (b) Result of resizing the image to 50% of its original size by pixel deletion. Aliasing is clearly visible. (c) Result of blurring the image in (a) with a 3×3 averaging filter prior to resizing. The image is slightly more blurred than (b), but aliasing is not longer objectionable. (Original image courtesy of the Signal Compression Laboratory, University of California, Santa Barbara.)

the sampling rate used initially was sufficient to avoid visible aliasing. In Fig. 4.17(b), the image was reduced to 50% of its original size using row-column deletion. The effects of aliasing are quite visible in this image (see, for example the areas around the subject's knees). The digital "equivalent" of anti-aliasing filtering of continuous images is to attenuate the high frequencies of a digital image by smoothing it before resampling. Figure 4.17(c) shows the result of smoothing the image in Fig. 4.17(a) with a 3 × 3 averaging filter (see Section 3.5) before reducing its size. The improvement over Fig. 4.17(b) is evident. Images (b) and (c) were resized up to their original dimension by pixel replication to simplify comparisons.   ■

When you work with images that have strong edge content, the effects of aliasing are seen as block-like image components, called *jaggies*. The following example illustrates this phenomenon.

EXAMPLE 4.8:
Illustration of jaggies in image shrinking.

■ Figure 4.18(a) shows a 1024 × 1024 digital image of a computer-generated scene in which aliasing is negligible. Figure 4.18(b) is the result of reducing the size of (a) by 75% to 256 × 256 pixels using bilinear interpolation and then using pixel replication to bring the image back to its original size in order to make the effects of aliasing (jaggies in this case) more visible. As in Example 4.7, the effects of aliasing can be made less objectionable by smoothing the image before resampling. Figure 4.18(c) is the result of using a 5 × 5 averaging filter prior to reducing the size of the image. As this figure shows, jaggies were reduced significantly. The size reduction and increase to the original size in Fig. 4.18(c) were done using the same approach used to generate Fig. 4.18(b).   ■

a b c

FIGURE 4.18 Illustration of jaggies. (a) A 1024 × 1024 digital image of a computer-generated scene with negligible visible aliasing. (b) Result of reducing (a) to 25% of its original size using bilinear interpolation. (c) Result of blurring the image in (a) with a 5 × 5 averaging filter prior to resizing it to 25% using bilinear interpolation. (Original image courtesy of D. P. Mitchell, Mental Landscape, LLC.)

■ In the previous two examples, we used pixel replication to zoom the small resampled images. This is not a preferred approach in general, as Fig. 4.19 illustrates. Figure 4.19(a) shows a 1024 × 1024 zoomed image generated by pixel replication from a 256 × 256 section out of the center of the image in Fig. 4.18(a). Note the "blocky" edges. The zoomed image in Fig. 4.19(b) was generated from the same 256 × 256 section, but using bilinear interpolation. The edges in this result are considerably smoother. For example, the edges of the bottle neck and the large checkerboard squares are not nearly as blocky in (b) as they are in (a). ■

EXAMPLE 4.9:
Illustration of jaggies in image zooming.

Moiré patterns

Before leaving this section, we examine another type of artifact, called *moiré patterns*,[†] that sometimes result from sampling scenes with periodic or nearly periodic components. In optics, moiré patterns refer to beat patterns produced between two gratings of approximately equal spacing. These patterns are a common everyday occurrence. We see them, for example, in overlapping insect window screens and on the interference between TV raster lines and striped materials. In digital image processing, the problem arises routinely when scanning media print, such as newspapers and magazines, or in images with periodic components whose spacing is comparable to the spacing between samples. It is important to note that moiré patterns are more general than sampling artifacts. For instance, Fig. 4.20 shows the moiré effect using ink drawings that have not been digitized. Separately, the patterns are clean and void of interference. However, superimposing one pattern on the other creates

a b

FIGURE 4.19 Image zooming. (a) A 1024 × 1024 digital image generated by pixel replication from a 256 × 256 image extracted from the middle of Fig. 4.18(a). (b) Image generated using bi-linear interpolation, showing a significant reduction in jaggies.

[†]The term *moiré* is a French word (not the name of a person) that appears to have originated with weavers who first noticed interference patterns visible on some fabrics; the term is rooted on the word *mohair*, a cloth made from Angola goat hairs.

FIGURE 4.20
Examples of the moiré effect. These are ink drawings, not digitized patterns. Superimposing one pattern on the other is equivalent mathematically to multiplying the patterns.

a beat pattern that has frequencies not present in either of the original patterns. Note in particular the moiré effect produced by two patterns of dots, as this is the effect of interest in the following discussion.

Newspapers and other printed materials make use of so called *halftone dots*, which are black dots or ellipses whose sizes and various joining schemes are used to simulate gray tones. As a rule, the following numbers are typical: newspapers are printed using 75 halftone dots per inch (*dpi* for short), magazines use 133 dpi, and high-quality brochures use 175 dpi. Figure 4.21 shows

Color printing uses red, green, and blue dots to produce the sensation in the eye of continuous color.

FIGURE 4.21
A newspaper image of size 246 × 168 pixels sampled at 75 dpi showing a moiré pattern. The moiré pattern in this image is the interference pattern created between the ±45° orientation of the halftone dots and the north–south orientation of the sampling grid used to digitize the image.

what happens when a newspaper image is sampled at 75 dpi. The sampling lattice (which is oriented vertically and horizontally) and dot patterns on the newspaper image (oriented at ±45°) interact to create a uniform moiré pattern that makes the image look blotchy. (We discuss a technique in Section 4.10.2 for reducing moiré interference patterns.)

As a related point of interest, Fig. 4.22 shows a newspaper image sampled at 400 dpi to avoid moiré effects. The enlargement of the region surrounding the subject's left eye illustrates how halftone dots are used to create shades of gray. The dot size is inversely proportional to image intensity. In light areas, the dots are small or totally absent (see, for example, the white part of the eye). In light gray areas, the dots are larger, as shown below the eye. In darker areas, when dot size exceeds a specified value (typically 50%), dots are allowed to join along two specified directions to form an interconnected mesh (see, for example, the left part of the eye). In some cases the dots join along only one direction, as in the top right area below the eyebrow.

4.5.5 The 2-D Discrete Fourier Transform and Its Inverse

A development similar to the material in Sections 4.3 and 4.4 would yield the following 2-D *discrete Fourier transform* (DFT):

$$F(u, v) = \sum_{x=0}^{M-1} \sum_{y=0}^{N-1} f(x, y) e^{-j2\pi(ux/M + vy/N)} \qquad (4.5\text{-}15)$$

where $f(x, y)$ is a digital image of size $M \times N$. As in the 1-D case, Eq. (4.5-15) must be evaluated for values of the discrete variables u and v in the ranges $u = 0, 1, 2, \ldots, M - 1$ and $v = 0, 1, 2, \ldots, N - 1$.[†]

Sometimes you will find in the literature the $1/MN$ constant in front of DFT instead of the IDFT. At times, the constant is expressed as $1/\sqrt{MN}$ and is included in front of the forward and inverse transforms, thus creating a more symmetric pair. Any of these formulations is correct, provided that you are consistent.

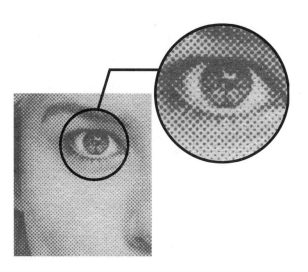

FIGURE 4.22
A newspaper image and an enlargement showing how halftone dots are arranged to render shades of gray.

[†]As mentioned in Section 4.4.1, keep in mind that in this chapter we use (t, z) and (μ, ν) to denote 2-D *continuous* spatial and frequency-domain variables. In the 2-D *discrete* case, we use (x, y) for spatial variables and (u, v) for frequency-domain variables.

Given the transform $F(u, v)$, we can obtain $f(x, y)$ by using the *inverse discrete Fourier transform* (IDFT):

$$f(x, y) = \frac{1}{MN} \sum_{u=0}^{M-1} \sum_{v=0}^{N-1} F(u, v) e^{j2\pi(ux/M+vy/N)} \qquad (4.5\text{-}16)$$

for $x = 0, 1, 2, \ldots, M - 1$ and $y = 0, 1, 2, \ldots, N - 1$. Equations (4.5-15) and (4.5-16) constitute the 2-D *discrete Fourier transform pair*. The rest of this chapter is based on properties of these two equations and their use for image filtering in the frequency domain.

4.6   Some Properties of the 2-D Discrete Fourier Transform

In this section, we introduce several properties of the 2-D discrete Fourier transform and its inverse.

4.6.1 Relationships Between Spatial and Frequency Intervals

The relationships between spatial sampling and the corresponding frequency-domain intervals are as explained in Section 4.4.2. Suppose that a continuous function $f(t, z)$ is sampled to form a digital image, $f(x, y)$, consisting of $M \times N$ samples taken in the t- and z-directions, respectively. Let ΔT and ΔZ denote the separations between samples (see Fig. 4.14). Then, the separations between the corresponding discrete, frequency domain variables are given by

$$\Delta u = \frac{1}{M\Delta T} \qquad (4.6\text{-}1)$$

and

$$\Delta v = \frac{1}{N\Delta Z} \qquad (4.6\text{-}2)$$

respectively. Note that the separations between samples in the frequency domain are inversely proportional both to the spacing between spatial samples and the number of samples.

4.6.2 Translation and Rotation

It can be shown by direct substitution into Eqs. (4.5-15) and (4.5-16) that the Fourier transform pair satisfies the following translation properties (Problem 4.16):

$$f(x, y) e^{j2\pi(u_0 x/M + v_0 y/N)} \Leftrightarrow F(u - u_0, v - v_0) \qquad (4.6\text{-}3)$$

and

$$f(x - x_0, y - y_0) \Leftrightarrow F(u, v) e^{-j2\pi(x_0 u/M + y_0 v/N)} \qquad (4.6\text{-}4)$$

That is, multiplying $f(x, y)$ by the exponential shown shifts the origin of the DFT to (u_0, v_0) and, conversely, multiplying $F(u, v)$ by the negative of that exponential shifts the origin of $f(x, y)$ to (x_0, y_0). As we illustrate in Example 4.13, translation has no effect on the magnitude (spectrum) of $F(u, v)$.

Using the polar coordinates

$$x = r \cos \theta \quad y = r \sin \theta \quad u = \omega \cos \varphi \quad v = \omega \sin \varphi$$

results in the following transform pair:

$$f(r, \theta + \theta_0) \Leftrightarrow F(\omega, \varphi + \theta_0) \tag{4.6-5}$$

which indicates that rotating $f(x, y)$ by an angle θ_0 rotates $F(u, v)$ by the same angle. Conversely, rotating $F(u, v)$ rotates $f(x, y)$ by the same angle.

4.6.3 Periodicity

As in the 1-D case, the 2-D Fourier transform and its inverse are infinitely periodic in the u and v directions; that is,

$$F(u, v) = F(u + k_1 M, v) = F(u, v + k_2 N) = F(u + k_1 M, v + k_2 N) \tag{4.6-6}$$

and

$$f(x, y) = f(x + k_1 M, y) = f(x, y + k_2 N) = f(x + k_1 M, y + k_2 N) \tag{4.6-7}$$

where k_1 and k_2 are integers.

The periodicities of the transform and its inverse are important issues in the implementation of DFT-based algorithms. Consider the 1-D spectrum in Fig. 4.23(a). As explained in Section 4.4.1, the transform data in the interval from 0 to $M - 1$ consists of two back-to-back half periods meeting at point $M/2$. For display and filtering purposes, it is more convenient to have in this interval a complete period of the transform in which the data are contiguous, as in Fig. 4.23(b). It follows from Eq. (4.6-3) that

$$f(x) e^{j2\pi(u_0 x/M)} \Leftrightarrow F(u - u_0)$$

In other words, multiplying $f(x)$ by the exponential term shown shifts the data so that the origin, $F(0)$, is located at u_0. If we let $u_0 = M/2$, the exponential term becomes $e^{j\pi x}$ which is equal to $(-1)^x$ because x is an integer. In this case,

$$f(x)(-1)^x \Leftrightarrow F(u - M/2)$$

That is, multiplying $f(x)$ by $(-1)^x$ shifts the data so that $F(0)$ is at the *center* of the interval $[0, M - 1]$, which corresponds to Fig. 4.23(b), as desired.

In 2-D the situation is more difficult to graph, but the principle is the same, as Fig. 4.23(c) shows. Instead of two half periods, there are now four quarter periods meeting at the point $(M/2, N/2)$. The dashed rectangles correspond to

a
b
c d

FIGURE 4.23
Centering the
Fourier transform.
(a) A 1-D DFT
showing an infinite
number of periods.
(b) Shifted DFT
obtained by
multiplying $f(x)$
by $(-1)^x$ before
computing $F(u)$.
(c) A 2-D DFT
showing an infinite
number of periods.
The solid area is
the $M \times N$ data
array, $F(u, v)$,
obtained with Eq.
(4.5-15). This array
consists of four
quarter periods.
(d) A Shifted DFT
obtained by
multiplying $f(x, y)$
by $(-1)^{x+y}$
before computing
$F(u, v)$. The data
now contains one
complete, centered
period, as in (b).

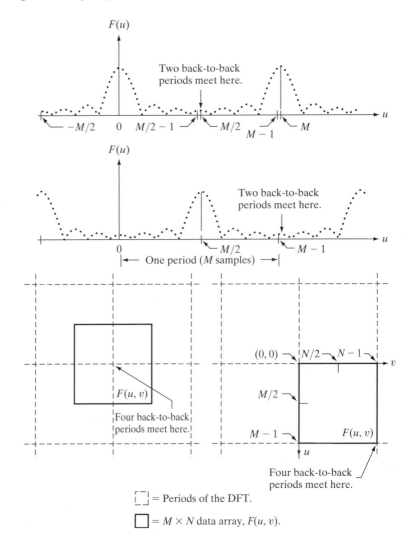

the infinite number of periods of the 2-D DFT. As in the 1-D case, visualization is simplified if we shift the data so that $F(0, 0)$ is at $(M/2, N/2)$. Letting $(u_0, v_0) = (M/2, N/2)$ in Eq. (4.6-3) results in the expression

$$f(x, y)(-1)^{x+y} \Leftrightarrow F(u - M/2, v - N/2) \qquad (4.6\text{-}8)$$

Using this equation shifts the data so that $F(0, 0)$ is at the center of the *frequency rectangle* defined by the intervals $[0, M - 1]$ and $[0, N - 1]$, as desired. Figure 4.23(d) shows the result. We illustrate these concepts later in this section as part of Example 4.11 and Fig. 4.24.

4.6.4 Symmetry Properties

An important result from functional analysis is that any real *or* complex function, $w(x, y)$, can be expressed as the sum of an even and an odd part (*each of which can be real or complex*):

$$w(x, y) = w_e(x, y) + w_o(x, y) \tag{4.6-9}$$

where the even and odd parts are defined as

$$w_e(x, y) \triangleq \frac{w(x, y) + w(-x, -y)}{2} \tag{4.6-10a}$$

and

$$w_o(x, y) \triangleq \frac{w(x, y) - w(-x, -y)}{2} \tag{4.6-10b}$$

Substituting Eqs. (4.6-10a) and (4.6-10b) into Eq. (4.6-9) gives the identity $w(x, y) \equiv w(x, y)$, thus proving the validity of the latter equation. It follows from the preceding definitions that

$$w_e(x, y) = w_e(-x, -y) \tag{4.6-11a}$$

and that

$$w_o(x, y) = -w_o(-x, -y) \tag{4.6-11b}$$

Even functions are said to be *symmetric* and odd functions are *antisymmetric*. Because all indices in the DFT and IDFT are positive, when we talk about symmetry (antisymmetry) we are referring to symmetry (antisymmetry) about the *center point* of a sequence. In terms of Eq. (4.6-11), indices to the right of the center point of a 1-D array are considered positive, and those to the left are considered negative (similarly in 2-D). In our work, it is more convenient to think only in terms of nonnegative indices, in which case the definitions of evenness and oddness become:

$$w_e(x, y) = w_e(M - x, N - y) \tag{4.6-12a}$$

and

$$w_o(x, y) = -w_o(M - x, N - y) \tag{4.6-12b}$$

where, as usual, M and N are the number of rows and columns of a 2-D array.

We know from elementary mathematical analysis that the product of two even or two odd functions is even, and that the product of an even and an odd function is odd. In addition, the only way that a discrete function can be odd is if all its samples sum to zero. These properties lead to the important result that

To convince yourself that the samples of an odd function sum to zero, sketch one period of a 1-D sine wave about the origin or any other interval spanning one period.

$$\sum_{x=0}^{M-1} \sum_{y=0}^{N-1} w_e(x, y)\, w_o(x, y) = 0 \qquad (4.6\text{-}13)$$

for any two discrete even and odd functions w_e and w_o. In other words, because the argument of Eq. (4.6-13) is odd, the result of the summations is 0. The functions can be real or complex.

EXAMPLE 4.10:
Even and odd functions.

■ Although evenness and oddness are visualized easily for continuous functions, these concepts are not as intuitive when dealing with discrete sequences. The following illustrations will help clarify the preceding ideas. Consider the 1-D sequence

$$f = \left\{ f(0) \quad f(1) \quad f(2) \quad f(3) \right\}$$

$$= \left\{ 2 \quad 1 \quad 1 \quad 1 \right\}$$

in which $M = 4$. To test for evenness, the condition $f(x) = f(4 - x)$ must be satisfied; that is, we require that

$$f(0) = f(4), \quad f(2) = f(2), \quad f(1) = f(3), \quad f(3) = f(1)$$

Because $f(4)$ is outside the range being examined, and it can be any value, the value of $f(0)$ is immaterial in the test for evenness. We see that the next three conditions are satisfied by the values in the array, so the sequence is even. In fact, we conclude that *any* 4-point even sequence has to have the form

$$\{a \quad b \quad c \quad b\}$$

That is, only the second and last points must be equal in a 4-point even sequence.

An odd sequence has the interesting property that its first term, $w_o(0, 0)$, is always 0, a fact that follows directly from Eq. (4.6-10b). Consider the 1-D sequence

$$g = \left\{ g(0) \quad g(1) \quad g(2) \quad g(3) \right\}$$

$$= \{0 \quad -1 \quad 0 \quad 1\}$$

We easily can confirm that this is an odd sequence by noting that the terms in the sequence satisfy the condition $g(x) = -g(4 - x)$. For example, $g(1) = -g(3)$. Any 4-point odd sequence has the form

$$\{0 \quad -b \quad 0 \quad b\}$$

That is, when M is an even number, a 1-D odd sequence has the property that the points at locations 0 and $M/2$ always are zero. When M is odd, the first term still has to be 0, but the remaining terms form pairs with equal value but opposite sign.

The preceding discussion indicates that evenness and oddness of sequences depend also on the length of the sequences. For example, we already showed that the sequence $\{0 \quad -1 \quad 0 \quad 1\}$ is odd. However, the sequence $\{0 \quad -1 \quad 0 \quad 1 \quad 0\}$ is neither odd nor even, although the "basic" structure appears to be odd. This is an important issue in interpreting DFT results. We show later in this section that the DFTs of even and odd functions have some very important characteristics. Thus, it often is the case that understanding when a function is odd or even plays a key role in our ability to interpret image results based on DFTs.

The same basic considerations hold in 2-D. For example, the 6×6 2-D sequence

$$
\begin{array}{cccccc}
0 & 0 & 0 & 0 & 0 & 0 \\
0 & 0 & 0 & 0 & 0 & 0 \\
0 & 0 & -1 & 0 & 1 & 0 \\
0 & 0 & -2 & 0 & 2 & 0 \\
0 & 0 & -1 & 0 & 1 & 0 \\
0 & 0 & 0 & 0 & 0 & 0
\end{array}
$$

As an exercise, you should use Eq. (4.6-12b) to convince yourself that this 2-D sequence is odd.

is odd. However, adding another row and column of 0s would give a result that is neither odd nor even. Note that the inner structure of this array is a Sobel mask, as discussed in Section 3.6.4. We return to this mask in Example 4.15. ■

Armed with the preceding concepts, we can establish a number of important symmetry properties of the DFT and its inverse. A property used frequently is that the Fourier transform of a *real* function, $f(x, y)$, is *conjugate symmetric*:

$$F^*(u, v) = F(-u, -v) \tag{4.6-14}$$

If $f(x, y)$ is *imaginary*, its Fourier transform is *conjugate antisymmetric*: $F^*(-u, -v) = -F(u, v)$. The proof of Eq. (4.6-14) is as follows:

Conjugate symmetry also is called *hermitian symmetry*. The term *antihermitian* is used sometimes to refer to conjugate antisymmetry.

$$F^*(u, v) = \left[\sum_{x=0}^{M-1} \sum_{y=0}^{N-1} f(x, y) e^{-j2\pi(ux/M + vy/N)} \right]^*$$

$$= \sum_{x=0}^{M-1} \sum_{y=0}^{N-1} f^*(x, y) e^{j2\pi(ux/M + vy/N)}$$

$$= \sum_{x=0}^{M-1} \sum_{y=0}^{N-1} f(x, y) e^{-j2\pi([-u]x/M + [-v]y/N)}$$

$$= F(-u, -v)$$

where the third step follows from the fact that $f(x, y)$ is real. A similar approach can be used to prove the conjugate antisymmetry exhibited by the transform of imaginary functions.

Table 4.1 lists symmetries and related properties of the DFT that are useful in digital image processing. Recall that the double arrows indicate Fourier transform pairs; that is, for any row in the table, the properties on the right are satisfied by the Fourier transform of the function having the properties listed on the left, and vice versa. For example, entry 5 reads: The DFT of a real function $f(x, y)$, in which (x, y) is replaced by $(-x, -y)$, is $F^*(u, v)$, where $F(u, v)$, the DFT of $f(x, y)$, is a complex function, and vice versa.

TABLE 4.1 Some symmetry properties of the 2-D DFT and its inverse. $R(u, v)$ and $I(u, v)$ are the real and imaginary parts of $F(u, v)$, respectively. The term *complex* indicates that a function has nonzero real and imaginary parts.

	Spatial Domain†		Frequency Domain†
1)	$f(x, y)$ real	⇔	$F^*(u, v) = F(-u, -v)$
2)	$f(x, y)$ imaginary	⇔	$F^*(-u, -v) = -F(u, v)$
3)	$f(x, y)$ real	⇔	$R(u, v)$ even; $I(u, v)$ odd
4)	$f(x, y)$ imaginary	⇔	$R(u, v)$ odd; $I(u, v)$ even
5)	$f(-x, -y)$ real	⇔	$F^*(u, v)$ complex
6)	$f(-x, -y)$ complex	⇔	$F(-u, -v)$ complex
7)	$f^*(x, y)$ complex	⇔	$F^*(-u - v)$ complex
8)	$f(x, y)$ real and even	⇔	$F(u, v)$ real and even
9)	$f(x, y)$ real and odd	⇔	$F(u, v)$ imaginary and odd
10)	$f(x, y)$ imaginary and even	⇔	$F(u, v)$ imaginary and even
11)	$f(x, y)$ imaginary and odd	⇔	$F(u, v)$ real and odd
12)	$f(x, y)$ complex and even	⇔	$F(u, v)$ complex and even
13)	$f(x, y)$ complex and odd	⇔	$F(u, v)$ complex and odd

†Recall that $x, y, u,$ and v are *discrete* (integer) variables, with x and u in the range $[0, M - 1]$, and y and v in the range $[0, N - 1]$. To say that a complex function is *even* means that its real *and* imaginary parts are even, and similarly for an odd complex function.

■ With reference to the even and odd concepts discussed earlier and illustrated in Example 4.10, the following 1-D sequences and their transforms are short examples of the properties listed in Table 4.1. The numbers in parentheses on the right are the individual elements of $F(u)$, and similarly for $f(x)$ in the last two properties.

EXAMPLE 4.11:
1-D illustrations
of properties from
Table 4.1.

Property	$f(x)$		$F(u)$
3	$\{1 \quad 2 \quad 3 \quad 4\}$	$\Leftrightarrow$	$\{(10)\,(-2+2j)\,(-2)\,(-2-2j)\}$
4	$j\{1 \quad 2 \quad 3 \quad 4\}$	$\Leftrightarrow$	$\{(2.5j)\,(.5-.5j)\,(-.5j)\,(-.5-.5j)\}$
8	$\{2 \quad 1 \quad 1 \quad 1\}$	$\Leftrightarrow$	$\{(5)\,(1)\,(1)\,(1)\}$
9	$\{0 \quad -1 \quad 0 \quad 1\}$	$\Leftrightarrow$	$\{(0)\,(2j)\,(0)\,(-2j)\}$
10	$j\{2 \quad 1 \quad 1 \quad 1\}$	$\Leftrightarrow$	$\{(5j)\,(j)\,(j)\,(j)\}$
11	$j\{0 \quad -1 \quad 0 \quad 1\}$	$\Leftrightarrow$	$\{(0)\,(-2)\,(0)\,(2)\}$
12	$\{(4+4j)\,(3+2j)\,(0+2j)\,(3+2j)\}$	$\Leftrightarrow$	$\{(10+10j)\,(4+2j)\,(-2+2j)\,(4+2j)\}$
13	$\{(0+0j)\,(1+1j)\,(0+0j)\,(-1-j)\}$	$\Leftrightarrow$	$\{(0+0j)\,(2-2j)\,(0+0j)\,(-2+2j)\}$

For example, in property 3 we see that a real function with elements $\{1 \ 2 \ 3 \ 4\}$ has Fourier transform whose real part, $\{10 \ -2 \ -2 \ -2\}$, is even and whose imaginary part, $\{0 \ 2 \ 0 \ -2\}$, is odd. Property 8 tells us that a real even function has a transform that is real and even also. Property 12 shows that an even complex function has a transform that is also complex and even. The other property examples are analyzed in a similar manner.   ■

■ In this example, we prove several of the properties in Table 4.1 to develop familiarity with manipulating these important properties, and to establish a basis for solving some of the problems at the end of the chapter. We prove only the properties on the right given the properties on the left. The converse is proved in a manner similar to the proofs we give here.

EXAMPLE 4.12:
Proving several
symmetry
properties of the
DFT from Table
4.1.

Consider property 3, which reads: If $f(x, y)$ is a real function, the real part of its DFT is even and the odd part is odd; similarly, if a DFT has real and imaginary parts that are even and odd, respectively, then its IDFT is a real function. We prove this property formally as follows. $F(u, v)$ is complex in general, so it can be expressed as the sum of a real and an imaginary part: $F(u, v) = R(u, v) + jI(u, v)$.   Then, $F^*(u, v) = R(u, v) - jI(u, v)$.   Also, $F(-u, -v) = R(-u, -v) + jI(-u, -v)$. But, as proved earlier, if $f(x, y)$ is real then $F^*(u, v) = F(-u, -v)$, which, based on the preceding two equations, means that $R(u, v) = R(-u, -v)$ and $I(u, v) = -I(-u, -v)$. In view of Eqs. (4.6-11a) and (4.6-11b), this proves that R is an even function and I is an odd function.

Next, we prove property 8. If $f(x, y)$ is real we know from property 3 that the real part of $F(u, v)$ is even, so to prove property 8 all we have to do is show that if $f(x, y)$ is real *and* even then the imaginary part of $F(u, v)$ is 0 (i.e., F is real). The steps are as follows:

$$F(u, v) = \sum_{x=0}^{M-1} \sum_{y=0}^{N-1} f(x, y) e^{-j2\pi(ux/M+vy/N)}$$

which we can write as

$$F(u, v) = \sum_{x=0}^{M-1} \sum_{y=0}^{N-1} [f_r(x, y)] e^{-j2\pi(ux/M + vy/N)}$$

$$= \sum_{x=0}^{M-1} \sum_{y=0}^{N-1} [f_r(x, y)] e^{-j2\pi(ux/M)} e^{-j2\pi(vy/N)}$$

$$= \sum_{x=0}^{M-1} \sum_{y=0}^{N-1} [\text{even}][\text{even} - j\text{odd}][\text{even} - j\text{odd}]$$

$$= \sum_{x=0}^{M-1} \sum_{y=0}^{N-1} [\text{even}][\text{even} \cdot \text{even} - 2j\text{even} \cdot \text{odd} - \text{odd} \cdot \text{odd}]$$

$$= \sum_{x=0}^{M-1} \sum_{y=0}^{N-1} [\text{even} \cdot \text{even}] - 2j \sum_{x=0}^{M-1} \sum_{y=0}^{N-1} [\text{even} \cdot \text{odd}]$$

$$- \sum_{x=0}^{M-1} \sum_{y=0}^{N-1} [\text{even} \cdot \text{even}]$$

$$= \text{real}$$

The fourth step follows from Euler's equation and the fact that the cos and sin are even and odd functions, respectively. We also know from property 8 that, in addition to being real, f is an even function. The only term in the penultimate line containing imaginary components is the second term, which is 0 according to Eq. (4.6-14). Thus, if f is real and even then F is real. As noted earlier, F is also even because f is real. This concludes the proof.

Finally, we prove the validity of property 6. From the definition of the DFT,

Note that we are not making a change of variable here. We are evaluating the DFT of $f(-x, -y)$, so we simply insert this function into the equation, as we would any other function.

$$\Im\{f(-x, -y)\} = \sum_{x=0}^{M-1} \sum_{y=0}^{N-1} f(-x, -y) e^{-j2\pi(ux/M + vy/N)}$$

Because of periodicity, $f(-x, -y) = f(M - x, N - y)$. If we now define $m = M - x$ and $n = N - y$, then

$$\Im\{f(-x, -y)\} = \sum_{m=0}^{M-1} \sum_{n=0}^{N-1} f(m, n) e^{-j2\pi(u[M-m]/M + v[N-n]/N)}$$

(To convince yourself that the summations are correct, try a 1-D transform and expand a few terms by hand.) Because $\exp[-j2\pi(\text{integer})] = 1$, it follows that

$$\Im\{f(-x, -y)\} = \sum_{m=0}^{M-1} \sum_{n=0}^{N-1} f(m, n) e^{j2\pi(um/M + vn/N)}$$

$$= F(-u, -v)$$

This concludes the proof.   ■

4.6.5 Fourier Spectrum and Phase Angle

Because the 2-D DFT is complex in general, it can be expressed in polar form:

$$F(u, v) = |F(u, v)| e^{j\phi(u,v)} \qquad (4.6\text{-}15)$$

where the magnitude

$$|F(u, v)| = \left[R^2(u, v) + I^2(u, v)\right]^{1/2} \qquad (4.6\text{-}16)$$

is called the *Fourier* (or *frequency*) *spectrum*, and

$$\phi(u, v) = \arctan\left[\frac{I(u, v)}{R(u, v)}\right] \qquad (4.6\text{-}17)$$

is the *phase angle*. Recall from the discussion in Section 4.2.1 that the arctan must be computed using a four-quadrant arctangent, such as MATLAB's atan2(Imag, Real) function.

Finally, the *power spectrum* is defined as

$$P(u, v) = |F(u, v)|^2$$
$$= R^2(u, v) + I^2(u, v) \qquad (4.6\text{-}18)$$

As before, R and I are the real and imaginary parts of $F(u, v)$ and all computations are carried out for the discrete variables $u = 0, 1, 2, \ldots, M - 1$ and $v = 0, 1, 2, \ldots, N - 1$. Therefore, $|F(u, v)|$, $\phi(u, v)$, and $P(u, v)$ are arrays of size $M \times N$.

The Fourier transform of a real function is conjugate symmetric [Eq. (4.6-14)], which implies that the spectrum has *even* symmetry about the origin:

$$|F(u, v)| = |F(-u, -v)| \qquad (4.6\text{-}19)$$

The phase angle exhibits the following *odd* symmetry about the origin:

$$\phi(u, v) = -\phi(-u, -v) \qquad (4.6\text{-}20)$$

It follows from Eq. (4.5-15) that

$$F(0, 0) = \sum_{x=0}^{M-1} \sum_{y=0}^{N-1} f(x, y)$$

a b
c d

FIGURE 4.24
(a) Image.
(b) Spectrum
showing bright spots
in the four corners.
(c) Centered
spectrum. (d) Result
showing increased
detail after a log
transformation. The
zero crossings of the
spectrum are closer in
the vertical direction
because the rectangle
in (a) is longer in that
direction. The
coordinate
convention used
throughout the book
places the origin of
the spatial and
frequency domains at
the top left.

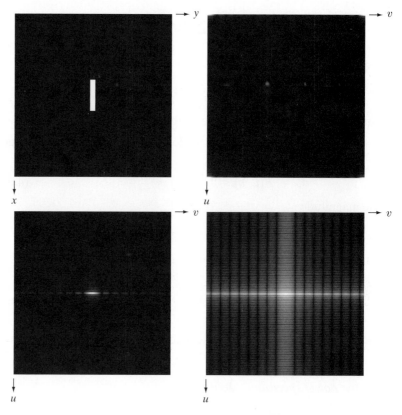

which indicates that the zero-frequency term is proportional to the average value of $f(x, y)$. That is,

$$F(0, 0) = MN\frac{1}{MN}\sum_{x=0}^{M-1}\sum_{y=0}^{N-1}f(x, y)$$

$$= MN\overline{f}(x, y) \tag{4.6-21}$$

where $\overline{f}$ denotes the average value of f. Then,

$$|F(0, 0)| = MN|\overline{f}(x, y)| \tag{4.6-22}$$

Because the proportionality constant MN usually is large, $|F(0, 0)|$ typically is the largest component of the spectrum by a factor that can be several orders of magnitude larger than other terms. Because frequency components u and v are zero at the origin, $F(0, 0)$ sometimes is called the *dc component* of the transform. This terminology is from electrical engineering, where "dc" signifies direct current (i.e., current of zero frequency).

EXAMPLE 4.13:
The 2-D Fourier
spectrum of a
simple function.

■ Figure 4.24(a) shows a simple image and Fig. 4.24(b) shows its spectrum, whose values were scaled to the range $[0, 255]$ and displayed in image form. The origins of both the spatial and frequency domains are at the top left. Two things are apparent in Fig. 4.22(b). As expected, the area around the origin of the transform contains the highest values (and thus appears brighter in the image). However, note that the four corners of the spectrum contain similarly high values. The reason is the periodicity property discussed in the previous section. To center the spectrum, we simply multiply the image in (a) by $(-1)^{x+y}$ before computing the DFT, as indicated in Eq. (4.6-8). Figure 4.22(c) shows the result, which clearly is much easier to visualize (note the symmetry about the center point). Because the dc term dominates the values of the spectrum, the dynamic range of other intensities in the displayed image are compressed. To bring out those details, we perform a log transformation, as described in Section 3.2.2. Figure 4.24(d) shows the display of $(1 + \log|F(u, v)|)$. The increased rendition of detail is evident. Most spectra shown in this and subsequent chapters are scaled in this manner.

It follows from Eqs. (4.6-4) and (4.6-5) that the spectrum is insensitive to image translation (the absolute value of the exponential term is 1), but it rotates by the same angle of a rotated image. Figure 4.25 illustrates these properties. The spectrum in Fig. 4.25(b) is identical to the spectrum in Fig. 4.24(d). Clearly, the images in Figs. 4.24(a) and 4.25(a) are different, so if their Fourier spectra are the same then, based on Eq. (4.6-15), their phase angles must be different. Figure 4.26 confirms this. Figures 4.26(a) and (b) are the phase angle arrays (shown as images) of the DFTs of Figs. 4.24(a) and 4.25(a). Note the lack of similarity between the phase images, in spite of the fact that the only differences between their corresponding images is simple translation. In general, visual analysis of phase angle images yields little intuitive information. For instance, due to its 45° orientation, one would expect intuitively that the phase angle in

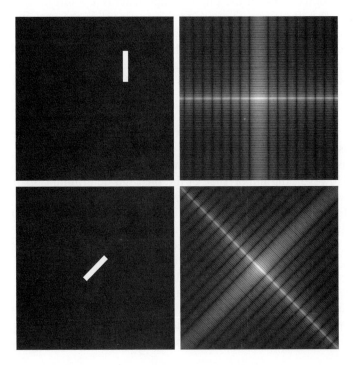

a b
c d

FIGURE 4.25
(a) The rectangle in Fig. 4.24(a) translated, and (b) the corresponding spectrum. (c) Rotated rectangle, and (d) the corresponding spectrum. The spectrum corresponding to the translated rectangle is identical to the spectrum corresponding to the original image in Fig. 4.24(a).

a b c

FIGURE 4.26 Phase angle array corresponding (a) to the image of the centered rectangle in Fig. 4.24(a), (b) to the translated image in Fig. 4.25(a), and (c) to the rotated image in Fig. 4.25(c).

Fig. 4.26(a) should correspond to the rotated image in Fig. 4.25(c), rather than to the image in Fig. 4.24(a). In fact, as Fig. 4.26(c) shows, the phase angle of the rotated image has a strong orientation that is much less than 45°.   ■

The components of the spectrum of the DFT determine the amplitudes of the sinusoids that combine to form the resulting image. At any given frequency in the DFT of an image, a large amplitude implies a greater prominence of a sinusoid of that frequency in the image. Conversely, a small amplitude implies that less of that sinusoid is present in the image. Although, as Fig. 4.26 shows, the contribution of the phase components is less intuitive, it is just as important. The phase is a measure of displacement of the various sinusoids with respect to their origin. Thus, while the magnitude of the 2-D DFT is an array whose components determine the intensities in the image, the corresponding phase is an array of angles that carry much of the information about where discernable objects are located in the image. The following example clarifies these concepts further.

EXAMPLE 4.14:
Further illustration of the properties of the Fourier spectrum and phase angle.

■ Figure 4.27(b) is the phase angle of the DFT of Fig. 4.27(a). There is no detail in this array that would lead us by visual analysis to associate it with features in its corresponding image (not even the symmetry of the phase angle is visible). However, the importance of the phase in determining shape characteristics is evident in Fig. 4.27(c), which was obtained by computing the inverse DFT of Eq. (4.6-15) using only phase information (i.e., with $|F(u, v)| = 1$ in the equation). Although the intensity information has been lost (remember, that information is carried by the spectrum) the key shape features in this image are unmistakably from Fig. 4.27(a).

Figure 4.27(d) was obtained using only the spectrum in Eq. (4.6-15) and computing the inverse DFT. This means setting the exponential term to 1, which in turn implies setting the phase angle to 0. The result is not unexpected. It contains only intensity information, with the dc term being the most dominant. There is no shape information in the image because the phase was set to zero.

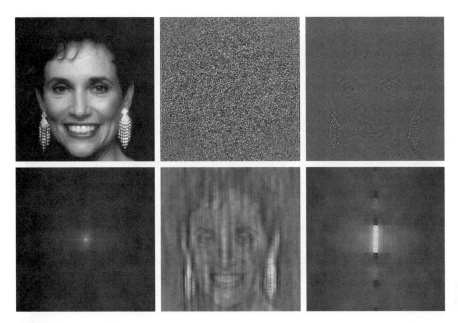

a b c
d e f

FIGURE 4.27 (a) Woman. (b) Phase angle. (c) Woman reconstructed using only the phase angle. (d) Woman reconstructed using only the spectrum. (e) Reconstruction using the phase angle corresponding to the woman and the spectrum corresponding to the rectangle in Fig. 4.24(a). (f) Reconstruction using the phase of the rectangle and the spectrum of the woman.

Finally, Figs. 4.27(e) and (f) show yet again the dominance of the phase in determining the feature content of an image. Figure 4.27(e) was obtained by computing the IDFT of Eq. (4.6-15) using the spectrum of the rectangle in Fig. 4.24(a) and the phase angle corresponding to the woman. The shape of the woman clearly dominates this result. Conversely, the rectangle dominates Fig. 4.27(f), which was computed using the spectrum of the woman and the phase angle of the rectangle. ■

4.6.6 The 2-D Convolution Theorem

Extending Eq. (4.4-10) to two variables results in the following expression for 2-D *circular convolution*:

$$f(x, y) \star h(x, y) = \sum_{m=0}^{M-1} \sum_{n=0}^{N-1} f(m, n)h(x - m, y - n) \qquad (4.6\text{-}23)$$

for $x = 0, 1, 2, \ldots, M - 1$ and $y = 0, 1, 2, \ldots, N - 1$. As in Eq. (4.4-10), Eq. (4.6-23) gives one period of a 2-D periodic sequence. The 2-D convolution theorem is given by the expressions

$$f(x, y) \star h(x, y) \Leftrightarrow F(u, v)H(u, v) \qquad (4.6\text{-}24)$$

and, conversely,

$$f(x, y)h(x, y) \Leftrightarrow F(u, v) \star H(u, v) \qquad (4.6\text{-}25)$$

where F and H are obtained using Eq. (4.5-15) and, as before, the double arrow is used to indicate that the left and right sides of the expressions constitute a Fourier transform pair. Our interest in the remainder of this chapter is in Eq. (4.6-24), which states that the inverse DFT of the product $F(u, v)H(u, v)$ yields $f(x, y) \star h(x, y)$, the 2-D spatial convolution of f and h. Similarly, the DFT of the spatial convolution yields the product of the transforms in the frequency domain. Equation (4.6-24) is the foundation of linear filtering and, as explained in Section 4.7, is the basis for all the filtering techniques discussed in this chapter.

We discuss efficient ways to compute the DFT in Section 4.11.

Because we are dealing here with discrete quantities, computation of the Fourier transforms is carried out with a DFT algorithm. If we elect to compute the spatial convolution using the IDFT of the product of the two transforms, then the periodicity issues discussed in Section 4.6.3 must be taken into account. We give a 1-D example of this and then extend the conclusions to two variables. The left column of Fig. 4.28 implements convolution of two functions, f and h, using the 1-D equivalent of Eq. (3.4-2) which, because the two functions are of same size, is written as

$$f(x) \star h(x) = \sum_{m=0}^{399} f(x)h(x - m)$$

This equation is identical to Eq. (4.4-10), *but* the requirement on the displacement x is that it be sufficiently large to cause the flipped (rotated) version of h to slide completely past f. In other words, the procedure consists of (1) mirroring h about the origin (i.e., rotating it by 180°) [Fig. 4.28(c)], (2) translating the mirrored function by an amount x [Fig. 4.28(d)], and (3) for *each* value x of translation, computing the *entire* sum of products in the right side of the preceding equation. In terms of Fig. 4.28 this means multiplying the function in Fig. 4.28(a) by the function in Fig. 4.28(d) for *each* value of x. The displacement x ranges over all values required to completely slide h across f. Figure 4.28(e) shows the convolution of these two functions. Note that convolution is a function of the displacement variable, x, and that the range of x required in this example to completely slide h past f is from 0 to 799.

If we use the DFT and the convolution theorem to obtain the same result as in the left column of Fig. 4.28, we must take into account the periodicity inherent in the expression for the DFT. This is equivalent to convolving the two periodic functions in Figs. 4.28(f) and (g). The convolution procedure is the same as we just discussed, but the two functions now are periodic. Proceeding with these two functions as in the previous paragraph would yield the result in Fig. 4.28(j) which obviously is incorrect. Because we are convolving two periodic functions, the convolution itself is periodic. The closeness of the periods in Fig. 4.28 is such that they interfere with each other to cause what is commonly referred to as *wraparound error*. According to the convolution theorem, if we had computed the DFT of the two 400-point functions, f and h, multiplied the

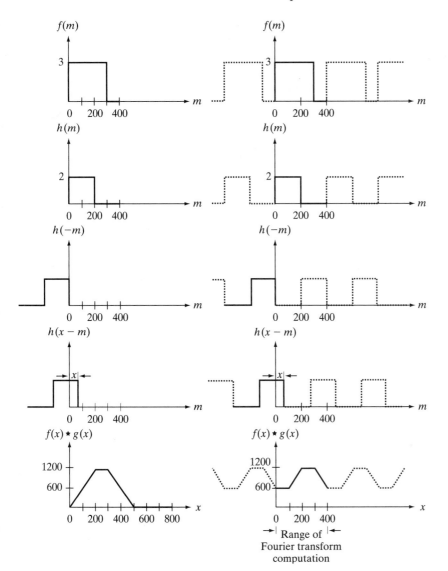

a f
b g
c h
d i
e j

FIGURE 4.28 Left column: convolution of two discrete functions obtained using the approach discussed in Section 3.4.2. The result in (e) is correct. Right column: Convolution of the same functions, but taking into account the periodicity implied by the DFT. Note in (j) how data from adjacent periods produce wraparound error, yielding an incorrect convolution result. To obtain the correct result, function padding must be used.

two transforms, and then computed the inverse DFT, we would have obtained the erroneous 400-point segment of the convolution shown in Fig. 4.28(j).

Fortunately, the solution to the wraparound error problem is simple. Consider two functions, $f(x)$ and $h(x)$ composed of A and B samples, respectively. It can be shown (Brigham [1988]) that if we append zeros to both functions so that they have the same length, denoted by P, then wraparound is avoided by choosing

$$P \geq A + B - 1 \qquad (4.6\text{-}26)$$

In our example, each function has 400 points, so the minimum value we could use is $P = 799$, which implies that we would append 399 zeros to the trailing edge of each function. This process is called *zero padding*. As an exercise, you

The zeros could be appended also to the beginning of the functions, or they could be divided between the beginning and end of the functions. It is simpler to append them at the end.

should convince yourself that if the periods of the functions in Figs. 4.28(f) and (g) were lengthened by appending to each period at least 399 zeros, the result would be a periodic convolution in which *each* period is identical to the correct result in Fig. 4.28(e). Using the DFT via the convolution theorem would result in a 799-point spatial function identical to Fig. 4.28(e). The conclusion, then, is that to obtain the same convolution result between the "straight" representation of the convolution equation approach in Chapter 3, and the DFT approach, functions in the latter must be padded prior to computing their transforms.

Visualizing a similar example in 2-D would be more difficult, but we would arrive at the same conclusion regarding wraparound error and the need for appending zeros to the functions. Let $f(x, y)$ and $h(x, y)$ be two image arrays of sizes $A \times B$ and $C \times D$ pixels, respectively. Wraparound error in their circular convolution can be avoided by padding these functions with zeros, as follows:

$$f_p(x, y) = \begin{cases} f(x, y) & 0 \le x \le A - 1 \quad \text{and} \quad 0 \le y \le B - 1 \\ 0 & A \le x \le P \quad \text{or} \quad B \le y \le Q \end{cases} \qquad (4.6\text{-}27)$$

and

$$h_p(x, y) = \begin{cases} h(x, y) & 0 \le x \le C - 1 \quad \text{and} \quad 0 \le y \le D - 1 \\ 0 & C \le x \le P \quad \text{or} \quad D \le y \le Q \end{cases} \qquad (4.6\text{-}28)$$

with

$$P \ge A + C - 1 \qquad (4.6\text{-}29)$$

and

$$Q \ge B + D - 1 \qquad (4.6\text{-}30)$$

The resulting padded images are of size $P \times Q$. If both arrays are of the same size, $M \times N$, then we require that

$$P \ge 2M - 1 \qquad (4.6\text{-}31)$$

and

$$Q \ge 2N - 1 \qquad (4.6\text{-}32)$$

We give an example in Section 4.7.2 showing the effects of wraparound error on images. As rule, DFT algorithms tend to execute faster with arrays of even size, so it is good practice to select P and Q as the smallest even integers that satisfy the preceding equations. If the two arrays are of the same size, this means that P and Q are selected as twice the array size.

The two functions in Figs. 4.28(a) and (b) conveniently become zero before the end of the sampling interval. If one or both of the functions were not zero at the end of the interval, then a discontinuity would be created when zeros were appended to the function to eliminate wraparound error. This is analogous to multiplying a function by a box, which in the frequency domain would imply convolution of the original transform with a sinc function (see Example 4.1). This, in turn, would create so-called *frequency leakage*, caused by the high-frequency components of the sinc function. Leakage produces a blocky effect on images. Although leakage never can be totally eliminated, it can be reduced significantly by multiplying the sampled function by another function that tapers smoothly to near zero at both ends of the sampled record to dampen the sharp transitions (and thus the high frequency components) of the box. This approach, called *windowing* or *apodizing*, is an important consideration when fidelity in image reconstruction (as in high-definition graphics) is desired. If you are faced with the need for windowing, a good approach is to use a 2-D Gaussian function (see Section 4.8.3). One advantage of this function is that its Fourier transform is Gaussian also, thus producing low leakage.

A simple apodizing function is a triangle, centered on the data record, which tapers to 0 at both ends of the record. This is called the *Bartlett* window. Other common windows are the *Hamming* and the *Hann* windows. We can even use a Gaussian function. We return to the issue of windowing in Section 5.11.5.

4.6.7 Summary of 2-D Discrete Fourier Transform Properties

Table 4.2 summarizes the principal DFT definitions introduced in this chapter. Separability is discussed in Section 4.11.1 and obtaining the inverse using a forward transform algorithm is discussed in Section 4.11.2. Correlation is discussed in Chapter 12.

Name	Expression(s)
1) Discrete Fourier transform (DFT) of $f(x, y)$	$F(u, v) = \sum_{x=0}^{M-1} \sum_{y=0}^{N-1} f(x, y) e^{-j2\pi(ux/M + vy/N)}$
2) Inverse discrete Fourier transform (IDFT) of $F(u, v)$	$f(x, y) = \dfrac{1}{MN} \sum_{u=0}^{M-1} \sum_{v=0}^{N-1} F(u, v) e^{j2\pi(ux/M + vy/N)}$
3) Polar representation	$F(u, v) = |F(u, v)| e^{j\phi(u,v)}$
4) Spectrum	$|F(u, v)| = \left[R^2(u, v) + I^2(u, v) \right]^{1/2}$ $R = \text{Real}(F); \quad I = \text{Imag}(F)$
5) Phase angle	$\phi(u, v) = \tan^{-1}\left[\dfrac{I(u, v)}{R(u, v)} \right]$
6) Power spectrum	$P(u, v) = |F(u, v)|^2$
7) Average value	$\bar{f}(x, y) = \dfrac{1}{MN} \sum_{x=0}^{M-1} \sum_{y=0}^{N-1} f(x, y) = \dfrac{1}{MN} F(0, 0)$

TABLE 4.2
Summary of DFT definitions and corresponding expressions.

(Continued)

TABLE 4.2
(*Continued*)

Name	Expression(s)
8) Periodicity (k_1 and k_2 are integers)	$F(u, v) = F(u + k_1M, v) = F(u, v + k_2N)$ $= F(u + k_1M, v + k_2N)$ $f(x, y) = f(x + k_1M, y) = f(x, y + k_2N)$ $= f(x + k_1M, y + k_2N)$
9) Convolution	$f(x, y) \star h(x, y) = \displaystyle\sum_{m=0}^{M-1}\sum_{n=0}^{N-1} f(m, n)h(x - m, y - n)$
10) Correlation	$f(x, y) \star h(x, y) = \displaystyle\sum_{m=0}^{M-1}\sum_{n=0}^{N-1} f^*(m, n)h(x + m, y + n)$
11) Separability	The 2-D DFT can be computed by computing 1-D DFT transforms along the rows (columns) of the image, followed by 1-D transforms along the columns (rows) of the result. See Section 4.11.1.
12) Obtaining the inverse Fourier transform using a forward transform algorithm.	$MNf^*(x, y) = \displaystyle\sum_{u=0}^{M-1}\sum_{v=0}^{N-1} F^*(u, v)e^{-j2\pi(ux/M+vy/N)}$ This equation indicates that inputting $F^*(u, v)$ into an algorithm that computes the forward transform (right side of above equation) yields $MNf^*(x, y)$. Taking the complex conjugate and dividing by MN gives the desired inverse. See Section 4.11.2.

Table 4.3 summarizes some important DFT pairs. Although our focus is on discrete functions, the last two entries in the table are Fourier transform pairs that can be derived only for continuous variables (note the use of continuous variable notation). We include them here because, with proper interpretation, they are quite useful in digital image processing. The differentiation pair can

TABLE 4.3
Summary of DFT pairs. The closed-form expressions in 12 and 13 are valid only for continuous variables. They can be used with discrete variables by sampling the closed-form, continuous expressions.

Name	DFT Pairs
1) Symmetry properties	See Table 4.1
2) Linearity	$af_1(x, y) + bf_2(x, y) \Leftrightarrow aF_1(u, v) + bF_2(u, v)$
3) Translation (general)	$f(x, y)e^{j2\pi(u_0x/M+v_0y/N)} \Leftrightarrow F(u - u_0, v - v_0)$ $f(x - x_0, y - y_0) \Leftrightarrow F(u, v)e^{-j2\pi(ux_0/M+vy_0/N)}$
4) Translation to center of the frequency rectangle, ($M/2, N/2$)	$f(x, y)(-1)^{x+y} \Leftrightarrow F(u - M/2, v - N/2)$ $f(x - M/2, y - N/2) \Leftrightarrow F(u, v)(-1)^{u+v}$
5) Rotation	$f(r, \theta + \theta_0) \Leftrightarrow F(\omega, \varphi + \theta_0)$ $x = r\cos\theta \quad y = r\sin\theta \quad u = \omega\cos\varphi \quad v = \omega\sin\varphi$
6) Convolution theorem[†]	$f(x, y) \star h(x, y) \Leftrightarrow F(u, v)H(u, v)$ $f(x, y)h(x, y) \Leftrightarrow F(u, v) \star H(u, v)$

(*Continued*)

TABLE 4.3
(*Continued*)

Name	DFT Pairs
7) Correlation theorem[†]	$f(x, y) \star h(x, y) \Leftrightarrow F^*(u, v)H(u, v)$ $f^*(x, y)h(x, y) \Leftrightarrow F(u, v) \star H(u, v)$
8) Discrete unit impulse	$\delta(x, y) \Leftrightarrow 1$
9) Rectangle	$\text{rect}[a, b] \Leftrightarrow ab \dfrac{\sin(\pi ua)}{(\pi ua)} \dfrac{\sin(\pi vb)}{(\pi vb)} e^{-j\pi(ua+vb)}$
10) Sine	$\sin(2\pi u_0 x + 2\pi v_0 y) \Leftrightarrow$ $j\dfrac{1}{2}\big[\delta(u + Mu_0, v + Nv_0) - \delta(u - Mu_0, v - Nv_0)\big]$
11) Cosine	$\cos(2\pi u_0 x + 2\pi v_0 y) \Leftrightarrow$ $\dfrac{1}{2}\big[\delta(u + Mu_0, v + Nv_0) + \delta(u - Mu_0, v - Nv_0)\big]$

The following Fourier transform pairs are derivable only for continuous variables, denoted as before by t and z for spatial variables and by μ and ν for frequency variables. These results can be used for DFT work by sampling the continuous forms.

12) *Differentiation* (The expressions on the right assume that $f(\pm\infty, \pm\infty) = 0$.)	$\left(\dfrac{\partial}{\partial t}\right)^m \left(\dfrac{\partial}{\partial z}\right)^n f(t, z) \Leftrightarrow (j2\pi\mu)^m (j2\pi\nu)^n F(\mu, \nu)$ $\dfrac{\partial^m f(t, z)}{\partial t^m} \Leftrightarrow (j2\pi\mu)^m F(\mu, \nu); \ \dfrac{\partial^n f(t, z)}{\partial z^n} \Leftrightarrow (j2\pi\nu)^n F(\mu, \nu)$
13) *Gaussian*	$A2\pi\sigma^2 e^{-2\pi^2\sigma^2(t^2+z^2)} \Leftrightarrow Ae^{-(\mu^2+\nu^2)/2\sigma^2}$ (*A* is a constant)

[†]Assumes that the functions have been extended by zero padding. Convolution and correlation are associative, commutative, and distributive.

be used to derive the frequency-domain equivalent of the Laplacian defined in Eq. (3.6-3) (Problem 4.26). The Gaussian pair is discussed in Section 4.7.4.

Tables 4.1 through 4.3 provide a summary of properties useful when working with the DFT. Many of these properties are key elements in the development of the material in the rest of this chapter, and some are used in subsequent chapters.

4.7 The Basics of Filtering in the Frequency Domain

In this section, we lay the groundwork for all the filtering techniques discussed in the remainder of the chapter.

4.7.1 Additional Characteristics of the Frequency Domain

We begin by observing in Eq. (4.5-15) that *each* term of $F(u, v)$ contains *all* values of $f(x, y)$, modified by the values of the exponential terms. Thus, with the exception of trivial cases, it usually is impossible to make direct associations between specific components of an image and its transform. However, some general statements can be made about the relationship between the frequency components of the Fourier transform and spatial features of an image. For instance, because frequency is directly related to spatial rates of change, it is not difficult intuitively to associate frequencies in the Fourier transform with patterns of intensity variations in an image. We showed in Section 4.6.5 that the slowest varying frequency component ($u = v = 0$) is proportional to the average intensity of an image. As we move away from the origin of the transform, the low frequencies correspond to the slowly varying intensity components of an image. In an image of a room, for example, these might correspond to smooth intensity variations on the walls and floor. As we move further away from the origin, the higher frequencies begin to correspond to faster and faster intensity changes in the image. These are the edges of objects and other components of an image characterized by abrupt changes in intensity.

Filtering techniques in the frequency domain are based on modifying the Fourier transform to achieve a specific objective and then computing the inverse DFT to get us back to the image domain, as introduced in Section 2.6.7. It follows from Eq. (4.6-15) that the two components of the transform to which we have access are the transform magnitude (spectrum) and the phase angle. Section 4.6.5 covered the basic properties of these two components of the transform. We learned there that visual analysis of the phase component generally is not very useful. The spectrum, however, provides some useful guidelines as to gross characteristics of the image from which the spectrum was generated. For example, consider Fig. 4.29(a), which is a scanning electron microscope image of an integrated circuit, magnified approximately 2500 times. Aside from the interesting construction of the device itself, we note two principal features: strong edges that run approximately at ±45° and two white, oxide protrusions resulting from thermally-induced failure. The Fourier spectrum in Fig. 4.29(b) shows prominent components along the ±45° directions that correspond to the edges just mentioned. Looking carefully along the vertical axis, we see a vertical component

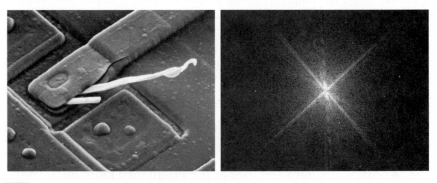

a b

FIGURE 4.29 (a) SEM image of a damaged integrated circuit. (b) Fourier spectrum of (a). (Original image courtesy of Dr. J. M. Hudak, Brockhouse Institute for Materials Research, McMaster University, Hamilton, Ontario, Canada.)

that is off-axis slightly to the left. This component was caused by the edges of the oxide protrusions. Note how the angle of the frequency component with respect to the vertical axis corresponds to the inclination (with respect to the horizontal axis) of the long white element, and note also the zeros in the vertical frequency component, corresponding to the narrow vertical span of the oxide protrusions.

These are typical of the types of associations that can be made in general between the frequency and spatial domains. As we show later in this chapter, even these types of gross associations, coupled with the relationships mentioned previously between frequency content and rate of change of intensity levels in an image, can lead to some very useful results. In the next section, we show the effects of modifying various frequency ranges in the transform of Fig. 4.29(a).

4.7.2 Frequency Domain Filtering Fundamentals

Filtering in the frequency domain consists of modifying the Fourier transform of an image and then computing the inverse transform to obtain the processed result. Thus, given a digital image, $f(x, y)$, of size $M \times N$, the basic filtering equation in which we are interested has the form:

$$g(x, y) = \mathfrak{I}^{-1}[H(u, v)F(u, v)] \tag{4.7-1}$$

where $\mathfrak{I}^{-1}$ is the IDFT, $F(u, v)$ is the DFT of the input image, $f(x, y)$, $H(u, v)$ is a *filter function* (also called simply the *filter*, or the *filter transfer function*), and $g(x, y)$ is the filtered (output) image. Functions F, H, and g are arrays of size $M \times N$, the same as the input image. The product $H(u, v)F(u, v)$ is formed using array multiplication, as defined in Section 2.6.1. The filter function modifies the transform of the input image to yield a processed output, $g(x, y)$. Specification of $H(u, v)$ is simplified considerably by using functions that are symmetric about their center, which requires that $F(u, v)$ be centered also. As explained in Section 4.6.3, this is accomplished by multiplying the input image by $(-1)^{x+y}$ prior to computing its transform.[†]

We are now in a position to consider the filtering process in some detail. One of the simplest filters we can construct is a filter $H(u, v)$ that is 0 at the center of the transform and 1 elsewhere. This filter would reject the dc term and "pass" (i.e., leave unchanged) all other terms of $F(u, v)$ when we form the product $H(u, v)F(u, v)$. We know from Eq. (4.6-21) that the dc term is responsible for the average intensity of an image, so setting it to zero will reduce the average intensity of the output image to zero. Figure 4.30 shows the result of this operation using Eq. (4.7-1). As expected, the image became much darker. (An average of zero

If H is real and symmetric and f is real (as is typically the case), then the IDFT in Eq. (4.7-1) should yield real quantities in theory. In practice, the inverse generally contains parasitic complex terms from round-off and other computational inaccuracies. Thus, it is customary to take the real part of the IDFT to form g.

[†]Many software implementations of the 2-D DFT (e.g., MATLAB) do not center the transform. This implies that filter functions must be arranged to correspond to the same data format as the uncentered transform (i.e., with the origin at the top left). The net result is that filters are more difficult to generate and display. We use centering in our discussions to aid in visualization, which is crucial in developing a clear understanding of filtering concepts. Either method can be used practice, as long as consistency is maintained.

FIGURE 4.30
Result of filtering the image in Fig. 4.29(a) by setting to 0 the term $F(M/2, N/2)$ in the Fourier transform.

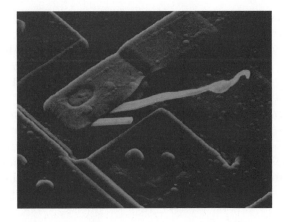

implies the existence of negative intensities. Therefore, although it illustrates the principle, Fig. 4.30 is not a true representation of the original, as all negative intensities were clipped (set to 0) for display purposes.)

As noted earlier, low frequencies in the transform are related to slowly varying intensity components in an image, such as the walls of a room or a cloudless sky in an outdoor scene. On the other hand, high frequencies are caused by sharp transitions in intensity, such as edges and noise. Therefore, we would expect that a filter $H(u, v)$ that attenuates high frequencies while passing low frequencies (appropriately called a *lowpass filter*) would blur an image, while a filter with the opposite property (called a *highpass filter*) would enhance sharp detail, but cause a reduction in contrast in the image. Figure 4.31 illustrates these effects. Note the similarity between Figs. 4.31(e) and Fig. 4.30. The reason is that the highpass filter shown eliminates the dc term, resulting in the same basic effect that led to Fig. 4.30. Adding a small constant to the filter does not affect sharpening appreciably, but it does prevent elimination of the dc term and thus preserves tonality, as Fig. 4.31(f) shows.

Equation (4.7-1) involves the product of two functions in the frequency domain which, by the convolution theorem, implies convolution in the spatial domain. We know from the discussion in Section 4.6.6 that if the functions in question are not padded we can expect wraparound error. Consider what happens when we apply Eq. (4.7-1) without padding. Figure 4.32(a) shows a simple image, and Fig. 4.32(b) is the result of lowpass filtering the image with a Gaussian lowpass filter of the form shown in Fig. 4.31(a). As expected, the image is blurred. However, the blurring is not uniform; the top white edge is blurred, but the side white edges are not. Padding the input image according to Eqs. (4.6-31) and (4.6-32) before applying Eq. (4.7-1) results in the filtered image in Fig. 4.32(c). This result is as expected.

Figure 4.33 illustrates the reason for the discrepancy between Figs. 4.32(b) and (c). The dashed areas in Fig. 4.33 correspond to the image in Fig. 4.32(a). Figure 4.33(a) shows the periodicity implicit in the use of the DFT, as explained in Section 4.6.3. Imagine convolving the *spatial* representation of the blurring filter with this image. When the filter is passing through the top of the

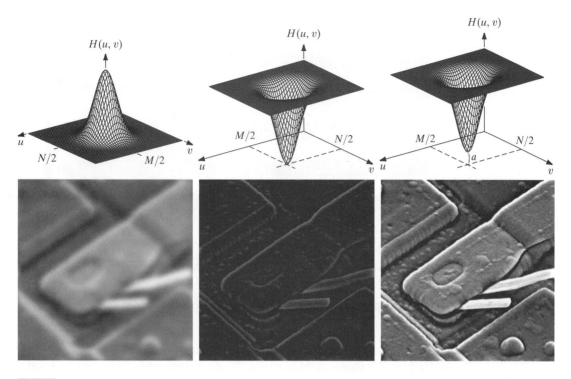

a b c
d e f

FIGURE 4.31 Top row: frequency domain filters. Bottom row: corresponding filtered images obtained using Eq. (4.7-1). We used $a = 0.85$ in (c) to obtain (f) (the height of the filter itself is 1). Compare (f) with Fig. 4.29(a).

a b c

FIGURE 4.32 (a) A simple image. (b) Result of blurring with a Gaussian lowpass filter without padding. (c) Result of lowpass filtering with padding. Compare the light area of the vertical edges in (b) and (c).

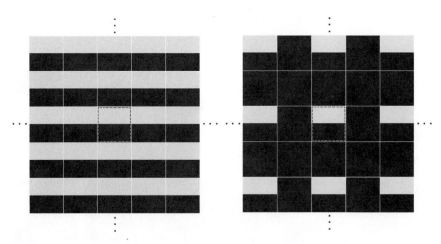

a b

FIGURE 4.33 2-D image periodicity inherent in using the DFT. (a) Periodicity without image padding. (b) Periodicity after padding with 0s (black). The dashed areas in the center correspond to the image in Fig. 4.32(a). (The thin white lines in both images are superimposed for clarity; they are not part of the data.)

dashed image, it will encompass part of the image and also part of the bottom of the periodic image right above it. When a dark and a light region reside under the filter, the result is a mid-gray, blurred output. However, when the filter is passing through the top right side of the image, the filter will encompass only light areas in the image and its right neighbor. The average of a constant is the same constant, so filtering will have no effect in this area, giving the result in Fig. 4.32(b). Padding the image with 0s creates a uniform border around the periodic sequence, as Fig. 4.33(b) shows. Convolving the blurring function with the padded "mosaic" of Fig. 4.33(b) gives the correct result in Fig. 4.32(c). You can see from this example that failure to pad an image can lead to erroneous results. If the purpose of filtering is only for rough visual analysis, the padding step is skipped sometimes.

Thus far, the discussion has centered on padding the input image, but Eq. (4.7-1) also involves a filter that can be specified either in the spatial or in the frequency domain. However, padding is done in the spatial domain, which raises an important question about the relationship between *spatial* padding and filters specified directly in the frequency domain.

At first glance, one could conclude that the way to handle padding of a frequency domain filter is to construct the filter to be of the same size as the image, compute the IDFT of the filter to obtain the corresponding spatial filter, pad that filter in the spatial domain, and then compute its DFT to return to the frequency domain. The 1-D example in Fig. 4.34 illustrates the pitfalls in this approach. Figure 4.34(a) shows a 1-D ideal lowpass filter in the frequency domain. The filter is real and has even symmetry, so we know from property 8 in Table 4.1 that its IDFT will be real and symmetric also. Figure 4.34(b) shows the result of multiplying the elements of the frequency domain filter

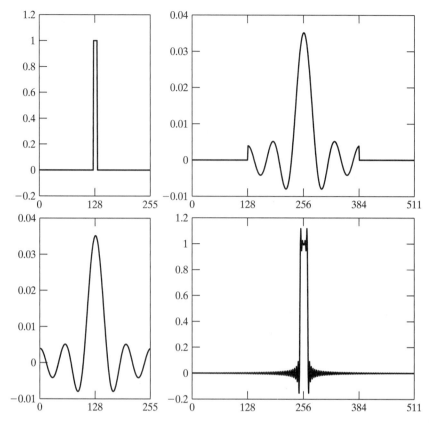

a c
b d

FIGURE 4.34
(a) Original filter specified in the (centered) frequency domain.
(b) Spatial representation obtained by computing the IDFT of (a).
(c) Result of padding (b) to twice its length (note the discontinuities).
(d) Corresponding filter in the frequency domain obtained by computing the DFT of (c). Note the ringing caused by the discontinuities in (c). (The curves appear continuous because the points were joined to simplify visual analysis.)

by $(-1)^u$ and computing its IDFT to obtain the corresponding spatial filter. The extremes of this spatial function are not zero so, as Fig. 4.34(c) shows, zero-padding the function created two discontinuities (padding the two ends of the function is the same as padding one end, as long as the total number of zeros used is the same).

To get back to the frequency domain, we compute the DFT of the spatial, padded filter. Figure 4.34(d) shows the result. The discontinuities in the spatial filter created ringing in its frequency domain counterpart, as you would expect from the results in Example 4.1. Viewed another way, we know from that example that the Fourier transform of a box function is a sinc function with frequency components extending to infinity, and we would expect the same behavior from the inverse transform of a box. That is, the spatial representation of an ideal (box) frequency domain filter has components extending to infinity. Therefore, any spatial truncation of the filter to implement zero-padding will introduce discontinuities, which will then in general result in ringing in the frequency domain (truncation can be avoided in this case if it is done at zero crossings, but we are interested in general procedures, and not all filters have zero crossings).

What the preceding results tell us is that, because we cannot work with an infinite number of components, we cannot use an ideal frequency domain filter [as in

See the end of Section 4.3.3 regarding the definition of an ideal filter.

Fig. 4.34(a)] and simultaneously use zero padding to avoid wraparound error. A decision on which limitation to accept is required. Our objective is to work with specified filter shapes in the frequency domain (including ideal filters) without having to be concerned with truncation issues. One approach is to zero-pad images and then create filters in the frequency domain to be of the same size as the padded images (remember, images and filters must be of the same size when using the DFT). Of course, this will result in wraparound error because no padding is used for the filter, but in practice this error is mitigated significantly by the separation provided by the padding of the image, and it is preferable to ringing. Smooth filters (such as those in Fig. 4.31) present even less of a problem. Specifically, then, the approach we will follow in this chapter in order to work with filters of a specified shape directly in the frequency domain is to pad images to size $P \times Q$ and construct filters of the same dimensions. As explained earlier, P and Q are given by Eqs. (4.6-29) and (4.6-30).

We conclude this section by analyzing the phase angle of the filtered transform. Because the DFT is a complex array, we can express it in terms of its real and imaginary parts:

$$F(u, v) = R(u, v) + jI(u, v) \tag{4.7-2}$$

Equation (4.7-1) then becomes

$$g(x, y) = \Im^{-1}\left[H(u, v)R(u, v) + jH(u, v)I(u, v)\right] \tag{4.7-3}$$

The phase angle is not altered by filtering in the manner just described because $H(u, v)$ cancels out when the ratio of the imaginary and real parts is formed in Eq. (4.6-17). Filters that affect the real and imaginary parts equally, and thus have no effect on the phase, are appropriately called *zero-phase-shift* filters. These are the only types of filters considered in this chapter.

Even small changes in the phase angle can have dramatic (usually undesirable) effects on the filtered output. Figure 4.35 illustrates the effect of something as simple as a scalar change. Figure 4.35(a) shows an image resulting from multiplying the angle array in Eq. (4.6-15) by 0.5, without changing

a b

FIGURE 4.35
(a) Image resulting from multiplying by 0.5 the phase angle in Eq. (4.6-15) and then computing the IDFT. (b) The result of multiplying the phase by 0.25. The spectrum was not changed in either of the two cases.

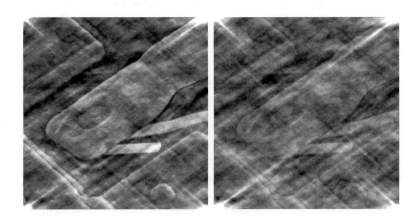

$|F(u, v)|$, and then computing the IDFT. The basic shapes remain unchanged, but the intensity distribution is quite distorted. Figure 4.35(b) shows the result of multiplying the phase by 0.25. The image is almost unrecognizable.

4.7.3 Summary of Steps for Filtering in the Frequency Domain

The material in the previous two sections can be summarized as follows:

1. Given an input image $f(x, y)$ of size $M \times N$, obtain the padding parameters P and Q from Eqs. (4.6-31) and (4.6-32). Typically, we select $P = 2M$ and $Q = 2N$.
2. Form a padded image, $f_p(x, y)$, of size $P \times Q$ by appending the necessary number of zeros to $f(x, y)$.
3. Multiply $f_p(x, y)$ by $(-1)^{x+y}$ to center its transform.
4. Compute the DFT, $F(u, v)$, of the image from step 3.
5. Generate a real, symmetric filter function, $H(u, v)$, of size $P \times Q$ with center at coordinates $(P/2, Q/2)$.† Form the product $G(u, v) = H(u, v)F(u, v)$ using array multiplication; that is, $G(i, k) = H(i, k)F(i, k)$.
6. Obtain the processed image:

$$g_p(x, y) = \left\{ \text{real} \left[\Im^{-1}[G(u, v)] \right] \right\}(-1)^{x+y}$$

where the real part is selected in order to ignore parasitic complex components resulting from computational inaccuracies, and the subscript p indicates that we are dealing with padded arrays.
7. Obtain the final processed result, $g(x, y)$, by extracting the $M \times N$ region from the top, left quadrant of $g_p(x, y)$.

As noted earlier, centering helps in visualizing the filtering process and in generating the filter functions themselves, but centering is not a fundamental requirement.

Figure 4.36 illustrates the preceding steps. The legend in the figure explains the source of each image. If it were enlarged, Fig. 4.36(c) would show black dots interleaved in the image because negative intensities are clipped to 0 for display. Note in Fig. 4.36(h) the characteristic dark border exhibited by lowpass filtered images processed using zero padding.

4.7.4 Correspondence Between Filtering in the Spatial and Frequency Domains

The link between filtering in the spatial and frequency domains is the convolution theorem. In Section 4.7.2, we defined filtering in the frequency domain as the multiplication of a filter function, $H(u, v)$, times $F(u, v)$, the Fourier transform of the input image. Given a filter $H(u, v)$, suppose that we want to find its equivalent representation in the spatial domain. If we let $f(x, y) = \delta(x, y)$, it follows from Table 4.3 that $F(u, v) = 1$. Then, from Eq. (4.7-1), the filtered output is $\Im^{-1}\{H(u, v)\}$. But this is the inverse transform of the frequency domain filter, which is the corresponding filter in the

---

†If $H(u, v)$ is to be generated from a *given* spatial filter, $h(x, y)$, then we form $h_p(x, y)$ by padding the spatial filter to size $P \times Q$, multiply the expanded array by $(-1)^{x+y}$, and compute the DFT of the result to obtain a centered $H(u, v)$. Example 4.15 illustrates this procedure.

FIGURE 4.36
(a) An $M \times N$ image, f.
(b) Padded image, f_p of size $P \times Q$.
(c) Result of multiplying f_p by $(-1)^{x+y}$.
(d) Spectrum of F_p. (e) Centered Gaussian lowpass filter, H, of size $P \times Q$.
(f) Spectrum of the product HF_p.
(g) g_p, the product of $(-1)^{x+y}$ and the real part of the IDFT of HF_p.
(h) Final result, g, obtained by cropping the first M rows and N columns of g_p.

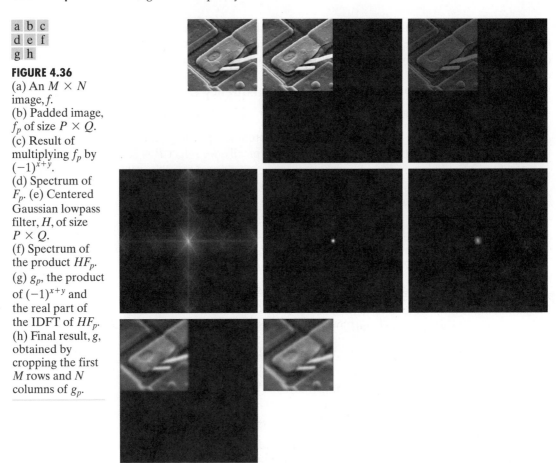

spatial domain. Conversely, it follows from a similar analysis and the convolution theorem that, given a spatial filter, we obtain its frequency domain representation by taking the forward Fourier transform of the spatial filter. Therefore, the two filters form a Fourier transform pair:

$$h(x, y) \Leftrightarrow H(u, v) \qquad (4.7\text{-}4)$$

where $h(x, y)$ is a spatial filter. Because this filter can be obtained from the response of a frequency domain filter to an impulse, $h(x, y)$ sometimes is referred to as the *impulse response* of $H(u, v)$. Also, because all quantities in a discrete implementation of Eq. (4.7-4) are finite, such filters are called *finite impulse response* (FIR) filters. These are the only types of linear spatial filters considered in this book.

We introduced spatial convolution in Section 3.4.1 and discussed its implementation in connection with Eq. (3.4-2), which involved convolving functions of different sizes. When we speak of spatial convolution in terms of the convolution theorem and the DFT, it is implied that we are convolving periodic functions, as explained in Fig. 4.28. For this reason, as explained earlier, Eq. (4.6-23) is referred to as *circular convolution*. Furthermore, convolution in the context of the DFT involves functions of the same size, whereas in Eq. (3.4-2) the functions typically are of different sizes.

In practice, we prefer to implement convolution filtering using Eq. (3.4-2) with small filter masks because of speed and ease of implementation in hardware and/or firmware. However, filtering concepts are more intuitive in the frequency domain. One way to take advantage of the properties of both domains is to specify a filter in the frequency domain, compute its IDFT, and then use the resulting, full-size spatial filter as a *guide* for constructing smaller spatial filter masks (more formal approaches are mentioned in Section 4.11.4). This is illustrated next. Later in this section, we illustrate also the converse, in which a small spatial filter is given and we obtain its full-size frequency domain representation. This approach is useful for analyzing the behavior of small spatial filters in the frequency domain. Keep in mind during the following discussion that the Fourier transform and its inverse are linear processes (Problem 4.14), so the discussion is limited to linear filtering.

In the following discussion, we use Gaussian filters to illustrate how frequency domain filters can be used as guides for specifying the coefficients of some of the small masks discussed in Chapter 3. Filters based on Gaussian functions are of particular interest because, as noted in Table 4.3, both the forward and inverse Fourier transforms of a Gaussian function are real Gaussian functions. We limit the discussion to 1-D to illustrate the underlying principles. Two-dimensional Gaussian filters are discussed later in this chapter.

Let $H(u)$ denote the 1-D frequency domain Gaussian filter:

$$H(u) = A e^{-u^2/2\sigma^2} \tag{4.7-5}$$

where σ is the standard deviation of the Gaussian curve. The corresponding filter in the spatial domain is obtained by taking the inverse Fourier transform of $H(u)$ (Problem 4.31):

$$h(x) = \sqrt{2\pi}\sigma A e^{-2\pi^2\sigma^2 x^2} \tag{4.7-6}$$

These equations[†] are important for two reasons: (1) They are a Fourier transform pair, both components of which are Gaussian and *real*. This facilitates analysis because we do not have to be concerned with complex numbers. In addition, Gaussian curves are intuitive and easy to manipulate. (2) The functions behave reciprocally. When $H(u)$ has a broad profile (large value of σ),

---

[†]As mentioned in Table 4.3, closed forms for the forward and inverse Fourier transforms of Gaussians are valid only for continuous functions. To use discrete formulations we simply sample the continuous Gaussian transforms. Our use of discrete variables here implies that we are dealing with sampled transforms.

$h(x)$ has a narrow profile, and vice versa. In fact, as σ approaches infinity, $H(u)$ tends toward a constant function and $h(x)$ tends toward an impulse, which implies no filtering in the frequency and spatial domains, respectively.

Figures 4.37(a) and (b) show plots of a Gaussian lowpass filter in the frequency domain and the corresponding lowpass filter in the spatial domain. Suppose that we want to use the shape of $h(x)$ in Fig. 4.37(b) as a *guide* for specifying the coefficients of a small spatial mask. The key similarity between the two filters is that all their values are positive. Thus, we conclude that we can implement lowpass filtering in the spatial domain by using a mask with all positive coefficients (as we did in Section 3.5.1). For reference, Fig. 4.37(b) shows two of the masks discussed in that section. Note the reciprocal relationship between the width of the filters, as discussed in the previous paragraph. The narrower the frequency domain filter, the more it will attenuate the low frequencies, resulting in increased blurring. In the spatial domain, this means that a larger mask must be used to increase blurring, as illustrated in Example 3.13.

More complex filters can be constructed using the basic Gaussian function of Eq. (4.7-5). For example, we can construct a highpass filter as the *difference* of Gaussians:

$$H(u) = Ae^{-u^2/2\sigma_1^2} - Be^{-u^2/2\sigma_2^2} \tag{4.7-7}$$

with $A \geq B$ and $\sigma_1 > \sigma_2$. The corresponding filter in the spatial domain is

$$h(x) = \sqrt{2\pi}\sigma_1 Ae^{-2\pi^2\sigma_1^2 x^2} - \sqrt{2\pi}\sigma_2 Be^{-2\pi^2\sigma_2^2 x^2} \tag{4.7-8}$$

Figures 4.37(c) and (d) show plots of these two equations. We note again the reciprocity in width, but the most important feature here is that $h(x)$ has a positive center term with negative terms on either side. The small masks shown in

FIGURE 4.37
(a) A 1-D Gaussian lowpass filter in the frequency domain. (b) Spatial lowpass filter corresponding to (a). (c) Gaussian highpass filter in the frequency domain. (d) Spatial highpass filter corresponding to (c). The small 2-D masks shown are spatial filters we used in Chapter 3.

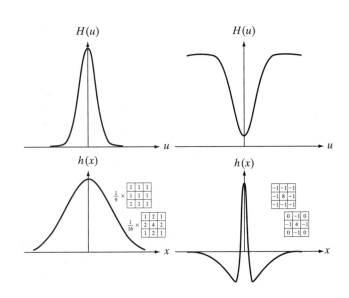

Fig. 4.37(d) "capture" this property. These two masks were used in Chapter 3 as sharpening filters, which we now know are highpass filters.

Although we have gone through significant effort to get here, be assured that it is impossible to truly understand filtering in the frequency domain without the foundation we have just established. In practice, the frequency domain can be viewed as a "laboratory" in which we take advantage of the correspondence between frequency content and image appearance. As is demonstrated numerous times later in this chapter, some tasks that would be exceptionally difficult or impossible to formulate directly in the spatial domain become almost trivial in the frequency domain. Once we have selected a specific filter via experimentation in the frequency domain, the actual implementation of the method usually is done in the spatial domain. One approach is to specify small spatial masks that attempt to capture the "essence" of the full filter function in the spatial domain, as we explained in Fig. 4.37. A more formal approach is to design a 2-D digital filter by using approximations based on mathematical or statistical criteria. We touch on this point again in Section 4.11.4.

■ In this example, we start with a spatial mask and show how to generate its corresponding filter in the frequency domain. Then, we compare the filtering results obtained using frequency domain and spatial techniques. This type of analysis is useful when one wishes to compare the performance of given spatial masks against one or more "full" filter candidates in the frequency domain, or to gain deeper understanding about the performance of a mask. To keep matters simple, we use the 3×3 Sobel vertical edge detector from Fig. 3.41(e). Figure 4.38(a) shows a 600×600 pixel image, $f(x, y)$, that we wish to filter, and Fig. 4.38(b) shows its spectrum.

Figure 4.39(a) shows the Sobel mask, $h(x, y)$ (the perspective plot is explained below). Because the input image is of size 600×600 pixels and the filter is of size 3×3 we avoid wraparound error by padding f and h to size

EXAMPLE 4.15:
Obtaining a frequency domain filter from a small spatial mask.

a b
FIGURE 4.38
(a) Image of a building, and
(b) its spectrum.

a b
c d

FIGURE 4.39
(a) A spatial mask and perspective plot of its corresponding frequency domain filter. (b) Filter shown as an image. (c) Result of filtering Fig. 4.38(a) in the frequency domain with the filter in (b). (d) Result of filtering the same image with the spatial filter in (a). The results are identical.

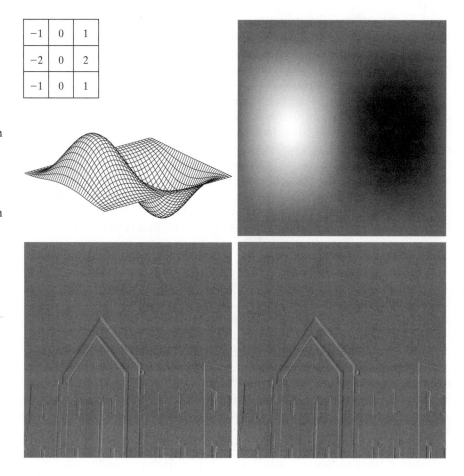

−1	0	1
−2	0	2
−1	0	1

602×602 pixels, according to Eqs. (4.6-29) and (4.6-30). The Sobel mask exhibits odd symmetry, provided that it is embedded in an array of zeros of even size (see Example 4.10). To maintain this symmetry, we place $h(x, y)$ so that its center is at the center of the 602×602 padded array. This is an important aspect of filter generation. If we preserve the odd symmetry with respect to the padded array in forming $h_p(x, y)$, we know from property 9 in Table 4.1 that $H(u, v)$ will be purely imaginary. As we show at the end of this example, this will yield results that are identical to filtering the image spatially using $h(x, y)$. If the symmetry were not preserved, the results would no longer be same.

The procedure used to generate $H(u, v)$ is: (1) multiply $h_p(x, y)$ by $(-1)^{x+y}$ to center the frequency domain filter; (2) compute the forward DFT of the result in (1); (3) set the real part of the resulting DFT to 0 to account for parasitic real parts (we know that $H(u, v)$ has to be purely imaginary); and (4) multiply the result by $(-1)^{u+v}$. This last step reverses the multiplication of $H(u, v)$ by $(-1)^{u+v}$, which is implicit when $h(x, y)$ was moved to the center of $h_p(x, y)$. Figure 4.39(a) shows a perspective plot of $H(u, v)$, and Fig. 4.39(b) shows

$H(u, v)$ as an image. As, expected, the function is odd, thus the antisymmetry about its center. Function $H(u, v)$ is used as any other frequency domain filter in the procedure outlined in Section 4.7.3.

Figure 4.39(c) is the result of using the filter just obtained in the procedure outlined in Section 4.7.3 to filter the image in Fig. 4.38(a). As expected from a derivative filter, edges are enhanced and all the constant intensity areas are reduced to zero (the grayish tone is due to scaling for display). Figure 4.39(d) shows the result of filtering the same image in the spatial domain directly, using $h(x, y)$ in the procedure outlined in Section 3.6.4. The results are identical.   ■

4.8 Image Smoothing Using Frequency Domain Filters

The remainder of this chapter deals with various filtering techniques in the frequency domain. We begin with lowpass filters. Edges and other sharp intensity transitions (such as noise) in an image contribute significantly to the high-frequency content of its Fourier transform. Hence, smoothing (blurring) is achieved in the frequency domain by high-frequency attenuation; that is, by *lowpass* filtering. In this section, we consider three types of lowpass filters: ideal, Butterworth, and Gaussian. These three categories cover the range from very sharp (ideal) to very smooth (Gaussian) filtering. The Butterworth filter has a parameter called the *filter order*. For high order values, the Butterworth filter approaches the ideal filter. For lower order values, the Butterworth filter is more like a Gaussian filter. Thus, the Butterworth filter may be viewed as providing a transition between two "extremes." All filtering in this section follows the procedure outlined in Section 4.7.3, so all filter functions, $H(u, v)$, are understood to be discrete functions of size $P \times Q$; that is, the discrete frequency variables are in the range $u = 0, 1, 2, \ldots, P - 1$ and $v = 0, 1, 2, \ldots, Q - 1$.

4.8.1 Ideal Lowpass Filters

A 2-D lowpass filter that passes without attenuation all frequencies within a circle of radius D_0 from the origin and "cuts off" all frequencies outside this circle is called an *ideal lowpass filter* (ILPF); it is specified by the function

$$H(u, v) = \begin{cases} 1 & \text{if } D(u, v) \leq D_0 \\ 0 & \text{if } D(u, v) > D_0 \end{cases} \tag{4.8-1}$$

where D_0 is a positive constant and $D(u, v)$ is the distance between a point (u, v) in the frequency domain and the center of the frequency rectangle; that is,

$$D(u, v) = \left[(u - P/2)^2 + (v - Q/2)^2 \right]^{1/2} \tag{4.8-2}$$

where, as before, P and Q are the padded sizes from Eqs. (4.6-31) and (4.6-32). Figure 4.40(a) shows a perspective plot of $H(u, v)$ and Fig. 4.40(b) shows the filter displayed as an image. As mentioned in Section 4.3.3, the name *ideal* indicates that all frequencies on or inside a circle of radius D_0 are passed

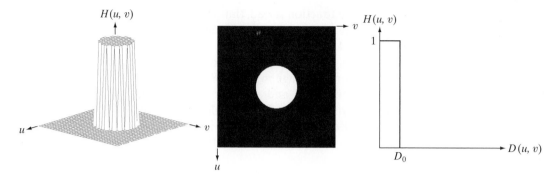

a b c

FIGURE 4.40 (a) Perspective plot of an ideal lowpass-filter transfer function. (b) Filter displayed as an image. (c) Filter radial cross section.

without attenuation, whereas all frequencies outside the circle are completely attenuated (filtered out). The ideal lowpass filter is radially symmetric about the origin, which means that the filter is completely defined by a radial cross section, as Fig. 4.40(c) shows. Rotating the cross section by 360° yields the filter in 2-D.

For an ILPF cross section, the point of transition between $H(u, v) = 1$ and $H(u, v) = 0$ is called the *cutoff frequency*. In the case of Fig. 4.40, for example, the cutoff frequency is D_0. The sharp cutoff frequencies of an ILPF cannot be realized with electronic components, although they certainly can be simulated in a computer. The effects of using these "nonphysical" filters on a digital image are discussed later in this section.

The lowpass filters introduced in this chapter are compared by studying their behavior as a function of the same cutoff frequencies. One way to establish a set of standard cutoff frequency loci is to compute circles that enclose specified amounts of total image power P_T. This quantity is obtained by summing the components of the power spectrum of the padded images at each point (u, v), for $u = 0, 1, \ldots, P - 1$ and $v = 0, 1, \ldots, Q - 1$; that is,

$$P_T = \sum_{u=0}^{P-1} \sum_{v=0}^{Q-1} P(u, v) \tag{4.8-3}$$

where $P(u, v)$ is given in Eq. (4.6-18). If the DFT has been centered, a circle of radius D_0 with origin at the center of the frequency rectangle encloses α percent of the power, where

$$\alpha = 100 \left[\sum_u \sum_v P(u, v)/P_T \right] \tag{4.8-4}$$

and the summation is taken over values of (u, v) that lie inside the circle or on its boundary.

Figures 4.41(a) and (b) show a test pattern image and its spectrum. The circles superimposed on the spectrum have radii of 10, 30, 60, 160, and 460 pixels, respectively. These circles enclose α percent of the image power, for $\alpha = 87.0$, 93.1, 95.7, 97.8, and 99.2%, respectively. The spectrum falls off rapidly, with 87% of the total power being enclosed by a relatively small circle of radius 10.

■ Figure 4.42 shows the results of applying ILPFs with cutoff frequencies at the radii shown in Fig. 4.41(b). Figure 4.42(b) is useless for all practical purposes, unless the objective of blurring is to eliminate all detail in the image, except the "blobs" representing the largest objects. The severe blurring in this image is a clear indication that most of the sharp detail information in the picture is contained in the 13% power removed by the filter. As the filter radius increases, less and less power is removed, resulting in less blurring. Note that the images in Figs. 4.42(c) through (e) are characterized by "ringing," which becomes finer in texture as the amount of high frequency content removed decreases. Ringing is visible even in the image [Fig. 4.42(e)] in which only 2% of the total power was removed. This ringing behavior is a characteristic of ideal filters, as you will see shortly. Finally, the result for $\alpha = 99.2$ shows very slight blurring in the noisy squares but, for the most part, this image is quite close to the original. This indicates that little edge information is contained in the upper 0.8% of the spectrum power in this particular case.

It is clear from this example that ideal lowpass filtering is not very practical. However, it is useful to study their behavior as part of our development of

EXAMPLE 4.16:
Image smoothing using an ILPF.

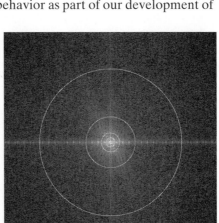

a b

FIGURE 4.41 (a) Test pattern of size 688 $\times$ 688 pixels, and (b) its Fourier spectrum. The spectrum is double the image size due to padding but is shown in half size so that it fits in the page. The superimposed circles have radii equal to 10, 30, 60, 160, and 460 with respect to the full-size spectrum image. These radii enclose 87.0, 93.1, 95.7, 97.8, and 99.2% of the padded image power, respectively.

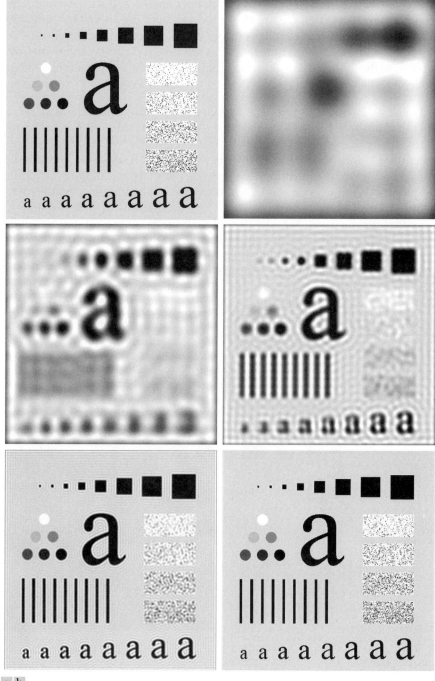

a b
c d
e f

FIGURE 4.42 (a) Original image. (b)–(f) Results of filtering using ILPFs with cutoff frequencies set at radii values 10, 30, 60, 160, and 460, as shown in Fig. 4.41(b). The power removed by these filters was 13, 6.9, 4.3, 2.2, and 0.8% of the total, respectively.

filtering concepts. Also, as shown in the discussion that follows, some interesting insight is gained by attempting to explain the ringing property of ILPFs in the spatial domain.   ■

The blurring and ringing properties of ILPFs can be explained using the convolution theorem. Figure 4.43(a) shows the spatial representation, $h(x, y)$, of an ILPF of radius 10, and Fig. 4.43(b) shows the intensity profile of a line passing through the center of the image. Because a cross section of the ILPF in the frequency domain looks like a box filter, it is not unexpected that a cross section of the corresponding spatial filter has the shape of a sinc function. Filtering in the spatial domain is done by convolving $h(x, y)$ with the image. Imagine each pixel in the image being a discrete impulse whose strength is proportional to the intensity of the image at that location. Convolving a sinc with an impulse copies the sinc at the location of the impulse. The center lobe of the sinc is the principal cause of blurring, while the outer, smaller lobes are mainly responsible for ringing. Convolving the sinc with every pixel in the image provides a nice model for explaining the behavior of ILPFs. Because the "spread" of the sinc function is inversely proportional to the radius of $H(u, v)$, the larger D_0 becomes, the more the spatial sinc approaches an impulse which, in the limit, causes no blurring at all when convolved with the image. This type of reciprocal behavior should be routine to you by now. In the next two sections, we show that it is possible to achieve blurring with little or no ringing, which is an important objective in lowpass filtering.

4.8.2 Butterworth Lowpass Filters

The transfer function of a Butterworth lowpass filter (BLPF) of order n, and with cutoff frequency at a distance D_0 from the origin, is defined as

$$H(u, v) = \frac{1}{1 + [D(u, v)/D_0]^{2n}} \qquad (4.8\text{-}5)$$

where $D(u, v)$ is given by Eq. (4.8-2). Figure 4.44 shows a perspective plot, image display, and radial cross sections of the BLPF function.

The transfer function of the Butterworth lowpass filter normally is written as the square root of our expression. However, our interest here is in the basic *form* of the filter, so we exclude the square root for computational convenience.

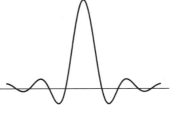

a b

FIGURE 4.43
(a) Representation in the spatial domain of an ILPF of radius 5 and size 1000 × 1000. (b) Intensity profile of a horizontal line passing through the center of the image.

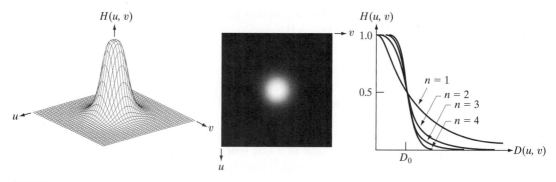

a b c

FIGURE 4.44 (a) Perspective plot of a Butterworth lowpass-filter transfer function. (b) Filter displayed as an image. (c) Filter radial cross sections of orders 1 through 4.

Unlike the ILPF, the BLPF transfer function does not have a sharp discontinuity that gives a clear cutoff between passed and filtered frequencies. For filters with smooth transfer functions, defining a cutoff frequency locus at points for which $H(u, v)$ is down to a certain fraction of its maximum value is customary. In Eq. (4.8-5), (down 50% from its maximum value of 1) when $D(u, v) = D_0$.

EXAMPLE 4.17:
Image smoothing with a Butterworth lowpass filter.

■ Figure 4.45 shows the results of applying the BLPF of Eq. (4.8-5) to Fig. 4.45(a), with $n = 2$ and D_0 equal to the five radii in Fig. 4.41(b). Unlike the results in Fig. 4.42 for the ILPF, we note here a smooth transition in blurring as a function of increasing cutoff frequency. Moreover, no ringing is visible in any of the images processed with this particular BLPF, a fact attributed to the filter's smooth transition between low and high frequencies. ■

A BLPF of order 1 has no ringing in the spatial domain. Ringing generally is imperceptible in filters of order 2, but can become significant in filters of higher order. Figure 4.46 shows a comparison between the *spatial* representation of BLPFs of various orders (using a cutoff frequency of 5 in all cases). Shown also is the intensity profile along a horizontal scan line through the center of each filter. These filters were obtained and displayed using the same procedure used to generate Fig. 4.43. To facilitate comparisons, additional enhancing with a gamma transformation [see Eq. (3.2-3)] was applied to the images of Fig. 4.46. The BLPF of order 1 [Fig. 4.46(a)] has neither ringing nor negative values. The filter of order 2 does show mild ringing and small negative values, but they certainly are less pronounced than in the ILPF. As the remaining images show, ringing in the BLPF becomes significant for higher-order filters. A Butterworth filter of order 20 exhibits characteristics similar to those of the ILPF (in the limit, both filters are identical). BLPFs of order 2 are a good compromise between effective lowpass filtering and acceptable ringing.

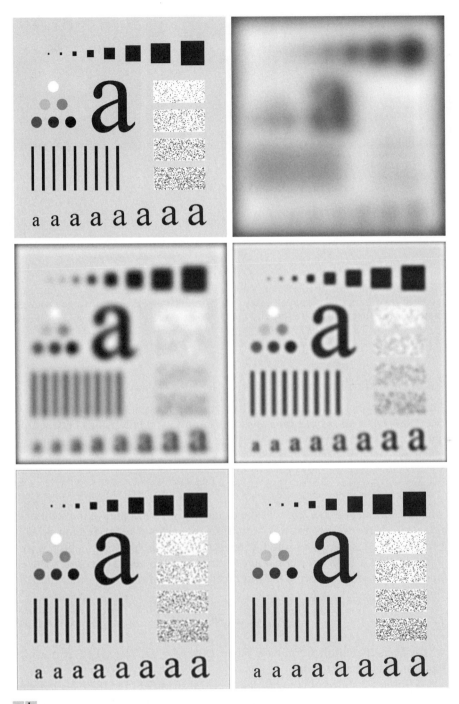

a b
c d
e f

FIGURE 4.45 (a) Original image. (b)–(f) Results of filtering using BLPFs of order 2, with cutoff frequencies at the radii shown in Fig. 4.41. Compare with Fig. 4.42.

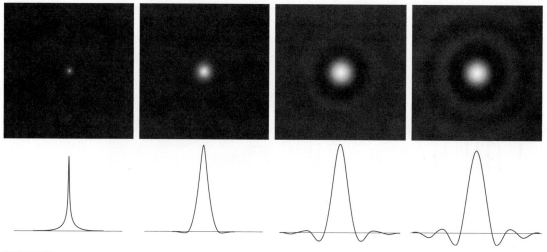

a b c d

FIGURE 4.46 (a)–(d) Spatial representation of BLPFs of order 1, 2, 5, and 20, and corresponding intensity profiles through the center of the filters (the size in all cases is 1000×1000 and the cutoff frequency is 5). Observe how ringing increases as a function of filter order.

4.8.3 Gaussian Lowpass Filters

Gaussian lowpass filters (GLPFs) of one dimension were introduced in Section 4.7.4 as an aid in exploring some important relationships between the spatial and frequency domains. The form of these filters in two dimensions is given by

$$H(u, v) = e^{-D^2(u, v)/2\sigma^2} \tag{4.8-6}$$

where, as in Eq. (4.8-2), $D(u, v)$ is the distance from the center of the frequency rectangle. Here we do not use a multiplying constant as in Section 4.7.4 in order to be consistent with the filters discussed in the present section, whose highest value is 1. As before, σ is a measure of spread about the center. By letting $\sigma = D_0$, we can express the filter using the notation of the other filters in this section:

$$H(u, v) = e^{-D^2(u, v)/2D_0^2} \tag{4.8-7}$$

where D_0 is the cutoff frequency. When $D(u, v) = D_0$, the GLPF is down to 0.607 of its maximum value.

As Table 4.3 shows, the inverse Fourier transform of the GLPF is Gaussian also. This means that a spatial Gaussian filter, obtained by computing the IDFT of Eq. (4.8-6) or (4.8-7), will have no ringing. Figure 4.47 shows a perspective plot, image display, and radial cross sections of a GLPF function, and Table 4.4 summarizes the lowpass filters discussed in this section.

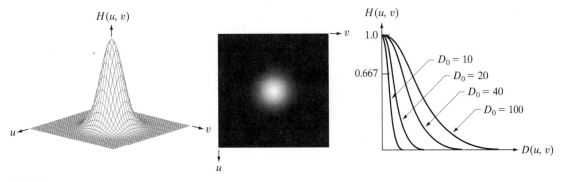

FIGURE 4.47 (a) Perspective plot of a GLPF transfer function. (b) Filter displayed as an image. (c) Filter radial cross sections for various values of D_0.

TABLE 4.4
Lowpass filters. D_0 is the cutoff frequency and n is the order of the Butterworth filter.

Ideal	Butterworth	Gaussian
$H(u, v) = \begin{cases} 1 & \text{if } D(u, v) \leq D_0 \\ 0 & \text{if } D(u, v) > D_0 \end{cases}$	$H(u, v) = \dfrac{1}{1 + [D(u, v)/D_0]^{2n}}$	$H(u, v) = e^{-D^2(u,v)/2D_0^2}$

■ Figure 4.48 shows the results of applying the GLPF of Eq. (4.8-7) to Fig. 4.48(a), with D_0 equal to the five radii in Fig. 4.41(b). As in the case of the BLPF of order 2 (Fig. 4.45), we note a smooth transition in blurring as a function of increasing cutoff frequency. The GLPF achieved slightly less smoothing than the BLPF of order 2 for the same value of cutoff frequency, as can be seen, for example, by comparing Figs. 4.45(c) and 4.48(c). This is expected, because the profile of the GLPF is not as "tight" as the profile of the BLPF of order 2. However, the results are quite comparable, and we are assured of no ringing in the case of the GLPF. This is an important characteristic in practice, especially in situations (e.g., medical imaging) in which any type of artifact is unacceptable. In cases where tight control of the transition between low and high frequencies about the cutoff frequency are needed, then the BLPF presents a more suitable choice. The price of this additional control over the filter profile is the possibility of ringing.   ■

EXAMPLE 4.18:
Image smoothing with a Gaussian lowpass filter.

4.8.4 Additional Examples of Lowpass Filtering

In the following discussion, we show several practical applications of lowpass filtering in the frequency domain. The first example is from the field of machine perception with application to character recognition; the second is from the printing and publishing industry; and the third is related to processing

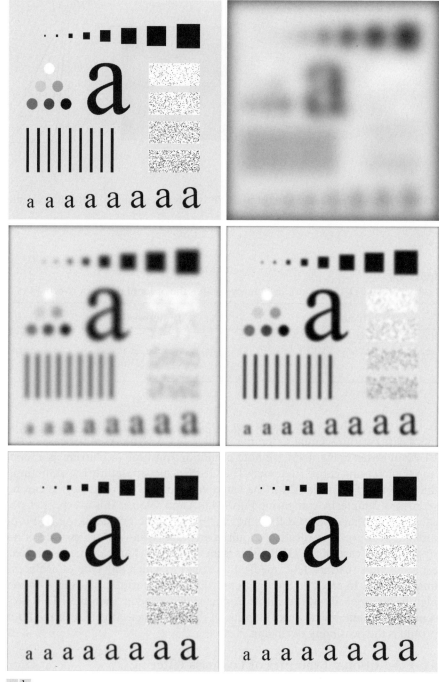

FIGURE 4.48 (a) Original image. (b)–(f) Results of filtering using GLPFs with cutoff frequencies at the radii shown in Fig. 4.41. Compare with Figs. 4.42 and 4.45.

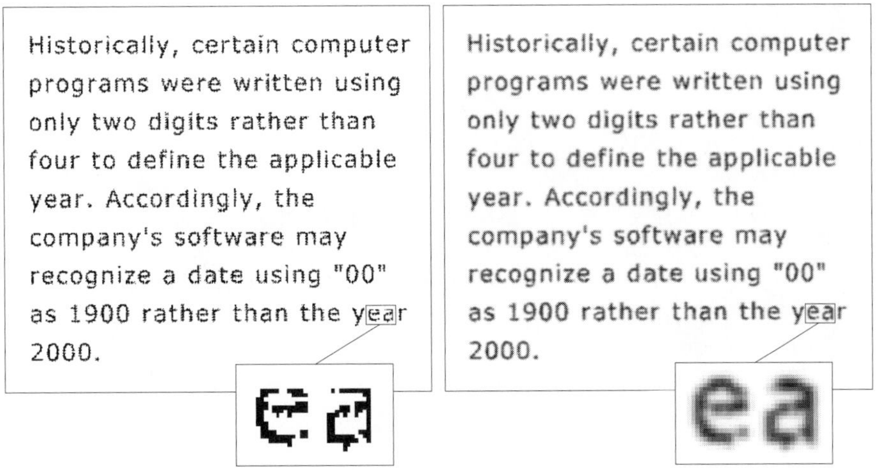

Historically, certain computer programs were written using only two digits rather than four to define the applicable year. Accordingly, the company's software may recognize a date using "00" as 1900 rather than the year 2000.

Historically, certain computer programs were written using only two digits rather than four to define the applicable year. Accordingly, the company's software may recognize a date using "00" as 1900 rather than the year 2000.

a b

FIGURE 4.49
(a) Sample text of low resolution (note broken characters in magnified view). (b) Result of filtering with a GLPF (broken character segments were joined).

satellite and aerial images. Similar results can be obtained using the lowpass spatial filtering techniques discussed in Section 3.5.

Figure 4.49 shows a sample of text of poor resolution. One encounters text like this, for example, in fax transmissions, duplicated material, and historical records. This particular sample is free of additional difficulties like smudges, creases, and torn sections. The magnified section in Fig. 4.49(a) shows that the characters in this document have distorted shapes due to lack of resolution, and many of the characters are broken. Although humans fill these gaps visually without difficulty, machine recognition systems have real difficulties reading broken characters. One approach for handling this problem is to bridge small gaps in the input image by blurring it. Figure 4.49(b) shows how well characters can be "repaired" by this simple process using a Gaussian lowpass filter with $D_0 = 80$. The images are of size 444×508 pixels.

Lowpass filtering is a staple in the printing and publishing industry, where it is used for numerous preprocessing functions, including unsharp masking, as discussed in Section 3.6.3. "Cosmetic" processing is another use of lowpass filtering prior to printing. Figure 4.50 shows an application of lowpass filtering for producing a smoother, softer-looking result from a sharp original. For human faces, the typical objective is to reduce the sharpness of fine skin lines and small blemishes. The magnified sections in Figs. 4.50(b) and (c) clearly show a significant reduction in fine skin lines around the eyes in this case. In fact, the smoothed images look quite soft and pleasing.

We discuss unsharp masking in the frequency domain in Section 4.9.5

Figure 4.51 shows two applications of lowpass filtering on the same image, but with totally different objectives. Figure 4.51(a) is an 808×754 very high resolution radiometer (VHRR) image showing part of the Gulf of Mexico (dark) and Florida (light), taken from a NOAA satellite (note the horizontal sensor scan lines). The boundaries between bodies of water were caused by loop currents. This image is illustrative of remotely sensed images in which sensors have the tendency to produce pronounced scan lines along the direction in which the scene is being scanned (see Example 4.24 for an illustration of a

a b c

FIGURE 4.50 (a) Original image (784 × 732 pixels). (b) Result of filtering using a GLPF with $D_0 = 100$. (c) Result of filtering using a GLPF with $D_0 = 80$. Note the reduction in fine skin lines in the magnified sections in (b) and (c).

physical cause). Lowpass filtering is a crude but simple way to reduce the effect of these lines, as Fig. 4.51(b) shows (we consider more effective approaches in Sections 4.10 and 5.4.1). This image was obtained using a GLFP with $D_0 = 50$. The reduction in the effect of the scan lines can simplify the detection of features such as the interface boundaries between ocean currents.

Figure 4.51(c) shows the result of significantly more aggressive Gaussian lowpass filtering with $D_0 = 20$. Here, the objective is to blur out as much detail as possible while leaving large features recognizable. For instance, this type of filtering could be part of a preprocessing stage for an image analysis system that searches for features in an image bank. An example of such features could be lakes of a given size, such as Lake Okeechobee in the lower eastern region of Florida, shown as a nearly round dark region in Fig. 4.51(c). Lowpass filtering helps simplify the analysis by averaging out features smaller than the ones of interest.

4.9   Image Sharpening Using Frequency Domain Filters

In the previous section, we showed that an image can be smoothed by attenuating the high-frequency components of its Fourier transform. Because edges and other abrupt changes in intensities are associated with high-frequency components, image sharpening can be achieved in the frequency domain by highpass filtering, which attenuates the low-frequency components without disturbing high-frequency information in the Fourier transform. As in Section

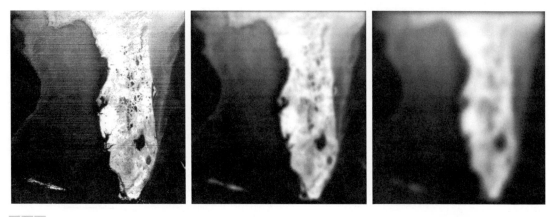

a b c

FIGURE 4.51 (a) Image showing prominent horizontal scan lines. (b) Result of filtering using a GLPF with $D_0 = 50$. (c) Result of using a GLPF with $D_0 = 20$. (Original image courtesy of NOAA.)

4.8, we consider only zero-phase-shift filters that are radially symmetric. All filtering in this section is based on the procedure outlined in Section 4.7.3, so all filter functions, $H(u, v)$, are understood to be discrete functions of size $P \times Q$; that is, the discrete frequency variables are in the range $u = 0, 1, 2, \ldots, P - 1$ and $v = 0, 1, 2, \ldots, Q - 1$.

A highpass filter is obtained from a given lowpass filter using the equation

$$H_{\text{HP}}(u, v) = 1 - H_{\text{LP}}(u, v) \tag{4.9-1}$$

where $H_{\text{LP}}(u, v)$ is the transfer function of the lowpass filter. That is, when the lowpass filter attenuates frequencies, the highpass filter passes them, and vice versa.

In this section, we consider ideal, Butterworth, and Gaussian highpass filters. As in the previous section, we illustrate the characteristics of these filters in both the frequency and spatial domains. Figure 4.52 shows typical 3-D plots, image representations, and cross sections for these filters. As before, we see that the Butterworth filter represents a transition between the sharpness of the ideal filter and the broad smoothness of the Gaussian filter. Figure 4.53, discussed in the sections that follow, illustrates what these filters look like in the spatial domain. The spatial filters were obtained and displayed by using the procedure used to generate Figs. 4.43 and 4.46.

4.9.1 Ideal Highpass Filters

A 2-D *ideal highpass filter* (IHPF) is defined as

$$H(u, v) = \begin{cases} 0 & \text{if } D(u, v) \leq D_0 \\ 1 & \text{if } D(u, v) > D_0 \end{cases} \tag{4.9-2}$$

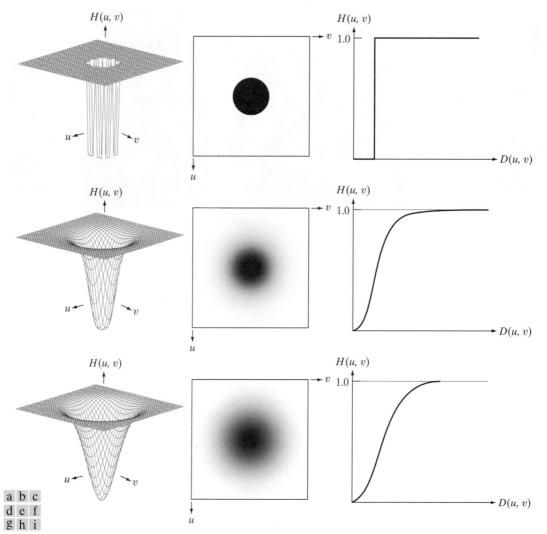

a b c
d e f
g h i

FIGURE 4.52 Top row: Perspective plot, image representation, and cross section of a typical ideal highpass filter. Middle and bottom rows: The same sequence for typical Butterworth and Gaussian highpass filters.

where D_0 is the cutoff frequency and $D(u, v)$ is given by Eq. (4.8-2). This expression follows directly from Eqs. (4.8-1) and (4.9-1). As intended, the IHPF is the opposite of the ILPF in the sense that it sets to zero all frequencies inside a circle of radius D_0 while passing, without attenuation, all frequencies outside the circle. As in the case of the ILPF, the IHPF is not physically realizable. However, we consider it here for completeness and, as before, because its properties can be used to explain phenomena such as ringing in the spatial domain. The discussion will be brief.

Because of the way in which they are related [Eq. (4.9-1)], we can expect IHPFs to have the same ringing properties as ILPFs. This is demonstrated

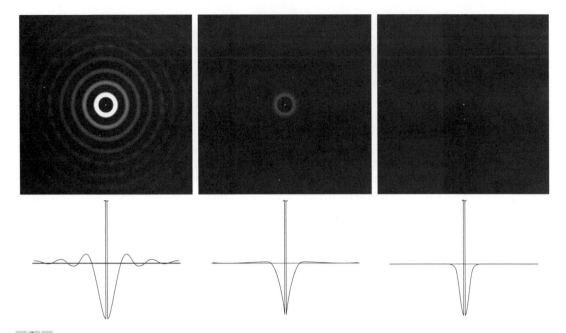

a b c

FIGURE 4.53 Spatial representation of typical (a) ideal, (b) Butterworth, and (c) Gaussian frequency domain highpass filters, and corresponding intensity profiles through their centers.

clearly in Fig. 4.54, which consists of various IHPF results using the original image in Fig. 4.41(a) with D_0 set to 30, 60, and 160 pixels, respectively. The ringing in Fig. 4.54(a) is so severe that it produced distorted, thickened object boundaries (e.g., look at the large letter "a"). Edges of the top three circles do not show well because they are not as strong as the other edges in the image (the intensity of these three objects is much closer to the background intensity,

a b c

FIGURE 4.54 Results of highpass filtering the image in Fig. 4.41(a) using an IHPF with $D_0 = 30$, 60, and 160.

giving discontinuities of smaller magnitude). Looking at the "spot" size of the spatial representation of the IHPF in Fig. 4.53(a) and keeping in mind that filtering in the spatial domain is convolution of the spatial filter with the image helps explain why the smaller objects and lines appear almost solid white. Look in particular at the three small squares in the top row and the thin, vertical bars in Fig. 4.54(a). The situation improved somewhat with $D_0 = 60$. Edge distortion is quite evident still, but now we begin to see filtering on the smaller objects. Due to the now familiar inverse relationship between the frequency and spatial domains, we know that the spot size of this filter is smaller than the spot of the filter with $D_0 = 30$. The result for $D_0 = 160$ is closer to what a highpass-filtered image should look like. Here, the edges are much cleaner and less distorted, and the smaller objects have been filtered properly. Of course, the constant background in all images is zero in these highpass-filtered images because highpass filtering is analogous to differentiation in the spatial domain.

4.9.2 Butterworth Highpass Filters

A 2-D *Butterworth highpass filter* (BHPF) of order n and cutoff frequency D_0 is defined as

$$H(u, v) = \frac{1}{1 + [D_0/D(u, v)]^{2n}} \tag{4.9-3}$$

where $D(u, v)$ is given by Eq. (4.8-2). This expression follows directly from Eqs. (4.8-5) and (4.9-1). The middle row of Fig. 4.52 shows an image and cross section of the BHPF function.

As with lowpass filters, we can expect Butterworth highpass filters to behave smoother than IHPFs. Figure 4.55 shows the performance of a BHPF, of

a b c

FIGURE 4.55 Results of highpass filtering the image in Fig. 4.41(a) using a BHPF of order 2 with $D_0 = 30, 60$, and 160, corresponding to the circles in Fig. 4.41(b). These results are much smoother than those obtained with an IHPF.

a b c

FIGURE 4.56 Results of highpass filtering the image in Fig. 4.41(a) using a GHPF with $D_0 = 30, 60,$ and 160, corresponding to the circles in Fig. 4.41(b). Compare with Figs. 4.54 and 4.55.

order 2 and with D_0 set to the same values as in Fig. 4.54. The boundaries are much less distorted than in Fig. 4.54, even for the smallest value of cutoff frequency. Because the spot sizes in the center areas of the IHPF and the BHPF are similar [see Figs. 4.53(a) and (b)], the performance of the two filters on the smaller objects is comparable. The transition into higher values of cutoff frequencies is much smoother with the BHPF.

4.9.3 Gaussian Highpass Filters

The transfer function of the Gaussian highpass filter (GHPF) with cutoff frequency locus at a distance D_0 from the center of the frequency rectangle is given by

$$H(u, v) = 1 - e^{-D^2(u,v)/2D_0^2} \tag{4.9-4}$$

where $D(u, v)$ is given by Eq. (4.8-2). This expression follows directly from Eqs. (4.8-7) and (4.9-1). The third row in Fig. 4.52 shows a perspective plot, image, and cross section of the GHPF function. Following the same format as for the BHPF, we show in Fig. 4.56 comparable results using GHPFs. As expected, the results obtained are more gradual than with the previous two filters. Even the filtering of the smaller objects and thin bars is cleaner with the Gaussian filter. Table 4.5 contains a summary of the highpass filters discussed in this section.

TABLE 4.5
Highpass filters. D_0 is the cutoff frequency and n is the order of the Butterworth filter.

Ideal	Butterworth	Gaussian
$H(u, v) = \begin{cases} 1 & \text{if } D(u, v) \leq D_0 \\ 0 & \text{if } D(u, v) > D_0 \end{cases}$	$H(u, v) = \dfrac{1}{1 + [D_0/D(u, v)]^{2n}}$	$H(u, v) = 1 - e^{-D^2(u,v)/2D_0^2}$

EXAMPLE 4.19:
Using highpass
filtering and
thresholding for
image
enhancement.

■ Figure 4.57(a) is a 1026 × 962 image of a thumb print in which smudges (a typical problem) are evident. A key step in automated fingerprint recognition is enhancement of print ridges and the reduction of smudges. Enhancement is useful also in human interpretation of prints. In this example, we use highpass filtering to enhance the ridges and reduce the effects of smudging. Enhancement of the ridges is accomplished by the fact that they contain high frequencies, which are unchanged by a highpass filter. On the other hand, the filter reduces low frequency components, which correspond to slowly varying intensities in the image, such as the background and smudges. Thus, enhancement is achieved by reducing the effect of all features except those with high frequencies, which are the features of interest in this case.

The value $D_0 = 50$ is approximately 2.5% of the short dimension of the padded image. The idea is for D_0 to be close to the origin so low frequencies are attenuated, but not completely eliminated. A range of 2% to 5% of the short dimension is a good starting point.

Figure 4.57(b) is the result of using a Butterworth highpass filter of order 4 with a cutoff frequency of 50. As expected, the highpass-filtered image lost its gray tones because the dc term was reduced to 0. The net result is that dark tones typically predominate in highpass-filtered images, thus requiring additional processing to enhance details of interest. A simple approach is to threshold the filtered image. Figure 4.57(c) shows the result of setting to black all negative values and to white all positive values in the filtered image. Note how the ridges are clear and the effect of the smudges has been reduced considerably. In fact, ridges that are barely visible in the top, right section of the image in Fig. 4.57(a) are nicely enhanced in Fig. 4.57(c). ■

4.9.4 The Laplacian in the Frequency Domain

In Section 3.6.2, we used the Laplacian for image enhancement in the spatial domain. In this section, we revisit the Laplacian and show that it yields equivalent results using frequency domain techniques. It can be shown (Problem 4.26) that the Laplacian can be implemented in the frequency domain using the filter

$$H(u, v) = -4\pi^2(u^2 + v^2) \tag{4.9-5}$$

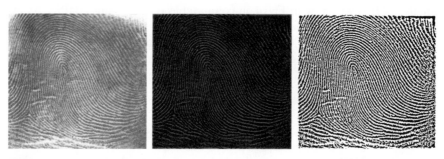

a b c

FIGURE 4.57 (a) Thumb print. (b) Result of highpass filtering (a). (c) Result of thresholding (b). (Original image courtesy of the U.S. National Institute of Standards and Technology.)

or, with respect to the center of the frequency rectangle, using the filter

$$H(u, v) = -4\pi^2\left[(u - P/2)^2 + (v - Q/2)^2\right]$$

$$= -4\pi^2 D^2(u, v)$$

(4.9-6)

where $D(u, v)$ is the distance function given in Eq. (4.8-2). Then, the Laplacian image is obtained as:

$$\nabla^2 f(x, y) = \Im^{-1}\{H(u, v)F(u, v)\}$$

(4.9-7)

where $F(u, v)$ is the DFT of $f(x, y)$. As explained in Section 3.6.2, enhancement is achieved using the equation:

$$g(x, y) = f(x, y) + c\nabla^2 f(x, y)$$

(4.9-8)

Here, $c = -1$ because $H(u, v)$ is negative. In Chapter 3, $f(x, y)$ and $\nabla^2 f(x, y)$ had comparable values. However, computing $\nabla^2 f(x, y)$ with Eq. (4.9-7) introduces DFT scaling factors that can be several orders of magnitude larger than the maximum value of f. Thus, the differences between f and its Laplacian must be brought into comparable ranges. The easiest way to handle this problem is to normalize the values of $f(x, y)$ to the range $[0, 1]$ (before computing its DFT) and divide $\nabla^2 f(x, y)$ by its maximum value, which will bring it to the approximate range $[-1, 1]$ (recall that the Laplacian has negative values). Equation (4.9-8) can then be applied.

In the frequency domain, Eq. (4.9-8) is written as

$$g(x, y) = \Im^{-1}\{F(u, v) - H(u, v)F(u, v)\}$$

$$= \Im^{-1}\{[1 - H(u, v)]F(u, v)\}$$

(4.9-9)

$$= \Im^{-1}\{[1 + 4\pi^2 D^2(u, v)]F(u, v)\}$$

Although this result is elegant, it has the same scaling issues just mentioned, compounded by the fact that the normalizing factor is not as easily computed. For this reason, Eq. (4.9-8) is the preferred implementation in the frequency domain, with $\nabla^2 f(x, y)$ computed using Eq. (4.9-7) and scaled using the approach mentioned in the previous paragraph.

■ Figure 4.58(a) is the same as Fig. 3.38(a), and Fig. 4.58(b) shows the result of using Eq. (4.9-8), in which the Laplacian was computed in the frequency domain using Eq. (4.9-7). Scaling was done as described in connection with that equation. We see by comparing Figs. 4.58(b) and 3.38(e) that the frequency domain and spatial results are identical visually. Observe that the results in these two figures correspond to the Laplacian mask in Fig. 3.37(b), which has a -8 in the center (Problem 4.26). ■

EXAMPLE 4.20:
Image sharpening in the frequency domain using the Laplacian.

a b

FIGURE 4.58
(a) Original,
blurry image.
(b) Image
enhanced using
the Laplacian in
the frequency
domain. Compare
with Fig. 3.38(e).

4.9.5 Unsharp Masking, Highboost Filtering, and High-Frequency-Emphasis Filtering

In this section, we discuss frequency domain formulations of the unsharp masking and high-boost filtering image sharpening techniques introduced in Section 3.6.3. Using frequency domain methods, the mask defined in Eq. (3.6-8) is given by

$$g_{\text{mask}}(x, y) = f(x, y) - f_{\text{LP}}(x, y) \qquad (4.9\text{-}10)$$

with

$$f_{\text{LP}}(x, y) = \Im^{-1}\big[H_{\text{LP}}(u, v)F(u, v)\big] \qquad (4.9\text{-}11)$$

where $H_{\text{LP}}(u, v)$ is a lowpass filter and $F(u, v)$ is the Fourier transform of $f(x, y)$. Here, $f_{\text{LP}}(x, y)$ is a smoothed image analogous to $\bar{f}(x, y)$ in Eq. (3.6-8). Then, as in Eq. (3.6-9),

$$g(x, y) = f(x, y) + k * g_{\text{mask}}(x, y) \qquad (4.9\text{-}12)$$

This expression defines unsharp masking when $k = 1$ and highboost filtering when $k > 1$. Using the preceding results, we can express Eq. (4.9-12) entirely in terms of frequency domain computations involving a lowpass filter:

$$g(x, y) = \Im^{-1}\Big\{\big[1 + k * [1 - H_{\text{LP}}(u, v)]\big]F(u, v)\Big\} \qquad (4.9\text{-}13)$$

Using Eq. (4.9-1), we can express this result in terms of a highpass filter:

$$g(x, y) = \Im^{-1}\Big\{[1 + k * H_{\text{HP}}(u, v)]F(u, v)\Big\} \qquad (4.9\text{-}14)$$

The expression contained within the square brackets is called a *high-frequency-emphasis filter*. As noted earlier, highpass filters set the dc term to zero, thus reducing the average intensity in the filtered image to 0. The high-frequency-emphasis filter does not have this problem because of the 1 that is added to the highpass filter. The constant, k, gives control over the proportion of high frequencies that influence the final result. A slightly more general formulation of high-frequency-emphasis filtering is the expression

$$g(x, y) = \mathfrak{I}^{-1}\{[k_1 + k_2 * H_{HP}(u, v)]F(u, v)\} \qquad (4.9\text{-}15)$$

where $k_1 \geq 0$ gives controls of the offset from the origin [see Fig. 4.31(c)] and $k_2 \geq 0$ controls the contribution of high frequencies.

■ Figure 4.59(a) shows a 416×596 chest X-ray with a narrow range of intensity levels. The objective of this example is to enhance the image using high-frequency-emphasis filtering. X-rays cannot be focused in the same manner that optical lenses are focused, and the resulting images generally tend to be slightly blurred. Because the intensities in this particular image are biased toward the dark end of the gray scale, we also take this opportunity to give an example of how spatial domain processing can be used to complement frequency-domain filtering.

EXAMPLE 4.21: Image enhancement using high-frequency-emphasis filtering.

Figure 4.59(b) shows the result of highpass filtering using a Gaussian filter with $D_0 = 40$ (approximately 5% of the short dimension of the padded image). As expected, the filtered result is rather featureless, but it shows faintly the principal edges in the image. Figure 4.59(c) shows the advantage of high-emphasis filtering, where we used Eq. (4.9-15) with $k_1 = 0.5$ and $k_2 = 0.75$. Although the image is still dark, the gray-level tonality due to the low-frequency components was not lost.

As discussed in Section 3.3.1, an image characterized by intensity levels in a narrow range of the gray scale is an ideal candidate for histogram equalization. As Fig. 4.59(d) shows, this was indeed an appropriate method to further enhance the image. Note the clarity of the bone structure and other details that simply are not visible in any of the other three images. The final enhanced image is a little noisy, but this is typical of X-ray images when their gray scale is expanded. The result obtained using a combination of high-frequency emphasis and histogram equalization is superior to the result that would be obtained by using either method alone. ■

Artifacts such as ringing are unacceptable in medical imaging. Thus, it is good practice to avoid using filters that have the potential for introducing artifacts in the processed image. Because spatial and frequency domain Gaussian filters are Fourier transform pairs, these filters produce smooth results that are void of artifacts.

4.9.6 Homomorphic Filtering

The illumination-reflectance model introduced in Section 2.3.4 can be used to develop a frequency domain procedure for improving the appearance of an image by simultaneous intensity range compression and contrast enhancement. From the discussion in that section, an image $f(x, y)$ can be expressed as the product of its illumination, $i(x, y)$, and reflectance, $r(x, y)$, components:

$$f(x, y) = i(x, y)r(x, y) \qquad (4.9\text{-}16)$$

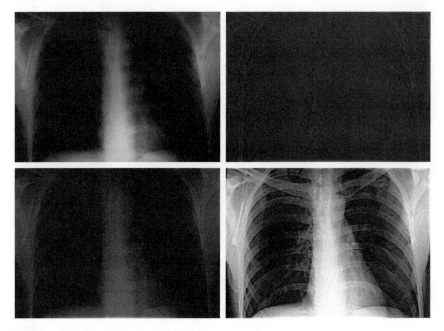

a b
c d

FIGURE 4.59 (a) A chest X-ray image. (b) Result of highpass filtering with a Gaussian filter. (c) Result of high-frequency-emphasis filtering using the same filter. (d) Result of performing histogram equalization on (c). (Original image courtesy of Dr. Thomas R. Gest, Division of Anatomical Sciences, University of Michigan Medical School.)

This equation cannot be used directly to operate on the frequency components of illumination and reflectance because the Fourier transform of a product is not the product of the transforms:

$$\Im[f(x, y)] \neq \Im[i(x, y)]\,\Im[r(x, y)] \qquad (4.9\text{-}17)$$

However, suppose that we define

If an image $f(x, y)$ with intensities in the range $[0, L - 1]$ has any 0 values, a 1 must be added to every element of the image to avoid having to deal with $\ln(0)$. The 1 is then subtracted at the end of the filtering process.

$$z(x, y) = \ln f(x, y)$$
$$\qquad\qquad\qquad\; = \ln i(x, y) + \ln r(x, y) \qquad (4.9\text{-}18)$$

Then,

$$\Im\{z(x, y)\} = \Im\{\ln f(x, y)\}$$
$$\qquad\qquad\qquad\qquad\;\; = \Im\{\ln i(x, y)\} + \Im\{\ln r(x, y)\} \qquad (4.9\text{-}19)$$

or

$$Z(u, v) = F_i(u, v) + F_r(u, v) \qquad (4.9\text{-}20)$$

where $F_i(u, v)$ and $F_r(u, v)$ are the Fourier transforms of $\ln i(x, y)$ and $\ln r(x, y)$, respectively.

We can filter $Z(u, v)$ using a filter $H(u, v)$ so that

$$S(u, v) = H(u, v)Z(u, v)$$
$$= H(u, v)F_i(u, v) + H(u, v)F_r(u, v) \tag{4.9-21}$$

The filtered image in the spatial domain is

$$s(x, y) = \Im^{-1}\{S(u, v)\}$$
$$= \Im^{-1}\{H(u, v)F_i(u, v)\} + \Im^{-1}\{H(u, v)F_r(u, v)\} \tag{4.9-22}$$

By defining

$$i'(x, y) = \Im^{-1}\{H(u, v)F_i(u, v)\} \tag{4.9-23}$$

and

$$r'(x, y) = \Im^{-1}\{H(u, v)F_r(u, v)\} \tag{4.9-24}$$

we can express Eq. (4.9-23) in the form

$$s(x, y) = i'(x, y) + r'(x, y) \tag{4.9-25}$$

Finally, because $z(x, y)$ was formed by taking the natural logarithm of the input image, we reverse the process by taking the exponential of the filtered result to form the output image:

$$g(x, y) = e^{s(x,y)}$$
$$= e^{i'(x,y)}e^{r'(x,y)} \tag{4.9-26}$$
$$= i_0(x, y)r_0(x, y)$$

where

$$i_0(x, y) = e^{i'(x,y)} \tag{4.9-27}$$

and

$$r_0(x, y) = e^{r'(x,y)} \tag{4.9-28}$$

are the illumination and reflectance components of the output (processed) image.

FIGURE 4.60
Summary of steps in homomorphic filtering.

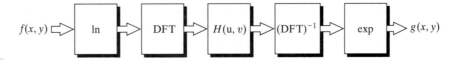

The filtering approach just derived is summarized in Fig. 4.60. This method is based on a special case of a class of systems known as *homomorphic systems*. In this particular application, the key to the approach is the separation of the illumination and reflectance components achieved in the form shown in Eq. (4.9-20). The *homomorphic filter function* $H(u, v)$ then can operate on these components separately, as indicated by Eq. (4.9-21).

The illumination component of an image generally is characterized by slow spatial variations, while the reflectance component tends to vary abruptly, particularly at the junctions of dissimilar objects. These characteristics lead to associating the low frequencies of the Fourier transform of the logarithm of an image with illumination and the high frequencies with reflectance. Although these associations are rough approximations, they can be used to advantage in image filtering, as illustrated in Example 4.22.

A good deal of control can be gained over the illumination and reflectance components with a homomorphic filter. This control requires specification of a filter function $H(u, v)$ that affects the low- and high-frequency components of the Fourier transform in different, controllable ways. Figure 4.61 shows a cross section of such a filter. If the parameters γ_L and γ_H are chosen so that $\gamma_L < 1$ and $\gamma_H > 1$, the filter function in Fig. 4.61 tends to attenuate the contribution made by the low frequencies (illumination) and amplify the contribution made by high frequencies (reflectance). The net result is simultaneous dynamic range compression and contrast enhancement.

The shape of the function in Fig. 4.61 can be approximated using the basic form of a highpass filter. For example, using a slightly modified form of the Gaussian highpass filter yields the function

$$H(u, v) = (\gamma_H - \gamma_L)\left[1 - e^{-c[D^2(u, v)/D_0^2]}\right] + \gamma_L \tag{4.9-29}$$

FIGURE 4.61
Radial cross section of a circularly symmetric homomorphic filter function. The vertical axis is at the center of the frequency rectangle and $D(u, v)$ is the distance from the center.

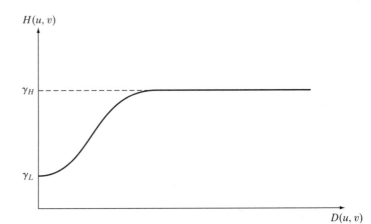

where $D(u, v)$ is defined in Eq. (4.8-2) and the constant c controls the sharpness of the slope of the function as it transitions between γ_L and γ_H. This filter is similar to the high-emphasis filter discussed in the previous section.

■ Figure 4.62(a) shows a full body PET (Positron Emission Tomography) scan of size 1162×746 pixels. The image is slightly blurry and many of its low-intensity features are obscured by the high intensity of the "hot spots" dominating the dynamic range of the display. (These hot spots were caused by a tumor in the brain and one in the lungs.) Figure 4.62(b) was obtained by homomorphic filtering Fig. 4.62(a) using the filter in Eq. (4.9-29) with $\gamma_L = 0.25$, $\gamma_H = 2$, $c = 1$, and $D_0 = 80$. A cross section of this filter looks just like Fig. 4.61, with a slightly steeper slope.

Note in Fig. 4.62(b) how much sharper the hot spots, the brain, and the skeleton are in the processed image, and how much more detail is visible in this image. By reducing the effects of the dominant illumination components (the hot spots), it became possible for the dynamic range of the display to allow lower intensities to become much more visible. Similarly, because the high frequencies are enhanced by homomorphic filtering, the reflectance components of the image (edge information) were sharpened considerably. The enhanced image in Fig. 4.62(b) is a significant improvement over the original. ∎

EXAMPLE 4.22:
Image enhancement using homomorphic filtering.

Recall that filtering uses image padding, so the filter is of size $P \times Q$.

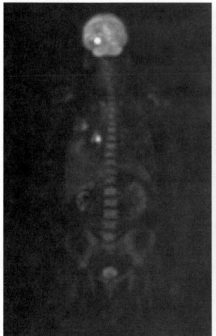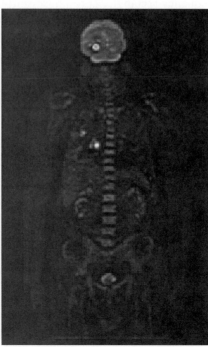

a b

FIGURE 4.62
(a) Full body PET scan. (b) Image enhanced using homomorphic filtering. (Original image courtesy of Dr. Michael E. Casey, CTI PET Systems.)

4.10 Selective Filtering

The filters discussed in the previous two sections operate over the entire frequency rectangle. There are applications in which it is of interest to process specific bands of frequencies or small regions of the frequency rectangle. Filters in the first category are called *bandreject* or *bandpass filters*, respectively. Filters in the second category are called *notch filters*.

4.10.1 Bandreject and Bandpass Filters

These types of filters are easy to construct using the concepts from the previous two sections. Table 4.6 shows expressions for ideal, Butterworth, and Gaussian bandreject filters, where $D(u, v)$ is the distance from the center of the frequency rectangle, as given in Eq. (4.8-2), D_0 is the radial center of the band, and W is the width of the band. Figure 4.63(a) shows a Gaussian bandreject filter in image form, where black is 0 and white is 1.

A bandpass filter is obtained from a bandreject filter in the same manner that we obtained a highpass filter from a lowpass filter:

$$H_{\mathrm{BP}}(u, v) = 1 - H_{\mathrm{BR}}(u, v) \tag{4.10-1}$$

Figure 4.63(b) shows a Gaussian bandpass filter in image form.

4.10.2 Notch Filters

Notch filters are the most useful of the selective filters. A notch filter rejects (or passes) frequencies in a predefined neighborhood about the center of the frequency rectangle. Zero-phase-shift filters must be symmetric about the origin, so a notch with center at (u_0, v_0) must have a corresponding notch at location $(-u_0, -v_0)$. Notch reject filters are constructed as products of highpass filters whose centers have been translated to the centers of the notches. The general form is:

$$H_{\mathrm{NR}}(u, v) = \prod_{k=1}^{Q} H_k(u, v) H_{-k}(u, v) \tag{4.10-2}$$

where $H_k(u, v)$ and $H_{-k}(u, v)$ are highpass filters whose centers are at (u_k, v_k) and $(-u_k, -v_k)$, respectively. These centers are specified with respect to the

TABLE 4.6
Bandreject filters. W is the width of the band, D is the distance $D(u, v)$ from the center of the filter, D_0 is the cutoff frequency, and n is the order of the Butterworth filter. We show D instead of $D(u, v)$ to simplify the notation in the table.

Ideal	Butterworth	Gaussian
$H(u, v) = \begin{cases} 0 & \text{if } D_0 - \dfrac{W}{2} \le D \le D_0 + \dfrac{W}{2} \\ 1 & \text{otherwise} \end{cases}$	$H(u, v) = \dfrac{1}{1 + \left[\dfrac{DW}{D^2 - D_0^2}\right]^{2n}}$	$H(u, v) = 1 - e^{-\left[\frac{D^2 - D_0^2}{DW}\right]^2}$

a b

FIGURE 4.63
(a) Bandreject
Gaussian filter.
(b) Corresponding
bandpass filter.
The thin black
border in (a) was
added for clarity; it
is not part of the
data.

center of the frequency rectangle, $(M/2, N/2)$. The distance computations for each filter are thus carried out using the expressions

$$D_k(u, v) = \left[(u - M/2 - u_k)^2 + (v - N/2 - v_k)^2\right]^{1/2} \qquad (4.10\text{-}3)$$

and

$$D_{-k}(u, v) = \left[(u - M/2 + u_k)^2 + (v - N/2 + v_k)^2\right]^{1/2} \qquad (4.10\text{-}4)$$

For example, the following is a Butterworth notch reject filter of order n, containing three notch pairs:

$$H_{\text{NR}}(u, v) = \prod_{k=1}^{3}\left[\frac{1}{1 + [D_{0k}/D_k(u, v)]^{2n}}\right]\left[\frac{1}{1 + [D_{0k}/D_{-k}(u, v)]^{2n}}\right] \qquad (4.10\text{-}5)$$

where D_k and D_{-k} are given by Eqs. (4.10-3) and (4.10-4). The constant D_{0k} is the same for each pair of notches, but it can be different for different pairs. Other notch reject filters are constructed in the same manner, depending on the highpass filter chosen. As with the filters discussed earlier, a *notch pass filter* is obtained from a notch reject filter using the expression

$$H_{\text{NP}}(u, v) = 1 - H_{\text{NR}}(u, v) \qquad (4.10\text{-}6)$$

As the next three examples show, one of the principal applications of notch filtering is for selectively modifying local regions of the DFT. This type of processing typically is done interactively, working directly on DFTs obtained without padding. The advantages of working interactively with actual DFTs (as opposed to having to "translate" from padded to actual frequency values) outweigh any wraparound errors that may result from not using padding in the filtering process. Also, as we show in Section 5.4.4, even more powerful notch filtering techniques than those discussed here are based on unpadded DFTs. To get an idea of how DFT values change as a function of padding, see Problem 4.22.

EXAMPLE 4.23:
Reduction of
moiré patterns
using notch
filtering.

■ Figure 4.64(a) is the scanned newspaper image from Fig. 4.21, showing a prominent moiré pattern, and Fig. 4.64(b) is its spectrum. We know from Table 4.3 that the Fourier transform of a pure sine, which is a periodic function, is a pair of conjugate symmetric impulses. The symmetric "impulse-like" bursts in Fig. 4.64(b) are a result of the near periodicity of the moiré pattern. We can attenuate these bursts by using notch filtering.

a b
c d

FIGURE 4.64
(a) Sampled
newspaper image
showing a
moiré pattern.
(b) Spectrum.
(c) Butterworth
notch reject filter
multiplied by the
Fourier
transform.
(d) Filtered
image.

Figure 4.64(c) shows the result of multiplying the DFT of Fig. 4.64(a) by a Butterworth notch reject filter with $D_0 = 3$ and $n = 4$ for all notch pairs. The value of the radius was selected (by visual inspection of the spectrum) to encompass the energy bursts completely, and the value of n was selected to give notches with mildly sharp transitions. The locations of the center of the notches were determined interactively from the spectrum. Figure 4.64(d) shows the result obtained with this filter using the procedure outlined in Section 4.7.3. The improvement is significant, considering the low resolution and degradation of the original image.   ■

■ Figure 4.65(a) shows an image of part of the rings surrounding the planet Saturn. This image was captured by *Cassini*, the first spacecraft to enter the planet's orbit. The vertical sinusoidal pattern was caused by an AC signal superimposed on the camera video signal just prior to digitizing the image. This was an unexpected problem that corrupted some images from the mission. Fortunately, this type of interference is fairly easy to correct by postprocessing. One approach is to use notch filtering.

Figure 4.65(b) shows the DFT spectrum. Careful analysis of the vertical axis reveals a series of small bursts of energy which correspond to the nearly sinusoidal

EXAMPLE 4.24:
Enhancement of corrupted *Cassini* Saturn image by notch filtering.

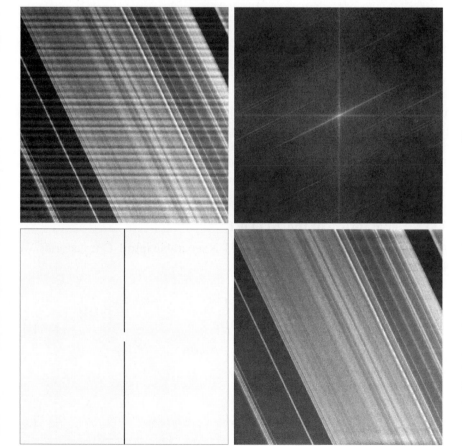

a b
c d

FIGURE 4.65
(a) 674 × 674 image of the Saturn rings showing nearly periodic interference. (b) Spectrum: The bursts of energy in the vertical axis near the origin correspond to the interference pattern. (c) A vertical notch reject filter. (d) Result of filtering. The thin black border in (c) was added for clarity; it is not part of the data. (Original image courtesy of Dr. Robert A. West, NASA/JPL.)

FIGURE 4.66
(a) Result
(spectrum) of
applying a notch
pass filter to
the DFT of
Fig. 4.65(a).
(b) Spatial
pattern obtained
by computing the
IDFT of (a).

interference. A simple approach is to use a narrow notch rectangle filter starting with the lowest frequency burst and extending for the remaining of the vertical axis. Figure 4.65(c) shows such a filter (white represents 1 and black 0). Figure 4.65(d) shows the result of filtering the corrupted image with this filter. This result is a significant improvement over the original image.

We isolated the frequencies in the vertical axis using a notch *pass* version of the same filter [Fig. 4.66(a)]. Then, as Fig. 4.66(b) shows, the IDFT of these frequencies yielded the spatial interference pattern itself. ■

4.11 Implementation

We have focused attention thus far on theoretical concepts and on examples of filtering in the frequency domain. One thing that should be clear by now is that computational requirements in this area of image processing are not trivial. Thus, it is important to develop a basic understanding of methods by which Fourier transform computations can be simplified and speeded up. This section deals with these issues.

4.11.1 Separability of the 2-D DFT

As mentioned in Table 4.2, the 2-D DFT is separable into 1-D transforms. We can write Eq. (4.5-15) as

$$F(u, v) = \sum_{x=0}^{M-1} e^{-j2\pi ux/M} \sum_{y=0}^{N-1} f(x, y) e^{-j2\pi vy/N}$$

$$= \sum_{x=0}^{M-1} F(x, v) e^{-j2\pi ux/M} \tag{4.11-1}$$

where

$$F(x, v) = \sum_{y=0}^{N-1} f(x, y) e^{-j2\pi vy/N} \tag{4.11-2}$$

For each value of x and for $v = 0, 1, 2, \ldots, N - 1$, we see that $F(x, v)$ is simply the 1-D DFT of a row of $f(x, y)$. By varying x from 0 to $M - 1$ in Eq. (4.11-2), we compute a set of 1-D DFTs for all rows of $f(x, y)$. The computations in Eq. (4.11-1) similarly are 1-D transforms of the columns of $F(x, v)$.

Thus, we conclude that the 2-D DFT of $f(x, y)$ can be obtained by computing the 1-D transform of each row of $f(x, y)$ and then computing the 1-D transform along each column of the result. This is an important simplification because we have to deal only with one variable at a time. A similar development applies to computing the 2-D IDFT using the 1-D IDFT. However, as we show in the following section, we can compute the IDFT using an algorithm designed to compute the DFT.

We could have expressed Eq. (4.11-1) and (4.11-2) in the form of 1-D column transforms followed by row transforms. The final result would have been the same.

4.11.2 Computing the IDFT Using a DFT Algorithm

Taking the complex conjugate of both sides of Eq. (4.5-16) and multiplying the results by MN yields

$$MNf^{*}(x, y) = \sum_{u=0}^{M-1}\sum_{v=0}^{N-1} F^{*}(u, v)\, e^{-j2\pi(ux/M + vy/N)} \qquad (4.11\text{-}3)$$

Multiplication by MN in this development assumes the forms in Eqs. (4.5-15) and (4.5-16). A different constant multiplication scheme is required if the constants are distributed differently between the forward and inverse transforms.

But, we recognize the form of the right side of this result as the DFT of $F^{*}(u, v)$. Therefore, Eq. (4.11-3) indicates that if we substitute $F^{*}(u, v)$ into an algorithm designed to compute the 2-D forward Fourier transform, the result will be $MNf^{*}(x, y)$. Taking the complex conjugate and multiplying this result by MN yields $f(x, y)$, which is the inverse of $F(u, v)$.

Computing the 2-D inverse from a 2-D forward DFT algorithm that is based on successive passes of 1-D transforms (as in the previous section) is a frequent source of confusion involving the complex conjugates and multiplication by a constant, neither of which is done in the 1-D algorithms. The key concept to keep in mind is that we simply input $F^{*}(u, v)$ into whatever forward algorithm we have. The result will be $MNf^{*}(x, y)$. All we have to do with this result to obtain $f(x, y)$ is to take its complex conjugate and multiply it by the constant MN. Of course, when $f(x, y)$ is real, as typically is the case, $f^{*}(x, y) = f(x, y)$.

4.11.3 The Fast Fourier Transform (FFT)

Work in the frequency domain would not be practical if we had to implement Eqs. (4.5-15) and (4.5-16) directly. Brute-force implementation of these equations requires on the order of $(MN)^2$ summations and additions. For images of moderate size (say, 1024×1024 pixels), this means on the order of a trillion multiplications and additions for just one DFT, excluding the exponentials, which could be computed once and stored in a look-up table. This would be a challenge even for super computers. Without the discovery of the *fast Fourier transform* (FFT), which reduces computations to the order of $MN \log_2 MN$ multiplications and additions, it is safe to say that the material presented in this chapter would be of little practical value. The computational reductions afforded by the FFT are impressive indeed. For example, computing the 2-D FFT of a 1024×1024 image would require on the order of 20 million multiplication and additions, which is a significant reduction from the one trillion computations mentioned above.

Although the FFT is a topic covered extensively in the literature on signal processing, this subject matter is of such significance in our work that this chapter would be incomplete if we did not provide at least an introduction explaining why the FFT works as it does. The algorithm we selected to accomplish this objective is the so-called *successive-doubling method*, which was the original algorithm that led to the birth of an entire industry. This particular algorithm assumes that the number of samples is an integer power of 2, but this is not a general requirement of other approaches (Brigham [1988]). We know from Section 4.11.1 that 2-D DFTs can be implemented by successive passes of the 1-D transform, so we need to focus only on the FFT of one variable.

When dealing with derivations of the FFT, it is customary to express Eq. (4.4-6) in the form

$$F(u) = \sum_{x=0}^{M-1} f(x)W_M^{ux} \tag{4.11-4}$$

$u = 0, 1, \ldots, M - 1$, where

$$W_M = e^{-j2\pi/M} \tag{4.11-5}$$

and M is assumed to be of the form

$$M = 2^n \tag{4.11-6}$$

with n being a positive integer. Hence, M can be expressed as

$$M = 2K \tag{4.11-7}$$

with K being a positive integer also. Substituting Eq. (4.11-7) into Eq. (4.11-4) yields

$$\begin{aligned} F(u) &= \sum_{x=0}^{2K-1} f(x)W_{2K}^{ux} \\ &= \sum_{x=0}^{K-1} f(2x)W_{2K}^{u(2x)} + \sum_{x=0}^{K-1} f(2x + 1)W_{2K}^{u(2x+1)} \end{aligned} \tag{4.11-8}$$

However, it can be shown using Eq. (4.11-5) that $W_{2K}^{2ux} = W_K^{ux}$, so Eq. (4.11-8) can be expressed as

$$F(u) = \sum_{x=0}^{K-1} f(2x)W_K^{ux} + \sum_{x=0}^{K-1} f(2x + 1)W_K^{ux}W_{2K}^{u} \tag{4.11-9}$$

Defining

$$F_{\text{even}}(u) = \sum_{x=0}^{K-1} f(2x)W_K^{ux} \tag{4.11-10}$$

for $u = 0, 1, 2, \ldots, K - 1$, and

$$F_{\text{odd}}(u) = \sum_{x=0}^{K-1} f(2x + 1)W_K^{ux} \qquad (4.11\text{-}11)$$

for $u = 0, 1, 2, \ldots, K - 1$, reduces Eq. (4.11-9) to

$$F(u) = F_{\text{even}}(u) + F_{\text{odd}}(u)W_{2K}^u \qquad (4.11\text{-}12)$$

Also, because $W_M^{u+M} = W_M^u$ and $W_{2M}^{u+M} = -W_{2M}^u$, Eqs. (4.11-10) through (4.11-12) give

$$F(u + K) = F_{\text{even}}(u) - F_{\text{odd}}(u)W_{2K}^u \qquad (4.11\text{-}13)$$

Analysis of Eqs. (4.11-10) through (4.11-13) reveals some interesting properties of these expressions. An M-point transform can be computed by dividing the original expression into two parts, as indicated in Eqs. (4.11-12) and (4.11-13). Computing the first half of $F(u)$ requires evaluation of the two $(M/2)$-point transforms given in Eqs. (4.11-10) and (4.11-11). The resulting values of $F_{\text{even}}(u)$ and $F_{\text{odd}}(u)$ are then substituted into Eq. (4.11-12) to obtain $F(u)$ for $u = 0, 1, 2, \ldots, (M/2 - 1)$. The other half then follows directly from Eq. (4.11-13) *without* additional transform evaluations.

In order to examine the computational implications of this procedure, let $m(n)$ and $a(n)$ represent the number of complex multiplications and additions, respectively, required to implement it. As before, the number of samples is 2^n with n a positive integer. Suppose first that $n = 1$. A two-point transform requires the evaluation of $F(0)$; then $F(1)$ follows from Eq. (4.11-13). To obtain $F(0)$ requires computing $F_{\text{even}}(0)$ and $F_{\text{odd}}(0)$. In this case $K = 1$ and Eqs. (4.11-10) and (4.11-11) are one-point transforms. However, because the DFT of a single sample point is the sample itself, no multiplications or additions are required to obtain $F_{\text{even}}(0)$ and $F_{\text{odd}}(0)$. One multiplication of $F_{\text{odd}}(0)$ by W_2^0 and one addition yield $F(0)$ from Eq. (4.11-12). Then $F(1)$ follows from (4.11-13) with one more addition (subtraction is considered to be the same as addition). Because $F_{\text{odd}}(0)W_2^0$ has already been computed, the total number of operations required for a two-point transform consists of $m(1) = 1$ multiplication and $a(1) = 2$ additions.

The next allowed value for n is 2. According to the above development, a four-point transform can be divided into two parts. The first half of $F(u)$ requires evaluation of two, two-point transforms, as given in Eqs. (4.11-10) and (4.11-11) for $K = 2$. As noted in the preceding paragraph, a two-point transform requires $m(1)$ multiplications and $a(1)$ additions, so evaluation of these two equations requires a total of $2m(1)$ multiplications and $2a(1)$ additions. Two further multiplications and additions are necessary to obtain $F(0)$ and $F(1)$ from Eq. (4.11-12). Because $F_{\text{odd}}(u)W_{2K}^u$ already has been computed for $u = \{0, 1\}$, two more additions give $F(2)$ and $F(3)$. The total is then $m(2) = 2m(1) + 2$ and $a(2) = 2a(1) + 4$.

When n is equal to 3, two four-point transforms are considered in the evaluation of $F_{\text{even}}(u)$ and $F_{\text{odd}}(u)$. They require $2m(2)$ multiplications and $2a(2)$ additions. Four more multiplications and eight more additions yield the complete transform. The total then is $m(3) = 2m(2) + 4$ and $a(3) = 2a(2) + 8$.

Continuing this argument for any positive integer value of n leads to recursive expressions for the number of multiplications and additions required to implement the FFT:

$$m(n) = 2m(n-1) + 2^{n-1} \quad n \geq 1 \tag{4.11-14}$$

and

$$a(n) = 2a(n-1) + 2^n \quad n \geq 1 \tag{4.11-15}$$

where $m(0) = 0$ and $a(0) = 0$ because the transform of a single point does not require any additions or multiplications.

Implementation of Eqs. (4.11-10) through (4.11-13) constitutes the successive doubling FFT algorithm. This name comes from the method of computing a two-point transform from two one-point transforms, a four-point transform from two two-point transforms, and so on, for any M equal to an integer power of 2. It is left as an exercise (Problem 4.41) to show that

$$m(n) = \frac{1}{2}M \log_2 M \tag{4.11-16}$$

and

$$a(n) = M \log_2 M \tag{4.11-17}$$

The computational advantage of the FFT over a direct implementation of the 1-D DFT is defined as

$$c(M) = \frac{M^2}{M \log_2 M}$$

$$= \frac{M}{\log_2 M} \tag{4.11-18}$$

Because it is assumed that $M = 2^n$, we can write Eq. (4.11-18) in terms of n:

$$c(n) = \frac{2^n}{n} \tag{4.11-19}$$

Figure 4.67 shows a plot of this function. It is evident that the computational advantage increases rapidly as a function of n. For instance, when $n = 15$ (32,768 points), the FFT has nearly a 2,200 to 1 advantage over the DFT. Thus, we would expect that the FFT can be computed nearly 2,200 times faster than the DFT on the same machine.

There are so many excellent sources that cover details of the FFT that we will not dwell on this topic further (see, for example, Brigham [1988]). Virtually all comprehensive signal and image processing software packages have generalized implementations of the FFT that handle cases in which the number of points is not an integer power of 2 (at the expense of less efficient computation). Free FFT programs also are readily available, principally over the Internet.

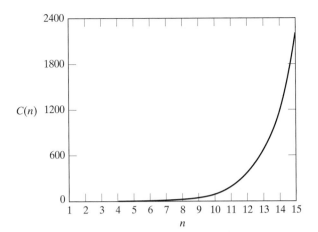

FIGURE 4.67
Computational
advantage of the
FFT over a direct
implementation
of the 1-D DFT.
Note that the
advantage
increases rapidly
as a function of n.

4.11.4 Some Comments on Filter Design

The approach to filtering discussed in this chapter is based strictly on fundamentals, the focus being specifically to explain the effects of filtering in the frequency domain as clearly as possible. We know of no better way to do that than to treat filtering the way we did here. One can view this development as the basis for "prototyping" a filter. In other words, given a problem for which we want to find a filter, the frequency domain approach is an ideal tool for experimenting, quickly and with full control over filter parameters.

Once a filter for a specific application has been found, it often is of interest to implement the filter directly in the spatial domain, using firmware and/or hardware. This topic is outside the scope of this book. Petrou and Bosdogianni [1999] present a nice tie between two-dimensional frequency domain filters and the corresponding digital filters. On the design of 2-D digital filters, see Lu and Antoniou [1992].

Summary

The material in this chapter is a progression from sampling to the Fourier transform, and then to filtering in the frequency domain. Some of the concepts, such as the sampling theorem, make very little sense if not explained in the context of the frequency domain. The same is true of effects such as aliasing. Thus, the material developed in the preceding sections is a solid foundation for understanding the fundamentals of digital signal processing. We took special care to develop the material starting with basic principles, so that any reader with a modest mathematical background would be in a position not only to absorb the material, but also to apply it.

A second major objective of this chapter was the development of the discrete Fourier transform and its use for filtering in the frequency domain. To get there, we had to introduce the convolution theorem. This result is the foundation of linear systems, and underlies many of the restoration techniques developed in Chapter 5. The types of filters we discussed are representative of what one finds in practice. The key point in presenting these filters, however, was to show how simple it is to formulate and implement filters in the frequency domain. While final implementation of a solution typically is based on spatial filters, the insight gained by working in the frequency domain as a guide in the selection of spatial filters cannot be overstated.

Although most filtering examples in this chapter are in the area of image enhancement, the procedures themselves are general and are utilized extensively in subsequent chapters.

References and Further Reading

For additional reading on the material in Section 4.1, see Hubbard [1998]. The books by Bracewell [2000, 1995] are good introductions to the continuous Fourier transform and its extensions to two dimensions for image processing. These two books, in addition to Lim [1990], Castleman [1996], Petrou and Bosdogianni [1999], Brigham [1988], and Smith [2003], provide comprehensive background for most of the discussion in Sections 4.2 through 4.6. For an overview of early work on the topic of moiré patterns, see Oster and Nishijima [1963]. Creath and Wyant [1992] discuss the state of the art in that field thirty years later. The sampling, aliasing, and image reconstruction issues discussed in Section 4.5 are also topics of significant interest in computer graphics, as exemplified by Shirley [2002].

For additional background on the material in Sections 4.7 through 4.11, see Castleman [1996], Pratt [2001], and Hall [1979]. To learn more about the imaging sensors in the Cassini spacecraft (Section 4.10.2), see Porco, West, et al. [2004]. Effective handling of issues on filter implementation (like ringing) still is a topic of interest, as exemplified by Bakir and Reeves [2000]. For unsharp masking and high-frequency-emphasis filtering, see Schowengerdt [1983]. The material on homomorphic filtering (Section 4.9.5) is based on a paper by Stockham [1972]; see also the books by Oppenheim and Schafer [1975] and Pitas and Venetsanopoulos [1990]. Brinkman et al. [1998] combine unsharp masking and homomorphic filtering for the enhancement of magnetic resonance images.

As noted in Section 4.1.1, "discovery" of the Fast Fourier transform (Section 4.11.3) was a major milestone in the popularization of the DFT as a fundamental signal processing tool. Our presentation of the FFT in Section 4.11.3 is based on a paper by Cooley and Tuckey [1965] and on the book by Brigham [1988], who also discusses several implementations of the FFT, including bases other than 2. Formulation of the fast Fourier transform is often credited to Cooley and Tukey [1965]. However, the FFT has an interesting history worth sketching here. In response to the Cooley–Tukey paper, Rudnick [1966] reported that he was using a similar technique, whose number of operations also was proportional to $N\log_2 N$ and which was based on a method published by Danielson and Lanczos [1942]. These authors, in turn, referenced Runge [1903, 1905] as the source of their technique. The latter two papers, together with the lecture notes of Runge and König [1924], contain the essential computational advantages of present FFT algorithms. Similar techniques also were published by Yates [1937], Stumpff [1939], Good [1958], and Thomas [1963]. A paper by Cooley, Lewis, and Welch [1967a] presents a historical summary and an interesting comparison of results prior to the 1965 Cooley–Tukey paper.

The FFT algorithm in Section 4.11.3 is from the original paper by Cooley and Tukey [1965]. See Brigham [1988] and Smith [2003] for complementary reading. For the design of digital filters (Section 4.11.4) based on the frequency domain formulations discussed in this chapter, see Lu and Antoniou [1992] and Petrou and Bosdogianni [1999]. For software implementation of many of the approaches discussed in Sections 4.7 through 4.11, see Gonzalez, Woods, and Eddins [2004].

Problems

Detailed solutions to the problems marked with a star can be found in the book Web site. The site also contains suggested projects based on the material in this chapter.

4.1 Repeat Example 4.1, but using the function $f(t) = A$ for $0 \leq t \leq W$ and $f(t) = 0$ for all other values of t. Explain the reason for any differences between your results and the results in the example.

★**4.2** Show that $\tilde{F}(\mu)$ in Eq. (4.4-2) is infinitely periodic in both directions, with period $1/\Delta T$.

★**4.3** It can be shown (Bracewell [2000]) that $1 \Leftrightarrow \delta(\mu)$ and $\delta(t) \Leftrightarrow 1$. Use the first of these properties and the translation property from Table 4.3 to show that the Fourier transform of the continuous function $f(t) = \sin(2\pi nt)$, where n is a real number, is $F(\mu) = (j/2)[\delta(\mu + n) - \delta(\mu - n)]$.

4.4   Consider the continuous function $f(t) = \sin(2\pi nt)$.

    ★**(a)** What is the period of $f(t)$?

    ★**(b)** What is the frequency of $f(t)$?

The Fourier transform, $F(\mu)$, of $f(t)$ is purely imaginary (Problem 4.3), and because the transform of the sampled data consists of periodic copies of $F(\mu)$, the transform of the sampled data, $\tilde{F}(\mu)$, will also be purely imaginary. Draw a diagram similar to Fig. 4.6, and answer the following questions based on your diagram (assume that sampling starts at $t = 0$).

    ★**(c)** What would the sampled function and its Fourier transform look like in general if $f(t)$ is sampled at a rate higher than the Nyquist rate?

    (d) What would the sampled function look like in general if $f(t)$ is sampled at a rate lower than the Nyquist rate?

    (e) What would the sampled function look like if $f(t)$ is sampled at the Nyquist rate with samples taken at $t = 0, \Delta T, 2\,\Delta T, \ldots$?

★**4.5**   Prove the validity of the 1-D convolution theorem of a continuous variable, as given in Eqs. (4.2-21) and (4.2-22).

4.6   Complete the steps that led from Eq. (4.3-11) to Eq. (4.3-12).

4.7   As the figure below shows, the Fourier transform of a "tent" function (on the left) is a squared sinc function (on the right). Advance an argument that shows that the Fourier transform of a tent function can be obtained from the Fourier transform of a box function. (*Hint:* The tent itself can be generated by convolving two equal boxes.)

4.8   **(a)** Show that Eqs. (4.4-4) and (4.4-5) constitute a Fourier transform pair.

    ★**(b)** Repeat (a) for Eqs. (4.4-6) and (4.4-7). You will need the following orthogonality property of exponentials in both parts of this problem:

$$\sum_{x=0}^{M-1} e^{j2\pi rx/M} e^{-j2\pi ux/M} = \begin{cases} M & \text{if } r = u \\ 0 & \text{otherwise} \end{cases}$$

4.9   Prove the validity of Eqs. (4.4-8) and (4.4-9).

★**4.10**   Prove the validity of the discrete convolution theorem of one variable [see Eqs. (4.2-21), (4.2-22), and (4.4-10)]. You will need to use the translation properties $f(x)e^{j2\pi u_0 x/M} \Leftrightarrow F(u - u_0)$ and conversely, $f(x - x_0) \Leftrightarrow F(u)e^{-j2\pi x_0 u/M}$.

★**4.11**   Write an expression for 2-D continuous convolution.

4.12   Consider a checkerboard image in which each square is 1×1 mm. Assuming that the image extends infinitely in both coordinate directions, what is the minimum sampling rate (in samples/mm) required to avoid aliasing?

4.13   We know from the discussion in Section 4.5.4 that shrinking an image can cause aliasing. Is this true also of zooming? Explain.

★**4.14**   Prove that both the 2-D continuous and discrete Fourier transforms are linear operations (see Section 2.6.2 for a definition of linearity).

4.15   You are given a "canned" program that computes the 2-D, DFT pair. However, it is not known in which of the two equations the $1/MN$ term is included or if it was split as two constants $1/\sqrt{MN}$ in front of both the forward and inverse transforms. How can you find where the term(s) is (are) included if this information is not available in the documentation?

4.16 ★**(a)** Prove the validity of the translation property in Eq. (4.6-3).

(b) Prove the validity of Eq. (4.6-4).

4.17 You can infer from Problem 4.3 that $1 \Leftrightarrow \delta(\mu, \nu)$ and $\delta(t, z) \Leftrightarrow 1$. Use the first of these properties and the translation property in Table 4.3 to show that the Fourier transform of the continuous function $f(t, z) = A\sin(2\pi\mu_0 t + 2\pi\nu_0 z)$ is

$$F(\mu, \nu) = \frac{j}{2}\left[\delta(\mu + \mu_0, \nu + \nu_0) - \delta(\mu - \mu_0, \nu - \nu_0)\right]$$

4.18 Show that the DFT of the discrete function $f(x, y) = 1$ is

$$\Im\{1\} = \delta(u, v) = \begin{cases} 1 & \text{if } u = v = 0 \\ 0 & \text{otherwise} \end{cases}$$

4.19 Show that the DFT of the discrete function $f(x, y) = \sin(2\pi u_0 x + 2\pi v_0 y)$ is

$$F(u, v) = \frac{j}{2}\left[\delta(u + Mu_0, v + Nv_0) - \delta(u - Mu_0, v - Nv_0)\right]$$

4.20 The following problems are related to the properties in Table 4.1.

★**(a)** Prove the validity of property 2.

★**(b)** Prove the validity of property 4.

(c) Prove the validity of property 5.

★**(d)** Prove the validity of property 7.

(e) Prove the validity of property 9.

(f) Prove the validity of property 10.

★**(g)** Prove the validity of property 11.

(h) Prove the validity of property 12.

(i) Prove the validity of property 13.

★**4.21** The need for image padding when filtering in the frequency domain was discussed in Section 4.6.6. We showed in that section that images needed to be padded by appending zeros to the ends of rows and columns in the image (see the following image on the left). Do you think it would make a difference if we

centered the image and surrounded it by a border of zeros instead (see image on the right), but without changing the total number of zeros used? Explain.

★**4.22** The two Fourier spectra shown are of the same image. The spectrum on the left corresponds to the original image, and the spectrum on the right was obtained after the image was padded with zeros. Explain the significant increase in signal strength along the vertical and horizontal axes of the spectrum shown on the right.

4.23 You know from Table 4.2 that the dc term, $F(0,0)$, of a DFT is proportional to the average value of its corresponding spatial image. Assume that the image is of size $M \times N$. Suppose that you pad the image with zeros to size $P \times Q$, where P and Q are given in Eqs. (4.6-31) and (4.6-32). Let $F_p(0,0)$ denote the dc term of the DFT of the padded function.

   ★**(a)** What is the ratio of the average values of the original and padded images?

   (b) Is $F_p(0,0) = F(0,0)$? Support your answer mathematically.

4.24 Prove the periodicity properties (entry 8) in Table 4.2.

4.25 The following problems are related to the entries in Table 4.3.

   ★**(a)** Prove the validity of the discrete convolution theorem (entry 6) for the 1-D case.

   (b) Repeat (a) for 2-D.

   ★**(c)** Prove the validity of entry 7.

   ★**(d)** Prove the validity of entry 12.

   (*Note:* Problems 4.18, 4.19, and 4.31 are related to Table 4.3 also.)

4.26 **(a)** Show that the Laplacian of a continuous function $f(t, z)$ of continuous variables t and z satisfies the following Fourier transform pair [see Eq. (3.6-3) for a definition of the Laplacian]:

$$\nabla^2 f(t, z) \Longleftrightarrow -4\pi^2(\mu^2 + \nu^2)F(\mu, \nu)$$

[*Hint:* Study entry 12 in Table 4.3 and see Problem 4.25(d).]

   ★**(b)** The preceding closed form expression is valid only for continuous variables. However, it can be the basis for implementing the Laplacian in the discrete frequency domain using the $M \times N$ filter

$$H(u, v) = -4\pi^2(u^2 + v^2)$$

for $u = 0, 1, 2, \ldots, M - 1$ and $v = 0, 1, 2, \ldots, N - 1$. Explain how you would implement this filter.

(c) As you saw in Example 4.20, the Laplacian result in the frequency domain was similar to the result of using a spatial mask with a center coefficient equal to -8. Explain the reason why the frequency domain result was not similar instead to the result of using a spatial mask with a center coefficient of -4. See Section 3.6.2 regarding the Laplacian in the spatial domain.

★**4.27** Consider a 3×3 spatial mask that averages the four closest neighbors of a point (x, y), but excludes the point itself from the average.

(a) Find the equivalent filter, $H(u, v)$, in the frequency domain.

(b) Show that your result is a lowpass filter.

4.28 Based on Eq. (3.6-1), one approach for approximating a discrete derivative in 2-D is based on computing differences of the form $f(x + 1, y) - f(x, y)$ and $f(x, y + 1) - f(x, y)$.

(a) Find the equivalent filter, $H(u, v)$, in the frequency domain.

(b) Show that your result is a highpass filter.

4.29 Find the equivalent filter, $H(u, v)$, that implements in the frequency domain the spatial operation performed by the Laplacian mask in Fig. 3.37(a).

★**4.30** Can you think of a way to use the Fourier transform to compute (or partially compute) the magnitude of the gradient [Eq. (3.6-11)] for use in image differentiation? If your answer is yes, give a method to do it. If your answer is no, explain why.

★**4.31** A continuous Gaussian lowpass filter in the continuous frequency domain has the transfer function

$$H(\mu, v) = Ae^{-(\mu^2 + v^2)/2\sigma^2}$$

Show that the corresponding filter in the spatial domain is

$$h(t, z) = A2\pi\sigma^2 e^{-2\pi^2\sigma^2(t^2 + z^2)}$$

4.32 As explained in Eq. (4.9-1), it is possible to obtain the transfer function, H_{HP}, of a highpass filter from the transfer function of a lowpass filter as

$$H_{HP} = 1 - H_{LP}$$

Using the information given in Problem 4.31, what is the form of the *spatial domain* Gaussian highpass filter?

4.33 Consider the images shown. The image on the right was obtained by: (a) multiplying the image on the left by $(-1)^{x+y}$; (b) computing the DFT; (c) taking the complex conjugate of the transform; (d) computing the inverse DFT; and (e) multiplying the real part of the result by $(-1)^{x+y}$. Explain (mathematically) why the image on the right appears as it does.

4.34 What is the source of the nearly periodic bright points in the horizontal axis of Fig. 4.41(b)?

★**4.35** Each filter in Fig. 4.53 has a strong spike in its center. Explain the source of these spikes.

4.36 Consider the images shown. The image on the right was obtained by lowpass filtering the image on the left with a Gaussian lowpass filter and then highpass filtering the result with a Gaussian highpass filter. The dimension of the images is 420×344, and $D_0 = 25$ was used for both filters.

   (a) Explain why the center part of the finger ring in the figure on the right appears so bright and solid, considering that the dominant characteristic of the filtered image consists of edges on the outer boundary of objects (e.g., fingers, wrist bones) with a darker area in between. In other words, would you not expect the highpass filter to render the constant area inside the ring dark, since a highpass filter eliminates the dc term?

   (b) Do you think the result would have been different if the order of the filtering process had been reversed?

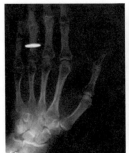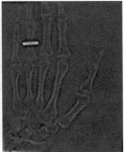

(Original image courtesy of Dr. Thomas R. Gest, Division of Anatomical Sciences, University of Michigan Medical School.)

4.37 Given an image of size $M \times N$, you are asked to perform an experiment that consists of repeatedly lowpass filtering the image using a Gaussian lowpass filter with a given cutoff frequency D_0. You may ignore computational round-off errors. Let c_{min} denote the smallest positive number representable in the machine in which the proposed experiment will be conducted.

   ★**(a)** Let K denote the number of applications of the filter. Can you predict (without doing the experiment) what the result (image) will be for a sufficiently large value of K? If so, what is that result?

   (b) Derive an expression for the *minimum* value of K that will guarantee the result that you predicted.

4.38 Consider the sequence of images shown. The image on the left is a segment of an X-ray image of a commercial printed circuit board. The images following it are, respectively, the results of subjecting the image to 1, 10, and 100 passes of a Gaussian highpass filter with $D_0 = 30$. The images are of size 330×334 pixels, with each pixel being represented by 8 bits of gray. The images were scaled for display, but this has no effect on the problem statement.

(a) It appears from the images that changes will cease to take place after some finite number of passes. Show whether or not this in fact is the case. You may ignore computational round-off errors. Let c_{min} denote the smallest positive number representable in the machine in which the proposed experiment will be conducted.

(b) If you determined in (a) that changes would cease after a finite number of iterations, determine the minimum value of that number.

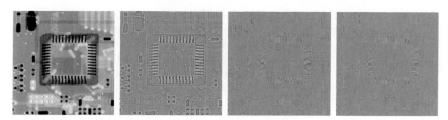

Original image courtesy of Mr. Joseph E. Pascente, Lixi, Inc.

4.39   As illustrated in Fig. 4.59, combining high-frequency emphasis and histogram equalization is an effective method for achieving edge sharpening and contrast enhancement.

(a) Show whether or not it matters which process is applied first.

(b) If the order does matter, give a rationale for using one or the other method first.

4.40   Use a Butterworth highpass filter to construct a homomorphic filter that has the same general shape as the filter in Fig. 4.61.

★**4.41**   Show the validity of Eqs. (4.11-16) and (4.11-17). (*Hint:* Use proof by induction.)

4.42   Suppose that you are given a set of images generated by an experiment dealing with the analysis of stellar events. Each image contains a set of bright, widely scattered dots corresponding to stars in a sparsely occupied section of the universe. The problem is that the stars are barely visible, due to superimposed illumination resulting from atmospheric dispersion. If these images are modeled as the product of a constant illumination component with a set of impulses, give an enhancement procedure based on homomorphic filtering designed to bring out the image components due to the stars themselves.

4.43   A skilled medical technician is assigned the job of inspecting a certain class of images generated by an electron microscope. In order to simplify the inspection task, the technician decides to use digital image enhancement and, to this end, examines a set of representative images and finds the following problems: (1) bright, isolated dots that are of no interest; (2) lack of sharpness; (3) not enough contrast in some images; and (4) shifts in the average intensity, when this value should be V to perform correctly certain intensity measurements. The technician wants to correct these problems and then display in white all intensities in a band between I_1 and I_2, while keeping normal tonality in the remaining intensities. Propose a sequence of processing steps that the technician can follow to achieve the desired goal. You may use techniques from both Chapters 3 and 4.

5 Image Restoration and Reconstruction

> Things which we see are not by themselves what we see. . . .
> It remains completely unknown to us what the objects may be by
> themselves and apart from the receptivity of our senses. We know
> nothing but our manner of perceiving them.
>
> *Immanuel Kant*

Preview

As in image enhancement, the principal goal of restoration techniques is to improve an image in some predefined sense. Although there are areas of overlap, image enhancement is largely a subjective process, while image restoration is for the most part an objective process. Restoration attempts to recover an image that has been degraded by using a priori knowledge of the degradation phenomenon. Thus, restoration techniques are oriented toward modeling the degradation and applying the inverse process in order to recover the original image.

This approach usually involves formulating a criterion of goodness that will yield an optimal estimate of the desired result. By contrast, enhancement techniques basically are heuristic procedures designed to manipulate an image in order to take advantage of the psychophysical aspects of the human visual system. For example, contrast stretching is considered an enhancement technique because it is based primarily on the pleasing aspects it might present to the viewer, whereas removal of image blur by applying a deblurring function is considered a restoration technique.

The material developed in this chapter is strictly introductory. We consider the restoration problem only from the point where a degraded, *digital* image is given; thus we consider topics dealing with sensor, digitizer, and display degradations only superficially. These subjects, although of importance in the overall treatment of image restoration applications, are beyond the scope of the present discussion.

As discussed in Chapters 3 and 4, some restoration techniques are best formulated in the spatial domain, while others are better suited for the frequency domain. For example, spatial processing is applicable when the only degradation is additive noise. On the other hand, degradations such as image blur are difficult to approach in the spatial domain using small filter masks. In this case, frequency domain filters based on various criteria of optimality are the approaches of choice. These filters also take into account the presence of noise. As in Chapter 4, a restoration filter that solves a given application in the frequency domain often is used as the basis for generating a digital filter that will be more suitable for routine operation using a hardware/firmware implementation.

Section 5.1 introduces a linear model of the image degradation/restoration process. Section 5.2 deals with various noise models encountered frequently in practice. In Section 5.3, we develop several spatial filtering techniques for reducing the noise content of an image, a process often referred to as *image denoising*. Section 5.4 is devoted to techniques for noise reduction using frequency-domain techniques. Section 5.5 introduces linear, position-invariant models of image degradation, and Section 5.6 deals with methods for estimating degradation functions. Sections 5.7 through 5.10 include the development of fundamental image-restoration approaches. We conclude the chapter (Section 5.11) with an introduction to image reconstruction from projections. The principal application of this concept is computed tomography (CT), one of the most important commercial applications of image processing, especially in health care.

5.1 A Model of the Image Degradation/Restoration Process

As Fig. 5.1 shows, the degradation process is modeled in this chapter as a degradation function that, together with an additive noise term, operates on an input image $f(x, y)$ to produce a degraded image $g(x, y)$. Given $g(x, y)$, some knowledge about the degradation function H, and some knowledge about the additive noise term $\eta(x, y)$, the objective of restoration is to obtain an estimate $\hat{f}(x, y)$ of the original image. We want the estimate to be as close as possible to the original input image and, in general, the more we know about H and η, the closer $\hat{f}(x, y)$ will be to $f(x, y)$. The restoration approach used throughout most of this chapter is based on various types of image restoration filters.

FIGURE 5.1
A model of the image degradation/ restoration process.

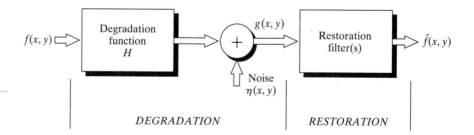

It is shown in Section 5.5 that if H is a linear, position-invariant process, then the degraded image is given in the *spatial domain* by

$$g(x, y) = h(x, y) \star f(x, y) + \eta(x, y) \qquad (5.1\text{-}1)$$

where $h(x, y)$ is the spatial representation of the degradation function and, as in Chapter 4, the symbol "$\star$" indicates convolution. We know from the discussion in Section 4.6.6 that convolution in the spatial domain is analogous to multiplication in the frequency domain, so we may write the model in Eq. (5.1-1) in an equivalent *frequency domain* representation:

$$G(u, v) = H(u, v)F(u, v) + N(u, v) \qquad (5.1\text{-}2)$$

where the terms in capital letters are the Fourier transforms of the corresponding terms in Eq. (5.1-1). These two equations are the bases for most of the restoration material in this chapter.

In the following three sections, we assume that H is the identity operator, and we deal only with degradations due to noise. Beginning in Section 5.6 we consider a number of important image degradations functions and look at several methods for image restoration in the presence of both H and η.

5.2 Noise Models

The principal sources of noise in digital images arise during image acquisition and/or transmission. The performance of imaging sensors is affected by a variety of factors, such as environmental conditions during image acquisition, and by the quality of the sensing elements themselves. For instance, in acquiring images with a CCD camera, light levels and sensor temperature are major factors affecting the amount of noise in the resulting image. Images are corrupted during transmission principally due to interference in the channel used for transmission. For example, an image transmitted using a wireless network might be corrupted as a result of lightning or other atmospheric disturbance.

5.2.1 Spatial and Frequency Properties of Noise

Relevant to our discussion are parameters that define the spatial characteristics of noise, and whether the noise is correlated with the image. Frequency properties refer to the frequency content of noise in the Fourier sense (i.e., as opposed to frequencies of the electromagnetic spectrum). For example, when the Fourier spectrum of noise is constant, the noise usually is called *white noise*. This terminology is a carryover from the physical properties of white light, which contains nearly all frequencies in the visible spectrum in equal proportions. From the discussion in Chapter 4, it is not difficult to show that the Fourier spectrum of a function containing all frequencies in equal proportions is a constant.

With the exception of spatially periodic noise (Section 5.2.3), we assume in this chapter that noise is independent of spatial coordinates, and that it is uncorrelated with respect to the image itself (that is, there is no correlation between pixel values and the values of noise components). Although these assumptions are at least partially invalid in some applications (quantum-limited imaging, such as in X-ray and nuclear-medicine imaging, is a good example), the complexities of dealing with spatially dependent and correlated noise are beyond the scope of our discussion.

5.2.2 Some Important Noise Probability Density Functions

Consult the book Web site for a brief review of probability theory.

Based on the assumptions in the previous section, the *spatial* noise descriptor with which we shall be concerned is the statistical behavior of the intensity values in the noise component of the model in Fig. 5.1. These may be considered random variables, characterized by a probability density function (PDF). The following are among the most common PDFs found in image processing applications.

Gaussian noise

Because of its mathematical tractability in both the spatial and frequency domains, Gaussian (also called *normal*) noise models are used frequently in practice. In fact, this tractability is so convenient that it often results in Gaussian models being used in situations in which they are marginally applicable at best.

The PDF of a Gaussian random variable, z, is given by

$$p(z) = \frac{1}{\sqrt{2\pi}\sigma} e^{-(z-\bar{z})^2/2\sigma^2} \qquad (5.2\text{-}1)$$

where z represents intensity, $\bar{z}$ is the mean[†] (average) value of z, and σ is its standard deviation. The standard deviation squared, σ^2, is called the *variance* of z. A plot of this function is shown in Fig. 5.2(a). When z is described by Eq. (5.2-1), approximately 70% of its values will be in the range $[(\bar{z} - \sigma), (\bar{z} + \sigma)]$, and about 95% will be in the range $[(\bar{z} - 2\sigma), (\bar{z} + 2\sigma)]$.

Rayleigh noise

The PDF of Rayleigh noise is given by

$$p(z) = \begin{cases} \dfrac{2}{b}(z - a)e^{-(z-a)^2/b} & \text{for } z \geq a \\ 0 & \text{for } z < a \end{cases} \qquad (5.2\text{-}2)$$

The mean and variance of this density are given by

$$\bar{z} = a + \sqrt{\pi b/4} \qquad (5.2\text{-}3)$$

[†]We use $\bar{z}$ instead of m to denote the mean in this section to avoid confusion when we use m and n later to denote neighborhood size.

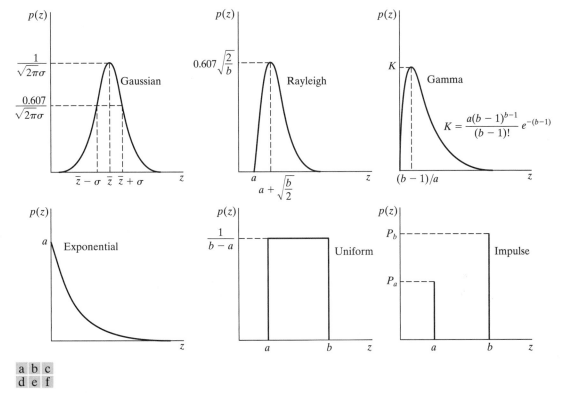

a	b	c
d	e	f

FIGURE 5.2 Some important probability density functions.

and

$$\sigma^2 = \frac{b(4 - \pi)}{4} \tag{5.2-4}$$

Figure 5.2(b) shows a plot of the Rayleigh density. Note the displacement from the origin and the fact that the basic shape of this density is skewed to the right. The Rayleigh density can be quite useful for approximating skewed histograms.

Erlang (gamma) noise

The PDF of Erlang noise is given by

$$p(z) = \begin{cases} \dfrac{a^b z^{b-1}}{(b-1)!} e^{-az} & \text{for } z \geq 0 \\ 0 & \text{for } z < 0 \end{cases} \tag{5.2-5}$$

where the parameters are such that $a > 0$, b is a positive integer, and "!" indicates factorial. The mean and variance of this density are given by

$$\bar{z} = \frac{b}{a} \tag{5.2-6}$$

and

$$\sigma^2 = \frac{b}{a^2} \tag{5.2-7}$$

Figure 5.2(c) shows a plot of this density. Although Eq. (5.2-5) often is referred to as the *gamma density*, strictly speaking this is correct only when the denominator is the gamma function, $\Gamma(b)$. When the denominator is as shown, the density is more appropriately called the *Erlang density*.

Exponential noise

The PDF of *exponential* noise is given by

$$p(z) = \begin{cases} ae^{-az} & \text{for } z \geq 0 \\ 0 & \text{for } z < 0 \end{cases} \tag{5.2-8}$$

where $a > 0$. The mean and variance of this density function are

$$\bar{z} = \frac{1}{a} \tag{5.2-9}$$

and

$$\sigma^2 = \frac{1}{a^2} \tag{5.2-10}$$

Note that this PDF is a special case of the Erlang PDF, with $b = 1$. Figure 5.2(d) shows a plot of this density function.

Uniform noise

The PDF of *uniform* noise is given by

$$p(z) = \begin{cases} \dfrac{1}{b - a} & \text{if } a \leq z \leq b \\ 0 & \text{otherwise} \end{cases} \tag{5.2-11}$$

The mean of this density function is given by

$$\bar{z} = \frac{a + b}{2} \tag{5.2-12}$$

and its variance by

$$\sigma^2 = \frac{(b - a)^2}{12} \tag{5.2-13}$$

Figure 5.2(e) shows a plot of the uniform density.

Impulse (salt-and-pepper) noise

The PDF of (*bipolar*) *impulse* noise is given by

$$p(z) = \begin{cases} P_a & \text{for } z = a \\ P_b & \text{for } z = b \\ 0 & \text{otherwise} \end{cases} \tag{5.2-14}$$

If $b > a$, intensity b will appear as a light dot in the image. Conversely, level a will appear like a dark dot. If either P_a or P_b is zero, the impulse noise is called *unipolar*. If neither probability is zero, and especially if they are approximately equal, impulse noise values will resemble salt-and-pepper granules randomly distributed over the image. For this reason, bipolar impulse noise also is called *salt-and-pepper* noise. *Data-drop-out* and *spike* noise also are terms used to refer to this type of noise. We use the terms *impulse* or *salt-and-pepper* noise interchangeably.

Noise impulses can be negative or positive. Scaling usually is part of the image digitizing process. Because impulse corruption usually is large compared with the strength of the image signal, impulse noise generally is digitized as extreme (pure black or white) values in an image. Thus, the assumption usually is that a and b are "saturated" values, in the sense that they are equal to the minimum and maximum allowed values in the digitized image. As a result, negative impulses appear as black (pepper) points in an image. For the same reason, positive impulses appear as white (salt) noise. For an 8-bit image this means typically that $a = 0$ (black) and $b = 255$ (white). Figure 5.2(f) shows the PDF of impulse noise.

As a group, the preceding PDFs provide useful tools for modeling a broad range of noise corruption situations found in practice. For example, Gaussian noise arises in an image due to factors such as electronic circuit noise and sensor noise due to poor illumination and/or high temperature. The Rayleigh density is helpful in characterizing noise phenomena in range imaging. The exponential and gamma densities find application in laser imaging. Impulse noise is found in situations where quick transients, such as faulty switching, take place during imaging, as mentioned in the previous paragraph. The uniform density is perhaps the least descriptive of practical situations. However, the uniform density is quite useful as the basis for numerous random number generators that are used in simulations (Peebles [1993] and Gonzalez, Woods, and Eddins [2004]).

■ Figure 5.3 shows a test pattern well suited for illustrating the noise models just discussed. This is a suitable pattern to use because it is composed of simple, constant areas that span the gray scale from black to near white in only three increments. This facilitates visual analysis of the characteristics of the various noise components added to the image.

Figure 5.4 shows the test pattern after addition of the six types of noise discussed thus far in this section. Shown below each image is the histogram computed directly from that image. The parameters of the noise were chosen in each case so that the histogram corresponding to the three intensity levels in the test pattern would start to merge. This made the noise quite visible, without obscuring the basic structure of the underlying image.

EXAMPLE 5.1:
Noisy images and their histograms.

FIGURE 5.3 Test pattern used to illustrate the characteristics of the noise PDFs shown in Fig. 5.2.

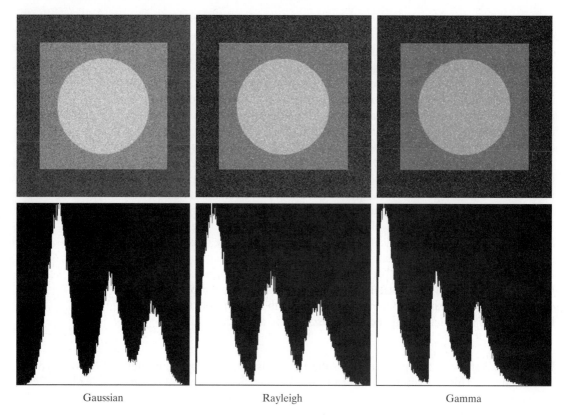

Gaussian   Rayleigh   Gamma

a	b	c
d	e	f

FIGURE 5.4 Images and histograms resulting from adding Gaussian, Rayleigh, and gamma noise to the image in Fig. 5.3.

We see a close correspondence in comparing the histograms in Fig. 5.4 with the PDFs in Fig. 5.2. The histogram for the salt-and-pepper example has an extra peak at the white end of the intensity scale because the noise components were pure black and white, and the lightest component of the test pattern (the circle) is light gray. With the exception of slightly different overall intensity, it is difficult to differentiate visually between the first five images in Fig. 5.4, even though their histograms are significantly different. The salt-and-pepper appearance of the image corrupted by impulse noise is the only one that is visually indicative of the type of noise causing the degradation. ■

5.2.3 Periodic Noise

Periodic noise in an image arises typically from electrical or electromechanical interference during image acquisition. This is the only type of spatially dependent noise that will be considered in this chapter. As discussed in Section 5.4, periodic noise can be reduced significantly via frequency domain filtering. For example, consider the image in Fig. 5.5(a). This image is severely corrupted by (spatial) sinusoidal noise of various frequencies. The Fourier transform of a pure

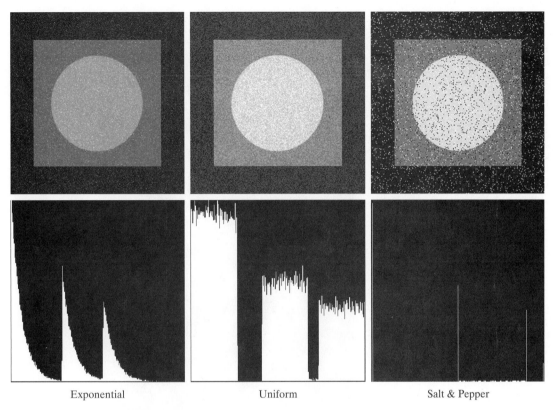

| Exponential | Uniform | Salt & Pepper |

g h i
j k l

FIGURE 5.4 *(Continued)* Images and histograms resulting from adding exponential, uniform, and salt-and-pepper noise to the image in Fig. 5.3.

sinusoid is a pair of conjugate impulses[†] located at the conjugate frequencies of the sine wave (Table 4.3). Thus, if the amplitude of a sine wave in the spatial domain is strong enough, we would expect to see in the spectrum of the image a pair of impulses for each sine wave in the image. As shown in Fig. 5.5(b), this is indeed the case, with the impulses appearing in an approximate circle because the frequency values in this particular case are so arranged. We will have much more to say in Section 5.4 about this and other examples of periodic noise.

5.2.4 Estimation of Noise Parameters

The parameters of periodic noise typically are estimated by inspection of the Fourier spectrum of the image. As noted in the previous section, periodic noise tends to produce frequency spikes that often can be detected even by visual analysis. Another approach is to attempt to infer the periodicity of noise components directly from the image, but this is possible only in simplistic cases.

[†]Be careful not to confuse the term *impulse* in the frequency domain with the use of the same term in impulse noise.

FIGURE 5.5
(a) Image
corrupted by
sinusoidal noise.
(b) Spectrum
(each pair of
conjugate
impulses
corresponds to
one sine wave).
(Original image
courtesy of
NASA.)

Automated analysis is possible in situations in which the noise spikes are either exceptionally pronounced, or when knowledge is available about the general location of the frequency components of the interference.

The parameters of noise PDFs may be known partially from sensor specifications, but it is often necessary to estimate them for a particular imaging arrangement. If the imaging system is available, one simple way to study the characteristics of system noise is to capture a set of images of "flat" environments. For example, in the case of an optical sensor, this is as simple as imaging a solid gray board that is illuminated uniformly. The resulting images typically are good indicators of system noise.

When only images already generated by a sensor are available, frequently it is possible to estimate the parameters of the PDF from small patches of reasonably constant background intensity. For example, the vertical strips (of 150×20 pixels) shown in Fig. 5.6 were cropped from the Gaussian, Rayleigh, and uniform images in Fig. 5.4. The histograms shown were calculated using image data from these small strips. The histograms in Fig. 5.4 that correspond to the histograms in Fig. 5.6 are the ones in the middle of the group of three in

Figs. 5.4(d), (e), and (k). We see that the shapes of these histograms correspond quite closely to the shapes of the histograms in Fig. 5.6. Their heights are different due to scaling, but the shapes are unmistakably similar.

The simplest use of the data from the image strips is for calculating the mean and variance of intensity levels. Consider a strip (subimage) denoted by S, and let $p_S(z_i)$, $i = 0, 1, 2, \ldots, L - 1$, denote the probability estimates (normalized histogram values) of the intensities of the pixels in S, where L is the number of possible intensities in the entire image (e.g., 256 for an 8-bit image). As in Chapter 3, we estimate the mean and variance of the pixels in S as follows:

$$\bar{z} = \sum_{i=0}^{L-1} z_i p_S(z_i) \qquad (5.2\text{-}15)$$

and

$$\sigma^2 = \sum_{i=0}^{L-1} (z_i - \bar{z})^2 p_S(z_i) \qquad (5.2\text{-}16)$$

The shape of the histogram identifies the closest PDF match. If the shape is approximately Gaussian, then the mean and variance are all we need because the Gaussian PDF is completely specified by these two parameters. For the other shapes discussed in Section 5.2.2, we use the mean and variance to solve for the parameters a and b. Impulse noise is handled differently because the estimate needed is of the actual probability of occurrence of white and black pixels. Obtaining this estimate requires that both black and white pixels be visible, so a midgray, relatively constant area is needed in the image in order to be able to compute a histogram. The heights of the peaks corresponding to black and white pixels are the estimates of P_a and P_b in Eq. (5.2-14).

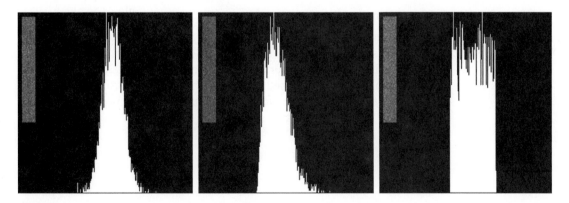

a b c

FIGURE 5.6 Histograms computed using small strips (shown as inserts) from (a) the Gaussian, (b) the Rayleigh, and (c) the uniform noisy images in Fig. 5.4.

Restoration in the Presence of Noise Only—Spatial Filtering

When the only degradation present in an image is noise, Eqs. (5.1-1) and (5.1-2) become

$$g(x, y) = f(x, y) + \eta(x, y) \tag{5.3-1}$$

and

$$G(u, v) = F(u, v) + N(u, v) \tag{5.3-2}$$

The noise terms are unknown, so subtracting them from $g(x, y)$ or $G(u, v)$ is not a realistic option. In the case of periodic noise, it usually is possible to estimate $N(u, v)$ from the spectrum of $G(u, v)$, as noted in Section 5.2.3. In this case $N(u, v)$ can be subtracted from $G(u, v)$ to obtain an estimate of the original image. In general, however, this type of knowledge is the exception, rather than the rule.

Spatial filtering is the method of choice in situations when only additive random noise is present. Spatial filtering is discussed in detail in Chapter 3. With the exception of the nature of the computation performed by a specific filter, the mechanics for implementing all the filters that follow are exactly as discussed in Sections 3.4 through 3.6.

5.3.1 Mean Filters

In this section we discuss briefly the noise-reduction capabilities of the spatial filters introduced in Section 3.5 and develop several other filters whose performance is in many cases superior to the filters discussed in that section.

Arithmetic mean filter

This is the simplest of the mean filters. Let S_{xy} represent the set of coordinates in a rectangular subimage window (neighborhood) of size $m \times n$, centered at point (x, y). The arithmetic mean filter computes the average value of the corrupted image $g(x, y)$ in the area defined by S_{xy}. The value of the restored image $\hat{f}$ at point (x, y) is simply the arithmetic mean computed using the pixels in the region defined by S_{xy}. In other words,

We assume that m and n are odd integers.

$$\hat{f}(x, y) = \frac{1}{mn} \sum_{(s, t) \in S_{xy}} g(s, t) \tag{5.3-3}$$

This operation can be implemented using a spatial filter of size $m \times n$ in which all coefficients have value $1/mn$. A mean filter smooths local variations in an image, and noise is reduced as a result of blurring.

Geometric mean filter

An image restored using a *geometric mean* filter is given by the expression

$$\hat{f}(x, y) = \left[\prod_{(s,t) \in S_{xy}} g(s, t) \right]^{\frac{1}{mn}} \tag{5.3-4}$$

Here, each restored pixel is given by the product of the pixels in the subimage window, raised to the power $1/mn$. As shown in Example 5.2, a geometric mean filter achieves smoothing comparable to the arithmetic mean filter, but it tends to lose less image detail in the process.

Harmonic mean filter

The *harmonic mean* filtering operation is given by the expression

$$\hat{f}(x, y) = \frac{mn}{\displaystyle\sum_{(s,t) \in S_{xy}} \frac{1}{g(s, t)}} \tag{5.3-5}$$

The harmonic mean filter works well for salt noise, but fails for pepper noise. It does well also with other types of noise like Gaussian noise.

Contraharmonic mean filter

The *contraharmonic* mean filter yields a restored image based on the expression

$$\hat{f}(x, y) = \frac{\displaystyle\sum_{(s,t) \in S_{xy}} g(s, t)^{Q+1}}{\displaystyle\sum_{(s,t) \in S_{xy}} g(s, t)^{Q}} \tag{5.3-6}$$

where Q is called the *order* of the filter. This filter is well suited for reducing or virtually eliminating the effects of salt-and-pepper noise. For positive values of Q, the filter eliminates pepper noise. For negative values of Q it eliminates salt noise. It cannot do both simultaneously. Note that the contraharmonic filter reduces to the arithmetic mean filter if $Q = 0$, and to the harmonic mean filter if $Q = -1$.

■ Figure 5.7(a) shows an 8-bit X-ray image of a circuit board, and Fig. 5.7(b) shows the same image, but corrupted with additive Gaussian noise of zero mean and variance of 400. For this type of image this is a significant level of noise. Figures 5.7(c) and (d) show, respectively, the result of filtering the noisy

EXAMPLE 5.2:
Illustration of mean filters.

a b
c d

FIGURE 5.7
(a) X-ray image.
(b) Image
corrupted by
additive Gaussian
noise. (c) Result
of filtering with
an arithmetic
mean filter of size
3×3. (d) Result
of filtering with a
geometric mean
filter of the same
size.
(Original image
courtesy of Mr.
Joseph E.
Pascente, Lixi,
Inc.)

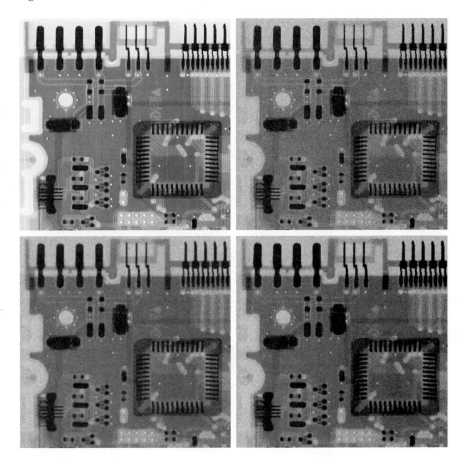

image with an arithmetic mean filter of size 3×3 and a geometric mean filter of the same size. Although both filters did a reasonable job of attenuating the contribution due to noise, the geometric mean filter did not blur the image as much as the arithmetic filter. For instance, the connector fingers at the top of the image are sharper in Fig. 5.7(d) than in (c). The same is true in other parts of the image.

Figure 5.8(a) shows the same circuit image, but corrupted now by pepper noise with probability of 0.1. Similarly, Fig. 5.8(b) shows the image corrupted by salt noise with the same probability. Figure 5.8(c) shows the result of filtering Fig. 5.8(a) using a contraharmonic mean filter with $Q = 1.5$, and Fig. 5.8(d) shows the result of filtering Fig. 5.8(b) with $Q = -1.5$. Both filters did a good job in reducing the effect of the noise. The positive-order filter did a better job of cleaning the background, at the expense of slightly thinning and blurring the dark areas. The opposite was true of the negative-order filter.

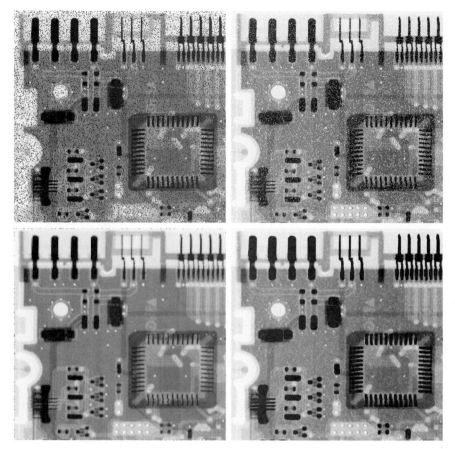

a b
c d

FIGURE 5.8
(a) Image corrupted by pepper noise with a probability of 0.1. (b) Image corrupted by salt noise with the same probability. (c) Result of filtering (a) with a 3×3 contra-harmonic filter of order 1.5. (d) Result of filtering (b) with $Q = -1.5$.

In general, the arithmetic and geometric mean filters (particularly the latter) are well suited for random noise like Gaussian or uniform noise. The contraharmonic filter is well suited for impulse noise, but it has the disadvantage that it must be known whether the noise is dark or light in order to select the proper sign for Q. The results of choosing the wrong sign for Q can be disastrous, as Fig. 5.9 shows. Some of the filters discussed in the following sections eliminate this shortcoming.   ■

5.3.2 Order-Statistic Filters

Order-statistic filters were introduced in Section 3.5.2. We now expand the discussion in that section and introduce some additional order-statistic filters. As noted in Section 3.5.2, order-statistic filters are spatial filters whose response is based on ordering (ranking) the values of the pixels contained in the image area encompassed by the filter. The ranking result determines the response of the filter.

a b

FIGURE 5.9
Results of selecting the wrong sign in contraharmonic filtering.
(a) Result of filtering
Fig. 5.8(a) with a contraharmonic filter of size 3×3 and $Q = -1.5$.
(b) Result of filtering 5.8(b) with $Q = 1.5$.

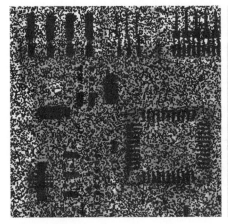
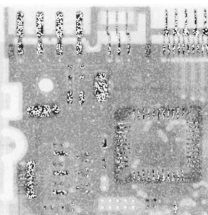

Median filter

The best-known order-statistic filter is the *median filter*, which, as its name implies, replaces the value of a pixel by the median of the intensity levels in the neighborhood of that pixel:

$$\hat{f}(x, y) = \underset{(s,t)\in S_{xy}}{\text{median}}\{g(s, t)\} \tag{5.3-7}$$

The value of the pixel at (x, y) is included in the computation of the median. Median filters are quite popular because, for certain types of random noise, they provide excellent noise-reduction capabilities, with considerably less blurring than linear smoothing filters of similar size. Median filters are particularly effective in the presence of both bipolar and unipolar impulse noise. In fact, as Example 5.3 below shows, the median filter yields excellent results for images corrupted by this type of noise. Computation of the median and implementation of this filter are discussed in Section 3.5.2.

Max and min filters

See the second margin note in Section 10.3.5 regarding percentiles.

Although the median filter is by far the order-statistic filter most used in image processing, it is by no means the only one. The median represents the 50th percentile of a ranked set of numbers, but you will recall from basic statistics that ranking lends itself to many other possibilities. For example, using the 100th percentile results in the so-called *max filter*, given by

$$\hat{f}(x, y) = \underset{(s,t)\in S_{xy}}{\text{max}} \{g(s, t)\} \tag{5.3-8}$$

This filter is useful for finding the brightest points in an image. Also, because pepper noise has very low values, it is reduced by this filter as a result of the max selection process in the subimage area S_{xy}.

The 0th percentile filter is the *min filter*:

$$\hat{f}(x, y) = \min_{(s,t)\in S_{xy}} \{g(s, t)\} \tag{5.3-9}$$

This filter is useful for finding the darkest points in an image. Also, it reduces salt noise as a result of the min operation.

Midpoint filter

The midpoint filter simply computes the midpoint between the maximum and minimum values in the area encompassed by the filter:

$$\hat{f}(x, y) = \frac{1}{2}\left[\max_{(s,t)\in S_{xy}} \{g(s, t)\} + \min_{(s,t)\in S_{xy}} \{g(s, t)\}\right] \tag{5.3-10}$$

Note that this filter combines order statistics and averaging. It works best for randomly distributed noise, like Gaussian or uniform noise.

Alpha-trimmed mean filter

Suppose that we delete the $d/2$ lowest and the $d/2$ highest intensity values of $g(s, t)$ in the neighborhood S_{xy}. Let $g_r(s, t)$ represent the remaining $mn - d$ pixels. A filter formed by averaging these remaining pixels is called an *alpha-trimmed mean* filter:

$$\hat{f}(x, y) = \frac{1}{mn - d} \sum_{(s,t)\in S_{xy}} g_r(s, t) \tag{5.3-11}$$

where the value of d can range from 0 to $mn - 1$. When $d = 0$, the alpha-trimmed filter reduces to the arithmetic mean filter discussed in the previous section. If we choose $d = mn - 1$, the filter becomes a median filter. For other values of d, the alpha-trimmed filter is useful in situations involving multiple types of noise, such as a combination of salt-and-pepper and Gaussian noise.

■ Figure 5.10(a) shows the circuit board image corrupted by salt-and-pepper noise with probabilities $P_a = P_b = 0.1$. Figure 5.10(b) shows the result of median filtering with a filter of size 3 × 3. The improvement over Fig. 5.10(a) is significant, but several noise points still are visible. A second pass [on the image in Fig. 5.10(b)] with the median filter removed most of these points, leaving only few, barely visible noise points. These were removed with a third pass of the filter. These results are good examples of the power of median filtering in handling impulse-like additive noise. Keep in mind that repeated passes of a median filter will blur the image, so it is desirable to keep the number of passes as low as possible.

Figure 5.11(a) shows the result of applying the max filter to the pepper noise image of Fig. 5.8(a). The filter did a reasonable job of removing the pepper noise, but we note that it also removed (set to a light intensity level) some dark pixels from the borders of the dark objects. Figure 5.11(b) shows the result of applying the min filter to the image in Fig. 5.8(b). In this case, the min filter did a better job than the max filter on noise removal, but it removed some white points around the border of light objects. These made the light objects smaller and

EXAMPLE 5.3:
Illustration of order-statistic filters.

a b
c d

FIGURE 5.10
(a) Image
corrupted by salt-
and-pepper noise
with probabilities
$P_a = P_b = 0.1$.
(b) Result of one
pass with a
median filter of
size 3×3.
(c) Result of
processing (b)
with this filter.
(d) Result of
processing (c)
with the same
filter.

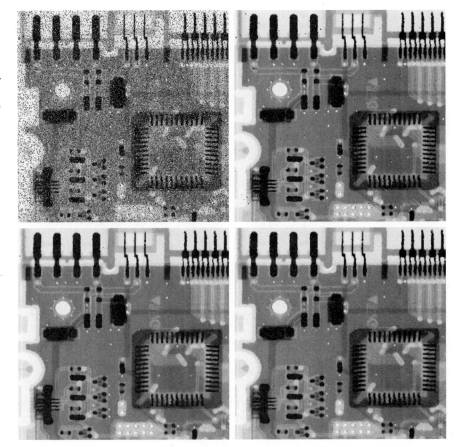

a b

FIGURE 5.11
(a) Result of
filtering
Fig. 5.8(a) with a
max filter of size
3×3. (b) Result
of filtering 5.8(b)
with a min filter
of the same size.

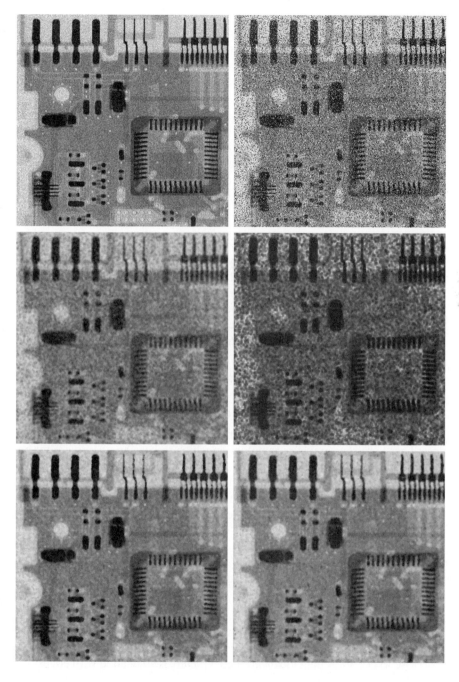

a b
c d
e f

FIGURE 5.12
(a) Image
corrupted
by additive
uniform noise.
(b) Image
additionally
corrupted by
additive salt-and-
pepper noise.
Image (b) filtered
with a 5×5:
(c) arithmetic
mean filter;
(d) geometric
mean filter;
(e) median filter;
and (f) alpha-
trimmed mean
filter with $d = 5$.

some of the dark objects larger (like the connector fingers in the top of the image) because white points around these objects were set to a dark level.

The alpha-trimmed filter is illustrated next. Figure 5.12(a) shows the circuit board image corrupted this time by additive, uniform noise of variance 800 and zero mean. This is a high level of noise corruption that is made worse by further addition of salt-and-pepper noise with $P_a = P_b = 0.1$, as Fig. 5.12(b) shows. The high level of noise in this image warrants use of larger filters. Figures 5.12(c) through (f) show the results obtained using arithmetic mean, geometric mean, median, and alpha-trimmed mean (with $d = 5$) filters of size 5×5. As expected, the arithmetic and geometric mean filters (especially the latter) did not do well because of the presence of impulse noise. The median and alpha-trimmed filters performed much better, with the alpha-trimmed filter giving slightly better noise reduction. Note, for example, that the fourth connector finger from the top left is slightly smoother in the alpha-trimmed result. This is not unexpected because, for a high value of d, the alpha-trimmed filter approaches the performance of the median filter, but still retains some smoothing capabilities.   ■

5.3.3 Adaptive Filters

Once selected, the filters discussed thus far are applied to an image without regard for how image characteristics vary from one point to another. In this section we take a look at two *adaptive* filters whose behavior changes based on statistical characteristics of the image inside the filter region defined by the $m \times n$ rectangular window S_{xy}. As the following discussion shows, adaptive filters are capable of performance superior to that of the filters discussed thus far. The price paid for improved filtering power is an increase in filter complexity. Keep in mind that we still are dealing with the case in which the degraded image is equal to the original image plus noise. No other types of degradations are being considered yet.

Adaptive, local noise reduction filter

The simplest statistical measures of a random variable are its mean and variance. These are reasonable parameters on which to base an adaptive filter because they are quantities closely related to the appearance of an image. The mean gives a measure of average intensity in the region over which the mean is computed, and the variance gives a measure of contrast in that region.

Our filter is to operate on a local region, S_{xy}. The response of the filter at any point (x, y) on which the region is centered is to be based on four quantities: (a) $g(x, y)$, the value of the noisy image at (x, y); (b) σ_η^2, the variance of the noise corrupting $f(x, y)$ to form $g(x, y)$; (c) m_L, the local mean of the pixels in S_{xy}; and (d) σ_L^2, the local variance of the pixels in S_{xy}. We want the behavior of the filter to be as follows:

1. If σ_η^2 is zero, the filter should return simply the value of $g(x, y)$. This is the trivial, zero-noise case in which $g(x, y)$ is equal to $f(x, y)$.
2. If the local variance is high relative to σ_η^2, the filter should return a value close to $g(x, y)$. A high local variance typically is associated with edges, and these should be preserved.
3. If the two variances are equal, we want the filter to return the arithmetic mean value of the pixels in S_{xy}. This condition occurs when the local area has the same properties as the overall image, and local noise is to be reduced simply by averaging.

An adaptive expression for obtaining $\hat{f}(x, y)$ based on these assumptions may be written as

$$\hat{f}(x, y) = g(x, y) - \frac{\sigma_\eta^2}{\sigma_L^2}\left[g(x, y) - m_L\right] \tag{5.3-12}$$

The only quantity that needs to be known or estimated is the variance of the overall noise, σ_η^2. The other parameters are computed from the pixels in S_{xy} at each location (x, y) on which the filter window is centered. A tacit assumption in Eq. (5.3-12) is that $\sigma_\eta^2 \leq \sigma_L^2$. The noise in our model is additive and position independent, so this is a reasonable assumption to make because S_{xy} is a subset of $g(x, y)$. However, we seldom have exact knowledge of σ_η^2. Therefore, it is possible for this condition to be violated in practice. For that reason, a test should be built into an implementation of Eq. (5.3-12) so that the ratio is set to 1 if the condition $\sigma_\eta^2 > \sigma_L^2$ occurs. This makes this filter nonlinear. However, it prevents nonsensical results (i.e., negative intensity levels, depending on the value of m_L) due to a potential lack of knowledge about the variance of the image noise. Another approach is to allow the negative values to occur, and then rescale the intensity values at the end. The result then would be a loss of dynamic range in the image.

■ Figure 5.13(a) shows the circuit-board image, corrupted this time by additive Gaussian noise of zero mean and a variance of 1000. This is a significant level of noise corruption, but it makes an ideal test bed on which to compare relative filter performance. Figure 5.13(b) is the result of processing the noisy image with an arithmetic mean filter of size 7×7. The noise was smoothed out, but at the cost of significant blurring in the image. Similar comments are applicable to Fig. 5.13(c), which shows the result of processing the noisy image with a geometric mean filter, also of size 7×7. The differences between these two filtered images are analogous to those we discussed in Example 5.2; only the degree of blurring is different.

EXAMPLE 5.4:
Illustration of adaptive, local noise-reduction filtering.

Figure 5.13(d) shows the result of using the adaptive filter of Eq. (5.3-12) with $\sigma_\eta^2 = 1000$. The improvements in this result compared with the two previous filters are significant. In terms of overall noise reduction, the adaptive filter achieved results similar to the arithmetic and geometric mean filters. However, the image filtered with the adaptive filter is much sharper. For example, the connector fingers at the top of the image are significantly sharper in Fig. 5.13(d). Other features, such as holes and the eight legs of the dark component on the lower left-hand side of the image, are much clearer in Fig. 5.13(d). These results are typical of what can be achieved with an adaptive filter. As mentioned earlier, the price paid for the improved performance is additional filter complexity.

The preceding results used a value for σ_η^2 that matched the variance of the noise exactly. If this quantity is not known and an estimate is used that is too low, the algorithm will return an image that closely resembles the original because the corrections will be smaller than they should be. Estimates that are too high

a b
c d

FIGURE 5.13
(a) Image
corrupted by
additive Gaussian
noise of zero
mean and
variance 1000.
(b) Result of
arithmetic mean
filtering.
(c) Result of
geometric mean
filtering.
(d) Result of
adaptive noise
reduction
filtering. All filters
were of size
7×7.

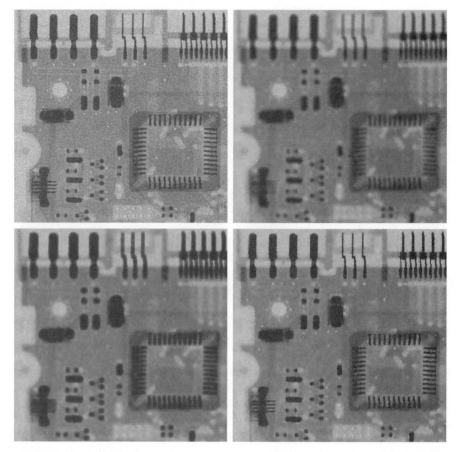

will cause the ratio of the variances to be clipped at 1.0, and the algorithm will subtract the mean from the image more frequently than it would normally. If negative values are allowed and the image is rescaled at the end, the result will be a loss of dynamic range, as mentioned previously. ■

Adaptive median filter

The median filter discussed in Section 5.3.2 performs well if the spatial density of the impulse noise is not large (as a rule of thumb, P_a and P_b less than 0.2). It is shown in this section that adaptive median filtering can handle impulse noise with probabilities larger than these. An additional benefit of the adaptive median filter is that it seeks to preserve detail while smoothing nonimpulse noise, something that the "traditional" median filter does not do. As in all the filters discussed in the preceding sections, the adaptive median filter also works in a rectangular window area S_{xy}. Unlike those filters, however, the adaptive median filter changes (increases) the size of S_{xy} during filter operation, depending on certain conditions listed in this section. Keep in mind that the output of the filter is a single value used to replace the value of the pixel at (x, y), the point on which the window S_{xy} is centered at a given time.

Consider the following notation:

$$z_{\min} = \text{minimum intensity value in } S_{xy}$$
$$z_{\max} = \text{maximum intensity value in } S_{xy}$$
$$z_{\text{med}} = \text{median of intensity values in } S_{xy}$$
$$z_{xy} = \text{intensity value at coordinates } (x, y)$$
$$S_{\max} = \text{maximum allowed size of } S_{xy}$$

The adaptive median-filtering algorithm works in two stages, denoted stage A and stage B, as follows:

Stage A: $A1 = z_{\text{med}} - z_{\min}$
$A2 = z_{\text{med}} - z_{\max}$
If $A1 > 0$ AND $A2 < 0$, go to stage B
Else increase the window size
If window size $\leq S_{\max}$ repeat stage A
Else output z_{med}

Stage B: $B1 = z_{xy} - z_{\min}$
$B2 = z_{xy} - z_{\max}$
If $B1 > 0$ AND $B2 < 0$, output z_{xy}
Else output z_{med}

The key to understanding the mechanics of this algorithm is to keep in mind that it has three main purposes: to remove salt-and-pepper (impulse) noise, to provide smoothing of other noise that may not be impulsive, and to reduce distortion, such as excessive thinning or thickening of object boundaries. The values $z_{\min}$ and $z_{\max}$ are considered statistically by the algorithm to be "impulse-like" noise components, even if these are not the lowest and highest possible pixel values in the image.

With these observations in mind, we see that the purpose of stage A is to determine if the median filter output, z_{med}, is an impulse (black *or* white) or not. If the condition $z_{\min} < z_{\text{med}} < z_{\max}$ holds, then z_{med} cannot be an impulse for the reason mentioned in the previous paragraph. In this case, we go to stage B and test to see if the point in the center of the window, z_{xy}, is itself an impulse (recall that z_{xy} is the point being processed). If the condition $B1 > 0$ AND $B2 < 0$ is true, then $z_{\min} < z_{xy} < z_{\max}$, and z_{xy} cannot be an impulse for the same reason that z_{med} was not. In this case, the algorithm outputs the unchanged pixel value, z_{xy}. By not changing these "intermediate-level" points, distortion is reduced in the image. If the condition $B1 > 0$ AND $B2 < 0$ is false, then either $z_{xy} = z_{\min}$ or $z_{xy} = z_{\max}$. In either case, the value of the pixel is an extreme value and the algorithm outputs the median value z_{med}, which we know from stage A is not a noise impulse. The last step is what the standard median filter does. The problem is that the standard median filter replaces every point in the image by the median of the corresponding neighborhood. This causes unnecessary loss of detail.

Continuing with the explanation, suppose that stage A *does* find an impulse (i.e., it fails the test that would cause it to branch to stage B). The algorithm then increases the size of the window and repeats stage A. This looping continues until the algorithm either finds a median value that is not an impulse (and branches to stage B), or the maximum window size is reached. If the maximum window size is reached, the algorithm returns the value of z_{med}. Note that there is no guarantee that this value is not an impulse. The smaller the noise probabilities P_a and/or P_b are, or the larger S_{max} is allowed to be, the less likely it is that a premature exit condition will occur. This is plausible. As the density of the impulses increases, it stands to reason that we would need a larger window to "clean up" the noise spikes.

Every time the algorithm outputs a value, the window S_{xy} is moved to the next location in the image. The algorithm then is reinitialized and applied to the pixels in the new location. As indicated in Problem 3.18, the median value can be updated iteratively using only the new pixels, thus reducing computational load.

EXAMPLE 5.5:
Illustration of adaptive median filtering.

■ Figure 5.14(a) shows the circuit-board image corrupted by salt-and-pepper noise with probabilities $P_a = P_b = 0.25$, which is 2.5 times the noise level used in Fig. 5.10(a). Here the noise level is high enough to obscure most of the detail in the image. As a basis for comparison, the image was filtered first using the smallest median filter required to remove most visible traces of impulse noise. A 7×7 median filter was required to do this, and the result is shown in Fig. 5.14(b). Although the noise was effectively removed, the filter caused significant loss of detail in the image. For instance, some of the connector fingers at the top of the image appear distorted or broken. Other image details are similarly distorted.

Figure 5.14(c) shows the result of using the adaptive median filter with $S_{\text{max}} = 7$. Noise removal performance was similar to the median filter. However, the adaptive filter did a better job of preserving sharpness and detail. The connector fingers are less distorted, and some other features that were either obscured or distorted beyond recognition by the median filter appear sharper and better defined in Fig. 5.14(c). Two notable examples are the feed-through small white holes throughout the board, and the dark component with eight legs in the bottom, left quadrant of the image.

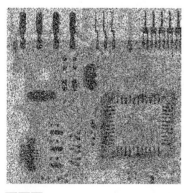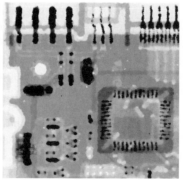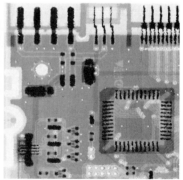

a b c

FIGURE 5.14 (a) Image corrupted by salt-and-pepper noise with probabilities $P_a = P_b = 0.25$. (b) Result of filtering with a 7×7 median filter. (c) Result of adaptive median filtering with $S_{\text{max}} = 7$.

Considering the high level of noise in Fig. 5.14(a), the adaptive algorithm performed quite well. The choice of maximum allowed window size depends on the application, but a reasonable starting value can be estimated by experimenting with various sizes of the standard median filter first. This will establish a visual baseline regarding expectations on the performance of the adaptive algorithm. ■

5.4 Periodic Noise Reduction by Frequency Domain Filtering

Periodic noise can be analyzed and filtered quite effectively using frequency domain techniques. The basic idea is that periodic noise appears as concentrated bursts of energy in the Fourier transform, at locations corresponding to the frequencies of the periodic interference. The approach is to use a selective filter (see Section 4.10) to isolate the noise. The three types of selective filters (bandreject, bandpass, and notch, introduced in Section 4.10) are used in Sections 5.4.1 through 5.4.3 for basic periodic noise reduction. We also develop an optimum notch filtering approach in Section 5.4.4.

5.4.1 Bandreject Filters

The transfer functions of ideal, Butterworth, and Gaussian bandreject filters, introduced in Section 4.10.1, are summarized in Table 4.6. Figure 5.15 shows perspective plots of these filters, and the following example illustrates using a bandreject filter for reducing the effects of periodic noise.

■ One of the principal applications of bandreject filtering is for noise removal in applications where the general location of the noise component(s) in the frequency domain is approximately known. A good example is an image corrupted by additive periodic noise that can be approximated as two-dimensional sinusoidal functions. It is not difficult to show that the Fourier transform of a sine consists of two impulses that are mirror images of each other about the origin of the transform. Their locations are given in Table 4.3. The impulses are both imaginary (the real part of the Fourier transform of a sine is zero) and are complex conjugates of each other. We will have more to say about this topic in Sections 5.4.3 and 5.4.4. Our purpose at the moment is to illustrate bandreject filtering.

EXAMPLE 5.6:
Use of bandreject filtering for periodic noise removal.

a b c

FIGURE 5.15 From left to right, perspective plots of ideal, Butterworth (of order 1), and Gaussian bandreject filters.

Figure 5.16(a), which is the same as Fig. 5.5(a), shows an image heavily corrupted by sinusoidal noise of various frequencies. The noise components are easily seen as symmetric pairs of bright dots in the Fourier spectrum shown in
Fig. 5.16(b). In this example, the components lie on an approximate circle about
the origin of the transform, so a circularly symmetric bandreject filter is a good
choice. Figure 5.16(c) shows a Butterworth bandreject filter of order 4, with the
appropriate radius and width to enclose completely the noise impulses. Since it is
desirable in general to remove as little as possible from the transform, sharp, narrow filters are common in bandreject filtering. The result of filtering Fig. 5.16(a)
with this filter is shown in Fig. 5.16(d). The improvement is quite evident. Even
small details and textures were restored effectively by this simple filtering approach. It is worth noting also that it would not be possible to get equivalent results
by a direct spatial domain filtering approach using small convolution masks.   ■

5.4.2 Bandpass Filters

A *bandpass* filter performs the opposite operation of a bandreject filter. We
showed in Section 4.10.1 how the transfer function $H_{\text{BP}}(u, v)$ of a bandpass filter is obtained from a corresponding bandreject filter with transfer function
$H_{\text{BR}}(u, v)$ by using the equation

$$H_{\text{BP}}(u, v) = 1 - H_{\text{BR}}(u, v) \qquad (5.4\text{-}1)$$

It is left as an exercise (Problem 5.12) to derive expressions for the bandpass
filters corresponding to the bandreject equations in Table 4.6.

a b
c d

FIGURE 5.16
(a) Image
corrupted by
sinusoidal noise.
(b) Spectrum of (a).
(c) Butterworth
bandreject filter
(white represents
1). (d) Result of
filtering.
(Original image
courtesy of
NASA.)

FIGURE 5.17
Noise pattern of
the image in
Fig. 5.16(a)
obtained by
bandpass filtering.

■ Performing straight bandpass filtering on an image is not a common procedure because it generally removes too much image detail. However, bandpass filtering is quite useful in isolating the effects on an image caused by selected frequency bands. This is illustrated in Fig. 5.17. This image was generated by (1) using Eq. (5.4-1) to obtain the bandpass filter corresponding to the band-reject filter used in Fig. 5.16; and (2) taking the inverse transform of the bandpass-filtered transform. Most image detail was lost, but the information that remains is most useful, as it is clear that the noise pattern recovered using this method is quite close to the noise that corrupted the image in Fig. 5.16(a). In other words, bandpass filtering helped isolate the noise pattern. This is a useful result because it simplifies analysis of the noise, reasonably independently of image content.   ■

EXAMPLE 5.7:
Bandpass filtering for extracting noise patterns.

5.4.3 Notch Filters

A *notch* filter rejects (or passes) frequencies in predefined neighborhoods about a center frequency. Equations for notch filtering are detailed in Section 4.10.2. Figure 5.18 shows 3-D plots of ideal, Butterworth, and Gaussian notch (reject) filters. Due to the symmetry of the Fourier transform, notch filters must appear in symmetric pairs about the origin in order to obtain meaningful results. The one exception to this rule is if the notch filter is located at the origin, in which case it appears by itself. Although we show only one pair for illustrative purposes, the number of pairs of notch filters that can be implemented is arbitrary. The shape of the notch areas also can be arbitrary (e.g., rectangular).

As explained in Section 4.10.2, we can obtain notch filters that *pass*, rather than suppress, the frequencies contained in the notch areas. Since these filters perform exactly the opposite function as the notch reject filters, their transfer functions are given by

$$H_{NP}(u, v) = 1 - H_{NR}(u, v) \qquad (5.4\text{-}2)$$

where $H_{NP}(u, v)$ is the transfer function of the notch pass filter corresponding to the notch reject filter with transfer function $H_{NR}(u, v)$.

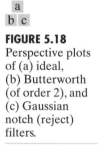

FIGURE 5.18
Perspective plots
of (a) ideal,
(b) Butterworth
(of order 2), and
(c) Gaussian
notch (reject)
filters.

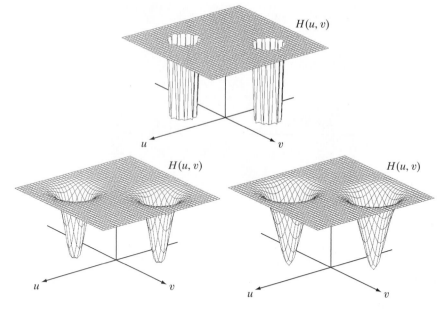

EXAMPLE 5.8:
Removal of
periodic noise by
notch filtering.

■ Figure 5.19(a) shows the same image as Fig. 4.51(a). The notch filtering approach that follows reduces the noise in this image, without introducing the appreciable blurring we saw in Section 4.8.4. Unless blurring is desirable for reasons we discussed in that section, notch filtering is preferable if a suitable filter can be found.

Just by looking at the nearly horizontal lines of the noise pattern in Fig. 5.19(a), we expect its contribution in the frequency domain to be concentrated along the vertical axis. However, the noise is not dominant enough to have a clear pattern along this axis, as is evident from the spectrum shown in Fig. 5.19(b). We can get an idea of what the noise contribution looks like by constructing a simple ideal notch pass filter along the vertical axis of the Fourier transform, as shown in Fig. 5.19(c). The spatial representation of the noise pattern (inverse transform of the notch-pass–filtered result) is shown in Fig. 5.19(d). This noise pattern corresponds closely to the pattern in Fig. 5.19(a). Having thus constructed a suitable notch pass filter that isolates the noise to a reasonable degree, we can obtain the corresponding notch reject filter from Eq. (5.4-2). The result of processing the image with the notch reject filter is shown in Fig. 5.19(e). This image contains significantly fewer visible noise scan lines than Fig. 5.19(a). ■

5.4.4 Optimum Notch Filtering

Figure 5.20(a), another example of periodic image degradation, shows a digital image of the Martian terrain taken by the *Mariner 6* spacecraft. The interference pattern is somewhat similar to the one in Fig. 5.16(a), but the former pattern is considerably more subtle and, consequently, harder to detect in the frequency plane. Figure 5.20(b) shows the Fourier spectrum of the image in

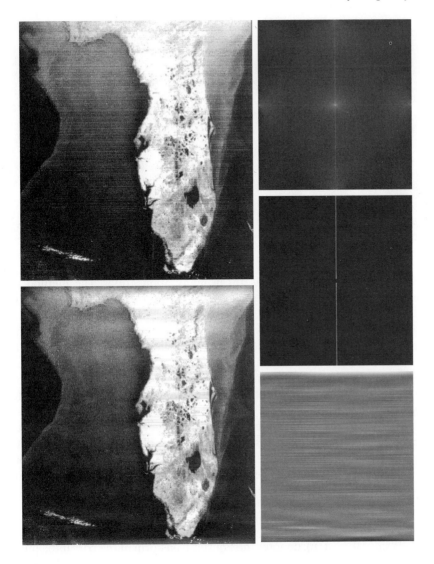

a b
c
e d

FIGURE 5.19
(a) Satellite image
of Florida and the
Gulf of Mexico
showing horizontal
scan lines.
(b) Spectrum.
(c) Notch
pass filter
superimposed on
(b). (d) Spatial
noise pattern.
(e) Result of notch
reject filtering.
(Original image
courtesy of
NOAA.)

question. The starlike components were caused by the interference, and several pairs of components are present, indicating that the pattern contains more than just one sinusoidal component.

When several interference components are present, the methods discussed in the preceding sections are not always acceptable because they may remove too much image information in the filtering process (a highly undesirable feature when images are unique and/or expensive to acquire). In addition, the interference components generally are not single-frequency bursts. Instead, they tend to have broad skirts that carry information about the interference pattern. These skirts are not always easily detectable from the normal transform background. Alternative filtering methods that reduce the effect of

FIGURE 5.20
(a) Image of the
Martian terrain
taken by *Mariner 6.*
(b) Fourier
spectrum showing
periodic
interference.
(Courtesy of
NASA.)

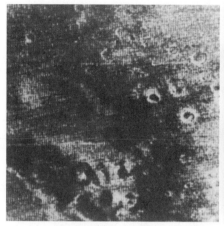

these degradations are quite useful in many applications. The method discussed here is optimum, in the sense that it minimizes local variances of the restored estimate $\hat{f}(x, y)$.

The procedure consists of first isolating the principal contributions of the interference pattern and then subtracting a variable, weighted portion of the pattern from the corrupted image. Although we develop the procedure in the context of a specific application, the basic approach is quite general and can be applied to other restoration tasks in which multiple periodic interference is a problem.

The first step is to extract the principal frequency components of the interference pattern. As before, this can be done by placing a notch pass filter, $H_{NP}(u, v)$, at the location of each spike. If the filter is constructed to pass only components associated with the interference pattern, then the Fourier transform of the interference noise pattern is given by the expression

$$N(u, v) = H_{NP}(u, v)G(u, v) \qquad (5.4\text{-}3)$$

where, as usual, $G(u, v)$, denotes the Fourier transform of the corrupted image.

Formation of $H_{NP}(u, v)$ requires considerable judgment about what is or is not an interference spike. For this reason, the notch pass filter generally is constructed interactively by observing the spectrum of $G(u, v)$ on a display. After a particular filter has been selected, the corresponding pattern in the spatial domain is obtained from the expression

$$\eta(x, y) = \Im^{-1}\{H_{NP}(u, v)G(u, v)\} \qquad (5.4\text{-}4)$$

Because the corrupted image is assumed to be formed by the addition of the uncorrupted image $f(x, y)$ and the interference, if $\eta(x, y)$ were known completely, subtracting the pattern from $g(x, y)$ to obtain $f(x, y)$ would be a simple matter. The problem, of course, is that this filtering procedure usually yields only an approximation of the true pattern. The effect of components not present in the estimate of $\eta(x, y)$ can be minimized instead by subtracting from $g(x, y)$ a *weighted* portion of $\eta(x, y)$ to obtain an estimate of $f(x, y)$:

$$\hat{f}(x, y) = g(x, y) - w(x, y)\eta(x, y) \tag{5.4-5}$$

where, as before, $\hat{f}(x, y)$ is the estimate of $f(x, y)$ and $w(x, y)$ is to be determined. The function $w(x, y)$ is called a *weighting* or *modulation* function, and the objective of the procedure is to select this function so that the result is optimized in some meaningful way. One approach is to select $w(x, y)$ so that the variance of the estimate $\hat{f}(x, y)$ is minimized over a specified neighborhood of every point (x, y).

Consider a neighborhood of size $(2a + 1)$ by $(2b + 1)$ about a point (x, y). The "local" variance of $\hat{f}(x, y)$ at coordinates (x, y) can be estimated from the samples, as follows:

$$\sigma^2(x, y) = \frac{1}{(2a + 1)(2b + 1)} \sum_{s=-a}^{a} \sum_{t=-b}^{b} \left[\hat{f}(x + s, y + t) - \bar{\hat{f}}(x, y)\right]^2 \tag{5.4-6}$$

where $\bar{\hat{f}}(x, y)$ is the average value of $\hat{f}$ in the neighborhood; that is,

$$\bar{\hat{f}}(x, y) = \frac{1}{(2a + 1)(2b + 1)} \sum_{s=-a}^{a} \sum_{t=-b}^{b} \hat{f}(x + s, y + t) \tag{5.4-7}$$

Points on or near the edge of the image can be treated by considering partial neighborhoods or by padding the border with 0s.

Substituting Eq. (5.4-5) into Eq. (5.4-6) yields

$$\sigma^2(x, y) = \frac{1}{(2a + 1)(2b + 1)} \sum_{s=-a}^{a} \sum_{t=-b}^{b} \{[g(x + s, y + t)$$

$$- w(x + s, y + t)\eta(x + s, y + t)] \tag{5.4-8}$$

$$- [\bar{g}(x, y) - \overline{w(x, y)\eta(x, y)}]\}^2$$

Assuming that $w(x, y)$ remains essentially constant over the neighborhood gives the approximation

$$w(x + s, y + t) = w(x, y) \tag{5.4-9}$$

for $-a \le s \le a$ and $-b \le t \le b$. This assumption also results in the expression

$$\overline{w(x, y)\eta(x, y)} = w(x, y)\bar{\eta}(x, y) \tag{5.4-10}$$

in the neighborhood. With these approximations, Eq. (5.4-8) becomes

$$\sigma^2(x, y) = \frac{1}{(2a + 1)(2b + 1)} \sum_{s=-a}^{a} \sum_{t=-b}^{b} \{[g(x + s, y + t)$$

$$- w(x, y)\eta(x + s, y + t)] \tag{5.4-11}$$

$$- [\bar{g}(x, y) - w(x, y)\bar{\eta}(x, y)]\}^2$$

To minimize $\sigma^2(x, y)$, we solve

$$\frac{\partial \sigma^2(x, y)}{\partial w(x, y)} = 0 \tag{5.4-12}$$

for $w(x, y)$. The result is

$$w(x, y) = \frac{\overline{g(x, y)\eta(x, y)} - \overline{g}(x, y)\overline{\eta}(x, y)}{\overline{\eta^2}(x, y) - \overline{\eta}^2(x, y)} \tag{5.4-13}$$

To obtain the restored image $\hat{f}(x, y)$, we compute $w(x, y)$ from Eq. (5.4-13) and then use Eq. (5.4-5). As $w(x, y)$ is assumed to be constant in a neighborhood, computing this function for every value of x and y in the image is unnecessary. Instead, $w(x, y)$ is computed for *one* point in each nonoverlapping neighborhood (preferably the center point) and then used to process all the image points contained in that neighborhood.

EXAMPLE 5.9:
Illustration of optimum notch filtering.

■ Figures 5.21 through 5.23 show the result of applying the preceding technique to the image in Fig. 5.20(a). This image is of size 512×512 pixels, and a neighborhood with $a = b = 15$ was selected. Figure 5.21 shows the Fourier spectrum of the corrupted image. The origin was not shifted to the center of the frequency plane in this particular case, so $u = v = 0$ is at the top left corner of the transform image in Fig. 5.21. Figure 5.22(a) shows the spectrum of $N(u, v)$, where only the noise spikes are present. Figure 5.22(b) shows the interference pattern $\eta(x, y)$ obtained by taking the inverse Fourier transform of $N(u, v)$. Note the similarity between this pattern and the structure of the noise present in Fig. 5.20(a). Finally, Fig. 5.23 shows the processed image obtained by using Eq. (5.4-5). The periodic interference was removed for all practical purposes. ■

FIGURE 5.21
Fourier spectrum (without shifting) of the image shown in Fig. 5.20(a). (Courtesy of NASA.)

FIGURE 5.22
(a) Fourier
spectrum of
$N(u, v)$, and
(b) corresponding
noise interference
pattern $\eta(x, y)$.
(Courtesy of
NASA.)

5.5 Linear, Position-Invariant Degradations

The input-output relationship in Fig. 5.1 before the restoration stage is expressed as

$$g(x, y) = H[f(x, y)] + \eta(x, y) \qquad (5.5\text{-}1)$$

For the moment, let us assume that $\eta(x, y) = 0$ so that $g(x, y) = H[f(x, y)]$. Based on the discussion in Section 2.6.2, H is *linear* if

$$H[af_1(x, y) + bf_2(x, y)] = aH[f_1(x, y)] + bH[f_2(x, y)] \qquad (5.5\text{-}2)$$

where a and b are scalars and $f_1(x, y)$ and $f_2(x, y)$ are any two input images. If $a = b = 1$, Eq. (5.5-2) becomes

Consult the book Web site
for a brief review of linear
system theory.

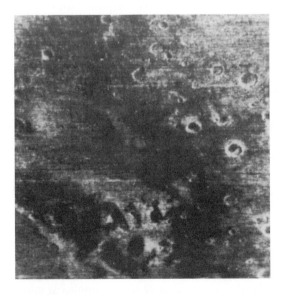

FIGURE 5.23
Processed image.
(Courtesy of
NASA.)

$$H[f_1(x, y) + f_2(x, y)] = H[f_1(x, y)] + H[f_2(x, y)] \tag{5.5-3}$$

which is called the property of *additivity*. This property simply says that, if H is a linear operator, the response to a sum of two inputs is equal to the sum of the two responses.

With $f_2(x, y) = 0$, Eq. (5.5-2) becomes

$$H[af_1(x, y)] = aH[f_1(x, y)] \tag{5.5-4}$$

which is called the property of *homogeneity*. It says that the response to a constant multiple of any input is equal to the response to that input multiplied by the same constant. Thus a linear operator possesses both the property of additivity and the property of homogeneity.

An operator having the input-output relationship $g(x, y) = H[f(x, y)]$ is said to be *position* (or *space*) *invariant* if

$$H[f(x - \alpha, y - \beta)] = g(x - \alpha, y - \beta) \tag{5.5-5}$$

for any $f(x, y)$ and any α and β. This definition indicates that the response at any point in the image depends only on the *value* of the input at that point, not on its *position*.

With a slight (but equivalent) change in notation in the definition of the impulse in Eq. (4.5-3), $f(x, y)$ can be expressed as:

See the footnote in page 369 regarding continuous and discrete variables.

$$f(x, y) = \int_{-\infty}^{\infty} \int_{-\infty}^{\infty} f(\alpha, \beta)\delta(x - \alpha, y - \beta) \, d\alpha \, d\beta \tag{5.5-6}$$

Assume again for a moment that $\eta(x, y) = 0$. Then, substitution of Eq. (5.5-6) into Eq. (5.5-1) results in the expression

$$g(x, y) = H[f(x, y)] = H\left[\int_{-\infty}^{\infty} \int_{-\infty}^{\infty} f(\alpha, \beta)\delta(x - \alpha, y - \beta) \, d\alpha \, d\beta\right] \tag{5.5-7}$$

If H is a linear operator and we extend the additivity property to integrals, then

$$g(x, y) = \int_{-\infty}^{\infty} \int_{-\infty}^{\infty} H[f(\alpha, \beta)\delta(x - \alpha, y - \beta)] \, d\alpha \, d\beta \tag{5.5-8}$$

Because $f(\alpha, \beta)$ is independent of x and y, and using the homogeneity property, it follows that

$$g(x, y) = \int_{-\infty}^{\infty} \int_{-\infty}^{\infty} f(\alpha, \beta)H[\delta(x - \alpha, y - \beta)] \, d\alpha \, d\beta \tag{5.5-9}$$

The term

$$h(x, \alpha, y, \beta) = H[\delta(x - \alpha, y - \beta)] \tag{5.5-10}$$

is called the *impulse response* of H. In other words, if $\eta(x, y) = 0$ in Eq. (5.5-1), then $h(x, \alpha, y, \beta)$ is the response of H to an impulse at coordinates (x, y). In optics, the impulse becomes a point of light and $h(x, \alpha, y, \beta)$ is commonly referred to as the *point spread function* (PSF). This name arises from the fact that all physical optical systems blur (spread) a point of light to some degree, with the amount of blurring being determined by the quality of the optical components.

Substituting Eq. (5.5-10) into Eq. (5.5-9) yields the expression

$$g(x, y) = \int_{-\infty}^{\infty} \int_{-\infty}^{\infty} f(\alpha, \beta) h(x, \alpha, y, \beta) \, d\alpha \, d\beta \qquad (5.5\text{-}11)$$

which is called the *superposition* (or *Fredholm*) *integral of the first kind*. This expression is a fundamental result that is at the core of linear system theory. It states that if the response of H to an impulse is known, the response to *any* input $f(\alpha, \beta)$ can be calculated by means of Eq. (5.5-11). In other words, a linear system H is completely characterized by its impulse response.

If H is position invariant, then, from Eq. (5.5-5),

$$H[\delta(x - \alpha, y - \beta)] = h(x - \alpha, y - \beta) \qquad (5.5\text{-}12)$$

Equation (5.5-11) reduces in this case to

$$g(x, y) = \int_{-\infty}^{\infty} \int_{-\infty}^{\infty} f(\alpha, \beta) h(x - \alpha, y - \beta) \, d\alpha \, d\beta \qquad (5.5\text{-}13)$$

This expression is the *convolution integral* introduced for one variable in Eq. (4.2-20) and extended to 2-D in Problem 4.11. This integral tells us that knowing the impulse response of a linear system allows us to compute its response, g, to *any* input f. The result is simply the convolution of the impulse response and the input function.

In the presence of additive noise, the expression of the linear degradation model [Eq. (5.5-11)] becomes

$$g(x, y) = \int_{-\infty}^{\infty} \int_{-\infty}^{\infty} f(\alpha, \beta) h(x, \alpha, y, \beta) \, d\alpha \, d\beta + \eta(x, y) \qquad (5.5\text{-}14)$$

If H is position invariant, Eq. (5.5-14) becomes

$$g(x, y) = \int_{-\infty}^{\infty} \int_{-\infty}^{\infty} f(\alpha, \beta) h(x - \alpha, y - \beta) \, d\alpha \, d\beta + \eta(x, y) \qquad (5.5\text{-}15)$$

The values of the noise term $\eta(x, y)$ are random, and are assumed to be independent of position. Using the familiar notation for convolution, we can write Eq. (5.5-15) as

$$g(x, y) = h(x, y) \star f(x, y) + \eta(x, y) \qquad (5.5\text{-}16)$$

or, based on the convolution theorem (see Section 4.6.6), we can express it in the frequency domain as

$$G(u, v) = H(u, v)F(u, v) + N(u, v) \qquad (5.5\text{-}17)$$

These two expressions agree with Eqs. (5.1-1) and (5.1-2). Keep in mind that, for discrete quantities, all products are term by term. For example, term ij of $H(u, v)F(u, v)$ is the product of term ij of $H(u, v)$ and term ij of $F(u, v)$.

In summary, the preceding discussion indicates that a linear, spatially-invariant degradation system with additive noise can be modeled in the spatial domain as the convolution of the degradation (point spread) function with an image, followed by the addition of noise. Based on the convolution theorem, the same process can be expressed in the frequency domain as the product of the transforms of the image and degradation, followed by the addition of the transform of the noise. When working in the frequency domain, we make use of an FFT algorithm, as discussed in Section 4.11. Keep in mind also the need for function padding in the implementation of discrete Fourier transforms, as outlined in Section 4.6.6.

Many types of degradations can be approximated by linear, position-invariant processes. The advantage of this approach is that the extensive tools of linear system theory then become available for the solution of image restoration problems. Nonlinear and position-dependent techniques, although more general (and usually more accurate), introduce difficulties that often have no known solution or are very difficult to solve computationally. This chapter focuses on linear, space-invariant restoration techniques. Because degradations are modeled as being the result of convolution, and restoration seeks to find filters that apply the process in reverse, the term *image deconvolution* is used frequently to signify linear image restoration. Similarly, the filters used in the restoration process often are called *deconvolution filters*.

5.6   Estimating the Degradation Function

There are three principal ways to estimate the degradation function for use in image restoration: (1) observation, (2) experimentation, and (3) mathematical modeling. These methods are discussed in the following sections. The process of restoring an image by using a degradation function that has been estimated in some way sometimes is called *blind deconvolution*, due to the fact that the true degradation function is seldom known completely.

5.6.1   Estimation by Image Observation

Suppose that we are given a degraded image without any knowledge about the degradation function H. Based on the assumption that the image was degraded by a linear, position-invariant process, one way to estimate H is to gather information from the image itself. For example, if the image is blurred, we can look at a small rectangular section of the image containing sample structures, like part of an object and the background. In order to reduce the effect of noise, we would look for an area in which the signal content is strong (e.g., an area of high contrast). The next step would be to process the subimage to arrive at a result that is as unblurred as possible. For example, we can do this by sharpening the subimage with a sharpening filter and even by processing small areas by hand.

Let the observed subimage be denoted by $g_s(x, y)$, and let the processed subimage (which in reality is our estimate of the original image in that area) be denoted by $\hat{f}_s(x, y)$. Then, assuming that the effect of noise is negligible because of our choice of a strong-signal area, it follows from Eq. (5.5-17) that

$$H_s(u, v) = \frac{G_s(u, v)}{\hat{F}_s(u, v)} \tag{5.6-1}$$

From the characteristics of this function, we then deduce the complete degradation function $H(u, v)$ based on our assumption of position invariance. For example, suppose that a radial plot of $H_s(u, v)$ has the approximate shape of a Gaussian curve. We can use that information to construct a function $H(u, v)$ on a larger scale, but having the same basic shape. We then use $H(u, v)$ in one of the restoration approaches to be discussed in the following sections. Clearly, this is a laborious process used only in very specific circumstances such as, for example, restoring an old photograph of historical value.

5.6.2 Estimation by Experimentation

If equipment similar to the equipment used to acquire the degraded image is available, it is possible in principle to obtain an accurate estimate of the degradation. Images similar to the degraded image can be acquired with various system settings until they are degraded as closely as possible to the image we wish to restore. Then the idea is to obtain the impulse response of the degradation by imaging an impulse (small dot of light) using the same system settings. As noted in Section 5.5, a linear, space-invariant system is characterized completely by its impulse response.

An impulse is simulated by a bright dot of light, as bright as possible to reduce the effect of noise to negligible values. Then, recalling that the Fourier transform of an impulse is a constant, it follows from Eq. (5.5-17) that

$$H(u, v) = \frac{G(u, v)}{A} \tag{5.6-2}$$

where, as before, $G(u, v)$ is the Fourier transform of the observed image and A is a constant describing the strength of the impulse. Figure 5.24 shows an example.

5.6.3 Estimation by Modeling

Degradation modeling has been used for many years because of the insight it affords into the image restoration problem. In some cases, the model can even take into account environmental conditions that cause degradations. For example, a degradation model proposed by Hufnagel and Stanley [1964] is based on the physical characteristics of atmospheric turbulence. This model has a familiar form:

$$H(u, v) = e^{-k(u^2+v^2)^{5/6}} \tag{5.6-3}$$

where k is a constant that depends on the nature of the turbulence. With the exception of the 5/6 power on the exponent, this equation has the same form as the Gaussian lowpass filter discussed in Section 4.8.3. In fact, the Gaussian LPF is used sometimes to model mild, uniform blurring. Figure 5.25 shows examples

a b

FIGURE 5.24
Degradation
estimation by
impulse
characterization.
(a) An impulse of
light (shown
magnified).
(b) Imaged
(degraded)
impulse.

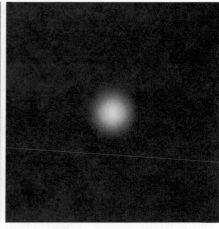

a b
c d

FIGURE 5.25
Illustration of the
atmospheric
turbulence model.
(a) Negligible
turbulence.
(b) Severe
turbulence,
$k = 0.0025$.
(c) Mild
turbulence,
$k = 0.001$.
(d) Low
turbulence,
$k = 0.00025$.
(Original image
courtesy of
NASA.)

obtained by simulating blurring an image using Eq. (5.6-3) with values $k = 0.0025$ (severe turbulence), $k = 0.001$ (mild turbulence), and $k = 0.00025$ (low turbulence). All images are of size 480×480 pixels.

Another major approach in modeling is to derive a mathematical model starting from basic principles. We illustrate this procedure by treating in some detail the case in which an image has been blurred by uniform linear motion between the image and the sensor during image acquisition. Suppose that an image $f(x, y)$ undergoes planar motion and that $x_0(t)$ and $y_0(t)$ are the time-varying components of motion in the x- and y-directions, respectively. The total exposure at any point of the recording medium (say, film or digital memory) is obtained by integrating the instantaneous exposure over the time interval during which the imaging system shutter is open.

Assuming that shutter opening and closing takes place instantaneously, and that the optical imaging process is perfect, isolates the effect of image motion. Then, if T is the duration of the exposure, it follows that

$$g(x, y) = \int_0^T f[x - x_0(t), y - y_0(t)] \, dt \qquad (5.6\text{-}4)$$

where $g(x, y)$ is the blurred image.

From Eq. (4.5-7), the Fourier transform of Eq. (5.6-4) is

$$
\begin{aligned}
G(u, v) &= \int_{-\infty}^{\infty} \int_{-\infty}^{\infty} g(x, y) e^{-j2\pi(ux+vy)} \, dx \, dy \\[6pt]
&= \int_{-\infty}^{\infty} \int_{-\infty}^{\infty} \left[\int_0^T f[x - x_0(t), y - y_0(t)] \, dt \right] e^{-j2\pi(ux+vy)} \, dx \, dy
\end{aligned}
\qquad (5.6\text{-}5)
$$

Reversing the order of integration allows Eq. (5.6-5) to be expressed in the form

$$G(u, v) = \int_0^T \left[\int_{-\infty}^{\infty} \int_{-\infty}^{\infty} f[x - x_0(t), y - y_0(t)] e^{-j2\pi(ux+vy)} \, dx \, dy \right] dt \qquad (5.6\text{-}6)$$

The term inside the outer brackets is the Fourier transform of the displaced function $f[x - x_0(t), y - y_0(t)]$. Using Eq. (4.6-4) then yields the expression

$$
\begin{aligned}
G(u, v) &= \int_0^T F(u, v) e^{-j2\pi[ux_0(t)+vy_0(t)]} \, dt \\[6pt]
&= F(u, v) \int_0^T e^{-j2\pi[ux_0(t)+vy_0(t)]} \, dt
\end{aligned}
\qquad (5.6\text{-}7)
$$

where the last step follows from the fact that $F(u, v)$ is independent of t.

By defining

$$H(u, v) = \int_0^T e^{-j2\pi[ux_0(t)+vy_0(t)]} \, dt \qquad (5.6\text{-}8)$$

Eq. (5.6-7) can be expressed in the familiar form

$$G(u, v) = H(u, v)F(u, v) \qquad (5.6\text{-}9)$$

If the motion variables $x_0(t)$ and $y_0(t)$ are known, the transfer function $H(u, v)$ can be obtained directly from Eq. (5.6-8). As an illustration, suppose that the image in question undergoes uniform linear motion in the x-direction only, at a rate given by $x_0(t) = at/T$. When $t = T$, the image has been displaced by a total distance a. With $y_0(t) = 0$, Eq. (5.6-8) yields

$$
\begin{aligned}
H(u, v) &= \int_0^T e^{-j2\pi u x_0(t)} \, dt \\
&= \int_0^T e^{-j2\pi u at/T} \, dt \\
&= \frac{T}{\pi u a} \sin(\pi u a) e^{-j\pi u a}
\end{aligned}
\qquad (5.6\text{-}10)
$$

Observe that H vanishes at values of u given by $u = n/a$, where n is an integer. If we allow the y-component to vary as well, with the motion given by $y_0 = bt/T$, then the degradation function becomes

As explained at the end of Table 4.3, we sample Eg (5.6-11) in u and v to generate a discrete filter.

$$H(u, v) = \frac{T}{\pi(ua + vb)} \sin[\pi(ua + vb)]e^{-j\pi(ua+vb)} \qquad (5.6\text{-}11)$$

EXAMPLE 5.10:
Image blurring due to motion.

■ Figure 5.26(b) is an image blurred by computing the Fourier transform of the image in Fig. 5.26(a), multiplying the transform by $H(u, v)$ from Eq. (5.6-11), and taking the inverse transform. The images are of size 688×688 pixels, and the parameters used in Eq. (5.6-11) were $a = b = 0.1$ and $T = 1$. As discussed in Sections 5.8 and 5.9, recovery of the original image from its blurred counterpart presents some interesting challenges, particularly when noise is present in the degraded image. ■

a b

FIGURE 5.26
(a) Original image.
(b) Result of blurring using the function in Eq. (5.6-11) with $a = b = 0.1$ and $T = 1$.

5.7 Inverse Filtering

The material in this section is our first step in studying restoration of images degraded by a degradation function H, which is given or obtained by a method such as those discussed in the previous section. The simplest approach to restoration is direct inverse filtering, where we compute an estimate, $\hat{F}(u, v)$, of the transform of the original image simply by dividing the transform of the degraded image, $G(u, v)$, by the degradation function:

$$\hat{F}(u, v) = \frac{G(u, v)}{H(u, v)} \qquad (5.7\text{-}1)$$

The division is an array operation, as defined in Section 2.6.1 and in connection with Eq. (5.5-17). Substituting the right side of Eq. (5.1-2) for $G(u, v)$ in Eq. (5.7-1) yields

$$\hat{F}(u, v) = F(u, v) + \frac{N(u, v)}{H(u, v)} \qquad (5.7\text{-}2)$$

This is an interesting expression. It tells us that even if we know the degradation function we cannot recover the undegraded image [the inverse Fourier transform of $F(u, v)$] exactly because $N(u, v)$ is not known. There is more bad news. If the degradation function has zero or very small values, then the ratio $N(u, v)/H(u, v)$ could easily dominate the estimate $\hat{F}(u, v)$. This, in fact, is frequently the case, as will be demonstrated shortly.

One approach to get around the zero or small-value problem is to limit the filter frequencies to values near the origin. From the discussion of Eq. (4.6-21) we know that $H(0, 0)$ is usually the highest value of $H(u, v)$ in the frequency domain. Thus, by limiting the analysis to frequencies near the origin, we reduce the probability of encountering zero values. This approach is illustrated in the following example.

■ The image in Fig. 5.25(b) was inverse filtered with Eq. (5.7-1) using the exact inverse of the degradation function that generated that image. That is, the degradation function used was

EXAMPLE 5.11:
Inverse filtering.

$$H(u, v) = e^{-k[(u-M/2)^2 + (v-N/2)^2]^{5/6}}$$

with $k = 0.0025$. The $M/2$ and $N/2$ constants are offset values; they center the function so that it will correspond with the centered Fourier transform, as discussed on numerous occasions in the previous chapter. In this case, $M = N = 480$. We know that a Gaussian-shape function has no zeros, so that will not be a concern here. However, in spite of this, the degradation values became so small that the result of full inverse filtering [Fig. 5.27(a)] is useless. The reasons for this poor result are as discussed in connection with Eq. (5.7-2).

Figures 5.27(b) through (d) show the results of cutting off values of the ratio $G(u, v)/H(u, v)$ outside a radius of 40, 70, and 85, respectively. The cut off was implemented by applying to the ratio a Butterworth lowpass function of order 10. This provided a sharp (but smooth) transition at the

FIGURE 5.27
Restoring
Fig. 5.25(b) with
Eq. (5.7-1).
(a) Result of
using the full
filter. (b) Result
with *H* cut off
outside a radius of
40; (c) outside a
radius of 70; and
(d) outside a
radius of 85.

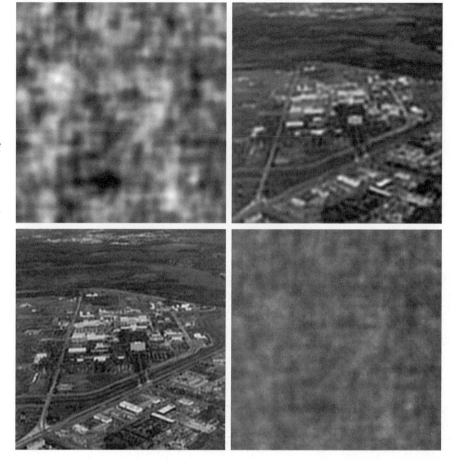

desired radius. Radii near 70 yielded the best visual results [Fig. 5.27(c)].
Radius values below that tended toward blurred images, as illustrated in
Fig. 5.27(b), which was obtained using a radius of 40. Values above 70 started
to produce degraded images, as illustrated in Fig. 5.27(d), which was obtained using a radius of 85. The image content is almost visible in this image
behind a "curtain" of noise, but the noise definitely dominates the result.
Further increases in radius values produced images that looked more and
more like Fig. 5.27(a).   ■

The results in the preceding example are illustrative of the poor performance of direct inverse filtering in general. The basic theme of the three sections that follow is how to improve on direct inverse filtering.

5.8 Minimum Mean Square Error (Wiener) Filtering

The inverse filtering approach discussed in the previous section makes no explicit provision for handling noise. In this section, we discuss an approach that
incorporates both the degradation function and statistical characteristics of

noise into the restoration process. The method is founded on considering images and noise as random variables, and the objective is to find an estimate $\hat{f}$ of the uncorrupted image f such that the mean square error between them is minimized. This error measure is given by

$$e^2 = E\{(f - \hat{f})^2\} \tag{5.8-1}$$

Note that entire images are being considered random variables, as discussed at the end of Section 2.6.8.

where $E\{\cdot\}$ is the expected value of the argument. It is assumed that the noise and the image are uncorrelated; that one or the other has zero mean; and that the intensity levels in the estimate are a linear function of the levels in the degraded image. Based on these conditions, the minimum of the error function in Eq. (5.8-1) is given in the frequency domain by the expression

$$
\begin{aligned}
\hat{F}(u, v) &= \left[\frac{H^*(u, v)S_f(u, v)}{S_f(u, v)|H(u, v)|^2 + S_\eta(u, v)} \right] G(u, v) \\[2mm]
&= \left[\frac{H^*(u, v)}{|H(u, v)|^2 + S_\eta(u, v)/S_f(u, v)} \right] G(u, v) \\[2mm]
&= \left[\frac{1}{H(u, v)} \frac{|H(u, v)|^2}{|H(u, v)|^2 + S_\eta(u, v)/S_f(u, v)} \right] G(u, v)
\end{aligned}
\tag{5.8-2}
$$

where we used the fact that the product of a complex quantity with its conjugate is equal to the magnitude of the complex quantity squared. This result is known as the *Wiener filter*, after N. Wiener [1942], who first proposed the concept in the year shown. The filter, which consists of the terms inside the brackets, also is commonly referred to as the *minimum mean square error filter* or the *least square error filter*. We include references at the end of the chapter to sources containing detailed derivations of the Wiener filter. Note from the first line in Eq. (5.8-2) that the Wiener filter does not have the same problem as the inverse filter with zeros in the degradation function, unless the entire denominator is zero for the same value(s) of u and v.

The terms in Eq. (5.8-2) are as follows:

$H(u, v) = $ degradation function

$H^*(u, v) = $ complex conjugate of $H(u, v)$

$|H(u, v)|^2 = H^*(u, v)H(u, v)$

$S_\eta(u, v) = |N(u, v)|^2 = $ power spectrum of the noise [see Eq. (4.6–18)][†]

$S_f(u, v) = |F(u, v)|^2 = $ power spectrum of the undegraded image

---

[†]The term $|N(u, v)|^2$ also is referred to as the *autocorrelation* of the noise. This terminology comes from the correlation theorem (first line of entry 7 in Table 4.3). When the two functions are the same, correlation becomes *autocorrelation* and the right side of that entry becomes $N^*(u, v)N(u, v)$, which is equal to $|N(u, v)|^2$. Similar comments apply to $|F(u, v)|^2$, which is the autocorrelation of the image. We discuss correlation in more detail in Chapter 12.

As before, $H(u, v)$ is the transform of the degradation function and $G(u, v)$ is the transform of the degraded image. The restored image in the spatial domain is given by the inverse Fourier transform of the frequency-domain estimate $\hat{F}(u, v)$. Note that if the noise is zero, then the noise power spectrum vanishes and the Wiener filter reduces to the inverse filter.

A number of useful measures are based on the power spectra of noise and of the undegraded image. One of the most important is the *signal-to-noise ratio*, approximated using frequency domain quantities such as

$$\text{SNR} = \frac{\sum_{u=0}^{M-1}\sum_{v=0}^{N-1}|F(u, v)|^2}{\sum_{u=0}^{M-1}\sum_{v=0}^{N-1}|N(u, v)|^2} \tag{5.8-3}$$

This ratio gives a measure of the level of information bearing signal power (i.e., of the original, undegraded image) to the level of noise power. Images with low noise tend to have a high SNR and, conversely, the same image with a higher level of noise has a lower SNR. This ratio by itself is of limited value, but it is an important metric used in characterizing the performance of restoration algorithms.

The *mean square error* given in statistical form in Eq. (5.8-1) can be approximated also in terms a summation involving the original and restored images:

$$\text{MSE} = \frac{1}{MN}\sum_{x=0}^{M-1}\sum_{y=0}^{N-1}[f(x, y) - \hat{f}(x, y)]^2 \tag{5.8-4}$$

In fact, if one considers the restored image to be "signal" and the difference between this image and the original to be noise, we can define a signal-to-noise ratio in the spatial domain as

$$\text{SNR} = \frac{\sum_{x=0}^{M-1}\sum_{y=0}^{N-1}\hat{f}(x, y)^2}{\sum_{x=0}^{M-1}\sum_{y=0}^{N-1}[f(x, y) - \hat{f}(x, y)]^2} \tag{5.8-5}$$

The closer f and $\hat{f}$ are, the larger this ratio will be. Sometimes the square root of these measures is used instead, in which case they are referred to as the *root-mean-square-signal-to-noise ratio* and the *root-mean-square-error*, respectively. As we have mentioned several times before, keep in mind that quantitative metrics do not necessarily relate well to perceived image quality.

When we are dealing with spectrally white noise, the spectrum $|N(u, v)|^2$ is a constant, which simplifies things considerably. However, the power spectrum of the undegraded image seldom is known. An approach used frequently when these quantities are not known or cannot be estimated is to approximate Eq. (5.8-2) by the expression

$$\hat{F}(u, v) = \left[\frac{1}{H(u, v)} \frac{|H(u, v)|^2}{|H(u, v)|^2 + K} \right] G(u, v) \qquad (5.8\text{-}6)$$

where K is a specified constant that is added to all terms of $|H(u, v)|^2$. The following examples illustrate the use of this expression.

■ Figure 5.28 illustrates the advantage of Wiener filtering over direct inverse filtering. Figure 5.28(a) is the full inverse-filtered result from Fig. 5.27(a). Similarly, Fig. 5.28(b) is the radially limited inverse filter result of Fig, 5.27(c). These images are duplicated here for convenience in making comparisons. Figure 5.28(c) shows the result obtained using Eq. (5.8-6) with the degradation function used in Example 5.11. The value of K was chosen interactively to yield the best visual results. The advantage of Wiener filtering over the direct inverse approach is evident in this example. By comparing Figs. 5.25(a) and 5.28(c), we see that the Wiener filter yielded a result very close in appearance to the original image.   ■

EXAMPLE 5.12:
Comparison of inverse and Wiener filtering.

■ The first row of Fig. 5.29 shows, from left to right, the blurred image of Fig. 5.26(b) heavily corrupted by additive Gaussian noise of zero mean and variance of 650; the result of direct inverse filtering; and the result of Wiener filtering. The Wiener filter of Eq. (5.8-6) was used, with $H(u, v)$ from Example 5.10, and with K chosen interactively to give the best possible visual result. As expected, the inverse filter produced an unusable image. Note that the noise in the inverse filtered image is so strong that its structure is in the direction of the *deblurring* filter. The Wiener filter result is by no means perfect, but it does give us a hint as to image content. With some difficulty, the text is readable.

The second row of Fig. 5.29 shows the same sequence, but with the level of noise variance reduced by one order of magnitude. This reduction had little effect on the inverse filter, but the Wiener results are considerably improved. The text

EXAMPLE 5.13:
Further comparisons of Wiener filtering.

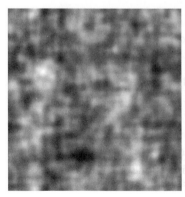

a b c

FIGURE 5.28 Comparison of inverse and Wiener filtering. (a) Result of full inverse filtering of Fig. 5.25(b). (b) Radially limited inverse filter result. (c) Wiener filter result.

262,144 × 1, and matrix **H** would be of dimensions 262,144 × 262,144. Manipulating vectors and matrices of such sizes is not a trivial task. The problem is complicated further by the fact **H** is highly sensitive to noise (after the experiences we had with the effect of noise in the previous two sections, this should not be a surprise). However, formulating the restoration problem in matrix form does facilitate derivation of restoration techniques.

Although we do not fully derive the method of constrained least squares that we are about to present, this method has its roots in a matrix formulation. We give references at the end of the chapter to sources where derivations are covered in detail. Central to the method is the issue of the sensitivity of **H** to noise. One way to alleviate the noise sensitivity problem is to base optimality of restoration on a measure of smoothness, such as the second derivative of an image (our old friend the Laplacian). To be meaningful, the restoration must be constrained by the parameters of the problems at hand. Thus, what is desired is to find the minimum of a criterion function, C, defined as

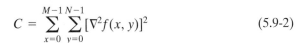

Consult the Tutorials section in the book Web site for an entire chapter devoted to the topic of algebraic techniques for image restoration.

$$C = \sum_{x=0}^{M-1} \sum_{y=0}^{N-1} [\nabla^2 f(x, y)]^2 \tag{5.9-2}$$

subject to the constraint

$$\|\mathbf{g} - \mathbf{H}\hat{\mathbf{f}}\|^2 = \|\boldsymbol{\eta}\|^2 \tag{5.9-3}$$

where $\|\mathbf{w}\|^2 \triangleq \mathbf{w}^T \mathbf{w}$ is the Euclidean vector norm,[†] and $\hat{\mathbf{f}}$ is the estimate of the undegraded image. The Laplacian operator ∇^2 is defined in Eq. (3.6-3).

The frequency domain solution to this optimization problem is given by the expression

$$\hat{F}(u, v) = \left[\frac{H^*(u, v)}{|H(u, v)|^2 + \gamma |P(u, v)|^2} \right] G(u, v) \tag{5.9-4}$$

where γ is a parameter that must be adjusted so that the constraint in Eq. (5.9-3) is satisfied, and $P(u, v)$ is the Fourier transform of the function

$$p(x, y) = \begin{bmatrix} 0 & -1 & 0 \\ -1 & 4 & -1 \\ 0 & -1 & 0 \end{bmatrix} \tag{5.9-5}$$

We recognize this function as the Laplacian operator introduced in Section 3.6.2. As noted earlier, it is important to keep in mind that $p(x, y)$, as well as all other relevant spatial domain functions, must be properly padded with zeros prior to computing their Fourier transforms for use in Eq. (5.9-4), as discussed in Section 4.6.6. Note that Eq. (5.9-4) reduces to inverse filtering if γ is zero.

[†]Recall that, for a vector $\mathbf{w}$ with n components, $\mathbf{w}^T \mathbf{w} = \sum_{k=1}^{n} w_k^2$, where w_k is the kth component of $\mathbf{w}$.

■ Figure 5.30 shows the result of processing Figs. 5.29(a), (d), and (g) with constrained least squares filters, in which the values of γ were selected manually to yield the best visual results. This is the same procedure we used to generate the Wiener filtered results in Fig. 5.29(c), (f), and (i). By comparing the constrained least squares and Wiener results, it is noted that the former yielded slightly better results for the high- and medium-noise cases, with both filters generating essentially equal results for the low-noise case. It is not unexpected that the constrained least squares filter would outperform the Wiener filter when selecting the parameters manually for better visual results. The parameter γ in Eq. (5.9-4) is a scalar, while the value of K in Eq. (5.8-6) is an approximation to the ratio of two unknown frequency domain functions; this ratio seldom is constant. Thus, it stands to reason that a result based on manually selecting γ would be a more accurate estimate of the undegraded image.   ■

EXAMPLE 5.14:
Comparison of Wiener and constrained least squares filtering.

As shown in the preceding example, it is possible to adjust the parameter γ interactively until acceptable results are achieved. If we are interested in optimality, however, then the parameter γ must be adjusted so that the constraint in Eq. (5.9-3) is satisfied. A procedure for computing γ by iteration is as follows.

Define a "residual" vector **r** as

$$\mathbf{r} = \mathbf{g} - \mathbf{H}\hat{\mathbf{f}} \qquad (5.9\text{-}6)$$

Since, from the solution in Eq. (5.9-4), $\hat{F}(u, v)$ (and by implication $\hat{\mathbf{f}}$) is a function of γ, then **r** also is a function of this parameter. It can be shown (Hunt [1973]) that

$$\phi(\gamma) = \mathbf{r}^T\mathbf{r}$$
$$= \|\mathbf{r}\|^2 \qquad (5.9\text{-}7)$$

a b c

FIGURE 5.30 Results of constrained least squares filtering. Compare (a), (b), and (c) with the Wiener filtering results in Figs. 5.29(c), (f), and (i), respectively.

is a monotonically increasing function of γ. What we want to do is adjust γ so that

$$\|\mathbf{r}\|^2 = \|\boldsymbol{\eta}\|^2 \pm a \tag{5.9-8}$$

where a is an accuracy factor. In view of Eq. (5.9-6), if $\|\mathbf{r}\|^2 = \|\boldsymbol{\eta}\|^2$, the constraint in Eq. (5.9-3) will be strictly satisfied.

Because $\phi(\gamma)$ is monotonic, finding the desired value of γ is not difficult. One approach is to

1. Specify an initial value of γ.
2. Compute $\|\mathbf{r}\|^2$.
3. Stop if Eq. (5.9-8) is satisfied; otherwise return to step 2 after increasing γ if $\|\mathbf{r}\|^2 < \|\boldsymbol{\eta}\|^2 - a$ or decreasing γ if $\|\mathbf{r}\|^2 > \|\boldsymbol{\eta}\|^2 + a$. Use the new value of γ in Eq. (5.9-4) to recompute the optimum estimate $\hat{F}(u, v)$.

Other procedures, such as a Newton–Raphson algorithm, can be used to improve the speed of convergence.

In order to use this algorithm, we need the quantities $\|\mathbf{r}\|^2$ and $\|\boldsymbol{\eta}\|^2$. To compute $\|\mathbf{r}\|^2$, we note from Eq. (5.9-6) that

$$R(u, v) = G(u, v) - H(u, v)\hat{F}(u, v) \tag{5.9-9}$$

from which we obtain $r(x, y)$ by computing the inverse transform of $R(u, v)$. Then

$$\|\mathbf{r}\|^2 = \sum_{x=0}^{M-1} \sum_{y=0}^{N-1} r^2(x, y) \tag{5.9-10}$$

Computation of $\|\boldsymbol{\eta}\|^2$ leads to an interesting result. First, consider the variance of the noise over the entire image, which we estimate by the sample-average method, as discussed in Section 3.3.4:

$$\sigma_\eta^2 = \frac{1}{MN} \sum_{x=0}^{M-1} \sum_{y=0}^{N-1} [\eta(x, y) - m_\eta]^2 \tag{5.9-11}$$

where

$$m_\eta = \frac{1}{MN} \sum_{x=0}^{M-1} \sum_{y=0}^{N-1} \eta(x, y) \tag{5.9-12}$$

is the sample mean. With reference to the *form* of Eq. (5.9-10), we note that the double summation in Eq. (5.9-11) is equal to $\|\boldsymbol{\eta}\|^2$. This gives us the expression

$$\|\boldsymbol{\eta}\|^2 = MN[\sigma_\eta^2 + m_\eta^2] \tag{5.9-13}$$

This is a most useful result. It tells us that we can implement an optimum restoration algorithm by having knowledge of only the mean and variance of the noise. These quantities are not difficult to estimate (Section 5.2.4), assuming that the noise and image intensity values are not correlated. This is a basic assumption of all the methods discussed in this chapter.

FIGURE 5.31
(a) Iteratively determined constrained least squares restoration of Fig. 5.16(b), using correct noise parameters. (b) Result obtained with wrong noise parameters.

■ Figure 5.31(a) shows the result obtained by using the algorithm just described to estimate the optimum filter for restoring Fig. 5.25(b). The initial value used for γ was 10^{-5}, the correction factor for adjusting γ was 10^{-6}, and the value for a was 0.25. The noise parameters specified were the same used to generate Fig. 5.25(a): a noise variance of 10^{-5}, and zero mean. The restored result is almost as good as Fig. 5.28(c), which was obtained by Wiener filtering with K manually specified for best visual results. Figure 5.31(b) shows what can happen if the wrong estimate of noise parameters are used. In this case, the noise variance specified was 10^{-2} and the mean was left at a value of 0. The result in this case is considerably more blurred. ■

EXAMPLE 5.15:
Iterative estimation of the optimum constrained least squares filter.

As stated at the beginning of this section, it is important to keep in mind that optimum restoration in the sense of constrained least squares does not necessarily imply "best" in the visual sense. Depending on the nature and magnitude of the degradation and noise, the other parameters in the algorithm for iteratively determining the optimum estimate also play a role in the final result. In general, automatically determined restoration filters yield inferior results to manual adjustment of filter parameters. This is particularly true of the constrained least squares filter, which is completely specified by a single, scalar parameter.

5.10 Geometric Mean Filter

It is possible to generalize slightly the Wiener filter discussed in Section 5.8. The generalization is in the form of the so-called *geometric mean filter*:

$$\hat{F}(u, v) = \left[\frac{H^*(u, v)}{|H(u, v)|^2} \right]^{\alpha} \left[\frac{H^*(u, v)}{|H(u, v)|^2 + \beta \left[\dfrac{S_\eta(u, v)}{S_f(u, v)} \right]} \right]^{1-\alpha} G(u, v) \quad (5.10\text{-}1)$$

with α and β being positive, real constants. The geometric mean filter consists of the two expressions in brackets raised to the powers α and $1 - \alpha$, respectively.

When $\alpha = 1$ this filter reduces to the inverse filter. With $\alpha = 0$ the filter becomes the so-called *parametric Wiener filter*, which reduces to the standard Wiener filter when $\beta = 1$. If $\alpha = 1/2$, the filter becomes a product of the two quantities raised to the same power, which is the definition of the geometric mean, thus giving the filter its name. With $\beta = 1$, as α decreases below $1/2$, the filter performance will tend more toward the inverse filter. Similarly, when α increases above $1/2$, the filter will behave more like the Wiener filter. When $\alpha = 1/2$ and $\beta = 1$, the filter also is commonly referred to as the *spectrum equalization filter*. Equation (5.10-1) is quite useful when implementing restoration filters because it represents a family of filters combined into a single expression.

5.11 Image Reconstruction from Projections

In the previous sections of this chapter, we dealt with techniques for restoring a degraded version of an image. In this section, we examine the problem of *reconstructing* an image from a series of projections, with a focus on X-ray *computed tomography* (CT). This is the earliest and still the most widely used type of CT and is currently one of the principal applications of digital image processing in medicine.

As noted in Chapter 1, the term *computerized axial tomography* (CAT) is used interchangeably to denote CT.

5.11.1 Introduction

The reconstruction problem is simple in principle and can be explained qualitatively in a straightforward, intuitive manner. To begin, consider Fig. 5.32(a), which consists of a single object on a uniform background. To bring physical

FIGURE 5.32
(a) Flat region showing a simple object, an input parallel beam, and a detector strip.
(b) Result of back-projecting the sensed strip data (i.e., the 1-D absorption profile). (c) The beam and detectors rotated by 90°.
(d) Back-projection.
(e) The sum of (b) and (d). The intensity where the back-projections intersect is twice the intensity of the individual back-projections.

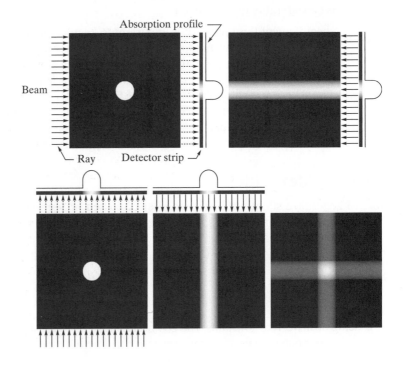

meaning to the following explanation, suppose that this image is a cross section of a 3-D region of a human body. Assume also that the background in the image represents soft, uniform tissue, while the round object is a tumor, also uniform, but with higher absorption characteristics.

Suppose next that we pass a thin, flat beam of X-rays from left to right (though the plane of the image), as Fig. 5.32(a) shows, and assume that the energy of the beam is absorbed more by the object than by the background, as typically is the case. Using a strip of X-ray absorption detectors on the other side of the region will yield the signal (absorption profile) shown, whose amplitude (intensity) is proportional to absorption.[†] We may view any point in the signal as the sum of the absorption values across the single ray in the beam corresponding spatially to that point (such a sum often is referred to as a *raysum*). At this juncture, all the information we have about the object is this 1-D absorption signal.

We have no way of determining from a single projection whether we are dealing with a single object or a multitude of objects along the path of the beam, but we begin the reconstruction by creating an *image* based on just this information. The approach is to project the 1-D signal back across the direction from which the beam came, as Fig. 5.32(b) shows. The process of backprojecting a 1-D signal across a 2-D area sometimes is referred to as *smearing* the projection back across the area. In terms of digital images, this means duplicating the same 1-D signal across the image perpendicularly to the direction of the beam. For example, Fig. 5.32(b) was created by duplicating the 1-D signal in all columns of the reconstructed image. For obvious reasons, the approach just described is called *backprojection*.

Next, suppose that we rotate the position of the source-detector pair by 90°, as in Fig. 5.32(c). Repeating the procedure explained in the previous paragraph yields a backprojection image in the vertical direction, as Fig. 5.32(d) shows. We continue the reconstruction by *adding* this result to the previous backprojection, resulting in Fig. 5.32(e). Now, we can tell that the object of interest is contained in the square shown, whose amplitude is twice the amplitude of the individual backprojections. A little thought will reveal that we should be able to learn more about the shape of the object in question by taking more views in the manner just described. In fact, this is exactly what happens, as Fig. 5.33 shows. As the number of projections increases, the strength of non-intersecting backprojections decreases relative to the strength of regions in which multiple backprojections intersect. The net effect is that brighter regions will dominate the result, and backprojections with few or no intersections will fade into the background as the image is scaled for display.

Figure 5.33(f), formed from 32 projections, illustrates this concept. Note, however, that while this reconstructed image is a reasonably good approximation to the shape of the original object, the image is blurred by a "halo" effect,

---

[†]A treatment of the physics of X-ray sources and detectors is beyond the scope of our discussion, which focuses on the image processing aspects of CT. See Prince and Links [2006] for an excellent introduction to the physics of X-ray image formation.

a b c
d e f

FIGURE 5.33
(a) Same as Fig.
5.32(a).
(b)–(e) Reconstruction using 1, 2, 3, and 4 backprojections 45° apart.
(f) Reconstruction with 32 backprojections 5.625° apart (note the blurring).

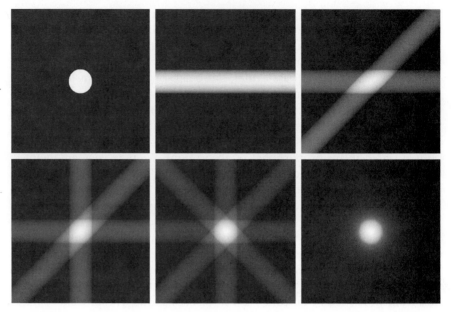

the formation of which can be seen in progressive stages in Fig. 5.33. For example, the halo in Fig. 5.33(e) appears as a "star" whose intensity is lower than that of the object, but higher than the background. As the number of views increases, the shape of the halo becomes circular, as in Fig. 5.33(f). Blurring in CT reconstruction is an important issue, whose solution is addressed in Section 5.11.5. Finally, we conclude from the discussion of Figs. 5.32 and 5.33 that projections 180° apart are mirror images of each other, so we only have to consider angle increments halfway around a circle in order to generate all the projections required for reconstruction.

EXAMPLE 5.16:
Backprojection of a simple planar region containing two objects.

■ Figure 5.34 illustrates reconstruction using backprojections on a slightly more complicated region that contains two objects with different absorption properties. Figure 5.34(b) shows the result of using one backprojection. We note three principal features in this figure, from bottom to top: a thin horizontal gray band corresponding to the unconcluded portion of the small object, a brighter (more absorption) band above it corresponding to the area shared by both objects, and an upper band corresponding to the rest of the elliptical object. Figures 5.34(c) and (d) show reconstruction using two projections 90° apart and four projections 45° apart, respectively. The explanation of these figures is similar to the discussion of Figs. 5.33(c) through (e). Figures 5.34(e) and (f) show more accurate reconstructions using 32 and 64 backprojections, respectively. These two results are quite close visually, and they both show the blurring problem mentioned earlier, whose solution we address in Section 5.11.5.   ■

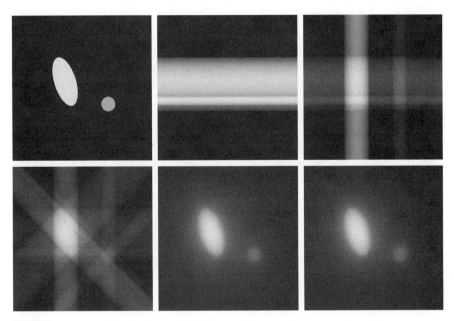

a b c
d e f

FIGURE 5.34 (a) A region with two objects. (b)–(d) Reconstruction using 1, 2, and 4 backprojections 45° apart. (e) Reconstruction with 32 backprojections 5.625° apart. (f) Reconstruction with 64 backprojections 2.8125° apart.

5.11.2 Principles of Computed Tomography (CT)

The goal of X-ray computed tomography is to obtain a 3-D representation of the internal structure of an object by X-raying the object from many different directions. Imagine a traditional chest X-ray, obtained by placing the subject against an X-ray sensitive plate and "illuminating" the individual with an X-ray beam in the form of a cone. The X-ray plate produces an image whose intensity at a point is proportional to the X-ray energy impinging on that point after it has passed through the subject. This image is the 2-D equivalent of the projections we discussed in the previous section. We could back-project this entire image and create a 3-D volume. Repeating this process through many angles and adding the backprojections would result in 3-D rendition of the structure of the chest cavity. Computed tomography attempts to get that same information (or localized parts of it) by generating *slices* through the body. A 3-D representation then can be obtained by stacking the slices. A CT implementation is much more economical, because the number of detectors required to obtain a high resolution slice is much smaller than the number of detectors needed to generate a complete 2-D projection of the same resolution. Computational burden and X-ray dosages are similarly reduced, making the 1-D projection CT a more practical approach.

As with the Fourier transform discussed in the last chapter, the basic mathematical concepts required for CT were in place years before the availability of digital computers made them practical. The theoretical foundation of CT dates back to Johann Radon, a mathematician from Vienna who derived a method in 1917 for projecting a 2-D object along parallel rays as part of his work on line integrals. The method now is referred to commonly as the *Radon transform*, a topic we discuss in the following section. Forty-five years later, Allan M. Cormack, a physicist at Tufts University, partially "rediscovered" these concepts and applied them to CT. Cormack published his initial findings in 1963 and 1964 and showed how they could be used to reconstruct cross-sectional images of the body from X-ray images taken at different angular directions. He gave the mathematical formulae needed for the reconstruction and built a CT prototype to show the practicality of his ideas. Working independently, electrical engineer Godfrey N. Hounsfield and his colleagues at EMI in London formulated a similar solution and built the first medical CT machine. Cormack and Hounsfield shared the 1979 Nobel Prize in Medicine for their contributions to medical tomography.

First-generation (G1) CT scanners employ a "pencil" X-ray beam and a single detector, as Fig. 5.35(a) shows. For a given angle of rotation, the source/detector

a b
c d

FIGURE 5.35 Four generations of CT scanners. The dotted arrow lines indicate incremental linear motion. The dotted arrow arcs indicate incremental rotation. The cross-mark on the subject's head indicates linear motion perpendicular to the plane of the paper. The double arrows in (a) and (b) indicate that the source/detector unit is translated and then brought back into its original position.

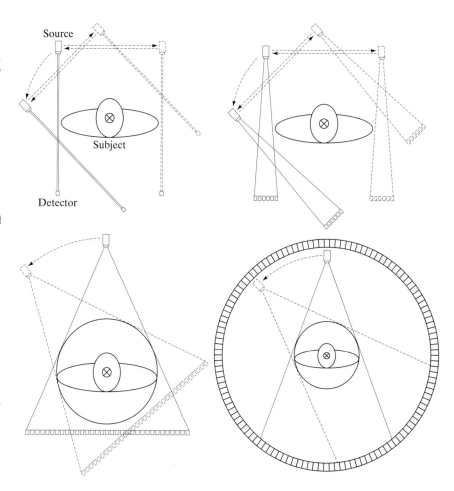

pair is translated incrementally along the linear direction shown. A projection (like the ones in Fig. 5.32), is generated by measuring the output of the detector at each increment of translation. After a complete linear translation, the source/detector assembly is rotated and the procedure is repeated to generate another projection at a different angle. The procedure is repeated for all desired angles in the range $[0°, 180°]$ to generate a complete set of projections, from which one image is generated by backprojection, as explained in the previous section. The cross-mark on the head of the subject indicates motion in a direction perpendicular to the plane of the source/detector pair. A set of cross sectional images (slices) is generated by incrementally moving the subject (after each complete scan) past the source/detector plane. Stacking these images computationally produces a 3-D volume of a section of the body. G1 scanners are no longer manufactured for medical imaging but, because they produce a parallel-ray beam (as in Fig. 5.32), their geometry is the one used predominantly for introducing the fundamentals of CT imaging. As discussed in the following section, this geometry is the starting point for deriving the equations necessary to implement image reconstruction from projections.

Second-generation (G2) CT scanners [Fig. 5.35(b)] operate on the same principle as G1 scanners, but the beam used is in the shape of a fan. This allows the use of multiple detectors, thus requiring fewer translations of the source/detector pair. *Third-generation (G3) scanners* are a significant improvement over the earlier two generations of CT geometries. As Fig. 5.35(c) shows, G3 scanners employ a bank of detectors long enough (on the order of 1000 individual detectors) to cover the entire field of view of a wider beam. Consequently, each increment of angle produces an entire projection, eliminating the need to translate the source/detector pair, as in the geometry of G1 and G2 scanners. *Fourth-generation (G4) scanners* go a step further. By employing a circular ring of detectors (on the order of 5000 individual detectors), only the source has to rotate. The key advantage of G3 and G4 scanners is speed. Key disadvantages are cost and greater X-ray scatter, which requires higher doses than G1 and G2 scanners to achieve comparable signal-to-noise characteristics.

Newer scanning modalities are beginning to be adopted. For example, *fifth-generation (G5) CT scanners*, also known as *electron beam computed tomography* (EBCT) *scanners*, eliminate all mechanical motion by employing electron beams controlled electromagnetically. By striking tungsten anodes that encircle the patient, these beams generate X-rays that are then shaped into a fan beam that passes through the patient and excites a ring of detectors, as in G4 scanners.

The conventional manner in which CT images are obtained is to keep the patient stationary during the scanning time required to generate one image. Scanning is then halted while the position of the patient is incremented in the direction perpendicular to the imaging plane using a motorized table. The next image is then obtained and the procedure is repeated for the number of increments required to cover a specified section of the body. Although an image may be obtained in less than one second, there are procedures (e.g., abdominal and chest scans) that require the patient to hold his/her breath during image acquisition. Completing these procedures for, say, 30 images, may require several minutes. An approach whose use is increasing is *helical CT*, sometimes referred to as *sixth-generation (G6) CT*. In this approach, a G3 or G4 scanner is configured using so-called *slip rings* that eliminate the need for electrical and signal cabling between the source/detectors and the processing unit. The source/detector pair then rotates continuously through 360° while the patient is moved at a constant speed along the axis perpendicular to the scan. The result is a continuous helical volume of data that is then processed to obtain individual slice images.

Seventh-generation (G7) scanners (also called *multislice CT scanners*) are emerging in which "thick" fan beams are used in conjunction with parallel banks of detectors to collect volumetric CT data simultaneously. That is, 3-D cross-sectional "slabs," rather than single cross-sectional images are generated per X-ray burst. In addition to a significant increase in detail, this approach has the advantage that it utilizes X-ray tubes more economically, thus reducing cost and potentially reducing dosage.

Beginning in the next section, we develop the mathematical tools necessary for formulating image projection and reconstruction algorithms. Our focus is on the image-processing fundamentals that underpin all the CT approaches just discussed. Information regarding the mechanical and source/detector characteristics of CT systems is provided in the references cited at the end of the chapter.

5.11.3 Projections and the Radon Transform

Throughout this section, we follow CT convention and place the origin of the *xy*-plane in the center, instead of at our customary top left corner (see Section 2.4.2). Note, however, that both are right-handed coordinate systems, the only difference being that our image coordinate system has no negative axes. We can account for the difference with a simple translation of the origin, so both representations are interchangeable.

In what follows, we develop in detail the mathematics needed for image reconstruction in the context of X-ray computed tomography, but the same basic principles are applicable in other CT imaging modalities, such as SPECT (single photon emission tomography), PET (positron emission tomography), MRI (magnetic resonance imaging), and some modalities of ultrasound imaging.

A straight line in Cartesian coordinates can be described either by its *slope-intercept* form, $y = ax + b$, or, as in Fig. 5.36, by its *normal* representation:

$$x \cos \theta + y \sin \theta = \rho \qquad (5.11-1)$$

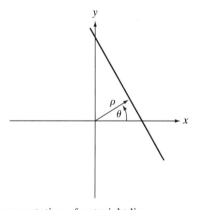

FIGURE 5.36 Normal representation of a straight line.

FIGURE 5.37
Geometry of a
parallel-ray beam.

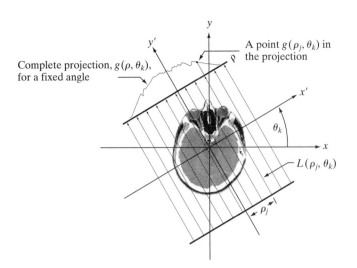

The projection of a parallel-ray beam may be modeled by a set of such lines, as Fig. 5.37 shows. An arbitrary point in the projection signal is given by the ray-sum along the line $x \cos \theta_k + y \sin \theta_k = \rho_j$. Working with continuous quantities[†] for the moment, the raysum is a line integral, given by

$$g(\rho_j, \theta_k) = \int_{-\infty}^{\infty} \int_{-\infty}^{\infty} f(x, y)\delta(x \cos \theta_k + y \sin \theta_k - \rho_j)\, dx\, dy \qquad (5.11\text{-}2)$$

where we used the properties of the impulse, δ, discussed in Section 4.5.1. In other words, the right side of Eq. (5.11-2) is zero unless the argument of δ is zero, indicating that the integral is computed only along the line $x \cos \theta_k + y \sin \theta_k = \rho_j$. If we consider all values of ρ and θ, the preceding equation generalizes to

$$g(\rho, \theta) = \int_{-\infty}^{\infty} \int_{-\infty}^{\infty} f(x, y)\delta(x \cos \theta + y \sin \theta - \rho)\, dx\, dy \qquad (5.11\text{-}3)$$

This equation, which gives the projection (line integral) of $f(x, y)$ along an arbitrary line in the xy-plane, is the *Radon transform* mentioned in the previous section. The notation $\Re\{f(x, y)\}$ or $\Re\{f\}$ is used sometimes in place of $g(\rho, \theta)$ in Eq. (5.11-3) to denote the Radon transform of f, but the type of notation used in Eq. (5.11-3) is more customary. As will become evident in the discussion that follows, the Radon transform is the cornerstone of reconstruction from projections, with computed tomography being its principal application in the field of image processing.

[†]In Chapter 4, we exercised great care in denoting continuous image coordinates by (t, z) and discrete coordinates by (x, y). At that time, the distinction was important because we were developing basic concepts to take us from continuous to sampled quantities. In the present discussion, we go back and forth so many times between continuous and discrete coordinates that adhering to this convention is likely to generate unnecessary confusion. For this reason, and also to follow the published literature in this field (e.g., see Prince and Links [2006]), we let the context determine whether coordinates (x, y) are continuous or discrete. When they are continuous, you will see integrals; otherwise you will see summations.

In the discrete case, Eq. (5.11-3) becomes

$$g(\rho, \theta) = \sum_{x=0}^{M-1} \sum_{y=0}^{N-1} f(x, y)\delta(x \cos \theta + y \sin \theta - \rho) \qquad (5.11\text{-}4)$$

where x, y, ρ, and θ are now discrete variables. If we fix θ and allow ρ to vary, we see that (5.11-4) simply sums the pixels of $f(x, y)$ along the line defined by the specified values of these two parameters. Incrementing through all values of ρ required to span the image (with θ fixed) yields *one* projection. Changing θ and repeating the foregoing procedure yields another projection, and so forth. This is precisely how the projections in Section 5.11.1 were generated.

EXAMPLE 5.17:
Using the Radon transform to obtain the projection of a circular region.

■ Before proceeding, we illustrate how to use the Radon transform to obtain an analytical expression for the projection of the circular object in Fig. 5.38(a):

$$f(x, y) = \begin{cases} A & x^2 + y^2 \le r^2 \\ 0 & \text{otherwise} \end{cases}$$

where A is a constant and r is the radius of the object. We assume that the circle is centered on the origin of the xy-plane. Because the object is circularly symmetric, its projections are the same for all angles, so all we have to do is obtain the projection for $\theta = 0°$. Equation (5.11-3) then becomes

$$g(\rho, \theta) = \int_{-\infty}^{\infty} \int_{-\infty}^{\infty} f(x, y)\delta(x - \rho) \, dx \, dy$$

$$= \int_{-\infty}^{\infty} f(\rho, y) \, dy$$

a
b

FIGURE 5.38 (a) A disk and (b) a plot of its Radon transform, derived analytically. Here we were able to plot the transform because it depends only on one variable. When g depends on both ρ and θ, the Radon transform becomes an image whose axes are ρ and θ, and the intensity of a pixel is proportional to the value of g at the location of that pixel.

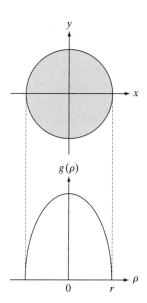

where the second line follows from Eq. (4.2-10). As noted earlier, this is a line integral (along the line $L(\rho, 0)$ in this case). Also, note that $g(\rho, \theta) = 0$ when $|\rho| > r$. When $|\rho| \leq r$ the integral is evaluated from $y = -\sqrt{r^2 - \rho^2}$ to $y = \sqrt{r^2 - \rho^2}$. Therefore,

$$g(\rho, \theta) = \int_{-\sqrt{r^2-\rho^2}}^{\sqrt{r^2-\rho^2}} f(\rho, y)dy$$

$$= \int_{-\sqrt{r^2-\rho^2}}^{\sqrt{r^2-\rho^2}} A\,dy$$

Carrying out the integration yields

$$g(\rho, \theta) = g(\rho) = \begin{cases} 2A\sqrt{r^2 - \rho^2} & |\rho| \leq r \\ 0 & \text{otherwise} \end{cases}$$

where we used the fact mentioned above that $g(\rho, \theta) = 0$ when $|\rho| > r$. Figure 5.38(b) shows the result, which agrees with the projections illustrated in Figs. 5.32 and 5.33. Note that $g(\rho, \theta) = g(\rho)$; that is, g is independent of θ because the object is symmetric about the origin. ∎

When the Radon transform, $g(\rho, \theta)$, is displayed as an image with ρ and θ as rectilinear coordinates, the result is called a *sinogram*, similar in concept to displaying the Fourier spectrum (unlike the Fourier transform, however, $g(\rho, \theta)$ is always a real function). Like the Fourier transform, a sinogram contains the data necessary to reconstruct $f(x,y)$. As is the case with displays of the Fourier spectrum, sinograms can be readily interpreted for simple regions, but become increasingly difficult to "read" as the region being projected becomes more complex. For example, Fig. 5.39(b) is the sinogram of the rectangle on the left. The vertical and horizontal axes correspond to θ and ρ, respectively. Thus, the bottom row is the projection of the rectangle in the horizontal direction (i.e., $\theta = 0°$), and the middle row is the projection in the vertical direction ($\theta = 90°$). The fact that the nonzero portion of the bottom row is smaller than the nonzero portion of the middle row tells us that the object is narrower in the horizontal direction. The fact that the sinogram is symmetric in both directions about the center of the image tells us that we are dealing with an object that is symmetric and parallel to the x and y axes. Finally, the sinogram is smooth, indicating that the object has a uniform intensity. Other than these types of general observations, we cannot say much more about this sinogram.

Figure 5.39(c) shows an image of the *Shepp-Logan phantom*, a widely used synthetic image designed to simulate the absorption of major areas of the brain, including small tumors. The sinogram of this image is considerably more difficult to interpret, as Fig. 5.39(d) shows. We still can infer some symmetry properties, but that is about all we can say. Visual analysis of sinograms is of limited practical use, but sometimes it is helpful in algorithm development.

To generate arrays with rows of the same size, the minimum dimension of the ρ-axis in sinograms corresponds to the largest dimension encountered during projection. For example, the minimum size of a sinogram of a square of size $M \times M$ obtained using increments of 1° is $180 \times Q$, where Q is the smallest integer greater than $\sqrt{2}M$.

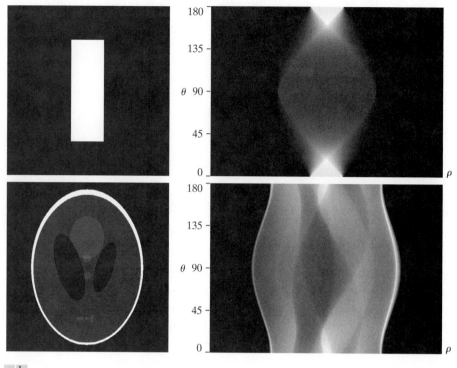

a b
c d

FIGURE 5.39 Two images and their sinograms (Radon transforms). Each row of a sinogram is a projection along the corresponding angle on the vertical axis. Image (c) is called the *Shepp-Logan phantom.* In its original form, the contrast of the phantom is quite low. It is shown enhanced here to facilitate viewing.

The key objective of CT is to obtain a 3-D representation of a volume from its projections. As introduced intuitively in Section 5.11.1, the approach is to back-project each projection and then sum all the backprojections to generate one image (slice). Stacking all the resulting images produces a 3-D rendition of the volume. To obtain a formal expression for a back-projected image from the Radon transform, let us begin with a *single* point, $g(\rho_j, \theta_k)$, of the complete projection, $g(\rho, \theta_k)$, for a *fixed* value of rotation, θ_k (see Fig. 5.37). Forming part of an image by back-projecting this *single* point is nothing more than copying the line $L(\rho_j, \theta_k)$ onto the image, where the value of each point in that line is $g(\rho_j, \theta_k)$. Repeating this process of all values of ρ_j in the projected signal (but keeping the value of θ fixed at θ_k) results in the following expression:

$$f_{\theta_k}(x, y) = g(\rho, \theta_k)$$
$$= g(x \cos \theta_k + y \sin \theta_k, \theta_k)$$

for the image due to back-projecting the projection obtained with a fixed angle, θ_k, as in Fig. 5.32(b). This equation holds for an arbitrary value of θ_k, so we may write in general that the image formed from a *single* backprojection obtained at an angle θ is given by

$$f_\theta(x, y) = g(x \cos \theta + y \sin \theta, \theta) \tag{5.11-5}$$

We form the final image by integrating over all the back-projected images:

$$f(x, y) = \int_0^\pi f_\theta(x, y) \, d\theta \tag{5.11-6}$$

In the discrete case, the integral becomes a sum of all the back-projected images:

$$f(x, y) = \sum_{\theta=0}^\pi f_\theta(x, y) \tag{5.11-7}$$

where, x, y, and θ are now discrete quantities. Recall from the discussion in Section 5.11.1 that the projections at $0°$ and $180°$ are mirror images of each other, so the summations are carried out to the last angle increment before $180°$. For example, if $0.5°$ increments are being used, the summation is from 0 to 179.5 in half-degree increments. A back-projected image formed in the manner just described sometimes is referred to as a *laminogram*. It is understood implicitly that a laminogram is only an approximation to the image from which the projections were generated, a fact that is illustrated clearly in the following example.

■ Equation (5.11-7) was used to generate the back-projected images in Figs. 5.32 through 5.34, from projections obtained with Eq. (5.11-4). Similarly, these equations were used to generate Figs. 5.40(a) and (b), which show the back-projected images corresponding to the sinograms in Fig. 5.39(b) and (d), respectively. As with the earlier figures, we note a significant amount of blurring, so it is obvious that a straight use of Eqs. (5.11-4) and (5.11-7) will not yield acceptable results. Early, experimental CT systems were based on these equations. However, as you will see in Section 5.11.5, significant improvements in reconstruction are possible by reformulating the backprojection approach. ■

EXAMPLE 5.18:
Obtaining back-projected images from sinograms.

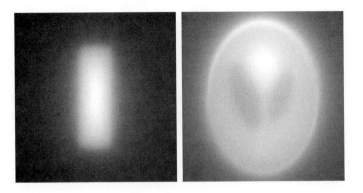

a b
FIGURE 5.40
Backprojections of the sinograms in Fig. 5.39.

5.11.4 The Fourier-Slice Theorem

In this section, we derive a fundamental result relating the 1-D Fourier transform of a projection and the 2-D Fourier transform of the region from which the projection was obtained. This relationship is the basis for reconstruction methods capable of dealing with the blurring problem just discussed.

The 1-D Fourier transform of a projection with respect to ρ is

$$G(\omega, \theta) = \int_{-\infty}^{\infty} g(\rho, \theta)e^{-j2\pi\omega\rho} \, d\rho \qquad (5.11\text{-}8)$$

where, as in Eq. (4.2-16), ω is the frequency variable, and it is understood that this expression is for a given value of θ. Substituting Eq. (5.11-3) for $g(\rho, \theta)$ results in the expression

$$
\begin{aligned}
G(\omega, \theta) &= \int_{-\infty}^{\infty}\int_{-\infty}^{\infty}\int_{-\infty}^{\infty} f(x, y)\delta(x\cos\theta + y\sin - \rho)e^{-j2\pi\omega\rho} \, dx \, dy \, d\rho \\[4pt]
&= \int_{-\infty}^{\infty}\int_{-\infty}^{\infty} f(x, y)\left[\int_{-\infty}^{\infty}\delta(x\cos\theta + y\sin\theta - \rho)e^{-j2\pi\omega\rho} \, d\rho\right] dx \, dy \\[4pt]
&= \int_{-\infty}^{\infty}\int_{-\infty}^{\infty} f(x, y)e^{-j2\pi\omega(x\cos\theta + y\sin\theta)} \, dx \, dy
\end{aligned}
$$

$$(5.11\text{-}9)$$

where the last step follows from the property of the impulse mentioned earlier in this section. By letting $u = \omega\cos\theta$ and $v = \omega\sin\theta$, Eq. (5.11-9) becomes

$$G(\omega, \theta) = \left[\int_{-\infty}^{\infty}\int_{-\infty}^{\infty} f(x, y)e^{-j2\pi(ux+vy)} \, dx \, dy\right]_{u=\omega\cos\theta;\, v=\omega\sin\theta} \qquad (5.11\text{-}10)$$

We recognize this expression as the 2-D Fourier transform of $f(x, y)$ [see Eq. (4.5-7)] evaluated at the values of u and v indicated. That is,

$$
\begin{aligned}
G(\omega, \theta) &= [F(u, v)]_{u=\omega\cos\theta;\, v=\omega\sin\theta} \\
&= F(\omega\cos\theta, \omega\sin\theta)
\end{aligned}
$$

$$(5.11\text{-}11)$$

where, as usual, $F(u, v)$ denotes the 2-D Fourier transform of $f(x, y)$.

Equation (5.11-11) is known as the *Fourier-slice theorem* (or the *projection-slice theorem*). It states that the Fourier transform of a projection is a *slice* of the 2-D Fourier transform of the region from which the projection was obtained. The reason for this terminology can be explained with the aid of Fig. 5.41. As this figure shows, the 1-D Fourier transform of an arbitrary projection is obtained by extracting the values of $F(u, v)$ along a line oriented at the same angle as the angle used in generating the projection. In principle, we could obtain $f(x, y)$ simply by obtaining the inverse

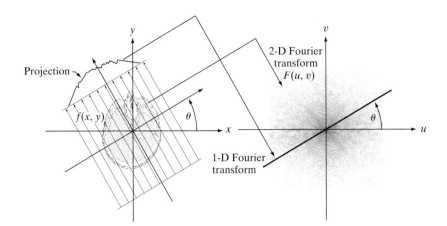

FIGURE 5.41
Illustration of the Fourier-slice theorem. The 1-D Fourier transform of a projection is a slice of the 2-D Fourier transform of the region from which the projection was obtained. Note the correspondence of the angle θ.

Fourier transform of $F(u, v)$.[†] However, this is expensive computationally, as it involves inverting a 2-D transform. The approach discussed in the following section is much more efficient.

5.11.5 Reconstruction Using Parallel-Beam Filtered Backprojections

As we saw in Section 5.11.1 and in Example 5.18, obtaining backprojections directly yields unacceptably blurred results. Fortunately, there is a straightforward solution to this problem based simply on filtering the projections before computing the backprojections. From Eq. (4.5-8), the 2-D inverse Fourier transform of $F(u, v)$ is

$$f(x, y) = \int_{-\infty}^{\infty} \int_{-\infty}^{\infty} F(u, v)e^{j2\pi(ux+vy)} \, du \, dv \qquad (5.11\text{-}12)$$

If, as in Eqs. (5.11-10) and (5.11-11), we let $u = \omega \cos \theta$ and $v = \omega \sin \theta$, then the differentials become $du \, dv = \omega \, d\omega \, d\theta$, and we can express Eq. (5.11-12) in polar coordinates:

The relationship $du \, dv = \omega \, d\omega \, d\theta$ is from basic integral calculus, where the Jacobian is used as the basis for a change of variables.

$$f(x, y) = \int_{0}^{2\pi} \int_{0}^{\infty} F(\omega \cos \theta, \omega \sin \theta)e^{j2\pi\omega(x \cos \theta + y \sin \theta)} \omega \, d\omega \, d\theta \qquad (5.11\text{-}13)$$

Then, using the Fourier-slice theorem,

$$f(x, y) = \int_{0}^{2\pi} \int_{0}^{\infty} G(\omega, \theta)e^{j2\pi\omega(x \cos \theta + y \sin \theta)} \omega \, d\omega \, d\theta \qquad (5.11\text{-}14)$$

[†]Keep in mind that blurring will still be present in an image recovered using the inverse Fourier transform, because the result is equivalent to the result obtained using the approach discussed in the previous section.

By splitting this integral into two expressions, one for θ in the range $0°$ to $180°$ and the other in the range $180°$ to $360°$, and using the fact that $G(\omega, \theta + 180°) = G(-\omega, \theta)$ (see Problem 5.32), we can express Eq. (5.11-14) as

$$f(x, y) = \int_0^\pi \int_{-\infty}^\infty |\omega| G(\omega, \theta) e^{j2\pi\omega(x \cos \theta + y \sin \theta)} \, d\omega \, d\theta \qquad (5.11\text{-}15)$$

In terms of integration with respect to ω, the term $x \cos \theta + y \sin \theta$ is a constant, which we recognize as ρ from Eq. (5.11-1). Thus, Eq. (5.11-15) can be written as:

$$f(x, y) = \int_0^\pi \left[\int_{-\infty}^\infty |\omega| G(\omega, \theta) e^{j2\pi\omega\rho} \, d\omega \right]_{\rho=x \cos \theta + y \sin \rho} d\theta \qquad (5.11\text{-}16)$$

The inner expression is in the form of an *inverse* 1-D Fourier transform [see Eq. (4.2-17)], with the added term $|\omega|$ which, based on the discussion in Section 4.7, we recognize as a *one-dimensional* filter function. Observe that $|\omega|$ is a *ramp* filter [see Fig. 5.42(a)].[†] This function is not integrable because its amplitude extends to $+\infty$ in both directions, so the inverse Fourier transform is undefined. Theoretically, this is handled by methods such as using so-called *generalized delta functions*. In practice, the approach is to window the ramp so it becomes zero outside of a defined frequency interval. That is, a window *band-limits* the ramp filter.

a b
c d e

FIGURE 5.42
(a) Frequency domain plot of the filter $|\omega|$ after band-limiting it with a box filter. (b) Spatial domain representation. (c) Hamming windowing function. (d) Windowed ramp filter, formed as the product of (a) and (c). (e) Spatial representation of the product (note the decrease in ringing).

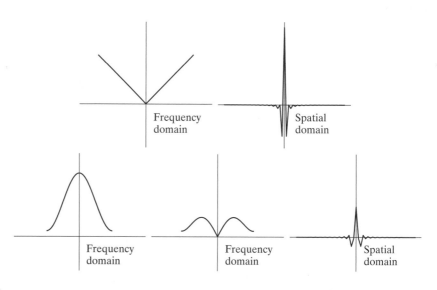

†The ramp filter often is referred to as the *Ram-Lak filter*, after Ramachandran and Lakshminarayanan [1971] who generally are credited with having been first to suggest it.

The simplest approach to band-limit a function is to use a box in the frequency domain. However, as we saw in Fig. 4.4, a box has undesirable ringing properties, so a smooth window is used instead. Figure 5.42(a) shows a plot of the ramp filter after it was band-limited by a box window, and Fig. 5.42(b) shows its spatial domain representation, obtained by computing its inverse Fourier transform. As expected, the resulting windowed filter exhibits noticeable ringing in the spatial domain. We know from Chapter 4 that filtering in the frequency domain is equivalent to convolution in the spatial domain, so spatial filtering with a function that exhibits ringing will produce a result corrupted by ringing also. Windowing with a smooth function helps this situation. An M-point discrete window function used frequently for implementation with the 1-D FFT is given by

$$
h(\omega) = \begin{cases} c + (c - 1) \cos \dfrac{2\pi\omega}{M - 1} & 0 \le \omega \le (M - 1) \\ 0 & \text{otherwise} \end{cases}
\qquad (5.11\text{-}17)
$$

When $c = 0.54$, this function is called the *Hamming window* (named after Richard Hamming) and, when $c = 0.5$, it is called the *Hann window* (named after Julius von Hann).[†] The key difference between the Hamming and Hann windows is that in the latter the end points are zero. The difference between the two generally is imperceptible in image processing applications.

Figure 5.42(c) is a plot of the Hamming window, and Fig. 5.42(d) shows the product of this window and the band-limited ramp filter in Fig. 5.42(a). Figure 5.42(e) shows the representation of the product in the spatial domain, obtained as usual by computing the inverse FFT. It is evident by comparing this figure and Fig. 5.42(b) that ringing was reduced in the windowed ramp (the ratios of the peak to trough in Figs. 5.42(b) and (e) are 2.5 and 3.4, respectively). On the other hand, because the width of the central lobe in Fig. 5.42(e) is slightly wider than in Fig. 5.42(b), we would expect backprojections based on using a Hamming window to have less ringing but be slightly more blurred. As Example 5.19 shows, this indeed is the case.

Recall from Eq. (5.11-8) that $G(\omega, \theta)$ is the 1-D Fourier transform of $g(\rho, \theta)$, which is a *single* projection obtained at a fixed angle, θ. Equation (5.11-16) states that the *complete*, back-projected image $f(x,y)$ is obtained as follows:

1. Compute the 1-D Fourier transform of each projection.
2. Multiply each Fourier transform by the filter function $|\omega|$ which, as explained above, has been multiplied by a suitable (e.g., Hamming) window.
3. Obtain the inverse 1-D Fourier transform of each resulting filtered transform.
4. Integrate (sum) all the 1-D inverse transforms from step 3.

[†]Sometimes the Hann window is referred to as the *Hanning window* in analogy to the Hamming window. However, this terminology is incorrect and is a frequent source of confusion.

Because a filter function is used, this image reconstruction approach is appropriately called *filtered backprojection*. In practice, the data are discrete, so all frequency domain computations are carried out using a 1-D FFT algorithm, and filtering is implemented using the same basic procedure explained in Chapter 4 for 2-D functions. Alternatively, we can implement filtering in the spatial domain using convolution, as explained later in this section.

The preceding discussion addresses the windowing aspects of filtered backprojections. As with any sampled data system, we also need to be concerned about sampling rates. We know from Chapter 4 that the selection of sampling rates has a profound influence on image processing results. In the present discussion, there are two sampling considerations. The first is the number of rays used, which determines the number of samples in each projection. The second is the number of rotation angle increments, which determines the number of reconstructed images (whose sum yields the final image). Under-sampling results in aliasing which, as we saw in Chapter 4, can manifest itself as artifacts in the image, such as streaks. We discuss CT sampling issues in more detail in Section 5.11.6.

EXAMPLE 5.19:
Image reconstruction using filtered backprojections.

∎ The focus of this example is to show reconstruction using filtered backprojections, first with a ramp filter and then using a ramp filter modified by a Hamming window. These filtered backprojections are compared against the results of "raw" backprojections in Fig. 5.40. In order to focus on the difference due only to filtering, the results in this example were generated with 0.5° increments of rotation, which is the increment we used to generate Fig. 5.40. The separation between rays was one pixel in both cases. The images in both examples are of size 600×600 pixels, so the length of the diagonal is $\sqrt{2} \times 600 \approx 849$. Consequently, 849 rays were used to provide coverage of the entire region when the angle of rotation was 45° and 135°.

Figure 5.43(a) shows the rectangle reconstructed using a ramp filter. The most vivid feature of this result is the absence of any visually detectable blurring. As expected, however, ringing is present, visible as faint lines, especially around the corners of the rectangle. These lines are more visible in the zoomed section in Fig. 5.43(c). Using a Hamming window on the ramp filter helped considerably with the ringing problem, at the expense of slight blurring, as Figs. 5.43(b) and (d) show. The improvements (even with the ramp filter without windowing) over Fig. 5.40(a) are evident. The phantom image does not have transitions that are as sharp and prominent as the rectangle so ringing, even with the un-windowed ramp filter, is imperceptible in this case, as you can see in Fig. 5.44(a). Using a Hamming window resulted in a slightly smoother image, as Fig. 5.44(b) shows. Both of these results are considerable improvements over Fig. 5.40(b), illustrating again the significant advantage inherent in the filtered-backprojection approach.

In most applications of CT (especially in medicine), artifacts such as ringing are a serious concern, so significant effort is devoted to minimizing them. Tuning the filtering algorithms and, as explained in Section 5.11.2,

a b
c d

FIGURE 5.43
Filtered back-projections of the rectangle using (a) a ramp filter, and (b) a Hamming-windowed ramp filter. The second row shows zoomed details of the images in the first row. Compare with Fig. 5.40(a).

using a large number of detectors are among the design considerations that help reduce these effects. ■

The preceding discussion is based on obtaining filtered backprojections via an FFT implementation. However, we know from the convolution theorem in Chapter 4 that equivalent results can be obtained using spatial convolution. In particular, note that the term inside the brackets in Eq. (5.11-16) is the inverse Fourier transform of the product of two frequency domain functions

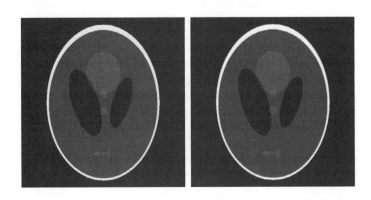

a b

FIGURE 5.44
Filtered backprojections of the head phantom using (a) a ramp filter, and (b) a Hamming-windowed ramp filter. Compare with Fig. 5.40(b).

which, according to the convolution theorem, we know to be equal to the convolution of the spatial representations (inverse Fourier transforms) of these two functions. In other words, letting $s(\rho)$ denote the inverse Fourier transform of $|\omega|$,[†] we write Eq. (5.11-16) as

$$f(x, y) = \int_0^\pi \left[\int_{-\infty}^\infty |\omega| G(\omega, \theta) e^{j2\pi\omega\rho} \, d\omega \right]_{\rho = x \cos\theta + y \sin\rho} d\theta$$

$$= \int_0^\pi [s(\rho) \star g(\rho, \theta)]_{\rho = x \cos\theta + y \sin\theta} \, d\theta \tag{5.11-18}$$

$$= \int_0^\pi \left[\int_{-\infty}^\infty g(\rho, \theta) s(x \cos\theta + y \sin\theta - \rho) \, d\rho \right] d\theta$$

where, as in Chapter 4, "$\star$" denotes convolution. The second line follows from the first for the reasons explained in the previous paragraph. The third line follows from the actual definition of convolution given in Eq. (4.2-20).

The last two lines of Eq. (5.11-18) say the same thing: Individual backprojections at an angle θ can be obtained by convolving the corresponding projection, $g(\rho, \theta)$, and the inverse Fourier transform of the ramp filter, $s(\rho)$. As before, the complete back-projected image is obtained by integrating (summing) all the individual back-projected images. With the exception of round-off differences in computation, the results of using convolution will be identical to the results using the FFT. In practical CT implementations, convolution generally turns out to be more efficient computationally, so most modern CT systems use this approach. The Fourier transform does play a central role in theoretical formulations and algorithm development (for example, CT image processing in MATLAB is based on the FFT). Also, we note that there is no need to store all the back-projected images during reconstruction. Instead, a single running sum is updated with the latest back-projected image. At the end of the procedure, the running sum will equal the sum total of all the backprojections.

Finally, we point out that, because the ramp filter (even when it is windowed) zeros the dc term in the frequency domain, each backprojection image will have a zero average value (see Fig. 4.30). This means that each backprojection image will have negative and positive pixels. When all the backprojections are added to form the final image, some negative locations may become positive and the average value may not be zero, but typically, the final image will still have negative pixels.

There are several ways to handle this problem. The simplest approach, when there is no knowledge regarding what the average values should be, is to accept the fact that negative values are inherent in the approach and scale the

---

[†]If a windowing function, such as a Hamming window, is used, then the inverse Fourier transform is performed on the windowed ramp. Also, we again ignore the issue mentioned earlier regarding the existence of the continuous inverse Fourier transform because all implementations are carried out using discrete quantities of finite length.

result using the procedure described in Eqs. (2.6-10) and (2.6-11). This is the approach followed in this section. When knowledge about what a "typical" average value should be is available, that value can be added to the filter in the frequency domain, thus offsetting the ramp and preventing zeroing the dc term [see Fig. 4.31(c)]. When working in the spatial domain with convolution, the very act of truncating the length of the spatial filter (inverse Fourier transform of the ramp) prevents it from having a zero average value, thus avoiding the zeroing problem altogether.

5.11.6 Reconstruction Using Fan-Beam Filtered Backprojections

The discussion thus far has centered on parallel beams. Because of its simplicity and intuitiveness, this is the imaging geometry used traditionally to introduce computed tomography. However, modern CT systems use a fan-beam geometry (see Fig. 5.35), the topic of discussion for the remainder of this section.

Figure 5.45 shows a basic fan-beam imaging geometry in which the detectors are arranged on a circular arc and the angular increments of the source are assumed to be equal. Let $p(\alpha, \beta)$ denote a fan-beam projection, where α is the angular position of a particular detector measured with respect to the *center ray*, and β is the angular displacement of the source, measured with respect to the y-axis, as shown in the figure. We also note in Fig. 5.45 that a ray in the fan beam can be represented as a line, $L(\rho, \theta)$, in normal form, which is the approach we used to represent a ray in the parallel-beam imaging geometry discussed in the previous sections. This allows us to utilize parallel-beam results as

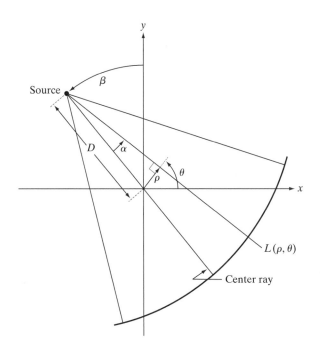

FIGURE 5.45
Basic fan-beam geometry. The line passing through the center of the source and the origin (assumed here to be the center of rotation of the source) is called the *center ray*.

the starting point for deriving the corresponding equations for the fan-beam geometry. We proceed to show this by deriving the fan-beam filtered backprojection based on convolution.[†]

We begin by noticing in Fig. 5.45 that the parameters of line $L(\rho, \theta)$ are related to the parameters of a fan-beam ray by

$$\theta = \beta + \alpha \tag{5.11-19}$$

and

$$\rho = D \sin \alpha \tag{5.11-20}$$

where D is the distance from the center of the source to the origin of the xy-plane.

The convolution backprojection formula for the parallel-beam imaging geometry is given by Eq. (5.11-18). Without loss of generality, suppose that we focus attention on objects that are encompassed within a circular area of radius T about the origin of the plane. Then $g(\rho, \theta) = 0$ for $|\rho| > T$ and Eq. (5.11-18) becomes

$$f(x, y) = \frac{1}{2} \int_0^{2\pi} \int_{-T}^{T} g(\rho, \theta) s(x \cos \theta + y \sin \theta - \rho) \, d\rho \, d\theta \tag{5.11-21}$$

where we used the fact stated in Section 5.11.1 that projections 180° apart are mirror images of each other. In this way, the limits of the outer integral in Eq. (5.11-21) are made to span a full circle, as required by a fan-beam arrangement in which the detectors are arranged in a circle.

We are interested in integrating with respect to α and β. To do this, we start by changing to polar coordinates (r, φ). That is, we let $x = r \cos \varphi$ and $y = r \sin \varphi$, from which it follows that

$$x \cos \theta + y \sin \theta = r \cos \varphi \cos \theta + r \sin \varphi \sin \theta$$
$$= r \cos(\theta - \varphi) \tag{5.11-22}$$

Using this result, we can express Eq. (5.11-21) as

$$f(x, y) = \frac{1}{2} \int_0^{2\pi} \int_{-T}^{T} g(\rho, \theta) s[r \cos(\theta - \alpha) - \rho] \, d\rho \, d\theta$$

This expression is nothing more than the parallel-beam reconstruction formula written in polar coordinates. However, integration still is with respect to ρ and θ. To integrate with respect to α and β requires a transformation of coordinates using Eqs. (5.11-19) and (5.11-20):

[†]The Fourier-slice theorem was derived for a parallel-beam geometry and is not directly applicable to fan beams. However, Eqs. (5.11-19) and (5.11-20) provide the basis for converting a fan-beam geometry to a parallel-beam geometry, thus allowing us to use the filtered parallel backprojection approach developed in the previous section, for which the slice theorem is applicable. We discuss this in more detail at the end of this section.

FIGURE 5.46
Maximum value
of α needed to
encompass a
region of interest.

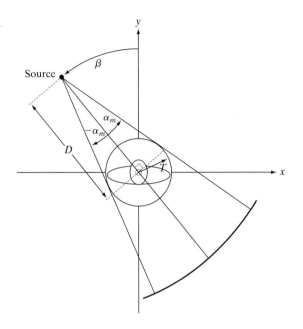

$$f(r, \varphi) = \frac{1}{2} \int_{-\alpha}^{2\pi - \alpha} \int_{\sin^{-1}(-T/D)}^{\sin^{-1}(T/D)} g(D \sin \alpha, \alpha + \beta)$$

$$s[r \cos(\beta + \alpha - \varphi) - D \sin \alpha]D \cos \alpha \, d\alpha \, d\beta \qquad (5.11\text{-}23)$$

where we used $d\rho \, d\theta = D \cos \alpha \, d\alpha \, d\beta$ [see the explanation of Eq. (5.11-13)].

This equation can be simplified further. First, note that the limits $-\alpha$ to $2\pi - \alpha$ for β span the entire range of 360°. Because all functions of β are periodic, with period 2π, the limits of the outer integral can be replaced by 0 and 2π, respectively. The term $\sin^{-1}(T/D)$ has a maximum value, α_m, corresponding to $|\rho| > T$, beyond which $g = 0$ (see Fig. 5.46), so we can replace the limits of the inner integral by $-\alpha_m$ and α_m, respectively. Finally, consider the line $L(\rho, \theta)$ in Fig. 5.45. A raysum of a fan beam along this line must equal the raysum of a parallel beam along the same line (a raysum is a sum of all values along a line, so the result must be the same for a given ray, regardless of the coordinate system is which it is expressed). This is true of any raysum for corresponding values of (α, β) and (ρ, θ). Thus, letting $p(\alpha, \beta)$ denote a fan-beam projection, it follows that $p(\alpha, \beta) = g(\rho, \theta)$ and, from Eqs. (5.11-19) and (5.11-20), that $p(\alpha, \beta) = g(D \sin \alpha, \alpha + \beta)$. Incorporating these observations into Eq. (5.11-23) results in the expression

$$f(r, \varphi) = \frac{1}{2} \int_{0}^{2\pi} \int_{-\alpha_m}^{\alpha_m} p(\alpha, \beta)s[r \cos(\beta + \alpha - \varphi) - D \sin \alpha] \qquad (5.11\text{-}24)$$

$$D \cos \alpha \, d\alpha \, d\beta$$

FIGURE 5.47
Polar represen-
tation of an arbi-
trary point on a
ray of a fan beam.

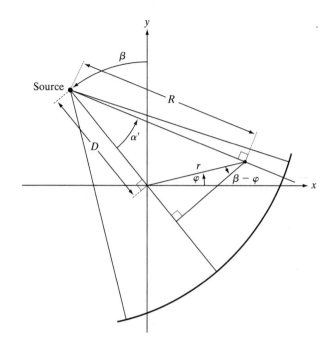

This is the fundamental fan-beam reconstruction formula based on filtered backprojections.

Equation (5.11-24) can be manipulated further to put it in a more familiar convolution form. With reference to Fig. 5.47, it can be shown (Problem 5.33) that

$$r \cos(\beta + \alpha - \varphi) - D \sin \alpha = R \sin(\alpha' - \alpha) \qquad (5.11\text{-}25)$$

where R is the distance from the source to an arbitrary point in a fan ray, and α' is the angle between this ray and the center ray. Note that R and α' are determined by the values of r, φ, and β. Substituting Eq. (5.11-25) into Eq. (5.11-24) yields

$$f(r, \varphi) = \frac{1}{2} \int_0^{2\pi} \int_{-\alpha_m}^{\alpha_m} p(\alpha, \beta) s[R \sin(\alpha' - \alpha)] D \cos \alpha \, d\alpha \, d\beta \qquad (5.11\text{-}26)$$

It can be shown (Problem 5.34) that

$$s(R \sin \alpha) = \left(\frac{\alpha}{R \sin \alpha} \right)^2 s(\alpha) \qquad (5.11\text{-}27)$$

Using this expression, we can write Eq. (5.11-26) as

$$f(r, \varphi) = \int_0^{2\pi} \frac{1}{R^2} \left[\int_{-\alpha_m}^{\alpha_m} q(\alpha, \beta) h(\alpha' - \alpha) \, d\alpha \right] d\beta \qquad (5.11\text{-}28)$$

where

$$h(\alpha) = \frac{1}{2}\left(\frac{\alpha}{\sin \alpha}\right)^2 s(\alpha) \qquad (5.11\text{-}29)$$

and

$$q(\alpha, \beta) = p(\alpha, \beta) \, D \cos \alpha \qquad (5.11\text{-}30)$$

We recognize the inner integral in Eq. (5.11-28) as a convolution expression, thus showing that the image reconstruction formula in Eq. (5.11-24) can be implemented as the convolution of functions $q(\alpha, \beta)$ and $h(\alpha)$. Unlike the reconstruction formula for parallel projections, reconstruction based on fan-beam projections involves a term $1/R^2$, which is a weighting factor inversely proportional to the distance from the source. The computational details of implementing Eq. (5.11-28) are beyond the scope of the present discussion (see Kak and Slaney [2001] for a detailed treatment of this subject).

Instead of implementing Eq. (5.11-28) directly, an approach used often, particularly in software simulations, is (1) to convert a fan-beam geometry to a parallel-beam geometry using Eqs. (5.11-19) and (5.11-20), and (2) use the parallel-beam reconstruction approach developed in Section 5.11.5. We conclude this section with an example of how this is done. As noted earlier, a fan-beam projection, p, taken at angle β has a corresponding parallel-beam projection, g, taken at a corresponding angle θ and, therefore,

$$p(\alpha, \beta) = g(\rho, \theta)$$
$$= g(D \sin \alpha, \alpha + \beta) \qquad (5.11\text{-}31)$$

where the second line follows from Eqs. (5.11-19) and (5.11-20).

Let $\Delta\beta$ denote the angular increment between successive fan-beam projections and let $\Delta\alpha$ be the angular increment between rays, which determines the number of samples in each projection. We impose the restriction that

$$\Delta\beta = \Delta\alpha = \gamma \qquad (5.11\text{-}32)$$

Then, $\beta = m\gamma$ and $\alpha = n\gamma$ for some integer values of m and n, and we can write Eq. (5.11-31) as

$$p(n\gamma, m\gamma) = g[D \sin n\gamma, (m + n)\gamma] \qquad (5.11\text{-}33)$$

This equation indicates that the nth ray in the mth radial projection is equal to the nth ray in the $(m + n)$th parallel projection. The $D \sin \gamma$ term on the right side of (5.11-33) implies that parallel projections converted from fan-beam projections are not sampled uniformly, an issue that can lead to blurring, ringing, and aliasing artifacts if the sampling intervals $\Delta\alpha$ and $\Delta\beta$ are too coarse, as the following example illustrates.

EXAMPLE 5.20:
Image
reconstruction
using filtered fan
backprojections.

■ Figure 5.48(a) shows the results of (1) generating fan projections of the rectangle image with $\Delta\alpha = \Delta\beta = 1°$, (2) converting each fan ray to the corresponding parallel ray using Eq. (5.11-33), and (3) using the filtered backprojection approach developed in Section 5.11.5 for parallel rays. Figures 5.48(b) through (d) show the results using $0.5°, 0.25°$, and $0.125°$ increments. A Hamming window was used in all cases. This variety of angle increments was used to illustrate the effects of under-sampling.

The result in Fig. 5.48(a) is a clear indication that $1°$ increments are too coarse, as blurring and ringing are quite evident. The result in (b) is interesting, in the sense that it compares poorly with Fig. 5.43(b), which was generated using the same angle increment of $0.5°$. In fact, as Fig. 5.48(c) shows, even with angle increments of $0.25°$ the reconstruction still is not as good as in Fig. 5.43(b). We have to use angle increments on the order of $0.125°$ before the two results become comparable, as Fig. 5.48(d) shows. This angle increment results in projections with $180 \times (1/0.25) = 720$ samples, which is close to the 849 rays used in the parallel projections of Example 5.19. Thus, it is not unexpected that the results are close in appearance when using $\Delta\alpha = 0.125°$.

Similar results were obtained with the head phantom, except that aliasing is much more visible as sinusoidal interference. We see in Fig. 5.49(c) that even with $\Delta\alpha = \Delta\beta = 0.25$ significant distortion still is present, especially in the periphery of the ellipse. As with the rectangle, using increments of $0.125°$ finally

a b
c d

FIGURE 5.48
Reconstruction of
the rectangle
image from
filtered fan
backprojections.
(a) $1°$ increments
of α and β. (b) $0.5°$
increments.
(c) $0.25°$ incre-
ments. (d) $0.125°$
increments.
Compare (d) with
Fig. 5.43(b).

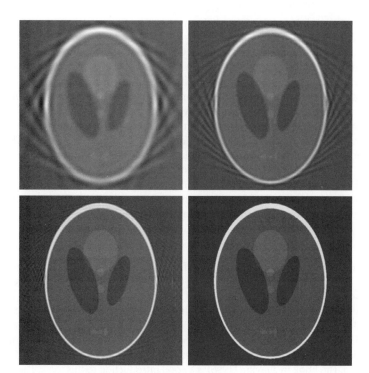

a b
c d

FIGURE 5.49
Reconstruction of the head phantom image from filtered fan backprojections. (a) 1° increments of α and β. (b) 0.5° increments. (c) 0.25° increments. (d) 0.125° increments. Compare (d) with Fig. 5.44(b).

produced results that are comparable with the back-projected image of the head phantom in Fig. 5.44(b). These results illustrate one of the principal reasons why thousands of detectors have to be used in the fan-beam geometry of modern CT systems in order to reduce aliasing artifacts.   ■

Summary

The restoration results in this chapter are based on the assumption that image degradation can be modeled as a linear, position invariant process followed by additive noise that is not correlated with image values. Even when these assumptions are not entirely valid, it often is possible to obtain useful results by using the methods developed in the preceding sections.

Some of the restoration techniques derived in this chapter are based on various criteria of optimality. Use of the word "optimal" in this context refers strictly to a mathematical concept, not to optimal response of the human visual system. In fact, the present lack of knowledge about visual perception precludes a general formulation of the image restoration problem that takes into account observer preferences and capabilities. In view of these limitations, the advantage of the concepts introduced in this chapter is the development of fundamental approaches that have reasonably predictable behavior and are supported by a solid body of knowledge.

As in Chapters 3 and 4, certain restoration tasks, such as random-noise reduction, are carried out in the spatial domain using convolution masks. The frequency domain was found ideal for reducing periodic noise and for modeling some important degradations, such as blur caused by motion during image acquisition. We also found the frequency domain to be a useful tool for formulating restoration filters, such as the Wiener and constrained least-squares filters.

As mentioned in Chapter 4, the frequency domain offers an intuitive, solid base for experimentation. Once an approach (filter) has been found to perform satisfactorily for a given application, implementation usually is carried out via the design of a digital filter that approximates the frequency domain solution, but runs much faster in a computer or in a dedicated hardware/firmware system, as indicated at the end of Chapter 4.

Our treatment of image reconstruction from projections, though introductory, is the foundation for the image-processing aspects of this field. As noted in Section 5.11, computed tomography (CT) is the main application area of image reconstruction from projections. Although we focused on X-ray tomography, the principles established in Section 5.11 are applicable in other CT imaging modalities, such as SPECT (single photon emission tomography), PET (positron emission tomography), MRI (magnetic resonance imaging), and some modalities of ultrasound imaging.

References and Further Reading

For additional reading on the linear model of degradation in Section 5.1, see Castleman [1996] and Pratt [1991]. The book by Peebles [1993] provides an intermediate-level coverage of noise probability density functions and their properties (Section 5.2). The book by Papoulis [1991] is more advanced and covers these concepts in more detail. References for Section 5.3 are Umbaugh [2005], Boie and Cox [1992], Hwang and Haddad [1995], and Wilburn [1998]. See Eng and Ma [2001, 2006] regarding adaptive median filtering. The general area of adaptive filter design is good background for the adaptive filters discussed in Section 5.3. The book by Haykin [1996] is a good introduction to this topic. The filters in Section 5.4 are direct extensions of the material in Chapter 4. For additional reading on the material of Section 5.5, see Rosenfeld and Kak [1982] and Pratt [1991].

The topic of estimating the degradation function (Section 5.6) is an area of considerable current interest. Some of the early techniques for estimating the degradation function are given in Andrews and Hunt [1977], Rosenfeld and Kak [1982], Bates and McDonnell [1986], and Stark [1987]. Since the degradation function seldom is known exactly, a number of techniques have been proposed over the years, in which specific aspects of restoration are emphasized. For example, Geman and Reynolds [1992] and Hurn and Jennison [1996] deal with issues of preserving sharp intensity transitions in an attempt to emphasize sharpness, while Boyd and Meloche [1998] are concerned with restoring thin objects in degraded images. Examples of techniques that deal with image blur are Yitzhaky et al. [1998], Harikumar and Bresler [1999], Mesarovic [2000], and Giannakis and Heath [2000]. Restoration of sequences of images also is of considerable interest. The book by Kokaram [1998] provides a good foundation in this area.

The filtering approaches discussed in Sections 5.7 through 5.10 have been explained in various ways over the years in numerous books and articles on image processing. There are two major approaches underpinning the development of these filters. One is based on a general formulation using matrix theory, as introduced by Andrews and Hunt [1977]. This approach is elegant and general, but it is difficult for newcomers to the field because it lacks intuitiveness. Approaches based directly on frequency domain filtering (the approach we followed in this chapter) usually are easier to follow by those who first encounter restoration, but lack the unifying mathematical rigor of the matrix approach. Both approaches arrive at the same results, but our experience in teaching this material in a variety of settings indicates that students first entering this field favor the latter approach by a significant margin. Complementary readings for our coverage of the filtering concepts presented in Sections 5.7 through 5.10 are Castleman [1996],

Umbaugh [2005], and Petrou and Bosdogianni [1999]. This last reference also presents a nice tie between two-dimensional frequency domain filters and the corresponding digital filters. On the design of 2-D digital filters, see Lu and Antoniou [1992].

Basic references for computed tomography are Rosenfeld and Kak [1982], Kak and Slaney [2001], and Prince and Links [2006]. For further reading on the Shepp-Logan phantom see Shepp and Logan [1974], and for additional details on the origin of the Ram-Lak filter see Ramachandran and Lakshminarayanan [1971]. The paper by O'Connor and Fessler [2006] is representative of current research in the signal and image processing aspects of computed tomography.

For software techniques to implement most of the material discussed in this chapter see Gonzalez, Woods, and Eddins [2004].

Problems

★ **5.1**   The white bars in the test pattern shown are 7 pixels wide and 210 pixels high. The separation between bars is 17 pixels. What would this image look like after application of

Detailed solutions to the problems marked with a star can be found in the book Web site. The site also contains suggested projects based on the material in this chapter.

(a) A 3 × 3 arithmetic mean filter?

(b) A 7 × 7 arithmetic mean filter?

(c) A 9 × 9 arithmetic mean filter?

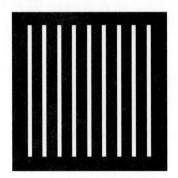

Note: This problem and the ones that follow it, related to filtering this image, may seem a bit tedious. However, they are worth the effort, as they help develop a real understanding of how these filters work. After you understand how a particular filter affects the image, your answer can be a brief verbal description of the result. For example, "the resulting image will consist of vertical bars 3 pixels wide and 206 pixels high." Be sure to describe any deformation of the bars, such as rounded corners. You may ignore image border effects, in which the masks only partially contain image pixels.

5.2   Repeat Problem 5.1 using a geometric mean filter.

★ **5.3**   Repeat Problem 5.1 using a harmonic mean filter.

5.4   Repeat Problem 5.1 using a contraharmonic mean filter with $Q = 1$.

★ **5.5**   Repeat Problem 5.1 using a contraharmonic mean filter with $Q = -1$.

5.6   Repeat Problem 5.1 using a median filter.

★ **5.7**   Repeat Problem 5.1 using a max filter.

5.8 Repeat Problem 5.1 using a min filter.

★5.9 Repeat Problem 5.1 using a midpoint filter.

5.10 The two subimages shown were extracted from the top right corners of Figs. 5.7(c) and (d), respectively. Thus, the subimage on the left is the result of using an arithmetic mean filter of size 3 × 3; the other subimage is the result of using a geometric mean filter of the same size.

  ★(a) Explain why the subimage obtained with geometric mean filtering is less blurred. (*Hint:* Start your analysis by examining a 1-D step transition in intensity.)

  (b) Explain why the black components in the right image are thicker.

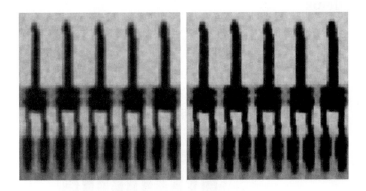

5.11 Refer to the contraharmonic filter given in Eq. (5.3-6).

  (a) Explain why the filter is effective in elimination pepper noise when Q is positive.

  (b) Explain why the filter is effective in eliminating salt noise when Q is negative.

  (c) Explain why the filter gives poor results (such as the results shown in Fig. 5.9) when the wrong polarity is chosen for Q.

  (d) Discuss the behavior of the filter when $Q = -1$.

  (e) Discuss (for positive and negative Q) the behavior of the filter in areas of constant intensity levels.

★5.12 Obtain equations for the bandpass filters corresponding to the bandreject filters in Table 4.6.

5.13 Obtain equations for Gaussian and ideal notch reject filters in the form of Eq. (4.10-5).

★5.14 Show that the Fourier transform of the 2-D continuous sine function

$$f(x, y) = A \sin(u_0 x + v_0 y)$$

is the pair of conjugate impulses

$$F(u, v) = -j\frac{A}{2}\left[\delta\left(u - \frac{u_0}{2\pi}, v - \frac{v_0}{2\pi}\right) - \delta\left(u + \frac{u_0}{2\pi}, v + \frac{v_0}{2\pi}\right)\right]$$

[*Hint:* Use the continuous version of the Fourier transform in Eq. (4.5-7), and express the sine in terms of exponentials.]

5.15   Start with Eq. (5.4-11) and derive Eq. (5.4-13).

★**5.16**   Consider a linear, position-invariant image degradation system with impulse response

$$h(x - \alpha, y - \beta) = e^{-[(x-\alpha)^2 + (y-\beta)^2]}$$

Suppose that the input to the system is an image consisting of a line of infinitesimal width located at $x = a$, and modeled by $f(x, y) = \delta(x - a)$, where δ is an impulse. Assuming no noise, what is the output image $g(x, y)$?

5.17   During acquisition, an image undergoes uniform linear motion in the vertical direction for a time T_1. The direction of motion then switches to the horizontal direction for a time interval T_2. Assuming that the time it takes the image to change directions is negligible, and that shutter opening and closing times are negligible also, give an expression for the blurring function, $H(u, v)$.

★**5.18**   Consider the problem of image blurring caused by uniform acceleration in the x-direction. If the image is at rest at time $t = 0$ and accelerates with a uniform acceleration $x_0(t) = at^2/2$ for a time T, find the blurring function $H(u, v)$. You may assume that shutter opening and closing times are negligible.

5.19   A space probe is designed to transmit images from a planet as it approaches it for landing. During the last stages of landing, one of the control thrusters fails, resulting in rapid rotation of the craft about its vertical axis. The images sent during the last two seconds prior to landing are blurred as a consequence of this circular motion. The camera is located in the bottom of the probe, along its vertical axis, and pointing down. Fortunately, the rotation of the craft is also about its vertical axis, so the images are blurred by uniform rotational motion. During the acquisition time of each image the craft rotation was limited to $\pi/8$ radians. The image acquisition process can be modeled as an ideal shutter that is open only during the time the craft rotated the $\pi/8$ radians. You may assume that vertical motion was negligible during image acquisition. Formulate a solution for restoring the images.

★**5.20**   The image shown is a blurred, 2-D projection of a volumetric rendition of a heart. It is known that each of the cross hairs on the right bottom part of the image was 3 pixels wide, 30 pixels long, and had an intensity value of 255 before blurring. Provide a step-by-step procedure indicating how you would use the information just given to obtain the blurring function $H(u, v)$.

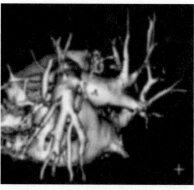

(Original image courtesy of G.E. Medical Systems.)

5.21   A certain X-ray imaging geometry produces a blurring degradation that can be modeled as the convolution of the sensed image with the spatial, circularly symmetric function

$$h(x, y) = \frac{x^2 + y^2 - 2\sigma^2}{\sigma^4} e^{-\frac{x^2+y^2}{2\sigma^2}}$$

Assuming continuous variables, show that the degradation in the frequency domain is given by the expression

$$H(u, v) = -8\pi^2\sigma^2(u^2 + v^2)e^{-2\pi^2\sigma^2(u^2+v^2)}$$

(*Hint:* Refer to Section 4.9.4, entry 13 in Table 4.3, and Problem 4.26.)

★**5.22**   Using the transfer function in Problem 5.21, give the expression for a Wiener filter, assuming that the ratio of power spectra of the noise and undegraded signal is a constant.

5.23   Using the transfer function in Problem 5.21, give the resulting expression for the constrained least squares filter.

5.24   Assume that the model in Fig. 5.1 is linear and position invariant and that the noise and image are uncorrelated. Show that the power spectrum of the output is

$$|G(u, v)|^2 = |H(u, v)|^2 |F(u, v)|^2 + |N(u, v)|^2$$

Refer to Eqs. (5.5-17) and (4.6-18).

5.25   Cannon [1974] suggested a restoration filter $R(u, v)$ satisfying the condition

$$|\hat{F}(u, v)|^2 = |R(u, v)|^2 |G(u, v)|^2$$

and based on the premise of forcing the power spectrum of the restored image, $|\hat{F}(u, v)|^2$, to equal the power spectrum of the original image, $|F(u, v)|^2$. Assume that the image and noise are uncorrelated.

★**(a)**   Find $R(u, v)$ in terms of $|F(u, v)|^2$, $|H(u, v)|^2$, and $|N(u, v)|^2$. [*Hint:* Refer to Fig. 5.1, Eq. (5.5-17), and Problem 5.24.]

(b)   Use your result in (a) to state a result in the form of Eq. (5.8-2).

5.26   An astronomer working with a large-scale telescope observes that her images are a little blurry. The manufacturer tells the astronomer that the unit is operating within specifications. The telescope lenses focus images onto a high-resolution, CCD imaging array, and the images are then converted by the telescope electronics into digital images. Trying to improve the situation by conducting controlled lab experiments with the lenses and imaging sensors is not possible due to the size and weight of the telescope components. The astronomer, having heard about your success as an image processing expert, calls you to help her formulate a digital image processing solution for sharpening the images a little more. How would you go about solving this problem, given that the only images you can obtain are images of stellar bodies?

★**5.27**   A professor of archeology doing research on currency exchange practices during the Roman Empire recently became aware that four Roman coins crucial to his research are listed in the holdings of the British Museum in London. Unfortunately, he was told after arriving there that the coins recently had been stolen. Further research on his part revealed that the museum keeps photographs of every item for which it is responsible. Unfortunately, the photos of the coins in question are blurred to the point where the date and other small markings are not readable. The cause of the blurring was the camera being out of focus when the pictures were taken. As an image processing expert and friend of the professor, you are asked as a favor to determine whether computer processing can be utilized to restore the images to the point where the professor can read the markings. You are told that the original camera used to take the photos is still available, as are other representative coins of the same era. Propose a step-by-step solution to this problem.

5.28 Sketch the Radon transform of the following square images. Label quantitatively all the important features of your sketches. Figure (a) consists of one dot in the center, and (b) has two dots along the diagonal. Describe your solution to (c) by an intensity profile. Assume a parallel-beam geometry.

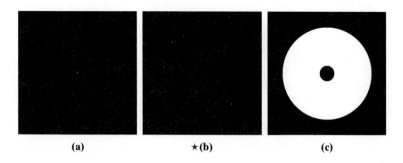

(a)   ★(b)   (c)

5.29 Show that the Radon transform [Eq. (5.11-3)] of the Gaussian shape $f(x, y) = A \exp(-x^2 - y^2)$ is $g(\rho, \theta) = A\sqrt{\pi} \exp(-\rho^2)$. (*Hint:* Refer to Example 5.17, where we used symmetry to simplify integration.)

5.30 ★(a) Show that the Radon transform [Eq. (5.11-3)] of the unit impulse $\delta(x, y)$ is a straight vertical line in the $\rho\theta$-plane passing through the origin.

    (b) Show that the radon transform of the impulse $\delta(x - x_0, y - y_0)$ is a sinusoidal curve in the $\rho\theta$-plane.

5.31 Prove the validity of the following properties of the Radon transform [Eq. (5.11-3)]:

    ★(a) Linearity: The Radon transform is a linear operator. (See Section 2.6.2 regarding the definition of linear operators.)

    (b) Translation property: The radon transform of $f(x - x_0, y - y_0)$ is $g(\rho - x_0 \cos_\theta - y_0 \sin_\theta, \theta)$.

    ★(c) Convolution property: Show that the Radon transform of the convolution of two functions is equal to the convolution of the Radon transforms of the two functions.

5.32 Provide the steps leading from Eq. (5.11-14) to (5.11-15). You will need to use the property $G(\omega, \theta + 180°) = G(-\omega, \theta)$.

★**5.33** Prove the validity of Eq. (5.11-25).

5.34 Prove the validity of Eq. (5.11-27).

6 *Color Image Processing*

It is only after years of preparation that the young artist should touch
color—not color used descriptively, that is, but as a means of
personal expression.

Henri Matisse

For a long time I limited myself to one color—as a form of discipline.

Pablo Picasso

Preview

The use of color in image processing is motivated by two principal factors.
First, color is a powerful descriptor that often simplifies object identification
and extraction from a scene. Second, humans can discern thousands of color
shades and intensities, compared to about only two dozen shades of gray. This
second factor is particularly important in manual (i.e., when performed by humans) image analysis.

Color image processing is divided into two major areas: *full-color* and
pseudocolor processing. In the first category, the images in question typically
are acquired with a full-color sensor, such as a color TV camera or color scanner. In the second category, the problem is one of assigning a color to a particular monochrome intensity or range of intensities. Until relatively recently,
most digital color image processing was done at the pseudocolor level. However, in the past decade, color sensors and hardware for processing color images have become available at reasonable prices. The result is that full-color
image processing techniques are now used in a broad range of applications, including publishing, visualization, and the Internet.

It will become evident in the discussions that follow that some of the gray-scale
methods covered in previous chapters are directly applicable to color images.

Others require reformulation to be consistent with the properties of the color spaces developed in this chapter. The techniques described here are far from exhaustive; they illustrate the range of methods available for color image processing.

6.1 Color Fundamentals

Although the process followed by the human brain in perceiving and interpreting color is a physiopsychological phenomenon that is not fully understood, the physical nature of color can be expressed on a formal basis supported by experimental and theoretical results.

In 1666, Sir Isaac Newton discovered that when a beam of sunlight passes through a glass prism, the emerging beam of light is not white but consists instead of a continuous spectrum of colors ranging from violet at one end to red at the other. As Fig. 6.1 shows, the color spectrum may be divided into six broad regions: violet, blue, green, yellow, orange, and red. When viewed in full color (Fig. 6.2), no color in the spectrum ends abruptly, but rather each color blends smoothly into the next.

Basically, the colors that humans and some other animals perceive in an object are determined by the nature of the light reflected from the object. As illustrated in Fig. 6.2, visible light is composed of a relatively narrow band of frequencies in the electromagnetic spectrum. A body that reflects light that is balanced in all visible wavelengths appears white to the observer. However, a body that favors reflectance in a limited range of the visible spectrum exhibits some shades of color. For example, green objects reflect light with wavelengths primarily in the 500 to 570 nm range while absorbing most of the energy at other wavelengths.

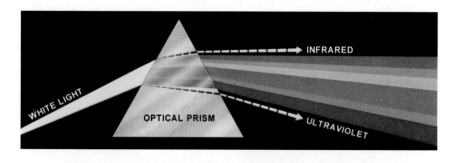

FIGURE 6.1 Color spectrum seen by passing white light through a prism. (Courtesy of the General Electric Co., Lamp Business Division.)

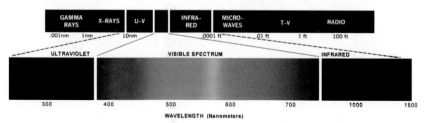

FIGURE 6.2 Wavelengths comprising the visible range of the electromagnetic spectrum. (Courtesy of the General Electric Co., Lamp Business Division.)

Characterization of light is central to the science of color. If the light is achromatic (void of color), its only attribute is its *intensity*, or amount. Achromatic light is what viewers see on a black and white television set, and it has been an implicit component of our discussion of image processing thus far. As defined in Chapter 2, and used numerous times since, the term *gray level* refers to a scalar measure of intensity that ranges from black, to grays, and finally to white.

Chromatic light spans the electromagnetic spectrum from approximately 400 to 700 nm. Three basic quantities are used to describe the quality of a chromatic light source: radiance, luminance, and brightness. *Radiance* is the total amount of energy that flows from the light source, and it is usually measured in watts (W). *Luminance*, measured in lumens (lm), gives a measure of the amount of energy an observer *perceives* from a light source. For example, light emitted from a source operating in the far infrared region of the spectrum could have significant energy (radiance), but an observer would hardly perceive it; its luminance would be almost zero. Finally, *brightness* is a subjective descriptor that is practically impossible to measure. It embodies the achromatic notion of intensity and is one of the key factors in describing color sensation.

As noted in Section 2.1.1, cones are the sensors in the eye responsible for color vision. Detailed experimental evidence has established that the 6 to 7 million cones in the human eye can be divided into three principal sensing categories, corresponding roughly to red, green, and blue. Approximately 65% of all cones are sensitive to red light, 33% are sensitive to green light, and only about 2% are sensitive to blue (but the blue cones are the most sensitive). Figure 6.3 shows average experimental curves detailing the absorption of light by the red, green, and blue cones in the eye. Due to these absorption characteristics of the

FIGURE 6.3
Absorption of light by the red, green, and blue cones in the human eye as a function of wavelength.

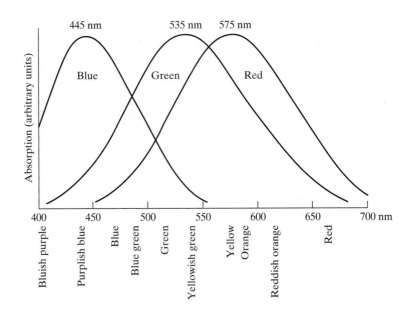

human eye, colors are seen as variable combinations of the so-called *primary colors* red (*R*), green (*G*), and blue (*B*). For the purpose of standardization, the CIE (Commission Internationale de l'Eclairage—the International Commission on Illumination) designated in 1931 the following specific wavelength values to the three primary colors: blue = 435.8 nm, green = 546.1 nm, and red = 700 nm. This standard was set before the detailed experimental curves shown in Fig. 6.3 became available in 1965. Thus, the CIE standards correspond only approximately with experimental data. We note from Figs. 6.2 and 6.3 that no single color may be called red, green, or blue. Also, it is important to keep in mind that having three specific primary color wavelengths for the purpose of standardization does not mean that these three fixed RGB components acting alone can generate all spectrum colors. Use of the word *primary* has been widely misinterpreted to mean that the three standard primaries, when mixed in various intensity proportions, can produce *all* visible colors. As you will see shortly, this interpretation is not correct unless the wavelength also is allowed to vary, in which case we would no longer have three fixed, standard primary colors.

The primary colors can be added to produce the *secondary* colors of light—magenta (red plus blue), cyan (green plus blue), and yellow (red plus green). Mixing the three primaries, or a secondary with its opposite primary color, in the right intensities produces white light. This result is shown in Fig. 6.4(a), which also illustrates the three primary colors and their combinations to produce the secondary colors.

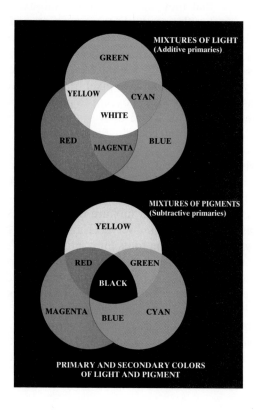

a
b

FIGURE 6.4
Primary and secondary colors of light and pigments. (Courtesy of the General Electric Co., Lamp Business Division.)

Differentiating between the primary colors of light and the primary colors of pigments or colorants is important. In the latter, a primary color is defined as one that subtracts or absorbs a primary color of light and reflects or transmits the other two. Therefore, the primary colors of pigments are magenta, cyan, and yellow, and the secondary colors are red, green, and blue. These colors are shown in Fig. 6.4(b). A proper combination of the three pigment primaries, or a secondary with its opposite primary, produces black.

Color television reception is an example of the additive nature of light colors. The interior of CRT (cathode ray tube) color TV screens is composed of a large array of triangular dot patterns of electron-sensitive phosphor. When excited, each dot in a triad produces light in one of the primary colors. The intensity of the red-emitting phosphor dots is modulated by an electron gun inside the tube, which generates pulses corresponding to the "red energy" seen by the TV camera. The green and blue phosphor dots in each triad are modulated in the same manner. The effect, viewed on the television receiver, is that the three primary colors from each phosphor triad are "added" together and received by the color-sensitive cones in the eye as a full-color image. Thirty successive image changes per second in all three colors complete the illusion of a continuous image display on the screen.

CRT displays are being replaced by "flat panel" digital technologies, such as *liquid crystal displays* (LCDs) and *plasma* devices. Although they are fundamentally different from CRTs, these and similar technologies use the same principle in the sense that they all require three subpixels (red, green, and blue) to generate a single color pixel. LCDs use properties of polarized light to block or pass light through the LCD screen and, in the case of active matrix display technology, thin film transistors (TFTs) are used to provide the proper signals to address each pixel on the screen. Light filters are used to produce the three primary colors of light at each pixel triad location. In plasma units, pixels are tiny gas cells coated with phosphor to produce one of the three primary colors. The individual cells are addressed in a manner analogous to LCDs. This individual pixel triad coordinate addressing capability is the foundation of digital displays.

The characteristics generally used to distinguish one color from another are *brightness, hue*, and *saturation*. As indicated earlier in this section, brightness embodies the achromatic notion of intensity. Hue is an attribute associated with the dominant wavelength in a mixture of light waves. Hue represents dominant color as perceived by an observer. Thus, when we call an object red, orange, or yellow, we are referring to its hue. Saturation refers to the relative purity or the amount of white light mixed with a hue. The pure spectrum colors are fully saturated. Colors such as pink (red and white) and lavender (violet and white) are less saturated, with the degree of saturation being inversely proportional to the amount of white light added.

Hue and saturation taken together are called *chromaticity*, and, therefore, a color may be characterized by its brightness and chromaticity. The amounts of red, green, and blue needed to form any particular color are called the *tristimulus* values and are denoted, X, Y, and Z, respectively. A color is then specified by its *trichromatic coefficients*, defined as

$$x = \frac{X}{X + Y + Z} \tag{6.1-1}$$

$$y = \frac{Y}{X + Y + Z} \tag{6.1-2}$$

and

$$z = \frac{Z}{X + Y + Z} \tag{6.1-3}$$

It is noted from these equations that[†]

$$x + y + z = 1 \tag{6.1-4}$$

For any wavelength of light in the visible spectrum, the tristimulus values needed to produce the color corresponding to that wavelength can be obtained directly from curves or tables that have been compiled from extensive experimental results (Poynton [1996]. See also the early references by Walsh [1958] and by Kiver [1965]).

Another approach for specifying colors is to use the CIE *chromaticity diagram* (Fig. 6.5), which shows color composition as a function of x (red) and y (green). For any value of x and y, the corresponding value of z (blue) is obtained from Eq. (6.1-4) by noting that $z = 1 - (x + y)$. The point marked green in Fig. 6.5, for example, has approximately 62% green and 25% red content. From Eq. (6.1-4), the composition of blue is approximately 13%.

The positions of the various spectrum colors—from violet at 380 nm to red at 780 nm—are indicated around the boundary of the tongue-shaped chromaticity diagram. These are the pure colors shown in the spectrum of Fig. 6.2. Any point not actually on the boundary but within the diagram represents some mixture of spectrum colors. The point of equal energy shown in Fig. 6.5 corresponds to equal fractions of the three primary colors; it represents the CIE standard for white light. Any point located on the boundary of the chromaticity chart is fully saturated. As a point leaves the boundary and approaches the point of equal energy, more white light is added to the color and it becomes less saturated. The saturation at the point of equal energy is zero.

The chromaticity diagram is useful for color mixing because a straight-line segment joining any two points in the diagram defines all the different color variations that can be obtained by combining these two colors additively. Consider, for example, a straight line drawn from the red to the green points shown in Fig. 6.5. If there is more red light than green light, the exact point representing the new color will be on the line segment, but it will be closer to the red point than to the green point. Similarly, a line drawn from the point of equal

---

[†]The use of x, y, z in this context follows notational convention. These should not be confused with the use of (x, y) to denote spatial coordinates in other sections of the book.

FIGURE 6.5
Chromaticity
diagram.
(Courtesy of the
General Electric
Co., Lamp
Business
Division.)

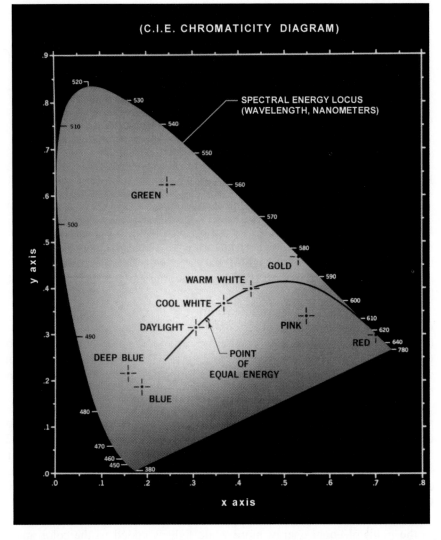

energy to any point on the boundary of the chart will define all the shades of
that particular spectrum color.

Extension of this procedure to three colors is straightforward. To determine
the range of colors that can be obtained from any three given colors in the
chromaticity diagram, we simply draw connecting lines to each of the three
color points. The result is a triangle, and any color on the boundary or inside
the triangle can be produced by various combinations of the three initial colors. A triangle with vertices at any three *fixed* colors cannot enclose the entire
color region in Fig. 6.5. This observation supports graphically the remark made
earlier that not all colors can be obtained with three single, fixed primaries.

The triangle in Figure 6.6 shows a typical range of colors (called the *color
gamut*) produced by RGB monitors. The irregular region inside the triangle
is representative of the color gamut of today's high-quality color printing

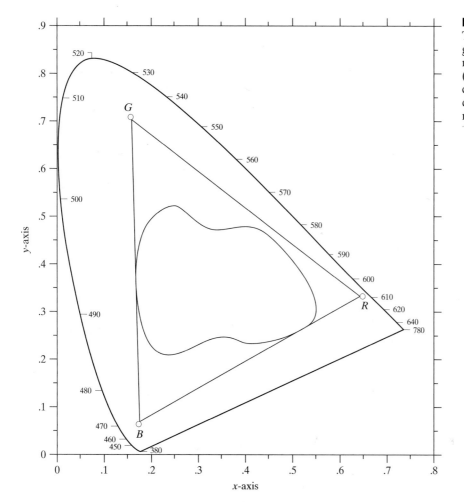

FIGURE 6.6
Typical color
gamut of color
monitors
(triangle) and
color printing
devices (irregular
region).

devices. The boundary of the color printing gamut is irregular because color printing is a combination of additive and subtractive color mixing, a process that is much more difficult to control than that of displaying colors on a monitor, which is based on the addition of three highly controllable light primaries.

6.2   Color Models

The purpose of a color model (also called *color space* or *color system*) is to facilitate the specification of colors in some standard, generally accepted way. In essence, a color model is a specification of a coordinate system and a subspace within that system where each color is represented by a single point.

Most color models in use today are oriented either toward hardware (such as for color monitors and printers) or toward applications where color manipulation is a goal (such as in the creation of color graphics for animation). In terms of digital image processing, the hardware-oriented models most commonly used in practice are the RGB (red, green, blue) model for color monitors and a broad class of color video cameras; the CMY (cyan, magenta, yellow) and CMYK (cyan, magenta, yellow, black) models for color printing; and the HSI (hue, saturation, intensity) model, which corresponds closely with the way humans describe and interpret color. The HSI model also has the advantage that it decouples the color and gray-scale information in an image, making it suitable for many of the gray-scale techniques developed in this book. There are numerous color models in use today due to the fact that color science is a broad field that encompasses many areas of application. It is tempting to dwell on some of these models here simply because they are interesting and informative. However, keeping to the task at hand, the models discussed in this chapter are leading models for image processing. Having mastered the material in this chapter, you will have no difficulty in understanding additional color models in use today.

6.2.1 The RGB Color Model

In the RGB model, each color appears in its primary spectral components of red, green, and blue. This model is based on a Cartesian coordinate system. The color subspace of interest is the cube shown in Fig. 6.7, in which RGB primary values are at three corners; the secondary colors cyan, magenta, and yellow are at three other corners; black is at the origin; and white is at the corner farthest from the origin. In this model, the gray scale (points of equal RGB values) extends from black to white along the line joining these two points. The different colors in this model are points on or inside the cube, and are defined by vectors extending from the origin. For convenience, the assumption is that all color values have been normalized so that the cube shown in Fig. 6.7 is the unit cube. That is, all values of R, G, and B are assumed to be in the range $[0, 1]$.

FIGURE 6.7
Schematic of the RGB color cube. Points along the main diagonal have gray values, from black at the origin to white at point $(1, 1, 1)$.

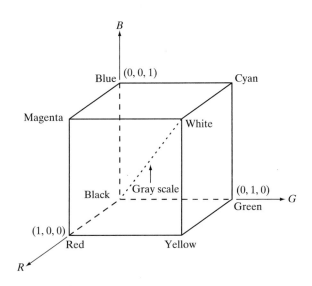

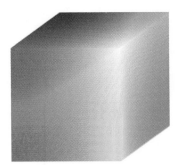

FIGURE 6.8 RGB 24-bit color cube.

Images represented in the RGB color model consist of three component images, one for each primary color. When fed into an RGB monitor, these three images combine on the screen to produce a composite color image, as explained in Section 6.1. The number of bits used to represent each pixel in RGB space is called the *pixel depth*. Consider an RGB image in which each of the red, green, and blue images is an 8-bit image. Under these conditions each RGB *color* pixel [that is, a triplet of values (R, G, B)] is said to have a depth of 24 bits (3 image planes times the number of bits per plane). The term *full-color* image is used often to denote a 24-bit RGB color image. The total number of colors in a 24-bit RGB image is $(2^8)^3 = 16,777,216$. Figure 6.8 shows the 24-bit RGB color cube corresponding to the diagram in Fig. 6.7.

■ The cube shown in Fig. 6.8 is a solid, composed of the $(2^8)^3 = 16,777,216$ colors mentioned in the preceding paragraph. A convenient way to view these colors is to generate color planes (faces or cross sections of the cube). This is accomplished simply by fixing one of the three colors and allowing the other two to vary. For instance, a cross-sectional plane through the center of the cube and parallel to the *GB*-plane in Fig. 6.8 is the plane $(127, G, B)$ for $G, B = 0, 1, 2, \ldots, 255$. Here we used the actual pixel values rather than the mathematically convenient normalized values in the range [0, 1] because the former values are the ones actually used in a computer to generate colors. Figure 6.9(a) shows that an image of the cross-sectional plane is viewed simply by feeding the three individual component images into a color monitor. In the component images, 0 represents black and 255 represents white (note that these are gray-scale images). Finally, Fig. 6.9(b) shows the three hidden surface planes of the cube in Fig. 6.8, generated in the same manner.

It is of interest to note that *acquiring* a color image is basically the process shown in Fig. 6.9 in reverse. A color image can be acquired by using three filters, sensitive to red, green, and blue, respectively. When we view a color scene with a monochrome camera equipped with one of these filters, the result is a monochrome image whose intensity is proportional to the response of that filter. Repeating this process with each filter produces three monochrome images that are the RGB component images of the color scene. (In practice, RGB color image sensors usually integrate this process into a single device.) Clearly, displaying these three RGB component images in the form shown in Fig. 6.9(a) would yield an RGB color rendition of the original color scene. ■

EXAMPLE 6.1: Generating the hidden face planes and a cross section of the RGB color cube.

a
b

FIGURE 6.9
(a) Generating
the RGB image of
the cross-sectional
color plane (127,
G, B). (b) The
three hidden
surface planes in
the color cube of
Fig. 6.8.

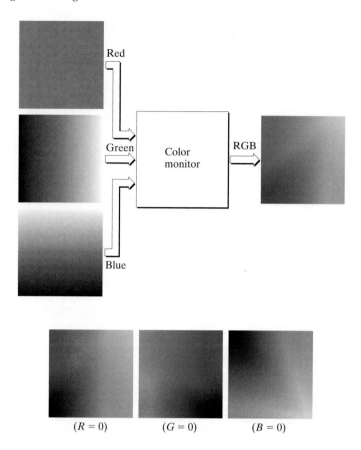

$(R = 0)$ $(G = 0)$ $(B = 0)$

While high-end display cards and monitors provide a reasonable rendition of the colors in a 24-bit RGB image, many systems in use today are limited to 256 colors. Also, there are numerous applications in which it simply makes no sense to use more than a few hundred, and sometimes fewer, colors. A good example of this is provided by the pseudocolor image processing techniques discussed in Section 6.3. Given the variety of systems in current use, it is of considerable interest to have a subset of colors that are likely to be reproduced faithfully, reasonably independently of viewer hardware capabilities. This subset of colors is called the set of *safe RGB colors*, or the set of *all-systems-safe colors*. In Internet applications, they are called *safe Web colors* or *safe browser colors*.

On the assumption that 256 colors is the minimum number of colors that can be reproduced faithfully by any system in which a desired result is likely to be displayed, it is useful to have an accepted standard notation to refer to these colors. Forty of these 256 colors are known to be processed differently by various operating systems, leaving only 216 colors that are common to most systems. These 216 colors have become the de facto standard for safe colors, especially in Internet applications. They are used whenever it is desired that the colors viewed by most people appear the same.

Number System	Color Equivalents					
Hex	00	33	66	99	CC	FF
Decimal	0	51	102	153	204	255

TABLE 6.1
Valid values of each RGB component in a safe color.

Each of the 216 safe colors is formed from three RGB values as before, but each value can only be 0, 51, 102, 153, 204, or 255. Thus, RGB triplets of these values give us $(6)^3 = 216$ possible values (note that all values are divisible by 3). It is customary to express these values in the hexagonal number system, as shown in Table 6.1. Recall that hex numbers $0, 1, 2, \ldots, 9, A, B, C, D, E, F$ correspond to decimal numbers $0, 1, 2, \ldots, 9, 10, 11, 12, 13, 14, 15$. Recall also that $(0)_{16} = (0000)_2$ and $(F)_{16} = (1111)_2$. Thus, for example, $(FF)_{16} = (255)_{10} = (11111111)_2$ and we see that a grouping of two hex numbers forms an 8-bit byte.

Since it takes three numbers to form an RGB color, each safe color is formed from three of the two digit hex numbers in Table 6.1. For example, the purest red is FF0000. The values 000000 and FFFFFF represent black and white, respectively. Keep in mind that the same result is obtained by using the more familiar decimal notation. For instance, the brightest red in decimal notation has $R = 255$ (FF) and $G = B = 0$.

Figure 6.10(a) shows the 216 safe colors, organized in descending RGB values. The square in the top left array has value FFFFFF (white), the second square to its right has value FFFFCC, the third square has value FFFF99, and

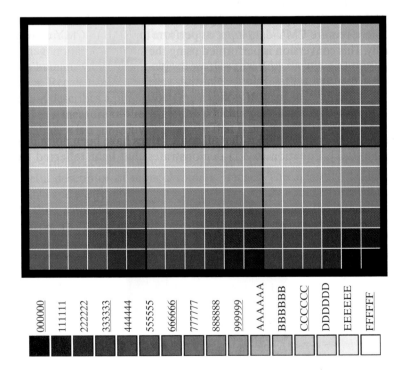

a
b

FIGURE 6.10
(a) The 216 safe RGB colors.
(b) All the grays in the 256-color RGB system (grays that are part of the safe color group are shown underlined).

Then the RGB components are

$$R = I(1 - S) \tag{6.2-9}$$

$$G = I\left[1 + \frac{S \cos H}{\cos(60° - H)}\right] \tag{6.2-10}$$

and

$$B = 3I - (R + G) \tag{6.2-11}$$

BR sector $(240° \le H \le 360°)$: Finally, if H is in this range, we subtract $240°$ from it:

$$H = H - 240° \tag{6.2-12}$$

Then the RGB components are

$$G = I(1 - S) \tag{6.2-13}$$

$$B = I\left[1 + \frac{S \cos H}{\cos(60° - H)}\right] \tag{6.2-14}$$

and

$$R = 3I - (G + B) \tag{6.2-15}$$

Uses of these equations for image processing are discussed in several of the following sections.

EXAMPLE 6.2:
The HSI values corresponding to the image of the RGB color cube.

■ Figure 6.15 shows the hue, saturation, and intensity images for the RGB values shown in Fig. 6.8. Figure 6.15(a) is the hue image. Its most distinguishing feature is the discontinuity in value along a 45° line in the front (red) plane of the cube. To understand the reason for this discontinuity, refer to Fig. 6.8, draw a line from the red to the white vertices of the cube, and select a point in the middle of this line. Starting at that point, draw a path to the right, following the cube around until you return to the starting point. The major colors encountered in this path are yellow, green, cyan, blue, magenta, and back to red. According to Fig. 6.13, the values of hue along this path should increase from 0°

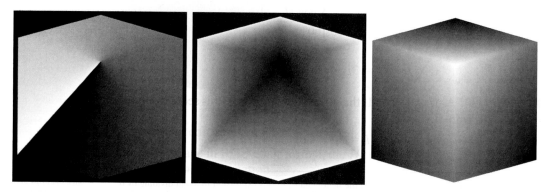

a b c

FIGURE 6.15 HSI components of the image in Fig. 6.8. (a) Hue, (b) saturation, and (c) intensity images.

to 360° (i.e., from the lowest to highest possible values of hue). This is precisely what Fig. 6.15(a) shows because the lowest value is represented as black and the highest value as white in the gray scale. In fact, the hue image was originally normalized to the range [0, 1] and then scaled to 8 bits; that is, it was converted to the range [0, 255], for display.

The saturation image in Fig. 6.15(b) shows progressively darker values toward the white vertex of the RGB cube, indicating that colors become less and less saturated as they approach white. Finally, every pixel in the intensity image shown in Fig. 6.15(c) is the average of the RGB values at the corresponding pixel in Fig. 6.8. ■

Manipulating HSI component images

In the following discussion, we take a look at some simple techniques for manipulating HSI component images. This will help you develop familiarity with these components and also help you deepen your understanding of the HSI color model. Figure 6.16(a) shows an image composed of the primary and secondary RGB colors. Figures 6.16(b) through (d) show the *H*, *S*, and *I* components of this image, generated using Eqs. (6.2-2) through (6.2-4). Recall from the discussion earlier in this section that the gray-level values in Fig. 6.16(b) correspond to angles; thus, for example, because red corresponds to 0°, the red region in Fig. 6.16(a) is mapped to a black region in the hue image. Similarly, the gray levels in Fig. 6.16(c) correspond to saturation (they were scaled to [0, 255] for display), and the gray levels in Fig. 6.16(d) are average intensities.

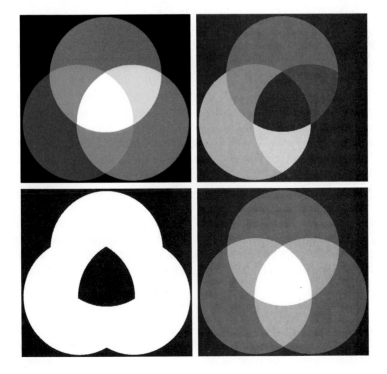

a b
c d

FIGURE 6.16
(a) RGB image and the components of its corresponding HSI image:
(b) hue,
(c) saturation, and
(d) intensity.

FIGURE 6.17
(a)–(c) Modified
HSI component
images.
(d) Resulting
RGB image. (See
Fig. 6.16 for the
original HSI
images.)

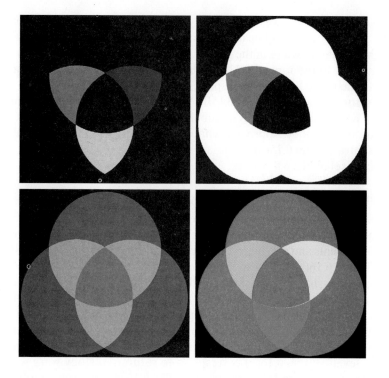

To change the individual color of any region in the RGB image, we change the values of the corresponding region in the hue image of Fig. 6.16(b). Then we convert the new H image, along with the unchanged S and I images, back to RGB using the procedure explained in connection with Eqs. (6.2-5) through (6.2-15). To change the saturation (purity) of the color in any region, we follow the same procedure, except that we make the changes in the saturation image in HSI space. Similar comments apply to changing the average intensity of any region. Of course, these changes can be made simultaneously. For example, the image in Fig. 6.17(a) was obtained by changing to 0 pixels corresponding to the blue and green regions in Fig. 6.16(b). In Fig. 6.17(b) we reduced by half the saturation of the cyan region in component image S from Fig. 6.16(c). In Fig. 6.17(c) we reduced by half the intensity of the central white region in the intensity image of Fig. 6.16(d). The result of converting this modified HSI image back to RGB is shown in Fig. 6.17(d). As expected, we see in this figure that the outer portions of all circles are now red; the purity of the cyan region was diminished, and the central region became gray rather than white. Although these results are simple, they illustrate clearly the power of the HSI color model in allowing *independent* control over hue, saturation, and intensity, quantities with which we are quite familiar when describing colors.

6.3 Pseudocolor Image Processing

Pseudocolor (also called *false color*) image processing consists of assigning colors to gray values based on a specified criterion. The term *pseudo* or *false* color is used to differentiate the process of assigning colors to monochrome images from the processes associated with true color images, a topic discussed starting in Section 6.4. The principal use of pseudocolor is for human visualization and interpretation of gray-scale events in an image or sequence of images. As noted at the beginning of this chapter, one of the principal motivations for using color is the fact that humans can discern thousands of color shades and intensities, compared to only two dozen or so shades of gray.

6.3.1 Intensity Slicing

The technique of *intensity* (sometimes called *density*) *slicing* and color coding is one of the simplest examples of pseudocolor image processing. If an image is interpreted as a 3-D function [see Fig. 2.18(a)], the method can be viewed as one of placing planes parallel to the coordinate plane of the image; each plane then "slices" the function in the area of intersection. Figure 6.18 shows an example of using a plane at $f(x, y) = l_i$ to slice the image function into two levels.

If a different color is assigned to each side of the plane shown in Fig. 6.18, any pixel whose intensity level is above the plane will be coded with one color, and any pixel below the plane will be coded with the other. Levels that lie on the plane itself may be arbitrarily assigned one of the two colors. The result is a two-color image whose relative appearance can be controlled by moving the slicing plane up and down the intensity axis.

In general, the technique may be summarized as follows. Let $[0, L - 1]$ represent the gray scale, let level l_0 represent black $[f(x, y) = 0]$, and level l_{L-1} represent white $[f(x, y) = L - 1]$. Suppose that P planes perpendicular to the intensity axis are defined at levels $l_1, l_2, \ldots, l_P$. Then, assuming that $0 < P < L - 1$, the P planes partition the gray scale into $P + 1$ intervals, $V_1, V_2, \ldots, V_{P+1}$. Intensity to color assignments are made according to the relation

$$f(x, y) = c_k \quad \text{if } f(x, y) \in V_k \quad (6.3\text{-}1)$$

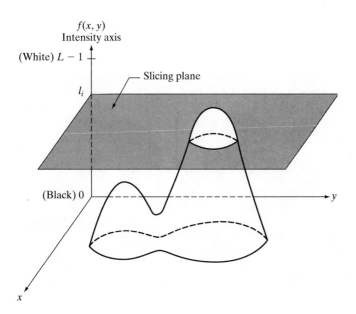

FIGURE 6.18
Geometric interpretation of the intensity-slicing technique.

FIGURE 6.19 An alternative representation of the intensity-slicing technique.

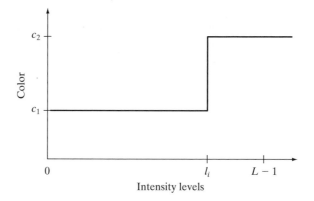

where c_k is the color associated with the kth intensity interval V_k defined by the partitioning planes at $l = k - 1$ and $l = k$.

The idea of planes is useful primarily for a geometric interpretation of the intensity-slicing technique. Figure 6.19 shows an alternative representation that defines the same mapping as in Fig. 6.18. According to the mapping function shown in Fig. 6.19, any input intensity level is assigned one of two colors, depending on whether it is above or below the value of l_i. When more levels are used, the mapping function takes on a staircase form.

EXAMPLE 6.3:
Intensity slicing.

■ A simple, but practical, use of intensity slicing is shown in Fig. 6.20. Figure 6.20(a) is a monochrome image of the Picker Thyroid Phantom (a radiation test pattern), and Fig. 6.20(b) is the result of intensity slicing this image into eight color regions. Regions that appear of constant intensity in the monochrome image are really quite variable, as shown by the various colors in the sliced image. The left lobe, for instance, is a dull gray in the monochrome image, and picking out variations in intensity is difficult. By contrast, the color image clearly shows eight different regions of constant intensity, one for each of the colors used.   ■

a b

FIGURE 6.20
(a) Monochrome image of the Picker Thyroid Phantom.
(b) Result of density slicing into eight colors.
(Courtesy of Dr. J. L. Blankenship, Instrumentation and Controls Division, Oak Ridge National Laboratory.)

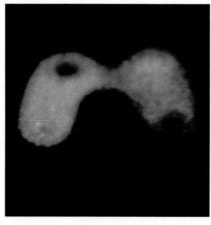 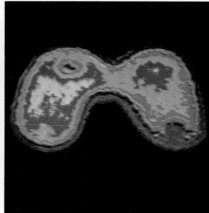

In the preceding simple example, the gray scale was divided into intervals and a different color was assigned to each region, without regard for the meaning of the gray levels in the image. Interest in that case was simply to view the different gray levels constituting the image. Intensity slicing assumes a much more meaningful and useful role when subdivision of the gray scale is based on physical characteristics of the image. For instance, Fig. 6.21(a) shows an X-ray image of a weld (the horizontal dark region) containing several cracks and porosities (the bright, white streaks running horizontally through the middle of the image). It is known that when there is a porosity or crack in a weld, the full strength of the X-rays going through the object saturates the imaging sensor on the other side of the object. Thus, intensity values of 255 in an 8-bit image coming from such a system automatically imply a problem with the weld. If a human were to be the ultimate judge of the analysis, and manual processes were employed to inspect welds (still a common procedure today), a simple color coding that assigns one color to

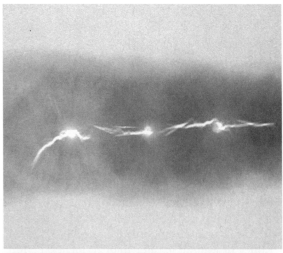

a
b

FIGURE 6.21
(a) Monochrome X-ray image of a weld. (b) Result of color coding. (Original image courtesy of X-TEK Systems, Ltd.)

level 255 and another to all other intensity levels would simplify the inspector's job considerably. Figure 6.21(b) shows the result. No explanation is required to arrive at the conclusion that human error rates would be lower if images were displayed in the form of Fig. 6.21(b), instead of the form shown in Fig. 6.21(a). In other words, if the exact intensity value or range of values one is looking for is known, intensity slicing is a simple but powerful aid in visualization, especially if numerous images are involved. The following is a more complex example.

EXAMPLE 6.4:
Use of color to highlight rainfall levels.

■ Measurement of rainfall levels, especially in the tropical regions of the Earth, is of interest in diverse applications dealing with the environment. Accurate measurements using ground-based sensors are difficult and expensive to acquire, and total rainfall figures are even more difficult to obtain because a significant portion of precipitation occurs over the ocean. One approach for obtaining rainfall figures is to use a satellite. The TRMM (Tropical Rainfall Measuring Mission) satellite utilizes, among others, three sensors specially designed to detect rain: a precipitation radar, a microwave imager, and a visible and infrared scanner (see Sections 1.3 and 2.3 regarding image sensing modalities).

The results from the various rain sensors are processed, resulting in estimates of average rainfall over a given time period in the area monitored by the sensors. From these estimates, it is not difficult to generate gray-scale images whose intensity values correspond directly to rainfall, with each pixel representing a physical land area whose size depends on the resolution of the sensors. Such an intensity image is shown in Fig. 6.22(a), where the area monitored by the satellite is the slightly lighter horizontal band in the middle one-third of the picture (these are the tropical regions). In this particular example, the rainfall values are average monthly values (in inches) over a three-year period.

Visual examination of this picture for rainfall patterns is quite difficult, if not impossible. However, suppose that we code intensity levels from 0 to 255 using the colors shown in Fig. 6.22(b). Values toward the blues signify low values of rainfall, with the opposite being true for red. Note that the scale tops out at pure red for values of rainfall greater than 20 inches. Figure 6.22(c) shows the result of color coding the gray image with the color map just discussed. The results are much easier to interpret, as shown in this figure and in the zoomed area of Fig. 6.22(d). In addition to providing global coverage, this type of data allows meteorologists to calibrate ground-based rain monitoring systems with greater precision than ever before. ■

6.3.2 Intensity to Color Transformations

Other types of transformations are more general and thus are capable of achieving a wider range of pseudocolor enhancement results than the simple slicing technique discussed in the preceding section. An approach that is particularly attractive is shown in Fig. 6.23. Basically, the idea underlying this approach is to perform three independent transformations on the intensity of any input pixel. The three results are then fed separately into the red, green, and blue channels of a color television monitor. This method produces a composite image whose color content is modulated by the nature of the transformation

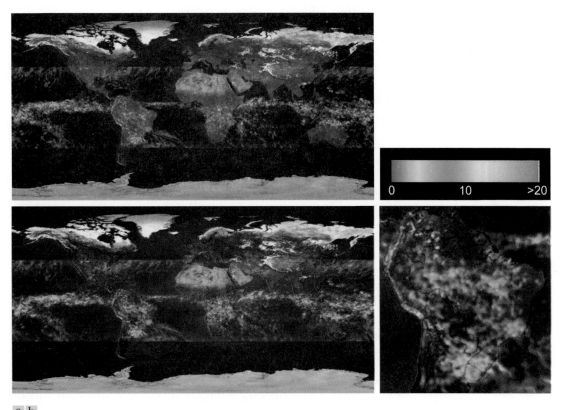

a b
c d

FIGURE 6.22 (a) Gray-scale image in which intensity (in the lighter horizontal band shown) corresponds to average monthly rainfall. (b) Colors assigned to intensity values. (c) Color-coded image. (d) Zoom of the South American region. (Courtesy of NASA.)

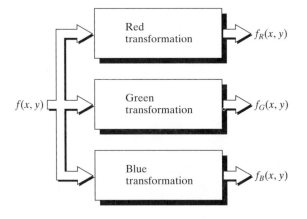

FIGURE 6.23
Functional block diagram for pseudocolor image processing. f_R, f_G, and f_B are fed into the corresponding red, green, and blue inputs of an RGB color monitor.

functions. Note that these are transformations on the intensity values of an image and are not functions of position.

The method discussed in the previous section is a special case of the technique just described. There, piecewise linear functions of the intensity levels (Fig. 6.19) are used to generate colors. The method discussed in this section, on the other hand, can be based on smooth, nonlinear functions, which, as might be expected, gives the technique considerable flexibility.

EXAMPLE 6.5:
Use of pseudocolor for highlighting explosives contained in luggage.

■ Figure 6.24(a) shows two monochrome images of luggage obtained from an airport X-ray scanning system. The image on the left contains ordinary articles. The image on the right contains the same articles, as well as a block of simulated plastic explosives. The purpose of this example is to illustrate the use of intensity level to color transformations to obtain various degrees of enhancement.

Figure 6.25 shows the transformation functions used. These sinusoidal functions contain regions of relatively constant value around the peaks as well as regions that change rapidly near the valleys. Changing the phase and frequency of each sinusoid can emphasize (in color) ranges in the gray scale. For instance, if all three transformations have the same phase and frequency, the output image will be monochrome. A small change in the phase between the three transformations produces little change in pixels whose intensities correspond to peaks in the sinusoids, especially if the sinusoids have broad profiles (low frequencies). Pixels with intensity values in the steep section of the sinusoids are assigned a much stronger color content as a result of significant differences between the amplitudes of the three sinusoids caused by the phase displacement between them.

FIGURE 6.24
Pseudocolor enhancement by using the gray level to color transformations in Fig. 6.25. (Original image courtesy of Dr. Mike Hurwitz, Westinghouse.)

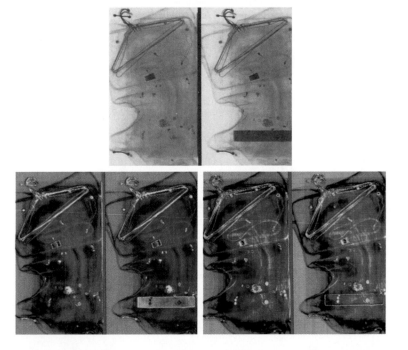

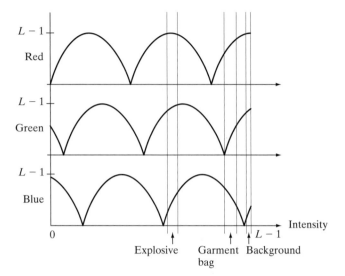

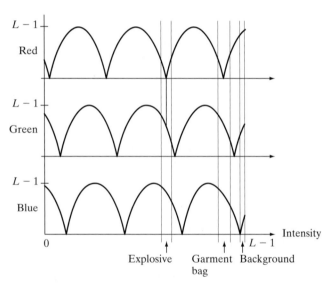

FIGURE 6.25
Transformation functions used to obtain the images in Fig. 6.24.

The image shown in Fig. 6.24(b) was obtained with the transformation functions in Fig. 6.25(a), which shows the gray-level bands corresponding to the explosive, garment bag, and background, respectively. Note that the explosive and background have quite different intensity levels, but they were both coded with approximately the same color as a result of the periodicity of the sine waves. The image shown in Fig. 6.24(c) was obtained with the transformation functions in Fig. 6.25(b). In this case the explosives and garment bag intensity bands were mapped by similar transformations and thus received essentially the same color assignments. Note that this mapping allows an observer to "see" through the explosives. The background mappings were about the same as those used for Fig. 6.24(b), producing almost identical color assignments. ■

FIGURE 6.26 A pseudocolor coding approach used when several monochrome images are available.

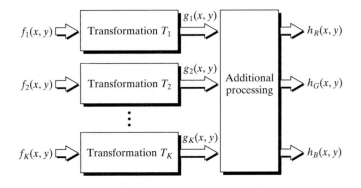

The approach shown in Fig. 6.23 is based on a single monochrome image. Often, it is of interest to combine several monochrome images into a single color composite, as shown in Fig. 6.26. A frequent use of this approach (illustrated in Example 6.6) is in multispectral image processing, where different sensors produce individual monochrome images, each in a different spectral band. The types of additional processes shown in Fig. 6.26 can be techniques such as color balancing (see Section 6.5.4), combining images, and selecting the three images for display based on knowledge about response characteristics of the sensors used to generate the images.

EXAMPLE 6.6:
Color coding of multispectral images.

■ Figures 6.27(a) through (d) show four spectral satellite images of Washington, D.C., including part of the Potomac River. The first three images are in the visible red, green, and blue, and the fourth is in the near infrared (see Table 1.1 and Fig. 1.10). Figure 6.27(e) is the full-color image obtained by combining the first three images into an RGB image. Full-color images of dense areas are difficult to interpret, but one notable feature of this image is the difference in color in various parts of the Potomac River. Figure 6.27(f) is a little more interesting. This image was formed by replacing the red component of Fig. 6.27(e) with the near-infrared image. From Table 1.1, we know that this band is strongly responsive to the biomass components of a scene. Figure 6.27(f) shows quite clearly the difference between biomass (in red) and the human-made features in the scene, composed primarily of concrete and asphalt, which appear bluish in the image.

The type of processing just illustrated is quite powerful in helping visualize events of interest in complex images, especially when those events are beyond our normal sensing capabilities. Figure 6.28 is an excellent illustration of this. These are images of the Jupiter moon Io, shown in pseudocolor by combining several of the sensor images from the *Galileo* spacecraft, some of which are in spectral regions not visible to the eye. However, by understanding the physical and chemical processes likely to affect sensor response, it is possible to combine the sensed images into a meaningful pseudocolor map. One way to combine the sensed image data is by how they show either differences in surface chemical composition or changes in the way the surface reflects sunlight. For example, in the pseudocolor image in Fig. 6.28(b), bright red depicts material newly ejected

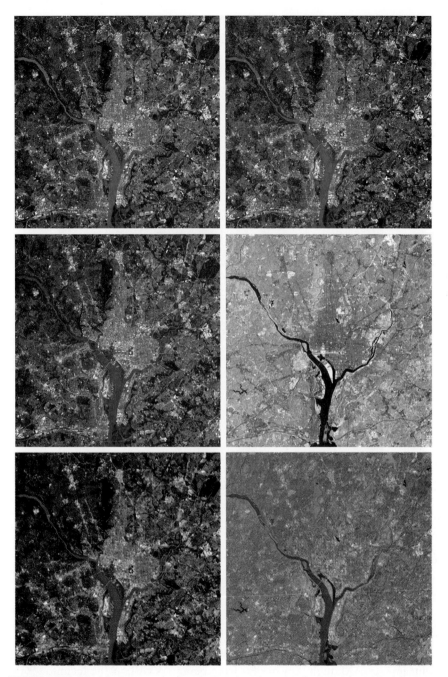

FIGURE 6.27 (a)–(d) Images in bands 1–4 in Fig. 1.10 (see Table 1.1). (e) Color composite image obtained by treating (a), (b), and (c) as the red, green, blue components of an RGB image. (f) Image obtained in the same manner, but using in the red channel the near-infrared image in (d). (Original multispectral images courtesy of NASA.)

a b
c d
e f

a
b

FIGURE 6.28
(a) Pseudocolor
rendition of
Jupiter Moon Io.
(b) A close-up.
(Courtesy of
NASA.)

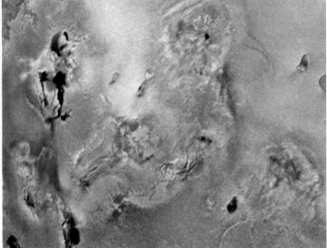

from an active volcano on Io, and the surrounding yellow materials are older sulfur deposits. This image conveys these characteristics much more readily than would be possible by analyzing the component images individually.   ■

6.4   Basics of Full-Color Image Processing

In this section, we begin the study of processing techniques applicable to full-color images. Although they are far from being exhaustive, the techniques developed in the sections that follow are illustrative of how full-color images are handled for a variety of image processing tasks. Full-color image processing approaches fall into two major categories. In the first category, we process each component image individually and then form a composite processed color image from the individually processed components. In the second category, we work with color pixels directly. Because full-color images have at least three components, color pixels are vectors. For example, in the RGB system, each color point can be interpreted as a vector extending from the origin to that point in the RGB coordinate system (see Fig. 6.7).

Let **c** represent an arbitrary vector in RGB color space:

$$\mathbf{c} = \begin{bmatrix} c_R \\ c_G \\ c_B \end{bmatrix} = \begin{bmatrix} R \\ G \\ B \end{bmatrix} \tag{6.4-1}$$

This equation indicates that the components of **c** are simply the RGB components of a color image at a point. We take into account the fact that the color components are a function of coordinates (x, y) by using the notation

$$\mathbf{c}(x, y) = \begin{bmatrix} c_R(x, y) \\ c_G(x, y) \\ c_B(x, y) \end{bmatrix} = \begin{bmatrix} R(x, y) \\ G(x, y) \\ B(x, y) \end{bmatrix} \tag{6.4-2}$$

For an image of size $M \times N$, there are MN such vectors, $\mathbf{c}(x, y)$, for $x = 0, 1, 2, \ldots, M - 1; y = 0, 1, 2, \ldots, N - 1$.

It is important to keep in mind that Eq. (6.4-2) depicts a vector whose components are *spatial* variables in x and y. This is a frequent source of confusion that can be avoided by focusing on the fact that our interest lies in spatial processes. That is, we are interested in image processing techniques formulated in x and y. The fact that the pixels are now color pixels introduces a factor that, in its easiest formulation, allows us to process a color image by processing each of its component images separately, using standard gray-scale image processing methods. However, the results of individual color component processing are not always equivalent to direct processing in color vector space, in which case we must formulate new approaches.

In order for per-color-component and vector-based processing to be equivalent, two conditions have to be satisfied: First, the process has to be applicable to both vectors and scalars. Second, the operation on each component of a vector must be independent of the other components. As an illustration, Fig. 6.29 shows neighborhood spatial processing of gray-scale and full-color images.

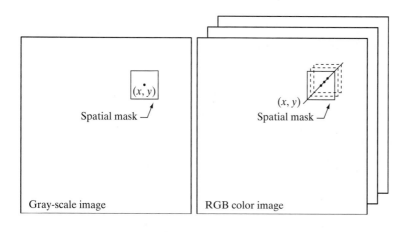

a b
FIGURE 6.29
Spatial masks for gray-scale and RGB color images.

Suppose that the process is neighborhood averaging. In Fig. 6.29(a), averaging would be accomplished by summing the intensities of all the pixels in the neighborhood and dividing by the total number of pixels in the neighborhood. In Fig. 6.29(b), averaging would be done by summing all the vectors in the neighborhood and dividing each component by the total number of vectors in the neighborhood. But each component of the average vector is the sum of the pixels in the image corresponding to that component, which is the same as the result that would be obtained if the averaging were done on a per-color-component basis and then the vector was formed. We show this in more detail in the following sections. We also show methods in which the results of the two approaches are not the same.

6.5 Color Transformations

The techniques described in this section, collectively called *color transformations*, deal with processing the components of a color image within the context of a *single* color model, as opposed to the conversion of those components between models (like the RGB-to-HSI and HSI-to-RGB conversion transformations of Section 6.2.3).

6.5.1 Formulation

As with the intensity transformation techniques of Chapter 3, we model color transformations using the expression

$$g(x, y) = T[f(x, y)] \tag{6.5-1}$$

where $f(x, y)$ is a color input image, $g(x, y)$ is the transformed or processed color output image, and T is an operator on f over a spatial neighborhood of (x, y). The principal difference between this equation and Eq. (3.1-1) is in its interpretation. The pixel values here are triplets or quartets (i.e., groups of three or four values) from the color space chosen to represent the images, as illustrated in Fig. 6.29(b).

Analogous to the approach we used to introduce the basic intensity transformations in Section 3.2, we will restrict attention in this section to color transformations of the form

$$s_i = T_i(r_1, r_2, \ldots, r_n), \qquad i = 1, 2, \ldots, n \tag{6.5-2}$$

where, for notational simplicity, r_i and s_i are variables denoting the color components of $f(x, y)$ and $g(x, y)$ at any point (x, y), n is the number of color components, and $\{T_1, T_2, \ldots, T_n\}$ is a set of *transformation* or *color mapping functions* that operate on r_i to produce s_i. Note that n transformations, T_i, combine to implement the single transformation function, T, in Eq. (6.5-1). The color space chosen to describe the pixels of f and g determines the value of n. If the RGB color space is selected, for example, $n = 3$ and r_1, r_2, and r_3 denote the red, green, and blue components of the input image, respectively. If the CMYK or HSI color spaces are chosen, $n = 4$ or $n = 3$.

FIGURE 6.30 A full-color image and its various color-space components. (Original image courtesy of MedData Interactive.)

The full-color image in Fig. 6.30 shows a high-resolution color image of a bowl of strawberries and a cup of coffee that was digitized from a large format (4″ × 5″) color negative. The second row of the figure contains the components of the initial CMYK scan. In these images, black represents 0 and white represents 1 in each CMYK color component. Thus, we see that the strawberries are composed of large amounts of magenta and yellow because the images corresponding to these two CMYK components are the brightest. Black is used sparingly and is generally confined to the coffee and shadows within the bowl of strawberries. When the CMYK image is converted to RGB, as shown in the third row of the figure, the strawberries are seen to contain a large amount of red and very little (although some) green and blue. The last row of Fig. 6.30 shows the HSI components of the full-color image—computed using Eqs. (6.2-2) through (6.2-4). As expected, the intensity component is a monochrome rendition of the full-color original. In addition, the strawberries are relatively pure in color; they possess the highest saturation or least dilution by white light of any of the hues in the image. Finally, we note some difficulty in interpreting the hue component. The problem is compounded by the fact that (1) there is a discontinuity in the HSI model where $0°$ and $360°$ meet (see Fig. 6.15), and (2) hue is undefined for a saturation of 0 (i.e., for white, black, and pure grays). The discontinuity of the model is most apparent around the strawberries, which are depicted in gray level values near both black (0) and white (1). The result is an unexpected mixture of highly contrasting gray levels to represent a single color—red.

Any of the color-space components in Fig. 6.30 can be used in conjunction with Eq. (6.5-2). In theory, any transformation can be performed in any color model. In practice, however, some operations are better suited to specific models. For a given transformation, the cost of converting between representations must be factored into the decision regarding the color space in which to implement it. Suppose, for example, that we wish to modify the intensity of the full-color image in Fig. 6.30 using

$$g(x, y) = kf(x, y) \qquad (6.5\text{-}3)$$

where $0 < k < 1$. In the HSI color space, this can be done with the simple transformation

$$s_3 = kr_3 \qquad (6.5\text{-}4)$$

where $s_1 = r_1$ and $s_2 = r_2$. Only HSI intensity component r_3 is modified. In the RGB color space, three components must be transformed:

$$s_i = kr_i \qquad i = 1, 2, 3 \qquad (6.5\text{-}5)$$

The CMY space requires a similar set of linear transformations:

$$s_i = kr_i + (1 - k) \qquad i = 1, 2, 3 \qquad (6.5\text{-}6)$$

Although the HSI transformation involves the fewest number of operations, the computations required to convert an RGB or CMY(K) image to the HSI space more than offsets (in this case) the advantages of the simpler transformation—the conversion calculations are more computationally intense than the intensity transformation itself. Regardless of the color space selected, however, the output is the same. Figure 6.31(b) shows the result of applying any of the transformations in Eqs. (6.5-4) through (6.5-6) to the full-color image of Fig. 6.30 using $k = 0.7$. The mapping functions themselves are depicted graphically in Figs. 6.31(c) through (e).

It is important to note that each transformation defined in Eqs. (6.5-4) through (6.5-6) depends only on one component within its color space. For example, the red output component, s_1, in Eq. (6.5-5) is independent of the green (r_2) and blue (r_3) inputs; it depends only on the red (r_1) input. Transformations of this type are among the simplest and most used color processing tools and can be carried out on a per-color-component basis, as mentioned at the beginning of our discussion. In the remainder of this section we examine several such transformations and discuss a case in which the component transformation functions are dependent on all the color components of the input image and, therefore, cannot be done on an individual color-component basis.

a b
c d e

FIGURE 6.31 Adjusting the intensity of an image using color transformations. (a) Original image. (b) Result of decreasing its intensity by 30% (i.e., letting $k = 0.7$). (c)–(e) The required RGB, CMY, and HSI transformation functions. (Original image courtesy of MedData Interactive.)

FIGURE 6.32
Complements on
the color circle.

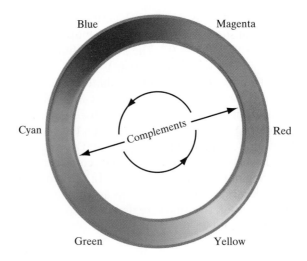

6.5.2 Color Complements

The hues directly opposite one another on the *color circle*[†] of Fig. 6.32 are called *complements*. Our interest in complements stems from the fact that they are analogous to the gray-scale negatives of Section 3.2.1. As in the gray-scale case, color complements are useful for enhancing detail that is embedded in dark regions of a color image—particularly when the regions are dominant in size.

EXAMPLE 6.7:
Computing color
image
complements.

■ Figures 6.33(a) and (c) show the full-color image from Fig. 6.30 and its color complement. The RGB transformations used to compute the complement are plotted in Fig. 6.33(b). They are identical to the gray-scale negative transformation defined in Section 3.2.1. Note that the computed complement is reminiscent of conventional photographic color film negatives. Reds of the original image are replaced by cyans in the complement. When the original image is black, the complement is white, and so on. Each of the hues in the complement image can be predicted from the original image using the color circle of Fig. 6.32, and each of the RGB component transforms involved in the computation of the complement is a function of *only* the corresponding input color component.

Unlike the intensity transformations of Fig. 6.31, the RGB complement transformation functions used in this example do not have a straightforward HSI space equivalent. It is left as an exercise for the reader (see Problem 6.18) to show that the saturation component of the complement cannot be computed from the saturation component of the input image alone. Figure 6.33(d) provides an approximation of the complement using the hue, saturation, and intensity transformations given in Fig. 6.33(b). Note that the saturation component of the input image is unaltered; it is responsible for the visual differences between Figs. 6.33(c) and (d). ■

[†]The color circle originated with Sir Isaac Newton, who in the seventeenth century joined the ends of the color spectrum to form the first color circle.

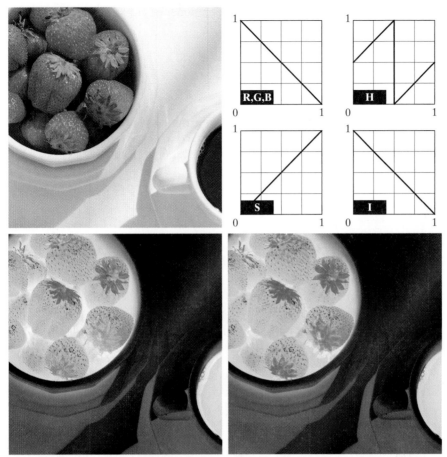

FIGURE 6.33
Color
complement
transformations.
(a) Original
image.
(b) Complement
transformation
functions.
(c) Complement
of (a) based on
the RGB mapping
functions. (d) An
approximation
of the RGB
complement
using HSI
transformations.

6.5.3 Color Slicing

Highlighting a specific range of colors in an image is useful for separating objects from their surroundings. The basic idea is either to (1) display the colors
of interest so that they stand out from the background or (2) use the region defined by the colors as a mask for further processing. The most straightforward
approach is to extend the intensity slicing techniques of Section 3.2.4. Because
a color pixel is an n-dimensional quantity, however, the resulting color transformation functions are more complicated than their gray-scale counterparts
in Fig. 3.11. In fact, the required transformations are more complex than the
color component transforms considered thus far. This is because all practical
color-slicing approaches require each pixel's transformed color components to
be a function of all n original pixel's color components.

One of the simplest ways to "slice" a color image is to map the colors outside
some range of interest to a nonprominent neutral color. If the colors of interest
are enclosed by a cube (or *hypercube* for $n > 3$) of width W and centered at a

prototypical (e.g., average) color with components $(a_1, a_2, \ldots, a_n)$, the necessary set of transformations is

$$
s_i =
\begin{cases}
0.5 & \text{if } \left[|r_j - a_j| > \dfrac{W}{2} \right]_{\text{any } 1 \leq j \leq n} \\
r_i & \text{otherwise}
\end{cases}
\qquad i = 1, 2, \ldots, n \qquad (6.5\text{-}7)
$$

These transformations highlight the colors around the prototype by forcing all other colors to the midpoint of the reference color space (an arbitrarily chosen neutral point). For the RGB color space, for example, a suitable neutral point is middle gray or color $(0.5, 0.5, 0.5)$.

If a sphere is used to specify the colors of interest, Eq. (6.5-7) becomes

$$
s_i =
\begin{cases}
0.5 & \text{if } \displaystyle\sum_{j=1}^{n} (r_j - a_j)^2 > R_0^2 \\
r_i & \text{otherwise}
\end{cases}
\qquad i = 1, 2, \ldots, n \qquad (6.5\text{-}8)
$$

Here, R_0 is the radius of the enclosing sphere (or hypersphere for $n > 3$) and $(a_1, a_2, \ldots, a_n)$ are the components of its center (i.e., the prototypical color). Other useful variations of Eqs. (6.5-7) and (6.5-8) include implementing multiple color prototypes and reducing the intensity of the colors outside the region of interest—rather than setting them to a neutral constant.

EXAMPLE 6.8:
An illustration of color slicing.

■ Equations (6.5-7) and (6.5-8) can be used to separate the edible part of the strawberries in Fig. 6.31(a) from the background cups, bowl, coffee, and table. Figures 6.34(a) and (b) show the results of applying both transformations. In

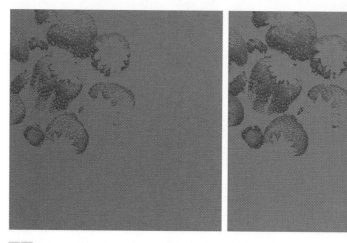

a b

FIGURE 6.34 Color-slicing transformations that detect (a) reds within an RGB cube of width $W = 0.2549$ centered at $(0.6863, 0.1608, 0.1922)$, and (b) reds within an RGB sphere of radius 0.1765 centered at the same point. Pixels outside the cube and sphere were replaced by color $(0.5, 0.5, 0.5)$.

each case, a prototype red with RGB color coordinate $(0.6863, 0.1608, 0.1922)$ was selected from the most prominent strawberry; W and R_0 were chosen so that the highlighted region would not expand to undesirable portions of the image. The actual values, $W = 0.2549$ and $R_0 = 0.1765$, were determined interactively. Note that the sphere-based transformation of Eq. (6.5-8) is slightly better, in the sense that it includes more of the strawberries' red areas. A sphere of radius 0.1765 does not completely enclose a cube of width 0.2549 but is itself not completely enclosed by the cube.   ■

6.5.4 Tone and Color Corrections

Color transformations can be performed on most desktop computers. In conjunction with digital cameras, flatbed scanners, and inkjet printers, they turn a personal computer into a *digital darkroom*—allowing tonal adjustments and color corrections, the mainstays of high-end color reproduction systems, to be performed without the need for traditionally outfitted wet processing (i.e., darkroom) facilities. Although tone and color corrections are useful in other areas of imaging, the focus of the current discussion is on the most common uses—photo enhancement and color reproduction.

The effectiveness of the transformations examined in this section is judged ultimately in print. Because these transformations are developed, refined, and evaluated on monitors, it is necessary to maintain a high degree of color consistency between the monitors used and the eventual output devices. In fact, the colors of the monitor should represent accurately any digitally scanned source images, as well as the final printed output. This is best accomplished with a *device-independent color model* that relates the color gamuts (see Section 6.1) of the monitors and output devices, as well as any other devices being used, to one another. The success of this approach is a function of the quality of the *color profiles* used to map each device to the model and the model itself. The model of choice for many *color management systems* (CMS) is the CIE $L*a*b*$ model, also called CIELAB (CIE [1978], Robertson [1977]). The $L*a*b*$ color components are given by the following equations:

$$L* = 116 \cdot h\left(\frac{Y}{Y_W}\right) - 16 \tag{6.5-9}$$

$$a* = 500\left[h\left(\frac{X}{X_W}\right) - h\left(\frac{Y}{Y_W}\right)\right] \tag{6.5-10}$$

$$b* = 200\left[h\left(\frac{Y}{Y_W}\right) - h\left(\frac{Z}{Z_W}\right)\right] \tag{6.5-11}$$

where

$$h(q) = \begin{cases} \sqrt[3]{q} & q > 0.008856 \\ 7.787q + 16/116 & q \le 0.008856 \end{cases} \tag{6.5-12}$$

and X_W, Y_W, and Z_W are reference white tristimulus values—typically the white of a perfectly reflecting diffuser under CIE standard $D65$ illumination (defined by $x = 0.3127$ and $y = 0.3290$ in the CIE chromaticity diagram of Fig. 6.5). The $L*a*b*$ color space is *colorimetric* (i.e., colors perceived as matching are encoded identically), *perceptually uniform* (i.e., color differences among various hues are perceived uniformly—see the classic paper by MacAdams [1942]), and *device independent*. While not a directly displayable format (conversion to another color space is required), its gamut encompasses the entire visible spectrum and can represent accurately the colors of any display, print, or input device. Like the HSI system, the $L*a*b*$ system is an excellent decoupler of intensity (represented by lightness $L*$) and color (represented by $a*$ for red minus green and $b*$ for green minus blue), making it useful in both image manipulation (tone and contrast editing) and image compression applications.[†]

The principal benefit of calibrated imaging systems is that they allow tonal and color imbalances to be corrected interactively and independently—that is, in two sequential operations. Before color irregularities, like over- and under-saturated colors, are resolved, problems involving the image's tonal range are corrected. The *tonal range* of an image, also called its *key type*, refers to its general distribution of color intensities. Most of the information in *high-key* images is concentrated at high (or light) intensities; the colors of *low-key* images are located predominantly at low intensities; *middle-key* images lie in between. As in the monochrome case, it is often desirable to distribute the intensities of a color image equally between the highlights and the shadows. The following examples demonstrate a variety of color transformations for the correction of tonal and color imbalances.

EXAMPLE 6.9:
Tonal
transformations.

■ Transformations for modifying image tones normally are selected interactively. The idea is to adjust experimentally the image's brightness and contrast to provide maximum detail over a suitable range of intensities. The colors themselves are not changed. In the RGB and CMY(K) spaces, this means mapping all three (or four) color components with the same transformation function; in the HSI color space, only the intensity component is modified.

Figure 6.35 shows typical transformations used for correcting three common tonal imbalances—flat, light, and dark images. The S-shaped curve in the

---

[†]Studies indicate that the degree to which the luminance (lightness) information is separated from the color information in $L*a*b*$ is greater than in other color models—such as CIELUV, YIQ, YUV, YCC, and XYZ (Kasson and Plouffe [1992]).

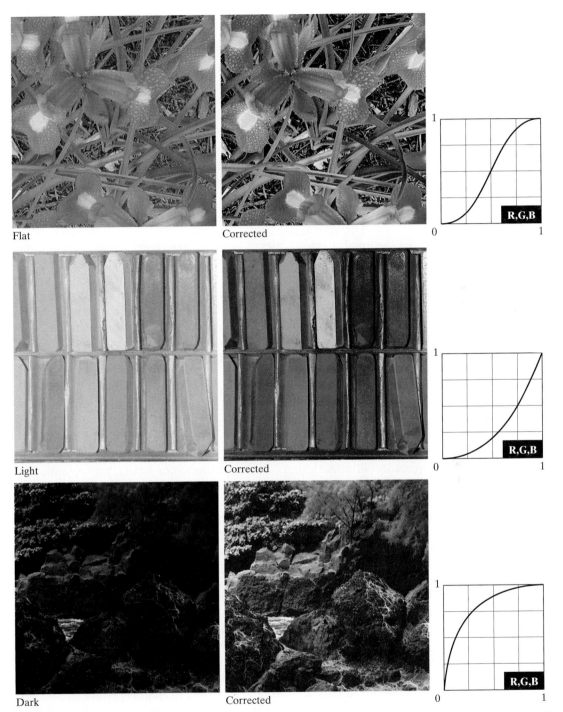

Flat   Corrected

Light   Corrected

Dark   Corrected

FIGURE 6.35 Tonal corrections for flat, light (high key), and dark (low key) color images. Adjusting the red, green, and blue components equally does not always alter the image hues significantly.

first row of the figure is ideal for boosting contrast [see Fig. 3.2(a)]. Its midpoint is anchored so that highlight and shadow areas can be lightened and darkened, respectively. (The inverse of this curve can be used to correct excessive contrast.) The transformations in the second and third rows of the figure correct light and dark images and are reminiscent of the power-law transformations in Fig. 3.6. Although the color components are discrete, as are the actual transformation functions, the transformation functions themselves are displayed and manipulated as continuous quantities—typically constructed from piecewise linear or higher order (for smoother mappings) polynomials. Note that the keys of the images in Fig. 6.35 are directly observable; they could also be determined using the histograms of the images' color components. ▪

EXAMPLE 6.10:
Color balancing.

▪ After the tonal characteristics of an image have been properly established, any color imbalances can be addressed. Although color imbalances can be determined objectively by analyzing—with a color spectrometer—a known color in an image, accurate visual assessments are possible when white areas, where the RGB or CMY(K) components should be equal, are present. As can be seen in Fig. 6.36, skin tones also are excellent subjects for visual color assessments because humans are highly perceptive of proper skin color. Vivid colors, such as bright red objects, are of little value when it comes to visual color assessment.

When a color imbalance is noted, there are a variety of ways to correct it. When adjusting the color components of an image, it is important to realize that every action affects the overall color balance of the image. That is, the perception of one color is affected by its surrounding colors. Nevertheless, the color wheel of Fig. 6.32 can be used to predict how one color component will affect others. Based on the color wheel, for example, the proportion of any color can be increased by decreasing the amount of the opposite (or complementary) color in the image. Similarly, it can be increased by raising the proportion of the two immediately adjacent colors or decreasing the percentage of the two colors adjacent to the complement. Suppose, for instance, that there is an abundance of magenta in an RGB image. It can be decreased by (1) removing both red and blue or (2) adding green.

Figure 6.36 shows the transformations used to correct simple CMYK output imbalances. Note that the transformations depicted are the functions required for correcting the images; the inverses of these functions were used to generate the associated color imbalances. Together, the images are analogous to a color ring-around print of a darkroom environment and are useful as a reference tool for identifying color printing problems. Note, for example, that too much red can be due to excessive magenta (per the bottom left image) or too little cyan (as shown in the rightmost image of the second row). ▪

Original/Corrected

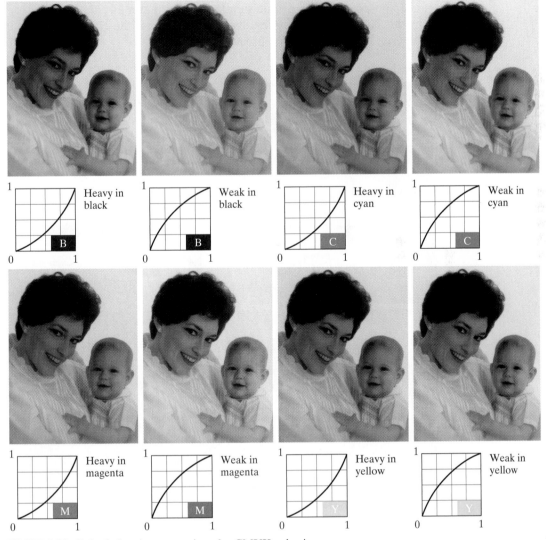

FIGURE 6.36 Color balancing corrections for CMYK color images.

6.5.5 Histogram Processing

Unlike the interactive enhancement approaches of the previous section, the gray-level histogram processing transformations of Section 3.3 can be applied to color images in an automated way. Recall that histogram equalization automatically determines a transformation that seeks to produce an image with a uniform histogram of intensity values. In the case of monochrome images, it was shown (see Fig. 3.20) to be reasonably successful at handling low-, high-, and middle-key images. Since color images are composed of multiple components, however, consideration must be given to adapting the gray-scale technique to more than one component and/or histogram. As might be expected, it is generally unwise to histogram equalize the components of a color image independently. This results in erroneous color. A more logical approach is to spread the color intensities uniformly, leaving the colors themselves (e.g., hues) unchanged. The following example shows that the HSI color space is ideally suited to this type of approach.

a b
c d

FIGURE 6.37
Histogram equalization (followed by saturation adjustment) in the HSI color space.

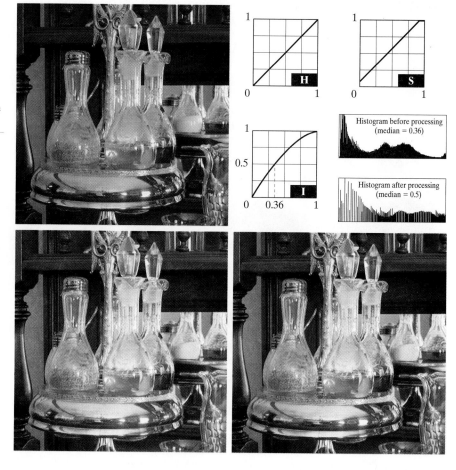

EXAMPLE 6.11:
Histogram
equalization in the
HSI color space.

■ Figure 6.37(a) shows a color image of a caster stand containing cruets and shakers whose intensity component spans the entire (normalized) range of possible values, [0, 1]. As can be seen in the histogram of its intensity component prior to processing [Fig. 6.37(b)], the image contains a large number of dark colors that reduce the median intensity to 0.36. Histogram equalizing the intensity component, without altering the hue and saturation, resulted in the image shown in Fig. 6.37(c). Note that the overall image is significantly brighter and that several moldings and the grain of the wooden table on which the caster is sitting are now visible. Figure 6.37(b) shows the intensity histogram of the new image, as well as the intensity transformation used to equalize the intensity component [see Eq. (3.3-8)].

Although the intensity equalization process did not alter the values of hue and saturation of the image, it did impact the overall color perception. Note, in particular, the loss of vibrancy in the oil and vinegar in the cruets. Figure 6.37(d) shows the result of correcting this partially by increasing the image's saturation component, subsequent to histogram equalization, using the transformation in Fig. 6.37(b). This type of adjustment is common when working with the intensity component in HSI space because changes in intensity usually affect the relative appearance of colors in an image.   ■

6.6 Smoothing and Sharpening

The next step beyond transforming each pixel of a color image without regard to its neighbors (as in the previous section) is to modify its value based on the characteristics of the surrounding pixels. In this section, the basics of this type of neighborhood processing are illustrated within the context of color image smoothing and sharpening.

6.6.1 Color Image Smoothing

With reference to Fig. 6.29(a) and the discussion in Sections 3.4 and 3.5, gray-scale image smoothing can be viewed as a spatial filtering operation in which the coefficients of the filtering mask have the same value. As the mask is slid across the image to be smoothed, each pixel is replaced by the average of the pixels in the neighborhood defined by the mask. As can be seen in Fig. 6.29(b), this concept is easily extended to the processing of full-color images. The principal difference is that instead of scalar intensity values we must deal with component vectors of the form given in Eq. (6.4-2).

Let S_{xy} denote the set of coordinates defining a neighborhood centered at (x, y) in an RGB color image. The average of the RGB component vectors in this neighborhood is

$$\bar{\mathbf{c}}(x, y) = \frac{1}{K} \sum_{(s,t) \in S_{xy}} \mathbf{c}(s, t) \tag{6.6-1}$$

It follows from Eq. (6.4-2) and the properties of vector addition that

Consult the book Web site for a brief review of vectors and matrices.

$$\bar{c}\,(x,\,y) = \begin{bmatrix} \dfrac{1}{K} \sum_{(s,\,t)\in S_{xy}} R(s,\,t) \\[2mm] \dfrac{1}{K} \sum_{(s,\,t)\in S_{xy}} G(s,\,t) \\[2mm] \dfrac{1}{K} \sum_{(s,\,t)\in S_{xy}} B(s,\,t) \end{bmatrix} \qquad (6.6\text{-}2)$$

We recognize the components of this vector as the scalar images that would be obtained by independently smoothing each plane of the starting RGB image using conventional gray-scale neighborhood processing. Thus, we conclude that smoothing by neighborhood averaging can be carried out on a per-color-plane basis. The result is the same as when the averaging is performed using RGB color vectors.

EXAMPLE 6.12:
Color image smoothing by neighborhood averaging.

■ Consider the RGB color image in Fig. 6.38(a). Its red, green, and blue component images are shown in Figs. 6.38(b) through (d). Figures 6.39(a) through (c) show the HSI components of the image. Based on the discussion in the previous paragraph, we smoothed each component image of the RGB image in Fig. 6.38 independently using a 5×5 spatial averaging mask. We then combined the individually smoothed images to form the smoothed, full-color RGB result shown in Fig. 6.40(a). Note that this image appears as we would expect from performing a spatial smoothing operation, as in the examples given in Section 3.5.

In Section 6.2, we noted that an important advantage of the HSI color model is that it decouples intensity and color information. This makes it suitable for many gray-scale processing techniques and suggests that it might be more efficient to smooth only the intensity component of the HSI representation in Fig. 6.39. To illustrate the merits and/or consequences of this approach, we next smooth only the intensity component (leaving the hue and saturation components unmodified) and convert the processed result to an RGB image for display. The smoothed color image is shown in Fig. 6.40(b). Note that it is similar to Fig. 6.40(a), but, as you can see from the difference image in Fig. 6.40(c), the two smoothed images are not identical. This is because in Fig. 6.40(a) the color of each pixel is the average color of the pixels in the neighborhood. On the other hand, by smoothing only the intensity component image in Fig. 6.40(b), the hue and saturation of each pixel was not affected and, therefore, the pixel colors did not change. It follows from this observation that the difference between the two smoothing approaches would become more pronounced as a function of increasing filter size. ■

FIGURE 6.38
(a) RGB image.
(b) Red
component image.
(c) Green compo-
nent. (d) Blue
component.

a b c

FIGURE 6.39 HSI components of the RGB color image in Fig. 6.38(a). (a) Hue. (b) Saturation. (c) Intensity.

a b c

FIGURE 6.40 Image smoothing with a 5 × 5 averaging mask. (a) Result of processing each RGB component image. (b) Result of processing the intensity component of the HSI image and converting to RGB. (c) Difference between the two results.

6.6.2 Color Image Sharpening

In this section we consider image sharpening using the Laplacian (see Section 3.6.2). From vector analysis, we know that the Laplacian of a vector is defined as a vector whose components are equal to the Laplacian of the individual scalar components of the input vector. In the RGB color system, the Laplacian of vector **c** in Eq. (6.4-2) is

$$\nabla^2[\mathbf{c}(x, y)] = \begin{bmatrix} \nabla^2 R(x, y) \\ \nabla^2 G(x, y) \\ \nabla^2 B(x, y) \end{bmatrix} \tag{6.6-3}$$

which, as in the previous section, tells us that we can compute the Laplacian of a full-color image by computing the Laplacian of each component image separately.

a b c

FIGURE 6.41 Image sharpening with the Laplacian. (a) Result of processing each RGB channel. (b) Result of processing the HSI intensity component and converting to RGB. (c) Difference between the two results.

■ Figure 6.41(a) was obtained using Eq. (3.6-7) and the mask in Fig. 3.37(c) to compute the Laplacians of the RGB component images in Fig. 6.38. These results were combined to produce the sharpened full-color result. Figure 6.41(b) shows a similarly sharpened image based on the HSI components in Fig. 6.39. This result was generated by combining the Laplacian of the intensity component with the unchanged hue and saturation components. The difference between the RGB and HSI sharpened images is shown in Fig. 6.41(c). The reason for the discrepancies between the two images is as in Example 6.12.   ■

EXAMPLE 6.13:
Sharpening with the Laplacian.

6.7 ■ Image Segmentation Based on Color

Segmentation is a process that partitions an image into regions. Although segmentation is the topic of Chapter 10, we consider color segmentation briefly here for the sake of continuity. You will have no difficulty following the discussion.

6.7.1 Segmentation in HSI Color Space

If we wish to segment an image based on color, and, in addition, we want to carry out the process on individual planes, it is natural to think first of the HSI space because color is conveniently represented in the hue image. Typically, saturation is used as a masking image in order to isolate further regions of interest in the hue image. The intensity image is used less frequently for segmentation of color images because it carries no color information. The following example is typical of how segmentation is performed in the HSI color space.

■ Suppose that it is of interest to segment the reddish region in the lower left of the image in Fig. 6.42(a). Although it was generated by pseudocolor methods, this image can be processed (segmented) as a full-color image without loss of generality. Figures 6.42(b) through (d) are its HSI component images. Note by comparing Figs. 6.42(a) and (b) that the region in which we are interested has relatively high values of hue, indicating that the colors are on the blue-magenta side of red (see Fig. 6.13). Figure 6.42(e) shows a binary mask generated by thresholding the saturation image with a threshold equal to 10% of the maximum value in that image. Any pixel value greater than the threshold was set to 1 (white). All others were set to 0 (black).

Figure 6.42(f) is the product of the mask with the hue image, and Fig. 6.42(g) is the histogram of the product image (note that the gray scale is in the range [0, 1]). We see in the histogram that high values (which are the values of interest) are grouped at the very high end of the gray scale, near 1.0. The result of thresholding the product image with threshold value of 0.9 resulted in the binary image shown in Fig. 6.42(h). The spatial location of the white points in this image identifies the points in the original image that have the reddish hue of interest. This was far from a perfect segmentation because there are points in the original image that we certainly would say have a reddish hue, but that were not identified by this segmentation method. However, it can be determined

EXAMPLE 6.14:
Segmentation in HSI space.

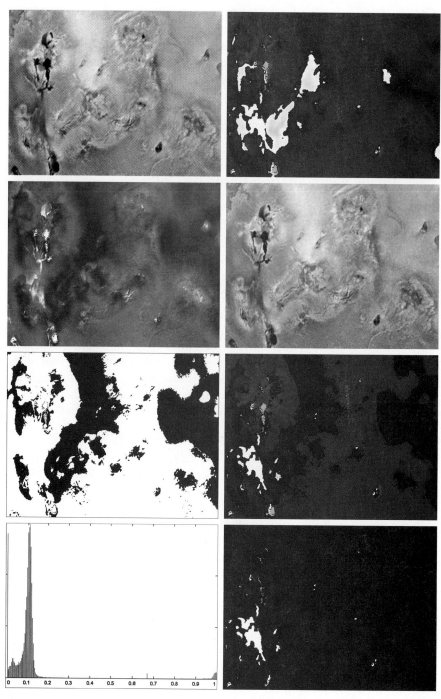

a b
c d
e f
g h

FIGURE 6.42 Image segmentation in HSI space. (a) Original. (b) Hue. (c) Saturation. (d) Intensity. (e) Binary saturation mask (black = 0). (f) Product of (b) and (e). (g) Histogram of (f). (h) Segmentation of red components in (a).

by experimentation that the regions shown in white in Fig. 6.42(h) are about the best this method can do in identifying the reddish components of the original image. The segmentation method discussed in the following section is capable of yielding considerably better results.   ■

6.7.2 Segmentation in RGB Vector Space

Although, as mentioned numerous times in this chapter, working in HSI space is more intuitive, segmentation is one area in which better results generally are obtained by using RGB color vectors. The approach is straightforward. Suppose that the objective is to segment objects of a specified color range in an RGB image. Given a set of sample color points representative of the colors of interest, we obtain an estimate of the "average" color that we wish to segment. Let this average color be denoted by the RGB vector **a**. The objective of segmentation is to classify each RGB pixel in a given image as having a color in the specified range or not. In order to perform this comparison, it is necessary to have a measure of similarity. One of the simplest measures is the Euclidean distance. Let **z** denote an arbitrary point in RGB space. We say that **z** is *similar* to **a** if the distance between them is less than a specified threshold, D_0. The Euclidean distance between **z** and **a** is given by

$$
\begin{aligned}
D(\mathbf{z}, \mathbf{a}) &= \|\mathbf{z} - \mathbf{a}\| \\
&= [(\mathbf{z} - \mathbf{a})^T (\mathbf{z} - \mathbf{a})]^{\frac{1}{2}} \\
&= [(z_R - a_R)^2 + (z_G - a_G)^2 + (z_B - a_B)^2]^{\frac{1}{2}}
\end{aligned}
\tag{6.7-1}
$$

where the subscripts $R, G,$ and B denote the RGB components of vectors **a** and **z**. The locus of points such that $D(\mathbf{z}, \mathbf{a}) \leq D_0$ is a solid sphere of radius D_0, as illustrated in Fig. 6.43(a). Points contained within the sphere satisfy the specified color criterion; points outside the sphere do not. Coding these two sets of points in the image with, say, black and white, produces a binary segmented image.

A useful generalization of Eq. (6.7-1) is a distance measure of the form

$$
D(\mathbf{z}, \mathbf{a}) = [(\mathbf{z} - \mathbf{a})^T \mathbf{C}^{-1} (\mathbf{z} - \mathbf{a})]^{\frac{1}{2}}
\tag{6.7-2}
$$

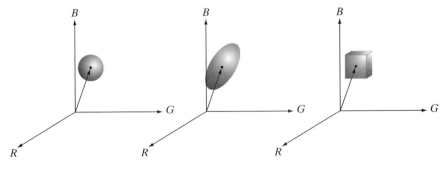

a b c

FIGURE 6.43
Three approaches for enclosing data regions for RGB vector segmentation.

where $\mathbf{C}$ is the covariance matrix[†] of the samples representative of the color we wish to segment. The locus of points such that $D(\mathbf{z}, \mathbf{a}) \leq D_0$ describes a solid 3-D elliptical body [Fig. 6.43(b)] with the important property that its principal axes are oriented in the direction of maximum data spread. When $\mathbf{C} = \mathbf{I}$, the 3×3 identity matrix, Eq. (6.7-2) reduces to Eq. (6.7-1). Segmentation is as described in the preceding paragraph.

Because distances are positive and monotonic, we can work with the distance squared instead, thus avoiding square root computations. However, implementing Eq. (6.7-1) or (6.7-2) is computationally expensive for images of practical size, even if the square roots are not computed. A compromise is to use a bounding box, as illustrated in Fig. 6.43(c). In this approach, the box is centered on $\mathbf{a}$, and its dimensions along each of the color axes is chosen proportional to the standard deviation of the samples along each of the axis. Computation of the standard deviations is done only once using sample color data.

Given an arbitrary color point, we segment it by determining whether or not it is on the surface or inside the box, as with the distance formulations. However, determining whether a color point is inside or outside a box is much simpler computationally when compared to a spherical or elliptical enclosure. Note that the preceding discussion is a generalization of the method introduced in Section 6.5.3 in connection with color slicing.

EXAMPLE 6.15:
Color image
segmentation in
RGB space.

■ The rectangular region shown Fig. 6.44(a) contains samples of reddish colors we wish to segment out of the color image. This is the same problem we considered in Example 6.14 using hue, but here we approach the problem using RGB color vectors. The approach followed was to compute the mean vector $\mathbf{a}$ using the color points contained within the rectangle in Fig. 6.44(a), and then to compute the standard deviation of the red, green, and blue values of those samples. A box was centered at $\mathbf{a}$, and its dimensions along each of the RGB axes were selected as 1.25 times the standard deviation of the data along the corresponding axis. For example, let σ_R denote the standard deviation of the red components of the sample points. Then the dimensions of the box along the R-axis extended from $(a_R - 1.25\sigma_R)$ to $(a_R + 1.25\sigma_R)$, where a_R denotes the red component of average vector $\mathbf{a}$. The result of coding each point in the entire color image as white if it was on the surface or inside the box, and as black otherwise, is shown in Fig. 6.44(b). Note how the segmented region was generalized from the color samples enclosed by the rectangle. In fact, by comparing Figs. 6.44(b) and. 6.42(h), we see that segmentation in the RGB vector space yielded results that are much more accurate, in the sense that they correspond much more closely with what we would define as "reddish" points in the original color image.   ■

[†]Computation of the covariance matrix of a set of vector samples is discussed in Section 11.4.

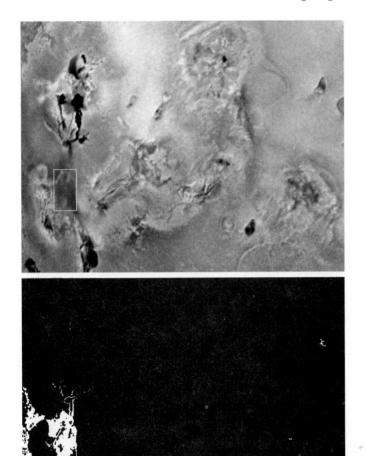

a
b

FIGURE 6.44
Segmentation in
RGB space.
(a) Original image
with colors of
interest shown
enclosed by a
rectangle.
(b) Result of
segmentation in
RGB vector
space. Compare
with Fig. 6.42(h).

6.7.3 Color Edge Detection

As discussed in Chapter 10, edge detection is an important tool for image segmentation. In this section, we are interested in the issue of computing edges on an individual-image basis versus computing edges directly in color vector space. The details of edge-based segmentation are given in Section 10.2.

Edge detection by gradient operators was introduced in Section 3.6.4 in connection with image sharpening. Unfortunately, the gradient discussed in Section 3.6.4 is not defined for vector quantities. Thus, we know immediately that computing the gradient on individual images and then using the results to form a color image will lead to erroneous results. A simple example will help illustrate the reason why.

Consider the two $M \times M$ color images (M odd) in Figs. 6.45(d) and (h), composed of the three component images in Figs. 6.45(a) through (c) and (e) through (g), respectively. If, for example, we compute the gradient image of each of the component images [see Eq. (3.6-11)] and add the results to form the two corresponding RGB gradient images, the value of the gradient at point $[(M + 1)/2, (M + 1)/2]$ would be the same in both cases. Intuitively, we would expect the gradient at that point to be stronger for the image in Fig. 6.45(d) because the edges of the R, G, and B images are in the same direction in that image, as opposed to the image in Fig. 6.45(h), in which only two of the edges are in the same direction. Thus we see from this simple example that processing the three individual planes to form a composite gradient image can yield erroneous results. If the problem is one of just detecting edges, then the individual-component approach usually yields acceptable results. If accuracy is an issue, however, then obviously we need a new definition of the gradient applicable to vector quantities. We discuss next a method proposed by Di Zenzo [1986] for doing this.

The problem at hand is to define the gradient (magnitude and direction) of the vector **c** in Eq. (6.4-2) at any point (x, y). As was just mentioned, the gradient we studied in Section 3.6.4 is applicable to a *scalar* function $f(x, y)$; it is not applicable to vector functions. The following is one of the various ways in which we can extend the concept of a gradient to vector functions. Recall that for a scalar function $f(x, y)$, the gradient is a vector pointing in the direction of maximum rate of change of f at coordinates (x, y).

a b c d
e f g h

FIGURE 6.45 (a)–(c) R, G, and B component images and (d) resulting RGB color image. (e)–(g) R, G, and B component images and (h) resulting RGB color image.

Let $\mathbf{r}, \mathbf{g}$, and $\mathbf{b}$ be unit vectors along the R, G, and B axis of RGB color space (Fig. 6.7), and define the vectors

$$\mathbf{u} = \frac{\partial R}{\partial x}\mathbf{r} + \frac{\partial G}{\partial x}\mathbf{g} + \frac{\partial B}{\partial x}\mathbf{b} \qquad (6.7\text{-}3)$$

and

$$\mathbf{v} = \frac{\partial R}{\partial y}\mathbf{r} + \frac{\partial G}{\partial y}\mathbf{g} + \frac{\partial B}{\partial y}\mathbf{b} \qquad (6.7\text{-}4)$$

Let the quantities g_{xx}, g_{yy}, and g_{xy} be defined in terms of the dot product of these vectors, as follows:

$$g_{xx} = \mathbf{u} \cdot \mathbf{u} = \mathbf{u}^T\mathbf{u} = \left|\frac{\partial R}{\partial x}\right|^2 + \left|\frac{\partial G}{\partial x}\right|^2 + \left|\frac{\partial B}{\partial x}\right|^2 \qquad (6.7\text{-}5)$$

$$g_{yy} = \mathbf{v} \cdot \mathbf{v} = \mathbf{v}^T\mathbf{v} = \left|\frac{\partial R}{\partial y}\right|^2 + \left|\frac{\partial G}{\partial y}\right|^2 + \left|\frac{\partial B}{\partial y}\right|^2 \qquad (6.7\text{-}6)$$

and

$$g_{xy} = \mathbf{u} \cdot \mathbf{v} = \mathbf{u}^T\mathbf{v} = \frac{\partial R}{\partial x}\frac{\partial R}{\partial y} + \frac{\partial G}{\partial x}\frac{\partial G}{\partial y} + \frac{\partial B}{\partial x}\frac{\partial B}{\partial y} \qquad (6.7\text{-}7)$$

Keep in mind that R, G, and B, and consequently the g's, are functions of x and y. Using this notation, it can be shown (Di Zenzo [1986]) that the direction of maximum rate of change of $\mathbf{c}(x, y)$ is given by the angle

$$\theta(x, y) = \frac{1}{2}\tan^{-1}\left[\frac{2g_{xy}}{g_{xx} - g_{yy}}\right] \qquad (6.7\text{-}8)$$

and that the value of the rate of change at (x, y), in the direction of $\theta(x, y)$, is given by

$$F_\theta(x, y) = \left\{\frac{1}{2}\left[(g_{xx} + g_{yy}) + (g_{xx} - g_{yy})\cos 2\theta(x, y) + 2g_{xy}\sin 2\theta(x, y)\right]\right\}^{\frac{1}{2}} \qquad (6.7\text{-}9)$$

Because $\tan(\alpha) = \tan(\alpha \pm \pi)$, if θ_0 is a solution to Eq. (6.7-8), so is $\theta_0 \pm \pi/2$. Furthermore, $F_\theta = F_{\theta+\pi}$, so F has to be computed only for values of θ in the half-open interval $[0, \pi)$. The fact that Eq. (6.7-8) provides two values 90° apart means that this equation associates with each point (x, y) a pair of orthogonal directions. Along one of those directions F is maximum, and it is minimum along the other. The derivation of these results is rather lengthy, and we would gain little in terms of the fundamental objective of our current discussion by detailing it here. Consult the paper by Di Zenzo [1986] for details. The partial derivatives required for implementing Eqs. (6.7-5) through (6.7-7) can be computed using, for example, the Sobel operators discussed in Section 3.6.4.

EXAMPLE 6.16:
Edge detection in
vector space.

■ Figure 6.46(b) is the gradient of the image in Fig. 6.46(a), obtained using
the vector method just discussed. Figure 6.46(c) shows the image obtained by
computing the gradient of each RGB component image and forming a composite gradient image by adding the corresponding values of the three component images at each coordinate (x, y). The edge detail of the vector
gradient image is more complete than the detail in the individual-plane gradient image in Fig. 6.46(c); for example, see the detail around the subject's right
eye. The image in Fig. 6.46(d) shows the difference between the two gradient
images at each point (x, y). It is important to note that both approaches yielded reasonable results. Whether the extra detail in Fig. 6.46(b) is worth the
added computational burden (as opposed to implementation of the Sobel operators, which were used to generate the gradient of the individual planes)
can only be determined by the requirements of a given problem. Figure 6.47
shows the three component gradient images, which, when added and scaled,
were used to obtain Fig. 6.46(c).   ■

a b
c d

FIGURE 6.46
(a) RGB image.
(b) Gradient
computed in RGB
color vector
space.
(c) Gradients
computed on a
per-image basis
and then added.
(d) Difference
between (b)
and (c).

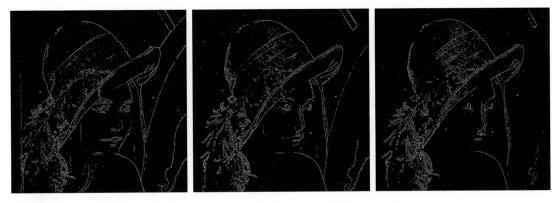

a b c

FIGURE 6.47 Component gradient images of the color image in Fig. 6.46. (a) Red component, (b) green component, and (c) blue component. These three images were added and scaled to produce the image in Fig. 6.46(c).

6.8 Noise in Color Images

The noise models discussed in Section 5.2 are applicable to color images. Usually, the noise content of a color image has the same characteristics in each color channel, but it is possible for color channels to be affected differently by noise. One possibility is for the electronics of a particular channel to malfunction. However, different noise levels are more likely to be caused by differences in the relative strength of illumination available to each of the color channels. For example, use of a red (reject) filter in a CCD camera will reduce the strength of illumination available to the red sensor. CCD sensors are noisier at lower levels of illumination, so the resulting red component of an RGB image would tend to be noisier than the other two component images in this situation.

■ In this example we take a brief look at noise in color images and how noise carries over when converting from one color model to another. Figures 6.48(a) through (c) show the three color planes of an RGB image corrupted by Gaussian noise, and Fig. 6.48(d) is the composite RGB image. Note that fine grain noise such as this tends to be less visually noticeable in a color image than it is in a monochrome image. Figures 6.49(a) through (c) show the result of converting the RGB image in Fig. 6.48(d) to HSI. Compare these results with the HSI components of the original image (Fig. 6.39) and note how significantly degraded the hue and saturation components of the noisy image are. This is due to the nonlinearity of the cos and min operations in Eqs. (6.2-2) and (6.2-3), respectively. On the other hand, the intensity component in Fig. 6.49(c) is slightly smoother than any of the three noisy RGB component images. This is due to the fact that the intensity image is the average of the RGB images, as indicated in Eq. (6.2-4). (Recall the discussion in Section 2.6.3 regarding the fact that image averaging reduces random noise.)

EXAMPLE 6.17:
Illustration of the effects of converting noisy RGB images to HSI.

FIGURE 6.48
(a)–(c) Red,
green, and blue
component
images corrupted
by additive
Gaussian noise of
mean 0 and
variance 800.
(d) Resulting
RGB image.
[Compare (d)
with Fig. 6.46(a).]

FIGURE 6.49 HSI components of the noisy color image in Fig. 6.48(d). (a) Hue. (b) Saturation. (c) Intensity.

a b
c d

FIGURE 6.50 (a) RGB image with green plane corrupted by salt-and-pepper noise. (b) Hue component of HSI image. (c) Saturation component. (d) Intensity component.

In cases when, say, only one RGB channel is affected by noise, conversion to HSI spreads the noise to all HSI component images. Figure 6.50 shows an example. Figure 6.50(a) shows an RGB image whose green image is corrupted by salt-and-pepper noise, in which the probability of either salt or pepper is 0.05. The HSI component images in Figs. 6.50(b) through (d) show clearly how the noise spread from the green RGB channel to all the HSI images. Of course, this is not unexpected because computation of the HSI components makes use of all RGB components, as shown in Section 6.2.3.   ■

As is true of the processes we have discussed thus far, filtering of full-color images can be carried out on a per-image basis or directly in color vector space, depending on the process. For example, noise reduction by using an averaging filter is the process discussed in Section 6.6.1, which we know gives the same result in vector space as it does if the component images are processed independently. Other filters, however, cannot be formulated in this manner. Examples include the class of order statistics filters discussed in Section 5.3.2. For instance, to implement a median filter in color vector space it is necessary to find a scheme for ordering vectors in a way that the median makes sense. While this was a simple process when dealing with scalars, the process is considerably more complex when dealing with vectors. A discussion of vector ordering is beyond the scope of our discussion here, but the book by Plataniotis and Venetsanopoulos [2000] is a good reference on vector ordering and some of the filters based on the ordering concept.

6.9 Color Image Compression

Because the number of bits required to represent color is typically three to four times greater than the number employed in the representation of gray levels, *data compression* plays a central role in the storage and transmission of color images. With respect to the RGB, CMY(K), and HSI images of the previous sections, the *data* that are the object of any compression are the components of each color pixel (e.g., the red, green, and blue components of the pixels in an RGB image); they are the means by which the color information is conveyed. *Compression* is the process of reducing or eliminating redundant and/or irrelevant data. Although compression is the topic of Chapter 8, we illustrate the concept briefly in the following example using a color image.

EXAMPLE 6.18:
A color image compression example.

■ Figure 6.51(a) shows a 24-bit RGB full-color image of an iris in which 8 bits each are used to represent the red, green, and blue components. Figure 6.51(b) was reconstructed from a compressed version of the image in (a) and is, in fact, a compressed and subsequently decompressed approximation of it. Although the compressed image is not directly displayable—it must be decompressed before input to a color monitor—the compressed image contains only 1 data bit (and thus 1 storage bit) for every 230 bits of data in the original image. Assuming that the compressed image could be transmitted over, say, the Internet, in 1 minute, transmission of the original image would require almost 4 hours. Of course, the transmitted data would have to be decompressed for viewing, but the decompression can be done in a matter of seconds. The JPEG 2000 compression algorithm used to generate Fig. 6.51(b) is a recently introduced standard that is described in detail in Section 8.2.10. Note that the reconstructed approximation image is slightly blurred. This is a characteristic of many *lossy* compression techniques; it can be reduced or eliminated by altering the level of compression.   ■

a
b

FIGURE 6.51
Color image
compression.
(a) Original RGB
image. (b) Result
of compressing
and decom-
pressing the
image in (a).

Summary

The material in this chapter is an introduction to color image processing and covers topics selected to provide a solid background in the techniques used in this branch of image processing. Our treatment of color fundamentals and color models was prepared as foundation material for a field that is wide in technical scope and areas of application. In particular, we focused on color models that we felt are not only useful in digital image processing but provide also the tools necessary for further study in this area of image processing. The discussion of pseudocolor and full-color processing on an individual image basis provides a tie to techniques that were covered in some detail in Chapters 3 through 5.

The material on color vector spaces is a departure from methods that we had studied before and highlights some important differences between gray-scale and full-color processing. In terms of techniques, the areas of direct color vector processing are numerous and include processes such as median and other order filters, adaptive and morphological filters, image restoration, image compression, and many others. These processes are not equivalent to color processing carried out on the individual component images of a color image. The references in the following section provide a pointer to further results in this field.

Our treatment of noise in color images also points out that the vector nature of the problem, along with the fact that color images are routinely transformed from one working space to another, has implications on the issue of how to reduce noise in these images. In some cases, noise filtering can be done on a per-image basis, but others, such as median filtering, require special treatment to reflect the fact that color pixels are vector quantities, as mentioned in the previous paragraph.

Although segmentation is the topic of Chapter 10 and image data compression is the topic of Chapter 8, we gained the advantage of continuity by introducing them here in the context of color image processing. As will become evident in subsequent discussions, many of the techniques developed in those chapters are applicable to the discussion in this chapter.

References and Further Reading

For a comprehensive reference on the science of color, see Malacara [2001]. Regarding the physiology of color, see Gegenfurtner and Sharpe [1999]. These two references, along with the early books by Walsh [1958] and by Kiver [1965], provide ample supplementary material for the discussion in Section 6.1. For further reading on color models (Section 6.2), see Fortner and Meyer [1997], Poynton [1996], and Fairchild [1998]. For a detailed derivation of the HSI model equations in Section 6.2.3 see the paper by Smith [1978] or consult the book Web site. The topic of pseudocolor (Section 6.3) is closely tied to the general area of data visualization. Wolff and Yaeger [1993] is a good basic reference on the use of pseudocolor. The book by Thorell and Smith [1990] also is of interest. For a discussion on the vector representation of color signals (Section 6.4), see Plataniotis and Venetsanopoulos [2000].

References for Section 6.5 are Benson [1985], Robertson [1977], and CIE [1978]. See also the classic paper by MacAdam [1942]. The material on color image filtering (Section 6.6) is based on the vector formulation introduced in Section 6.4 and on our discussion of spatial filtering in Chapter 3. Segmentation of color images (Section 6.7) has been a topic of much attention during the past ten years. The papers by Liu and Yang [1994] and by Shafarenko et al. [1998] are representative of work in this field. A special issue of the *IEEE Transactions on Image Processing* [1997] also is of interest. The discussion on color edge detection (Section 6.7.3) is from Di Zenzo [1986]. The book by Plataniotis and Venetsanopoulos [2000] does a good job of summarizing a variety of approaches to the segmentation of color images. The discussion in Section 6.8 is based on the noise models introduced in Section 5.2. References on image compression (Section 6.9) are listed at the end of Chapter 8. For details of software implementation of many of the techniques discussed in this chapter, see Gonzalez, Woods, and Eddins [2004].

Detailed solutions to the problems marked with a star can be found in the book Web site. The site also contains suggested projects based on the material in this chapter.

Problems

6.1   Give the percentages of red *(X)*, green *(Y)*, and blue *(Z)* light required to generate the point labeled "warm white" in Fig. 6.5.

★**6.2**   Consider any two valid colors c_1 and c_2 with coordinates (x_1, y_1) and (x_2, y_2) in the chromaticity diagram of Fig. 6.5. Derive the necessary general expression(s) for computing the relative percentages of colors c_1 and c_2 composing a given color that is known to lie on the straight line joining these two colors.

6.3 Consider any three valid colors c_1, c_2, and c_3 with coordinates (x_1, y_1), (x_2, y_2), and (x_3, y_3) in the chromaticity diagram of Fig. 6.5. Derive the necessary general expression(s) for computing the relative percentages of c_1, c_2, and c_3 composing a given color that is known to lie within the triangle whose vertices are at the coordinates of c_1, c_2, and c_3.

★**6.4** In an automated assembly application, three classes of parts are to be color coded in order to simplify detection. However, only a monochrome TV camera is available to acquire digital images. Propose a technique for using this camera to detect the three different colors.

6.5 In a simple RGB image, the R, G, and B component images have the horizontal intensity profiles shown in the following diagram. What color would a person see in the middle column of this image?

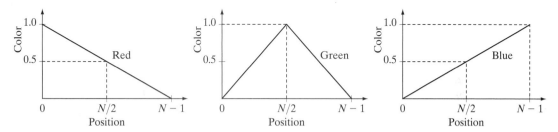

★**6.6** Sketch the RGB components of the following image as they would appear on a monochrome monitor. All colors are at maximum intensity and saturation. In working this problem, consider the middle gray border as part of the image.

6.7 How many different shades of gray are there in a color RGB system in which each RGB image is an 8-bit image?

6.8 Consider the RGB color cube shown in Fig 6.8, and answer each of the following:

★**(a)** Describe how the gray levels vary in the R, G, and B primary images that make up the front face of the color cube.

(b) Suppose that we replace every color in the RGB cube by its CMY color. This new cube is displayed on an RGB monitor. Label with a color name the eight vertices of the new cube that you would see on the screen.

    (c) What can you say about the colors on the edges of the RGB color cube regarding saturation?

6.9   **(a)** Sketch the CMY components of the image in Problem 6.6 as they would appear on a monochrome monitor.

    (b) If the CMY components sketched in (a) are fed into the red, green, and blue inputs of a color monitor, respectively, describe the resulting image.

★**6.10**   Derive the CMY intensity mapping function of Eq. (6.5-6) from its RGB counterpart in Eq. (6.5-5).

6.11   Consider the entire 216 safe-color array shown in Fig. 6.10(a). Label each cell by its (row, column) designation, so that the top left cell is (1, 1) and the rightmost bottom cell is (12, 18). At which cells will you find

    (a) The purest green?

    (b) The purest blue?

★**6.12**   Sketch the HSI components of the image in Problem 6.6 as they would appear on a monochrome monitor.

6.13   Propose a method for generating a color band similar to the one shown in the zoomed section entitled *Visible Spectrum* in Fig. 6.2. Note that the band starts at a dark purple on the left and proceeds toward pure red on the right. (*Hint:* Use the HSI color model.)

★**6.14**   Propose a method for generating a color version of the image shown diagrammatically in Fig. 6.13(c). Give your answer in the form of a flow chart. Assume that the intensity value is fixed and given. (*Hint:* Use the HSI color model.)

6.15   Consider the following image composed of solid color squares. For discussing your answer, choose a gray scale consisting of eight shades of gray, 0 through 7, where 0 is black and 7 is white. Suppose that the image is converted to HSI color space. In answering the following questions, use specific numbers for the gray shades if using numbers makes sense. Otherwise, the relationships "same as," "lighter than," or "darker than" are sufficient. If you cannot assign a specific gray level or one of these relationships to the image you are discussing, give the reason.

    (a) Sketch the hue image.

    (b) Sketch the saturation image.

    (c) Sketch the intensity image.

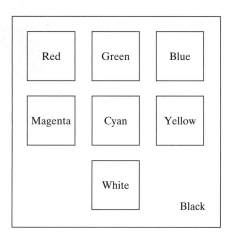

6.16 The following 8-bit images are (left to right) the *H*, *S*, and *I* component images from Fig. 6.16. The numbers indicate gray-level values. Answer the following questions, explaining the basis for your answer in each. If it is not possible to answer a question based on the given information, state why you cannot do so.

★**(a)** Give the gray-level values of all regions in the hue image.

(b) Give the gray-level value of all regions in the saturation image.

(c) Give the gray-level values of all regions in the intensity image.

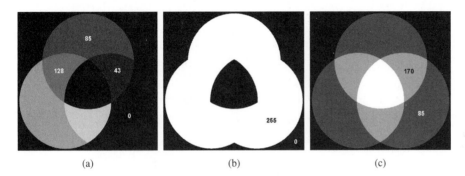

(a)   (b)   (c)

6.17 Refer to Fig. 6.27 in answering the following:

(a) Why does the image in Fig. 6.27(f) exhibit predominantly red tones?

(b) Suggest an automated procedure for coding the water in Fig. 6.27 in a bright-blue color.

(c) Suggest an automated procedure for coding the predominantly man-made components in a bright yellow color. [*Hint:* Work with Fig. 6.27(f).]

★**6.18** Show that the saturation component of the complement of a color image cannot be computed from the saturation component of the input image alone.

6.19 Explain the shape of the hue transformation function for the complement approximation in Fig. 6.33(b) using the HSI color model.

★**6.20** Derive the CMY transformations to generate the complement of a color image.

6.21 Draw the general shape of the transformation functions used to correct excessive contrast in the RGB color space.

★**6.22** Assume that the monitor and printer of an imaging system are imperfectly calibrated. An image that looks balanced on the monitor appears yellowish in print. Describe general transformations that might correct the imbalance.

6.23 Compute the $L^*a^*b^*$ components of the image in Problem 6.6 assuming

$$\begin{bmatrix} X \\ Y \\ Z \end{bmatrix} = \begin{bmatrix} 0.588 & 0.179 & 0.183 \\ 0.29 & 0.606 & 0.105 \\ 0 & 0.068 & 1.021 \end{bmatrix} \begin{bmatrix} R \\ G \\ B \end{bmatrix}$$

This matrix equation defines the tristimulus values of the colors generated by standard National Television System Committee (NTSC) color TV phosphors viewed under *D*65 standard illumination (Benson [1985]).

★**6.24** How would you implement the color equivalent of gray scale histogram matching (specification) from Section 3.3.2?

6.25 Consider the following 500×500 RGB image, in which the squares are fully saturated red, green, and blue, and each of the colors is at maximum intensity [e.g., $(1, 0, 0)$ for the red square]. An HSI image is generated from this image.

(a) Describe the appearance of each HSI component image.

(b) The saturation component of the HSI image is smoothed using an averaging mask of size 125×125. Describe the appearance of the result (you may ignore image border effects in the filtering operation).

(c) Repeat (b) for the hue image.

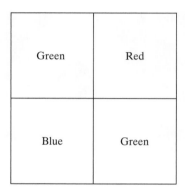

6.26 Show that Eq. (6.7-2) reduces to Eq. (6.7-1) when $\mathbf{C} = \mathbf{I}$, the identity matrix.

6.27 ★**(a)** With reference to the discussion in Section 6.7.2, give a procedure (in flow chart form) for determining whether a color vector (point) $\mathbf{z}$ is inside a cube with sides W, centered at an average color vector $\mathbf{a}$. Distance computations are not allowed.

(b) This process also can be implemented on an image-by-image basis if the box is lined up with the axes. Show how you would do it.

6.28 Sketch the surface in RGB space for the points that satisfy the equation

$$D(\mathbf{z}, \mathbf{a}) = [(\mathbf{z} - \mathbf{a})^T \mathbf{C}^{-1} (\mathbf{z} - \mathbf{a})]^{\frac{1}{2}} = D_0$$

where D_0 is a specified nonzero constant. Assume that $\mathbf{a} = \mathbf{0}$ and that

$$\mathbf{C} = \begin{bmatrix} 8 & 0 & 0 \\ 0 & 1 & 0 \\ 0 & 0 & 1 \end{bmatrix}$$

6.29 Refer to Section 6.7.3. One might think that a logical approach for defining the gradient of an RGB image at any point (x, y) would be to compute the gradient *vector* (see Section 3.6.4) of each component image and then form a gradient vector for the color image by summing the three individual gradient vectors. Unfortunately, this method can at times yield erroneous results. Specifically, it is possible for a color image with clearly defined edges to have a zero gradient if this method were used. Give an example of such an image. (*Hint:* Set one of the color planes to a constant value to simplify your analysis.)

7 Wavelets and Multiresolution Processing

> All this time, the guard was looking at her, first through a telescope, then through a microscope, and then through an opera glass.
>
> Lewis Carrol, *Through the Looking Glass*

Preview

Although the Fourier transform has been the mainstay of transform-based image processing since the late 1950s, a more recent transformation, called the *wavelet transform*, is now making it even easier to compress, transmit, and analyze many images. Unlike the Fourier transform, whose basis functions are sinusoids, wavelet transforms are based on small waves, called *wavelets*, of varying frequency *and limited duration*. This allows them to provide the equivalent of a musical score for an image, revealing not only what notes (or frequencies) to play but also when to play them. Fourier transforms, on the other hand, provide only the notes or frequency information; temporal information is lost in the transformation process.

In 1987, wavelets were first shown to be the foundation of a powerful new approach to signal processing and analysis called *multiresolution* theory (Mallat [1987]). Multiresolution theory incorporates and unifies techniques from a variety of disciplines, including subband coding from signal processing, quadrature mirror filtering from digital speech recognition, and pyramidal image processing. As its name implies, multiresolution theory is concerned with the representation and analysis of signals (or images) at more than one resolution. The appeal of such an approach is obvious—features that might go undetected at one resolution may be easy to detect at another. Although the imaging community's interest in multiresolution analysis was limited until the late 1980s, it is now difficult to keep up with the number of papers, theses, and books devoted to the subject.

In this chapter, we examine wavelet-based transformations from a multiresolution point of view. Although such transformations can be presented in other ways, this approach simplifies both their mathematical and physical interpretations. We begin with an overview of imaging techniques that influenced the formulation of multiresolution theory. Our objective is to introduce the theory's fundamental concepts within the context of image processing and simultaneously provide a brief historical perspective of the method and its application. The bulk of the chapter is focused on the development and use of the discrete wavelet transform. To demonstrate the usefulness of the transform, examples ranging from image coding to noise removal and edge detection are provided. In the next chapter, wavelets will be used for image compression, an application in which they have received considerable attention.

7.1 Background

When we look at images, generally we see connected regions of similar texture and intensity levels that combine to form objects. If the objects are small in size or low in contrast, we normally examine them at high resolutions; if they are large in size or high in contrast, a coarse view is all that is required. If both small and large objects—or low- and high-contrast objects—are present simultaneously, it can be advantageous to study them at several resolutions. This, of course, is the fundamental motivation for multiresolution processing.

Local histograms are histograms of the pixels in a neighborhood (see Section 3.3.3).

From a mathematical viewpoint, images are two-dimensional arrays of intensity values with locally varying statistics that result from different combinations of abrupt features like edges and contrasting homogeneous regions. As illustrated in Fig. 7.1—an image that will be examined repeatedly in the remainder of the

FIGURE 7.1
An image and its local histogram variations.

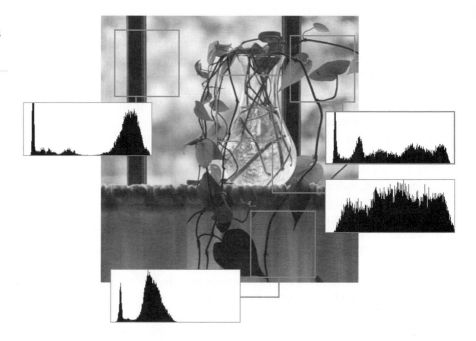

section—local histograms can vary significantly from one part of an image to another, making statistical modeling over the span of an entire image a difficult, or impossible task.

7.1.1 Image Pyramids

A powerful, yet conceptually simple structure for representing images at more than one resolution is the *image pyramid* (Burt and Adelson [1983]). Originally devised for machine vision and image compression applications, an image pyramid is a collection of decreasing resolution images arranged in the shape of a pyramid. As can be seen in Fig. 7.2(a), the base of the pyramid contains a high-resolution representation of the image being processed; the apex contains a low-resolution approximation. As you move up the pyramid, both size and resolution decrease. Base level J is of size $2^J \times 2^J$ or $N \times N$, where $J = \log_2 N$, apex level 0 is of size 1×1, and general level j is of size $2^j \times 2^j$, where $0 \le j \le J$. Although the pyramid shown in Fig. 7.2(a) is composed of $J + 1$ resolution levels from $2^J \times 2^J$ to $2^0 \times 2^0$, most image pyramids are truncated to $P + 1$ levels, where $1 \le P \le J$ and $j = J - P, \ldots, J - 2, J - 1, J$. That is, we normally limit ourselves to P reduced resolution approximations of the original image; a 1×1 (i.e., single pixel) approximation of a 512×512 image, for example, is of little value. The total number of pixels in a $P + 1$ level pyramid for $P > 0$ is

$$N^2\left(1 + \frac{1}{(4)^1} + \frac{1}{(4)^2} + \cdots + \frac{1}{(4)^P}\right) \le \frac{4}{3}N^2$$

Figure 7.2(b) shows a simple system for constructing two intimately related image pyramids. The *Level j − 1 approximation* output provides the images

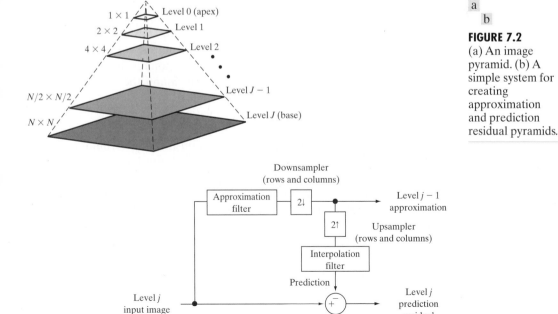

a
b

FIGURE 7.2
(a) An image pyramid. (b) A simple system for creating approximation and prediction residual pyramids.

needed to build an *approximation pyramid* (as described in the preceding paragraph), while the *Level j prediction residual* output is used to build a complementary *prediction residual pyramid*. Unlike approximation pyramids, prediction residual pyramids contain only one reduced-resolution approximation of the input image (at the top of the pyramid, level $J - P$). All other levels contain prediction residuals, where the level *j prediction residual* (for $J - P + 1 \le j \le J$) is defined as the difference between the level *j* approximation (the input to the block diagram) and an estimate of the level *j* approximation based on the level $j - 1$ approximation (the approximation output in the block diagram).

In general, a prediction residual can be defined as the difference between an image and a predicted version of the image. As will be seen in Section 8.2.9, prediction residuals can often be coded more efficiently than 2-D intensity arrays.

As Fig. 7.2(b) suggests, both approximation and prediction residual pyramids are computed in an iterative fashion. Before the first iteration, the image to be represented in pyramidal form is placed in level *J* of the approximation pyramid. The following three-step procedure is then executed *P* times—for $j = J, J - 1, \ldots,$ and $J - P + 1$ (in that order):

Step 1. Compute a reduced-resolution approximation of the *Level j input image* [the input on the left side of the block diagram in Fig. 7.2(b)]. This is done by filtering and downsampling the filtered result by a factor of 2. Both of these operations are described in the next paragraph. Place the resulting approximation at level $j - 1$ of the approximation pyramid.

Step 2. Create an estimate of the *Level j input image* from the reduced-resolution approximation generated in step 1. This is done by upsampling and filtering (see the next paragraph) the generated approximation. The resulting prediction image will have the same dimensions as the *Level j input image*.

Step 3. Compute the difference between the prediction image of step 2 and the input to step 1. Place this result in level *j* of the prediction residual pyramid.

At the conclusion of *P* iterations (i.e., following the iteration in which $j = J - P + 1$), the level $J - P$ approximation output is placed in the prediction residual pyramid at level $J - P$. If a prediction residual pyramid is not needed, this operation—along with steps 2 and 3 and the upsampler, interpolation filter, and summer of Fig. 7.2(b)—can be omitted.

A variety of approximation and interpolation filters can be incorporated into the system of Fig. 7.2(b). Typically, the filtering is performed in the spatial domain (see Section 3.4). Useful approximation filtering techniques include neighborhood averaging (see Section 3.5.1.), which produces *mean pyramids*; lowpass Gaussian filtering (see Sections 4.7.4 and 4.8.3), which produces *Gaussian pyramids*; and no filtering, which results in *subsampling pyramids*. Any of the interpolation methods described in Section 2.4.4, including nearest neighbor, bilinear, and bicubic, can be incorporated into the interpolation filter. Finally, we note that the upsampling and downsampling blocks of Fig. 7.2(b) are used to double and halve the spatial dimensions of the approximation and prediction images that are computed. Given an integer variable *n* and 1-D sequence of samples $f(n)$, *upsampled* sequence $f_{2\uparrow}(n)$ is defined as

$$f_{2\uparrow}(n) = \begin{cases} f(n/2) & \text{if } n \text{ is even} \\ 0 & \text{otherwise} \end{cases} \qquad (7.1\text{-}1)$$

In this chapter, we will be working with both continuous and discrete functions and variables. With the notable exception of 2-D image $f(x, y)$ and unless otherwise noted, $x, y, z, \ldots$ are continuous variables; $i, j, k, l, m, n, \ldots$ are discrete variables.

where, as is indicated by the subscript, the upsampling is by a factor of 2. The complementary operation of *downsampling* by 2 is defined as

$$f_{2\downarrow}(n) = f(2n) \qquad (7.1\text{-}2)$$

Upsampling can be thought of as inserting a 0 after every sample in a sequence; downsampling can be viewed as discarding every other sample. The upsampling and downsampling blocks in Fig. 7.2(b), which are labeled $2\uparrow$ and $2\downarrow$, respectively, are annotated to indicate that both the rows and columns of the 2-D inputs on which they operate are to be up- and downsampled. Like the separable 2-D DFT in Section 4.11.1, 2-D upsampling and downsampling can be performed by successive passes of the 1-D operations defined in Eqs. (7.1-1) and (7.1-2).

■ Figure 7.3 shows both an approximation pyramid and a prediction residual pyramid for the vase of Fig. 7.1. A lowpass Gaussian smoothing filter (see Section 4.7.4) was used to produce the four-level approximation pyramid in Fig. 7.3(a). As you can see, the resulting pyramid contains the original 512×512 resolution image (at its base) and three low-resolution approximations (of resolution 256×256, 128×128, and 64×64). Thus, P is 3 and levels 9, 8, 7, and 6 out of a possible $\log_2(512) + 1$ or 10 levels are present. Note the reduction in detail that accompanies the lower resolutions of the pyramid. The level 6 (i.e., 64×64) approximation image is suitable for locating the window stiles (i.e., the window pane framing), for example, but not for finding the stems of the plant. In general, the lower-resolution levels of a pyramid can be used for the analysis of large structures or overall image context; the high-resolution images are appropriate for analyzing individual object characteristics. Such a coarse-to-fine analysis strategy is particularly useful in pattern recognition.

A bilinear interpolation filter was used to produce the prediction residual pyramid in Fig. 7.3(b). In the absence of quantization error, the resulting prediction residual pyramid can be used to generate the complementary approximation pyramid in Fig. 7.3(a), including the original image, without error. To do so, we begin with the level 6 64×64 approximation image (the only approximation image in the prediction residual pyramid), predict the level 7 128×128 resolution approximation (by upsampling and filtering), and add the level 7 prediction residual. This process is repeated using successively computed approximation images until the original 512×512 image is generated. Note that the prediction residual histogram in Fig. 7.3(b) is highly peaked around zero; the approximation histogram in Fig. 7.3(a) is not. Unlike approximation images, prediction residual images can be highly compressed by assigning fewer bits to the more probable values (see the variable-length codes of Section 8.2.1). Finally, we note that the prediction residuals in Fig. 7.3(b) are scaled to make small prediction errors more visible; the prediction residual histogram, however, is based on the original residual values, with level 128 representing zero error. ■

EXAMPLE 7.1:
Approximation and prediction residual pyramids.

a
b

FIGURE 7.3
Two image
pyramids and
their histograms:
(a) an
approximation
pyramid;
(b) a prediction
residual pyramid.

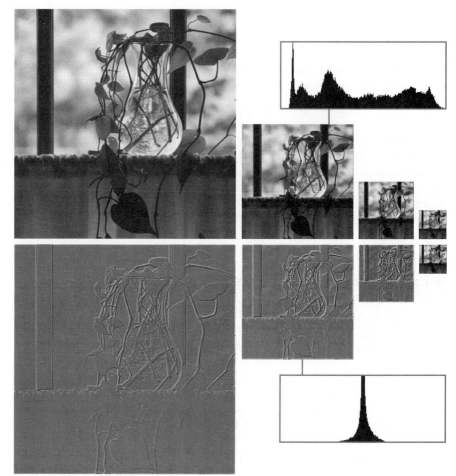

The approximation
pyramid in (a) is called a
Gaussian pyramid
because a Gaussian filter
was used to construct it.
The prediction residual
pyramid in (b) is often
called a Laplacian
pyramid; note the
similarity in appearance
with the Laplacian fil-
tered images in Chapter 3.

7.1.2 Subband Coding

Another important imaging technique with ties to multiresolution analysis is
subband coding. In subband coding, an image is decomposed into a set of
bandlimited components, called subbands. The decomposition is performed so
that the subbands can be reassembled to reconstruct the original image without error. Because the decomposition and reconstruction are performed by
means of digital filters, we begin our discussion with a brief introduction to
digital signal processing (DSP) and *digital signal filtering*.

The term "delay" implies
a time-based input
sequence and reflects the
fact that in digital signal
filtering, the input is
usually a sampled analog
signal.

Consider the simple *digital filter* in Fig. 7.4(a) and note that it is constructed
from three basic components—*unit delays*, *multipliers*, and *adders*. Along the
top of the filter, unit delays are connected in series to create $K - 1$ delayed
(i.e., right shifted) versions of the input sequence $f(n)$. Delayed sequence
$f(n - 2)$, for example, is

$$f(n-2) = \begin{cases} \vdots \\ f(0) & \text{for } n = 2 \\ f(1) & \text{for } n = 2 + 1 = 3 \\ \vdots \end{cases}$$

As the grayed annotations in Fig. 7.4(a) indicate, input sequence $f(n) = f(n-0)$ and the $K-1$ delayed sequences at the outputs of the unit delays, denoted $f(n-1), f(n-2), \ldots, f(n-K+1)$, are multiplied by constants $h(0), h(1), \ldots, h(K-1)$, respectively, and summed to produce the filtered output sequence

$$\hat{f}(n) = \sum_{k=-\infty}^{\infty} h(k)f(n-k)$$

$$= f(n) \star h(n) \tag{7.1-3}$$

where $\star$ denotes convolution. Note that—except for a change in variables— Eq. (7.1-3) is equivalent to the discrete convolution defined in Eq. (4.4-10) of Chapter 4. The K multiplication constants in Fig. 7.4(a) and Eq. (7.1-3) are

If the coefficients of the filter in Fig. 7.4(a) are indexed using values of n between 0 and $K-1$ (as we have done), the limits on the sum in Eq. (7.1-3) can be reduced to 0 to $K-1$ [like Eq. (4.4-10)].

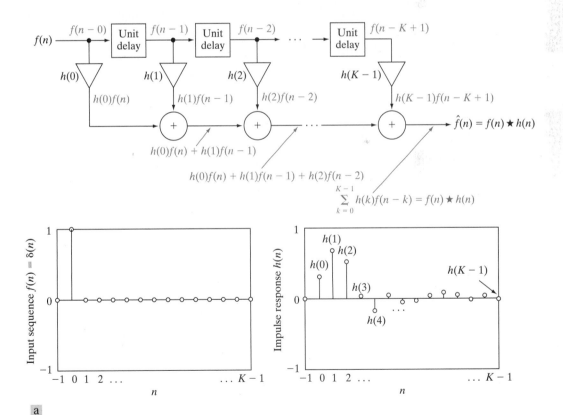

FIGURE 7.4 (a) A digital filter; (b) a unit discrete impulse sequence; and (c) the impulse response of the filter.

called *filter coefficients*. Each coefficient defines a *filter tap*, which can be thought of as the components needed to compute one term of the sum in Eq. (7.1-3), and the filter is said to be of *order K*.

If the input to the filter of Fig. 7.4(a) is the unit discrete impulse of Fig. 7.4(b) and Section 4.2.3, Eq. (7.1-3) becomes

$$\hat{f}(n) = \sum_{k=-\infty}^{\infty} h(k)\delta(n - k)$$
$$= h(n)$$

(7.1-4)

That is, by substituting $\delta(n)$ for input $f(n)$ in Eq. (7.1-3) and making use of the sifting property of the unit discrete impulse as defined in Eq. (4.2-13), we find that the *impulse response* of the filter in Fig. 7.4(a) is the K-element sequence of filter coefficients that define the filter. Physically, the unit impulse is shifted from left to right across the top of the filter (from one unit delay to the next), producing an output that assumes the value of the coefficient at the location of the delayed impulse. Because there are K coefficients, the impulse response is of length K and the filter is called a *finite impulse response* (FIR) filter.

In the remainder of the chapter, "filter $h(n)$" will be used to refer to the filter whose impulse response is $h(n)$.

Figure 7.5 shows the impulse responses of six functionally related filters. Filter $h_2(n)$ in Fig. 7.5(b) is a *sign-reversed* (i.e., reflected about the horizontal axis) version of $h_1(n)$ in Fig. 7.5(a). That is,

$$h_2(n) = -h_1(n)$$

(7.1-5)

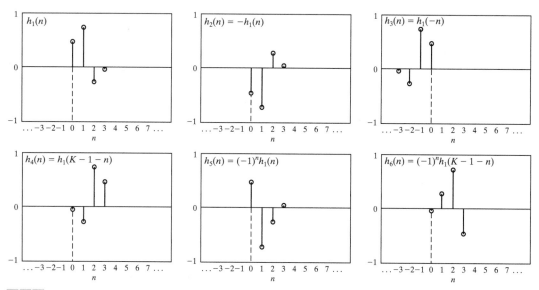

a b c
d e f

FIGURE 7.5 Six functionally related filter impulse responses: (a) reference response; (b) sign reversal; (c) and (d) order reversal (differing by the delay introduced); (e) modulation; and (f) order reversal and modulation.

Filters $h_3(n)$ and $h_4(n)$ in Figs. 7.5(c) and (d) are *order-reversed* versions of $h_1(n)$:

Order reversal is often called *time reversal* when the input sequence is a sampled analog signal.

$$h_3(n) = h_1(-n) \tag{7.1-6}$$

$$h_4(n) = h_1(K - 1 - n) \tag{7.1-7}$$

Filter $h_3(n)$ is a reflection of $h_1(n)$ about the vertical axis; filter $h_4(n)$ is a reflected and translated (i.e., shifted) version of $h_1(n)$. Neglecting translation, the responses of the two filters are identical. Filter $h_5(n)$ in Fig. 7.5(e), which is defined as

$$h_5(n) = (-1)^n h_1(n) \tag{7.1-8}$$

is called a *modulated* version of $h_1(n)$. Because modulation changes the signs of all odd-indexed coefficients [i.e., the coefficients for which n is odd in Fig. 7.5(e)], $h_5(1) = -h_1(1)$ and $h_5(3) = -h_1(3)$, while $h_5(0) = h_1(0)$ and $h_5(2) = h_1(2)$. Finally, the sequence shown in Fig. 7.5(f) is an order-reversed version of $h_1(n)$ that is also modulated:

$$h_6(n) = (-1)^n h_1(K - 1 - n) \tag{7.1-9}$$

This sequence is included to illustrate the fact that sign reversal, order reversal, and modulation are sometimes combined in the specification of the relationship between two filters.

With this brief introduction to digital signal filtering, consider the two-band subband coding and decoding system in Fig. 7.6(a). As indicated in the figure, the system is composed of two *filter banks*, each containing two FIR filters of the type shown in Fig. 7.4(a). Note that each of the four FIR filters is depicted

A *filter bank* is a collection of two or more filters.

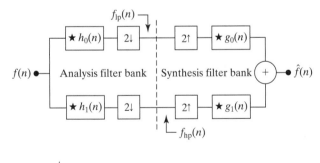

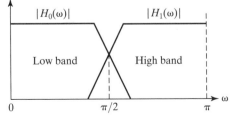

a
b

FIGURE 7.6
(a) A two-band subband coding and decoding system, and (b) its spectrum splitting properties.

as a single block in Fig. 7.6(a), with the impulse response of each filter (and the convolution symbol) written inside it. The *analysis* filter bank, which includes filters $h_0(n)$ and $h_1(n)$, is used to break input sequence $f(n)$ into two half-length sequences $f_{lp}(n)$ and $f_{hp}(n)$, the *subbands* that represent the input. Note that filters $h_0(n)$ and $h_1(n)$ are half-band filters whose idealized transfer characteristics, H_0 and H_1, are shown in Fig. 7.6(b). Filter $h_0(n)$ is a lowpass filter whose output, subband $f_{lp}(n)$, is called an *approximation* of $f(n)$; filter $h_1(n)$ is a highpass filter whose output, subband $f_{hp}(n)$, is called the high frequency or *detail* part of $f(n)$. *Synthesis* bank filters $g_0(n)$ and $g_1(n)$ combine $f_{lp}(n)$ and $f_{hp}(n)$ to produce $\hat{f}(n)$. The goal in subband coding is to select $h_0(n), h_1(n), g_0(n),$ and $g_1(n)$ so that $\hat{f}(n) = f(n)$. That is, so that the input and output of the subband coding and decoding system are identical. When this is accomplished, the resulting system is said to employ *perfect reconstruction filters*.

By *real-coefficient*, we mean that the filter coefficients are real (not complex) numbers.

There are many two-band, real-coefficient, FIR, perfect reconstruction filter banks described in the filter bank literature. In all of them, the synthesis filters are modulated versions of the analysis filters—with one (and only one) synthesis filter being sign reversed as well. For perfect reconstruction, the impulse responses of the synthesis and analysis filters must be related in one of the following two ways:

Equations (7.1-10) through (7.1-14) are described in detail in the filter bank literature (see, for example, Vetterli and Kovacevic [1995]).

$$g_0(n) = (-1)^n h_1(n)$$
$$g_1(n) = (-1)^{n+1} h_0(n)$$
(7.1-10)

or

$$g_0(n) = (-1)^{n+1} h_1(n)$$
$$g_1(n) = (-1)^n h_0(n)$$
(7.1-11)

Filters $h_0(n), h_1(n), g_0(n),$ and $g_1(n)$ in Eqs. (7.1-10) and (7.1-11) are said to be *cross-modulated* because diagonally opposed filters in the block diagram of Fig. 7.6(a) are related by modulation [and sign reversal when the modulation factor is $-(-1)^n$ or $(-1)^{n+1}$]. Moreover, they can be shown to satisfy the following *biorthogonality* condition:

$$\langle h_i(2n - k), g_j(k) \rangle = \delta(i - j)\delta(n), \quad i, j = \{0, 1\}$$
(7.1-12)

Here, $\langle h_i(2n - k), g_j(k) \rangle$ denotes the inner product of $h_i(2n - k)$ and $g_j(k)$.[†] When i is not equal to j, the inner product is 0; when i and j are equal, the product is the unit discrete impulse function, $\delta(n)$. Biorthogonality will be considered again in Section 7.2.1.

Of special interest in subband coding—and in the development of the fast wavelet transform of Section 7.4—are filters that move beyond biorthogonality and require

---

[†]The vector inner product of sequences $f_1(n)$ and $f_2(n)$ is $\langle f_1, f_2 \rangle = \sum_n f_1^*(n)f_2(n)$, where the * denotes the complex conjugate operation. If $f_1(n)$ and $f_2(n)$ are real, $\langle f_1, f_2 \rangle = \langle f_2, f_1 \rangle$.

$$\langle g_i(n), g_j(n + 2m) \rangle = \delta(i - j)\delta(m), \quad i, j = \{0, 1\} \qquad (7.1\text{-}13)$$

which defines *orthonormality* for perfect reconstruction filter banks. In addition to Eq. (7.1-13), orthonormal filters can be shown to satisfy the following two conditions:

$$g_1(n) = (-1)^n g_0(K_{\text{even}} - 1 - n)$$
$$h_i(n) = g_i(K_{\text{even}} - 1 - n), \quad i = \{0, 1\} \qquad (7.1\text{-}14)$$

where the subscript on K_{even} is used to indicate that the number of filter coefficients must be divisible by 2 (i.e., an even number). As Eq. (7.1-14) indicates, synthesis filter g_1 is related to g_0 by order reversal and modulation. In addition, both h_0 and h_1 are order-reversed versions of synthesis filters, g_0 and g_1, respectively. Thus, an orthonormal filter bank can be developed around the impulse response of a single filter, called the *prototype*; the remaining filters can be computed from the specified prototype's impulse response. For biorthogonal filter banks, two prototypes are required; the remaining filters can be computed via Eq. (7.1-10) or (7.1-11). The generation of useful prototype filters, whether orthonormal or biorthogonal, is beyond the scope of this chapter. We simply use filters that have been presented in the literature and provide references for further study.

Before concluding the section with a 2-D subband coding example, we note that 1-D orthonormal and biorthogonal filters can be used as 2-D separable filters for the processing of images. As can be seen in Fig. 7.7, the separable filters are first applied in one dimension (e.g., vertically) and then in the other (e.g., horizontally) in the manner introduced in Section 2.6.7. Moreover, downsampling is performed in two stages—once before the second filtering operation to reduce the overall number of computations. The resulting filtered

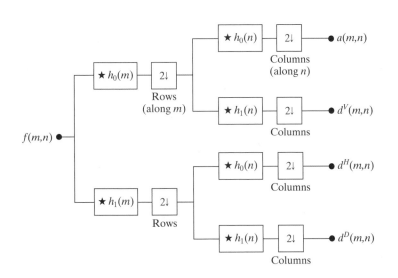

FIGURE 7.7

A two-dimensional, four-band filter bank for subband image coding.

outputs, denoted $a(m, n)$, $d^V(m, n)$, $d^H(m, n)$, and $d^D(m, n)$ in Fig. 7.7, are called the *approximation, vertical detail, horizontal detail,* and *diagonal detail* subbands of the input image, respectively. These subbands can be split into four smaller subbands, which can be split again, and so on—a property that will be described in greater detail in Section 7.4.

EXAMPLE 7.2:
A four-band subband coding of the vase in Fig. 7.1.

■ Figure 7.8 shows the impulse responses of four 8-tap orthonormal filters. The coefficients of prototype synthesis filter $g_0(n)$ for $0 \leq n \leq 7$ [in Fig. 7.8(c)] are defined in Table 7.1 (Daubechies [1992]). The coefficients of the remaining orthonormal filters can be computed using Eq. (7.1-14). With the help of Fig. 7.5, note (by visual inspection) the cross modulation of the analysis and synthesis filters in Fig. 7.8. It is relatively easy to show numerically that the filters are

TABLE 7.1
Daubechies 8-tap orthonormal filter coefficients for $g_0(n)$ (Daubechies [1992]).

n	$g_0(n)$
0	0.23037781
1	0.71484657
2	0.63088076
3	−0.02798376
4	−0.18703481
5	0.03084138
6	0.03288301
7	−0.01059740

a	b
c	d

FIGURE 7.8
The impulse responses of four 8-tap Daubechies orthonormal filters. See Table 7.1 for the values of $g_0(n)$ for $0 \leq n \leq 7$.

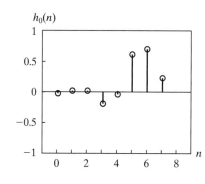

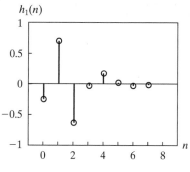

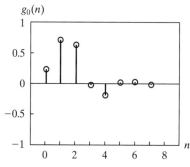

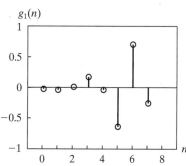

a b
c d

FIGURE 7.9
A four-band split
of the vase in
Fig. 7.1 using the
subband coding
system of Fig. 7.7.
The four
subbands that
result are the
(a) approximation,
(b) horizontal
detail, (c) vertical
detail, and
(d) diagonal detail
subbands.

both biorthogonal (they satisfy Eq. 7.1-12) and orthonormal (they satisfy Eq. 7.1-13). As a result, the Daubechies 8-tap filters in Fig. 7.8 support error-free reconstruction of the decomposed input.

A four-band split of the 512×512 image of a vase in Fig. 7.1, based on the filters in Fig. 7.8, is shown in Fig. 7.9. Each quadrant of this image is a subband of size 256×256. Beginning with the upper-left corner and proceeding in a clockwise manner, the four quadrants contain approximation subband a, horizontal detail subband d^H, diagonal detail subband d^D, and vertical detail subband d^V, respectively. All subbands, except the approximation subband in Fig. 7.9(a), have been scaled to make their underlying structure more visible. Note the visual effects of aliasing that are present in Figs. 7.9(b) and (c)—the d^H and d^V subbands. The wavy lines in the window area are due to the downsampling of a barely discernable window screen in Fig. 7.1. Despite the aliasing, the original image can be reconstructed from the subbands in Fig. 7.9 without error. The required synthesis filters, $g_0(n)$ and $g_1(n)$, are determined from Table 7.1 and Eq. (7.1-14), and incorporated into a filter bank that roughly mirrors the system in Fig. 7.7. In the new filter bank, filters $h_i(n)$ for $i = \{0, 1\}$ are replaced by their $g_i(n)$ counterparts, and upsamplers and summers are added. ■

See Section 4.5.4 for
more on aliasing.

7.1.3 The Haar Transform

The third and final imaging-related operation with ties to multiresolution analysis that we will look at is the Haar transform (Haar [1910]). Within the context of this chapter, its importance stems from the fact that its basis functions (defined below) are the oldest and simplest known orthonormal wavelets. They will be used in a number of examples in the sections that follow.

With reference to the discussion in Section 2.6.7, the Haar transform can be expressed in the following matrix form

$$\mathbf{T} = \mathbf{H}\mathbf{F}\mathbf{H}^T \tag{7.1-15}$$

where $\mathbf{F}$ is an $N \times N$ image matrix, $\mathbf{H}$ is an $N \times N$ Haar transformation matrix, and $\mathbf{T}$ is the resulting $N \times N$ transform. The transpose is required because $\mathbf{H}$ is not symmetric; in Eq. (2.6-38) of Section 2.6.7, the transformation matrix is assumed to be symmetric. For the Haar transform, $\mathbf{H}$ contains the Haar basis functions, $h_k(z)$. They are defined over the continuous, closed interval $z \in [0, 1]$ for $k = 0, 1, 2, \ldots, N - 1$, where $N = 2^n$. To generate $\mathbf{H}$, we define the integer k such that $k = 2^p + q - 1$, where $0 \le p \le n - 1$, $q = 0$ or 1 for $p = 0$, and $1 \le q \le 2^p$ for $p \ne 0$. Then the *Haar basis functions* are

$$h_0(z) = h_{00}(z) = \frac{1}{\sqrt{N}}, \quad z \in [0, 1] \tag{7.1-16}$$

and

$$h_k(z) = h_{pq}(z) = \frac{1}{\sqrt{N}} \begin{cases} 2^{p/2} & (q-1)/2^p \le z < (q-0.5)/2^p \\ -2^{p/2} & (q-0.5)/2^p \le z < q/2^p \\ 0 & \text{otherwise}, z \in [0, 1] \end{cases} \tag{7.1-17}$$

The ith row of an $N \times N$ Haar transformation matrix contains the elements of $h_i(z)$ for $z = 0/N, 1/N, 2/N, \ldots, (N-1)/N$. For instance, if $N = 2$, the first row of the 2×2 Haar matrix is computed using $h_0(z)$ with $z = 0/2, 1/2$. From Eq. (7.1-16), $h_0(z)$ is equal to $1/\sqrt{2}$, independent of z, so the first row of $\mathbf{H}_2$ has two identical $1/\sqrt{2}$ elements. The second row is obtained by computing $h_1(z)$ for $z = 0/2, 1/2$. Because $k = 2^p + q - 1$, when $k = 1, p = 0$ and $q = 1$. Thus, from Eq. (7.1-17), $h_1(0) = 2^0/\sqrt{2} = 1/\sqrt{2}, h_1(1/2) = -2^0/\sqrt{2} = -1/\sqrt{2}$, and the 2×2 Haar matrix is

$$\mathbf{H}_2 = \frac{1}{\sqrt{2}} \begin{bmatrix} 1 & 1 \\ 1 & -1 \end{bmatrix} \tag{7.1-18}$$

If $N = 4$, k, q, and p assume the values

k	p	q
0	0	0
1	0	1
2	1	1
3	1	2

and the 4×4 transformation matrix, $\mathbf{H}_4$, is

$$\mathbf{H}_4 = \frac{1}{\sqrt{4}} \begin{bmatrix} 1 & 1 & 1 & 1 \\ 1 & 1 & -1 & -1 \\ \sqrt{2} & -\sqrt{2} & 0 & 0 \\ 0 & 0 & \sqrt{2} & -\sqrt{2} \end{bmatrix} \tag{7.1-19}$$

Our principal interest in the Haar transform is that the rows of $\mathbf{H}_2$ can be used to define the analysis filters, $h_0(n)$ and $h_1(n)$, of a 2-tap perfect reconstruction filter bank (see the previous section), as well as the scaling and wavelet vectors (defined in Sections 7.2.2 and 7.2.3, respectively) of the simplest and oldest wavelet transform (see Example 7.10 in Section 7.4). Rather than concluding the section with the computation of a Haar transform, we close with an example that illustrates the influence of the decomposition methods that have been considered to this point on the methods that will be developed in the remainder of the chapter.

■ Figure 7.10(a) shows a decomposition of the 512×512 image in Fig. 7.1 that combines the key features of pyramid coding, subband coding, and the Haar transform (the three techniques we have discussed so far). Called the discrete wavelet transform (and developed later in the chapter), the representation is characterized by the following important features:

EXAMPLE 7.3:
Haar functions in a discrete wavelet transform.

1. With the exception of the subimage in the upper-left corner of Fig. 7.10(a), the local histograms are very similar. Many of the pixels are close to zero. Because the subimages (except for the subimage in the upper-left corner) have been scaled to make their underlying structure more visible, the displayed histograms are peaked at intensity 128 (the zeroes have been scaled to mid-gray). The large number of zeroes in the decomposition makes the image an excellent candidate for compression (see Chapter 8).
2. In a manner that is similar to the way in which the levels of the prediction residual pyramid of Fig. 7.3(b) were used to create approximation images of differing resolutions, the subimages in Fig. 7.10(a) can be used to construct both coarse and fine resolution approximations of the original vase image in Fig. 7.1. Figures 7.10(b) through (d), which are of size

FIGURE 7.10
(a) A discrete wavelet transform using Haar $\mathbf{H}_2$ basis functions. Its local histogram variations are also shown. (b)–(d) Several different approximations (64 × 64, 128 × 128, and 256 × 256) that can be obtained from (a).

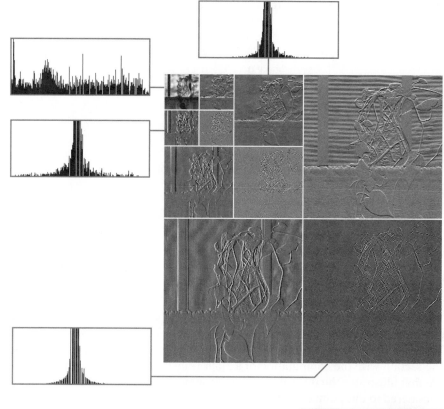

64 × 64, 128 × 128, and 256 × 256, respectively, were generated from the subimages in Fig. 7.10(a). A perfect 512 × 512 reconstruction of the original image is also possible.

3. Like the subband coding decomposition in Fig. 7.9, a simple real-coefficient, FIR filter bank of the form given in Fig. 7.7 was used to produce Fig. 7.10(a). After the generation of a four subband image like that of Fig. 7.9, the 256 × 256 approximation subband was decomposed and replaced by four 128 × 128 subbands (using the same filter bank), and the resulting approximation subband was again decomposed and replaced by four 64 × 64 subbands. This process produced the unique arrangement of subimages that

characterizes discrete wavelet transforms. The subimages in Fig. 7.10(a) become smaller in size as you move from the lower-right-hand to upper-left-hand corner of the image.

4. Figure 7.10(a) is not the Haar transform of the image in Fig. 7.1. Although the filter bank coefficients that were used to produce this decomposition were taken from Haar transformation matrix $\mathbf{H}_2$, a variety of othronormal and biorthogonal filter bank coefficients can be used in discrete wavelet transforms.

5. As will be shown in Section 7.4, each subimage in Fig. 7.10(a) represents a specific band of spatial frequencies in the original image. In addition, many of the subimages demonstrate directional sensitivity [e.g., the subimage in the upper-right corner of Fig. 7.10(a) captures horizontal edge information in the original image].

Considering this impressive list of features, it is remarkable that the discrete wavelet transform of Fig. 7.10(a) was generated using two 2-tap digital filters with a total of four filter coefficients.   ■

7.2 ■ Multiresolution Expansions

The previous section introduced three well-known imaging techniques that play an important role in a mathematical framework called *multiresolution analysis* (MRA). In MRA, a *scaling function* is used to create a series of approximations of a function or image, each differing by a factor of 2 in resolution from its nearest neighboring approximations. Additional functions, called *wavelets*, are then used to encode the difference in information between adjacent approximations.

7.2.1 Series Expansions

A signal or function $f(x)$ can often be better analyzed as a linear combination of expansion functions

$$f(x) = \sum_k \alpha_k \varphi_k(x) \qquad (7.2\text{-}1)$$

where k is an integer index of a finite or infinite sum, the α_k are real-valued *expansion coefficients*, and the $\varphi_k(x)$ are real-valued *expansion functions*. If the expansion is unique—that is, there is only one set of α_k for any given $f(x)$— the $\varphi_k(x)$ are called *basis functions*, and the *expansion set*, $\{\varphi_k(x)\}$, is called a *basis* for the class of functions that can be so expressed. The expressible functions form a *function space* that is referred to as the *closed span* of the expansion set, denoted

$$V = \overline{\text{Span}_k \{\varphi_k(x)\}} \qquad (7.2\text{-}2)$$

To say that $f(x) \in V$ means that $f(x)$ is in the closed span of $\{\varphi_k(x)\}$ and can be written in the form of Eq. (7.2-1).

For any function space V and corresponding expansion set $\{\varphi_k(x)\}$, there is a set of *dual* functions denoted $\{\tilde{\varphi}_k(x)\}$ that can be used to compute the α_k coefficients of Eq. (7.2-1) for any $f(x) \in V$. These coefficients are computed by taking the *integral inner products*[†] of the dual $\tilde{\varphi}_k(x)$ and function $f(x)$. That is,

$$\alpha_k = \langle \tilde{\varphi}_k(x), f(x) \rangle = \int \tilde{\varphi}_k^*(x) f(x)\, dx \tag{7.2-3}$$

where the $*$ denotes the complex conjugate operation. Depending on the orthogonality of the expansion set, this computation assumes one of three possible forms. Problem 7.10 at the end of the chapter illustrates the three cases using vectors in two-dimensional Euclidean space.

Case 1: If the expansion functions form an orthonormal basis for V, meaning that

$$\langle \varphi_j(x), \varphi_k(x) \rangle = \delta_{jk} = \begin{cases} 0 & j \neq k \\ 1 & j = k \end{cases} \tag{7.2-4}$$

the basis and its dual are equivalent. That is, $\varphi_k(x) = \tilde{\varphi}_k(x)$ and Eq. (7.2-3) becomes

$$\alpha_k = \langle \varphi_k(x), f(x) \rangle \tag{7.2-5}$$

The α_k are computed as the inner products of the basis functions and $f(x)$.

Case 2: If the expansion functions are not orthonormal, but are an orthogonal basis for V, then

$$\langle \varphi_j(x), \varphi_k(x) \rangle = 0 \quad j \neq k \tag{7.2-6}$$

and the basis functions and their duals are called *biorthogonal*. The α_k are computed using Eq. (7.2-3), and the biorthogonal basis and its dual are such that

$$\langle \varphi_j(x), \tilde{\varphi}_k(x) \rangle = \delta_{jk} = \begin{cases} 0 & j \neq k \\ 1 & j = k \end{cases} \tag{7.2-7}$$

Case 3: If the expansion set is not a basis for V, but supports the expansion defined in Eq. (7.2-1), it is a spanning set in which there is more than one set of α_k for any $f(x) \in V$. The expansion functions and their duals are said to be *overcomplete* or redundant. They form a *frame* in which[‡]

$$A\|f(x)\|^2 \leq \sum_k |\langle \varphi_k(x), f(x) \rangle|^2 \leq B\|f(x)\|^2 \tag{7.2-8}$$

[†]The integral inner product of two real or complex-valued functions $f(x)$ and $g(x)$ is $\langle f(x), g(x) \rangle = \int f^*(x)g(x)\, dx$. If $f(x)$ is real, $f^*(x) = f(x)$ and $\langle f(x), g(x) \rangle = \int f(x)g(x)\, dx$.

[‡]The norm of $f(x)$, denoted $\|f(x)\|$, is defined as the square root of the absolute value of the inner product of $f(x)$ with itself.

for some $A > 0$, $B < \infty$, and all $f(x) \in V$. Dividing this equation by the norm squared of $f(x)$, we see that A and B "frame" the normalized inner products of the expansion coefficients and the function. Equations similar to (7.2-3) and (7.2-5) can be used to find the expansion coefficients for frames. If $A = B$, the expansion set is called a *tight frame* and it can be shown that (Daubechies [1992])

$$f(x) = \frac{1}{A} \sum_k \langle \varphi_k(x), f(x) \rangle \varphi_k(x) \tag{7.2-9}$$

Except for the A^{-1} term, which is a measure of the frame's redundancy, this is identical to the expression obtained by substituting Eq. (7.2-5) (for orthonormal bases) into Eqs. (7.2-1).

7.2.2 Scaling Functions

Consider the set of expansion functions composed of integer translations and binary scalings of the real, square-integrable function $\varphi(x)$; this is the set $\{\varphi_{j,k}(x)\}$, where

$$\varphi_{j,k}(x) = 2^{j/2} \varphi(2^j x - k) \tag{7.2-10}$$

for all $j, k \in \mathbf{Z}$ and $\varphi(x) \in L^2(\mathbf{R})$.[†] Here, k determines the position of $\varphi_{j,k}(x)$ along the x-axis, and j determines the width of $\varphi_{j,k}(x)$—that is, how broad or narrow it is along the x-axis. The term $2^{j/2}$ controls the amplitude of the function. Because the shape of $\varphi_{j,k}(x)$ changes with j, $\varphi(x)$ is called a *scaling function*. By choosing $\varphi(x)$ properly, $\{\varphi_{j,k}(x)\}$ can be made to span $L^2(\mathbf{R})$, which is the set of all measurable, square-integrable functions.

If we restrict j in Eq. (7.2-10) to a specific value, say $j = j_0$, the resulting expansion set, $\{\varphi_{j_0,k}(x)\}$, is a subset of $\{\varphi_{j,k}(x)\}$ that spans a subspace of $L^2(\mathbf{R})$. Using the notation of the previous section, we can define that subspace as

$$V_{j_0} = \overline{\underset{k}{\text{Span}}\{\varphi_{j_0,k}(x)\}} \tag{7.2-11}$$

That is, V_{j_0} is the span of $\varphi_{j_0,k}(x)$ over k. If $f(x) \in V_{j_0}$, we can write

$$f(x) = \sum_k \alpha_k \varphi_{j_0,k}(x) \tag{7.2-12}$$

More generally, we will denote the subspace spanned over k for any j as

$$V_j = \overline{\underset{k}{\text{Span}}\{\varphi_{j,k}(x)\}} \tag{7.2-13}$$

As will be seen in the following example, increasing j increases the size of V_j, allowing functions with smaller variations or finer detail to be included in the subspace. This is a consequence of the fact that, as j increases, the $\varphi_{j,k}(x)$ that are used to represent the subspace functions become narrower and separated by smaller changes in x.

[†] The notation $L^2(\mathbf{R})$, where $\mathbf{R}$ is the set of real numbers, denotes the set of measurable, square-integrable, one-dimensional functions; $\mathbf{Z}$ is the set of integers.

EXAMPLE 7.4:
The Haar scaling
function.

■ Consider the unit-height, unit-width scaling function (Haar [1910])

$$\varphi(x) = \begin{cases} 1 & 0 \le x < 1 \\ 0 & \text{otherwise} \end{cases} \tag{7.2-14}$$

Figures 7.11(a) through (d) show four of the many expansion functions that can be generated by substituting this pulse-shaped scaling function into Eq. (7.2-10). Note that the expansion functions for $j = 1$ in Figs. 7.11(c) and (d) are half as wide as those for $j = 0$ in Figs. 7.11(a) and (b). For a given interval on x, we can define twice as many V_1 scaling functions as V_0 scaling functions (e.g., $\varphi_{1,0}$ and $\varphi_{1,1}$ of V_1 versus $\varphi_{0,0}$ of V_0 for the interval $0 \le x < 1$).

Figure 7.11(e) shows a member of subspace V_1. This function does not belong to V_0, because the V_0 expansion functions in 7.11(a) and (b) are too coarse to represent it. Higher-resolution functions like those in 7.11(c) and (d)

a b
c d
e f

FIGURE 7.11
Some Haar
scaling functions.

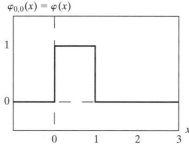

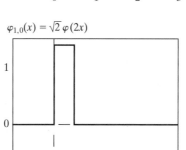

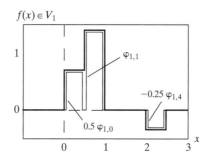

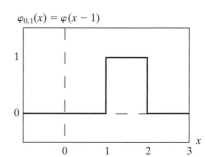

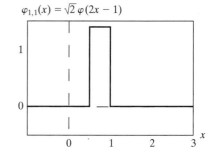

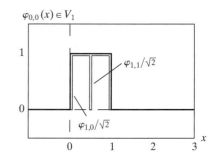

are required. They can be used, as shown in (e), to represent the function by the three-term expansion

$$f(x) = 0.5\varphi_{1,0}(x) + \varphi_{1,1}(x) - 0.25\varphi_{1,4}(x)$$

To conclude the example, Fig. 7.11(f) illustrates the decomposition of $\varphi_{0,0}(x)$ as a sum of V_1 expansion functions. In a similar manner, any V_0 expansion function can be decomposed using

$$\varphi_{0,k}(x) = \frac{1}{\sqrt{2}}\varphi_{1,2k}(x) + \frac{1}{\sqrt{2}}\varphi_{1,2k+1}(x)$$

Thus, if $f(x)$ is an element of V_0, it is also an element of V_1. This is because all V_0 expansion functions are contained in V_1. Mathematically, we write that V_0 is a subspace of V_1, denoted $V_0 \subset V_1$. ■

The simple scaling function in the preceding example obeys the four fundamental requirements of multiresolution analysis (Mallat [1989a]):

MRA Requirement 1: The scaling function is orthogonal to its integer translates.

This is easy to see in the case of the Haar function, because whenever it has a value of 1, its integer translates are 0, so that the product of the two is 0. The Haar scaling function is said to have *compact support*, which means that it is 0 everywhere outside a finite interval called the *support*. In fact, the width of the support is 1; it is 0 outside the half open interval [0, 1). It should be noted that the requirement for orthogonal integer translates becomes harder to satisfy as the width of support of the scaling function becomes larger than 1.

MRA Requirement 2: The subspaces spanned by the scaling function at low scales are nested within those spanned at higher scales.

As can be seen in Fig. 7.12, subspaces containing high-resolution functions must also contain all lower resolution functions. That is,

$$V_{-\infty} \subset \cdots \subset V_{-1} \subset V_0 \subset V_1 \subset V_2 \subset \cdots \subset V_{\infty} \qquad (7.2\text{-}15)$$

Moreover, the subspaces satisfy the intuitive condition that if $f(x) \in V_j$, then $f(2x) \in V_{j+1}$. The fact that the Haar scaling function meets this requirement

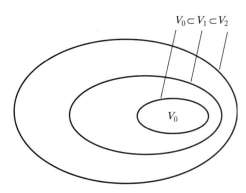

$$V_0 \subset V_1 \subset V_2$$

V_0

FIGURE 7.12
The nested function spaces spanned by a scaling function.

should not be taken to indicate that any function with a support width of 1 automatically satisfies the condition. It is left as an exercise for the reader to show that the equally simple function

$$
\varphi(x) = \begin{cases} 1 & 0.25 \le x < 0.75 \\ 0 & \text{elsewhere} \end{cases}
$$

is not a valid scaling function for a multiresolution analysis (see Problem 7.11).

MRA Requirement 3: The only function that is common to all V_j is $f(x) = 0$. If we consider the coarsest possible expansion functions (i.e., $j = -\infty$), the only representable function is the function of no information. That is,

$$
V_{-\infty} = \{0\} \tag{7.2-16}
$$

MRA Requirement 4: Any function can be represented with arbitrary precision. Though it may not be possible to expand a particular $f(x)$ at an arbitrarily coarse resolution, as was the case for the function in Fig. 7.11(e), all measurable, square-integrable functions can be represented by the scaling functions in the limit as $j \to \infty$. That is,

$$
V_{\infty} = \{L^2(\mathbf{R})\} \tag{7.2-17}
$$

Under these conditions, the expansion functions of subspace V_j can be expressed as a weighted sum of the expansion functions of subspace V_{j+1}. Using Eq. (7.2-12), we let

$$
\varphi_{j,k}(x) = \sum_n \alpha_n \varphi_{j+1,n}(x)
$$

where the index of summation has been changed to n for clarity. Substituting for $\varphi_{j+1,n}(x)$ from Eq. (7.2-10) and changing variable α_n to $h_\varphi(n)$, this becomes

The α_n are changed to $h_\varphi(n)$ because they are used later (see Section 7.4) as filter bank coefficients.

$$
\varphi_{j,k}(x) = \sum_n h_\varphi(n) 2^{(j+1)/2} \varphi(2^{j+1}x - n)
$$

Because $\varphi(x) = \varphi_{0,0}(x)$, both j and k can be set to 0 to obtain the simpler non-subscripted expression

$$
\varphi(x) = \sum_n h_\varphi(n) \sqrt{2} \varphi(2x - n) \tag{7.2-18}
$$

The $h_\varphi(n)$ coefficients in this recursive equation are called *scaling function coefficients*; h_φ is referred to as a *scaling vector*. Equation (7.2-18) is fundamental to multiresolution analysis and is called the *refinement equation*, the *MRA equation*, or the *dilation equation*. It states that the expansion functions of any subspace can be built from double-resolution copies of themselves—that is, from expansion functions of the next higher resolution space. The choice of a reference subspace, V_0, is arbitrary.

EXAMPLE 7.5:
Haar scaling
function
coefficients.

■ The scaling function coefficients for the Haar function of Eq. (7.2-14) are $h_\varphi(0) = h_\varphi(1) = 1/\sqrt{2}$, the first row of matrix $\mathbf{H}_2$ in Eq. (7.1-18). Thus, Eq. (7.2-18) yields

$$\varphi(x) = \frac{1}{\sqrt{2}}\left[\sqrt{2}\varphi(2x)\right] + \frac{1}{\sqrt{2}}\left[\sqrt{2}\varphi(2x - 1)\right]$$

This decomposition was illustrated graphically for $\varphi_{0,0}(x)$ in Fig. 7.11(f), where the bracketed terms of the preceding expression are seen to be $\varphi_{1,0}(x)$ and $\varphi_{1,1}(x)$. Additional simplification yields $\varphi(x) = \varphi(2x) + \varphi(2x - 1)$.   ■

7.2.3 Wavelet Functions

Given a scaling function that meets the MRA requirements of the previous section, we can define a *wavelet function* $\psi(x)$ that, together with its integer translates and binary scalings, spans the difference between any two adjacent scaling subspaces, V_j and V_{j+1}. The situation is illustrated graphically in Fig. 7.13. We define the set $\{\psi_{j,k}(x)\}$ of wavelets

$$\psi_{j,k}(x) = 2^{j/2}\psi(2^j x - k) \tag{7.2-19}$$

for all $k \in \mathbf{Z}$ that span the W_j spaces in the figure. As with scaling functions, we write

$$W_j = \overline{\underset{k}{\mathrm{Span}}\{\psi_{j,k}(x)\}} \tag{7.2-20}$$

and note that if $f(x) \in W_j$,

$$f(x) = \sum_k \alpha_k \psi_{j,k}(x) \tag{7.2-21}$$

The scaling and wavelet function subspaces in Fig. 7.13 are related by

$$V_{j+1} = V_j \oplus W_j \tag{7.2-22}$$

where $\oplus$ denotes the union of spaces (like the union of sets). The orthogonal complement of V_j in V_{j+1} is W_j, and all members of V_j are orthogonal to the members of W_j. Thus,

$$\langle \varphi_{j,k}(x), \psi_{j,l}(x)\rangle = 0 \tag{7.2-23}$$

for all appropriate $j, k, l \in \mathbf{Z}$.

$V_2 = V_1 \oplus W_1 = V_0 \oplus W_0 \oplus W_1$

$V_1 = V_0 \oplus W_0$

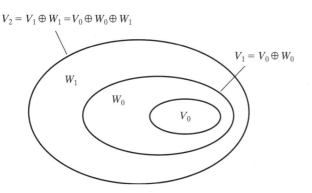

FIGURE 7.13
The relationship between scaling and wavelet function spaces.

We can now express the space of all measurable, square-integrable functions as

$$L^2(\mathbf{R}) = V_0 \oplus W_0 \oplus W_1 \oplus \dots \qquad (7.2\text{-}24)$$

or

$$L^2(\mathbf{R}) = V_1 \oplus W_1 \oplus W_2 \oplus \dots \qquad (7.2\text{-}25)$$

or even

$$L^2(\mathbf{R}) = \dots \oplus W_{-2} \oplus W_{-1} \oplus W_0 \oplus W_1 \oplus W_2 \oplus \dots \qquad (7.2\text{-}26)$$

which eliminates the scaling function, and represents a function in terms of wavelets alone [i.e., there are only wavelet function spaces in Eq. (7.2-26)]. Note that if $f(x)$ is an element of V_1, but not V_0, an expansion using Eq. (7.2-24) contains an *approximation* of $f(x)$ using V_0 scaling functions. Wavelets from W_0 would encode the *difference* between this approximation and the actual function. Equations (7.2-24) through (7.2-26) can be generalized to yield

$$L^2(\mathbf{R}) = V_{j_0} \oplus W_{j_0} \oplus W_{j_0+1} \oplus \dots \qquad (7.2\text{-}27)$$

where j_0 is an arbitrary starting scale.

Since wavelet spaces reside within the spaces spanned by the next higher resolution scaling functions (see Fig. 7.13), any wavelet function—like its scaling function counterpart of Eq. (7.2-18)—can be expressed as a weighted sum of shifted, double-resolution scaling functions. That is, we can write

$$\psi(x) = \sum_n h_\psi(n) \sqrt{2}\varphi(2x - n) \qquad (7.2\text{-}28)$$

where the $h_\psi(n)$ are called the *wavelet function coefficients* and h_ψ is the *wavelet vector*. Using the condition that wavelets span the orthogonal complement spaces in Fig. 7.13 and that integer wavelet translates are orthogonal, it can be shown that $h_\psi(n)$ is related to $h_\varphi(n)$ by (see, for example, Burrus, Gopinath, and Guo [1998])

$$h_\psi(n) = (-1)^n h_\varphi(1 - n) \qquad (7.2\text{-}29)$$

Note the similarity of this result and Eq. (7.1-14), the relationship governing the impulse responses of orthonormal subband coding and decoding filters.

EXAMPLE 7.6:
The Haar wavelet function coefficients.

■ In the previous example, the Haar scaling vector was defined as $h_\varphi(0) = h_\varphi(1) = 1/\sqrt{2}$. Using Eq. (7.2-29), the corresponding wavelet vector is $h_\psi(0) = (-1)^0 h_\varphi(1 - 0) = 1/\sqrt{2}$ and $h_\psi(1) = (-1)^1 h_\varphi(1 - 1) = -1/\sqrt{2}$. Note that these coefficients correspond to the second row of matrix $\mathbf{H}_2$ in Eq. (7.1-18). Substituting these values into Eq. (7.2-28), we get

$\psi(x) = \varphi(2x) - \varphi(2x - 1)$, which is plotted in Fig. 7.14(a). Thus, the Haar wavelet function is

$$\psi(x) = \begin{cases} 1 & 0 \le x < 0.5 \\ -1 & 0.5 \le x < 1 \\ 0 & \text{elsewhere} \end{cases} \qquad (7.2\text{-}30)$$

Using Eq. (7.2-19), we can now generate the universe of scaled and translated Haar wavelets. Two such wavelets, $\psi_{0,2}(x)$ and $\psi_{1,0}(x)$, are plotted in Figs. 7.14(b) and (c), respectively. Note that wavelet $\psi_{1,0}(x)$ for space W_1 is narrower than $\psi_{0,2}(x)$ for W_0; it can be used to represent finer detail.

Figure 7.14(d) shows a function of subspace V_1 that is not in subspace V_0. This function was considered in an earlier example [see Fig. 7.11(e)]. Although the function cannot be represented accurately in V_0, Eq. (7.2-22) indicates that it can be expanded using V_0 and W_0 expansion functions. The resulting expansion is

$$f(x) = f_a(x) + f_d(x)$$

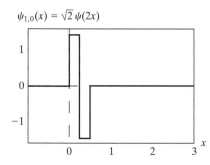

$\psi(x) = \psi_{0,0}(x)$

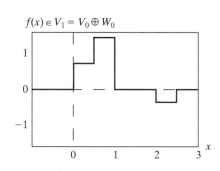

$\psi_{0,2}(x) = \psi(x - 2)$

a	b
c	d
e	f

FIGURE 7.14
Haar wavelet functions in W_0 and W_1.

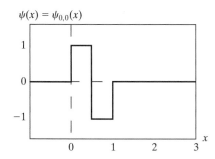

$\psi_{1,0}(x) = \sqrt{2}\,\psi(2x)$

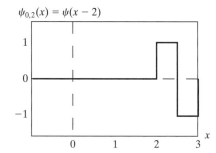

$f(x) \in V_1 = V_0 \oplus W_0$

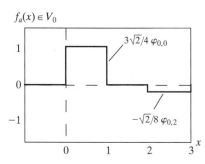

$f_a(x) \in V_0$

$\frac{3\sqrt{2}}{4}\,\varphi_{0,0}$

$-\sqrt{2}/8\,\varphi_{0,2}$

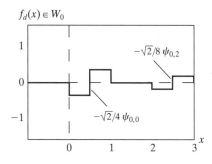

$f_d(x) \in W_0$

$-\sqrt{2}/8\,\psi_{0,2}$

$-\sqrt{2}/4\,\psi_{0,0}$

where

$$f_a(x) = \frac{3\sqrt{2}}{4}\varphi_{0,0}(x) - \frac{\sqrt{2}}{8}\varphi_{0,2}(x)$$

and

$$f_d(x) = \frac{-\sqrt{2}}{4}\psi_{0,0}(x) - \frac{\sqrt{2}}{8}\psi_{0,2}(x)$$

Here, $f_a(x)$ is an approximation of $f(x)$ using V_0 scaling functions, while $f_d(x)$ is the difference $f(x) - f_a(x)$ as a sum of W_0 wavelets. The two expansions, which are shown in Figs. 7.14(e) and (f), divide $f(x)$ in a manner similar to a lowpass and highpass filter as discussed in connection with Fig. 7.6. The low frequencies of $f(x)$ are captured in $f_a(x)$—it assumes the average value of $f(x)$ in each integer interval—while the high-frequency details are encoded in $f_d(x)$. ■

7.3   Wavelet Transforms in One Dimension

We can now formally define several closely related wavelet transformations: the generalized *wavelet series expansion*, the *discrete wavelet transform*, and the *continuous wavelet transform*. Their counterparts in the Fourier domain are the Fourier series expansion, the discrete Fourier transform, and the integral Fourier transform, respectively. In Section 7.4, we develop a computationally efficient implementation of the discrete wavelet transform called the *fast wavelet transform*.

7.3.1 The Wavelet Series Expansions

We begin by defining the *wavelet series expansion* of function $f(x) \in L^2(\mathbf{R})$ relative to wavelet $\psi(x)$ and scaling function $\varphi(x)$. In accordance with Eq. (7.2-27), $f(x)$ can be represented by a scaling function expansion in subspace V_{j_0} [Eq. (7.2-12) defines such an expansion] and some number of wavelet function expansions in subspaces $W_{j_0}, W_{j_0+1}, \dots$ [as defined in Eq. (7.2-21)]. Thus,

$$f(x) = \sum_k c_{j_0}(k)\varphi_{j_0,k}(x) + \sum_{j=j_0}^{\infty}\sum_k d_j(k)\psi_{j,k}(x) \qquad (7.3\text{-}1)$$

where j_0 is an arbitrary starting scale and the $c_{j_0}(k)$ and $d_j(k)$ are relabeled α_k from Eqs. (7.2-12) and (7.2-21), respectively. The $c_{j_0}(k)$ are normally called *approximation* and/or *scaling coefficients*; the $d_j(k)$ are referred to as *detail* and/or *wavelet coefficients*. This is because the first sum in Eq. (7.3-1) uses scaling functions to provide an approximation of $f(x)$ at scale j_0 [unless $f(x) \in V_{j_0}$ so that the sum of the scaling functions is equal to $f(x)$]. For each higher scale $j \geq j_0$ in the second sum, a finer resolution function—a sum of wavelets—is added to the approximation to provide increasing detail. If the expansion functions form an orthonormal basis or tight frame, which is often the case, the expansion coefficients are calculated—based on Eqs. (7.2-5) and (7.2-9)—as

$$c_{j_0}(k) = \langle f(x), \varphi_{j_0, k}(x) \rangle = \int f(x)\varphi_{j_0, k}(x) \, dx \qquad (7.3\text{-}2)$$

and

$$d_j(k) = \langle f(x), \psi_{j, k}(x) \rangle = \int f(x)\psi_{j, k}(x) \, dx \qquad (7.3\text{-}3)$$

Because f is real, no conjugates are needed in the inner products of Eqs. (7.3-2) and (7.3-3).

In Eqs. (7.2-5) and (7.2-9), the expansion coefficients (i.e., the α_k) are defined as inner products of the function being expanded and the expansion functions being used. In Eqs. (7.3-2) and (7.3-3), the expansion functions are the $\varphi_{j_0, k}$ and $\psi_{j, k}$; the expansion coefficients are the c_{j_0} and d_j. If the expansion functions are part of a biorthogonal basis, the φ and ψ terms in these equations must be replaced by their dual functions, $\tilde{\varphi}$ and $\tilde{\psi}$, respectively.

■ Consider the simple function

$$y = \begin{cases} x^2 & 0 \le x \le 1 \\ 0 & \text{otherwise} \end{cases}$$

EXAMPLE 7.7:
The Haar wavelet series expansion of $y = x^2$.

shown in Fig. 7.15(a). Using Haar wavelets—see Eqs. (7.2-14) and (7.2-30)—and a starting scale $j_0 = 0$, Eqs. (7.3-2) and (7.3-3) can be used to compute the following expansion coefficients:

$$c_0(0) = \int_0^1 x^2 \varphi_{0,0}(x) \, dx = \int_0^1 x^2 \, dx = \left. \frac{x^3}{3} \right|_0^1 = \frac{1}{3}$$

$$d_0(0) = \int_0^1 x^2 \psi_{0,0}(x) \, dx = \int_0^{0.5} x^2 \, dx - \int_{0.5}^1 x^2 \, dx = -\frac{1}{4}$$

$$d_1(0) = \int_0^1 x^2 \psi_{1,0}(x) \, dx = \int_0^{0.25} x^2\sqrt{2} \, dx - \int_{0.25}^{0.5} x^2\sqrt{2} \, dx = -\frac{\sqrt{2}}{32}$$

$$d_1(1) = \int_0^1 x^2 \psi_{1,1}(x) \, dx = \int_{0.5}^{0.75} x^2\sqrt{2} \, dx - \int_{0.75}^1 x^2\sqrt{2} \, dx = -\frac{3\sqrt{2}}{32}$$

Substituting these values into Eq. (7.3-1), we get the wavelet series expansion

$$y = \underbrace{\frac{1}{3}\varphi_{0,0}(x)}_{V_0} + \underbrace{\left[-\frac{1}{4}\psi_{0,0}(x) \right]}_{W_0} + \underbrace{\left[-\frac{\sqrt{2}}{32}\psi_{1,0}(x) - \frac{3\sqrt{2}}{32}\psi_{1,1}(x) \right]}_{W_1} + \cdots$$

$$\underbrace{\phantom{\frac{1}{3}\varphi_{0,0}(x) + \left[-\frac{1}{4}\psi_{0,0}(x) \right]}}_{V_1 = V_0 \oplus W_0}$$

$$\underbrace{\phantom{\frac{1}{3}\varphi_{0,0}(x) + \left[-\frac{1}{4}\psi_{0,0}(x) \right] + \left[-\frac{\sqrt{2}}{32}\psi_{1,0}(x) - \frac{3\sqrt{2}}{32}\psi_{1,1}(x) \right]}}_{V_2 = V_1 \oplus W_1 = V_0 \oplus W_0 \oplus W_1}$$

a b
c d
e f

FIGURE 7.15
A wavelet series
expansion of
$y = x^2$ using Haar
wavelets.

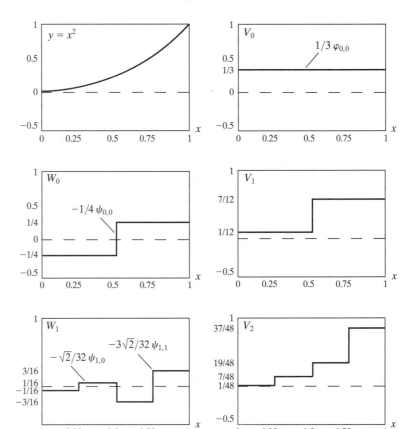

The first term in this expansion uses $c_0(0)$ to generate a subspace V_0 approximation of the function being expanded. This approximation is shown in Fig. 7.15(b) and is the average value of the original function. The second term uses $d_0(0)$ to refine the approximation by adding a level of detail from subspace W_0. The added detail and resulting V_1 approximation are shown in Figs. 7.15(c) and (d), respectively. Another level of detail is added by the subspace W_1 coefficients $d_1(0)$ and $d_1(1)$. This additional detail is shown in Fig. 7.15(e), and the resulting V_2 approximation is depicted in 7.15(f). Note that the expansion is now beginning to resemble the original function. As higher scales (greater levels of detail) are added, the approximation becomes a more precise representation of the function, realizing it in the limit as $j \to \infty$. ■

7.3.2 The Discrete Wavelet Transform

Like the Fourier series expansion, the wavelet series expansion of the previous section maps a function of a continuous variable into a sequence of coefficients. If the function being expanded is discrete (i.e., a sequence of numbers), the resulting coefficients are called the *discrete wavelet transform* (DWT). For example, if $f(n) = f(x_0 + n \Delta x)$ for some x_0, Δx, and $n = 0, 1, 2, \ldots, M - 1$, the wavelet series expansion coefficients for $f(x)$ [defined by Eqs. (7.3-2) and (7.3-3)] become the *forward* DWT coefficients for sequence $f(n)$:

$$W_\varphi(j_0, k) = \frac{1}{\sqrt{M}} \sum_n f(n)\varphi_{j_0, k}(n) \tag{7.3-5}$$

$$W_\psi(j, k) = \frac{1}{\sqrt{M}} \sum_n f(n)\psi_{j, k}(n) \quad \text{for } j \geq j_0 \tag{7.3-6}$$

The $\varphi_{j_0, k}(n)$ and $\psi_{j, k}(n)$ in these equations are sampled versions of basis functions $\varphi_{j_0, k}(x)$ and $\psi_{j, k}(x)$. For example, $\varphi_{j_0, k}(n) = \varphi_{j_0, k}(x_s + n\Delta x_s)$ for some x_s, Δx_s, and $n = 0, 1, 2, \ldots, M - 1$. Thus, we employ M equally spaced samples over the support of the basis functions (see Example 7.8 below). In accordance with Eq. (7.3-1), the complementary *inverse* DWT is

$$f(n) = \frac{1}{\sqrt{M}} \sum_k W_\varphi(j_0, k)\varphi_{j_0, k}(n) + \frac{1}{\sqrt{M}} \sum_{j=j_0}^\infty \sum_k W_\psi(j, k)\psi_{j, k}(n) \tag{7.3-7}$$

Normally, we let $j_0 = 0$ and select M to be a power of 2 (i.e., $M = 2^J$) so that the summations in Eqs. (7.3-5) through (7.3-7) are performed over $n = 0, 1, 2, \ldots, M - 1$, $j = 0, 1, 2, \ldots, J - 1$, and $k = 0, 1, 2, \ldots, 2^j - 1$. For Haar wavelets, the discretized scaling and wavelet functions employed in the transform (i.e., the basis functions) correspond to the rows of the $M \times M$ Haar transformation matrix of Section 7.1.3. The transform itself is composed of M coefficients, the minimum scale is 0, and the maximum scale is $J - 1$. For reasons noted in Section 7.3.1 and illustrated in Example 7.6, the coefficients defined in Eqs. (7.3-5) and (7.3-6) are usually called *approximation* and *detail coefficients*, respectively.

The $W_\varphi(j_0, k)$ and $W_\psi(j, k)$ in Eqs. (7.3-5) to (7.3-7) correspond to the $c_{j_0}(k)$ and $d_j(k)$ of the wavelet series expansion in the previous section. (This change of variables is not necessary but paves the way for the standard notation used for the continuous wavelet transform of the next section.) Note that the integrations in the series expansion have been replaced by summations, and a $1/\sqrt{M}$ normalizing factor, reminiscent of the DFT in Section 4.4.1, has been added to both the forward and inverse expressions. This factor alternately could be incorporated into the forward or inverse alone as $1/M$. Finally, it should be remembered that Eqs. (7.3-5) through (7.3-7) are valid for orthonormal bases and tight frames alone. For biorthogonal bases, the φ and ψ terms in Eqs. (7.3-5) and (7.3-6) must be replaced by their duals, $\widetilde{\varphi}$ and $\widetilde{\psi}$, respectively.

■ To illustrate the use of Eqs. (7.3-5) through (7.3-7), consider the discrete function of four points: $f(0) = 1$, $f(1) = 4$, $f(2) = -3$, and $f(3) = 0$. Because $M = 4$, $J = 2$ and, with $j_0 = 0$, the summations are performed over $x = 0, 1, 2, 3$, $j = 0, 1$, and $k = 0$ for $j = 0$ or $k = 0, 1$ for $j = 1$. We will use the Haar scaling and wavelet functions and assume that the four samples of

EXAMPLE 7.8:
Computing a one-dimensional discrete wavelet transform.

$f(x)$ are distributed over the support of the basis functions, which is 1 in width. Substituting the four samples into Eq. (7.3-5), we find that

$$
W_\varphi(0,0) = \frac{1}{2} \sum_{n=0}^{3} f(n) \varphi_{0,0}(n)
$$

$$
= \frac{1}{2} \left[1 \cdot 1 + 4 \cdot 1 - 3 \cdot 1 + 0 \cdot 1 \right] = 1
$$

because $\varphi_{0,0}(n) = 1$ for $n = 0, 1, 2, 3$. Note that we have employed uniformly spaced samples of the Haar scaling function for $j = 0$ and $k = 0$. The values correspond to the first row of Haar transformation matrix $\mathbf{H}_4$ of Section 7.1.3. Continuing with Eq. (7.3-6) and similarly spaced samples of $\psi_{j,k}(x)$, which correspond to rows 2, 3, and 4 of $\mathbf{H}_4$, we get

$$
W_\psi(0,0) = \frac{1}{2} \left[1 \cdot 1 + 4 \cdot 1 - 3 \cdot (-1) + 0 \cdot (-1) \right] = 4
$$

$$
W_\psi(1,0) = \frac{1}{2} \left[1 \cdot \sqrt{2} + 4 \cdot \left(-\sqrt{2} \right) - 3 \cdot 0 + 0 \cdot 0 \right] = -1.5\sqrt{2}
$$

$$
W_\psi(1,1) = \frac{1}{2} \left[1 \cdot 0 + 4 \cdot 0 - 3 \cdot \sqrt{2} + 0 \cdot \left(-\sqrt{2} \right) \right] = -1.5\sqrt{2}
$$

Thus, the discrete wavelet transform of our simple four-sample function relative to the Haar wavelet and scaling function is $\{1, 4, -1.5\sqrt{2}, -1.5\sqrt{2}\}$, where the transform coefficients have been arranged in the order in which they were computed.

Equation (7.3-7) lets us reconstruct the original function from its transform. Iterating through its summation indices, we get

$$
f(n) = \frac{1}{2} \Big[W_\varphi(0,0) \varphi_{0,0}(n) + W_\psi(0,0) \psi_{0,0}(n) + W_\psi(1,0) \psi_{1,0}(n)
$$

$$
+ W_\psi(1,1) \psi_{1,1}(n) \Big]
$$

for $n = 0, 1, 2, 3$. If $n = 0$, for instance,

$$
f(0) = \frac{1}{2} \left[1 \cdot 1 + 4 \cdot 1 - 1.5\sqrt{2} \cdot \left(\sqrt{2} \right) - 1.5\sqrt{2} \cdot 0 \right] = 1
$$

As in the forward case, uniformly spaced samples of the scaling and wavelet functions are used in the computation of the inverse.   ■

The four-point DWT in the preceding example is an illustration of a two-scale decomposition of $f(n)$—that is, $j = \{0, 1\}$. The underlying assumption was that starting scale j_0 was zero, but other starting scales are possible. It is left as an exercise for the reader (see Problem 7.16) to compute the single-scale transform $\{2.5\sqrt{2}, -1.5\sqrt{2}, -1.5\sqrt{2}, -1.5\sqrt{2}\}$, which results when the starting scale is 1. Thus, Eqs. (7.3-5) and (7.3-6) define a "family" of transforms that differ in starting scale j_0.

7.3.3 The Continuous Wavelet Transform

The natural extension of the discrete wavelet transform is the *continuous wavelet transform* (CWT), which transforms a continuous function into a highly redundant function of two continuous variables—translation and scale. The resulting transform is easy to interpret and valuable for time-frequency analysis. Although our interest is in discrete images, the continuous transform is covered here for completeness.

The continuous wavelet transform of a continuous, square-integrable function, $f(x)$, relative to a real-valued wavelet, $\psi(x)$, is defined as

$$W_{\psi}(s, \tau) = \int_{-\infty}^{\infty} f(x)\psi_{s,\tau}(x)\, dx \tag{7.3-8}$$

where

$$\psi_{s,\tau}(x) = \frac{1}{\sqrt{s}}\psi\!\left(\frac{x - \tau}{s}\right) \tag{7.3-9}$$

and s and τ are called *scale* and *translation* parameters, respectively. Given $W_{\psi}(s, \tau)$, $f(x)$ can be obtained using the *inverse continuous wavelet transform*

$$f(x) = \frac{1}{C_{\psi}} \int_{0}^{\infty} \int_{-\infty}^{\infty} W_{\psi}(s, \tau) \frac{\psi_{s,\tau}(x)}{s^2}\, d\tau\, ds \tag{7.3-10}$$

where

$$C_{\psi} = \int_{-\infty}^{\infty} \frac{|\Psi(\mu)|^2}{|\mu|}\, d\mu \tag{7.3-11}$$

and $\Psi(\mu)$ is the Fourier transform of $\psi(x)$. Equations (7.3-8) through (7.3-11) define a reversible transformation as long as the so-called *admissibility criterion*, $C_{\psi} < \infty$, is satisfied (Grossman and Morlet [1984]). In most cases, this simply means that $\Psi(0) = 0$ and $\Psi(\mu) \to 0$ as $\mu \to \infty$ fast enough to make $C_{\psi} < \infty$.

The preceding equations are reminiscent of their discrete counterparts— Eqs. (7.2-19), (7.3-1), (7.3-3), (7.3-6), and (7.3-7). The following similarities should be noted:

1. The continuous translation parameter, τ, takes the place of the integer translation parameter, k.
2. The continuous scale parameter, s, is inversely related to the binary scale parameter, 2^j. This is because s appears in the denominator of $\psi\big((x - \tau)/s\big)$ in Eq. (7.3-9). Thus, wavelets used in continuous transforms are compressed or reduced in width when $0 < s < 1$ and dilated or expanded when $s > 1$. Wavelet scale and our traditional notion of frequency are inversely related.

3. The continuous transform is similar to a series expansion [see Eq. (7.3-1)] or discrete transform [see Eq. (7.3-6)] in which the starting scale $j_0 = -\infty$. This—in accordance with Eq. (7.2-26)—eliminates explicit scaling function dependence, so that the function is represented in terms of wavelets alone.

4. Like the discrete transform, the continuous transform can be viewed as a set of transform coefficients, $\{W_\psi(s, \tau)\}$, that measure the similarity of $f(x)$ with a set of basis functions, $\{\psi_{s,\tau}(x)\}$. In the continuous case, however, both sets are infinite. Because $\psi_{s,\tau}(x)$ is real valued and $\psi_{s,\tau}(x) = \psi^*_{s,\tau}(x)$, each coefficient from Eq. (7.3-8) is the integral inner product, $\langle f(x), \psi_{s,\tau}(x) \rangle$, of $f(x)$ and $\psi_{s,\tau}(x)$.

EXAMPLE 7.9:
A one-dimensional continuous wavelet transform.

■ The *Mexican hat* wavelet,

$$\psi(x) = \left(\frac{2}{\sqrt{3}} \pi^{-1/4} \right)(1 - x^2)e^{-x^2/2} \tag{7.3-12}$$

gets its name from its distinctive shape [see Fig. 7.16(a)]. It is proportional to the second derivative of the Gaussian probability function, has an average value of 0, and is compactly supported (i.e., dies out rapidly as $|x| \to \infty$). Although it satisfies the admissibility requirement for the existence of continuous, reversible transforms, there is not an associated scaling function, and the computed transform does not result in an orthogonal analysis. Its most distinguishing features are its symmetry and the existence of the explicit expression of Eq. (7.3-12).

The continuous, one-dimensional function in Fig. 7.16(a) is the sum of two Mexican hat wavelets:

$$f(x) = \psi_{1, 10}(x) + \psi_{6, 80}(x)$$

Its Fourier spectrum, shown in Fig. 7.16(b), reveals the close connection between scaled wavelets and Fourier frequency bands. The spectrum contains two broad frequency bands (or peaks) that correspond to the two Gaussian-like perturbations of the function.

Figure 7.16(c) shows a portion ($1 \le s \le 10$ and $\tau \le 100$) of the CWT of the function in Fig. 7.16(a) relative to the Mexican hat wavelet. Unlike the Fourier spectrum in Fig. 7.16(b), it provides both spatial and frequency information. Note, for example, that when $s = 1$, the transform achieves a maximum at $\tau = 10$, which corresponds to the location of the $\psi_{1, 10}(x)$ component of $f(x)$. Because the transform provides an objective measure of the similarity between $f(x)$ and the wavelets for which it is computed, it is easy to see how it can be used for feature detection. We simply need wavelets that match the features of interest. Similar observations can be drawn from the intensity plot in Fig. 7.16(d), where the absolute value of the transform $|W_\psi(s, \tau)|$ is displayed as intensities between black and white. Note that the continuous wavelet transform turns a 1-D function into a 2-D result. ■

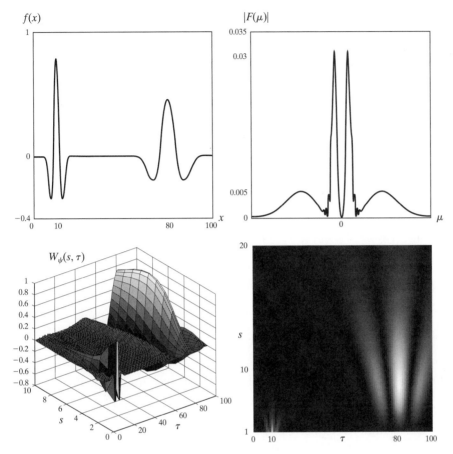

$f(x)$

$|F(\mu)|$

$W_\psi(s, \tau)$

FIGURE 7.16
The continuous
wavelet transform
(c and d) and
Fourier spectrum
(b) of a
continuous 1-D
function (a).

7.4 The Fast Wavelet Transform

The *fast wavelet transform* (FWT) is a computationally efficient implementation of the discrete wavelet transform (DWT) that exploits a surprising but fortunate relationship between the coefficients of the DWT at adjacent scales. Also called *Mallat's herringbone algorithm* (Mallat [1989a, 1989b]), the FWT resembles the two-band subband coding scheme of Section 7.1.2.

Consider again the multiresolution refinement equation

$$\varphi(x) = \sum_n h_\varphi(n)\sqrt{2}\varphi(2x - n) \qquad (7.4\text{-}1)$$

Equation (7.4-1) is Eq.
(7.2-18) of Section 7.2.2.

Scaling x by 2^j, translating it by k, and letting $m = 2k + n$ gives

$$\varphi(2^j x - k) = \sum_n h_\varphi(n)\sqrt{2}\varphi\left(2(2^j x - k) - n\right)$$

$$= \sum_m h_\varphi(n)\sqrt{2}\varphi(2^{j+1}x - 2k - n)$$

$$= \sum_m h_\varphi(m - 2k)\sqrt{2}\varphi(2^{j+1}x - m) \qquad (7.4\text{-}2)$$

Note that scaling vector h_φ can be thought of as the "weights" used to expand $\varphi(2^j x - k)$ as a sum of scale $j + 1$ scaling functions. A similar sequence of operations—beginning with Eq. (7.2-28)—provides an analogous result for $\psi(2^j x - k)$. That is,

$$\psi(2^j x - k) = \sum_m h_\psi(m - 2k)\sqrt{2}\varphi(2^{j+1}x - m) \tag{7.4-3}$$

where scaling vector $h_\varphi(n)$ in Eq. (7.4-2) corresponds to wavelet vector $h_\psi(n)$ in Eq. (7.4-3).

The wavelet series expansion coefficients become the DWT coefficient when f is discrete. Here, we begin with the series expansion coefficients to simplify the derivation; we will be able to substitute freely from earlier results (like the scaling and wavelet function definitions).

Now consider Eqs. (7.3-2) and (7.3-3) of Section 7.3.1. They define the wavelet series expansion coefficients of continuous function $f(x)$. Substituting Eq. (7.2-19)—the wavelet defining equation—into Eq. (7.3-3), we get

$$d_j(k) = \int f(x)2^{j/2}\psi(2^j x - k)\, dx \tag{7.4-4}$$

which, upon replacing $\psi(2^j x - k)$ with the right side of Eq. (7.4-3), becomes

$$d_j(k) = \int f(x)2^{j/2}\left[\sum_m h_\psi(m - 2k)\sqrt{2}\varphi(2^{j+1}x - m)\right] dx \tag{7.4-5}$$

Interchanging the sum and integral and rearranging terms then gives

$$d_j(k) = \sum_m h_\psi(m - 2k)\left[\int f(x)2^{(j+1)/2}\varphi(2^{j+1}x - m)\right] \tag{7.4-6}$$

where the bracketed quantity is $c_{j_0}(k)$ of Eq. (7.3-2) with $j_0 = j + 1$ and $k = m$. To see this, substitute Eq. (7.2-10) into Eq. (7.3-2) and replace j_0 and k with $j + 1$ and m, respectively. Therefore, we can write

$$d_j(k) = \sum_m h_\psi(m - 2k)c_{j+1}(m) \tag{7.4-7}$$

and note that the detail coefficients at scale j are a function of the approximation coefficients at scale $j + 1$. Using Eqs. (7.4-2) and (7.3-2) as the starting point of a similar derivation involving the wavelet series expansion (and DWT) approximation coefficients, we find similarly that

$$c_j(k) = \sum_m h_\varphi(m - 2k)c_{j+1}(m) \tag{7.4-8}$$

Because the $c_j(k)$ and $d_j(k)$ coefficients of the wavelet series expansion become the $W_\varphi(j, k)$ and $W_\psi(j, k)$ coefficients of the DWT when $f(x)$ is discrete (see Section 7.3.2), we can write

$$W_{\psi}(j, k) = \sum_{m} h_{\psi}(m - 2k) W_{\varphi}(j + 1, m) \qquad (7.4\text{-}9)$$

$$W_{\varphi}(j, k) = \sum_{m} h_{\varphi}(m - 2k) W_{\varphi}(j + 1, m) \qquad (7.4\text{-}10)$$

Equations (7.4-9) and (7.4-10) reveal a remarkable relationship between the DWT coefficients of adjacent scales. Comparing these results to Eq. (7.1-7), we see that both $W_{\varphi}(j, k)$ and $W_{\psi}(j, k)$, the scale j approximation and the detail coefficients, can be computed by convolving $W_{\varphi}(j + 1, k)$, the scale $j + 1$ approximation coefficients, with the order-reversed scaling and wavelet vectors, $h_{\varphi}(-n)$ and $h_{\psi}(-n)$, and subsampling the results. Figure 7.17 summarizes these operations in block diagram form. Note that this diagram is identical to the analysis portion of the two-band subband coding and decoding system of Fig. 7.6, with $h_0(n) = h_{\varphi}(-n)$ and $h_1(n) = h_{\psi}(-n)$. Therefore, we can write

$$W_{\psi}(j, k) = h_{\psi}(-n) \star W_{\varphi}(j + 1, n) \Big|_{n=2k, k \geq 0} \qquad (7.4\text{-}11)$$

and

$$W_{\varphi}(j, k) = h_{\varphi}(-n) \star W_{\varphi}(j + 1, n) \Big|_{n=2k, k \geq 0} \qquad (7.4\text{-}12)$$

If $h_{\varphi}(m - 2k)$ in Eq. (7.4-9) is rewritten as $h_{\varphi}(-(2k - m))$, we see that the first minus sign is responsible for the order reversal [see Eq. (7.1-6)], the $2k$ is responsible for the subsampling [see Eq. (7.1-2)], and m is the dummy variable for convolution [see Eq. (7.1-7)].

where the convolutions are evaluated at instants $n = 2k$ for $k \geq 0$. As will be shown in Example 7.10, evaluating convolutions at nonnegative, even indices is equivalent to filtering and downsampling by 2.

Equations (7.4-11) and (7.4-12) are the defining equations for the computation of the fast wavelet transform. For a sequence of length $M = 2^J$, the number of mathematical operations involved is on the order of $O(M)$. That is, the number of multiplications and additions is linear with respect to the length of the input sequence—because the number of multiplications and additions involved in the convolutions performed by the FWT analysis bank in Fig. 7.17 is proportional to the length of the sequences being convolved. Thus, the FWT compares favorably with the FFT algorithm, which requires on the order of $O(M \log_2 M)$ operations.

To conclude the development of the FWT, we simply note that the filter bank in Fig. 7.17 can be "iterated" to create multistage structures for computing DWT coefficients at two or more successive scales. For example, Fig. 7.18(a) shows a two-stage filter bank for generating the coefficients at the two highest scales of the transform. Note that the highest scale coefficients are assumed to be samples of the function itself. That is, $W_{\varphi}(J, n) = f(n)$, where J is the highest

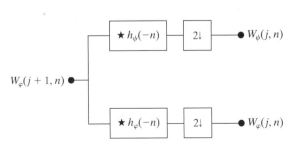

FIGURE 7.17
An FWT analysis bank.

a
b

FIGURE 7.18
(a) A two-stage or two-scale FWT analysis bank and (b) its frequency splitting characteristics.

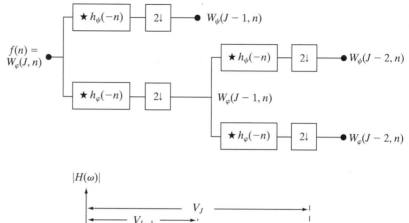

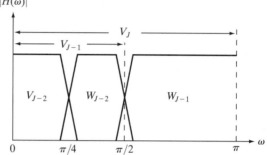

scale. [In accordance with Section 7.2.2, $f(x) \in V_J$, where V_J is the scaling space in which $f(x)$ resides.] The first filter bank in Fig. 7.18(a) splits the original function into a lowpass, approximation component, which corresponds to scaling coefficients $W_\varphi(J - 1, n)$; and a highpass, detail component, corresponding to coefficients $W_\psi(J - 1, n)$. This is illustrated graphically in Fig. 7.18(b), where scaling space V_J is split into wavelet subspace W_{J-1} and scaling subspace V_{J-1}. The spectrum of the original function is split into two half-band components. The second filter bank of Fig. 7.18(a) splits the spectrum and subspace V_{J-1}, the lower half-band, into quarter-band subspaces W_{J-2} and V_{J-2} with corresponding DWT coefficients $W_\psi(J - 2, n)$ and $W_\varphi(J - 2, n)$, respectively.

The two-stage filter bank of Fig. 7.18(a) is extended easily to any number of scales. A third filter bank, for example, would operate on the $W_\varphi(J - 2, n)$ coefficients, splitting scaling space V_{J-2} into two eighth-band subspaces W_{J-3} and V_{J-3}. Normally, we choose 2^J samples of $f(x)$ and employ P filter banks (as in Fig. 7.17) to generate a P-scale FWT at scales $J - 1, J - 2, \ldots, J - P$. The highest scale (i.e., $J - 1$) coefficients are computed first; the lowest scale (i.e., $J - P$) last. If function $f(x)$ is sampled above the Nyquist rate, as is usually the case, its samples are good approximations of the scaling coefficients at the sampling resolution and can be used as the starting high-resolution scaling coefficient inputs. In other words, no wavelet or detail coefficients are needed at the sampling scale. The highest-resolution scaling functions act as unit discrete impulse functions in Eqs. (7.3-5) and (7.3-6), allowing $f(n)$ to be used as the scaling (approximation) input to the first two-band filter bank (Odegard, Gopinath, and Burrus [1992]).

EXAMPLE 7.10:
Computing a 1-D
fast wavelet
transform.

■ To illustrate the preceding concepts, consider the discrete function $f(n)$ = $\{1, 4, -3, 0\}$ from Example 7.8. As in that example, we will compute the transform based on Haar scaling and wavelet functions. Here, however, we will not use the basis functions directly, as was done in the DWT of Example 7.8. Instead, we will use the corresponding scaling and wavelet vectors from Examples 7.5 and 7.6:

$$h_\varphi(n) = \begin{cases} 1/\sqrt{2} & n = 0, 1 \\ 0 & \text{otherwise} \end{cases} \tag{7.4-13}$$

and

$$h_\psi(n) = \begin{cases} 1/\sqrt{2} & n = 0 \\ -1/\sqrt{2} & n = 1 \\ 0 & \text{otherwise} \end{cases} \tag{7.4-14}$$

These are the functions used to build the FWT filter banks; they provide the filter coefficients. Note that because Haar scaling and wavelet functions are orthonormal, Eq. (7.1-14) can be used to generate the FWT filter coefficients from a single prototype filter—like $h_\varphi(n)$ in Table 7.2, which corresponds to $g_0(n)$ in Eq. (7.1-14):

Since the DWT computed in Example 7.8 was composed of elements $\{W_\varphi(0, 0), W_\psi(0, 0), W_\psi(1, 0), W_\psi(1, 1)\}$, we will compute the corresponding two-scale FWT for scales $j = \{0, 1\}$. That is, $J = 2$ (there are $2^J = 2^2$ samples) and $P = 2$ (we are working with scales $J - 1 = 2 - 1 = 1$ and $J - P = 2 - 2 = 0$ in that order). The transform will be computed using the two-stage filter bank of Fig. 7.18(a). Figure 7.19 shows the sequences that result from the required FWT convolutions and downsamplings. Note that function $f(n)$ itself is the scaling (approximation) input to the leftmost filter bank. To compute the $W_\psi(1, k)$ coefficients that appear at the end of the upper branch of Fig. 7.19, for example, we first convolve $f(n)$ with $h_\psi(-n)$. As explained in Section 3.4.2, this requires flipping one of the functions about the origin, sliding it past the other, and computing the sum of the point-wise product of the two functions. For sequences $\{1, 4, -3, 0\}$ and $\{-1/\sqrt{2}, 1/\sqrt{2}\}$, this produces

$$\{-1/\sqrt{2}, \underline{-3/\sqrt{2}}, 7/\sqrt{2}, -3/\sqrt{2}, 0\}$$

where the second term corresponds to index $k = 2n = 0$. (In Fig. 7.19, underlined values represent negative indices, i.e., $n < 0$.) When downsampled by

n	$h_\varphi(n)$
0	$1/\sqrt{2}$
1	$1/\sqrt{2}$

TABLE 7.2
Orthonormal
Haar filter
coefficients for
$h_\varphi(n)$.

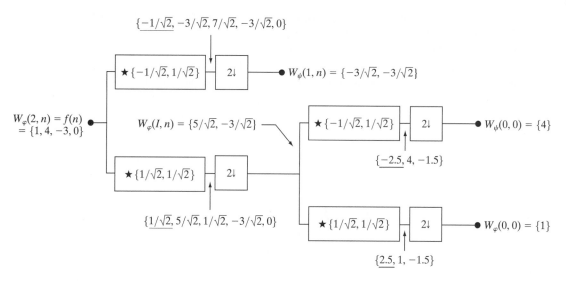

FIGURE 7.19 Computing a two-scale fast wavelet transform of sequence $\{1, 4, -3, 0\}$ using Haar scaling and wavelet vectors.

taking the even-indexed points, we get $W_\psi(1, k) = \{-3/\sqrt{2}, -3/\sqrt{2}\}$ for $k = \{0, 1\}$. Alternatively, we can use Eq. (7.4-12) to compute

$$W_\psi(1, k) = h_\psi(-n) \star W_\varphi(2, n)\Big|_{n=2k, k\geq 0} = h_\psi(-n) \star f(n)\Big|_{n=2k, k\geq 0}$$

$$= \sum_l h_\psi(l - 2k) x(l)\Big|_{k=0,1}$$

$$= \frac{1}{\sqrt{2}} x(2k) - \frac{1}{\sqrt{2}} x(2k + 1)\Big|_{k=0,1}$$

Here, we have substituted $2k$ for n in the convolution and employed l as a dummy variable of convolution (i.e., for displacing the two sequences relative to one another). There are only two terms in the expanded sum because there are only two nonzero values in the order-reversed wavelet vector $h_\psi(-n)$. Substituting $k = 0$, we find that $W_\psi(1, 0) = -3/\sqrt{2}$; for $k = 1$, we get $W_\psi(1, 1) = -3/\sqrt{2}$. Thus, the filtered and downsampled sequence is $\{-3/\sqrt{2}, -3/\sqrt{2}\}$, which matches the earlier result. The remaining convolutions and downsamplings are performed in a similar manner. ∎

As one might expect, a fast inverse transform for the reconstruction of $f(n)$ from the results of the forward transform can be formulated. Called the *inverse fast wavelet transform* (FWT^{-1}), it uses the scaling and wavelet vectors employed in the forward transform, together with the level j approximation

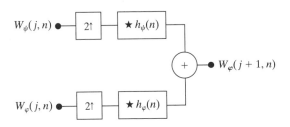

FIGURE 7.20
The FWT^{-1}
synthesis filter
bank.

and detail coefficients, to generate the level $j + 1$ approximation coefficients. Noting the similarity between the FWT analysis bank in Fig. 7.17 and the two-band subband analysis portion of Fig. 7.6(a), we can immediately postulate the required FWT^{-1} *synthesis filter bank*. Figure 7.20 details its structure, which is identical to the synthesis portion of the two-band subband coding and decoding system in Fig. 7.6(a). Equation (7.1-14) of Section 7.1.2 defines the relevant synthesis filters. As noted there, perfect reconstruction (for two-band orthonormal filters) requires $g_i(n) = h_i(-n)$ for $i = \{0, 1\}$. That is, the synthesis and analysis filters must be order-reversed versions of one another. Since the FWT analysis filters (see Fig. 7.17) are $h_0(n) = h_\varphi(-n)$ and $h_1(n) = h_\psi(-n)$, the required FWT^{-1} synthesis filters are $g_0(n) = h_0(-n) = h_\varphi(n)$ and $g_1(n) = h_1(-n) = h_\psi(n)$. It should be remembered, however, that it is possible also to use biorthogonal analysis and synthesis filters, which are not order-reversed versions of one another. Biorthogonal analysis and synthesis filters are cross-modulated per Eqs. (7.1-10) and (7.1-11).

The FWT^{-1} filter bank in Fig. 7.20 implements the computation

$$W_\varphi(j + 1, k) = h_\varphi(k) \star W_\varphi^{2\uparrow}(j, k) + h_\psi(k) \star W_\psi^{2\uparrow}(j, k) \Big|_{k \geq 0} \qquad (7.4\text{-}15)$$

where $W^{2\uparrow}$ signifies upsampling by 2 [i.e., inserting zeros in W as defined by Eq. (7.1-1) so that it is twice its original length]. The upsampled coefficients are filtered by convolution with $h_\varphi(n)$ and $h_\psi(n)$ and added to generate a higher scale approximation. In essence, a better approximation of sequence $f(n)$ with greater detail and resolution is created. As with the forward FWT, the inverse filter bank can be iterated as shown in Fig. 7.21, where a two-scale structure for computing the final two scales of a FWT^{-1} reconstruction is depicted. This coefficient combining process can be extended to any number of scales and guarantees perfect reconstruction of sequence $f(n)$.

Remember that like in pyramid coding (see Section 7.1.1), wavelet transforms can be computed at a user-specified number of scales. For a $2^J \times 2^J$ image, for example, there are $1 + \log_2 J$ possible scales.

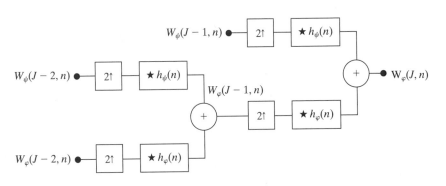

FIGURE 7.21
A two-stage or
two-scale FWT^{-1}
synthesis bank.

EXAMPLE 7.11:
Computing a 1-D inverse fast wavelet transform.

∎ Computation of the inverse fast wavelet transform mirrors its forward counterpart. Figure 7.22 illustrates the process for the sequence considered in Example 7.10. To begin the calculation, the level 0 approximation and detail coefficients are upsampled to yield $\{1, 0\}$ and $\{4, 0\}$, respectively. Convolution with filters $g_0(n) = h_\varphi(n) = \{1/\sqrt{2}, 1/\sqrt{2}\}$ and $g_1(n) = h_\psi(n) = \{1/\sqrt{2}, -1/\sqrt{2}\}$ produces $\{1/\sqrt{2}, 1/\sqrt{2}, 0\}$ and $\{4/\sqrt{2}, -4/\sqrt{2}, 0\}$, which when added give $W_\varphi(1, n) = \{5/\sqrt{2}, -3/\sqrt{2}\}$. Thus, the level 1 approximation of Fig. 7.22, which matches the computed approximation in Fig. 7.19, is reconstructed. Continuing in this manner, $f(n)$ is formed at the right of the second synthesis filter bank. ∎

We conclude our discussion of the fast wavelet transform by noting that while the Fourier basis functions (i.e., sinusoids) guarantee the existence of the FFT, the existence of the FWT depends upon the availability of a scaling function for the wavelets being used, as well as the orthogonality (or biorthogonality) of the scaling function and corresponding wavelets. Thus, the Mexican hat wavelet of Eq. (7.3-12), which does not have a companion scaling function, cannot be used in the computation of the FWT. In other words, we cannot construct a filter bank like that of Fig. 7.17 for the Mexican hat wavelet; it does not satisfy the underlying assumptions of the FWT approach.

Finally, we note that while time and frequency usually are viewed as different domains when representing functions, they are inextricably linked. When you try to analyze a function simultaneously in time and frequency, you run into the following problem: If you want precise information about time, you must accept some vagueness about frequency, and vice versa. This is the *Heisenberg uncertainty principle* applied to information processing. To illustrate the principle graphically, each basis function used in the representation of a function can be viewed schematically as a *tile* in a *time-frequency plane*. The tile, also called a *Heisenberg cell* or *Heisenberg box*, shows the frequency content of the basis function that it represents and where the basis function resides in time. Basis functions that are orthonormal are characterized by nonoverlapping tiles.

Figure 7.23 shows the time-frequency tiles for (a) an impulse function (i.e., conventional time domain) basis, (b) a sinusoidal (FFT) basis, and (c) an FWT

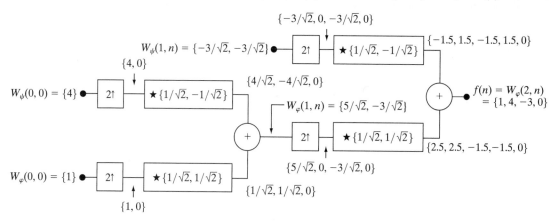

FIGURE 7.22 Computing a two-scale inverse fast wavelet transform of sequence $\{1, 4, -1.5\sqrt{2}, -1.5\sqrt{2}\}$ with Haar scaling and wavelet functions.

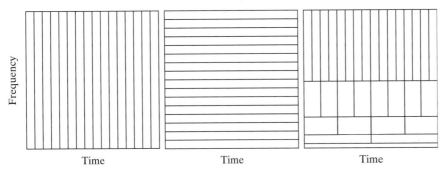

FIGURE 7.23 Time-frequency tilings for the basis functions associated with (a) sampled data, (b) the FFT, and (c) the FWT. Note that the horizontal strips of equal height rectangles in (c) represent FWT scales.

basis. Each tile is a rectangular region in Figs. 7.23(a) through (c); the height and width of the region defines the frequency and time characteristics of the functions that can be represented using the basis function. Note that the standard time domain basis in Fig. 7.23(a) pinpoints the instants when events occur but provides no frequency information [the width of each rectangle in Fig. 7.23(a) should be considered one instant in time]. Thus, to represent a single frequency sinusoid as an expansion using impulse basis functions, every basis function is required. The sinusoidal basis in Fig. 7.23(b), on the other hand, pinpoints the frequencies that are present in events that occur over long periods but provides no time resolution [the height of each rectangle in Fig. 7.23(b) should be considered a single frequency]. Thus, the single frequency sinusoid that was represented by an infinite number of impulse basis functions can be represented as an expansion involving one sinusoidal basis function. The time and frequency resolution of the FWT tiles in Fig. 7.23(c) vary, but the area of each tile (rectangle) is the same. At low frequencies, the tiles are shorter (i.e., have better frequency resolution or less ambiguity regarding frequency) but are wider (which corresponds to poorer time resolution or more ambiguity regarding time). At high frequencies, tile width is smaller (so the time resolution is improved) and tile height is greater (which means the frequency resolution is poorer). Thus, the FWT basis functions provide a compromise between the two limiting cases in Fig. 7.23(a) and (b). This fundamental difference between the FFT and FWT was noted in the introduction to the chapter and is important in the analysis of nonstationary functions whose frequencies vary in time.

7.5 Wavelet Transforms in Two Dimensions

The one-dimensional transforms of the previous sections are easily extended to two-dimensional functions like images. In two dimensions, a two-dimensional scaling function, $\varphi(x, y)$, and three two-dimensional wavelets, $\psi^H(x, y)$, $\psi^V(x, y)$, and $\psi^D(x, y)$, are required. Each is the product of two one-dimensional functions. Excluding products that produce one-dimensional results, like $\varphi(x)\psi(x)$, the four remaining products produce the *separable* scaling function

$$\varphi(x, y) = \varphi(x)\varphi(y) \tag{7.5-1}$$

and separable, "directionally sensitive" wavelets

$$\psi^H(x, y) = \psi(x)\varphi(y) \tag{7.5-2}$$

$$\psi^V(x, y) = \varphi(x)\psi(y) \tag{7.5-3}$$

$$\psi^D(x, y) = \psi(x)\psi(y) \tag{7.5-4}$$

These wavelets measure functional variations—intensity variations for images—along different directions: ψ^H measures variations along columns (for example, horizontal edges), ψ^V responds to variations along rows (like vertical edges), and ψ^D corresponds to variations along diagonals. The directional sensitivity is a natural consequence of the separability in Eqs. (7.5-2) to (7.5-4); it does not increase the computational complexity of the 2-D transform discussed in this section.

Given separable two-dimensional scaling and wavelet functions, extension of the 1-D DWT to two dimensions is straightforward. We first define the scaled and translated basis functions:

$$\varphi_{j,m,n}(x, y) = 2^{j/2}\varphi(2^j x - m, 2^j y - n) \tag{7.5-5}$$

$$\psi^i_{j,m,n}(x, y) = 2^{j/2}\psi^i(2^j x - m, 2^j y - n), \quad i = \{H, V, D\} \tag{7.5-6}$$

where index i identifies the directional wavelets in Eqs. (7.5-2) to (7.5-4). Rather than an exponent, i is a superscript that assumes the values $H, V,$ and D. The discrete wavelet transform of image $f(x, y)$ of size $M \times N$ is then

Now that we are dealing with 2-D images, $f(x, y)$ is a discrete function or sequence of values and x and y are discrete variables. The scaling and wavelet functions in Eq. (7.5-7) and (7.5-8) are sampled over their support (as was done in the 1-D case in Section 7.3.2).

$$W_\varphi(j_0, m, n) = \frac{1}{\sqrt{MN}} \sum_{x=0}^{M-1} \sum_{y=0}^{N-1} f(x, y)\varphi_{j_0,m,n}(x, y) \tag{7.5-7}$$

$$W^i_\psi(j, m, n) = \frac{1}{\sqrt{MN}} \sum_{x=0}^{M-1} \sum_{y=0}^{N-1} f(x, y)\psi^i_{j,m,n}(x, y), \quad i = \{H, V, D\} \tag{7.5-8}$$

As in the one-dimensional case, j_0 is an arbitrary starting scale and the $W_\varphi(j_0, m, n)$ coefficients define an approximation of $f(x, y)$ at scale j_0. The $W^i_\psi(j, m, n)$ coefficients add horizontal, vertical, and diagonal details for scales $j \geq j_0$. We normally let $j_0 = 0$ and select $N = M = 2^J$ so that $j = 0, 1, 2, \ldots,$ $J - 1$ and $m = n = 0, 1, 2, \ldots, 2^j - 1$. Given the W_φ and W^i_ψ of Eqs. (7.5-7) and (7.5-8), $f(x, y)$ is obtained via the inverse discrete wavelet transform

$$f(x, y) = \frac{1}{\sqrt{MN}} \sum_m \sum_n W_\varphi(j_0, m, n)\varphi_{j_0,m,n}(x, y)$$

$$+ \frac{1}{\sqrt{MN}} \sum_{i=H,V,D} \sum_{j=j_0}^{\infty} \sum_m \sum_n W^i_\psi(j, m, n)\psi^i_{j,m,n}(x, y) \tag{7.5-9}$$

Like the 1-D discrete wavelet transform, the 2-D DWT can be implemented using digital filters and downsamplers. With separable two-dimensional scaling and wavelet functions, we simply take the 1-D FWT of the rows of $f(x, y)$, followed by the 1-D FWT of the resulting columns. Figure 7.24(a) shows the

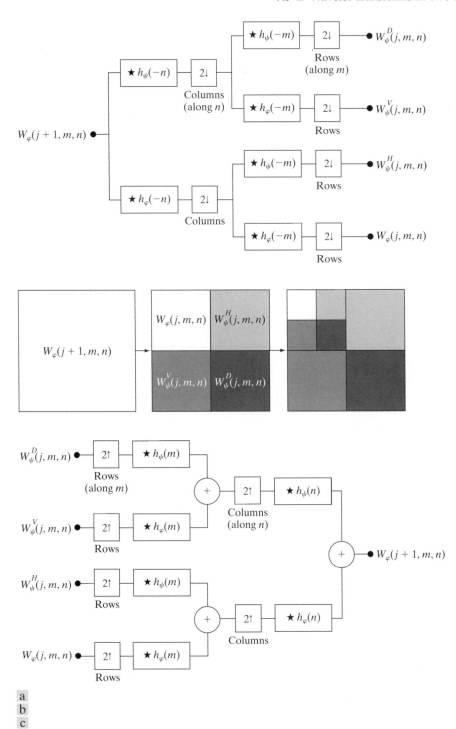

a
b
c

FIGURE 7.24 The 2-D fast wavelet transform: (a) the analysis filter bank; (b) the resulting decomposition; and (c) the synthesis filter bank.

process in block diagram form. Note that, like its one-dimensional counterpart in Fig. 7.17, the 2-D FWT "filters" the scale $j + 1$ approximation coefficients to construct the scale j approximation and detail coefficients. In the two-dimensional case, however, we get three sets of detail coefficients—the horizontal, vertical, and diagonal details.

The single-scale filter bank of Fig. 7.24(a) can be "iterated" (by tying the approximation output to the input of another filter bank) to produce a P scale transform in which scale j is equal to $J - 1, J - 2, \ldots, J - P$. As in the one-dimensional case, image $f(x, y)$ is used as the $W_\varphi(J, m, n)$ input. Convolving its rows with $h_\varphi(-n)$ and $h_\psi(-n)$ and downsampling its columns, we get two subimages whose horizontal resolutions are reduced by a factor of 2. The high-pass or detail component characterizes the image's high-frequency information with vertical orientation; the lowpass, approximation component contains its low-frequency, vertical information. Both subimages are then filtered columnwise and downsampled to yield four quarter-size output subimages—

Note how W_φ, W_ψ^H, W_ψ^V, and W_ψ^D are arranged in Fig. 7.24(b). For each scale that is computed, they replace the previous scale approximation on which they were based.

W_φ, W_ψ^H, W_ψ^V, and W_ψ^D. These subimages, which are shown in the middle of Fig. 7.24(b), are the inner products of $f(x, y)$ and the two-dimensional scaling and wavelet functions in Eqs. (7.5-1) through (7.5-4), followed by downsampling by two in each dimension. Two iterations of the filtering process produces the two-scale decomposition at the far right of Fig. 7.24(b).

Figure 7.24(c) shows the synthesis filter bank that reverses the process just described. As would be expected, the reconstruction algorithm is similar to the one-dimensional case. At each iteration, four scale j approximation and detail subimages are upsampled and convolved with two one-dimensional filters—one operating on the subimages' columns and the other on its rows. Addition of the results yields the scale $j + 1$ approximation, and the process is repeated until the original image is reconstructed.

EXAMPLE 7.12:
Computing a 2-D fast wavelet transform.

■ Figure 7.25(a) is a 128×128 computer-generated image consisting of 2-D sine-like pulses on a black background. The objective of this example is to illustrate the mechanics involved in computing the 2-D FWT of this image. Figures 7.25(b) through (d) show three FWTs of the image in Fig. 7.25(a). The 2-D filter bank of Fig. 7.24(a) and the decomposition filters shown in Figs. 7.26(a) and (b) were used to generate all three results.

The scaling and wavelet vectors used in this example are described later. Our focus here is on the mechanics of the transform computation, which are independent of the filter coefficients employed.

Figure 7.25(b) shows the one-scale FWT of the image in Fig. 7.25(a). To compute this transform, the original image was used as the input to the filter bank of Fig. 7.24(a). The four resulting quarter-size decomposition outputs (i.e., the approximation and horizontal, vertical, and diagonal details) were then arranged in accordance with Fig. 7.24(b) to produce the image in Fig. 7.25(b). A similar process was used to generate the two-scale FWT in Fig. 7.25(c), but the input to the filter bank was changed to the quarter-size approximation subimage from the upper-left-hand corner of Fig. 7.25(b). As can be seen in Fig. 7.25(c), that quarter-size subimage was then replaced by the four quarter-size

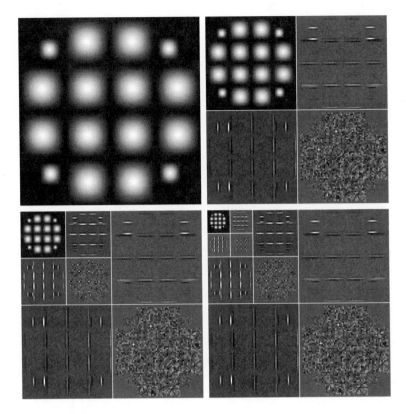

FIGURE 7.25
Computing a 2-D
three-scale FWT:
(a) the original
image; (b) a one-
scale FWT; (c) a
two-scale FWT;
and (d) a three-
scale FWT.

(now 1/16th of the size of the original image) decomposition results that were
generated in the second filtering pass. Finally, Fig. 7.25(d) is the three-scale
FWT that resulted when the subimage from the upper-left-hand corner of Fig.
7.25(c) was used as the filter bank input. Each pass through the filter bank produced four quarter-size output images that were substituted for the input from
which they were derived. Note the directional nature of the wavelet-based
subimages, W_ψ^H, W_ψ^V, and W_ψ^D, at each scale.   ■

The decomposition filters used in the preceding example are part of a well-
known family of wavelets called *symlets*, short for "symmetrical wavelets." Although they are not perfectly symmetrical, they are designed to have the least
asymmetry and highest number of vanishing moments[†] for a given compact
support (Daubechies [1992]). Figures 7.26(e) and (f) show the fourth-order

Recall that the compact
support of a function is
the interval in which the
function has non-zero
values.

[†]The kth moment of wavelet $\psi(x)$ is $m(k) = \int x^k \psi(x)\, dx$. Zero moments impact the smoothness of the
scaling and wavelet functions and our ability to represent them as polynomials. An order-N symlet has
N vanishing moments.

FIGURE 7.26
Fourth-order
symlets: (a)–(b)
decomposition
filters; (c)–(d)
reconstruction
filters; (e) the
one-dimensional
wavelet; (f) the
one-dimensional
scaling function;
and (g) one of
three two-
dimensional
wavelets, $\psi^V(x, y)$.
See Table 7.3 for
the values of
$h_\varphi(n)$ for
$0 \le n \le 7$.

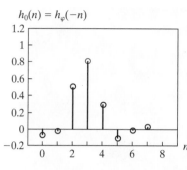

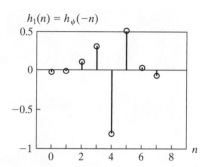

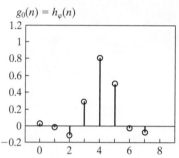

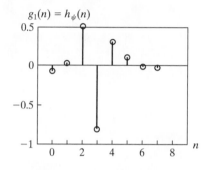

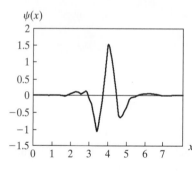

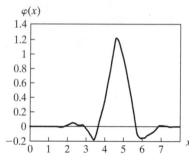

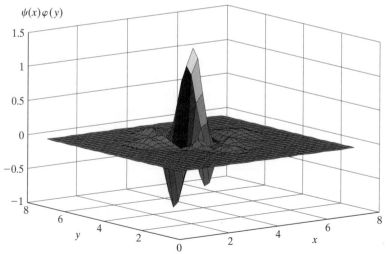

n	$h_\varphi(n)$
0	0.0322
1	−0.0126
2	−0.0992
3	0.2979
4	0.8037
5	0.4976
6	−0.0296
7	−0.0758

TABLE 7.3
Orthonormal
fourth-order
symlet filter
coefficients for
$h_\varphi(n)$.
(Daubechies
[1992].)

1-D symlets (i.e., wavelet and scaling functions). Figures 7.26(a) through (d) show the corresponding decomposition and reconstruction filters. The coefficients of lowpass reconstruction filter $g_0(n) = h_\varphi(n)$ for $0 \le n \le 7$ are given in Table 7.3. The coefficients of the remaining orthonormal filters are obtained using Eq. (7.1-14). Figure 7.26(g), a low-resolution graphic depiction of wavelet $\psi^V(x, y)$, is provided as an illustration of how a one-dimensional scaling and wavelet function can combine to form a separable, two-dimensional wavelet.

We conclude this section with two examples that demonstrate the usefulness of wavelets in image processing. As in the Fourier domain, the basic approach is to

Step 1. Compute a 2-D wavelet transform of an image.
Step 2. Alter the transform.
Step 3. Compute the inverse transform.

Because the DWT's scaling and wavelet vectors are used as lowpass and highpass filters, most Fourier-based filtering techniques have an equivalent "wavelet domain" counterpart.

■ Figure 7.27 provides a simple illustration of the preceding three steps. In Fig. 7.27(a), the lowest scale approximation component of the discrete wavelet transform shown in Fig. 7.25(c) has been eliminated by setting its values to zero. As Fig. 7.27(b) shows, the net effect of computing the inverse wavelet transform using these modified coefficients is edge enhancement, reminiscent of the Fourier-based image sharpening results discussed in Section 4.9. Note how well the transitions between signal and background are delineated, despite the fact that they are relatively soft, sinusoidal transitions. By zeroing the horizontal details as well—see Figs. 7.27(c) and (d)—we can isolate the vertical edges.   ■

EXAMPLE 7.13:
Wavelet-based
edge detection.

was generated by simply zeroing the highest-resolution detail coefficients (not thresholding the lower-resolution details) and reconstructing the image. Here, almost all of the background noise has been eliminated and the edges are only slightly disturbed. The difference image in Fig. 7.28(d) shows the information that is lost in the process. This result was generated by computing the inverse FWT of the two-scale transform with all but the highest-resolution detail coefficients zeroed. As can be seen, the resulting image contains most of the noise in the original image and some of the edge information. Figures 7.28(e) and (f) are included to show the negative effect of deleting all the detail coefficients. That is, Fig. 7.28(e) is a reconstruction of the DWT in which the details at both levels of the two-scale transform have been zeroed; Fig. 7.28(f) shows the information that is lost. Note the significant increase in edge information in Fig. 7.28(f) and the corresponding decrease in edge detail in Fig. 7.28(e). ■

> Because only the highest resolution detail coefficients were kept when generating Fig. 7.28(d), the inverse transform is their contribution to the image. In the same way, Fig. 7.28(f) is the contribution of all the detail coefficients.

7.6   Wavelet Packets

The fast wavelet transform decomposes a function into a sum of scaling and wavelet functions whose bandwidths are logarithmically related. That is, the low frequency content (of the function) is represented using (scaling and wavelet) functions with narrow bandwidths, while the high-frequency content is represented using functions with wider bandwidths. If you look along the frequency axis of the time-frequency plane in Fig. 7.23(c), this is immediately apparent. Each horizontal strip of constant height tiles, which contains the basis functions for a single FWT scale, increases logarithmically in height as you move up the frequency axis. If we want greater control over the partitioning of the time-frequency plane (e.g., smaller bands at the higher frequencies), the FWT must be generalized to yield a more flexible decomposition—called a *wavelet packet* (Coifman and Wickerhauser [1992]). The cost of this generalization is an increase in computational complexity from $O(M)$ for the FWT to $O(M \log_2 M)$ for a wavelet packet.

Consider again the two-scale filter bank of Fig. 7.18(a)—but imagine the decomposition as a *binary tree*. Figure 7.29(a) details the structure of the tree, and links the appropriate FWT scaling and wavelet coefficients [from Fig. 7.18(a)] to its *nodes*. The *root node* is assigned the highest-scale approximation coefficients,

a b

FIGURE 7.29
An (a) coefficient tree and (b) analysis tree for the two-scale FWT analysis bank of Fig. 7.18.

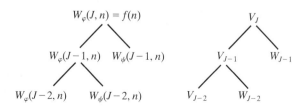

which are samples of the function itself, while the *leaves* inherit the transform's approximation and detail coefficient outputs. The lone intermediate node, $W_\varphi(J - 1, n)$, is a filter bank approximation that is ultimately filtered to become two leaf nodes. Note that the coefficients of each node are the weights of a linear expansion that produces a band-limited "piece" of root node $f(n)$. Because any such piece is an element of a known scaling or wavelet subspace (see Sections 7.2.2 and 7.2.3), we can replace the generating coefficients in Fig. 7.29(a) by the corresponding subspace. The result is the *subspace analysis tree* of Fig. 7.29(b). Although the variable W is used to denote both coefficients and subspaces, the two quantities are distinguishable by the format of their subscripts.

These concepts are further illustrated in Fig. 7.30, where a three-scale FWT analysis bank, analysis tree, and corresponding frequency spectrum are depicted. Unlike Fig. 7.18(a), the block diagram of Fig. 7.30(a) is labeled to resemble the analysis tree in Fig. 7.30(b)—as well as the spectrum in Fig. 7.30(c). Thus, while the output of the upper-left filter and subsampler is, to be accurate, $W_\psi(J - 1, n)$, it has been labeled W_{J-1}—the subspace of the function that is generated by the $W_\psi(J - 1, n)$ transform coefficients. This subspace corresponds to the upper-right leaf of the associated analysis tree, as well as the rightmost (widest bandwidth) segment of the corresponding frequency spectrum.

Analysis trees provide a compact and informative way of representing multiscale wavelet transforms. They are simple to draw, take less space than their corresponding filter and subsampler-based block diagrams, and make it relatively

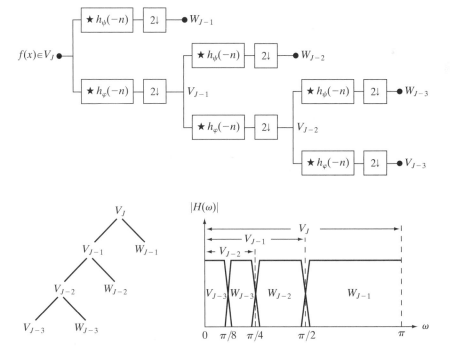

a
b c

FIGURE 7.30
A three-scale FWT filter bank: (a) block diagram; (b) decomposition space tree; and (c) spectrum splitting characteristics.

easy to detect valid decompositions. The three-scale analysis tree of Fig. 7.30(b), for example, makes possible the following three expansion options:

$$V_J = V_{J-1} \oplus W_{J-1} \tag{7.6-1}$$

$$V_J = V_{J-2} \oplus W_{J-2} \oplus W_{J-1} \tag{7.6-2}$$

$$V_J = V_{J-3} \oplus W_{J-3} \oplus W_{J-2} \oplus W_{J-1} \tag{7.6-3}$$

They correspond to the one-, two-, and three-scale FWT decompositions of Section 7.4 and may be obtained from Eq. (7.2-27) of Section 7.2.3 by letting $j_0 = J - P$ for $P = \{1, 2, 3\}$. In general, a P-scale FWT analysis tree supports P unique decompositions.

Analysis trees also are an efficient mechanism for representing *wavelet packets*, which are nothing more than *conventional wavelet transforms in which the details are filtered iteratively*. Thus, the three-scale FWT analysis tree of Fig. 7.30(b) becomes the three-scale *wavelet packet* tree of Fig. 7.31. Note the additional subscripting that is introduced. The first subscript of a double-subscripted node identifies the scale of the FWT *parent* node from which it descended. The second—a variable length string of As and Ds—encodes the path from the parent to the node. An A designates approximation filtering, while a D indicates detail filtering. Subspace $W_{J-1,DA}$, for example, is obtained by "filtering" the scale $J - 1$ FWT coefficients (i.e., parent W_{J-1} in Fig. 7.31) through an additional detail filter (yielding $W_{J-1,D}$), followed by an approximation filter (giving $W_{J-1,DA}$). Figures 7.32(a) and (b) are the filter bank and spectrum splitting characteristics of the analysis tree in Fig. 7.31. Note that the "naturally ordered" outputs of the filter bank in Fig. 7.32(a) have been reordered based on frequency content in Fig. 7.32(b) (see Problem 7.25 for more on "frequency ordered" wavelets).

The three-scale packet tree in Fig. 7.31 almost triples the number of decompositions (and associated time-frequency tilings) that are available from the three-scale FWT tree. Recall that in a normal FWT, we split, filter, and downsample the lowpass bands alone. This creates a fixed logarithmic (base 2) relationship between the bandwidths of the scaling and wavelet spaces used in the representation of a function [see Figure 7.30(c)]. Thus, while the three-scale FWT analysis tree of Fig. 7.30(a) offers three possible decompositions—defined by Eqs. (7.6-1) to (7.6-3)—the wavelet packet tree of Fig. 7.31 supports 26 different decompositions. For instance, V_J [and therefore function $f(n)$] can be expanded as

FIGURE 7.31
A three-scale wavelet packet analysis tree.

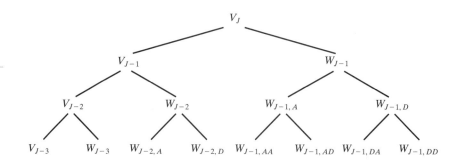

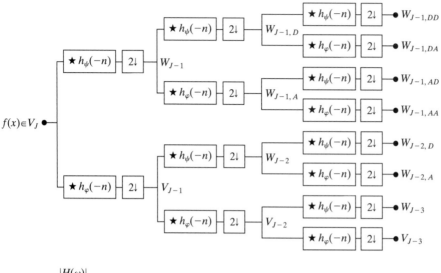

FIGURE 7.32
The (a) filter bank
and (b) spectrum
splitting
characteristics of
a three-scale full
wavelet packet
analysis tree.

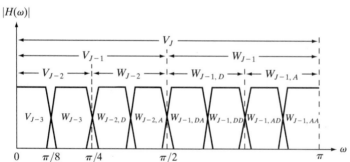

$$V_J = V_{J-3} \oplus W_{J-3} \oplus W_{J-2,A} \oplus W_{J-2,D} \oplus W_{J-1,AA}$$
$$\oplus W_{J-1,AD} \oplus W_{J-1,DA} \oplus W_{J-1,DD} \quad (7.6\text{-}4)$$

whose spectrum is shown in Fig. 7.32(b), or

$$V_J = V_{J-1} \oplus W_{J-1,A} \oplus W_{J-1,DA} \oplus W_{J-1,DD} \quad (7.6\text{-}5)$$

whose spectrum is depicted in Fig. 7.33. Note the difference between this last
spectrum and the full packet spectrum of Fig. 7.32(b), or the three-scale FWT

Recall that $\oplus$ denotes
the union of spaces (like
the union of sets). The 26
decompositions associated
with Fig. 7.31 are
determined by various
combinations of nodes
(spaces) that can be
combined to represent
the root node (space) at
the top of the tree.
Eqs. (7.6-4) and (7.6-5)
define two of them.

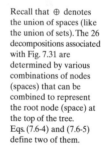

FIGURE 7.33
The spectrum of
the decomposi-
tion in Eq. (7.6-5).

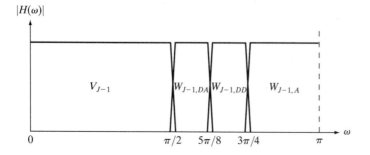

spectrum of Fig. 7.30(c). In general, P-scale, one-dimensional wavelet packet transforms (and associated $P + 1$-level analysis trees) support

$$D(P + 1) = \left[D(P)\right]^2 + 1 \qquad (7.6\text{-}6)$$

unique decompositions, where $D(1) = 1$. With such a large number of valid expansions, packet-based transforms provide improved control over partitioning the spectrum of the decomposed function. The cost of this control is an increase in computational complexity [compare the filter bank in Fig. 7.30(a) to that of Fig. 7.32(a)].

Now consider the two-dimensional, four-band filter bank of Fig. 7.24(a). As was noted in Section 7.5, it splits approximation $W_\varphi(j + 1, m, n)$ into outputs, $W_\varphi(j, m, n)$, $W_\psi^H(j, m, n)$, $W_\psi^V(j, m, n)$, and $W_\psi^D(j, m, n)$. As in the one-dimensional case, it can be "iterated" to generate P scale transforms for scales $j = J - 1, J - 2, \ldots, J - P$, with $W_\varphi(J, m, n) = f(m, n)$. The spectrum resulting from the first iteration [i.e., using $j + 1 = J$ in Fig. 7.24(a)] is shown in Fig. 7.34(a). Note that it divides the frequency plane into four equal areas. The low-frequency quarter-band in the center of the plane coincides with transform coefficients $W_\varphi(J - 1, m, n)$ and scaling space V_{J-1}. (This nomenclature is consistent with the one-dimensional case.) To accommodate the two-dimensional nature of the input, however, we now have three (rather than one) wavelet subspaces. They are denoted W_{J-1}^H, W_{J-1}^V, and W_{J-1}^D and correspond to coefficients $W_\psi^H(J - 1, m, n)$, $W_\psi^V(J - 1, m, n)$, and $W_\psi^D(J - 1, m, n)$, respectively. Figure 7.34(b) shows the resulting four-band, single-scale *quaternary FWT analysis tree*. Note the superscripts that link the wavelet subspace designations to their transform coefficient counterparts.

Figure 7.35 shows a portion of a three-scale, two-dimensional wavelet packet analysis tree. Like its one-dimensional counterpart in Fig. 7.31, the first subscript of every node that is a descendant of a conventional FWT detail node is the scale of that *parent* detail node. The second subscript—a variable length string of As, Hs, Vs, and Ds—encodes the path from the parent to the node under consideration. The node labeled $W_{J-1,VD}^H$, for example, is obtained by "row/column filtering" the

FIGURE 7.34
The first decomposition of a two-dimensional FWT: (a) the spectrum and (b) the subspace analysis tree.

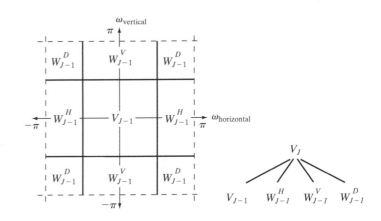

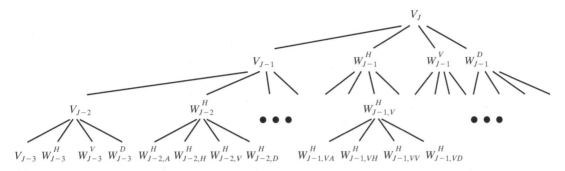

FIGURE 7.35 A three-scale, full wavelet packet decomposition tree. Only a portion of the tree is provided.

scale $J - 1$ FWT horizontal detail coefficients (i.e., parent W_{J-1}^{H} in Fig. 7.35) through an additional detail/approximation filter (yielding $W_{J-1,V}^{H}$), followed by a detail/detail filter (giving $W_{J-1,VD}^{H}$). A P-scale, two-dimensional wavelet packet tree supports

$$D(P + 1) = \left[D(P)\right]^4 + 1 \qquad (7.6\text{-}7)$$

unique expansions, where $D(1) = 1$. Thus, the three-scale tree of Fig. 7.35 offers 83,522 possible decompositions. The problem of selecting among them is the subject of the next example.

■ As noted in the above discussion, a single wavelet packet tree presents numerous decomposition options. In fact, the number of possible decompositions is often so large that it is impractical, if not impossible, to enumerate or examine them individually. An efficient algorithm for finding optimal decompositions with respect to application specific criteria is highly desirable. As will be seen, classical entropy- and energy-based cost functions are applicable in many situations and are well suited for use in binary and quaternary tree searching algorithms.

Consider the problem of reducing the amount of data needed to represent the 400 × 480 fingerprint image in Fig. 7.36(a). Image compression is discussed in detail in Chapter 8. In this example, we want to select the "best" three-scale wavelet packet decomposition as a starting point for the compression process. Using three-scale wavelet packet trees, there are 83,522 [see Eq. (7.6-7)] potential decompositions. Figure 7.36(b) shows one of them—a full wavelet packet, 64-leaf decomposition like the analysis tree of Fig. 7.35. Note that the leaves of the tree correspond to the subbands of the 8 × 8 array of decomposed subimages in Fig. 7.36(b). The probability that this particular 64-leaf decomposition is in some way optimal for the purpose of compression, however, is relatively low. In the absence of a suitable optimality criterion, we can neither confirm nor deny it.

One reasonable criterion for selecting a decomposition for the compression of the image of Fig. 7.36(a) is the additive cost function

$$E(f) = \sum_{m,n} |f(m, n)| \qquad (7.6\text{-}8)$$

EXAMPLE 7.15:
Two-dimensional wavelet packet decompositions.

The 64 leaf nodes in Fig. 7.35 correspond to the 8 × 8 array of 64 subimages in Fig. 7.36(b). Despite appearances, they are not square. The distortion (particularly noticeable in the approximation subimage) is due to the program used to produce the result.

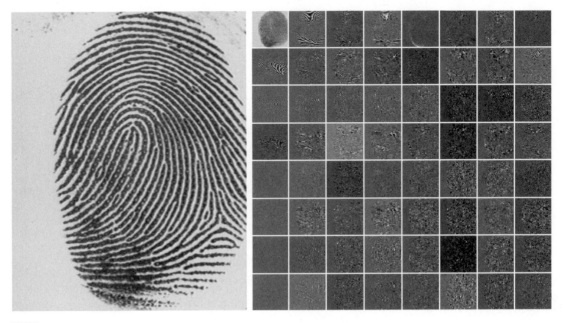

a b

FIGURE 7.36 (a) A scanned fingerprint and (b) its three-scale, full wavelet packet decomposition. (Original image courtesy of the National Institute of Standards and Technology.)

Other possible energy measures include the sum of the squares of $f(x, y)$, the sum of the log of the squares, etc. Problem 7.27 defines one possible entropy-based cost function.

This function provides one possible measure of the energy content of two-dimensional function f. Under this measure, the energy of function $f(m, n) = 0$ for all m and n is 0. High values of E, on the other hand, are indicative of functions with many nonzero values. Since most transform-based compression schemes work by truncating or thresholding the small coefficients to zero, a cost function that maximizes the number of near-zero values is a reasonable criterion for selecting a "good" decomposition from a compression point of view.

The cost function just described is both computationally simple and easily adapted to tree optimization routines. The optimization algorithm must use the function to minimize the "cost" of the leaf nodes in the decomposition tree. Minimal energy leaf nodes should be favored because they have more near-zero values, which leads to greater compression. Because the cost function of Eq. (7.6-8) is a local measure that uses only the information available at the node under consideration, an efficient algorithm for finding minimal energy solutions is easily constructed as follows:

For each node of the analysis tree, beginning with the root and proceeding level by level to the leaves:

Step 1. Compute both the energy of the node, denoted E_P (for parent energy), and the energy of its four offspring—denoted E_A, E_H, E_V, and E_D. For two-dimensional wavelet packet decompositions, the parent is a two-dimensional array of approximation or detail coefficients; the offspring are the filtered approximation, horizontal, vertical, and diagonal details.

Step 2. If the combined energy of the offspring is less than the energy of the parent—that is, $E_A + E_H + E_V + E_D < E_P$—include the offspring in the analysis tree. If the combined energy of the offspring is greater than or equal to that of the parent, prune the offspring, keeping only the parent. It is a leaf of the optimized analysis tree.

The preceding algorithm can be used to (1) prune wavelet packet trees or (2) design procedures for computing optimal trees from scratch. In the latter case, nonessential siblings—descendants of nodes that would be eliminated in step 2 of the algorithm—would not be computed. Figure 7.37 shows the optimized decomposition that results from applying the algorithm to the image of Fig. 7.36(a) with the cost function of Eq. (7.6-8). The corresponding analysis tree is given in Fig. 7.38. Note that many of the original full packet decomposition's 64 subbands in Fig. 7.36(b) (and corresponding 64 leaves of the analysis tree in Fig. 7.35) have been eliminated. In addition, the subimages that are not split (further decomposed) in Fig. 7.37 are relatively smooth and composed of pixels that are middle gray in value. Because all but the approximation subimage of this figure have been scaled so that gray level 128 indicates a zero-valued coefficient, these subimages contain little energy. There would be no overall decrease in energy realized by splitting them. ■

The preceding example is based on a real-world problem that was solved through the use of wavelets. The Federal Bureau of Investigation (FBI) currently maintains a large database of fingerprints and has established a wavelet-based national standard for the digitization and compression of fingerprint

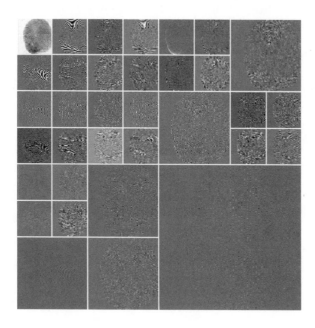

FIGURE 7.37
An optimal wavelet packet decomposition for the fingerprint of Fig. 7.36(a).

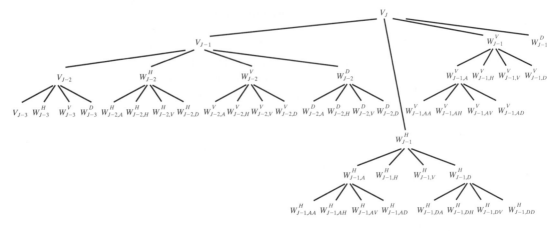

FIGURE 7.38 The optimal wavelet packet analysis tree for the decomposition in Fig. 7.37.

images (FBI [1993]). Using biorthogonal wavelets, the standard achieves a typical compression ratio of 15:1. The advantages of wavelet-based compression over the more traditional JPEG approach are examined in the next chapter.

The decomposition filters used in Example 7.15, as well as by the FBI, are part of a well-known family of wavelets called Cohen-Daubechies-Feauveau biorthogonal wavelets (Cohen, Daubechies, and Feauveau [1992]). Because the scaling and wavelet functions of the family are symmetrical and have similar lengths, they are among the most widely used biorthogonal wavelets. Figures 7.39(e) through (h) show the dual scaling and wavelet functions. Figures 7.39(a) through (d) are the corresponding decomposition and reconstruction filters. The coefficients of the lowpass and highpass decomposition filters, $h_0(n)$ and $h_1(n)$ for $0 \le n \le 17$ are shown in Table 7.4. The corresponding coefficients of the biorthogonal synthesis filters can be computed using $g_0(n) = (-1)^{n+1}h_1(n)$ and $g_1(n) = (-1)^n h_0(n)$ of Eq. (7.1-11). That is, they are cross-modulated versions of the decomposition filters. Note that zero padding is employed to make the filters the same length and that Table 7.4 and Fig. 7.39 define them with respect to the subband coding and decoding system of Fig. 7.6(a); with respect to the FWT, $h_\varphi(-n) = h_0(n)$ and $h_\psi(-n) = h_1(n)$.

TABLE 7.4
Biorthogonal
Cohen-
Daubechies-
Feauveau filter
coefficients
(Cohen,
Daubechies, and
Feauveau [1992]).

n	$h_0(n)$	$h_1(n)$	n	$h_0(n)$	$h_1(n)$
0	0	0	9	0.8259	0.4178
1	0.0019	0	10	0.4208	0.0404
2	−0.0019	0	11	−0.0941	−0.0787
3	−0.017	0.0144	12	−0.0773	−0.0145
4	0.0119	−0.0145	13	0.0497	0.0144
5	0.0497	−0.0787	14	0.0119	0
6	−0.0773	0.0404	15	−0.017	0
7	−0.0941	0.4178	16	−0.0019	0
8	0.4208	−0.7589	17	0.0010	0

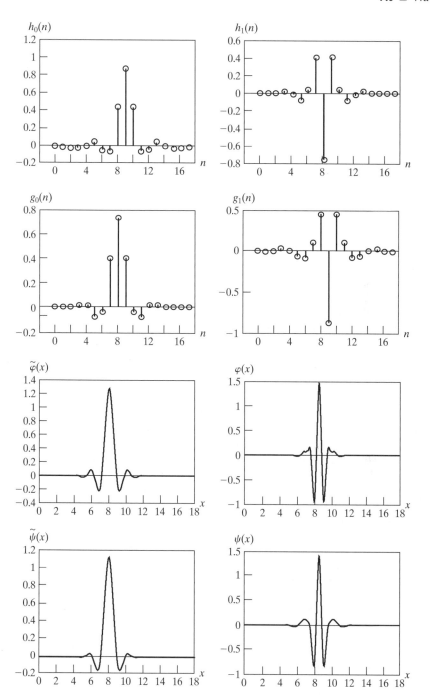

FIGURE 7.39
A member of the Cohen-Daubechies-Feauveau biorthogonal wavelet family: (a) and (b) decomposition filter coefficients; (c) and (d) reconstruction filter coefficients; (e)–(h) dual wavelet and scaling functions. See Table 7.3 for the values of $h_0(n)$ and $h_1(n)$ for $0 \leq n \leq 17$.

Summary

The material of this chapter establishes a solid mathematical foundation for under-standing and accessing the role of wavelets and multiresolution analysis in image pro-cessing. Wavelets and wavelet transforms are relatively new imaging tools that are being rapidly applied to a wide variety of image processing problems. Because of their similarity to the Fourier transform, many of the techniques in Chapter 4 have wavelet domain counterparts. A partial listing of the imaging applications that have been ap-proached from a wavelet point of view includes image matching, registration, segmen-tation, denoising, restoration, enhancement, compression, morphological filtering, and computed tomography. Since it is impractical to cover all of these applications in a sin-gle chapter, the topics included were chosen for their value in introducing or clarifying fundamental concepts and preparing the reader for further study in the field. In Chapter 8, we will apply wavelets to the compression of images.

References and Further Reading

There are many good texts on wavelets and their application. Several complement our treatment and were relied upon during the development of the core sections of the chapter. The material in Section 7.1.2 on subband coding and digital filtering follows the book by Vetterli and Kovacevic [1995], while Sections 7.2 and 7.4 on multiresolu-tion expansions and the fast wavelet transform follow the treatment of these subjects in Burrus, Gopinath, and Guo [1998]. The remainder of the material in the chapter is based on the references cited in the text. All of the examples in the chapter were done using MATLAB (see Gonzalez et al. [2004]).

The history of wavelet analysis is recorded in a book by Hubbard [1998]. The early predecessors of wavelets were developed simultaneously in different fields and unified in a paper by Mallat [1987]. It brought a mathematical framework to the field. Much of the history of wavelets can be traced through the works of Meyer [1987] [1990] [1992a, 1992b] [1993], Mallat [1987] [1989a–c] [1998], and Daubechies [1988] [1990] [1992] [1993] [1996]. The current interest in wavelets was stimulated by many of their publica-tions. The book by Daubechies [1992] is a classic source for the mathematical details of wavelet theory.

The application of wavelets to image processing is addressed in general image pro-cessing texts, like Castleman [1996], and many application specific books, some of which are conference proceedings. In this latter category, for example, are Rosenfeld [1984], Prasad and Iyengar [1997], and Topiwala [1998]. Recent articles that can serve as starting points for further research into specific imaging applications include Gao et al. [2007] on corner detection; Olkkonen and Olkkonen [2007] on lattice implemen-tations; Selesnick et al. [2005] and Kokare et al. [2005] on complex wavelets; Thévenaz and Unser [2000] for image registration; Chang and Kuo [1993] and Unser [1995] on texture-based classification; Heijmans and Goutsias [2000] on morphological wavelets; Banham et al. [1994], Wang, Zhang, and Pan [1995], and Banham and Kastaggelos [1996] on image restoration; Xu et al. [1994] and Chang, Yu, and Vetterli [2000] on image enhancement; Delaney and Bresler [1995] and Westenberg and Roerdink [2000] on computed tomography; and Lee, Sun, and Chen [1995], Liang and Kuo [1999], Wang, Lee, and Toraichi [1999], and You and Bhattacharya [2000] on image description and matching. One of the most important applications of wavelets is image compression— see, for example, Brechet et al. [2007], Demin Wang et al. [2006], Antonini et al. [1992], Wei et al. [1998], and the book by Topiwala [1998]. Finally, there have been a number of special issues devoted to wavelets, including a special issue on wavelet transforms and multiresolution signal anaysis in the *IEEE Transactions on Information Theory* [1992], a special issue on wavelets and signal processing in the *IEEE Transactions on Signal*

Processing [1993], and a special section on multiresolution representation in the *IEEE Transactions on Pattern Analysis and Machine Intelligence* [1989].

Although the chapter focuses on the fundamentals of wavelets and their application to image processing, there is considerable interest in the construction of wavelets themselves. The interested reader is referred to the work of Battle [1987] [1988], Daubechies [1988] [1992], Cohen and Daubechies [1992], Meyer [1990], Mallat [1989b], Unser, Aldroubi, and Eden [1993], and Gröchenig and Madych [1992]. This is not an exhaustive list but should serve as a starting point for further reading. See also the general references on subband coding and filter banks, including Strang and Nguyen [1996] and Vetterli and Kovacevic [1995], and the references included in the chapter with respect to the wavelets we used as examples.

Problems

7.1 Design a system for decoding the prediction residual pyramid generated by the encoder of Fig. 7.2(b) and draw its block diagram. Assume there is no quantization error introduced by the encoder.

★7.2 Construct a fully populated approximation pyramid and corresponding prediction residual pyramid for the image

$$f(x, y) = \begin{bmatrix} 1 & 2 & 3 & 4 \\ 5 & 6 & 7 & 8 \\ 9 & 10 & 11 & 12 \\ 13 & 14 & 15 & 16 \end{bmatrix}$$

Use 2×2 block neighborhood averaging for the approximation filter in Fig. 7.2(b) and assume the interpolation filter implements pixel replication.

★7.3 Given a $2^J \times 2^J$ image, does a $J + 1$-level pyramid reduce or expand the amount of data required to represent the image? What is the compression or expansion ratio?

7.4 Is the two-band subband coding filter bank containing filters $h_0(n) = \{1/\sqrt{2}, 1/\sqrt{2}\}$, $h_1(n) = \{-1/\sqrt{2}, 1/\sqrt{2}\}$, $g_0(n) = \{1/\sqrt{2}, 1/\sqrt{2}\}$, and $g_1(n) = \{1/\sqrt{2}, -1/\sqrt{2}\}$ orthonormal, biorthogonal, or both?

7.5 Given the sequence $f(n) = \{0, 0.5, 0.25, 1\}$ where $n = 0, 1, 2, 3$, compute:
  (a) The sign-reversed sequence.
  (b) The order-reversed sequence.
  (c) The modulated sequence.
  (d) The modulated and then order-reversed sequence.
  (e) The order-reversed and then modulated sequence.
  (f) Does the result from (d) or (e) correspond to Eq. (7.1-9)?

7.6 Compute the coefficients of the Daubechies synthesis filters $g_0(n)$ and $g_1(n)$ for Example 7.2. Using Eq. (7.1-13) with $m = 0$ only, show that the filters are orthonormal.

★7.7 Draw a two-dimensional four-band filter bank decoder to reconstruct input $f(m, n)$ in Fig. 7.7.

7.8 Obtain the Haar transformation matrix for $N = 8$.

7.9 **(a)** Compute the Haar transform of the 2×2 image

$$F = \begin{bmatrix} 3 & -1 \\ 6 & 2 \end{bmatrix}$$

(b) The inverse Haar transform is $\mathbf{F} = \mathbf{H}^T \mathbf{T} \mathbf{H}$, where $\mathbf{T}$ is the Haar transform of $\mathbf{F}$ and $\mathbf{H}^T$ is the matrix inverse of $\mathbf{H}$. Show that $\mathbf{H}_2^{-1} = \mathbf{H}_2^T$ and use it to compute the inverse Haar transform of the result in (a).

7.10 Compute the expansion coefficients of 2-tuple $[3, 2]^T$ for the following bases and write the corresponding expansions:

★ **(a)** Basis $\varphi_0 = [1/\sqrt{2}, 1/\sqrt{2},]^T$ and $\varphi_1 = [1/\sqrt{2}, -1/\sqrt{2},]^T$ on $\mathbf{R}^2$, the set of real 2-tuples.

(b) Basis $\varphi_0 = [1, 0]^T$ and $\varphi_1 = [1, 1]^T$, and its dual, $\tilde{\varphi}_0 = [1, -1]^T$ and $\tilde{\varphi}_1 = [0, 1]^T$, on $\mathbf{R}^2$.

(c) Basis $\varphi_0 = [1, 0]^T$, $\varphi_1 = [-1/2, \sqrt{3}/2]^T$, and $\varphi_2 = [-1/2, -\sqrt{3}/2]^T$, and their duals, $\tilde{\varphi}_i = 2\varphi_i/3$ for $i = \{0, 1, 2,\}$, on $\mathbf{R}^2$.

(*Hint:* Vector inner products must be used in place of the integral inner products of Section 7.2.1.)

7.11 Show that scaling function

$$\varphi(x) = \begin{cases} 1 & 0.25 \le x < 0.75 \\ 0 & \text{elsewhere} \end{cases}$$

does not satisfy the second requirement of a multiresolution analysis.

7.12 Write an expression for scaling space V_3 as a function of scaling function $\varphi(x)$. Use the Haar scaling function definition of Eq. (7.2-14) to draw the Haar V_3 scaling functions at translations $k = \{0, 1, 2\}$.

★ **7.13** Draw wavelet $\psi_{3,3}(x)$ for the Haar wavelet function. Write an expression for $\psi_{3,3}(x)$ in terms of Haar scaling functions.

7.14 Suppose function $f(x)$ is a member of Haar scaling space V_3—that is, $f(x) \in V_3$. Use Eq. (7.2-22) to express V_3 as a function of scaling space V_0 and any required wavelet spaces. If $f(x)$ is 0 outside the interval $[0, 1)$, sketch the scaling and wavelet functions required for a linear expansion of $f(x)$ based on your expression.

7.15 Compute the first four terms of the wavelet series expansion of the function used in Example 7.7 with starting scale $j_0 = 1$. Write the resulting expansion in terms of the scaling and wavelet functions involved. How does your result compare to the example, where the starting scale was $j_0 = 0$?

7.16 The DWT in Eqs. (7.3-5) and (7.3-6) is a function of starting scale j_0.

(a) Recompute the one-dimensional DWT of function $f(n) = \{1, 4, -3, 0\}$ for $0 \le n \le 3$ in Example 7.8 with $j_0 = 1$ (rather than 0).

(b) Use the result from (a) to compute $f(1)$ from the transform values.

★ **7.17** What does the following continuous wavelet transform reveal about the one-dimensional function upon which it was based?

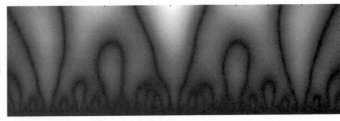

Time

7.18   **(a)** The continuous wavelet transform of Problem 7.17 is computer generated. The function upon which it is based was first sampled at discrete intervals. What is continuous about the transform—or what distinguishes it from the discrete wavelet transform of the function?

  ★ **(b)** Under what circumstances is the DWT a better choice than the CWT? Are there times when the CWT is better than the DWT?

★**7.19**   Draw the FWT filter bank required to compute the transform in Problem 7.16. Label all inputs and outputs with the appropriate sequences.

7.20   The computational complexity of an M-point fast wavelet transform is $O(M)$. That is, the number of operations is proportional M. What determines the constant of proportionality?

7.21 ★ **(a)** If the input to the three-scale FWT filter bank of Fig. 7.30(a) is the Haar scaling function $\varphi(n) = 1$ for $n = 0, 1, \ldots, 7$ and 0 elsewhere, what is the resulting transform with respect to Haar wavelets?

  (b) What is the transform if the input is the corresponding Haar wavelet function $\psi(n) = \{1, 1, 1, 1, -1, -1, -1, -1\}$ for $n = 0, 1, \ldots, 7$?

  (c) What input sequence produces transform $\{0, 0, 0, 0, 0, 0, B, 0\}$ with nonzero coefficient $W_\psi(2, 2) = B$?

★**7.22**   The two-dimensional fast wavelet transform is similar to the pyramidal coding scheme of Section 7.2.1. How are they similar? Given the three-scale wavelet transform in Fig. 7.10(a), how would you construct the corresponding approximation pyramid? How many levels would it have?

7.23   Compute the two-dimensional wavelet transform with respect to Haar wavelets of the 2×2 image in Problem 7.9. Draw the required filter bank and label all inputs and outputs with the proper arrays.

★**7.24**   In the Fourier domain

$$f(x - x_0, y - y_0) \Leftrightarrow F(\mu, v)e^{-2\pi(\mu x_0/M + v y_0/N)}$$

and translation does not affect the display of $|F(\mu, v)|$. Using the following sequence of images, explain the translation property of wavelet transforms. The leftmost image contains two 32×32 white squares centered on a 128×128 gray background. The second image (from the left) is its single-scale wavelet transform with respect to Haar wavelets. The third is the wavelet transform of the original image after shifting it 32 pixels to the right and downward, and the final (rightmost) image is the wavelet transform of the original image after it has been shifted one pixel to the right and downward.

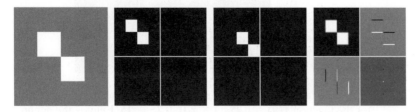

7.25   The following table shows the Haar wavelet and scaling functions for a four-scale fast wavelet transform. Sketch the additional basis functions needed for a full three-scale packet decomposition. Give the mathematical expression or expressions for determining them. Then order the basis functions according to frequency content and explain the results.

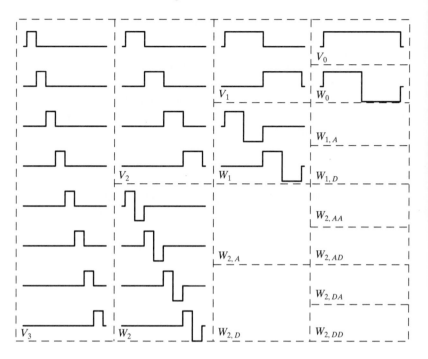

7.26 A wavelet packet decomposition of the vase from Fig. 7.1 is shown below.

    (a) Draw the corresponding decomposition analysis tree, labeling all nodes with the names of the proper scaling and wavelet spaces.

    (b) Draw and label the decomposition's frequency spectrum.

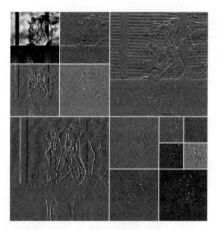

7.27 Using the Haar wavelet, determine the minimum entropy packet decomposition for the function $f(n) = 0.25$ for $n = 0, 1, 2, \ldots, 15$. Employ the nonnormalized Shannon entropy,

$$E\big[f(n)\big] = \sum_n f^2(n) \ln\big[f^2(n)\big]$$

as the minimization criterion. Draw the optimal tree, labeling the nodes with the computed entropy values.

8 Image Compression

> But life is short and information endless ...
> Abbreviation is a necessary evil and the abbreviator's
> business is to make the best of a job which, although
> intrinsically bad, is still better than nothing.
>
> *Aldous Huxley*

Preview

Image compression, the art and science of reducing the amount of data required to represent an image, is one of the most useful and commercially successful technologies in the field of digital image processing. The number of images that are compressed and decompressed daily is staggering, and the compressions and decompressions themselves are virtually invisible to the user. Anyone who owns a digital camera, surfs the web, or watches the latest Hollywood movies on *Digital Video Disks* (DVDs) benefits from the algorithms and standards discussed in this chapter.

To better understand the need for compact image representations, consider the amount of data required to represent a two-hour *standard definition* (SD) *television* movie using $720 \times 480 \times 24$ bit pixel arrays. A digital movie (or *video*) is a sequence of *video frames* in which each frame is a full-color still image. Because video players must display the frames sequentially at rates near 30 fps (frames per second), SD digital video data must be accessed at

$$30 \frac{\text{frames}}{\text{sec}} \times (720 \times 480) \frac{\text{pixels}}{\text{frame}} \times 3 \frac{\text{bytes}}{\text{pixel}} = 31{,}104{,}000 \text{ bytes/sec}$$

and a two-hour movie consists of

$$31{,}104{,}000 \frac{\text{bytes}}{\text{sec}} \times \left(60^2\right) \frac{\text{sec}}{\text{hr}} \times 2 \text{ hrs} \cong 2.24 \times 10^{11} \text{ bytes}$$

or 224 GB (gigabytes) of data. Twenty-seven 8.5 GB dual-layer DVDs (assuming conventional 12 cm disks) are needed to store it. To put a two-hour movie on a single DVD, each frame must be compressed—on average—by a factor of 26.3. The compression must be even higher for *high definition* (HD) *television*, where image resolutions reach $1920 \times 1080 \times 24$ bits/image.

Web page images and high-resolution digital camera photos also are compressed routinely to save storage space and reduce transmission time. For example, residential Internet connections deliver data at speeds ranging from 56 Kbps (kilobits per second) via conventional phone lines to more than 12 Mbps (megabits per second) for broadband. The time required to transmit a small $128 \times 128 \times 24$ bit full-color image over this range of speeds is from 7.0 to 0.03 seconds. Compression can reduce transmission time by a factor of 2 to 10 or more. In the same way, the number of uncompressed full-color images that an 8-megapixel digital camera can store on a 1-GB flash memory card [about forty-one 24 MB (megabyte) images] can be similarly increased. In addition to these applications, image compression plays an important role in many other areas, including televideo conferencing, remote sensing, document and medical imaging, and facsimile transmission (FAX). An increasing number of applications depend on the efficient manipulation, storage, and transmission of binary, gray-scale, and color images.

In this chapter, we introduce the theory and practice of digital image compression. We examine the most frequently used compression techniques and describe the industry standards that make them useful. The material is introductory in nature and applicable to both still image and video applications. The chapter concludes with an introduction to *digital image watermarking*, the process of inserting visible and invisible data (like copyright information) into images.

8.1 Fundamentals

The term *data compression* refers to the process of reducing the amount of data required to represent a given quantity of information. In this definition, *data* and *information* are not the same thing; data are the means by which information is conveyed. Because various amounts of data can be used to represent the same amount of information, representations that contain irrelevant or repeated information are said to contain *redundant data*. If we let b and b' denote the number of bits (or information-carrying units) in two representations of the same information, the *relative data redundancy* R of the representation with b bits is

$$R = 1 - \frac{1}{C} \qquad (8.1-1)$$

where C, commonly called the *compression ratio*, is defined as

$$C = \frac{b}{b'} \qquad (8.1-2)$$

If $C = 10$ (sometimes written 10:1), for instance, the larger representation has 10 bits of data for every 1 bit of data in the smaller representation.

The corresponding relative data redundancy of the larger representation is 0.9 ($R = 0.9$), indicating that 90% of its data is redundant.

In the context of digital image compression, b in Eq. (8.1-2) usually is the number of bits needed to represent an image as a 2-D array of intensity values. The 2-D intensity arrays introduced in Section 2.4.2 are the preferred formats for human viewing and interpretation—and the standard by which all other representations are judged. When it comes to compact image representation, however, these formats are far from optimal. Two-dimensional intensity arrays suffer from three principal types of data redundancies that can be identified and exploited:

1. *Coding redundancy.* A *code* is a system of symbols (letters, numbers, bits, and the like) used to represent a body of information or set of events. Each piece of information or event is assigned a sequence of *code symbols*, called a *code word*. The number of symbols in each code word is its *length*. The 8-bit codes that are used to represent the intensities in most 2-D intensity arrays contain more bits than are needed to represent the intensities.

2. *Spatial* and *temporal redundancy.* Because the pixels of most 2-D intensity arrays are correlated spatially (i.e., each pixel is similar to or dependent on neighboring pixels), information is unnecessarily replicated in the representations of the correlated pixels. In a video sequence, temporally correlated pixels (i.e., those similar to or dependent on pixels in nearby frames) also duplicate information.

3. *Irrelevant information.* Most 2-D intensity arrays contain information that is ignored by the human visual system and/or extraneous to the intended use of the image. It is redundant in the sense that it is not used.

The computer-generated images in Figs. 8.1(a) through (c) exhibit each of these fundamental redundancies. As will be seen in the next three sections, compression is achieved when one or more redundancy is reduced or eliminated.

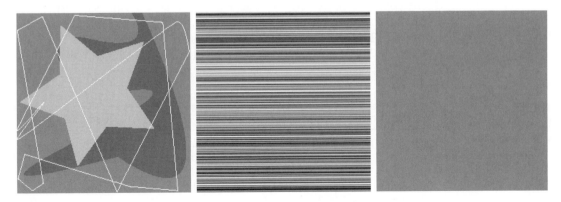

a b c

FIGURE 8.1 Computer generated $256 \times 256 \times 8$ bit images with (a) coding redundancy, (b) spatial redundancy, and (c) irrelevant information. (Each was designed to demonstrate one principal redundancy but may exhibit others as well.)

8.1.1 Coding Redundancy

In Chapter 3, we developed techniques for image enhancement by histogram processing, assuming that the intensity values of an image are random quantities. In this section, we use a similar formulation to introduce optimal information coding.

Assume that a discrete random variable r_k in the interval $[0, L - 1]$ is used to represent the intensities of an $M \times N$ image and that each r_k occurs with probability $p_r(r_k)$. As in Section 3.3,

$$p_r(r_k) = \frac{n_k}{MN} \qquad k = 0, 1, 2, \ldots, L - 1 \tag{8.1-3}$$

where L is the number of intensity values, and n_k is the number of times that the kth intensity appears in the image. If the number of bits used to represent each value of r_k is $l(r_k)$, then the average number of bits required to represent each pixel is

$$L_{\text{avg}} = \sum_{k=0}^{L-1} l(r_k) p_r(r_k) \tag{8.1-4}$$

That is, the average length of the code words assigned to the various intensity values is found by summing the products of the number of bits used to represent each intensity and the probability that the intensity occurs. The total number of bits required to represent an $M \times N$ image is MNL_{avg}. If the intensities are represented using a *natural m*-bit fixed-length code,[†] the right-hand side of Eq. (8.1-4) reduces to m bits. That is, $L_{\text{avg}} = m$ when m is substituted for $l(r_k)$. The constant m can be taken outside the summation, leaving only the sum of the $p_r(r_k)$ for $0 \leq k \leq L - 1$, which, of course, equals 1.

EXAMPLE 8.1:
A simple illustration of variable-length coding.

■ The computer-generated image in Fig. 8.1(a) has the intensity distribution shown in the second column of Table 8.1. If a natural 8-bit binary code (denoted as code 1 in Table 8.1) is used to represent its 4 possible intensities, L_{avg}—the average number of bits for code 1—is 8 bits, because $l_1(r_k) = 8$ bits for all r_k.

TABLE 8.1
Example of variable-length coding.

r_k	$p_r(r_k)$	Code 1	$l_1(r_k)$	Code 2	$l_2(r_k)$
$r_{87} = 87$	0.25	01010111	8	01	2
$r_{128} = 128$	0.47	10000000	8	1	1
$r_{186} = 186$	0.25	11000100	8	000	3
$r_{255} = 255$	0.03	11111111	8	001	3
r_k for $k \neq 87, 128, 186, 255$	0	—	8	—	0

[†]A *natural* binary code is one in which each event or piece of information to be encoded (such as intensity value) is assigned one of 2^m codes from an m-bit binary counting sequence.

On the other hand, if the scheme designated as code 2 in Table 8.1 is used, the average length of the encoded pixels is, in accordance with Eq. (8.1-4),

$$L_{avg} = 0.25(2) + 0.47(1) + 0.25(3) + 0.03(3) = 1.81 \text{ bits}$$

The total number of bits needed to represent the entire image is $MNL_{avg} = 256 \times 256 \times 1.81$ or 118,621. From Eqs. (8.1-2) and (8.1-1), the resulting compression and corresponding relative redundancy are

$$C = \frac{256 \times 256 \times 8}{118{,}621} = \frac{8}{1.81} \approx 4.42$$

and

$$R = 1 - \frac{1}{4.42} = 0.774$$

respectively. Thus 77.4% of the data in the original 8-bit 2-D intensity array is redundant.

The compression achieved by code 2 results from assigning fewer bits to the more probable intensity values than to the less probable ones. In the resulting *variable-length code*, r_{128}—the image's most probable intensity—is assigned the 1-bit code word 1 [of length $l_2(r_{128}) = 1$], while r_{255}—its least probable occurring intensity—is assigned the 3-bit code word 001 [of length $l_2(r_{255}) = 3$]. Note that the best *fixed-length code* that can be assigned to the intensities of the image in Fig. 8.1(a) is the natural 2-bit counting sequence {00, 01, 10, 11}, but the resulting compression is only 8/2 or 4:1—about 10% less than the 4.42:1 compression of the variable-length code. ▪

As the preceding example shows, *coding redundancy* is present when the codes assigned to a set of events (such as intensity values) do not take full advantage of the probabilities of the events. Coding redundancy is almost always present when the intensities of an image are represented using a natural binary code. The reason is that most images are composed of objects that have a regular and somewhat predictable morphology (shape) and reflectance, and are sampled so that the objects being depicted are much larger than the picture elements. The natural consequence is that, for most images, certain intensities are more probable than others (that is, the histograms of most images are not uniform). A natural binary encoding assigns the same number of bits to both the most and least probable values, failing to minimize Eq. (8.1-4) and resulting in coding redundancy.

8.1.2 Spatial and Temporal Redundancy

Consider the computer-generated collection of constant intensity lines in Fig. 8.1(b). In the corresponding 2-D intensity array:

1. All 256 intensities are equally probable. As Fig. 8.2 shows, the histogram of the image is uniform.

FIGURE 8.2 The intensity histogram of the image in Fig. 8.1(b).

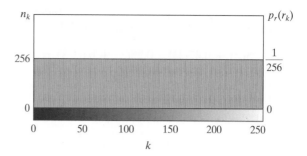

2. Because the intensity of each line was selected randomly, its pixels are independent of one another in the vertical direction.
3. Because the pixels along each line are identical, they are maximally correlated (completely dependent on one another) in the horizontal direction.

The first observation tells us that the image in Fig. 8.1(b)—when represented as a conventional 8-bit intensity array—cannot be compressed by variable-length coding alone. Unlike the image of Fig. 8.1(a) (and Example 8.1), whose histogram was *not* uniform, a fixed-length 8-bit code in this case minimizes Eq. (8.1-4). Observations 2 and 3 reveal a significant spatial redundancy that can be eliminated, for instance, by representing the image in Fig. 8.1(b) as a sequence of *run-length pairs*, where each run-length pair specifies the start of a new intensity and the number of consecutive pixels that have that intensity. A run-length based representation compresses the original 2-D, 8-bit intensity array by $(256 \times 256 \times 8)/[(256 + 256) \times 8]$ or 128:1. Each 256-pixel line of the original representation is replaced by a single 8-bit intensity value and length 256 in the run-length representation.

In most images, pixels are correlated spatially (in both x and y) and in time (when the image is part of a video sequence). Because most pixel intensities can be predicted reasonably well from neighboring intensities, the information carried by a single pixel is small. Much of its visual contribution is redundant in the sense that it can be inferred from its neighbors. To reduce the redundancy associated with spatially and temporally correlated pixels, a 2-D intensity array must be transformed into a more efficient but usually "non-visual" representation. For example, run-lengths or the differences between adjacent pixels can be used. Transformations of this type are called *mappings*. A mapping is said to be *reversible* if the pixels of the original 2-D intensity array can be reconstructed without error from the transformed data set; otherwise the mapping is said to be *irreversible*.

8.1.3 Irrelevant Information

One of the simplest ways to compress a set of data is to remove superfluous data from the set. In the context of digital image compression, information that is ignored by the human visual system or is extraneous to the intended use of an image are obvious candidates for omission. Thus, the computer-generated image in Fig. 8.1(c), because it appears to be a homogeneous field of gray, can be represented by its average intensity alone—a single 8-bit value. The original $256 \times 256 \times 8$ bit intensity array is reduced to a single byte; and the resulting compression is $(256 \times 256 \times 8)/8$ or $65,536{:}1$. Of course, the original $256 \times 256 \times 8$ bit image must be recreated to view and/or analyze it—but there would be little or no perceived decrease in reconstructed image quality.

Figure 8.3(a) shows the histogram of the image in Fig. 8.1(c). Note that there are several intensity values (intensities 125 through 131) actually present. The human visual system averages these intensities, perceives only the average value, and ignores the small changes in intensity that are present in this case. Figure 8.3(b), a histogram equalized version of the image in Fig. 8.1(c), makes the intensity changes visible *and* reveals two previously undetected regions of constant intensity—one oriented vertically and the other horizontally. If the image in Fig. 8.1(c) is represented by its average value alone, this "invisible" structure (i.e., the constant intensity regions) and the random intensity variations surrounding them—real information—is lost. Whether or not this information should be preserved is application dependent. If the information is important, as it might be in a medical application (like digital X-ray archival), it should not be omitted; otherwise, the information is redundant and can be excluded for the sake of compression performance.

We conclude the section by noting that the redundancy examined here is fundamentally different from the redundancies discussed in Sections 8.1.1 and 8.1.2. Its elimination is possible because the information itself is not essential for normal visual processing and/or the intended use of the image. Because its omission results in a loss of quantitative information, its removal is commonly referred to as *quantization*. This terminology is consistent with normal use of the word, which generally means the mapping of a broad range of input values to a limited number of output values (see Section 2.4). Because information is lost, quantization is an irreversible operation.

8.1.4 Measuring Image Information

In the previous sections, we introduced several ways to reduce the amount of data used to represent an image. The question that naturally arises is this: How

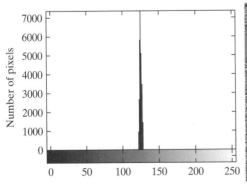

a b
FIGURE 8.3
(a) Histogram of the image in Fig. 8.1(c) and (b) a histogram equalized version of the image.

Consult the book Web site for a brief review of information and probability theory.

few bits are actually needed to represent the information in an image? That is, is there a minimum amount of data that is sufficient to describe an image without losing information? *Information theory* provides the mathematical framework to answer this and related questions. Its fundamental premise is that the generation of information can be modeled as a probabilistic process that can be measured in a manner that agrees with intuition. In accordance with this supposition, a random event E with probability $P(E)$ is said to contain

$$I(E) = \log \frac{1}{P(E)} = -\log P(E) \tag{8.1-5}$$

units of information. If $P(E) = 1$ (that is, the event always occurs), $I(E) = 0$ and no information is attributed to it. Because no uncertainty is associated with the event, no information would be transferred by communicating that the event has occurred [it *always* occurs if $P(E) = 1$].

The base of the logarithm in Eq. (8.1-5) determines the unit used to measure information. If the base m logarithm is used, the measurement is said to be in m-ary units. If the base 2 is selected, the unit of information is the *bit*. Note that if $P(E) = \frac{1}{2}$, $I(E) = -\log_2 \frac{1}{2}$, or 1 bit. That is, 1 bit is the amount of information conveyed when one of two possible equally likely events occurs. A simple example is flipping a coin and communicating the result.

Given a source of statistically independent random events from a discrete set of possible events $\{a_1, a_2, \ldots, a_J\}$ with associated probabilities $\{P(a_1), P(a_2), \ldots, P(a_J)\}$, the average information per source output, called the *entropy* of the source, is

$$H = -\sum_{j=1}^{J} P(a_j) \log P(a_j) \tag{8.1-6}$$

The a_j in this equation are called *source symbols*. Because they are statistically independent, the source itself is called a *zero-memory source*.

If an image is considered to be the output of an imaginary zero-memory "intensity source," we can use the histogram of the observed image to estimate the symbol probabilities of the source. Then the intensity source's entropy becomes

Equation (8.1-6) is for zero-memory sources with J source symbols; Eq. (8.1-7) uses probability estimates for the $L - 1$ intensity values in an image.

$$\tilde{H} = -\sum_{k=0}^{L-1} p_r(r_k) \log_2 p_r(r_k) \tag{8.1-7}$$

where variables L, r_k, and $p_r(r_k)$ are as defined in Sections 8.1.1 and 3.3. Because the base 2 logarithm is used, Eq. (8.1-7) is the average information per intensity output of the imaginary intensity source in bits. It is not possible to code the *intensity values* of the imaginary source (and thus the sample image) with fewer than $\tilde{H}$ bits/pixel.

■ The entropy of the image in Fig. 8.1(a) can be estimated by substituting the intensity probabilities from Table 8.1 into Eq. (8.1-7):

EXAMPLE 8.2:
Image entropy
estimates.

$$\tilde{H} = -[0.25 \log_2 0.25 + 0.47 \log_2 0.47 + 0.25 \log_2 0.25 + 0.03 \log_2 0.03]$$

$$\approx -[0.25(-2) + 0.47(-1.09) + 0.25(-2) + 0.03(-5.06)]$$

$$\approx 1.6614 \text{ bits/pixel}$$

In a similar manner, the entropies of the images in Fig. 8.1(b) and (c) can be shown to be 8 bits/pixel and 1.566 bits/pixel, respectively. Note that the image in Fig. 8.1(a) appears to have the most visual information, but has almost the lowest computed entropy—1.66 bits/pixel. The image in Fig. 8.1(b) has almost five times the entropy of the image in (a), but appears to have about the same (or less) visual information; and the image in Fig. 8.1(c), which seems to have little or no information, has almost the same entropy as the image in (a). The obvious conclusion is that the amount of entropy and thus information in an image is far from intuitive. ■

Shannon's first theorem

Recall that the variable-length code in Example 8.1 was able to represent the intensities of the image in Fig. 8.1(a) using only 1.81 bits/pixel. Although this is higher than the 1.6614 bits/pixel entropy estimate from Example 8.2, *Shannon's first theorem*—also called the *noiseless coding theorem* (Shannon [1948])—assures us that the image in Fig. 8.1(a) can be represented with as few as 1.6614 bits/pixel. To prove it in a general way, Shannon looked at representing groups of n consecutive source symbols with a single code word (rather than one code word per source symbol) and showed that

$$\lim_{n \to \infty} \left[\frac{L_{\text{avg},n}}{n} \right] = H \qquad (8.1\text{-}8)$$

where $L_{\text{avg},n}$ is the average number of code symbols required to represent all n-symbol groups. In the proof, he defined the *nth extension* of a zero-memory source to be the hypothetical source that produces n-symbol blocks[†] using the symbols of the original source; and computed $L_{\text{avg},n}$ by applying Eq. (8.1-4) to the code words used to represent the n-symbol blocks. Equation (8.1-8) tells us that $L_{\text{avg},n}/n$ can be made arbitrarily close to H by encoding infinitely long extensions of the single-symbol source. That is, it is possible to represent the output of a zero-memory source with an average of H information units per source symbol.

[†]The output of the nth extension is an n-tuple of symbols from the underlying *single-symbol* source. It was considered a *block random variable* in which the probability of each n-tuple is the product of the probabilities of its individual symbols. The entropy of the nth extension is then n times the entropy of the single-symbol source from which it is derived.

If we now return to the idea that an image is a "sample" of the intensity source that produced it, a block of n source symbols corresponds to a group of n adjacent pixels. To construct a variable-length code for n-pixel blocks, the relative frequencies of the blocks must be computed. But the nth extension of a hypothetical intensity source with 256 intensity values has 256^n possible n-pixel blocks. Even in the simple case of $n = 2$, a 65,536 element histogram and up to 65,536 variable-length code words must be generated. For $n = 3$, as many as 16,777,216 code words are needed. So even for small values of n, computational complexity limits the usefulness of the extension coding approach in practice.

Finally, we note that although Eq. (8.1-7) provides a lower bound on the compression that can be achieved when coding statistically independent pixels directly, it breaks down when the pixels of an image are correlated. Blocks of correlated pixels can be coded with fewer average bits per pixel than the equation predicts. Rather than using source extensions, less correlated descriptors (like intensity run-lengths) are normally selected and coded without extension. This was the approach used to compress Fig. 8.1(b) in Section 8.1.2. When the output of a source of information depends on a finite number of preceding outputs, the source is called a *Markov* or *finite memory source*.

8.1.5 Fidelity Criteria

In Section 8.1.3, it was noted that the removal of "irrelevant visual" information involves a loss of real or quantitative image information. Because information is lost, a means of quantifying the nature of the loss is needed. Two types of criteria can be used for such an assessment: (1) objective fidelity criteria and (2) subjective fidelity criteria.

When information loss can be expressed as a mathematical function of the input and output of a compression process, it is said to be based on an *objective fidelity criterion*. An example is the root-mean-square (rms) error between two images. Let $f(x, y)$ be an input image and $\hat{f}(x, y)$ be an approximation of $f(x, y)$ that results from compressing and subsequently decompressing the input. For any value of x and y, the error $e(x, y)$ between $f(x, y)$ and $\hat{f}(x, y)$ is

$$e(x, y) = \hat{f}(x, y) - f(x, y) \tag{8.1-9}$$

so that the total error between the two images is

$$\sum_{x=0}^{M-1} \sum_{y=0}^{N-1} \left[\hat{f}(x, y) - f(x, y) \right]$$

where the images are of size $M \times N$. The *root-mean-square error*, e_{rms}, between $f(x, y)$ and $\hat{f}(x, y)$ is then the square root of the squared error averaged over the $M \times N$ array, or

$$e_{rms} = \left[\frac{1}{MN} \sum_{x=0}^{M-1} \sum_{y=0}^{N-1} \left[\hat{f}(x, y) - f(x, y) \right]^2 \right]^{1/2} \tag{8.1-10}$$

If $\hat{f}(x,y)$ is considered [by a simple rearrangement of the terms in Eq. (8.1-9)] to be the sum of the original image $f(x,y)$ and an error or "noise" signal $e(x,y)$, the *mean-square signal-to-noise ratio* of the output image, denoted SNR_{ms}, can be defined as in Section 5.8:

$$SNR_{ms} = \frac{\sum_{x=0}^{M-1}\sum_{y=0}^{N-1}\hat{f}(x,y)^2}{\sum_{x=0}^{M-1}\sum_{y=0}^{N-1}\left[\hat{f}(x,y)-f(x,y)\right]^2} \tag{8.1-11}$$

The rms value of the signal-to-noise ratio, denoted SNR_{rms}, is obtained by taking the square root of Eq. (8.1-11).

While objective fidelity criteria offer a simple and convenient way to evaluate information loss, decompressed images are ultimately viewed by humans. So, measuring image quality by the subjective evaluations of people is often more appropriate. This can be done by presenting a decompressed image to a cross section of viewers and averaging their evaluations. The evaluations may be made using an absolute rating scale or by means of side-by-side comparisons of $f(x, y)$ and $\hat{f}(x, y)$. Table 8.2 shows one possible absolute rating scale. Side-by-side comparisons can be done with a scale such as $\{-3, -2, -1, 0, 1, 2, 3\}$ to represent the subjective evaluations *{much worse, worse, slightly worse, the same, slightly better, better, much better}*, respectively. In either case, the evaluations are based on *subjective fidelity criteria*.

■ Figure 8.4 shows three different approximations of the image in Fig. 8.1(a). Using Eq. (8.1-10) with Fig. 8.1(a) for $f(x,y)$ and the images in Figs. 8.4(a) through (c) as $\hat{f}(x,y)$, the computed rms errors are 5.17, 15.67, and 14.17 intensity levels, respectively. In terms of rms error—an objective fidelity criterion—the three images in Fig. 8.4 are ranked in order of decreasing quality as $\{(a), (c), (b)\}$.

EXAMPLE 8.3:
Image quality comparisons.

Value	Rating	Description
1	Excellent	An image of extremely high quality, as good as you could desire.
2	Fine	An image of high quality, providing enjoyable viewing. Interference is not objectionable.
3	Passable	An image of acceptable quality. Interference is not objectionable.
4	Marginal	An image of poor quality; you wish you could improve it. Interference is somewhat objectionable.
5	Inferior	A very poor image, but you could watch it. Objectionable interference is definitely present.
6	Unusable	An image so bad that you could not watch it.

TABLE 8.2
Rating scale of the Television Allocations Study Organization. (Frendendall and Behrend.)

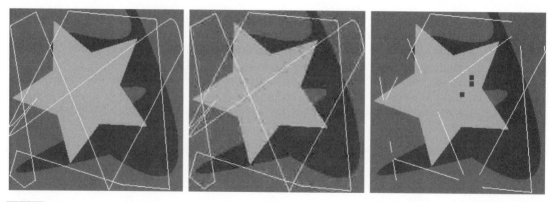

a b c

FIGURE 8.4 Three approximations of the image in Fig. 8.1(a).

Figures 8.4(a) and (b) are typical of images that have been compressed and subsequently reconstructed. Both retain the essential information of the original image—like the spatial and intensity characteristics of its objects. And their rms errors correspond roughly to perceived quality. Figure 8.4(a), which is practically as good as the original image, has the lowest rms error, while Fig. 8.4(b) has more error but noticeable degradation at the boundaries between objects. This is exactly as one would expect.

Figure 8.4(c) is an artificially generated image that demonstrates the limitations of objective fidelity criteria. Note that the image is missing large sections of several important lines (i.e., visual information), and has small dark squares (i.e., artifacts) in the upper right quadrant. The visual content of the image is misleading and certainly not as accurate as the image in (b), but it has less rms error—14.17 versus 15.67 intensity values. A subjective evaluation of the three images using Table 8.2 might yield an *excellent* rating for (a), a *passable* or *marginal* rating for (b), and an *inferior* of *unusable* rating for (c). The rms error measure, on the other hand, ranks (c) ahead of (b). ■

8.1.6 Image Compression Models

As Fig. 8.5 shows, an image compression system is composed of two distinct functional components: an *encoder* and a *decoder*. The encoder performs compression, and the decoder performs the complementary operation of decompression. Both operations can be performed in software, as is the case in Web browsers and many commercial image editing programs, or in a combination of hardware and firmware, as in commercial DVD players. A *codec* is a device or program that is capable of both encoding and decoding.

Here, the notation $f(x, \ldots)$ is used to denote both $f(x, y)$ and $f(x, y, t)$.

Input image $f(x, \ldots)$ is fed into the encoder, which creates a compressed representation of the input. This representation is stored for later use, or transmitted for storage and use at a remote location. When the compressed representation is presented to its complementary decoder, a reconstructed output image $\hat{f}(x, \ldots)$ is generated. In still-image applications, the encoded input and decoder output are $f(x, y)$ and $\hat{f}(x, y)$, respectively; in video applications, they

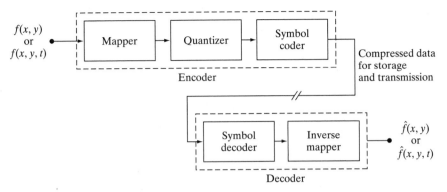

FIGURE 8.5
Functional block
diagram of a
general image
compression
system.

are $f(x, y, t)$ and $\hat{f}(x, y, t)$, where discrete parameter t specifies time. In general, $\hat{f}(x, \dots)$ may or may not be an exact replica of $f(x, \dots)$. If it is, the compression system is called *error free*, *lossless*, or *information preserving*. If not, the reconstructed output image is distorted and the compression system is referred to as *lossy*.

The encoding or compression process

The encoder of Fig. 8.5 is designed to remove the redundancies described in Sections 8.1.1–8.1.3 through a series of three independent operations. In the first stage of the encoding process, a *mapper* transforms $f(x, \dots)$ into a (usually nonvisual) format designed to reduce spatial and temporal redundancy. This operation generally is reversible and may or may not reduce directly the amount of data required to represent the image. Run-length coding (see Sections 8.1.2 and 8.2.5) is an example of a mapping that normally yields compression in the first step of the encoding process. The mapping of an image into a set of less correlated transform coefficients (see Section 8.2.8) is an example of the opposite case (the coefficients must be further processed to achieve compression). In video applications, the mapper uses previous (and in some cases future) video frames to facilitate the removal of temporal redundancy.

The *quantizer* in Fig. 8.5 reduces the accuracy of the mapper's output in accordance with a pre-established fidelity criterion. The goal is to keep irrelevant information out of the compressed representation. As noted in Section 8.1.3, this operation is irreversible. It must be omitted when error-free compression is desired. In video applications, the *bit rate* of the encoded output is often measured (in bits/second) and used to adjust the operation of the quantizer so that a predetermined average output rate is maintained. Thus, the visual quality of the output can vary from frame to frame as a function of image content.

In the third and final stage of the encoding process, the *symbol coder* of Fig. 8.5 generates a fixed- or variable-length code to represent the quantizer output and maps the output in accordance with the code. In many cases, a variable-length code is used. The shortest code words are assigned to the most frequently occurring quantizer output values—thus minimizing coding redundancy. This operation is reversible. Upon its completion, the input image has been processed for the removal of each of the three redundancies described in Sections 8.1.1 to 8.1.3.

The decoding or decompression process

The decoder of Fig. 8.5 contains only two components: a *symbol decoder* and an *inverse mapper*. They perform, in reverse order, the inverse operations of the encoder's symbol encoder and mapper. Because quantization results in irreversible information loss, an inverse quantizer block is not included in the general decoder model. In video applications, decoded output frames are maintained in an internal frame store (not shown) and used to reinsert the temporal redundancy that was removed at the encoder.

8.1.7 Image Formats, Containers, and Compression Standards

In the context of digital imaging, an *image file format* is a standard way to organize and store image data. It defines how the data is arranged and the type of compression—if any—that is used. An *image container* is similar to a file format but handles multiple types of image data. Image *compression standards*, on the other hand, define procedures for compressing and decompressing images—that is, for reducing the amount of data needed to represent an image. These standards are the underpinning of the widespread acceptance of image compression technology.

Figure 8.6 lists the most important image compression standards, file formats, and containers in use today, grouped by the type of image handled. The entries in black are international standards sanctioned by the *International Standards Organization* (ISO), the *International Electrotechnical Commission* (IEC), and/or the *International Telecommunications Union* (ITU-T)—a *United Nations* (UN) organization that was once called the *Consultative Committee of the International Telephone and Telegraph* (CCITT). Two video compression standards, VC-1 by the *Society of Motion Pictures and Television Engineers* (SMPTE) and AVS by the Chinese *Ministry of Information Industry* (MII), are

FIGURE 8.6 Some popular image compression standards, file formats, and containers. Internationally sanctioned entries are shown in black; all others are grayed.

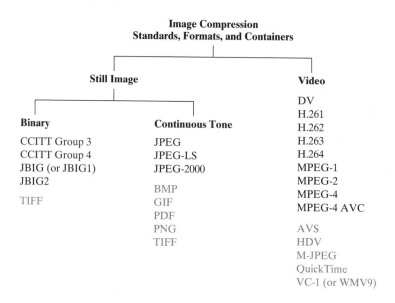

also included. Note that they are shown in gray, which is used in Fig. 8.6 to denote entries that are not sanctioned by an international standards organization.

Tables 8.3 and 8.4 summarize the standards, formats, and containers listed in Fig. 8.6. Responsible organizations, targeted applications, and key compression methods are identified. The compression methods themselves are the subject of the next section. In both tables, forward references to the relevant subsections of Section 8.2 are enclosed in square brackets.

TABLE 8.3
Internationally sanctioned image compression standards. The numbers in brackets refer to sections in this chapter.

Name	Organization	Description
Bi-Level Still Images		
CCITT Group 3	ITU-T	Designed as a facsimile (FAX) method for transmitting binary documents over telephone lines. Supports 1-D and 2-D run-length [8.2.5] and Huffman [8.2.1] coding.
CCITT Group 4	ITU-T	A simplified and streamlined version of the CCITT Group 3 standard supporting 2-D run-length coding only.
JBIG *or* JBIG1	ISO/IEC/ ITU-T	A *Joint Bi-level Image Experts Group* standard for progressive, lossless compression of bi-level images. Continuous-tone images of up to 6 bits/pixel can be coded on a bit-plane basis [8.2.7]. Context sensitive arithmetic coding [8.2.3] is used and an initial low resolution version of the image can be gradually enhanced with additional compressed data.
JBIG2	ISO/IEC/ ITU-T	A follow-on to JBIG1 for bi-level images in desktop, Internet, and FAX applications. The compression method used is content based, with dictionary based methods [8.2.6] for text and halftone regions, and Huffman [8.2.1] or arithmetic coding [8.2.3] for other image content. It can be lossy or lossless.
Continuous-Tone Still Images		
JPEG	ISO/IEC/ ITU-T	A *Joint Photographic Experts Group* standard for images of photographic quality. Its lossy *baseline coding system* (most commonly implemented) uses quantized discrete cosine transforms (DCT) on 8×8 image blocks [8.2.8], Huffman [8.2.1], and run-length [8.2.5] coding. It is one of the most popular methods for compressing images on the Internet.
JPEG-LS	ISO/IEC/ ITU-T	A lossless to near-lossless standard for continuous tone images based on adaptive prediction [8.2.9], context modeling [8.2.3], and Golomb coding [8.2.2].
JPEG-2000	ISO/IEC/ ITU-T	A follow-on to JPEG for increased compression of photographic quality images. Arithmetic coding [8.2.3] and quantized discrete wavelet transforms (DWT) [8.2.10] are used. The compression can be lossy or lossless.

(Continues)

TABLE 8.3
(*Continued*)

Name	Organization	Description
Video		
DV	IEC	*Digital Video.* A video standard tailored to home and semiprofessional video production applications and equipment—like electronic news gathering and camcorders. Frames are compressed independently for uncomplicated editing using a DCT-based approach [8.2.8] similar to JPEG.
H.261	ITU-T	A two-way videoconferencing standard for ISDN (*integrated services digital network*) lines. It supports non-interlaced 352×288 and 176×144 resolution images, called CIF (*Common Intermediate Format*) and QCIF (*Quarter CIF*), respectively. A DCT-based compression approach [8.2.8] similar to JPEG is used, with frame-to-frame prediction differencing [8.2.9] to reduce temporal redundancy. A block-based technique is used to compensate for motion between frames.
H.262	ITU-T	See MPEG-2 below.
H.263	ITU-T	An enhanced version of H.261 designed for ordinary telephone modems (i.e., 28.8 Kb/s) with additional resolutions: SQCIF (*Sub-Quarter* CIF 128×96), 4CIF (704×576), and 16CIF (1408×512).
H.264	ITU-T	An extension of H.261–H.263 for videoconferencing, Internet streaming, and television broadcasting. It supports prediction differences within frames [8.2.9], variable block size integer transforms (rather than the DCT), and context adaptive arithmetic coding [8.2.3].
MPEG-1	ISO/IEC	A *Motion Pictures Expert Group* standard for CD-ROM applications with non-interlaced video at up to 1.5 Mb/s. It is similar to H.261 but frame predictions can be based on the previous frame, next frame, or an interpolation of both. It is supported by almost all computers and DVD players.
MPEG-2	ISO/IEC	An extension of MPEG-1 designed for DVDs with transfer rates to 15 Mb/s. Supports interlaced video and HDTV. It is the most successful video standard to date.
MPEG-4	ISO/IEC	An extension of MPEG-2 that supports variable block sizes and prediction differencing [8.2.9] within frames.
MPEG-4 AVC	ISO/IEC	MPEG-4 Part 10 *Advanced Video Coding* (AVC). Identical to H.264 above.

Name	Organization	Description
Continuous-Tone Still Images		
BMP	Microsoft	*Windows Bitmap.* A file format used mainly for simple uncompressed images.
GIF	CompuServe	*Graphic Interchange Format.* A file format that uses lossless LZW coding [8.2.4] for 1- through 8-bit images. It is frequently used to make small animations and short low resolution films for the World Wide Web.
PDF	Adobe Systems	*Portable Document Format.* A format for representing 2-D documents in a device and resolution independent way. It can function as a container for JPEG, JPEG 2000, CCITT, and other compressed images. Some PDF versions have become ISO standards.
PNG	*World Wide Web Consortium* (W3C)	*Portable Network Graphics.* A file format that losslessly compresses full color images with transparency (up to 48 bits/pixel) by coding the difference between each pixel's value and a predicted value based on past pixels [8.2.9].
TIFF	Aldus	*Tagged Image File Format.* A flexible file format supporting a variety of image compression standards, including JPEG, JPEG-LS, JPEG-2000, JBIG2, and others.
Video		
AVS	MII	*Audio-Video Standard.* Similar to H.264 but uses exponential Golomb coding [8.2.2]. Developed in China.
HDV	Company consortium	*High Definition Video.* An extension of DV for HD television that uses MPEG-2 like compression, including temporal redundancy removal by prediction differencing [8.2.9].
M-JPEG	Various companies	*Motion JPEG.* A compression format in which each frame is compressed independently using JPEG.
Quick-Time	Apple Computer	A media container supporting DV, H.261, H.262, H.264, MPEG-1, MPEG-2, MPEG-4, and other video compression formats.
VC-1 WMV9	SMPTE Microsoft	The most used video format on the Internet. Adopted for HD and *Blu-ray* high-definition DVDs. It is similar to H.264/AVC, using an integer DCT with varying block sizes [8.2.8 and 8.2.9] and context dependent variable-length code tables [8.2.1]—but no predictions within frames.

TABLE 8.4
Popular image compression standards, file formats, and containers, not included in Table 8.3.

8.2   Some Basic Compression Methods

In this section, we describe the principal lossy and error-free compression methods in use today. Our focus is on methods that have proven useful in mainstream binary, continuous-tone still images, and video compression standards. The standards themselves are used to demonstrate the methods presented.

8.2.1 Huffman Coding

With reference to Tables 8.3 and 8.4, Huffman codes are used in

- CCITT
- JBIG2
- JPEG
- MPEG-1,2,4
- H.261, H.262, H.263, H.264

and other compression standards.

One of the most popular techniques for removing coding redundancy is due to Huffman (Huffman [1952]). When coding the symbols of an information source individually, *Huffman coding* yields the smallest possible number of code symbols per source symbol. In terms of Shannon's first theorem (see Section 8.1.4), the resulting code is optimal for a fixed value of n, subject to the constraint that the source symbols be coded *one at a time*. In practice, the source symbols may be either the intensities of an image or the output of an intensity mapping operation (pixel differences, run lengths, and so on).

The first step in Huffman's approach is to create a series of source reductions by ordering the probabilities of the symbols under consideration and combining the lowest probability symbols into a single symbol that replaces them in the next source reduction. Figure 8.7 illustrates this process for binary coding (K-ary Huffman codes can also be constructed). At the far left, a hypothetical set of source symbols and their probabilities are ordered from top to bottom in terms of decreasing probability values. To form the first source reduction, the bottom two probabilities, 0.06 and 0.04, are combined to form a "compound symbol" with probability 0.1. This compound symbol and its associated probability are placed in the first source reduction column so that the probabilities of the reduced source also are ordered from the most to the least probable. This process is then repeated until a reduced source with two symbols (at the far right) is reached.

The second step in Huffman's procedure is to code each reduced source, starting with the smallest source and working back to the original source. The minimal length binary code for a two-symbol source, of course, are the symbols 0 and 1. As Fig. 8.8 shows, these symbols are assigned to the two symbols on the right (the assignment is arbitrary; reversing the order of the 0 and 1 would work just as well). As the reduced source symbol with probability 0.6 was generated by combining two symbols in the reduced source to its left, the 0 used to code it is now assigned to *both* of these symbols, and a 0 and 1 are arbitrarily appended to each to distinguish them from each other. This operation is then repeated for

FIGURE 8.7
Huffman source reductions.

Original source		Source reduction			
Symbol	Probability	1	2	3	4
a_2	0.4	0.4	0.4	0.4	0.6
a_6	0.3	0.3	0.3	0.3	0.4
a_1	0.1	0.1	0.2	0.3	
a_4	0.1	0.1	0.1		
a_3	0.06	0.1			
a_5	0.04				

Original source			Source reduction			
Symbol	Probability	Code	1	2	3	4
a_2	0.4	1	0.4 1	0.4 1	0.4 1	┌─0.6 0
a_6	0.3	00	0.3 00	0.3 00	0.3 00 ◄┐	0.4 1
a_1	0.1	011	0.1 011	┌─0.2 010 ◄	┌─0.3 01 ◄┘	
a_4	0.1	0100	0.1 0100 ◄┐	0.1 011 ◄┘		
a_3	0.06	01010 ◄┐	┌─0.1 0101 ◄┘			
a_5	0.04	01011 ◄┘				

FIGURE 8.8
Huffman code assignment procedure.

each reduced source until the original source is reached. The final code appears at the far left in Fig. 8.8. The average length of this code is

$$L_{avg} = (0.4)(1) + (0.3)(2) + (0.1)(3) + (0.1)(4) + (0.06)(5) + (0.04)(5)$$
$$= 2.2 \text{ bits/pixel}$$

and the entropy of the source is 2.14 bits/symbol.

Huffman's procedure creates the optimal code for a set of symbols and probabilities *subject to the constraint* that the symbols be coded one at a time. After the code has been created, coding and/or error-free decoding is accomplished in a simple lookup table manner. The code itself is an instantaneous uniquely decodable block code. It is called a *block code* because each source symbol is mapped into a fixed sequence of code symbols. It is *instantaneous* because each code word in a string of code symbols can be decoded without referencing succeeding symbols. It is *uniquely decodable* because any string of code symbols can be decoded in only one way. Thus, any string of Huffman encoded symbols can be decoded by examining the individual symbols of the string in a left-to-right manner. For the binary code of Fig. 8.8, a left-to-right scan of the encoded string 010100111100 reveals that the first valid code word is 01010, which is the code for symbol a_3. The next valid code is 011, which corresponds to symbol a_1. Continuing in this manner reveals the completely decoded message to be $a_3 a_1 a_2 a_2 a_6$.

■ The $512 \times 512 \times 8$ bit monochrome image in Fig. 8.9(a) has the intensity histogram shown in Fig. 8.9(b). Because the intensities are not equally probable,

EXAMPLE 8.4:
Huffman coding.

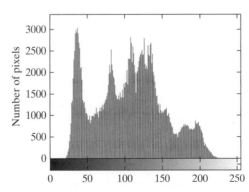

a b

FIGURE 8.9 (a) A 512×512 8-bit image, and (b) its histogram.

a MATLAB implementation of Huffman's procedure was used to encode them with 7.428 bits/pixel—including the Huffman code table that is required to reconstruct the original 8-bit image intensities. The compressed representation exceeds the estimated entropy of the image [7.3838 bits/pixel from Eq. (8.1-7)] by $512^2 \times (7.428 - 7.3838)$ or 11,587 bits—about 0.6%. The resulting compression ratio and corresponding relative redundancy are $C = 8/7.428 = 1.077$ and $R = 1 - (1/1.077) = 0.0715$, respectively. Thus 7.15% of the original 8-bit fixed-length intensity representation was removed as coding redundancy. ■

When a large number of symbols is to be coded, the construction of an optimal Huffman code is a nontrivial task. For the general case of J source symbols, J symbol probabilities, $J - 2$ source reductions, and $J - 2$ code assignments are required. When source symbol probabilities can be estimated in advance, "near optimal" coding can be achieved with pre-computed Huffman codes. Several popular image compression standards, including the JPEG and MPEG standards discussed in Sections 8.2.8 and 8.2.9, specify default Huffman coding tables that have been pre-computed based on experimental data.

8.2.2 Golomb Coding

With reference to Tables 8.3 and 8.4, Golomb codes are used in

- JPEG-LS
- AVS

compression.

In this section we consider the coding of nonnegative integer inputs with exponentially decaying probability distributions. Inputs of this type can be optimally encoded (in the sense of Shannon's first theorem) using a family of codes that are computationally simpler than Huffman codes. The codes themselves were first proposed for the representation of nonnegative run lengths (Golomb [1966]). In the discussion that follows, the notation $\lfloor x \rfloor$ denotes the largest integer less than or equal to x, $\lceil x \rceil$ means the smallest integer greater than or equal to x, and $x \bmod y$ is the remainder of x divided by y.

Given a nonnegative integer n and a positive integer *divisor* $m > 0$, the *Golomb code* of n with respect to m, denoted $G_m(n)$, is a combination of the unary code of *quotient* $\lfloor n/m \rfloor$ and the binary representation of *remainder* $n \bmod m$. $G_m(n)$ is constructed as follows:

Step 1. Form the unary code of quotient $\lfloor n/m \rfloor$. (The *unary code* of an integer q is defined as q 1s followed by a 0.)

Step 2. Let $k = \lceil \log_2 m \rceil, c = 2^k - m, r = n \bmod m$, and compute truncated remainder r' such that

$$r' = \begin{cases} r \text{ truncated to } k - 1 \text{ bits} & 0 \le r < c \\ r + c \text{ truncated to } k \text{ bits} & \text{otherwise} \end{cases} \tag{8.2-1}$$

Step 3. Concatenate the results of steps 1 and 2.

To compute $G_4(9)$, for example, begin by determining the unary code of the quotient $\lfloor 9/4 \rfloor = \lfloor 2.25 \rfloor = 2$, which is 110 (the result of step 1). Then let $k = \lceil \log_2 4 \rceil = 2, c = 2^2 - 4 = 0$, and $r = 9 \bmod 4$, which in binary is $1001 \bmod 0100$ or 0001. In accordance with Eq. (8.2-1), r' is then r (i.e., 0001) truncated to 2 bits, which is 01 (the result of step 2). Finally, concatenate 110 from step 1 and 01 from step 2 to get 11001, which is $G_4(9)$.

For the special case of $m = 2^k$, $c = 0$ and $r' = r = n \bmod m$ truncated to k bits in Eq. (8.2-1) for all n. The divisions required to generate the resulting Golomb codes become binary shift operations and the computationally simpler codes are called *Golomb-Rice* or *Rice codes* (Rice [1975]). Columns 2, 3, and 4 of Table 8.5 list the G_1, G_2, and G_4 codes of the first ten nonnegative integers. Because m is a power of 2 in each case (i.e., $1 = 2^0$, $2 = 2^1$, and $4 = 2^2$), they are the first three Golomb-Rice codes as well. Moreover, G_1 is the unary code of the nonnegative integers because $\lfloor n/1 \rfloor = n$ and $n \bmod 1 = 0$ for all n.

Keeping in mind that Golomb codes can only be used to represent nonnegative integers and that there are many Golomb codes to choose from, a key step in their effective application is the selection of divisor m. When the integers to be represented are *geometrically* distributed with *probability mass function* (PMF)[†]

$$P(n) = (1 - \rho)\rho^n \tag{8.2-2}$$

for some $0 < \rho < 1$, Golomb codes can be shown to be optimal—in the sense that $G_m(n)$ provides the shortest average code length of all uniquely decipherable codes—when (Gallager and Voorhis [1975])

$$m = \left\lceil \frac{\log_2(1 + \rho)}{\log_2(1/\rho)} \right\rceil \tag{8.2-3}$$

Figure 8.10(a) plots Eq. (8.2-2) for three values of ρ and illustrates graphically the symbol probabilities that Golomb codes handle well (that is, code efficiently). As is shown in the figure, small integers are much more probable than large ones.

Because the probabilities of the intensities in an image [see, for example, the histogram of Fig. 8.9(b)] are unlikely to match the probabilities specified in Eq. (8.2-2) and shown in Fig. 8.10(a), Golomb codes are seldom used for the coding of intensities. When intensity differences are to be coded, however, the

The discrete probability distribution defined by the PMF in Eq. (8.2-2) is called the geometric probability distribution. *Its continuous counterpart is the* exponential distribution.

The graphical representation of a PMF is a histogram.

n	$G_1(n)$	$G_2(n)$	$G_4(n)$	$G_{exp}^0(n)$
0	0	00	000	0
1	10	01	001	100
2	110	100	010	101
3	1110	101	011	11000
4	11110	1100	1000	11001
5	111110	1101	1001	11010
6	1111110	11100	1010	11011
7	11111110	11101	1011	1110000
8	111111110	111100	11000	1110001
9	1111111110	111101	11001	1110010

TABLE 8.5
Several Golomb codes for the integers 0 – 9.

[†]A *probability mass function* (PMF) is a function that defines the probability that a discrete random variable is exactly equal to some value. A PMF differs from a PDF in that a PDF's values are not probabilities; rather, the integral of a PDF over a specified interval is a probability.

FIGURE 8.10
(a) Three one-sided geometric distributions from Eq. (8.2-2); (b) a two-sided exponentially decaying distribution; and (c) a reordered version of (b) using Eq. (8.2-4).

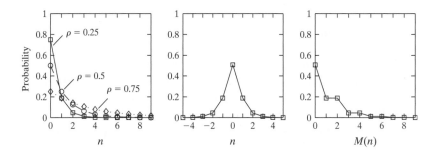

probabilities of the resulting "difference values" (see Section 8.2.9)—with the notable exception of the negative differences—often resemble those of Eq. (8.2-2) and Fig. 8.10(a). To handle negative differences in Golomb coding, which can only represent nonnegative integers, a mapping like

$$M(n) = \begin{cases} 2n & n \geq 0 \\ 2|n| - 1 & n < 0 \end{cases} \tag{8.2-4}$$

typically is used. Using this mapping, for example, the two-sided PMF shown in Fig. 8.10(b) can be transformed into the one-sided PMF in Fig. 8.10(c). Its integers are reordered, alternating the negative and positive integers so that the negative integers are mapped into the odd positive integer positions. If $P(n)$ is two-sided and centered at zero, $P(M(n))$ will be one-sided. The mapped integers, $M(n)$, can then be efficiently encoded using an appropriate Golomb-Rice code (Weinberger et al. [1996]).

EXAMPLE 8.5:
Golomb-Rice coding.

■ Consider again the image from Fig. 8.1(c) and note that its histogram—see Fig. 8.3(a)—is similar to the two-sided distribution in Fig. 8.10(b) above. If we let n be some nonnegative integer intensity in the image, where $0 \leq n \leq 255$, and μ be the mean intensity, $P(n - \mu)$ is the two-sided distribution shown in Fig. 8.11(a). This plot was generated by normalizing the histogram in Fig. 8.3(a) by the total number of pixels in the image and shifting the normalized values to the left by 128 (which in effect subtracts the mean intensity from the image). In accordance with Eq. (8.2-4), $P(M(n - \mu))$ is then the one-sided distribution shown in Fig. 8.11(b). If the reordered intensity values are Golomb coded using a MATLAB implementation of code G_1 in column 2 of Table 8.5, the encoded representation is 4.5 times smaller than the original image (i.e., $C = 4.5$). The G_1 code realizes 4.5/5.1 or 88% of the theoretical compression possible with variable-length coding. (Based on the entropy calculated in Example 8.2, the maximum possible compression ratio through variable-length coding is $C = 8/1.566 \approx 5.1$.) Moreover, Golomb coding achieves 96% of the compression provided by a MATLAB implementation of Huffman's approach—and doesn't require the computation of a custom Huffman coding table.

Now consider the image in Fig. 8.9(a). If its intensities are Golomb coded using the same G_1 code as above, $C = 0.0922$. That is, there is *data expansion*.

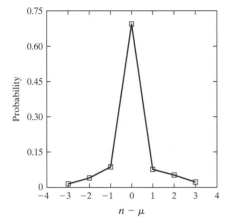

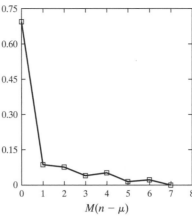

FIGURE 8.11
(a) The probability distribution of the image in Fig. 8.1(c) after subtracting the mean intensity from each pixel, and (b) a mapped version of (a) using Eq. (8.2-4).

This is due to the fact that the probabilities of the intensities of the image in Fig. 8.9(a) are much different than the probabilities defined in Eq. (8.2-2). In a similar manner, Huffman codes can produce data expansion when used to encode symbols whose probabilities are different from those for which the code was computed. In practice, the further you depart from the input probability assumptions for which a code is designed, the greater the risk of poor compression performance and data expansion.   ■

When C is less than 1 in Eq. (8.1-2), there is data expansion.

To conclude our coverage of Golomb codes, we note that Column 5 of Table 8.5 contains the first 10 codes of the zeroth order *exponential-Golomb code*, denoted $G_{\exp}^0(n)$. Exponential-Golomb codes are useful for the encoding of run lengths, because both short and long runs are encoded efficiently. An order-k exponential-Golomb code $G_{\exp}^k(n)$ is computed as follows:

Step 1. Find an integer $i \geq 0$ such that

$$\sum_{j=0}^{i-1} 2^{j+k} \leq n < \sum_{j=0}^{i} 2^{j+k} \qquad (8.2\text{-}5)$$

and form the unary code of i. If $k = 0$, $i = \lfloor \log_2(n + 1) \rfloor$ and the code is also known as the *Elias gamma code*.
Step 2. Truncate the binary representation of

$$n - \sum_{j=0}^{i-1} 2^{j+k} \qquad (8.2\text{-}6)$$

to $k + i$ least significant bits.
Step 3. Concatenate the results of steps 1 and 2.

To find $G_{exp}^0(8)$, for example, we let $i = \lfloor \log_2 9 \rfloor$ or 3 in step 1 because $k = 0$. Equation (8.2-5) is then satisfied because

$$\sum_{j=0}^{3-1} 2^{j+0} \leq 8 < \sum_{j=0}^{3} 2^{j+0}$$

$$\sum_{j=0}^{2} 2^j \leq 8 < \sum_{j=0}^{3} 2^j$$

$$2^0 + 2^1 + 2^2 \leq 8 < 2^0 + 2^1 + 2^2 + 2^3$$

$$7 \leq 8 < 15$$

The unary code of 3 is 1110 and Eq. (8.2-6) of step 2 yields

$$8 - \sum_{j=0}^{3-1} 2^{j+0} = 8 - \sum_{j=0}^{2} 2^j = 8 - (2^0 + 2^1 + 2^2) = 8 - 7 = 1 = 0001$$

which when truncated to its $3 + 0$ least significant bits becomes 001. The concatenation of the results from steps 1 and 2 then yields 1110001. Note that this is the entry in column 4 of Table 8.5 for $n = 8$. Finally, we note that like the Huffman codes of the last section, the Golomb codes of Table 8.5 are variable-length, instantaneous uniquely decodable block codes.

8.2.3 Arithmetic Coding

With reference to Tables 8.3 and 8.4, arithmetic coding is used in

- JBIG1
- JBIG2
- JPEG-2000
- H.264
- MPEG-4 AVC

and other compression standards.

Unlike the variable-length codes of the previous two sections, *arithmetic coding* generates nonblock codes. In arithmetic coding, which can be traced to the work of Elias (see Abramson [1963]), a one-to-one correspondence between source symbols and code words does not exist. Instead, an entire sequence of source symbols (or message) is assigned a single arithmetic code word. The code word itself defines an interval of real numbers between 0 and 1. As the number of symbols in the message increases, the interval used to represent it becomes smaller and the number of information units (say, bits) required to represent the interval becomes larger. Each symbol of the message reduces the size of the interval in accordance with its probability of occurrence. Because the technique does not require, as does Huffman's approach, that each source symbol translate into an integral number of code symbols (that is, that the symbols be coded one at a time), it achieves (but only in theory) the bound established by Shannon's first theorem of Section 8.1.4.

Figure 8.12 illustrates the basic arithmetic coding process. Here, a five-symbol sequence or message, $a_1a_2a_3a_3a_4$, from a four-symbol source is coded. At the start of the coding process, the message is assumed to occupy the entire half-open interval $[0, 1)$. As Table 8.6 shows, this interval is subdivided initially into four regions based on the probabilities of each source symbol. Symbol a_1, for example, is associated with subinterval $[0, 0.2)$. Because it is the first symbol of the message being coded, the message interval is initially narrowed to $[0, 0.2)$. Thus in Fig. 8.12

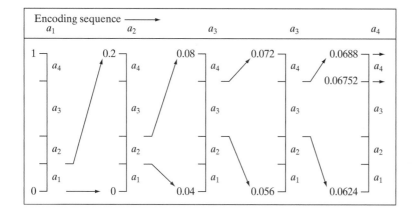

FIGURE 8.12
Arithmetic coding
procedure.

Source Symbol	Probability	Initial Subinterval
a_1	0.2	$[0.0, 0.2)$
a_2	0.2	$[0.2, 0.4)$
a_3	0.4	$[0.4, 0.8)$
a_4	0.2	$[0.8, 1.0)$

TABLE 8.6
Arithmetic coding
example.

$[0, 0.2)$ is expanded to the full height of the figure and its end points labeled by the values of the narrowed range. The narrowed range is then subdivided in accordance with the original source symbol probabilities and the process continues with the next message symbol. In this manner, symbol a_2 narrows the subinterval to $[0.04, 0.08)$, a_3 further narrows it to $[0.056, 0.072)$, and so on. The final message symbol, which must be reserved as a special end-of-message indicator, narrows the range to $[0.06752, 0.0688)$. Of course, any number within this subinterval—for example, 0.068—can be used to represent the message.

In the arithmetically-coded message of Fig. 8.12, three decimal digits are used to represent the five-symbol message. This translates into 0.6 decimal digits per source symbol and compares favorably with the entropy of the source, which, from Eq. (8.1-6), is 0.58 decimal digits per source symbol. As the length of the sequence being coded increases, the resulting arithmetic code approaches the bound established by Shannon's first theorem. In practice, two factors cause coding performance to fall short of the bound: (1) the addition of the end-of-message indicator that is needed to separate one message from another; and (2) the use of finite precision arithmetic. Practical implementations of arithmetic coding address the latter problem by introducing a scaling strategy and a rounding strategy (Langdon and Rissanen [1981]). The scaling strategy renormalizes each subinterval to the $[0, 1)$ range before subdividing it in accordance with the symbol probabilities. The rounding strategy guarantees that the truncations associated with finite precision arithmetic do not prevent the coding subintervals from being represented accurately.

Adaptive context dependent probability estimates

With accurate input symbol *probability models*, that is, models that provide the true probabilities of the symbols being coded, arithmetic coders are near optimal in the sense of minimizing the average number of code symbols required to represent the symbols being coded. Like in both Huffman and Golomb coding, however, inaccurate probability models can lead to non-optimal results. A simple way to improve the accuracy of the probabilities employed is to use an adaptive, context dependent probability model. *Adaptive* probability models update symbol probabilities as symbols are coded or become known. Thus, the probabilities adapt to the local statistics of the symbols being coded. *Context dependent* models provide probabilities that are based on a predefined neighborhood of pixels—called the *context*—around the symbols being coded. Normally, a *causal context*—one limited to symbols that have already been coded—is used. Both the Q-coder (Pennebaker et al. [1988]) and MQ-coder (ISO/IEC [2000]), two well-known arithmetic coding techniques that have been incorporated into the JBIG, JPEG-2000, and other important image compression standards, use probability models that are both adaptive and context dependent. The Q-coder dynamically updates symbol probabilities during the interval renormalizations that are part of the arithmetic coding process. Adaptive context dependent models also have been used in Golomb coding—for example, in the JPEG-LS compression standard.

Figure 8.13(a) diagrams the steps involved in adaptive, context-dependent arithmetic coding of *binary* source symbols. Arithmetic coding often is used when binary symbols are to be coded. As each symbol (or bit) begins the coding process, its context is formed in the *Context determination* block of Fig. 8.13(a). Figures 8.13(b) through (d) show three possible contexts that can be used: (1) the immediately preceding symbol, (2) a group of preceding symbols, and (3) some number of preceding symbols plus symbols on the previous scan line. For the three cases shown, the *Probability estimation* block must manage 2^1 (or 2), 2^8 (or 256), and 2^5 (or 32) contexts and their associated probabilities. For instance, if the context in Fig. 8.13(b) is used, conditional probabilities

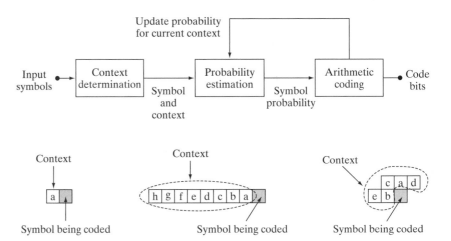

a
b c d

FIGURE 8.13
(a) An adaptive, context-based arithmetic coding approach (often used for binary source symbols). (b)–(d) Three possible context models.

$P(0|a = 0)$ (the probability that the symbol being coded is a 0 given that the preceding symbol is a 0), $P(1|a = 0)$, $P(0|a = 1)$, and $P(1|a = 1)$ must be tracked. The appropriate probabilities are then passed to the *Arithmetic coding* block as a function of the current context and drive the generation of the arithmetically coded output sequence in accordance with the process illustrated in Fig. 8.12. The probabilities associated with the context involved in the current coding step are then updated to reflect the fact that another symbol within that context has been processed.

Finally, we note that a variety of arithmetic coding techniques are protected by United States patents (and may in addition be protected in other jurisdictions). Because of these patents and the possibility of unfavorable monetary judgments for their infringement, most implementations of the JPEG compression standard, which contains options for both Huffman and arithmetic coding, typically support Huffman coding alone.

8.2.4 LZW Coding

The techniques covered in the previous sections are focused on the removal of coding redundancy. In this section, we consider an error-free compression approach that also addresses spatial redundancies in an image. The technique, called *Lempel-Ziv-Welch (LZW) coding*, assigns fixed-length code words to variable length sequences of source symbols. Recall from Section 8.1.4 that Shannon used the idea of coding sequences of source symbols, rather than individual source symbols, in the proof of his first theorem. A key feature of LZW coding is that it requires no a priori knowledge of the probability of occurrence of the symbols to be encoded. Despite the fact that until recently it was protected under a United States patent, LZW compression has been integrated into a variety of mainstream imaging file formats, including GIF, TIFF, and PDF. The PNG format was created to get around LZW licensing requirements.

With reference to Tables 8.3 and 8.4, LZW coding is used in the

- GIF
- TIFF
- PDF

formats, but not in any of the internationally sanctioned compression standards.

◾ Consider again the 512×512, 8-bit image from Fig. 8.9(a). Using Adobe Photoshop, an uncompressed TIFF version of this image requires 286,740 bytes of disk space — 262,144 bytes for the 512×512 8-bit pixels plus 24,596 bytes of overhead. Using TIFF's LZW compression option, however, the resulting file is 224,420 bytes. The compression ratio is $C = 1.28$. Recall that for the Huffman encoded representation of Fig. 8.9(a) in Example 8.4, $C = 1.077$. The additional compression realized by the LZW approach is due the removal of some of the image's spatial redundancy. ◾

EXAMPLE 8.6:
LZW coding
Fig. 8.9(a).

LZW coding is conceptually very simple (Welch [1984]). At the onset of the coding process, a codebook or *dictionary* containing the source symbols to be coded is constructed. For 8-bit monochrome images, the first 256 words of the dictionary are assigned to intensities 0, 1, 2, ..., 255. As the encoder sequentially examines image pixels, intensity sequences that are not in the dictionary are placed in algorithmically determined (e.g., the next unused) locations. If the first two pixels of the image are white, for instance, sequence "255–255" might be assigned to location 256, the address following the locations reserved for intensity levels 0 through 255. The next time that two consecutive white pixels are encountered, code word 256, the address of the location containing sequence 255–255, is used to represent them. If a 9-bit, 512-word dictionary is employed in the coding process, the original $(8 + 8)$ bits that were used to represent the two pixels are replaced by a single 9-bit code word. Clearly, the size of the dictionary is an important system parameter. If it is too small, the detection of matching intensity-level sequences will be less likely; if it is too large, the size of the code words will adversely affect compression performance.

EXAMPLE 8.7:
LZW coding.

■ Consider the following 4 × 4, 8-bit image of a vertical edge:

$$
\begin{array}{cccc}
39 & 39 & 126 & 126 \\
39 & 39 & 126 & 126 \\
39 & 39 & 126 & 126 \\
39 & 39 & 126 & 126 \\
\end{array}
$$

Table 8.7 details the steps involved in coding its 16 pixels. A 512-word dictionary with the following starting content is assumed:

Dictionary Location	Entry
0	0
1	1
⋮	⋮
255	255
256	—
⋮	⋮
511	—

Locations 256 through 511 initially are unused.

The image is encoded by processing its pixels in a left-to-right, top-to-bottom manner. Each successive intensity value is concatenated with a variable—column 1 of Table 8.7—called the "currently recognized sequence." As can be seen, this variable is initially null or empty. The dictionary is searched for each concatenated sequence and if found, as was the case in the first row of the table, is replaced by the newly concatenated and recognized (i.e., located in the dictionary) sequence. This was done in column 1 of row 2. No output codes are generated, nor is the dictionary altered. If the concatenated sequence is not found, however, the address of the currently recognized sequence is output as the next encoded value, the concatenated but unrecognized sequence is added to the dictionary, and the currently recognized sequence is initialized to the current pixel value. This occurred in row 2 of the table. The last two columns detail the intensity sequences that are added to the dictionary when scanning the entire 4 × 4 image. Nine additional code words are defined. At the conclusion of coding, the dictionary contains 265 code words and the LZW algorithm has successfully identified several repeating intensity sequences—leveraging them to reduce the original 128-bit image to 90 bits (i.e., 10 9-bit codes). The encoded output is obtained by reading the third column from top to bottom. The resulting compression ratio is 1.42:1. ■

TABLE 8.7
LZW coding example.

Currently Recognized Sequence	Pixel Being Processed	Encoded Output	Dictionary Location (Code Word)	Dictionary Entry
	39			
39	39	39	256	39-39
39	126	39	257	39-126
126	126	126	258	126-126
126	39	126	259	126-39
39	39			
39-39	126	256	260	39-39-126
126	126			
126-126	39	258	261	126-126-39
39	39			
39-39	126			
39-39-126	126	260	262	39-39-126-126
126	39			
126-39	39	259	263	126-39-39
39	126			
39-126	126	257	264	39-126-126
126		126		

A unique feature of the LZW coding just demonstrated is that the coding dictionary or code book is created while the data are being encoded. Remarkably, an LZW decoder builds an identical decompression dictionary as it decodes simultaneously the encoded data stream. It is left as an exercise to the reader (see Problem 8.20) to decode the output of the preceding example and reconstruct the code book. Although not needed in this example, most practical applications require a strategy for handling dictionary overflow. A simple solution is to flush or reinitialize the dictionary when it becomes full and continue coding with a new initialized dictionary. A more complex option is to monitor compression performance and flush the dictionary when it becomes poor or unacceptable. Alternatively, the least used dictionary entries can be tracked and replaced when necessary.

8.2.5 Run-Length Coding

As was noted in Section 8.1.2, images with repeating intensities along their rows (or columns) can often be compressed by representing runs of identical intensities as *run-length pairs*, where each run-length pair specifies the start of a new intensity and the number of consecutive pixels that have that intensity. The technique, referred to as *run-length encoding* (RLE), was developed in the 1950s and became, along with its 2-D extensions, the standard compression approach in facsimile (FAX) coding. Compression is achieved by eliminating a simple form of spatial redundancy—groups of identical intensities. When there are few (or no) runs of identical pixels, run-length encoding results in data expansion.

With reference to Tables 8.3 and 8.4, the coding of run-lengths is used in

- CCITT
- JBIG2
- JPEG
- M-JPEG
- MPEG-1,2,4
- BMP

and other compression standards and file formats.

EXAMPLE 8.8:
RLE in the BMP
file format.

■ The BMP file format uses a form of run-length encoding in which image data is represented in two different modes: encoded and absolute — and either mode can occur anywhere in the image. In *encoded* mode, a two byte RLE representation is used. The first byte specifies the number of consecutive pixels that have the color index contained in the second byte. The 8-bit color index selects the run's intensity (color or gray value) from a table of 256 possible intensities.

In *absolute* mode, the first byte is 0 and the second byte signals one of four possible conditions, as shown in Table 8.8. When the second byte is 0 or 1, the end of a line or the end of the image has been reached. If it is 2, the next two bytes contain unsigned horizontal and vertical offsets to a new spatial position (and pixel) in the image. If the second byte is between 3 and 255, it specifies the number of uncompressed pixels that follow — with each subsequent byte containing the color index of one pixel. The total number of bytes must be aligned on a 16-bit word boundary.

Note that due to differences in overhead, the uncompressed BMP file is smaller than the uncompressed TIFF file in Example 8.7.

An uncompressed BMP file (saved using Photoshop) of the $512 \times 512 \times 8$ bit image shown in Fig. 8.9(a) requires 263,244 bytes of memory. Compressed using BMP's RLE option, the file expands to 267,706 bytes — and the compression ratio is $C = 0.98$. There are not enough equal intensity runs to make run-length compression effective; a small amount of expansion occurs. For the image in Fig. 8.1(c), however, the BMP RLE option results in a compression ratio $C = 1.35$.   ■

Run-length encoding is particularly effective when compressing binary images. Because there are only two possible intensities (black and white), adjacent pixels are more likely to be identical. In addition, each image row can be represented by a sequence of lengths only — rather than length-intensity pairs as was used in Example 8.8. The basic idea is to code each contiguous group (i.e., run) of 0s or 1s encountered in a left to right scan of a row by its length *and* to establish a convention for determining the value of the run. The most common conventions are (1) to specify the value of the first run of each row, or (2) to assume that each row begins with a white run, whose run length may in fact be zero.

Although run-length encoding is in itself an effective method of compressing binary images, additional compression can be achieved by variable-length coding the run lengths themselves. The black and white run lengths can be coded separately using variable-length codes that are specifically tailored to their own statistics. For example, letting symbol a_j represent a black run of length j, we can estimate the probability that symbol a_j was emitted by an imaginary black run-length source by dividing the number of black run lengths

TABLE 8.8
BMP absolute coding mode options. In this mode, the first byte of the BMP pair is 0.

Second Byte Value	Condition
0	End of line
1	End of image
2	Move to a new position
3–255	Specify pixels individually

of length j in the entire image by the total number of black runs. An estimate of the entropy of this black run-length source, denoted H_0, follows by substituting these probabilities into Eq. (8.1-6). A similar argument holds for the entropy of the white runs, denoted H_1. The approximate run-length entropy of the image is then

$$H_{RL} = \frac{H_0 + H_1}{L_0 + L_1} \qquad (8.2\text{-}7)$$

where the variables L_0 and L_1 denote the average values of black and white run lengths, respectively. Equation (8.2-7) provides an estimate of the average number of bits per pixel required to code the run lengths in a binary image using a variable-length code.

Two of the oldest and most widely used image compression standards are the CCITT Group 3 and 4 standards for binary image compression. Although they have been used in a variety of computer applications, they were originally designed as facsimile (FAX) coding methods for transmitting documents over telephone networks. The Group 3 standard uses a 1-D run-length coding technique in which the last $K - 1$ lines of each group of K lines (for $K = 2$ or 4) can be optionally coded in a 2-D manner. The Group 4 standard is a simplified or streamlined version of the Group 3 standard in which only 2-D coding is allowed. Both standards use the same 2-D coding approach, which is two-dimensional in the sense that information from the previous line is used to encode the current line. Both 1-D and 2-D coding are discussed next.

One-dimensional CCITT compression

In the 1-D CCITT Group 3 compression standard, each line of an image[†] is encoded as a series of variable-length Huffman code words that represent the run lengths of alternating white and black runs in a left-to-right scan of the line. The compression method employed is commonly referred to as *Modified Huffman* (MH) coding. The code words themselves are of two types, which the standard refers to as *terminating codes* and *makeup codes*. If run length r is less than 63, a terminating code from Table A.1 in Appendix A is used to represent it. Note that the standard specifies different terminating codes for black and white runs. If $r > 63$, two codes are used—a makeup code for quotient $\lfloor r/64 \rfloor$ and terminating code for remainder $r \bmod 64$. Makeup codes are listed in Table A.2 and may or may not depend on the intensity (black or white) of the run being coded. If $\lfloor r/64 \rfloor < 1792$, separate black and white run makeup codes are specified; otherwise, makeup codes are independent of run intensity. The standard requires that each line begin with a white run-length code word, which may in fact be 00110101, the code for a white run of length zero. Finally, a unique end-of-line (EOL) code word 000000000001 is used to terminate each line, as well as to signal the first line of each new image. The end of a sequence of images is indicated by six consecutive EOLs.

Recall from Section 8.2.2 that the notation $\lfloor x \rfloor$ denotes the largest integer less than or equal to x.

[†]In the standard, images are referred to as *pages* and sequences of images are called *documents*.

Two-dimensional CCITT compression

The 2-D compression approach adopted for both the CCITT Group 3 and 4 standards is a line-by-line method in which the position of each black-to-white or white-to-black run transition is coded with respect to the position of a *reference element* a_0 that is situated on the current *coding line*. The previously coded line is called the *reference line*; the reference line for the first line of each new image is an imaginary white line. The 2-D coding technique that is used is called *Relative Element Address Designate* (READ) coding. In the Group 3 standard, one or three READ coded lines are allowed between successive MH coded lines and the technique is called *Modified READ* (MR) coding. In the Group 4 standard, a greater number of READ coded lines are allowed and the method is called *Modified Modified READ* (MMR) coding. As was previously noted, the coding is two-dimensional in the sense that information from the previous line is used to encode the current line. Two-dimensional transforms are not involved.

Figure 8.14 shows the basic 2-D coding process for a single scan line. Note that the initial steps of the procedure are directed at locating several key *changing elements*: a_0, a_1, a_2, b_1, and b_2. A changing element is defined by the standard as a pixel whose value is different from that of the previous pixel on the same line. The most important changing element is a_0 (the reference element), which is either set to the location of an imaginary white changing element to the left of the first pixel of each new coding line or determined from the previous coding mode. Coding modes are discussed in the following paragraph. After a_0 is located, a_1 is identified as the location of the next changing element to the right of a_0 on the current coding line, a_2 as the next changing element to the right of a_1 on the coding line, b_1 as the changing element of the opposite value (of a_0) and to the right of a_0 on the reference (or previous) line, and b_2 as the next changing element to the right of b_1 on the reference line. If any of these changing elements are not detected, they are set to the location of an imaginary pixel to the right of the last pixel on the appropriate line. Figure 8.15 provides two illustrations of the general relationships between the various changing elements.

After identification of the current reference element and associated changing elements, two simple tests are performed to select one of three possible coding modes: *pass mode*, *vertical mode*, or *horizontal mode*. The initial test, which corresponds to the first branch point in the flowchart in Fig. 8.14, compares the location of b_2 to that of a_1. The second test, which corresponds to the second branch point in Fig. 8.14, computes the distance (in pixels) between the locations of a_1 and b_1 and compares it against 3. Depending on the outcome of these tests, one of the three outlined coding blocks of Fig. 8.14 is entered and the appropriate coding procedure is executed. A new reference element is then established, as per the flowchart, in preparation for the next coding iteration.

Table 8.9 defines the specific codes utilized for each of the three possible coding modes. In pass mode, which specifically excludes the case in which b_2 is directly above a_1, only the pass mode code word 0001 is needed. As Fig. 8.15(a) shows, this mode identifies white or black reference line runs that do not overlap

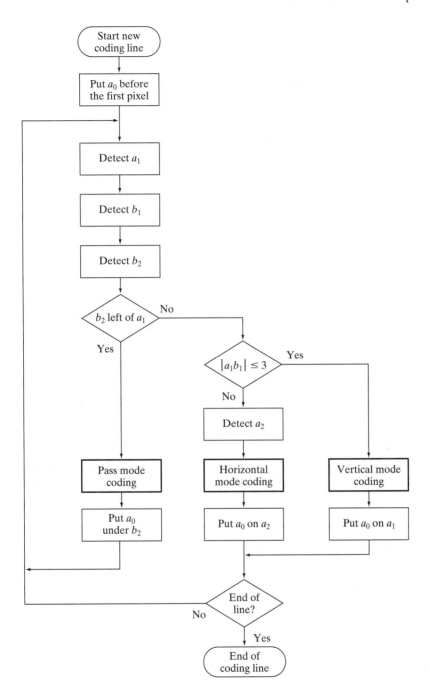

FIGURE 8.14
CCITT 2-D
READ coding
procedure. The
notation $|a_1b_1|$
denotes the
absolute value of
the distance
between changing
elements a_1
and b_1.

TABLE 8.9
CCITT two-
dimensional code
table.

Mode	Code Word
Pass	0001
Horizontal	$001 + M(a_0a_1) + M(a_1a_2)$
Vertical	
a_1 below b_1	1
a_1 one to the right of b_1	011
a_1 two to the right of b_1	000011
a_1 three to the right of b_1	0000011
a_1 one to the left of b_1	010
a_1 two to the left of b_1	000010
a_1 three to the left of b_1	0000010
Extension	0000001xxx

the current white or black coding line runs. In horizontal coding mode, the distances from a_0 to a_1 and a_1 to a_2 must be coded in accordance with the termination and makeup codes of Tables A.1 and A.2 of Appendix A and then appended to the horizontal mode code word 001. This is indicated in Table 8.9 by the notation $001 + M(a_0a_1) + M(a_1a_2)$, where a_0a_1 and a_1a_2 denote the distances from a_0 to a_1 and a_1 to a_2, respectively. Finally, in vertical coding mode, one of six special variable-length codes is assigned to the distance between a_1 and b_1. Figure 8.15(b) illustrates the parameters involved in both horizontal and vertical mode coding. The extension mode code word at the bottom of Table 8.9 is used to enter an optional facsimile coding mode. For example, the 0000001111 code is used to initiate an uncompressed mode of transmission.

EXAMPLE 8.9:
CCITT vertical
mode coding
example.

■ Although Fig. 8.15(b) is annotated with the parameters for both horizontal and vertical mode coding (to facilitate the discussion above), the depicted pattern of black and white pixels is a case for vertical mode coding. That is, because b_2 is to the right of a_1, the first (or pass mode) test in Fig. 8.14 fails. The second test, which determines whether the vertical or horizontal coding mode

a
b

FIGURE 8.15
CCITT (a) pass
mode and
(b) horizontal
and vertical mode
coding
parameters.

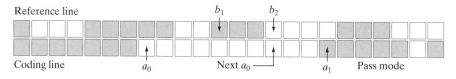

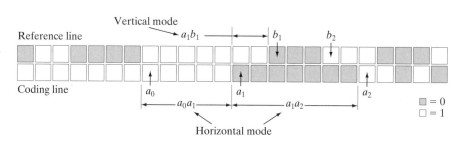

is entered, indicates that vertical mode coding should be used, because the distance from a_1 to b_1 is less than 3. In accordance with Table 8.9, the appropriate code word is 000010, implying that a_1 is two pixels left of b_1. In preparation for the next coding iteration, a_0 is moved to the location of a_1.   ■

■ Figure 8.16(a) is a 300 dpi scan of a 7×9.25 inch book page displayed at about 1/3 scale. Note that about half of the page contains text, around 9% is occupied by a halftone image, and the rest is white space. A section of the page is enlarged in Fig. 8.16(b). Keep in mind that we are dealing with a binary image; the illusion of gray tones is created, as was described in Section 4.5.4, by the halftoning process used in printing. If the binary pixels of the image in Fig. 8.16(a) are stored in groups of 8 pixels per byte, the 1952×2697 bit scanned image, commonly called a *document*, requires 658,068 bytes. An uncompressed PDF file of the document (created in Photoshop) requires 663,445 bytes. CCITT Group 3 compression reduces the file to 123,497 bytes—resulting in a compression ratio $C = 5.37$; CCITT Group 4 compression reduces the file to 110,456 bytes, increasing the compression ratio to about 6.   ■

EXAMPLE 8.10:
CCITT
compression
example.

Do not confuse the PDF used here, which stands for *Portable Document Format*, with the PDF used in previous sections and chapters for probability density function.

8.2.6 Symbol-Based Coding

In *symbol-* or *token-based* coding, an image is represented as a collection of frequently occurring sub-images, called *symbols*. Each such symbol is stored in a *symbol dictionary* and the image is coded as a set of triplets $\{(x_1, y_1, t_1), (x_2, y_2, t_2), \dots\}$, where each (x_i, y_i) pair specifies the location of a symbol in

With reference to Tables 8.3 and 8.4, symbol-based coding is used in

- JBIG2

compression.

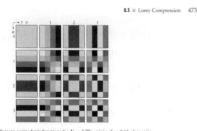

8.5 ■ Lossy Compression   473

FIGURE 8.30 Discrete-cosine basis functions for $N = 4$. The origin of each block is at its top left.

where

$$\alpha(u) = \begin{cases} \sqrt{\dfrac{1}{N}} & \text{for } u = 0 \\ \sqrt{\dfrac{2}{N}} & \text{for } u = 1, 2, \dots, N-1 \end{cases} \qquad (8.5\text{-}33)$$

and similarly for $\alpha(v)$. Figure 8.30 shows $g(x, y, u, v)$ for the case $N = 4$. The computation follows the same format as explained for Fig. 8.29, with the difference that the values of g are not integers. In Fig. 8.30, the lighter gray levels correspond to larger values of g.

■ Figures 8.31(a), (c), and (e) show three approximations of the 512×512 monochrome image in Fig. 8.23. These pictures were obtained by dividing the original image into subimages of size 8×8, representing each subimage using one of the transforms just described (i.e., the DFT, WHT, or DCT transform), truncating 50% of the resulting coefficients, and taking the inverse transform of the truncated coefficient arrays.

In each case, the 32 retained coefficients were selected on the basis of maximum magnitude. When we disregard any quantization or coding issues, this process amounts to compressing the original image by a factor of 2. Note that in all cases, the 32 discarded coefficients had little visual impact on reconstructed image quality. Their elimination, however, was accompanied by some mean-square error, which can be seen in the scaled error images of Figs. 8.31(b), (d), and (f). The actual rms errors were 1.28, 0.86, and 0.68 gray levels, respectively.   ■

EXAMPLE 8.19:
Transform coding with the DFT, WHT, and DCT.

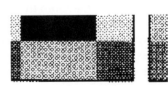

a b

FIGURE 8.16
A binary scan of a book page:
(a) scaled to show the general page content; (b) scaled to show the binary pixels used in dithering.

the image and *token t_i* is the address of the symbol or sub-image in the dictionary. That is, each triplet represents an instance of a dictionary symbol in the image. Storing repeated symbols only once can compress images significantly—particularly in document storage and retrieval applications, where the symbols are often character bitmaps that are repeated many times.

Consider the simple bilevel image in Fig. 8.17(a). It contains the single word, *banana*, which is composed of three unique symbols: a *b*, three *a*'s, and two *n*'s. Assuming that the *b* is the first symbol identified in the coding process, its 9 × 7 bitmap is stored in location 0 of the symbol dictionary. As Fig. 8.17(b) shows, the token identifying the *b* bitmap is 0. Thus, the first triplet in the encoded image's representation [see Fig. 8.17(c)] is (0, 2, 0)—indicating that the upper-left corner (an arbitrary convention) of the rectangular bitmap representing the *b* symbol is to be placed at location (0, 2) in the decoded image. After the bitmaps for the *a* and *n* symbols have been identified and added to the dictionary, the remainder of the image can be encoded with five additional triplets. As long as the six triplets required to locate the symbols in the image, together with the three bitmaps required to define them, are smaller than the original image, compression occurs. In this case, the starting image has 9 × 51 × 1 or 459 bits and, assuming that each triplet is composed of 3 bytes, the compressed representation has (6 × 3 × 8) + [(9 × 7) + (6 × 7) + (6 × 6)] or 285 bits; the resulting compression ratio $C = 1.61$. To decode the symbol-based representation in Fig. 8.17(c), you simply read the bitmaps of the symbols specified in the triplets from the symbol dictionary and place them at the spatial coordinates specified in each triplet.

Symbol-based compression was proposed in the early 1970s (Ascher and Nagy [1974]), but has become practical only recently. Advances in symbol matching algorithms (see Chapter 12) and increased CPU computer processing speeds have made it possible both to select dictionary symbols and to find where they occur in an image in a timely manner. And like many other compression methods, symbol-based decoding is significantly faster than encoding. Finally, we note that both the symbol bitmaps that are stored in the dictionary and the triplets used to reference them can themselves be encoded to further improve compression performance. If—as in Fig. 8.17—only exact symbol matches are allowed, the resulting compression is lossless; if small differences are permitted, some level of reconstruction error will be present.

a b c

FIGURE 8.17
(a) A bi-level document,
(b) symbol dictionary, and
(c) the triplets used to locate the symbols in the document.

Token	Symbol
0	
1	
2	

Triplet
(0, 2, 0)
(3, 10, 1)
(3, 18, 2)
(3, 26, 1)
(3, 34, 2)
(3, 42, 1)

JBIG2 compression

JBIG2 is an international standard for bilevel image compression. By segmenting an image into overlapping and/or non-overlapping regions of *text*, *halftone*, and *generic* content, compression techniques that are specifically optimized for each type of content are employed:

- *Text regions* are composed of characters that are ideally suited for a symbol-based coding approach. Typically, each symbol will correspond to a character bitmap—a subimage representing a character of text. There is normally only one character bitmap (or subimage) in the symbol dictionary for each upper- and lowercase character of the font being used. For example, there would be one "a" bitmap in the dictionary, one "A" bitmap, one "b" bitmap, and so on.

  In lossy JBIG2 compression, often called *perceptually lossless* or *visually lossless*, we neglect differences between dictionary bitmaps (i.e., the reference character bitmaps or character templates) and specific instances of the corresponding characters in the image. In lossless compression, the differences are stored and used in conjunction with the triplets encoding each character (by the decoder) to produce the actual image bitmaps. All bitmaps are encoded either arithmetically or using MMR (see Section 8.2.5); the triplets used to access dictionary entries are either arithmetically or Huffman encoded.

- *Halftone regions* are similar to text regions in that they are composed of patterns arranged in a regular grid. The symbols that are stored in the dictionary, however, are not character bitmaps but periodic patterns that represent intensities (e.g., of a photograph) that have been dithered to produce bilevel images for printing.

- *Generic regions* contain non-text, non-halftone information, like line art and noise, and are compressed using either arithmetic or MMR coding.

As is true of many image compression standards, JBIG2 defines decoder behavior. It does not explicitly define a standard encoder, but is flexible enough to allow various encoder designs. Although the design of the encoder is left unspecified, it is nevertheless important, because it determines the level of compression that is achieved. After all, the encoder must segment the image into regions, choose the text and halftone symbols that are stored in the dictionaries, and decide when those symbols are essentially the same as, or different from, potential instances of the symbols in the image. The decoder simply uses that information to recreate the original image.

■ Consider again the bilevel image in Fig. 8.16(a). Figure 8.18(a) shows a reconstructed section of the image after lossless JBIG2 encoding (by a commercially available document compression application). It is an exact replica of the original image. Note that the *d*s in the reconstructed text vary slightly, despite the fact that they were generated from the same *d* entry in the dictionary. The differences between that *d* and the *d*s in the image were used to refine the output of the dictionary. The standard defines an algorithm for accomplishing

EXAMPLE 8.11:
JBIG2 compression example.

a b c

FIGURE 8.18
JBIG2
compression
comparison:
(a) lossless
compression and
reconstruction;
(b) perceptually
lossless; and
(c) the scaled
difference
between the two.

...images of size   ...images of size
just described   just described
esulting coeffic   esulting coeffic
nt arrays.   nt arrays.
retained coeffi   retained coeffi
n we disregar   n we disregar

this during the decoding of the encoded dictionary bitmaps. For the purposes of our discussion, you can think of it as adding the difference between a dictionary bitmap and a specific instance of the corresponding character in the image to the bitmap read from the dictionary.

Figure 8.18(b) is another reconstruction of the area in (a) after perceptually lossless JBIG2 compression. Note that the ds in this figure are identical. They have been copied directly from the symbol dictionary. The reconstruction is called perceptually lossless because the text is readable and the font is even the same. The small differences—shown in Fig. 8.18(c)—between the ds in the original image and the d in the dictionary are considered unimportant because they do not affect readability. Remember that we are dealing with bilevel images, so there are only three intensities in Fig. 8.18(c). Intensity 128 indicates areas where there is no difference between the corresponding pixels of the images in Figs. 8.18(a) and (b); intensities 0 (black) and 255 (white) indicate pixels of opposite intensities in the two images—for example, a black pixel in one image that is white in the other, and vice versa.

The lossless JBIG2 compression that was used to generate Fig. 8.18(a) reduces the original 663,445 byte uncompressed PDF image to 32,705 bytes; the compression ratio is $C = 20.3$. Perceptually lossless JBIG2 compression reduces the image to 23,913 bytes, increasing the compression ratio to about 27.7. These compressions are 4 to 5 times greater than the CCITT Group 3 and 4 results from Example 8.10.   ■

8.2.7 Bit-Plane Coding

With reference to
Tables 8.3 and 8.4,
bit-plane coding is used
in the

• JBIG1
• JPEG-2000

compression standards.

The run-length and symbol-based techniques of the previous sections can be applied to images with more than two intensities by processing their bit planes individually. The technique, called *bit-plane coding*, is based on the concept of decomposing a multilevel (monochrome or color) image into a series of binary images (see Section 3.2.4) and compressing each binary image via one of several well-known binary compression methods. In this section, we describe the two most popular decomposition approaches.

The intensities of an m-bit monochrome image can be represented in the form of the base-2 polynomial

$$a_{m-1}2^{m-1} + a_{m-2}2^{m-2} + \ldots + a_1 2^1 + a_0 2^0 \qquad (8.2\text{-}8)$$

Based on this property, a simple method of decomposing the image into a collection of binary images is to separate the m coefficients of the polynomial into m 1-bit bit planes. As noted in Section 3.2.4, the lowest order bit plane (the plane corresponding to the least significant bit) is generated by collecting the a_0 bits of each pixel, while the highest order bit plane contains the a_{m-1} bits or coefficients. In general, each bit plane is constructed by setting its pixels equal to the values of the appropriate bits or polynomial coefficients from each pixel in the original image. The inherent disadvantage of this decomposition approach is that small changes in intensity can have a significant impact on the complexity of the bit planes. If a pixel of intensity 127 (01111111) is adjacent to a pixel of intensity 128 (10000000), for instance, every bit plane will contain a corresponding 0 to 1 (or 1 to 0) transition. For example, because the most significant bits of the binary codes for 127 and 128 are different, the highest bit plane will contain a zero-valued pixel next to a pixel of value 1, creating a 0 to 1 (or 1 to 0) transition at that point.

An alternative decomposition approach (which reduces the effect of small intensity variations) is to first represent the image by an m-bit *Gray code*. The m-bit Gray code $g_{m-1} \ldots g_2 g_1 g_0$ that corresponds to the polynomial in Eq. (8.2-8) can be computed from

$$g_i = a_i \oplus a_{i+1} \quad 0 \leq i \leq m - 2$$
$$g_{m-1} = a_{m-1}$$

$$(8.2\text{-}9)$$

Here, $\oplus$ denotes the exclusive OR operation. This code has the unique property that successive code words differ in only one bit position. Thus, small changes in intensity are less likely to affect all m bit planes. For instance, when intensity levels 127 and 128 are adjacent, only the highest order bit plane will contain a 0 to 1 transition, because the Gray codes that correspond to 127 and 128 are 01000000 and 11000000, respectively.

▪ Figures 8.19 and 8.20 show the eight binary and Gray-coded bit planes of the 8-bit monochrome image of the child in Fig. 8.19(a). Note that the high-order bit planes are far less complex than their low-order counterparts. That is, they contain large uniform areas of significantly less detail, busyness, or randomness. In addition, the Gray-coded bit planes are less complex than the corresponding binary bit planes. Both observations are reflected in the JBIG2 coding results of Table 8.10. Note, for instance, that the a_5 and g_5 results are

EXAMPLE 8.12:
Bit-plane coding.

Coefficient m	Binary Code (PDF bits)	Gray Code (PDF bits)	Compression Ratio
7	6,999	6,999	1.00
6	12,791	11,024	1.16
5	40,104	36,914	1.09
4	55,911	47,415	1.18
3	78,915	67,787	1.16
2	101,535	92,630	1.10
1	107,909	105,286	1.03
0	99,753	107,909	0.92

TABLE 8.10
JBIG2 lossless coding results for the binary and Gray-coded bit planes of Fig. 8.19(a). These results include the overhead of each bit plane's PDF representation.

FIGURE 8.19
(a) A 256-bit
monochrome
image. (b)–(h)
The four most
significant binary
and Gray-coded
bit planes of the
image in (a).

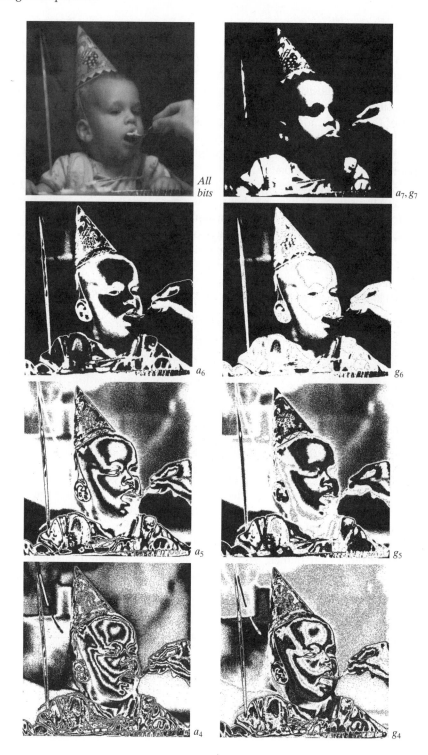

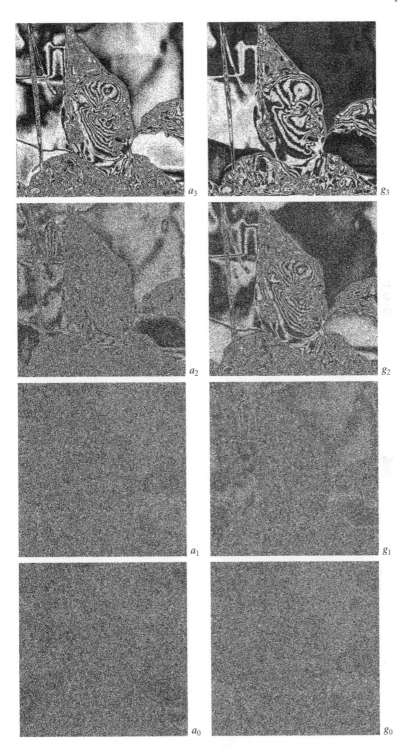

a b
c d
e f
g h

FIGURE 8.20
(a)–(h) The four
least significant
binary (left
column) and
Gray-coded
(right column)
bit planes of
the image in
Fig. 8.19(a).

significantly larger than the a_6 and g_6 compressions; and that both g_5 and g_6 are smaller than their a_5 and a_6 counterparts. This trend continues throughout the table, with the single exception of a_0. Gray-coding provides a compression advantage of about 1.06:1 on average. Combined together, the Gray-coded files compress the original monochrome image by 678,676/475,964 or 1.43:1; the non-Gray-coded files compress the image by 678,676/503,916 or 1.35:1.

Finally, we note that the two least significant bits in Fig. 8.20 have little apparent structure. Because this is typical of most 8-bit monochrome images, bit-plane coding is usually restricted to images of 6 bits/pixel or less. JBIG1, the predecessor to JBIG2, imposes such a limit. ■

8.2.8 Block Transform Coding

With reference to Tables 8.3 and 8.4, block transform coding is used in

- JPEG
- M-JPEG
- MPEG-1, 2, 4
- H.261, H.262, H.263, and H.264
- DV and HDV
- VC-1

and other compression standards.

In this section, we consider a compression technique that divides an image into small non-overlapping blocks of equal size (e.g., 8×8) and processes the blocks independently using a 2-D transform. In *block transform coding*, a reversible, linear transform (such as the Fourier transform) is used to map each *block* or *subimage* into a set of transform coefficients, which are then quantized and coded. For most images, a significant number of the coefficients have small magnitudes and can be coarsely quantized (or discarded entirely) with little image distortion. A variety of transformations, including the discrete Fourier transform (DFT) of Chapter 4, can be used to transform the image data.

Figure 8.21 shows a typical block transform coding system. The decoder implements the inverse sequence of steps (with the exception of the quantization function) of the encoder, which performs four relatively straightforward operations: subimage decomposition, transformation, quantization, and coding. An $M \times N$ input image is subdivided first into subimages of size $n \times n$, which are then transformed to generate MN/n^2 subimage transform arrays, each of size $n \times n$. The goal of the transformation process is to decorrelate the pixels of each subimage, or to pack as much information as possible into the smallest number of transform coefficients. The quantization stage then selectively eliminates or more coarsely quantizes the coefficients that carry the least amount of information in a predefined sense (several methods are discussed later in the section). These coefficients have the smallest impact on reconstructed subimage quality. The encoding process terminates by coding (normally using a variable-length code) the quantized coefficients. Any or all of the transform encoding steps can be adapted to local image content, called *adaptive transform coding*, or fixed for all subimages, called *nonadaptive transform coding*.

In this section, we restrict our attention to square subimages (the most commonly used). It is assumed that the input image is padded, if necessary, so that both M and N are multiples of n.

a
b

FIGURE 8.21
A block transform coding system:
(a) encoder;
(b) decoder.

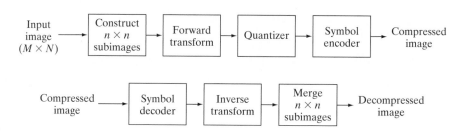

Transform selection

Block transform coding systems based on a variety of discrete 2-D transforms have been constructed and/or studied extensively. The choice of a particular transform in a given application depends on the amount of reconstruction error that can be tolerated and the computational resources available. Compression is achieved during the quantization of the transformed coefficients (not during the transformation step).

With reference to the discussion in Section 2.6.7, consider a subimage $g(x, y)$ of size $n \times n$ whose forward, discrete transform, $T(u, v)$, can be expressed in terms of the general relation

$$T(u, v) = \sum_{x=0}^{n-1} \sum_{y=0}^{n-1} g(x, y) r(x, y, u, v) \tag{8.2-10}$$

We use $g(x, y)$ to differentiate a subimage from the input image $f(x, y)$. Thus, the summation limits become n rather than M and N.

for $u, v = 0, 1, 2, \ldots, n - 1$. Given $T(u, v)$, $g(x, y)$ similarly can be obtained using the generalized inverse discrete transform

$$g(x, y) = \sum_{u=0}^{n-1} \sum_{v=0}^{n-1} T(u, v) s(x, y, u, v) \tag{8.2-11}$$

for $x, y = 0, 1, 2, \ldots, n - 1$. In these equations, $r(x, y, u, v)$ and $s(x, y, u, v)$ are called the *forward* and *inverse transformation kernels*, respectively. For reasons that will become clear later in the section, they also are referred to as *basis functions* or *basis images*. The $T(u, v)$ for $u, v = 0, 1, 2, \ldots, n - 1$ in Eq. (8.2-10) are called *transform coefficients*; they can be viewed as the expansion coefficients—see Section 7.2.1—of a series expansion of $g(x, y)$ with respect to basis functions $s(x, y, u, v)$.

As explained in Section 2.6.7, the kernel in Eq. (8.2-10) is separable if

$$r(x, y, u, v) = r_1(x, u) r_2(y, v) \tag{8.2-12}$$

In addition, the kernel is symmetric if r_1 is functionally equal to r_2. In this case, Eq. (8.2-12) can be expressed in the form

$$r(x, y, u, v) = r_1(x, u) r_1(y, v) \tag{8.2-13}$$

Identical comments apply to the inverse kernel if $r(x, y, u, v)$ is replaced by $s(x, y, u, v)$ in Eqs. (8.2-12) and (8.2-13). It is not difficult to show that a 2-D transform with a separable kernel can be computed using row-column or column-row passes of the corresponding 1-D transform, in the manner explained in Section 4.11.1.

The forward and inverse transformation kernels in Eqs. (8.2-10) and (8.2-11) determine the type of transform that is computed and the overall computational complexity and reconstruction error of the block transform coding system in which they are employed. The best known transformation kernel pair is

$$r(x, y, u, v) = e^{-j2\pi(ux+vy)/n} \tag{8.2-14}$$

and

$$s(x, y, u, v) = \frac{1}{n^2} e^{j2\pi(ux+vy)/n} \tag{8.2-15}$$

where $j = \sqrt{-1}$. These are the transformation kernels defined in Eqs. (2.6-34) and (2.6-35) of Chapter 2 with $M = N = n$. Substituting these kernels into Eqs. (8.2-10) and (8.2-11) yields a simplified version of the discrete Fourier transform pair introduced in Section 4.5.5.

A computationally simpler transformation that is also useful in transform coding, called the *Walsh-Hadamard transform* (WHT), is derived from the functionally identical kernels

To compute the WHT of an $N \times N$ input image $f(x, y)$, rather than a subimage, change n to N in Eq. (8.2-16).

$$r(x, y, u, v) = s(x, y, u, v) = \frac{1}{n}(-1)^{\sum_{i=0}^{m-1} \lfloor b_i(x)p_i(u)+b_i(y)p_i(v)\rfloor} \tag{8.2-16}$$

where $n = 2^m$. The summation in the exponent of this expression is performed in modulo 2 arithmetic and $b_k(z)$ is the kth bit (from right to left) in the binary representation of z. If $m = 3$ and $z = 6$ (110 in binary), for example, $b_0(z) = 0$, $b_1(z) = 1$, and $b_2(z) = 1$. The $p_i(u)$ in Eq. (8.2-16) are computed using:

$$
\begin{aligned}
p_0(u) &= b_{m-1}(u) \\
p_1(u) &= b_{m-1}(u) + b_{m-2}(u) \\
p_2(u) &= b_{m-2}(u) + b_{m-3}(u) \\
&\vdots \\
p_{m-1}(u) &= b_1(u) + b_0(u)
\end{aligned}
\tag{8.2-17}
$$

where the sums, as noted previously, are performed in modulo 2 arithmetic. Similar expressions apply to $p_i(v)$.

Unlike the kernels of the DFT, which are sums of sines and cosines [see Eqs. (8.2-14) and (8.2-15)], the Walsh-Hadamard kernels consist of alternating plus and minus 1s arranged in a checkerboard pattern. Figure 8.22 shows the kernel for $n = 4$. Each block consists of $4 \times 4 = 16$ elements (subsquares).

FIGURE 8.22
Walsh-Hadamard basis functions for $n = 4$. The origin of each block is at its top left.

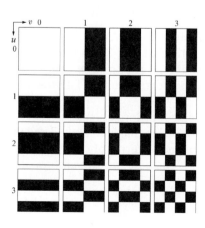

White denotes $+1$ and black denotes -1. To obtain the top left block, we let $u = v = 0$ and plot values of $r(x, y, 0, 0)$ for $x, y = 0, 1, 2, 3$. All values in this case are $+1$. The second block on the top row is a plot of values of $r(x, y, 0, 1)$ for $x, y = 0, 1, 2, 3$, and so on. As already noted, the importance of the Walsh-Hadamard transform is its simplicity of implementation—all kernel values are $+1$ or -1.

One of the transformations used most frequently for image compression is the *discrete cosine transform* (DCT). It is obtained by substituting the following (equal) kernels into Eqs. (8.2-10) and (8.2-11)

$$r(x, y, u, v) = s(x, y, u, v)$$

$$= \alpha(u)\alpha(v) \cos\left[\frac{(2x + 1)u\pi}{2n}\right] \cos\left[\frac{(2y + 1)v\pi}{2n}\right] \quad (8.2\text{-}18)$$

where

$$\alpha(u) = \begin{cases} \sqrt{\dfrac{1}{n}} & \text{for } u = 0 \\[2mm] \sqrt{\dfrac{2}{n}} & \text{for } u = 1, 2, \ldots, n - 1 \end{cases} \quad (8.2\text{-}19)$$

To compute the DCT of an $N \times N$ input image $f(x, y)$, rather than a subimage, change n to N in Eqs. (8.2-18) and (8.2-19).

and similarly for $\alpha(v)$. Figure 8.23 shows $r(x, y, u, v)$ for the case $n = 4$. The computation follows the same format as explained for Fig. 8.22, with the difference that the values of r are not integers. In Fig. 8.23, the lighter intensity values correspond to larger values of r.

■ Figures 8.24(a) through (c) show three approximations of the 512 × 512 monochrome image in Fig. 8.9(a). These pictures were obtained by dividing the original image into subimages of size 8 × 8, representing each subimage using one of the transforms just described (i.e., the DFT, WHT, or DCT transform), truncating 50% of the resulting coefficients, and taking the inverse transform of the truncated coefficient arrays.

EXAMPLE 8.13:
Block transform coding with the DFT, WHT, and DCT.

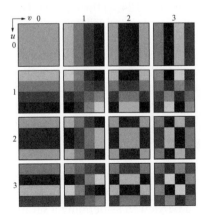

FIGURE 8.23
Discrete-cosine basis functions for $n = 4$. The origin of each block is at its top left.

a b c
d e f

FIGURE 8.24 Approximations of Fig. 8.9(a) using the (a) Fourier, (b) Walsh-Hadamard, and (c) cosine transforms, together with the corresponding scaled error images in (d)–(f).

In each case, the 32 retained coefficients were selected on the basis of maximum magnitude. Note that in all cases, the 32 discarded coefficients had little visual impact on the quality of the reconstructed image. Their elimination, however, was accompanied by some mean-square error, which can be seen in the scaled error images of Figs. 8.24(d) through (f). The actual rms errors were 2.32, 1.78, and 1.13 intensities, respectively. ■

The small differences in mean-square reconstruction error noted in the preceding example are related directly to the energy or information packing properties of the transforms employed. In accordance with Eq. (8.2-11), an $n \times n$ subimage $g(x, y)$ can be expressed as a function of its 2-D transform $T(u, v)$:

$$g(x, y) = \sum_{u=0}^{n-1}\sum_{v=0}^{n-1} T(u, v)\, s(x, y, u, v) \qquad (8.2\text{-}20)$$

for $x, y = 0, 1, 2, \ldots, n - 1$. Because the inverse kernel $s(x, y, u, v)$ in Eq. (8.2-20) depends only on the indices x, y, u, v, and not on the values of $g(x, y)$ or $T(u, v)$, it can be viewed as defining a set of *basis functions* or *basis images* for the series defined by Eq. (8.2-20). This interpretation becomes clearer if the notation used in Eq. (8.2-20) is modified to obtain

$$\mathbf{G} = \sum_{u=0}^{n-1}\sum_{v=0}^{n-1} T(u, v)\,\mathbf{S}_{uv} \tag{8.2-21}$$

where $\mathbf{G}$ is an $n \times n$ matrix containing the pixels of $g(x, y)$ and

$$\mathbf{S}_{uv} = \begin{bmatrix} s(0, 0, u, v) & s(0, 1, u, v) & \cdots & s(0, n-1, u, v) \\ s(1, 0, u, v) & & \cdots & \vdots \\ \vdots & \vdots & \cdots & \vdots \\ \vdots & \vdots & \cdots & \vdots \\ \vdots & & & \\ s(n-1, 0, u, v) & s(n-1, 1, u, v) & \cdots & s(n-1, n-1, u, v) \end{bmatrix} \tag{8.2-22}$$

Then $\mathbf{G}$, the matrix containing the pixels of the input subimage, is explicitly defined as a linear combination of n^2 matrices of size $n \times n$; that is, the $\mathbf{S}_{uv}$ for $u, v = 0, 1, 2, \ldots, n - 1$ in Eq. (8.2-22). These matrices in fact are the basis images (or functions) of the series expansion in Eq. (8.2-20); the associated $T(u, v)$ are the expansion coefficients. Figures 8.22 and 8.23 illustrate graphically the WHT and DCT basis images for the case of $n = 4$.

If we now define a transform coefficient *masking function*

$$\chi(u, v) = \begin{cases} 0 & \text{if } T(u, v) \text{ satisfies a specified truncation criterion} \\ 1 & \text{otherwise} \end{cases} \tag{8.2-23}$$

for $u, v = 0, 1, 2, \ldots, n - 1$, an approximation of $\mathbf{G}$ can be obtained from the truncated expansion

$$\hat{\mathbf{G}} = \sum_{u=0}^{n-1}\sum_{v=0}^{n-1} \chi(u, v)\,T(u, v)\,\mathbf{S}_{uv} \tag{8.2-24}$$

where $\chi(u, v)$ is constructed to eliminate the basis images that make the smallest contribution to the total sum in Eq. (8.2-21). The mean-square error between subimage $\mathbf{G}$ and approximation $\hat{\mathbf{G}}$ then is

$$\begin{aligned}
e_{ms} &= E\left\{\|\mathbf{G} - \hat{\mathbf{G}}\|^2\right\} \\
&= E\left\{\left\|\sum_{u=0}^{n-1}\sum_{v=0}^{n-1} T(u, v)\mathbf{S}_{uv} - \sum_{u=0}^{n-1}\sum_{v=0}^{n-1} \chi(u, v)T(u, v)\mathbf{S}_{uv}\right\|^2\right\} \\
&= E\left\{\left\|\sum_{u=0}^{n-1}\sum_{v=0}^{n-1} T(u, v)\mathbf{S}_{uv}\bigl[1 - \chi(u, v)\bigr]\right\|^2\right\} \\
&= \sum_{u=0}^{n-1}\sum_{v=0}^{n-1} \sigma^2_{T(u,v)}\bigl[1 - \chi(u, v)\bigr]
\end{aligned} \tag{8.2-25}$$

where $\|\mathbf{G} - \hat{\mathbf{G}}\|$ is the norm of matrix $\left(\mathbf{G} - \hat{\mathbf{G}}\right)$ and $\sigma^2_{T(u,v)}$ is the variance of the coefficient at transform location (u, v). The final simplification is based on the orthonormal nature of the basis images and the assumption that the pixels of $\mathbf{G}$ are generated by a random process with zero mean and known covariance. The total mean-square approximation error thus is the sum of the variances of the discarded transform coefficients; that is, the coefficients for which $\chi(u, v) = 0$, so that $\left[1 - \chi(u, v)\right]$ in Eq. (8.2-25) is 1. Transformations that redistribute or pack the most information into the fewest coefficients provide the best subimage approximations and, consequently, the smallest reconstruction errors. Finally, under the assumptions that led to Eq. (8.2-25), the mean-square error of the MN/n^2 subimages of an $M \times N$ image are identical. Thus the mean-square error (being a measure of *average* error) of the $M \times N$ image equals the mean-square error of a single subimage.

In Example 8.13, 50% of a DFT, WHT, and DCT block transform coded image's coefficients were discarded (using 8×8 blocks). After decoding, the DCT-based result had the smallest rms error, indicating that with respect to rms error the least amount of information was discarded.

The earlier example showed that the information packing ability of the DCT is superior to that of the DFT and WHT. Although this condition usually holds for most images, the Karhunen-Loève transform (see Chapter 11), not the DCT, is the optimal transform in an information packing sense. This is due to the fact that the KLT minimizes the mean-square error in Eq. (8.2-25) for any input image and any number of retained coefficients (Kramer and Mathews [1956]).[†] However, because the KLT is data dependent, obtaining the KLT basis images for each subimage, in general, is a nontrivial computational task. For this reason, the KLT is used infrequently in practice for image compression. Instead, a transform, such as the DFT, WHT, or DCT, whose basis images are fixed (input independent), normally is used. Of the possible input independent transforms, the nonsinusoidal transforms (such as the WHT transform) are the simplest to implement. The sinusoidal transforms (such as the DFT or DCT) more closely approximate the information packing ability of the optimal KLT.

Hence, most transform coding systems are based on the DCT, which provides a good compromise between information packing ability and computational complexity. In fact, the properties of the DCT have proved to be of such practical value that the DCT has become an international standard for transform coding systems. Compared to the other input independent transforms, it has the advantages of having been implemented in a single integrated circuit, packing the most information into the fewest coefficients[‡] (for most images), and minimizing the block-like appearance, called *blocking artifact*, that results when the boundaries between subimages become visible. This last property is particularly important in comparisons with the other sinusoidal transforms. As Fig. 8.25(a) shows, the implicit n-point periodicity (see Section 4.6.3) of the DFT gives rise to boundary discontinuities that result in substantial high-frequency transform

---

[†]An additional condition for optimality is that the masking function of Eq. (8.2-23) selects the KLT coefficients of maximum variance.

[‡]Ahmed et al. [1974] first noticed that the KLT basis images of a first-order Markov image source closely resemble the DCT's basis images. As the correlation between adjacent pixels approaches one, the input dependent KLT basis images become identical to the input independent DCT basis images (Clarke [1985]).

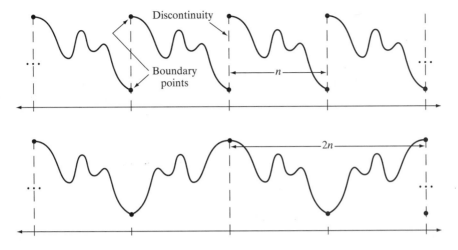

FIGURE 8.25 The periodicity implicit in the 1-D (a) DFT and (b) DCT.

content. When the DFT transform coefficients are truncated or quantized, the Gibbs phenomenon[†] causes the boundary points to take on erroneous values, which appear in an image as blocking artifact. That is, the boundaries between adjacent subimages become visible because the boundary pixels of the subimages assume the mean values of discontinuities formed at the boundary points [see Fig. 8.25(a)]. The DCT of Fig. 8.25(b) reduces this effect, because its implicit $2n$-point periodicity does not inherently produce boundary discontinuities.

Subimage size selection

Another significant factor affecting transform coding error and computational complexity is subimage size. In most applications, images are subdivided so that the correlation (redundancy) between adjacent subimages is reduced to some acceptable level and so that n is an integer power of 2 where, as before, n is the subimage dimension. The latter condition simplifies the computation of the subimage transforms (see the base-2 successive doubling method discussed in Section 4.11.3). In general, both the level of compression and computational complexity increase as the subimage size increases. The most popular subimage sizes are 8×8 and 16×16.

■ Figure 8.26 illustrates graphically the impact of subimage size on transform coding reconstruction error. The data plotted were obtained by dividing the monochrome image of Fig. 8.9(a) into subimages of size $n \times n$, for $n = 2, 4, 8, 16, \ldots, 256, 512$, computing the transform of each subimage, truncating 75% of the resulting coefficients, and taking the inverse transform of the truncated arrays. Note that the Hadamard and cosine curves flatten as the size of the subimage becomes greater than 8×8, whereas the Fourier reconstruction

EXAMPLE 8.14:
Effects of subimage size on transform coding.

[†]This phenomenon, described in most electrical engineering texts on circuit analysis, occurs because the Fourier transform fails to converge uniformly at discontinuities. At discontinuities, Fourier expansions take the mean values of the points of discontinuity.

FIGURE 8.26
Reconstruction
error versus
subimage size.

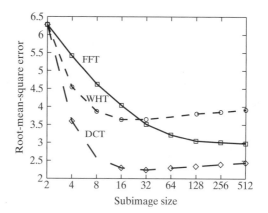

error continues to decrease in this region. As *n* further increases, the Fourier reconstruction error crosses the Walsh-Hadamard curve and approaches the cosine result. This result is consistent with the theoretical and experimental findings reported by Netravali and Limb [1980] and by Pratt [1991] for a 2-D Markov image source.

All three curves intersect when 2×2 subimages are used. In this case, only one of the four coefficients (25%) of each transformed array was retained. The coefficient in all cases was the dc component, so the inverse transform simply replaced the four subimage pixels by their average value [see Eq. (4.6-21)]. This condition is evident in Fig. 8.27(b), which shows a zoomed portion of the 2×2 DCT result. Note that the blocking artifact that is prevalent in this result decreases as the subimage size increases to 4×4 and 8×8 in Figs. 8.27(c) and (d). Figure 8.27(a) shows a zoomed portion of the original image for reference. ■

Bit allocation

The reconstruction error associated with the truncated series expansion of Eq. (8.2-24) is a function of the number and relative importance of the

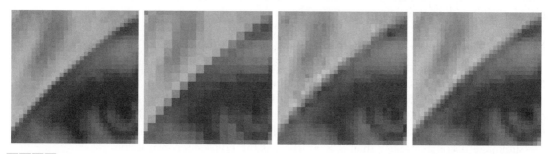

a b c d

FIGURE 8.27 Approximations of Fig. 8.27(a) using 25% of the DCT coefficients and (b) 2×2 subimages, (c) 4×4 subimages, and (d) 8×8 subimages. The original image in (a) is a zoomed section of Fig. 8.9(a).

transform coefficients that are discarded, as well as the precision that is used to represent the retained coefficients. In most transform coding systems, the retained coefficients are selected [that is, the masking function of Eq. (8.2-23) is constructed] on the basis of maximum variance, called *zonal coding*, or on the basis of maximum magnitude, called *threshold coding*. The overall process of truncating, quantizing, and coding the coefficients of a transformed subimage is commonly called *bit allocation*.

■ Figures 8.28(a) and (c) show two approximations of Fig. 8.9(a) in which 87.5% of the DCT coefficients of each 8×8 subimage were discarded. The first result was obtained via threshold coding by keeping the eight largest transform coefficients, and the second image was generated by using a zonal coding approach. In the latter case, each DCT coefficient was considered a random variable whose distribution could be computed over the ensemble of all transformed subimages. The 8 distributions of largest variance (12.5% of the 64 coefficients in the transformed 8×8 subimage) were located and used to determine the coordinates, u and v, of the coefficients, $T(u, v)$, that were retained for all subimages. Note that the threshold coding difference image of Fig. 8.28(b) contains less error than the zonal coding result in Fig. 8.28(d). Both images have been scaled to make the errors more visible. The corresponding rms errors are 4.5 and 6.5 intensities, respectively.   ■

EXAMPLE 8.15:
Bit allocation.

a b
c d

FIGURE 8.28
Approximations of Fig. 8.9(a) using 12.5% of the 8×8 DCT coefficients: (a)—(b) threshold coding results; (c)—(d) zonal coding results. The difference images are scaled by 4.

Zonal coding implementation Zonal coding is based on the information theory concept of viewing information as uncertainty. Therefore the transform coefficients of maximum variance carry the most image information and should be retained in the coding process. The variances themselves can be calculated directly from the ensemble of MN/n^2 transformed subimage arrays, as in the preceding example, or based on an assumed image model (say, a Markov autocorrelation function). In either case, the zonal sampling process can be viewed, in accordance with Eq. (8.2-24), as multiplying each $T(u, v)$ by the corresponding element in a *zonal mask*, which is constructed by placing a 1 in the locations of maximum variance and a 0 in all other locations. Coefficients of maximum variance usually are located around the origin of an image transform, resulting in the typical zonal mask shown in Fig. 8.29(a).

The coefficients retained during the zonal sampling process must be quantized and coded, so zonal masks are sometimes depicted showing the number of bits used to code each coefficient [Fig. 8.29(b)]. In most cases, the coefficients are allocated the same number of bits, or some fixed number of bits is distributed among them unequally. In the first case, the coefficients generally are normalized by their standard deviations and uniformly quantized. In the second case, a quantizer, such as an optimal Lloyd-Max quantizer (see Optimal quantizers in Section 8.2.9), is designed for each coefficient. To construct the required quantizers, the zeroth or dc coefficient normally is modeled by a Rayleigh density function, whereas the remaining coefficients are modeled by a Laplacian or

a b
c d

FIGURE 8.29
A typical
(a) zonal mask,
(b) zonal bit allocation,
(c) threshold mask, and
(d) thresholded coefficient ordering sequence. Shading highlights the coefficients that are retained.

1	1	1	1	1	0	0	0
1	1	1	1	0	0	0	0
1	1	1	0	0	0	0	0
1	1	0	0	0	0	0	0
1	0	0	0	0	0	0	0
0	0	0	0	0	0	0	0
0	0	0	0	0	0	0	0
0	0	0	0	0	0	0	0

8	7	6	4	3	2	1	0
7	6	5	4	3	2	1	0
6	5	4	3	3	1	1	0
4	4	3	3	2	1	0	0
3	3	3	2	1	1	0	0
2	2	1	1	1	0	0	0
1	1	1	0	0	0	0	0
0	0	0	0	0	0	0	0

1	1	0	1	1	0	0	0
1	1	1	1	0	0	0	0
1	1	0	0	0	0	0	0
1	0	0	0	0	0	0	0
0	0	0	0	0	0	0	0
0	1	0	0	0	0	0	0
0	0	0	0	0	0	0	0
0	0	0	0	0	0	0	0

0	1	5	6	14	15	27	28
2	4	7	13	16	26	29	42
3	8	12	17	25	30	41	43
9	11	18	24	31	40	44	53
10	19	23	32	39	45	52	54
20	22	33	38	46	51	55	60
21	34	37	47	50	56	59	61
35	36	48	49	57	58	62	63

Gaussian density.[†] The number of quantization levels (and thus the number of bits) allotted to each quantizer is made proportional to $\log_2 \sigma^2_{T(u, v)}$. Thus the retained coefficients in Eq. (8.2-24)—which (in the context of the current discussion) are selected on the basis of maximum variance—are assigned bits in proportion to the logarithm of the coefficient variances.

Threshold coding implementation   Zonal coding usually is implemented by using a single fixed mask for all subimages. Threshold coding, however, is inherently adaptive in the sense that the location of the transform coefficients retained for each subimage vary from one subimage to another. In fact, threshold coding is the adaptive transform coding approach most often used in practice because of its computational simplicity. The underlying concept is that, for any subimage, the transform coefficients of largest magnitude make the most significant contribution to reconstructed subimage quality, as demonstrated in the last example. Because the locations of the maximum coefficients vary from one subimage to another, the elements of $X(u, v)T(u, v)$ normally are reordered (in a predefined manner) to form a 1-D, run-length coded sequence. Figure 8.29(c) shows a typical *threshold mask* for one subimage of a hypothetical image. This mask provides a convenient way to visualize the threshold coding process for the corresponding subimage, as well as to mathematically describe the process using Eq. (8.2-24). When the mask is applied [via Eq. (8.2-24)] to the subimage for which it was derived, and the resulting $n \times n$ array is reordered to form an n^2-element coefficient sequence in accordance with the zigzag ordering pattern of Fig. 8.29(d), the reordered 1-D sequence contains several long runs of 0s [the zigzag pattern becomes evident by starting at 0 in Fig. 8.29(d) and following the numbers in sequence]. These runs normally are run-length coded. The nonzero or retained coefficients, corresponding to the mask locations that contain a 1, are represented using a variable-length code.

There are three basic ways to threshold a transformed subimage or, stated differently, to create a subimage threshold masking function of the form given in Eq. (8.2-23): (1) A single global threshold can be applied to all subimages; (2) a different threshold can be used for each subimage; or (3) the threshold can be varied as a function of the location of each coefficient within the subimage. In the first approach, the level of compression differs from image to image, depending on the number of coefficients that exceed the global threshold. In the second, called *N-largest coding*, the same number of coefficients is discarded for each subimage. As a result, the code rate is constant and known in advance. The third technique, like the first, results in a variable code rate, but offers the advantage that thresholding *and* quantization can be combined

The N in "N-largest coding" is not an image dimension, but refers to the number of coefficients that are kept.

[†]As each coefficient is a linear combination of the pixels in its subimage [see Eq. (8.2-10)], the central-limit theorem suggests that, as subimage size increases, the coefficients tend to become Gaussian. This result does not apply to the dc coefficient, however, because nonnegative images always have positive dc coefficients.

by replacing $X(u, v)T(u, v)$ in Eq. (8.2-24) with

$$\hat{T}(u, v) = \text{round}\left[\frac{T(u, v)}{Z(u, v)}\right] \tag{8.2-26}$$

where $\hat{T}(u, v)$ is a thresholded and quantized approximation of $T(u, v)$, and $Z(u, v)$ is an element of the transform normalization array

$$\mathbf{Z} = \begin{bmatrix} Z(0, 0) & Z(0, 1) & \cdots & Z(0, n-1) \\ Z(1, 0) & \vdots & \cdots & \vdots \\ \vdots & \vdots & \cdots & \vdots \\ \vdots & \vdots & \cdots & \vdots \\ \vdots & \vdots & \cdots & \vdots \\ Z(n-1, 0) & Z(n-1, 1) & \cdots & Z(n-1, n-1) \end{bmatrix} \tag{8.2-27}$$

Before a normalized (thresholded and quantized) subimage transform, $\hat{T}(u, v)$, can be inverse transformed to obtain an approximation of subimage $g(x, y)$, it must be multiplied by $Z(u, v)$. The resulting denormalized array, denoted $\dot{T}(u, v)$ is an approximation of $\hat{T}(u, v)$:

$$\dot{T}(u, v) = \hat{T}(u, v)Z(u, v) \tag{8.2-28}$$

The inverse transform of $\dot{T}(u, v)$ yields the decompressed subimage approximation.

Figure 8.30(a) depicts Eq. (8.2-26) graphically for the case in which $Z(u, v)$ is assigned a particular value c. Note that $\hat{T}(u, v)$ assumes integer value k if and only if

$$kc - \frac{c}{2} \le T(u, v) < kc + \frac{c}{2} \tag{8.2-29}$$

If $Z(u, v) > 2T(u, v)$, then $\hat{T}(u, v) = 0$ and the transform coefficient is completely truncated or discarded. When $\hat{T}(u, v)$ is represented with a variable-length code that increases in length as the magnitude of k increases, the number of bits used to represent $T(u, v)$ is controlled by the value of c. Thus the elements of $\mathbf{Z}$

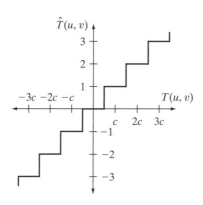

a b

FIGURE 8.30
(a) A threshold coding quantization curve [see Eq. (8.2-29)]. (b) A typical normalization matrix.

16	11	10	16	24	40	51	61
12	12	14	19	26	58	60	55
14	13	16	24	40	57	69	56
14	17	22	29	51	87	80	62
18	22	37	56	68	109	103	77
24	35	55	64	81	104	113	92
49	64	78	87	103	121	120	101
72	92	95	98	112	100	103	99

can be scaled to achieve a variety of compression levels. Figure 8.30(b) shows a typical normalization array. This array, which has been used extensively in the JPEG standardization efforts (see the next section), weighs each coefficient of a transformed subimage according to heuristically determined perceptual or psychovisual importance.

■ Figures 8.31(a) through (f) show six threshold-coded approximations of the monochrome image in Fig. 8.9(a). All images were generated using an 8×8 DCT and the normalization array of Fig. 8.30(b). The first result, which provides a compression ratio of about 12 to 1 (i.e., $C = 12$), was obtained by direct application of that normalization array. The remaining results, which compress the original image by 19, 30, 49, 85, and 182 to 1, were generated after multiplying (scaling) the normalization arrays by 2, 4, 8, 16, and 32, respectively. The corresponding rms errors are 3.83, 4.93, 6.62, 9.35, 13.94, and 22.46 intensity levels. ■

EXAMPLE 8.16:
Illustration of threshold coding.

JPEG

One of the most popular and comprehensive continuous tone, still frame compression standards is the JPEG standard. It defines three different coding systems: (1) a lossy baseline coding system, which is based on the DCT and is adequate for most compression applications; (2) an extended coding system for

a b c
d e f

FIGURE 8.31 Approximations of Fig. 8.9(a) using the DCT and normalization array of Fig. 8.30(b): (a) **Z**, (b) 2**Z**, (c) 4**Z**, (d) 8**Z**, (e) 16**Z**, and (f) 32**Z**.

greater compression, higher precision, or progressive reconstruction applications; and (3) a lossless independent coding system for reversible compression. To be JPEG compatible, a product or system must include support for the baseline system. No particular file format, spatial resolution, or color space model is specified.

In the baseline system, often called the *sequential baseline system*, the input and output data precision is limited to 8 bits, whereas the quantized DCT values are restricted to 11 bits. The compression itself is performed in three sequential steps: DCT computation, quantization, and variable-length code assignment. The image is first subdivided into pixel blocks of size 8×8, which are processed left to right, top to bottom. As each 8×8 block or subimage is encountered, its 64 pixels are level-shifted by subtracting the quantity 2^{k-1}, where 2^k is the maximum number of intensity levels. The 2-D discrete cosine transform of the block is then computed, quantized in accordance with Eq. (8.2-26), and reordered, using the zigzag pattern of Fig. 8.29(d), to form a 1-D sequence of quantized coefficients.

Because the one-dimensionally reordered array generated under the zigzag pattern of Fig. 8.29(d) is arranged qualitatively according to increasing spatial frequency, the JPEG coding procedure is designed to take advantage of the long runs of zeros that normally result from the reordering. In particular, the nonzero AC[†] coefficients are coded using a variable-length code that defines the coefficient values and number of preceding zeros. The DC coefficient is difference coded relative to the DC coefficient of the previous subimage. Tables A.3, A.4, and A.5 in Appendix A provide the default JPEG Huffman codes for the luminance component of a color image or intensity of a monochrome image. The JPEG recommended luminance quantization array is given in Fig. 8.30(b) and can be scaled to provide a variety of compression levels. The scaling of this array allows users to select the "quality" of JPEG compressions. Although default coding tables and quantization arrays are provided for both color and monochrome processing, the user is free to construct custom tables and/or arrays, which may in fact be adapted to the characteristics of the image(s) being compressed.

EXAMPLE 8.17:
JPEG baseline coding and decoding.

■ Consider compression and reconstruction of the following 8×8 subimage with the JPEG baseline standard:

52	55	61	66	70	61	64	73
63	59	66	90	109	85	69	72
62	59	68	113	144	104	66	73
63	58	71	122	154	106	70	69
67	61	68	104	126	88	68	70
79	65	60	70	77	63	58	75
85	71	64	59	55	61	65	83
87	79	69	68	65	76	78	94

[†]In the standard, the term AC denotes all transform coefficients with the exception of the zeroth or DC coefficient.

The original image consists of 256 or 2^8 possible intensities, so the coding process begins by level shifting the pixels of the original subimage by -2^7 or -128 intensity levels. The resulting shifted array is

-76	-73	-67	-62	-58	-67	-64	-55
-65	-69	-62	-38	-19	-43	-59	-56
-66	-69	-60	-15	16	-24	-62	-55
-65	-70	-57	-6	26	-22	-58	-59
-61	-67	-60	-24	-2	-40	-60	-58
-49	-63	-68	-58	-51	-65	-70	-53
-43	-57	-64	-69	-73	-67	-63	-45
-41	-49	-59	-60	-63	-52	-50	-34

which, when transformed in accordance with the forward DCT of Eqs. (8.2-10) and (8.2-18) for $n = 8$, becomes

-415	-29	-62	25	55	-20	-1	3
7	-21	-62	9	11	-7	-6	6
-46	8	77	-25	-30	10	7	-5
-50	13	35	-15	-9	6	0	3
11	-8	-13	-2	-1	1	-4	1
-10	1	3	-3	-1	0	2	-1
-4	-1	2	-1	2	-3	1	-2
-1	-1	-1	-2	-1	-1	0	-1

If the JPEG recommended normalization array of Fig. 8.30(b) is used to quantize the transformed array, the scaled and truncated [that is, normalized in accordance with Eq. (8.2-26)] coefficients are

-26	-3	-6	2	2	0	0	0
1	-2	-4	0	0	0	0	0
-3	1	5	-1	-1	0	0	0
-4	1	2	-1	0	0	0	0
1	0	0	0	0	0	0	0
0	0	0	0	0	0	0	0
0	0	0	0	0	0	0	0
0	0	0	0	0	0	0	0

where, for instance, the DC coefficient is computed as

$$\hat{T}(0, 0) = \text{round}\left[\frac{T(0, 0)}{Z(0, 0)}\right]$$

$$= \text{round}\left[\frac{-415}{16}\right] = -26$$

Note that the transformation and normalization process produces a large number of zero-valued coefficients. When the coefficients are reordered in accordance with the zigzag ordering pattern of Fig. 8.29(d), the resulting 1-D coefficient sequence is

$$[-26\ -3\ 1\ -3\ -2\ -6\ 2\ -4\ 1\ -4\ 1\ 1\ 5\ 0\ 2\ 0\ 0\ -1\ 2\ 0\ 0\ 0\ 0\ 0\ -1\ -1\ \text{EOB}]$$

where the EOB symbol denotes the end-of-block condition. A special EOB Huffman code word (see category 0 and run-length 0 in Table A.5) is provided to indicate that the remainder of the coefficients in a reordered sequence are zeros.

The construction of the default JPEG code for the reordered coefficient sequence begins with the computation of the difference between the current DC coefficient and that of the previously encoded subimage. Assuming the DC coefficient of the transformed and quantized subimage to its immediate left was 17, the resulting DPCM difference is $[-26 - (-17)]$ or -9, which lies in DC difference category 4 of Table A.3. In accordance with the default Huffman difference code of Table A.4, the proper base code for a category 4 difference is 101 (a 3-bit code), while the total length of a completely encoded category 4 coefficient is 7 bits. The remaining 4 bits must be generated from the least significant bits (LSBs) of the difference value. For a general DC difference category (say, category K), an additional K bits are needed and computed as either the K LSBs of the positive difference or the K LSBs of the negative difference minus 1. For a difference of -9, the appropriate LSBs are (0111) $-$ 1 or 0110, and the complete DPCM coded DC code word is 1010110.

The nonzero AC coefficients of the reordered array are coded similarly from Tables A.3 and A.5. The principal difference is that each default AC Huffman code word depends on the number of zero-valued coefficients preceding the nonzero coefficient to be coded, as well as the magnitude category of the nonzero coefficient. (See the column labeled Run/Category in Table A.5.) Thus the first nonzero AC coefficient of the reordered array (-3) is coded as 0100. The first 2 bits of this code indicate that the coefficient was in magnitude category 2 and preceded by no zero-valued coefficients (see Table A.3); the last 2 bits are generated by the same process used to arrive at the LSBs of the DC difference code. Continuing in this manner, the completely coded (reordered) array is

1010110 0100 001 0100 0101 100001 0110 100011 001 100011 001
001 100101 11100110 110110 0110 11110100 000 1010

where the spaces have been inserted solely for readability. Although it was not needed in this example, the default JPEG code contains a special code word for a run of 15 zeros followed by a zero (see category 0 and run-length F in Table A.5). The total number of bits in the completely coded reordered array (and thus the number of bits required to represent the entire 8×8, 8-bit subimage of this example) is 92. The resulting compression ratio is 512/92, or about 5.6:1.

To decompress a JPEG compressed subimage, the decoder must first recreate the normalized transform coefficients that led to the compressed bit stream. Because a Huffman-coded binary sequence is instantaneous and uniquely decodable, this step is easily accomplished in a simple lookup table manner.

Here the regenerated array of quantized coefficients is

−26	−3	−6	2	2	0	0	0
1	−2	−4	0	0	0	0	0
−3	1	5	−1	−1	0	0	0
−4	1	2	−1	0	0	0	0
1	0	0	0	0	0	0	0
0	0	0	0	0	0	0	0
0	0	0	0	0	0	0	0
0	0	0	0	0	0	0	0

After denormalization in accordance with Eq. (8.2-28), the array becomes

−416	−33	−60	32	48	0	0	0
12	−24	−56	0	0	0	0	0
−42	13	80	−24	−40	0	0	0
−56	17	44	−29	0	0	0	0
18	0	0	0	0	0	0	0
0	0	0	0	0	0	0	0
0	0	0	0	0	0	0	0
0	0	0	0	0	0	0	0

where, for example, the DC coefficient is computed as

$$\dot{T}(0, 0) = \hat{T}(0, 0)Z(0, 0) = (-26)(16) = -416$$

The completely reconstructed subimage is obtained by taking the inverse DCT of the denormalized array in accordance with Eqs. (8.2-11) and (8.2-18) to obtain

−70	−64	−61	−64	−69	−66	−58	−50
−72	−73	−61	−39	−30	−40	−54	−59
−68	−78	−58	−9	13	−12	−48	−64
−59	−77	−57	0	22	−13	−51	−60
−54	−75	−64	−23	−13	−44	−63	−56
−52	−71	−72	−54	−54	−71	−71	−54
−45	−59	−70	−68	−67	−67	−61	−50
−35	−47	−61	−66	−60	−48	−44	−44

and level shifting each inverse transformed pixel by 2^7 (or +128) to yield

58	64	67	64	59	62	70	78
56	55	67	89	98	88	74	69
60	50	70	119	141	116	80	64
69	51	71	128	149	115	77	68
74	53	64	105	115	84	65	72
76	57	56	74	75	57	57	74
83	69	59	60	61	61	67	78
93	81	67	62	69	80	84	84

Any differences between the original and reconstructed subimage are a result of the lossy nature of the JPEG compression and decompression process. In this example, the errors range from -14 to $+11$ and are distributed as follows:

-6	-9	-6	2	11	-1	-6	-5
7	4	-1	1	11	-3	-5	3
2	9	-2	-6	-3	-12	-14	9
-6	7	0	-4	-5	-9	-7	1
-7	8	4	-1	6	4	3	-2
3	8	4	-4	2	6	1	1
2	2	5	-1	-6	0	-2	5
-6	-2	2	6	-4	-4	-6	10

The root-mean-square error of the overall compression and reconstruction process is approximately 5.8 intensity levels. ■

EXAMPLE 8.18:
Illustration of
JPEG coding.

■ Figures 8.32(a) and (d) show two JPEG approximations of the monochrome image in Fig. 8.9(a). The first result provides a compression of 25:1; the second compresses the original image by 52:1. The differences between the original image and the reconstructed images in Figs. 8.30(a) and (d) are shown in Figs. 8.30(b) and (e), respectively. The corresponding rms errors are 5.4 and 10.7 intensities. The errors are clearly visible in the zoomed images in Figs. 8.32(c) and (f). These images show a magnified section of Figs. 8.32(a) and (d), respectively. Note that the JPEG blocking artifact increases with compression. ■

8.2.9 Predictive Coding

With reference to
Tables 8.3 and 8.4,
predictive coding is
used in

- JBIG2
- JPEG
- JPEG-LS
- MPEG-1,2,4
- H.261, H.262,
  H.263, and H.264
- HDV
- VC-1

and other compression
standards and file
formats.

We now turn to a simpler compression approach that achieves good compression without significant computational overhead *and* can be either error-free or lossy. The approach, commonly referred to as *predictive coding*, is based on eliminating the redundancies of closely spaced pixels—in space and/or time—by extracting and coding only the new information in each pixel. The *new information* of a pixel is defined as the difference between the actual and predicted value of the pixel.

Lossless predictive coding

Figure 8.33 shows the basic components of a *lossless predictive coding* system. The system consists of an encoder and a decoder, each containing an identical *predictor*. As successive samples of discrete time input signal, $f(n)$, are introduced to the encoder, the predictor generates the anticipated value of each sample based on a specified number of past samples. The output of the predictor is then rounded to the nearest integer, denoted $\hat{f}(n)$, and used to form the difference or *prediction error*

$$e(n) = f(n) - \hat{f}(n) \qquad (8.2\text{-}30)$$

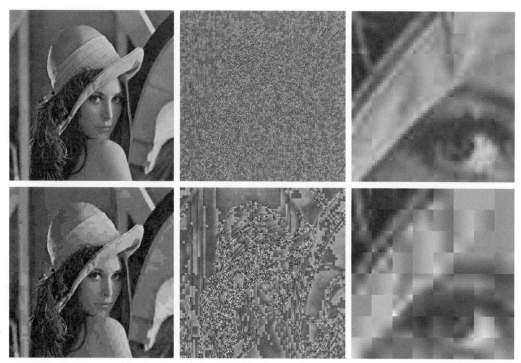

a b c
d e f

FIGURE 8.32 Two JPEG approximations of Fig. 8.9(a). Each row contains a result after compression and reconstruction, the scaled difference between the result and the original image, and a zoomed portion of the reconstructed image.

which is encoded using a variable-length code (by the symbol encoder) to generate the next element of the compressed data stream. The decoder in Fig. 8.33(b) reconstructs $e(n)$ from the received variable-length code words and performs the inverse operation

$$f(n) = e(n) + \hat{f}(n) \tag{8.2-31}$$

to decompress or recreate the original input sequence.

Various local, global, and adaptive methods (see the later subsection entitled Lossy predictive coding) can be used to generate $\hat{f}(n)$. In many cases, the prediction is formed as a linear combination of m previous samples. That is,

$$\hat{f}(n) = \text{round}\left[\sum_{i=1}^{m} \alpha_i f(n - i) \right] \tag{8.2-32}$$

where m is the *order* of the linear predictor, round is a function used to denote the rounding or nearest integer operation, and the α_i for $i = 1, 2, \ldots, m$ are prediction coefficients. If the input sequence in Fig. 8.33(a) is considered to be

a
b

FIGURE 8.33
A lossless
predictive coding
model:
(a) encoder;
(b) decoder.

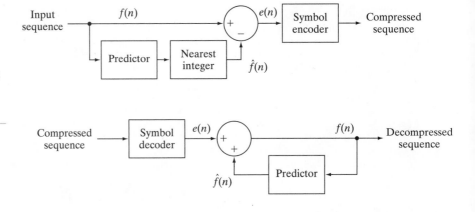

samples of an image, the $f(n)$ in Eqs. (8.2-30) through (8.2-32) are pixels—and the m samples used to predict the value of each pixel come from the current scan line (called 1-D linear predictive coding), from the current and previous scan lines (called 2-D linear predictive coding), or from the current image and previous images in a sequence of images (called 3-D linear predictive coding). Thus, for 1-D linear predictive image coding, Eq. (8.2-32) can be written as

$$\hat{f}(x, y) = \text{round}\left[\sum_{i=1}^{m}\alpha_i f(x, y - i)\right] \qquad (8.2\text{-}33)$$

where each sample is now expressed explicitly as a function of the input image's spatial coordinates, x and y. Note that Eq. (8.2-33) indicates that the 1-D linear prediction is a function of the previous pixels on the current line alone. In 2-D predictive coding, the prediction is a function of the previous pixels in a left-to-right, top-to-bottom scan of an image. In the 3-D case, it is based on these pixels and the previous pixels of preceding frames. Equation (8.2-33) cannot be evaluated for the first m pixels of each line, so those pixels must be coded by using other means (such as a Huffman code) and considered as an overhead of the predictive coding process. Similar comments apply to the higher-dimensional cases.

EXAMPLE 8.19:
Predictive coding
and spatial
redundancy.

■ Consider encoding the monochrome image of Fig. 8.34(a) using the simple first-order (i.e., $m = 1$) linear predictor from Eq. (8.2-33)

$$\hat{f}(x, y) = \text{round}\left[\alpha f(x, y - 1)\right] \qquad (8.2\text{-}34)$$

This equation is a simplification of Eq. (8.2-33) with $m = 1$ and the subscript of lone prediction coefficient α_1 dropped as unnecessary. A predictor of this general form is called a *previous pixel* predictor, and the corresponding predictive coding procedure is known as *differential coding* or *previous pixel coding*. Figure 8.34(c) shows the prediction error image, $e(x, y) = f(x, y) - \hat{f}(x, y)$, that results from Eq. (8.2-34) with $\alpha = 1$. The scaling of this image is such that intensity 128 represents a prediction error of zero, while all nonzero positive

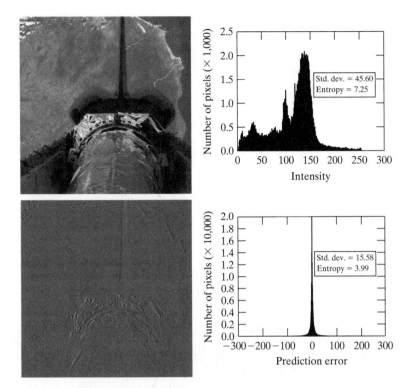

FIGURE 8.34
(a) A view of the
Earth from an
orbiting space
shuttle. (b) The
intensity
histogram of
(a). (c) The
prediction error
image resulting
from Eq. (8.2-34).
(d) A histogram
of the prediction
error.
(Original image
courtesy of
NASA.)

and negative prediction errors (under and over estimates) are displayed as lighter and darker shades of gray, respectively. The mean value of the prediction image is 128.26. Because intensity 128 corresponds to a prediction error of 0, the average prediction error is only 0.26 bits.

Figures 8.34(b) and (d) show the intensity histogram of the image in Fig. 8.34(a) and the histogram of prediction error $e(x, y)$, respectively. Note that the standard deviation of the prediction error in Fig. 8.34(d) is much smaller than the standard deviation of the intensities in the original image. Moreover, the entropy of the prediction error—as estimated using Eq. (8.1-7)—is significantly less than the estimated entropy of the original image (3.99 bits/pixel as opposed to 7.25 bits/pixel). This decrease in entropy reflects removal of a great deal of spatial redundancy, despite the fact that for k-bit images, $(k + 1)$-bit numbers are needed to represent accurately prediction error sequence $e(x, y)$. In general, the maximum compression of a predictive coding approach can be estimated by dividing the average number of bits used to represent each pixel in the original image by an estimate of the entropy of the prediction error. In this example, any variable-length coding procedure can be used to code $e(x, y)$, but the resulting compression will be limited to about 8/3.99 or 2:1. ■

Note that the variable-length encoded prediction error is the compressed image.

The preceding example illustrates that the compression achieved in predictive coding is related directly to the entropy reduction that results from mapping an input image into a prediction error sequence—often called a *prediction residual*. Because spatial redundancy is removed by the prediction and differencing process, the probability density function of the prediction residual is, in general, highly peaked at zero and characterized by a relatively small (in comparison to the input intensity distribution) variance. In fact, it is often modeled by a zero mean uncorrelated Laplacian PDF

$$p_e(e) = \frac{1}{\sqrt{2}\sigma_e} e^{\frac{-\sqrt{2}|e|}{\sigma_e}}$$

(8.2-35)

where σ_e is the standard deviation of e.

EXAMPLE 8.20:
Predictive coding and temporal redundancy.

■ The image in Fig. 8.34(a) is a portion of a frame of NASA video in which the Earth is moving from left to right with respect to a stationary camera attached to the space shuttle. It is repeated in Fig. 8.35(b)—along with its immediately preceding frame in Fig. 8.35(a). Using the first-order linear predictor

$$\hat{f}(x, y, t) = \text{round}\left[\alpha f(x, y, t - 1)\right]$$

(8.2-36)

a b
c d

FIGURE 8.35
(a) and (b) Two views of Earth from an orbiting space shuttle video. (c) The prediction error image resulting from Eq. (8.2-36). (d) A histogram of the prediction error. (Original images courtesy of NASA.)

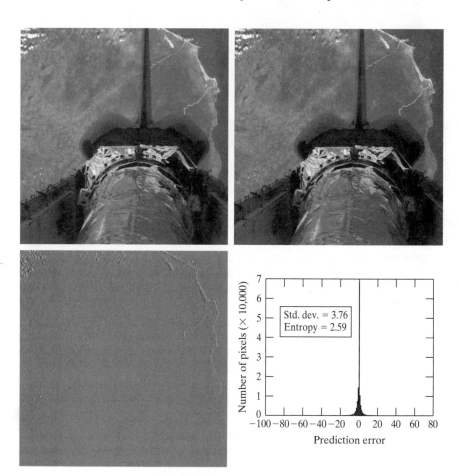

with $\alpha = 1$, the intensities of the pixels in Fig. 8.35(b) can be predicted from the corresponding pixels in (a). Figure 8.34(c) is the resulting prediction residual image, $e(x, y, t) = f(x, y, t) - \hat{f}(x, y, t)$. Figure 8.34(d) is the histogram of $e(x, y, t)$. Note that there is very little prediction error. The standard deviation of the error is much smaller than in the previous example—3.76 bits/pixel as opposed to 15.58 bits/pixel. In addition, the entropy of the prediction error [computed using Eq. (8.1-7)] has decreased from 3.99 to 2.59 bits/pixel. By variable-length coding the resulting prediction residual, the original image is compressed by approximately 8/2.59 or 3.1:1—a 50% improvement over the 2:1 compression obtained using the spatially-oriented previous pixel predictor in Example 8.19. ■

Recall again that the variable-length encoded prediction error is the compressed image.

Motion compensated prediction residuals

As you saw in Example 8.20, successive frames in a video sequence often are very similar. Coding their differences can reduce temporal redundancy and provide significant compression. However, when a sequence of frames contains rapidly moving objects—or involves camera zoom and pan, sudden scene changes, or fade-ins and fade-outs—the similarity between neighboring frames is reduced and compression is affected negatively. That is, like most compression techniques (see Example 8.5), temporally-based predictive coding works best with certain kinds of inputs—namely, a sequence of images with significant temporal redundancy. When used on images with little temporal redundancy, data expansion can occur. Video compression systems avoid the problem of data expansion in two ways:

1. By tracking object movement and compensating for it during the prediction and differencing process.
2. By switching to an alternate coding method when there is insufficient *interframe* correlation (similarity between frames) to make predictive coding advantageous.

The first of these—called *motion compensation*—is the subject of the remainder of this section. Before proceeding, however, we note that when there is insufficient interframe correlation to make predictive coding effective, the second problem is typically addressed using a block-oriented 2-D transform, like JPEG's DCT-based coding (see Section 8.2.8). Frames compressed in this way (i.e., without a prediction residual) are called *intraframes* or *Independent frames* (*I-frames*). They can be decoded without access to other frames in the video to which they belong. I-frames usually resemble JPEG encoded images and are ideal starting points for the generation of prediction residuals. Moreover, they provide a high degree of random access, ease of editing, and resistance to the propagation of transmission error. As a result, all standards require the periodic insertion of I-frames into the compressed video codestream.

Figure 8.36 illustrates the basics of motion compensated predictive coding. Each video frame is divided into non-overlapping rectangular regions—typically of size 4×4 to 16×16—called *macroblocks*. (Only one macroblock is shown in Fig. 8.36.) The "movement" of each macroblock with respect to its "most likely" position in the previous (or subsequent) video frame, called the *reference frame*,

FIGURE 8.36
Macroblock
motion
specification.

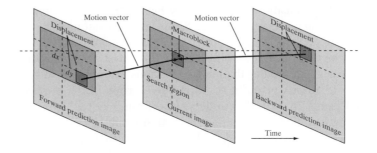

The "most likely" position is the one that minimizes an error measure between the reference macroblock and macroblock being encoded. The two blocks do not have to be representations of the same object, but they must minimize the error measure.

is encoded in a *motion vector*. The vector describes the motion by defining the horizontal and vertical *displacement* from the "most likely" position. The displacements typically are specified to the nearest pixel, $\frac{1}{2}$ pixel, or $\frac{1}{4}$ pixel precision. If sub-pixel precision is used, the predictions must be interpolated [e.g., using bilinear interpolation (see Section 2.4.4)] from a combination of pixels in the reference frame. An encoded frame that is based on the previous frame (a *forward prediction* in Fig. 8.36) is called a *Predictive frame* (*P-frame*); one that is based on the subsequent frame (a *backward prediction* in Fig. 8.36) is called a *Bidirectional frame* (*B-frame*). B-frames require the compressed codestream to be reordered so that frames are presented to the decoder in the proper decoding sequence—rather than the natural display order.

As you might expect, *motion estimation* is the key component of motion compensation. During motion estimation, the motion of objects is measured and encoded into motion vectors. The search for the "best" motion vector requires that a criterion of optimality be defined. For example, motion vectors may be selected on the basis of maximum correlation or minimum error between macroblock pixels and the predicted pixels (or interpolated pixels for sub-pixel motion vectors) from the chosen reference frame. One of the most commonly used error measures is *mean absolute distortion* (*MAD*)

$$MAD(x, y) = \frac{1}{mn} \sum_{i=1}^{m} \sum_{j=1}^{n} |f(x + i, y + j) - p(x + i + dx, y + j + dy)|$$

$$(8.2\text{-}37)$$

where x and y are the coordinates of the upper-left pixel of the $m \times n$ macroblock being coded, dx and dy are displacements from the reference frame as shown in Fig. 8.36, and p is an array of predicted macroblock pixel values. For sub-pixel motion vector estimation, p is interpolated from pixels in the reference frame. Typically, dx and dy must fall within a limited search region (see Fig. 8.36) around each macroblock. Values from ±8 to ±64 pixels are common, and the horizontal search area often is slightly larger than the vertical area. A more computationally efficient error measure, called the *sum of absolute distortions* (*SAD*), omits the $1/mn$ factor in Eq. (8.2-37).

Given a selection criterion like that of Eq. (8.2-37), motion estimation is performed by searching for the dx and dy that minimize $MAD(x, y)$ over the allowed range of motion vector displacements—including sub-pixel displacements. This process often is called *block matching*. An exhaustive search guarantees the best possible result, but is computationally expensive, because every possible motion must be tested over the entire displacement range. For 16×16 macroblocks and a ± 32 pixel displacement range (not out of the question for action films and sporting events), 4225 16×16 *MAD* calculations must be performed for each macroblock in a frame when integer displacement precision is used. If $\frac{1}{2}$ or $\frac{1}{4}$ pixel precision is desired, the number of calculations is multiplied by a factor of 4 or 16, respectively. Fast search algorithms can reduce the computational burden but may or may not yield optimal motion vectors. A number of fast block-based motion estimation algorithms have been proposed and studied in the literature (see, for example, Furht et al. [1997] or Mitchell et al. [1997]).

■ Figures 8.37(a) and (b) were taken from the same NASA video sequence used in Examples 8.19 and 8.20. Figure 8.37(b) is identical to Figs. 8.34(a) and 8.35(b); Fig. 8.37(a) is the corresponding section of a frame occurring thirteen frames earlier. Figure 8.37(c) is the difference between the two frames, scaled to the full intensity range. Note that the difference is 0 in the area of the stationary (with respect to the camera) space shuttle, but there are significant differences in the remainder of the image due to the relative motion of the Earth. The standard deviation of the prediction residual in Fig. 8.37(c) is 12.73 intensity levels; its entropy [using Eq. (8.1-7)] is 4.17 bits/pixel. The maximum compression achievable when variable-length coding the prediction residual is $C = 8/4.17 = 1.92$.

EXAMPLE 8.21:
Motion compensated prediction.

Figure 8.37(d) shows a motion compensated prediction residual with a much lower standard deviation (5.62 as opposed to 12.73 intensity levels) and slightly lower entropy (3.04 vs. 4.17 bits/pixel). The entropy was computed using Eq. (8.1-7). If the prediction residual in Fig. 8.37(d) is variable-length coded, the resulting compression ratio is $C = 8/3.04 = 2.63$. To generate this prediction residual, we divided Fig. 8.37(b) into non-overlapping 16×16 macroblocks and compared each macroblock against every 16×16 region in Fig. 8.37(a)—the reference frame—that fell within ± 16 pixels of the macroblock's position in (b). We used Eq. (8.2-37) to determine the best match by selecting displacement (dx, dy) with the lowest *MAD*. The resulting displacements are the x and y components of the motion vectors shown in Fig. 8.37(e). The white dots in the figure are the heads of the motion vectors; they indicate the upper-left-hand corner of the coded macroblocks. As you can see from the pattern of the vectors, the predominant motion in the image is from left to right. In the lower portion of the image, which corresponds to the area of the space shuttle in the original image, there is no motion and therefore no motion vectors displayed. Macroblocks in this area are predicted from similarly located (i.e., the corresponding) macroblocks in the reference frame. Because the motion vectors in Fig. 8.37(e) are highly correlated, they can be variable-length coded to reduce their storage and transmission requirements.   ■

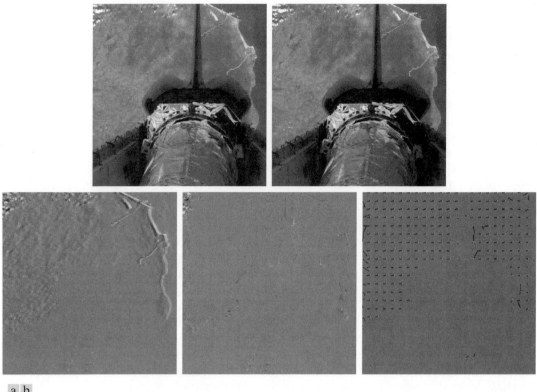

a b
c d e

FIGURE 8.37 (a) and (b) Two views of Earth that are thirteen frames apart in an orbiting space shuttle video. (c) A prediction error image without motion compensation. (d) The prediction residual with motion compensation. (e) The motion vectors associated with (d). The white dots in (d) represent the arrow heads of the motion vectors that are depicted. (Original images courtesy of NASA.)

The visual difference between Figs. 8.37(c) and 8.38(a) is due to scaling. The image in Fig. 8.38(a) has been scaled to match Figs. 8.38(b)–(d).

Figure 8.38 illustrates the increased prediction accuracy that is possible with sub-pixel motion compensation. Figure 8.38(a) is repeated from Fig. 8.37(c) and included as a point of reference; it shows the prediction error that results without motion compensation. The images in Figs. 8.38(b), (c), and (d) are motion compensated prediction residuals. They are based on the same two frames that were used in Example 8.21 and computed with macroblock displacements to $1, \frac{1}{2},$ and $\frac{1}{4}$ pixel resolution (i.e., precision), respectively. Macroblocks of size 8×8 were used; displacements were limited to ± 8 pixels.

The most significant visual difference between the prediction residuals in Fig. 8.38 is the number and size of intensity peaks and valleys—their darkest and lightest areas of intensity. The $\frac{1}{4}$ pixel residual in Fig. 8.38(d) is the "flattest" of the four images, with the fewest excursions to black or white. As would be expected, it has the narrowest histogram. The standard deviations of the prediction residuals in Figs. 8.38(a) through (d) decrease as motion vector precision increases—from 12.7 to 4.4, 4, and 3.8 pixels, respectively. The entropies of the

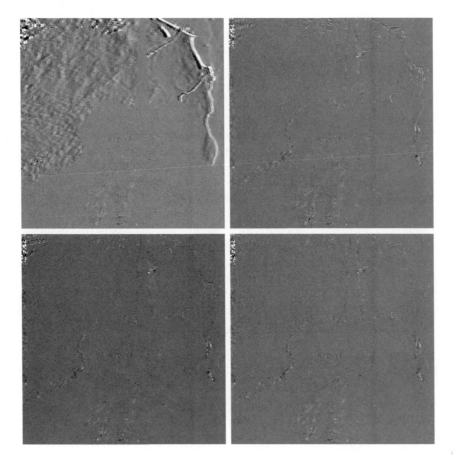

FIGURE 8.38
Sub-pixel motion compensated prediction residuals: (a) without motion compensation; (b) single pixel precision; (c) $\frac{1}{2}$ pixel precision; and (d) $\frac{1}{4}$ pixel precision. (All prediction errors have been scaled to the full intensity range and then multiplied by 2 to increase their visibility.)

residuals, as determined using Eq. (8.1-7), are 4.17, 3.34, 3.35, and 3.34 bits/pixel, respectively. Thus, the motion compensated residuals contain about the same amount of information, despite the fact that the residuals in Figs. 8.38(c) and (d) use additional bits to accommodate $\frac{1}{2}$ and $\frac{1}{4}$ pixel interpolation. Finally, we note that there is an obvious strip of increased prediction error on the left side of each motion compensated residual. This is due to the left-to-right motion of the Earth, which introduces new or previously unseen areas of the Earth's terrain into the left side of each image. Because these areas are absent from the previous frames, they cannot be accurately predicted, regardless of the precision used to compute motion vectors.

Motion estimation is a computationally demanding task. Fortunately, only the encoder must estimate macroblock motion. Given the motion vectors of the macroblocks, the decoder simply accesses the areas of the reference frames that were used in the encoder to form the prediction residuals. Because of this, motion estimation is not included in most video compression standards. Compression standards focus on the decoder—placing constraints on macroblock dimensions, motion vector precision, horizontal and vertical displacement ranges, and the like. Table 8.11 gives the key predictive coding parameters of some the most important video compression standards. Note that most of the standards use an 8×8 DCT for I-frame encoding, but specify a larger area (i.e., 16×16 macroblock) for motion compensation. In addition, even the P- and B-frame prediction residuals are transform coded due to the effectiveness of DCT coefficient quantization. Finally, we note that the H.264 and MPEG-4 AVC standards support intraframe predictive coding (in I-frames) to reduce spatial redundancy.

Figure 8.39 shows a typical motion compensated video encoder. It exploits redundancies within and between adjacent video frames, motion uniformity between frames, and the psychovisual properties of the human visual system. We can think of the input to the encoder as sequential macroblocks of video. For color video, each macroblock is composed of a luminance block and two chrominance blocks. Because the eye has far less spatial acuity for color than for luminance, the chrominance blocks often are sampled at half the horizontal and vertical resolution of the luminance block. The grayed elements in the figure parallel the transformation, quantization, and variable-length coding operations of a JPEG encoder. The principal difference is the input, which may be a conventional macroblock of image data (for I-frames)

TABLE 8.11

Predictive coding in video compression standards.

Parameter	H.261	MPEG-1	H.262 MPEG-2	H.263	MPEG-4	VC-1 WMV-9	H.264 MPEG-4 AVC
Motion vector precision	1	$1/2$	$1/2$	$1/2$	$1/4$	$1/4$	$1/4$
Macroblock sizes	16×16	16×16	16×16 16×8	16×16 8×8	16×16 8×8	16×16 8×8	16×16 16×8 8×16 8×8 8×4 4×8 4×4
Transform	8×8 DCT	8×8 DCT	8×8 DCT	8×8 DCT	8×8 DCT	8×8 8×4 4×8 4×4 Integer DCT	4×4 8×8 Integer
Interframe predictions	P	P, B	P, B	P, B	P, B	P, B	P, B
I-frame intra-predictions	No	No	No	No	No	No	Yes

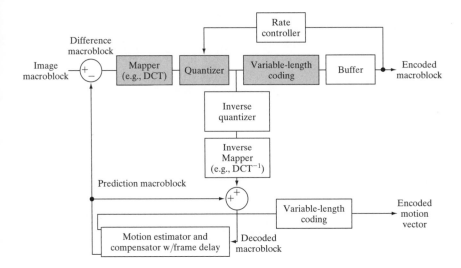

FIGURE 8.39
A typical motion compensated video encoder.

or the difference between a conventional macroblock and a prediction of it based on previous and/or subsequent video frames (for P- and B-frames). The encoder includes an *inverse quantizer* and *inverse mapper* (e.g., inverse DCT) so that its predictions match those of the complementary decoder. Also, it is designed to produce compressed bit streams that match the capacity of the intended video channel. To accomplish this, the quantization parameters are adjusted by a *rate controller* as a function of the occupancy of an output *buffer*. As the buffer becomes fuller, the quantization is made coarser, so that fewer bits stream into the buffer.

Quantization as defined earlier in the chapter is irreversible. The "inverse quantizer" in Fig. 8.39 does not prevent information loss.

■ We conclude our discussion of motion compensated predictive coding with an example illustrating the kind of compression that is possible with modern video compression methods. Figure 8.40 shows fifteen frames of a 1 minute HD (1280 × 720) full-color NASA video, parts of which have been used throughout this section. Although the images shown are monochrome, the video is a sequence of 1,829 full-color frames. Note that there are a variety of scenes, a great deal of motion, and multiple fade effects. For example, the video opens with a 150 frame fade-in from black, which includes frames 21 and 44 in Fig. 8.40, and concludes with a fade sequence containing frames 1595, 1609, and 1652 in Fig. 8.40, followed by a final fade to black. There are also several abrupt scene changes, like the change involving frames 1303 and 1304 in Fig. 8.40.

An H.264 compressed version of the NASA video stored as a Quicktime file (see Table 8.4) requires 44.56 MB of storage—plus another 1.39 MB for the associated audio. The video quality is excellent. About 5 GB of data would be needed to store the video frames as uncompressed full-color images. It should be noted that the video contains sequences involving both rotation and scale change (e.g., the sequence including frames 959, 1023, and 1088 in Fig. 840). The discussion in this section, however, has been limited to translation alone. ■

EXAMPLE 8.22:
Video compression example.

See the book Web site for the NASA video segment used in this section.

FIGURE 8.40 Fifteen frames from an 1829-frame, 1-minute NASA video. The original video is in HD full color. (Courtesy of NASA.)

Lossy predictive coding

In this section, we add a quantizer to the lossless predictive coding model introduced earlier and examine the trade-off between reconstruction accuracy and compression performance within the context of spatial predictors. As Fig. 8.41 shows, the quantizer, which replaces the nearest integer function of the error-free

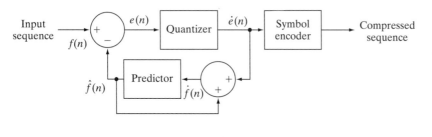

FIGURE 8.41
A lossless
predictive
coding model:
(a) encoder;
(b) decoder.

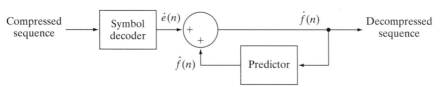

encoder, is inserted between the symbol encoder and the point at which the pre-diction error is formed. It maps the prediction error into a limited range of out-puts, denoted $\dot{e}(n)$, which establish the amount of compression and distortion that occurs.

In order to accommodate the insertion of the quantization step, the error-free encoder of Fig. 8.33(a) must be altered so that the predictions generated by the encoder and decoder are equivalent. As Fig. 8.41(a) shows, this is ac-complished by placing the lossy encoder's predictor within a feedback loop, where its input, denoted $\dot{f}(n)$, is generated as a function of past predictions and the corresponding quantized errors. That is,

$$\dot{f}(n) = \dot{e}(n) + \hat{f}(n) \tag{8.2-38}$$

where $\hat{f}(n)$ is as defined earlier. This closed loop configuration prevents error buildup at the decoder's output. Note in Fig. 8.41(b) that the output of the de-coder is given also by Eq. (8.2-38).

■ *Delta modulation* (DM) is a simple but well-known form of lossy predictive coding in which the predictor and quantizer are defined as

EXAMPLE 8.23:
Delta modulation.

$$\hat{f}(n) = \alpha \dot{f}(n - 1) \tag{8.2-39}$$

and

$$\dot{e}(n) = \begin{cases} +\zeta & \text{for } e(n) > 0 \\ -\zeta & \text{otherwise} \end{cases} \tag{8.2-40}$$

where α is a prediction coefficient (normally less than 1) and ζ is a positive constant. The output of the quantizer, $\dot{e}(n)$, can be represented by a single bit [Fig. 8.42(a)], so the symbol encoder of Fig. 8.41(a) can utilize a 1-bit fixed-length code. The resulting DM code rate is 1 bit/pixel.

Figure 8.42(c) illustrates the mechanics of the delta modulation process, where the calculations needed to compress and reconstruct input sequence

a b
c

FIGURE 8.42
An example of
delta modulation.

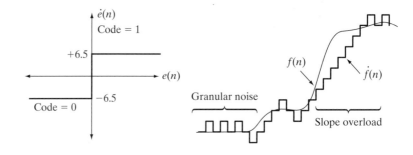

Input		Encoder				Decoder		Error
n	$f(n)$	$\hat{f}(n)$	$e(n)$	$\dot{e}(n)$	$\dot{f}(n)$	$\hat{f}(n)$	$\dot{f}(n)$	$f(n) - \dot{f}(n)$
0	14	—	—	—	14.0	—	14.0	0.0
1	15	14.0	1.0	6.5	20.5	14.0	20.5	−5.5
2	14	20.5	−6.5	−6.5	14.0	20.5	14.0	0.0
3	15	14.0	1.0	6.5	20.5	14.0	20.5	−5.5
⋅	⋅	⋅	⋅	⋅	⋅	⋅	⋅	⋅
⋅	⋅	⋅	⋅	⋅	⋅	⋅	⋅	⋅
14	29	20.5	8.5	6.5	27.0	20.5	27.0	2.0
15	37	27.0	10.0	6.5	33.5	27.0	33.5	3.5
16	47	33.5	13.5	6.5	40.0	33.5	40.0	7.0
17	62	40.0	22.0	6.5	46.5	40.0	46.5	15.5
18	75	46.5	28.5	6.5	53.0	46.5	53.0	22.0
19	77	53.0	24.0	6.5	59.6	53.0	59.6	17.5
⋅	⋅	⋅	⋅	⋅	⋅	⋅	⋅	⋅
⋅	⋅	⋅	⋅	⋅	⋅	⋅	⋅	⋅

$\{14, 15, 14, 15, 13, 15, 15, 14, 20, 26, 27, 28, 27, 27, 29, 37, 47, 62, 75, 77, 78,$
$79, 80, 81, 81, 82, 82\}$ with $\alpha = 1$ and $\zeta = 6.5$ are tabulated. The process begins with the error-free transfer of the first input sample to the decoder. With
the initial condition $\dot{f}(0) = f(0) = 14$ established at both the encoder and
decoder, the remaining outputs can be computed by repeatedly evaluating
Eqs. (8.2-39), (8.2-30), (8.2-40), and (8.2-38). Thus, when $n = 1$, for example,
$\hat{f}(1) = (1)(14) = 14$, $e(1) = 15 - 14 = 1$, $\dot{e}(1) = +6.5$ (because $e(1) > 0$),
$\dot{f}(1) = 6.4 + 14 = 20.5$, and the resulting reconstruction error is $(15 - 20.5)$,
or -5.5.

Figure 8.42(b) shows graphically the tabulated data in Fig. 8.42(c). Both the
input and completely decoded output $[f(n)$ and $\dot{f}(n)]$ are shown. Note that in
the rapidly changing area from $n = 14$ to 19, where ζ was too small to represent the input's largest changes, a distortion known as *slope overload* occurs.
Moreover, when ζ was too large to represent the input's smallest changes, as in
the relatively smooth region from $n = 0$ to $n = 7$, *granular noise* appears. In
images, these two phenomena lead to blurred object edges and grainy or noisy
surfaces (that is, distorted smooth areas).   ■

The distortions noted in the preceding example are common to all forms of
lossy predictive coding. The severity of these distortions depends on a complex set
of interactions between the quantization and prediction methods employed. Despite these interactions, the predictor normally is designed with the assumption of

no quantization error, and the quantizer is designed to minimize its own error. That is, the predictor and quantizer are designed independently of each other.

Optimal predictors

In many predictive coding applications, the predictor is chosen to minimize the encoder's mean-square prediction error[†]

$$E\{e^2(n)\} = E\{[f(n) - \hat{f}(n)]^2\} \tag{8.2-41}$$

subject to the constraint that

$$\dot{f}(n) = \dot{e}(n) + \hat{f}(n) \approx e(n) + \hat{f}(n) = f(n) \tag{8.2-42}$$

and

$$\hat{f}(n) = \sum_{i=1}^{m} \alpha_i f(n - i) \tag{8.2-43}$$

That is, the optimization criterion is minimal mean-square prediction error, the quantization error is assumed to be negligible [$\dot{e}(n) \approx e(n)$] and the prediction is constrained to a linear combination of m previous samples.[‡] These restrictions are not essential, but they simplify the analysis considerably and, at the same time, decrease the computational complexity of the predictor. The resulting predictive coding approach is referred to as *differential pulse code modulation* (DPCM).

Under these conditions, the optimal predictor design problem is reduced to the relatively straightforward exercise of selecting the m prediction coefficients that minimize the expression

$$E\{e^2(n)\} = E\left\{\left[f(n) - \sum_{i=1}^{m} \alpha_i f(n - i)\right]^2\right\} \tag{8.2-44}$$

Differentiating Eq. (8.2-44) with respect to each coefficient, equating the derivatives to zero, and solving the resulting set of simultaneous equations under the assumption that $f(n)$ has mean zero and variance σ^2 yields

$$\boldsymbol{\alpha} = \mathbf{R}^{-1}\mathbf{r} \tag{8.2-45}$$

where $\mathbf{R}^{-1}$ is the inverse of the $m \times m$ autocorrelation matrix

$$\mathbf{R} = \begin{bmatrix} E\{f(n-1)f(n-1)\} & E\{f(n-1)f(n-2)\} & \cdots & E\{f(n-1)f(n-m)\} \\ E\{f(n-2)f(n-1)\} & \vdots & \cdots & \vdots \\ \vdots & \vdots & \cdots & \vdots \\ \vdots & \vdots & \cdots & \vdots \\ E\{f(n-m)f(n-1)\} & E\{f(n-m)f(n-2)\} & \cdots & E\{f(n-m)f(n-m)\} \end{bmatrix}$$

$$\tag{8.2-46}$$

[†]The notation $E\{\cdot\}$ denotes the statistical expectation operator.

[‡]In general, the optimal predictor for a non-Gaussian sequence is a nonlinear function of the samples used to form the estimate.

and **r** and $\boldsymbol{\alpha}$ are the m-element vectors

$$\mathbf{r} = \begin{bmatrix} E\{f(n)f(n-1)\} \\ E\{f(n)f(n-2)\} \\ \vdots \\ E\{f(n)f(n-m)\} \end{bmatrix} \text{ and } \boldsymbol{\alpha} = \begin{bmatrix} \alpha_1 \\ \alpha_2 \\ \vdots \\ \alpha_m \end{bmatrix} \tag{8.2-47}$$

Thus for any input sequence, the coefficients that minimize Eq. (8.2-44) can be determined via a series of elementary matrix operations. Moreover, the coefficients depend only on the autocorrelations of the samples in the original sequence. The variance of the prediction error that results from the use of these optimal coefficients is

$$\sigma_e^2 = \sigma^2 - \boldsymbol{\alpha}^T \mathbf{r} = \sigma^2 - \sum_{i=1}^{m} E\{f(n)f(n-i)\}\alpha_i \tag{8.2-48}$$

Although the mechanics of evaluating Eq. (8.2-45) are quite simple, computation of the autocorrelations needed to form **R** and **r** is so difficult in practice that *local* predictions (those in which the prediction coefficients are computed for each input sequence) are almost never used. In most cases, a set of *global* coefficients is computed by assuming a simple input model and substituting the corresponding autocorrelations into Eqs. (8.2-46) and (8.2-47). For instance, when a 2-D Markov image source (see Section 8.1.4) with separable autocorrelation function

$$E\{f(x, y)f(x-i, y-j)\} = \sigma^2 \rho_v^i \rho_h^j \tag{8.2-49}$$

and generalized fourth-order linear predictor

$$\begin{aligned} \hat{f}(x, y) &= \alpha_1 f(x, y-1) + \alpha_2 f(x-1, y-1) \\ &\quad + \alpha_3 f(x-1, y) + \alpha_4 f(x-1, y+1) \end{aligned} \tag{8.2-50}$$

are assumed, the resulting optimal coefficients (Jain [1989]) are

$$\alpha_1 = \rho_h \quad \alpha_2 = -\rho_v \rho_h \quad \alpha_3 = \rho_v \quad \alpha_4 = 0 \tag{8.2-51}$$

where ρ_h and ρ_v are the horizontal and vertical correlation coefficients, respectively, of the image under consideration.

Finally, the sum of the prediction coefficients in Eq. (8.2-43) normally is required to be less than or equal to one. That is,

$$\sum_{i=1}^{m} \alpha_i \leq 1 \tag{8.2-52}$$

This restriction is made to ensure that the output of the predictor falls within the allowed range of the input and to reduce the impact of transmission noise [which generally is seen as horizontal streaks in reconstructed images when the input to Fig. 8.41(a) is an image]. Reducing the DPCM decoder's susceptibility to input noise is important, because a single error (under the right circumstances) can propagate to all future outputs. That is, the decoder's output may become unstable. By further restricting Eq. (8.2-52) to be strictly less than 1 confines the impact of an input error to a small number of outputs.

■ Consider the prediction error that results from DPCM coding the monochrome image of Fig. 8.9(a) under the assumption of zero quantization error and with each of four predictors:

EXAMPLE 8.24:
Comparison of
prediction
techniques.

$$\hat{f}(x, y) = 0.97f(x, y - 1) \qquad (8.2\text{-}53)$$

$$\hat{f}(x, y) = 0.5f(x, y - 1) + 0.5f(x - 1, y) \qquad (8.2\text{-}54)$$

$$\hat{f}(x, y) = 0.75f(x, y - 1) + 0.75f(x - 1, y) - 0.5f(x - 1, y - 1) \qquad (8.2\text{-}55)$$

$$\hat{f}(x, y) = \begin{cases} 0.97f(x, y - 1) & \text{if } \Delta h \le \Delta v \\ 0.97f(x - 1, y) & \text{otherwise} \end{cases} \qquad (8.2\text{-}56)$$

where $\Delta h = |f(x - 1, y) - f(x - 1, y - 1)|$ and $\Delta v = |f(x, y - 1) - f(x - 1, y - 1)|$ denote the horizontal and vertical gradients at point (x, y). Equations (8.2-53) through (8.2-56) define a relatively robust set of α_i that provide satisfactory performance over a wide range of images. The adaptive predictor of Eq. (8.2-56) is designed to improve edge rendition by computing a local measure of the directional properties of an image (Δh and Δv) and selecting a predictor specifically tailored to the measured behavior.

Figures 8.43(a) through (d) show the prediction error images that result from using the predictors of Eqs. (8.2-53) through (8.2-56). Note that the

a b
c d

FIGURE 8.43
A comparison of
four linear
prediction
techniques.

visually perceptible error decreases as the order of the predictor increases.[†]
The standard deviations of the prediction errors follow a similar pattern. They
are 11.1, 9.8, 9.1, and 9.7 intensity levels, respectively.   ■

Optimal quantization

The staircase quantization function $t = q(s)$ in Fig. 8.44 is an odd function of s
[that is, $q(-s) = -q(s)$] that can be described completely by the $L/2$ values of
s_i and t_i shown in the first quadrant of the graph. These break points define
function discontinuities and are called the *decision* and *reconstruction levels* of
the quantizer. As a matter of convention, s is considered to be mapped to t_i if
it lies in the half-open interval $(s_i, s_{i+1}]$.

The quantizer design problem is to select the best s_i and t_i for a particular optimization criterion and input probability density function $p(s)$. If the optimization criterion, which could be either a statistical or psychovisual measure,[‡] is the
minimization of the mean-square quantization error (that is, $E\{(s_i - t_i)^2\}$) and
$p(s)$ is an even function, the conditions for minimal error (Max [1960]) are

$$\int_{s_{i-1}}^{s_i} (s - t_i)p(s)\, ds \quad i = 1, 2, \ldots, \frac{L}{2} \tag{8.2-57}$$

$$s_i = \begin{cases} 0 & i = 0 \\ \dfrac{t_i + t_{i+1}}{2} & i = 1, 2, \ldots, \dfrac{L}{2} - 1 \\ \infty & i = \dfrac{L}{2} \end{cases} \tag{8.2-58}$$

FIGURE 8.44
A typical
quantization
function.

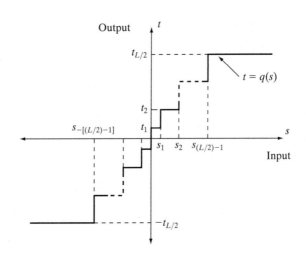

Output

$-tL/2$ ... (figure)

[†]Predictors that use more than three or four previous pixels provide little compression gain for the
added predictor complexity (Habibi [1971]).

[‡]See Netravali [1977] and Limb and Rubinstein [1978] for more on psychovisual measures.

and

$$s_{-i} = -s_i \quad t_{-i} = -t_i \tag{8.2-59}$$

Equation (8.2-57) indicates that the reconstruction levels are the centroids of the areas under $p(s)$ over the specified decision intervals, whereas Eq. (8.2-58) indicates that the decision levels are halfway between the reconstruction levels. Equation (8.2-59) is a consequence of the fact that q is an odd function. For any L, the s_i and t_i that satisfy Eqs. (8.2-57) through (8.2-59) are optimal in the mean-square error sense; the corresponding quantizer is called an L-level *Lloyd-Max* quantizer.

Table 8.12 lists the 2-, 4-, and 8-level Lloyd-Max decision and reconstruction levels for a unit variance Laplacian probability density function [see Eq. (8.2-35)]. Because obtaining an explicit or closed-form solution to Eqs. (8.2-57) through (8.2-59) for most nontrivial $p(s)$ is difficult, these values were generated numerically (Paez and Glisson [1972]). The three quantizers shown provide fixed output rates of 1, 2, and 3 bits/pixel, respectively. As Table 8.12 was constructed for a unit variance distribution, the reconstruction and decision levels for the case of $\sigma \neq 1$ are obtained by multiplying the tabulated values by the standard deviation of the probability density function under consideration. The final row of the table lists the step size, θ, that simultaneously satisfies Eqs. (8.2-57) through (8.5-59) *and* the additional constraint that

$$t_i - t_{i-1} = s_i - s_{i-1} = \theta \tag{8.2-60}$$

If a symbol encoder that utilizes a variable-length code is used in the general lossy predictive encoder of Fig. 8.41(a), an *optimum uniform quantizer* of step size θ will provide a lower code rate (for a Laplacian PDF) than a fixed-length coded Lloyd-Max quantizer with the same output fidelity (O'Neil [1971]).

Although the Lloyd-Max and optimum uniform quantizers are not adaptive, much can be gained from adjusting the quantization levels based on the local behavior of an image. In theory, slowly changing regions can be finely quantized, while the rapidly changing areas are quantized more coarsely. This approach simultaneously reduces both granular noise and slope overload, while requiring only a minimal increase in code rate. The trade-off is increased quantizer complexity.

Levels	2		4		8	
i	s_i	t_i	s_i	t_i	s_i	t_i
1	∞	0.707	1.102	0.395	0.504	0.222
2			∞	1.810	1.181	0.785
3					2.285	1.576
4					∞	2.994
θ	1.414		1.087		0.731	

TABLE 8.12
Lloyd-Max quantizers for a Laplacian probability density function of unit variance.

8.2.10 Wavelet Coding

With reference to Tables 8.3 and 8.4, wavelet coding is used in the

• JPEG-2000

compression standard.

As with the transform coding techniques of Section 8.2.8, wavelet coding is based on the idea that the coefficients of a transform that decorrelates the pixels of an image can be coded more efficiently than the original pixels themselves. If the basis functions of the transform—in this case wavelets—pack most of the important visual information into a small number of coefficients, the remaining coefficients can be quantized coarsely or truncated to zero with little image distortion.

Figure 8.45 shows a typical wavelet coding system. To encode a $2^J \times 2^J$ image, an analyzing wavelet, ψ, and minimum decomposition level, $J - P$, are selected and used to compute the discrete wavelet transform of the image. If the wavelet has a complementary scaling function φ, the fast wavelet transform (see Sections 7.4 and 7.5) can be used. In either case, the computed transform converts a large portion of the original image to horizontal, vertical, and diagonal decomposition coefficients with zero mean and Laplacian-like probabilities. Recall the image of Fig. 7.1 and the dramatically simpler statistics of its wavelet transform in Fig. 7.10(a). Because many of the computed coefficients carry little visual information, they can be quantized and coded to minimize intercoefficient and coding redundancy. Moreover, the quantization can be adapted to exploit any positional correlation across the P decomposition levels. One or more lossless coding methods, like run-length, Huffman, arithmetic, and bit-plane coding, can be incorporated into the final symbol coding step. Decoding is accomplished by inverting the encoding operations—with the exception of quantization, which cannot be reversed exactly.

The principal difference between the wavelet-based system of Fig. 8.45 and the transform coding system of Fig. 8.21 is the omission of the subimage processing stages of the transform coder. Because wavelet transforms are both computationally efficient and inherently local (i.e., their basis functions are limited in duration), subdivision of the original image is unnecessary. As you will see later in this section, the removal of the subdivision step eliminates the blocking artifact that characterizes DCT-based approximations at high compression ratios.

Wavelet selection

The wavelets chosen as the basis of the forward and inverse transforms in Fig. 8.45 affect all aspects of wavelet coding system design and performance. They impact directly the computational complexity of the transforms and, less directly, the system's ability to compress and reconstruct

a
b

FIGURE 8.45
A wavelet coding system:
(a) encoder;
(b) decoder.

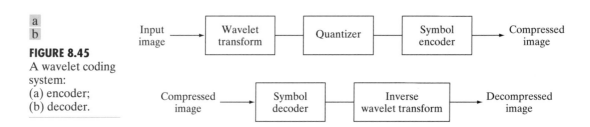

images of acceptable error. When the transforming wavelet has a companion scaling function, the transformation can be implemented as a sequence of digital filtering operations, with the number of filter taps equal to the number of nonzero wavelet and scaling vector coefficients. The ability of the wavelet to pack information into a small number of transform coefficients determines its compression and reconstruction performance.

> In digital filtering, each filter tap multiplies a filter coefficient by a delayed version of the signal being filtered.

The most widely used expansion functions for wavelet-based compression are the Daubechies wavelets and biorthogonal wavelets. The latter allow useful analysis properties, like the number of zero moments (see Section 7.5), to be incorporated into the decomposition filters, while important synthesis properties, like smoothness of reconstruction, are built into the reconstruction filters.

■ Figure 8.46 contains four discrete wavelet transforms of Fig. 8.9(a). Haar wavelets, the simplest and only discontinuous wavelets considered in this example, were used as the expansion or basis functions in Fig. 8.46(a). Daubechies wavelets, among the most popular imaging wavelets, were used in Fig. 8.46(b),

EXAMPLE 8.25:
Wavelet bases in wavelet coding.

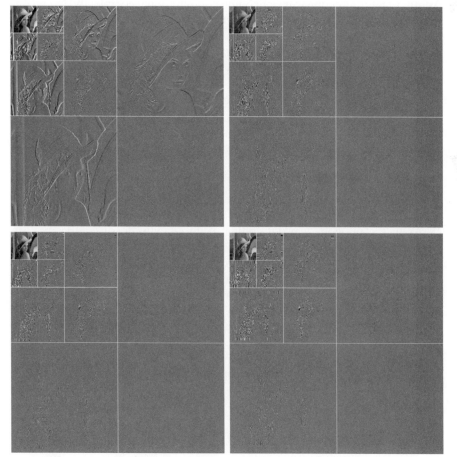

a b
c d

FIGURE 8.46
Three-scale wavelet transforms of Fig. 8.9(a) with respect to (a) Haar wavelets, (b) Daubechies wavelets, (c) symlets, and (d) Cohen-Daubechies Feauveau biorthogonal wavelets.

and symlets, which are an extension of the Daubechies wavelets with increased symmetry, were used in Fig. 8.46(c). The Cohen-Daubechies-Feauveau wavelets that were employed in Fig. 8.46(d) are included to illustrate the capabilities of biorthogonal wavelets. As in previous results of this type, all detail coefficients were scaled to make the underlying structure more visible—with intensity 128 corresponding to coefficient value 0.

DWT detail coefficients are discussed in Section 7.3.2.

As you can see in Table 8.13, the number of operations involved in the computation of the transforms in Fig. 8.46 increases from 4 to 28 multiplications and additions per coefficient (for each decomposition level) as you move from Fig. 8.46(a) to (d). All four transforms were computed using a fast wavelet transform (i.e., filter bank) formulation. Note that as the computational complexity (i.e., the number of filter taps) increases, the information packing performance does as well. When Haar wavelets are employed and the detail coefficients below 1.5 are truncated to zero, 33.8% of the total transform is zeroed. With the more complex biorthogonal wavelets, the number of zeroed coefficients rises to 42.1%, increasing the potential compression by almost 10%. ■

Decomposition level selection

Another factor affecting wavelet coding computational complexity and reconstruction error is the number of transform decomposition levels. Because a P-scale fast wavelet transform involves P-filter bank iterations, the number of operations in the computation of the forward and inverse transforms increases with the number of decomposition levels. Moreover, quantizing the increasingly lower-scale coefficients that result with more decomposition levels affects increasingly larger areas of the reconstructed image. In many applications, like searching image databases or transmitting images for progressive reconstruction, the resolution of the stored or transmitted images and the scale of the lowest useful approximations normally determine the number of transform levels.

EXAMPLE 8.26:
Decomposition levels in wavelet coding.

■ Table 8.14 illustrates the effect of decomposition level selection on the coding of Fig. 8.9(a) using biorthogonal wavelets and a fixed global threshold of 25. As in the previous wavelet coding example, only detail coefficients are truncated. The table lists both the percentage of zeroed coefficients and the resulting rms reconstruction errors from Eq. (8.1-10). Note that the initial decompositions are responsible for the majority of the data compression. There is little change in the number of truncated coefficients above three decomposition levels. ■

TABLE 8.13
Wavelet transform filter taps and zeroed coefficients when truncating the transforms in Fig. 8.46 below 1.5.

Wavelet	Filter Taps (Scaling + Wavelet)	Zeroed Coefficients
Haar (see Ex. 7.10)	2 + 2	33.8%
Daubechies (see Fig. 7.8)	8 + 8	40.9%
Symlet (see Fig. 7.26)	8 + 8	41.2%
Biorthogonal (see Fig. 7.39)	17 + 11	42.1%

Decomposition Level (Scales or Filter Bank Iterations)	Approximation Coefficient Image	Truncated Coefficients (%)	Reconstruction Error (rms)
1	256 × 256	74.7%	3.27
2	128 × 128	91.7%	4.23
3	64 × 64	95.1%	4.54
4	32 × 32	95.6%	4.61
5	16 × 16	95.5%	4.63

TABLE 8.14
Decomposition level impact on wavelet coding the 512 × 512 image of Fig. 8.9(a).

Quantizer design

The most important factor affecting wavelet coding compression and reconstruction error is coefficient quantization. Although the most widely used quantizers are uniform, the effectiveness of the quantization can be improved significantly by (1) introducing a larger quantization interval around zero, called a *dead zone*, or (2) adapting the size of the quantization interval from scale to scale. In either case, the selected quantization intervals must be transmitted to the decoder with the encoded image bit stream. The intervals themselves may be determined heuristically or computed automatically based on the image being compressed. For example, a global coefficient threshold could be computed as the median of the absolute values of the first-level detail coefficients or as a function of the number of zeroes that are truncated and the amount of energy that is retained in the reconstructed image.

One measure of the energy of a digital signal is the sum of the squared samples.

■ Figure 8.47 illustrates the impact of dead zone interval size on the percentage of truncated detail coefficients for a three-scale biorthogonal wavelet-based encoding of Fig. 8.9(a). As the size of the dead zone increases, the number of truncated coefficients does as well. Above the knee of the curve (i.e., beyond 5), there is little gain. This is due to the fact that the histogram of the detail coefficients is highly peaked around zero (see, for example, Fig. 7.10).

The rms reconstruction errors corresponding to the dead zone thresholds in Fig. 8.47 increase from 0 to 1.94 intensity levels at a threshold of 5 and to 3.83 intensity levels for a threshold of 18, where the number of zeroes reaches 93.85%. If every detail coefficient were eliminated, that percentage would increase to about 97.92% (about 4%), but the reconstruction error would grow to 12.3 intensity levels. ■

EXAMPLE 8.27:
Dead zone interval selection in wavelet coding.

JPEG-2000

JPEG-2000 extends the popular JPEG standard to provide increased flexibility in both the compression of continuous-tone still images and access to the compressed data. For example, portions of a JPEG-2000 compressed image can be extracted for retransmission, storage, display, and/or editing. The standard is based on the wavelet coding techniques just described. Coefficient quantization is adapted to individual scales and subbands and the quantized

FIGURE 8.47 The impact of dead zone interval selection on wavelet coding.

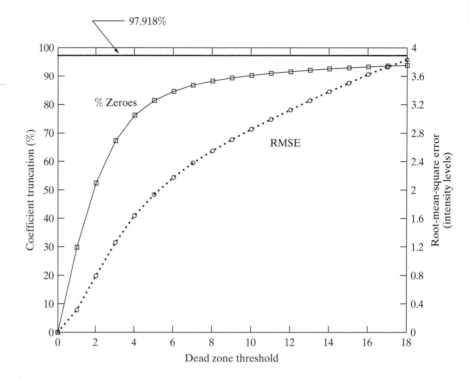

Ssiz is used in the standard to denote intensity resolution.

The irreversible component transform is the component transform used for lossy compression. The component transform itself is not irreversible. A different component transform is used for reversible compression.

coefficients are arithmetically coded on a bit-plane basis (see Sections 8.2.3 and 8.2.7). Using the notation of the standard, an image is encoded as follows (ISO/IEC [2000]).

The first step of the encoding process is to DC level shift the samples of the $Ssiz$-bit unsigned image to be coded by subtracting 2^{Ssiz-1}. If the image has more than one *component*—like the red, green, and blue planes of a color image—each component is shifted individually. If there are exactly three components, they may be optionally decorrelated using a reversible or nonreversible linear combination of the components. The *irreversible component transform* of the standard, for example, is

$$Y_0(x, y) = 0.299I_0(x, y) + 0.587I_1(x, y) + 0.114I_2(x, y)$$
$$Y_1(x, y) = -0.16875I_0(x, y) - 0.33126I_1(x, y) + 0.5I_2(x, y) \quad (8.2\text{-}61)$$
$$Y_2(x, y) = 0.5I_0(x, y) - 0.41869I_1(x, y) - 0.08131I_2(x, y)$$

where I_0, I_1, and I_2 are the level-shifted input components and Y_0, Y_1, and Y_2 are the corresponding decorrelated components. If the input components are the red, green, and blue planes of a color image, Eq. (8.2-61) approximates the $R'G'B'$ to $Y'C_bC_r$ color video transform (Poynton [1996]).[†] The goal of the transformation is to improve compression efficiency; transformed components Y_1 and Y_2 are difference images whose histograms are highly peaked around zero.

[†] $R'G'B'$ is a gamma corrected, nonlinear version of a linear CIE (International Commission on Illumination) RGB colorimetry value. Y' is luminance and C_b and C_r are color differences (i.e., scaled $B' - Y'$ and $R' - Y'$ values).

After the image has been level shifted and optionally decorrelated, its components can be divided into *tiles*. Tiles are rectangular arrays of pixels that are processed independently. Because an image can have more than one component (e.g., it could be made up of three color components), the tiling process creates *tile components*. Each tile component can be reconstructed independently, providing a simple mechanism for accessing and/or manipulating a limited region of a coded image. For example, an image having a 16:9 aspect ratio could be subdivided into tiles so that one of its tiles is a subimage with a 4:3 aspect ratio. That tile could then be reconstructed without accessing the other tiles in the compressed image. If the image is not subdivided into tiles, it is a single tile.

The 1-D discrete wavelet transform of the rows and columns of each tile component is then computed. For error-free compression, the transform is based on a biorthogonal, 5-3 coefficient scaling and wavelet vector (Le Gall and Tabatabai [1988]). A rounding procedure is defined for non-integer-valued transform coefficients. In lossy applications, a 9-7 coefficient scaling-wavelet vector (Antonini, Barlaud, Mathieu, and Daubechies [1992]) is employed. In either case, the transform is computed using the fast wavelet transform of Section 7.4 or via a complementary *lifting-based* approach (Mallat [1999]). For example, in lossy applications, the coefficients used to construct the 9-7 FWT analysis filter bank are given in Table 8.15. The complementary lifting-based implementation involves six sequential "lifting" and "scaling" operations:

> Lifting-based implementations are another way to compute wavelet transforms. The coefficients used in the approach are directly related to the FWT filter bank coefficients.

$$Y(2n+1) = X(2n+1) + \alpha[X(2n) + X(2n+2)], \quad i_0 - 3 \le 2n + 1 < i_1 + 3$$

$$Y(2n) = X(2n) + \beta[Y(2n-1) + Y(2n+1)], \quad i_0 - 2 \le 2n < i_1 + 2$$

$$Y(2n+1) = Y(2n+1) + \gamma[Y(2n) + Y(2n+2)], \quad i_0 - 1 \le 2n + 1 < i_1 + 1$$

$$Y(2n) = Y(2n) + \delta[Y(2n-1) + Y(2n+1)], \quad i_0 \le 2n < i_1$$

$$Y(2n+1) = -K \cdot Y(2n+1), \quad i_0 \le 2n + 1 < i_1$$

$$Y(2n) = Y(2n)/K, \quad i_0 \le 2n < i_1 \quad (8.2\text{-}62)$$

Here, X is the tile component being transformed, Y is the resulting transform, and i_0 and i_1 define the position of the tile component within a component. That is, they are the indices of the first sample of the tile-component row or column being transformed and the one immediately following the last sample. Variable n assumes values based on i_0, i_1, and which of the six operations is

Filter Tap	Highpass Wavelet Coefficient	Lowpass Scaling Coefficient
0	−1.115087052456994	0.6029490182363579
±1	0.5912717631142470	0.2668641184428723
±2	0.05754352622849957	−0.07822326652898785
±3	−0.09127176311424948	−0.01686411844287495
±4	0	0.02674875741080976

TABLE 8.15
Impulse responses of the low- and highpass analysis filters for an irreversible 9-7 wavelet transform.

being performed. If $n < i_0$ or $n \geq i_1$, $X(n)$ is obtained by symmetrically extending X. For example, $X(i_0 - 1) = X(i_0 + 1)$, $X(i_0 - 2) = X(i_0 + 2)$, $X(i_1) = X(i_1 - 2)$, and $X(i_1 + 1) = X(i_1 - 3)$. At the conclusion of the lifting and scaling operations, the even-indexed values of Y are equivalent to the FWT lowpass filtered output; the odd-indexed values of Y correspond to the highpass FWT filtered result. Lifting parameters α, β, γ, and δ are -1.586134342, -0.052980118, 0.882911075, and 0.433506852, respectively. Scaling factor K is 1.230174105.

These lifting-based coefficients are specified in the standard.

The transformation just described produces four subbands—a low-resolution approximation of the tile component and the component's horizontal, vertical, and diagonal frequency characteristics. Repeating the transformation N_L times, with subsequent iterations restricted to the previous decomposition's approximation coefficients, produces an N_L-scale wavelet transform. Adjacent scales are related spatially by powers of 2 and the lowest scale contains the only explicitly defined approximation of the original tile component. As can be surmised from Fig. 8.48, where the notation of the JPEG-2000 standard is summarized for the case of $N_L = 2$, a general N_L-scale transform contains $3N_L + 1$ subbands whose coefficients are denoted a_b, for $b = N_L LL$, $N_L HL, \ldots, 1HL, 1LH, 1HH$. The standard does not specify the number of scales to be computed.

Recall from Chapter 7 that the DWT decomposes an image into a set of band-limited components called subbands.

When each of the tile components has been processed, the total number of transform coefficients is equal to the number of samples in the original image—but the important visual information is concentrated in a few coefficients. To reduce the number of bits needed to represent the transform, coefficient $a_b(u, v)$ of subband b is quantized to value $q_b(u, v)$ using

$$q_b(u, v) = \text{sign}\big[a_b(u, v)\big] \cdot \text{floor}\left[\frac{|a_b(u, v)|}{\Delta_b}\right] \qquad (8.2\text{-}63)$$

FIGURE 8.48
JPEG 2000
two-scale wavelet
transform
tile-component
coefficient
notation and
analysis gain.

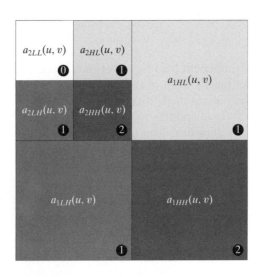

where the *quantiztion step size* Δ_b is

$$\Delta_b = 2^{R_b - \varepsilon_b}\left(1 + \frac{\mu_b}{2^{11}}\right) \tag{8.2-64}$$

R_b is the *nominal dynamic range* of subband b, and ε_b and μ_b are the number of bits allotted to the *exponent* and *mantissa* of the subband's coefficients. The nominal dynamic range of subband b is the sum of the number of bits used to represent the original image and the *analysis gain* bits for subband b. Subband analysis gain bits follow the simple pattern shown in Fig. 8.48. For example, there are two analysis gain bits for subband $b = 1HH$.

Do not confuse the standard's definition of nominal dynamic range with the closely related definition in Chapter 2.

For error-free compression, $\mu_b = 0$, $R_b = \varepsilon_b$, and $\Delta_b = 1$. For irreversible compression, no particular quantization step size is specified in the standard. Instead, the number of exponent and mantissa bits must be provided to the decoder on a subband basis, called *expounded quantization*, or for the $N_L LL$ subband only, called *derived quantization*. In the latter case, the remaining subbands are quantized using extrapolated $N_L LL$ subband parameters. Letting ε_0 and μ_0 be the number of bits allocated to the $N_L LL$ subband, the extrapolated parameters for subband b are

$$\mu_b = \mu_0$$
$$\varepsilon_b = \varepsilon_0 + n_b - N_L \tag{8.2-65}$$

where n_b denotes the number of subband decomposition levels from the original image tile component to subband b.

In the final steps of the encoding process, the coefficients of each transformed tile-component's subbands are arranged into rectangular blocks called *code blocks*, which are coded individually, one bit plane at a time. Starting from the most significant bit plane with a nonzero element, each bit plane is processed in three passes. Each bit (in a bit plane) is coded in only one of the three passes, which are called *significance propagation*, *magnitude refinement*, and *cleanup*. The outputs are then arithmetically coded and grouped with similar passes from other code blocks to form *layers*. A layer is an arbitrary number of groupings of coding passes from each code block. The resulting layers finally are partitioned into *packets*, providing an additional method of extracting a spatial region of interest from the total code stream. Packets are the fundamental unit of the encoded code stream.

JPEG-2000 decoders simply invert the operations described previously. After reconstructing the subbands of the tile-components from the arithmetically coded JPEG-2000 packets, a user-selected number of the subbands is decoded. Although the encoder may have encoded M_b bit planes for a particular subband, the user—due to the embedded nature of the code stream—may choose to decode only N_b bit planes. This amounts to quantizing the coefficients of the code block using a step size of $2^{M_b - N_b} \cdot \Delta_b$. Any nondecoded bits are set to zero and the resulting coefficients, denoted

Quantization as defined earlier in the chapter is irreversible. The term "inverse quantized" does not mean that there is no information loss. This process is lossy except for the case of reversible JPEG-2000 compression, where $\mu_b = 0$, $R_b = \varepsilon_b$, and $\Delta_b = 1$.

$\bar{q}_b(u, v)$, are inverse quantized using

$$
R_{q_b}(u, v) = \begin{cases}
(\bar{q}_b(u, v) + r \cdot 2^{M_b - N_b(u, v)}) \cdot \Delta_b & \bar{q}_b(u, v) > 0 \\
(\bar{q}_b(u, v) - r \cdot 2^{M_b - N_b(u, v)}) \cdot \Delta_b & \bar{q}_b(u, v) < 0 \quad (8.2\text{-}66) \\
0 & \bar{q}_b(u, v) = 0
\end{cases}
$$

where $R_{q_b}(u, v)$ denotes an inverse-quantized transform coefficient and $N_b(u, v)$ is the number of decoded bit planes for $\bar{q}_b(u, v)$. *Reconstruction parameter r is chosen by the decoder to produce the best visual or objective quality of reconstruction.* Generally $0 \le r < 1$, with a common value being $r = 1/2$. The inverse-quantized coefficients then are inverse-transformed by column and by row using an FWT^{-1} filter bank whose coefficients are obtained from Table 8.15 and Eq. (7.1-11), or via the following lifting-based operations:

$$
\begin{array}{ll}
X(2n) = K \cdot Y(2n), & i_0 - 3 \le 2n < i_1 + 3 \\
X(2n+1) = (-1/K) \cdot Y(2n+1), & i_0 - 2 \le 2n - 1 < i_1 + 2 \\
X(2n) = X(2n) - \delta[X(2n-1) + X(2n+1)], & i_0 - 3 \le 2n < i_1 + 3 \\
X(2n+1) = X(2n+1) - \gamma[X(2n) + X(2n+2)], & i_0 - 2 \le 2n+1 < i_1 + 2 \\
X(2n) = X(2n) - \beta[X(2n-1) + X(2n+1)], & i_0 - 1 \le 2n < i_1 + 1 \\
X(2n+1) = X(2n+1) - \alpha[X(2n) + X(2n+2)], & i_0 \le 2n+1 < i_1
\end{array}
$$

$$(8.2\text{-}67)$$

where parameters α, β, γ, δ, and K are as defined for Eq. (8.2-62). Inverse-quantized coefficient row or column element $Y(n)$ is symmetrically extended when necessary. The final decoding steps are the assembly of the component tiles, inverse component transformation (if required), and DC level shifting. For irreversible coding, the inverse component transformation is

$$
\begin{aligned}
I_0(x, y) &= Y_0(x, y) + 1.402 Y_2(x, y) \\
I_1(x, y) &= Y_0(x, y) - 0.34413 Y_1(x, y) - 0.71414 Y_2(x, y) \quad (8.2\text{-}68) \\
I_2(x, y) &= Y_0(x, y) + 1.772 Y_1(x, y)
\end{aligned}
$$

and the transformed pixels are shifted by $+2^{Ssiz-1}$.

EXAMPLE 8.28:
A comparison of JPEG-2000 wavelet-based coding and JPEG DCT-based compression.

■ Figure 8.49 shows four JPEG-2000 approximations of the monochrome image in Figure 8.9(a). Successive rows of the figure illustrate increasing levels of compression — including $C = 25, 52, 75$, and 105. The images in column 1 are decompressed JPEG-2000 encodings. The differences between these images and the original image [Fig. 8.9(a)] are shown in the second column, and the third column contains a zoomed portion of the reconstructions in column 1. Because the compression ratios for the first two rows are virtually identical to the compression ratios in Example 8.18, these results can be compared — both qualitatively and quantitatively — to the JPEG transform-based results in Figs. 8.32(a) through (f).

FIGURE 8.49 Four JPEG-2000 approximations of Fig. 8.9(a). Each row contains a result after compression and reconstruction, the scaled difference between the result and the original image, and a zoomed portion of the reconstructed image. (Compare the results in rows 1 and 2 with the JPEG results in Fig. 8.32.)

A visual comparison of the error images in rows 1 and 2 of Fig. 8.49 with the corresponding images in Figs. 8.32(b) and (e) reveals a noticeable decrease of error in the JPEG-2000 results—3.86 and 5.77 intensity levels as opposed to 5.4 and 10.7 intensity levels for the JPEG results. The computed errors favor the wavelet-based results at both compression levels. Besides decreasing reconstruction error, wavelet coding dramatically increases (in a subjective sense) image quality. Note that the blocking artifact that dominated the JPEG results [see Figs. 8.32(c) and (f)] is not present in Fig. 8.49. Finally, we note that the compression achieved in rows 3 and 4 of Fig. 8.49 is not practical with JPEG. JPEG-2000 provides useable images that are compressed by more than 100:1—with the most objectionable degradation being increased image blur. ■

8.3   Digital Image Watermarking

The methods and standards of Section 8.2 make the distribution of images (whether in photographs or videos) on digital media and over the Internet practical. Unfortunately, the images so distributed can be copied repeatedly and without error, putting the rights of their owners at risk. Even when encrypted for distribution, images are unprotected after decryption. One way to discourage illegal duplication is to insert one or more items of information, collectively called a *watermark*, into potentially vulnerable images in such a way that the watermarks are inseparable from the images themselves. As integral parts of the *watermarked images*, they protect the rights of their owners in a variety of ways, including:

1. *Copyright identification.* Watermarks can provide information that serves as proof of ownership when the rights of the owner have been infringed.
2. *User identification* or *fingerprinting.* The identity of legal users can be encoded in watermarks and used to identify sources of illegal copies.
3. *Authenticity determination.* The presence of a watermark can guarantee that an image has not been altered—assuming the watermark is designed to be destroyed by any modification of the image.
4. *Automated monitoring.* Watermarks can be monitored by systems that track when and where images are used (e.g., programs that search the Web for images placed on Web pages). Monitoring is useful for royalty collection and/or the location of illegal users.
5. *Copy protection.* Watermarks can specify rules of image usage and copying (e.g., to DVD players).

In this section, we provide a brief overview of *digital image watermarking*—the process of inserting data into an image in such a way that it can be used to make an assertion about the image. The methods described have little in common with the compression techniques presented in the previous sections—although they do involve the coding of information. In fact, watermarking and compression are in some ways opposites. While the objective in compression is to reduce the amount of data used to represent images, the goal in watermarking is to add information and thus data (i.e., watermarks) to them. As will be seen in the remainder of the section, the watermarks themselves can be either visible or invisible.

A *visible watermark* is an opaque or semi-transparent sub-image or image that is placed on top of another image (i.e., the image being watermarked) so that it is obvious to the viewer. Television networks often place visible watermarks (fashioned after their logos) in the upper- or lower-right hand corner of the television screen. As the following example illustrates, visible watermarking typically is performed in the spatial domain.

■ The image in Fig. 8.50(b) is the lower-right-hand quadrant of the image in Fig. 8.9(a) with a scaled version of the watermark in Fig. 8.50(a) overlaid on top of it. Letting f_w denote the watermarked image, we can express it as a linear combination of the unmarked image f and watermark w using

EXAMPLE 8.29:
A simple visible watermark.

$$f_w = (1 - \alpha)f + \alpha w \qquad (8.3\text{-}1)$$

where constant α controls the relative visibility of the watermark and the underlying image. If α is 1, the watermark is opaque and the underlying image is completely obscured. As α approaches 0, more of the underlying image and less of the watermark is seen. In general, $0 < \alpha \leq 1$; in Fig. 8.50(b), $\alpha = 0.3$. Figure 8.50(c) is the computed difference (scaled in intensity) between the watermarked image in (b) and the unmarked image in Fig. 8.9(a). Intensity 128 represents a difference of 0. Note that the underlying image is clearly visible through the "semi-transparent" watermark. This is evident in both Fig. 8.50(b) and the difference image in (c). ■

a
b c

FIGURE 8.50
A simple visible watermark:
(a) watermark;
(b) the watermarked image; and (c) the difference between the watermarked image and the original (non-watermarked) image.

Unlike the visible watermark of the previous example, *invisible watermarks* cannot be seen with the naked eye. They are imperceptible—but can be recovered with an appropriate decoding algorithm. Invisibility is assured by inserting them as visually redundant information—as information that the human visual system ignores or cannot perceive (see Section 8.1.3). Figure 8.51(a) provides a simple example. Because the least significant bits of an 8-bit image have virtually no effect on our perception of the image, the watermark from Fig. 8.50(a) was inserted or "hidden" in its two least significant bits. Using the notation introduced above, we let

$$f_w = 4\left(\frac{f}{4}\right) + \frac{w}{64} \qquad (8.3\text{-}2)$$

and use unsigned integer arithmetic to perform the calculations. Dividing and multiplying by 4 sets the two least significant bits of f to 0, dividing w by 64 shifts its two most significant bits into the two least significant bit positions, and adding the two results generates the *LSB watermarked image*. Note that the embedded watermark is not visible in Fig. 8.51(a). By zeroing the most significant 6 bits of this image and scaling the remaining values to the full intensity range, however, the watermark can be extracted as in Fig. 8.51(b).

a b
c d

FIGURE 8.51 A simple invisible watermark: (a) watermarked image; (b) the extracted watermark; (c) the watermarked image after high quality JPEG compression and decompression; and (d) the extracted watermark from (c).

An important property of invisible watermarks is their resistance to both accidental and intentional attempts to remove them. *Fragile invisible watermarks* are destroyed by any modification of the images in which they are embedded. In some applications, like image authentication, this is a desirable characteristic. As Figs. 8.51(c) and (d) show, the LSB watermarked image in Fig. 8.51(a) contains a fragile invisible watermark. If the image in (a) is compressed and decompressed using lossy JPEG, the watermark is destroyed. Figure 8.51(c) is the result after compressing and decompressing Fig. 8.51(a); the rms error is 2.1 bits. If we try to extract the watermark from this image using the same method as in (b), the result is unintelligible [see Fig. 8.51(d)]. Although lossy compression and decompression preserved the important visual information in the image, the fragile watermark was destroyed.

Robust invisible watermarks are designed to survive image modification, whether the so called *attacks* are inadvertent or intentional. Common inadvertent attacks include lossy compression, linear and non-linear filtering, cropping, rotation, resampling, and the like. Intentional attacks range from printing and rescanning to adding additional watermarks and/or noise. Of course, it is unnecessary to withstand attacks that leave the image itself unusable.

Figure 8.52 shows the basic components of a typical image watermarking system. The encoder in Fig. 8.52(a) inserts watermark w_i into image f_i, producing watermarked image f_{w_i}; the complementary decoder in (b) extracts and validates the presence of w_i in watermarked input f_{w_i} or unmarked input f_j. If w_i is visible, the decoder is not needed. If it is invisible, the decoder may or may not require a copy of f_i and w_i [shown in gray in Fig. 8.52(b)] to do its job. If f_i and/or w_i are used, the watermarking system is known as a *private* or *restricted-key* system; if not, it is a *public* or *unrestricted-key* system. Because the decoder must process both marked and unmarked images, $w_\varnothing$ is used in Fig. 8.52(b) to denote the absence of a mark. Finally, we note that to determine the presence of w_i in an image, the decoder must correlate extracted watermark w_j with w_i and compare the result to a predefined threshold. The threshold sets the degree of similarity that is acceptable for a "match."

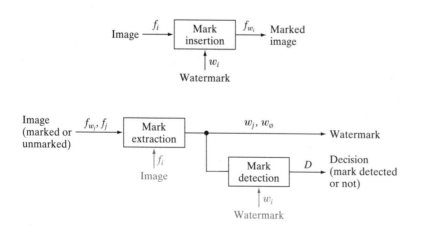

a
b

FIGURE 8.52
A typical image watermarking system:
(a) encoder;
(b) decoder.

EXAMPLE 8.30:
A DCT-based
invisible robust
watermark.

■ *Mark insertion* and *extraction* can be performed in the spatial domain, as in the previous examples, or in the transform domain. Figures 8.53(a) and (c) show two watermarked versions of the image in Fig. 8.9(a) using the DCT-based watermarking approach outlined below (Cox et al. [1997]):

Step 1. Compute the 2-D DCT of the image to be watermarked.

Step 2. Locate its K largest coefficients, $c_1, c_2, \ldots, c_K$, by magnitude.

a b
c d

FIGURE 8.53 (a) and (c) Two watermarked versions of Fig. 8.9(a); (b) and (d) the differences (scaled in intensity) between the watermarked versions and the unmarked image. These two images show the intensity contribution (although scaled dramatically) of the pseudo-random watermarks on the original image.

Step 3. Create a watermark by generating a K-element pseudo-random sequence of numbers, $\omega_1, \omega_2, \ldots, \omega_K$, taken from a Gaussian distribution with mean $\mu = 0$ and variance $\sigma^2 = 1$.

A pseudo-random number sequence approximates the properties of random numbers. It is not truly random because it depends on a predetermined initial value.

Step 4. Embed the watermark from step 3 into the K largest DCT coefficients from step 2 using the following equation

$$c_i' = c_i \cdot (1 + \alpha\omega_i) \quad 1 \le i \le K \tag{8.3-3}$$

for a specified constant $\alpha > 0$ (that controls the extent to which ω_i alters c_i). Replace the original c_i with the computed c_i' from Eq. (8.3-3).

For the images in Fig. 8.53, $\alpha = 0.1$ and $K = 1000$.

Step 5. Compute the inverse DCT of the result from step 4.

By employing watermarks made from pseudo-random numbers and spreading them across an image's perceptually significant frequency components, α can be made small, reducing watermark visibility. At the same time, watermark security is kept high because (1) the watermarks are composed of pseudo-random numbers with no obvious structure, (2) the watermarks are embedded in multiple frequency components with spatial impact over the entire 2-D image (so their location is not obvious) and (3) attacks against them tend to degrade the image as well (i.e., the image's most important frequency components must be altered to affect the watermarks).

Figures 8.53(b) and (d) make the changes in image intensity that result from the pseudo-random numbers that are embedded in the DCT coefficients of the watermarked images in Figs. 8.53(a) and (c) visible. Obviously, the pseudo-random numbers must have an effect—even if too small to see—on the watermarked images. To display the effect, the images in Figs. 8.53(a) and (c) were subtracted from the unmarked image in Fig. 8.9(a) and scaled in intensity to the range [0, 255]. Figures 8.53(b) and (d) are the resulting images; they show the 2-D spatial contributions of the pseudo-random numbers. Because they have been scaled, however, you cannot simply add these images to the image in Fig. 8.9(a) and get the watermarked images in Figs. 8.53(a) and (c). As can be seen in Figs. 8.53(a) and (c), their actual intensity perturbations are small to negligible.

To determine whether a particular image is a copy of a previously watermarked image with watermark $\omega_1, \omega_2, \ldots, \omega_K$ and DCT coefficients $c_1, c_2, \ldots, c_K$, we use the following procedure:

Step 1. Compute the 2-D DCT of the image in question.

Step 2. Extract the K DCT coefficients (in the positions corresponding to $c_1, c_2, \ldots, c_K$ of step 2 in the watermarking procedure) and denote the coefficients as $\hat{c}_1, \hat{c}_2, \ldots, \hat{c}_K$. If the image in question is the previously watermarked image (without modification), $\hat{c}_i = c_i'$ for $1 \le i \le K$. If it is a modified copy of the watermarked image (i.e., it has undergone some sort of attack), $\hat{c}_i \approx c_i'$ for $1 \le i \le K$ (the $\hat{c}_i$ will be approximations of the c_i'). Otherwise, the image in question will be an unmarked image or an image with a completely different watermark—and the $\hat{c}_i$ will bear no resemblance to the original c_i'.

Step 3. Compute watermark $\hat{\omega}_1, \hat{\omega}_2, \ldots, \hat{\omega}_K$ using

$$\hat{\omega}_i = \hat{c}_i - c_i \quad \text{for} \quad 1 \le i \le K \tag{8.3-4}$$

Recall that watermarks are a sequence of pseudo-random numbers.

Step 4. Measure the similarity of $\hat{\omega}_1, \hat{\omega}_2, \ldots, \hat{\omega}_K$ (from step 2) and $\omega_1, \omega_2, \ldots, \omega_K$ (from step 3 of the watermarking procedure) using a metric such as the correlation coefficient

We discuss the correlation coefficient in detail in Section 12.2.1.

$$\gamma = \frac{\sum\limits_{i=1}^{K} (\hat{\omega}_i - \overline{\hat{\omega}})(\omega_i - \overline{\omega})}{\sqrt{\sum\limits_{i=1}^{K} (\hat{\omega}_i - \overline{\hat{\omega}})^2 \cdot \sum\limits_{i=1}^{K} (\omega_i - \overline{\omega})^2}} \quad 1 \le i \le K \tag{8.3-5}$$

where $\overline{\omega}$ and $\overline{\hat{\omega}}$ are the means of the two K-element watermarks.

Step 5. Compare the measured similarity, γ, to a predefined threshold, T, and make a binary detection decision

$$D = \begin{cases} 1 & \text{if } \gamma \ge T \\ 0 & \text{otherwise} \end{cases} \tag{8.3-6}$$

In other words, $D = 1$ indicates that watermark $\omega_1, \omega_2, \ldots, \omega_K$ is present (with respect to the specified threshold, T); $D = 0$ indicates that it was not.

Using this procedure, the original watermarked image in Fig. 8.53(a)—measured against itself—yields a correlation coefficient of 0.9999, i.e., $\gamma = 0.9999$. It is an unmistakable match. In a similar manner, the image in Fig. 8.53(b), when measured against the image in Fig. 8.53(a), results in a γ of 0.0417—it could not be mistaken for the watermarked image in Fig. 8.53(a) because the correlation coefficient is so low. ■

To conclude the section, we note that the DCT-based watermarking approach of the previous example is fairly resistant to watermark attacks, partly because it is a private or restricted-key method. Restricted-key methods are always more resilient than their unrestricted-key counterparts. Using the watermarked image in Fig. 8.53(a), Fig. 8.54 illustrates the ability of the method to withstand a variety of common attacks. As can be seen in the figure, watermark detection is quite good over the range of attacks that were implemented—the resulting correlation coefficients (shown under each image in the figure) vary from 0.3113 to 0.9945. When subjected to a high quality but lossy (resulting in an rms error of 7 intensities) JPEG compression and decompression, $\gamma = 0.9945$. Even when the compression and reconstructed yields an rms error of 10 intensity levels, $\gamma = 0.7395$—and the usability of this image has been significantly degraded. Significant smoothing by spatial filtering and the addition of Gaussian noise do not reduce the correlation coefficient below 0.8230. However, histogram equalization reduces γ to 0.5210; and rotation has the largest effect—reducing γ to 0.3313. All attacks, except for the lossy JPEG

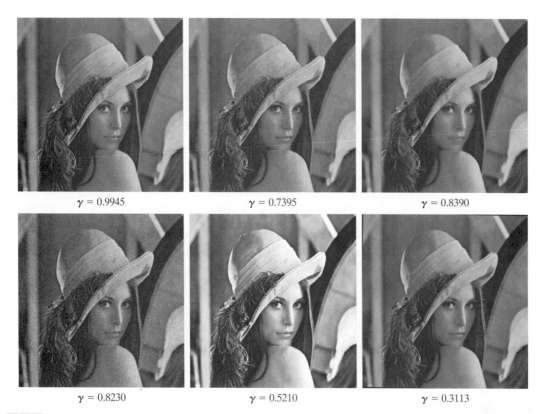

$\gamma = 0.9945$ $\gamma = 0.7395$ $\gamma = 0.8390$

$\gamma = 0.8230$ $\gamma = 0.5210$ $\gamma = 0.3113$

a b c
d e f

FIGURE 8.54 Attacks on the watermarked image in Fig. 8.53(a): (a) lossy JPEG compression and decompression with an rms error of 7 intensity levels; (b) lossy JPEG compression and decompression with an rms error of 10 intensity levels (note the blocking artifact); (c) smoothing by spatial filtering; (d) the addition of Gaussian noise; (e) histogram equalization; and (f) rotation. Each image is a modified version of the watermarked image in Fig. 8.53(a). After modification, they retain their watermarks to varying degrees, as indicated by the correlation coefficients below each image.

compression and reconstruction in (a), have significantly reduced the usability of the original watermarked image.

Summary

The principal objectives of this chapter were to present the theoretic foundation of digital image compression, to describe the most commonly used compression methods, and to introduce the related area of digital image watermarking. Although the level of the presentation is introductory in nature, the references provide an entry into the extensive body of literature dealing with the topics discussed. As evidenced by the international standards listed in Tables 8.3 and 8.4, compression plays a key role in document image storage and transmission, the Internet, and commercial video distribution (e.g., DVDs). It is one of the few areas of image processing that has received a sufficiently broad commercial appeal to warrant the adoption of widely accepted standards. And image watermarking is becoming increasingly important as more and more images are distributed in compressed digital form.

References and Further Reading

The introductory material of the chapter, which is generally confined to Section 8.1, is basic to image compression and may be found in one form or another in most of the general image processing books cited at the end of Chapter 1. For additional information on the human visual system, see Netravali and Limb [1980], as well as Huang [1966], Schreiber and Knapp [1958], and the references cited at the end of Chapter 2. For more on information theory, see the book Web site or Abramson [1963], Blahut [1987], and Berger [1971]. Shannon's classic paper, "A Mathematical Theory of Communication" [1948], lays the foundation for the area and is another excellent reference. Subjective fidelity criteria are discussed in Frendendall and Behrend [1960].

Throughout the chapter, a variety of compression standards are used in examples. Most of them were implemented using Adobe Photoshop (with freely available compression plug-ins) and/or MATLAB, which is described in Gonzalez et al. [2004]. Compression standards, as a rule, are lengthy and complex; we have not attempted to cover any of them in their entirety. For more information on a particular standard, see the published documents of the appropriate standards organization—the International Standards Organization, International Electrotechnical Commission, and/or the International Telecommunications Union. Additional references on standards include Hunter and Robinson [1980], Ang et al. [1991], Fox [1991], Pennebaker and Mitchell [1992], Bhatt et al. [1997], Sikora [1997], Bhaskaran and Konstantinos [1997], Ngan et al. [1999], Weinberger et al. [2000], Symes [2001], Mitchell et al. [1997], and Manjunath et al. [2001].

The lossy and error-free compression techniques described in Section 8.2 and watermarking techniques in Section 8.3 are, for the most part, based on the original papers cited in the text. The algorithms covered are representative of the work in this area, but are by no means exhaustive. The material on LZW coding has its origins in the work of Ziv and Lempel [1977, 1978]. The material on arithmetic coding follows the development in Witten, Neal, and Cleary [1987]. One of the more important implementations of arithmetic coding is summarized in Pennebaker et al. [1988]. For a good discussion of lossless predictive coding, see the tutorial by Rabbani and Jones [1991]. The adaptive predictor of Eq. (8.2-56) is from Graham [1958]. For more on motion compensation, see S. Solari [1997], which also contains an introduction to general video compression and compression standards, and Mitchell et al. [1997]. The DCT-based watermarking technique in Section 8.3 is based on the paper by Cox et al. [1997]. For more on watermarking, see the books by Cox et al. [2001] and Parhi and Nishitani [1999]. See also the paper by S. Mohanty [1999].

Many survey articles have been devoted to the field of image compression. Noteworthy are Netravali and Limb [1980], A. K. Jain [1981], a special issue on picture communication systems in the *IEEE Transactions on Communications* [1981], a special issue on the encoding of graphics in the *Proceedings of IEEE* [1980], a special issue on visual communication systems in the *Proceedings of the IEEE* [1985], a special issue on image sequence compression in the *IEEE Transactions on Image Processing* [1994], and a special issue on vector quantization in the *IEEE Transactions on Image Processing* [1996]. In addition, most issues of the *IEEE Transactions on Image Processing, IEEE Transactions on Circuits and Systems for Video Technology,* and *IEEE Transactions on Multimedia* include articles on video and still image compression, motion compensation, and watermarking. See, for example, Robinson [2006], Chandler and Hemami [2005], Yan and Cosman [2003], Boulgouris et al. [2001], Martin and Bell [2001], Chen and Wilson [2000], Hartenstein et al. [2000], Yang and Ramchandran [2000], Meyer et al. [2000], S. Mitra et al. [1998], Mukherjee and Mitra [2003], Xu et al. [2005], Rane and Sapiro [2001], Hu et al. [2006], Pi et al. [2006], Dugelay et al. [2006], and Kamstra and Heijmans [2005] as a starting point for further reading and references.

Problems

8.1 **(a)** Can variable-length coding procedures be used to compress a histogram equalized image with 2^n intensity levels? Explain.

  (b) Can such an image contain spatial or temporal redundancies that could be exploited for data compression?

8.2 One variation of run-length coding involves (1) coding only the runs of 0's or 1's (not both) and (2) assigning a special code to the start of each line to reduce the effect of transmission errors. One possible code pair is (x_k, r_k), where x_k and r_k represent the kth run's starting coordinate and run length, respectively. The code $(0, 0)$ is used to signal each new line.

  (a) Derive a general expression for the maximum average runs per scan line required to guarantee data compression when run-length coding a $2^n \times 2^n$ binary image.

  (b) Compute the maximum allowable value for $n = 10$.

8.3 Consider an 8-pixel line of intensity data, $\{108, 139, 135, 244, 172, 173, 56, 99\}$. If it is uniformly quantized with 4-bit accuracy, compute the rms error and rms signal-to-noise ratios for the quantized data.

★8.4 Although quantization results in information loss, it is sometimes invisible to the eye. For example, when 8-bit pixels are uniformly quantized to fewer bits/pixel, false contouring often occurs. It can be reduced or eliminated using *improved gray-scale* (IGS) *quantization*. A sum—initially set to zero—is formed from the current 8-bit intensity value and the four least significant bits of the previously generated sum. If the four most significant bits of the intensity value are 1111_2, however, 0000_2 is added instead. The four most significant bits of the resulting sum are used as the coded pixel value.

  (a) Construct the IGS code for the intensity data in Problem 8.3.

  (b) Compute the rms error and rms signal-to-noise ratios for the decoded IGS data.

8.5 A 1024×1024 8-bit image with 5.3 bits/pixel entropy [computed from its histogram using Eq. (8.1-7)] is to be Huffman coded.

  (a) What is the maximum compression that can be expected?

  (b) Will it be obtained?

  (c) If a greater level of lossless compression is required, what else can be done?

★8.6 The base e unit of information is commonly called a *nat*, and the base-10 information unit is called a *Hartley*. Compute the conversion factors needed to relate these units to the base-2 unit of information (the bit).

★**8.7**   Prove that, for a zero-memory source with q symbols, the maximum value of the entropy is $\log q$, which is achieved if and only if all source symbols are equiprobable. [*Hint*: Consider the quantity $\log q - H(z)$ and note the inequality $\ln x \leq x - 1$.]

8.8   **(a)** How many unique Huffman codes are there for a three-symbol source?

(b) Construct them.

8.9   Consider the simple 4×8, 8-bit image:

$$
\begin{array}{cccccccc}
21 & 21 & 21 & 95 & 169 & 243 & 243 & 243 \\
21 & 21 & 21 & 95 & 169 & 243 & 243 & 243 \\
21 & 21 & 21 & 95 & 169 & 243 & 243 & 243 \\
21 & 21 & 21 & 95 & 169 & 243 & 243 & 243
\end{array}
$$

(a) Compute the entropy of the image.

(b) Compress the image using Huffman coding.

(c) Compute the compression achieved and the effectiveness of the Huffman coding.

★**(d)** Consider Huffman encoding pairs of pixels rather than individual pixels. That is, consider the image to be produced by the second extension of the zero-memory source that produced the original image. What is the entropy of the image when looked at as pairs of pixels?

(e) Consider coding the differences between adjacent pixels. What is the entropy of the new difference image? What does this tell us about compressing the image?

(f) Explain the entropy differences in (a), (d) and (e).

8.10   Using the Huffman code in Fig. 8.8, decode the encoded string 0101000001010111110100.

8.11   Compute Golomb code $G_3(n)$ for $0 \leq n \leq 15$.

8.12   Write a general procedure for decoding Golomb code $G_m(n)$.

8.13   Why is it not possible to compute the Huffman code of the nonnegative integers, $n \geq 0$, with the geometric probability mass function of Eq. (8.2-2)?

8.14   Compute exponential Golomb code $G^2_{\exp}(n)$ for $0 \leq n \leq 15$.

★**8.15**   Write a general procedure for decoding exponential Golomb code $G^k_{\exp}(n)$.

8.16   Plot the optimal Golomb coding parameter m as a function of ρ for $0 < \rho < 1$ in Eq. (8.2-3).

8.17   Given a four-symbol source $\{a, b, c, d\}$ with source probabilities $\{0.1, 0.4, 0.3, 0.2\}$, arithmetically encode the sequence *bbadc*.

★**8.18**   The arithmetic decoding process is the reverse of the encoding procedure. Decode the message 0.23355 given the coding model

Symbol	Probability
a	0.2
e	0.3
i	0.1
o	0.2
u	0.1
!	0.1

8.19 Use the LZW coding algorithm of Section 8.2.4 to encode the 7-bit ASCII string "aaaaaaaaaaa".

★**8.20** Devise an algorithm for decoding the LZW encoded output of Example 8.7. Since the dictionary that was used during the encoding is not available, the code book must be reproduced as the output is decoded.

8.21 Decode the BMP encoded sequence $\{3, 4, 5, 6, 0, 3, 103, 125, 67, 0, 2, 47\}$.

8.22 **(a)** Construct the entire 4-bit Gray code.

(b) Create a general procedure for converting a Gray-coded number to its binary equivalent and use it to decode 0111010100111.

8.23 Use the CCITT Group 4 compression algorithm to code the second line of the following two-line segment:

$$011001110011111111100001$$
$$111111100011100001111111$$

Assume that the initial reference element a_0 is located on the first pixel of the second line segment.

★**8.24** **(a)** List all the members of JPEG DC coefficient difference category 3.

(b) Compute their default Huffman codes using Table A.4.

8.25 How many computations are required to find the optimal motion vector of an 8×8 macroblock using the MAD optimality criterion, single pixel precision, and a maximum allowable displacement of 8 pixels? What would it become for $\frac{1}{4}$ pixel precision?

8.26 What are the advantages of using B-frames for motion compensation?

★**8.27** Draw the block diagram of the companion motion compensated video decoder for the encoder in Fig. 8.39.

8.28 An image whose autocorrelation function is of the form of Eq. (8.2-49) with $\rho_h = 0$ is to be DPCM coded using a second-order predictor.

(a) Form the autocorrelation matrix $\mathbf{R}$ and vector $\mathbf{r}$.

(b) Find the optimal prediction coefficients.

(c) Compute the variance of the prediction error that would result from using the optimal coefficients.

★**8.29** Derive the Lloyd-Max decision and reconstruction levels for $L = 4$ and the uniform probability density function.

$$p(s) = \begin{cases} \dfrac{1}{2A} & -A \le s \le A \\ 0 & \text{otherwise} \end{cases}$$

8.30 A radiologist from a well-known research hospital recently attended a medical conference at which a system that could transmit 4096×4096 12-bit digitized X-ray images over standard T1 (1.544 Mb/s) phone lines was exhibited. The system transmitted the images in a compressed form using a progressive technique in which a reasonably good approximation of the X-ray was first reconstructed at the viewing station and then refined gradually to produce an error-free display. The transmission of the data needed to generate the first approximation took approximately 5 or 6 s. Refinements were made every 5 or 6 s (on the average) for the next 1 min, with the first and last refinements having the most and least significant impact on the reconstructed X-ray, respectively. The physician was favorably impressed with the system, because she could begin her diagnosis by using the first approximation of the X-ray and complete it as the error-free reconstruction of the X-ray was being generated. Upon returning to her office, she submitted a purchase request to the hospital administrator. Unfortunately, the hospital was on a relatively tight budget, which recently had been stretched thinner by the hiring of an aspiring young electrical engineering graduate. To appease the radiologist, the administrator gave the young engineer the task of designing such a system. (He thought it might be cheaper to design and build a similar system in-house. The hospital currently owned some of the elements of such a system, but the transmission of the raw X-ray data took more than 2 min.) The administrator asked the engineer to have an initial block diagram by the afternoon staff meeting. With little time and only a copy of *Digital Image Processing* from his recent school days in hand, the engineer was able to devise conceptually a system to satisfy the transmission and associated compression requirements. Construct a conceptual block diagram of such a system, specifying the compression techniques you would recommend.

8.31 Show that the lifting-based wavelet transform defined by Eq. (8.2-62) is equivalent to the traditional FWT filter bank implementation using the coefficients in Table 8.15. Define the filter coefficients in terms of α, β, γ, δ, and K.

8.32 Compute the quantization step sizes of the subbands for a JPEG-2000 encoded image in which derived quantization is used and 8 bits are allotted to the mantissa and exponent of the $2LL$ subband.

8.33 How would you add a visible watermark to an image in the frequency domain?

★**8.34** Design an invisible watermarking system based on the discrete Fourier transform.

8.35 Design an invisible watermarking system based on the discrete wavelet transform.

9 *Morphological Image Processing*

In form and feature, face and limb,
I grew so like my brother
That folks got taking me for him
And each for one another.

Henry Sambrooke Leigh, Carols of Cockayne, The Twins

Preview

The word *morphology* commonly denotes a branch of biology that deals with the form and structure of animals and plants. We use the same word here in the context of *mathematical morphology* as a tool for extracting image components that are useful in the representation and description of region shape, such as boundaries, skeletons, and the convex hull. We are interested also in morphological techniques for pre- or postprocessing, such as morphological filtering, thinning, and pruning.

In the following sections we develop and illustrate several important concepts in mathematical morphology. Many of the ideas introduced here can be formulated in terms of n-dimensional Euclidean space, E^n. However, our interest initially is on binary images whose components are elements of Z^2 (see Section 2.4.2). We discuss extensions to gray-scale images in Section 9.6.

The material in this chapter begins a transition from a focus on purely image processing methods, whose input and output are images, to processes in which the inputs are images, but the outputs are attributes extracted from those images, in the sense defined in Section 1.1. Tools such as morphology and related concepts are a cornerstone of the mathematical foundation that is utilized for extracting "meaning" from an image. Other approaches are developed and applied in the remaining chapters of the book.

You will find it helpful to review Sections 2.4.2 and 2.6.4 before proceeding.

9.1   Preliminaries

The language of mathematical morphology is set theory. As such, morphology offers a unified and powerful approach to numerous image processing problems. Sets in mathematical morphology represent objects in an image. For example, the set of all white pixels in a binary image is a complete morphological description of the image. In binary images, the sets in question are members of the 2-D integer space Z^2 (see Section 2.4.2), where each element of a set is a tuple (2-D vector) whose coordinates are the (x, y) coordinates of a white (or black, depending on convention) pixel in the image. Grayscale digital images of the form discussed in the previous chapters can be represented as sets whose components are in Z^3. In this case, two components of each element of the set refer to the coordinates of a pixel, and the third corresponds to its discrete intensity value. Sets in higher dimensional spaces can contain other image attributes, such as color and time varying components.

The set reflection operation is analogous to the flipping (rotating) operation performed in spatial convolution (Section 3.4.2).

In addition to the basic set definitions in Section 2.6.4, the concepts of set reflection and translation are used extensively in morphology. The *reflection* of a set B, denoted $\hat{B}$, is defined as

$$\hat{B} = \{w | w = -b, \quad \text{for} \quad b \in B\} \tag{9.1-1}$$

If B is the set of pixels (2-D points) representing an object in an image, then $\hat{B}$ is simply the set of points in B whose (x, y) coordinates have been replaced by $(-x, -y)$. Figures 9.1(a) and (b) show a simple set and its reflection.[†]

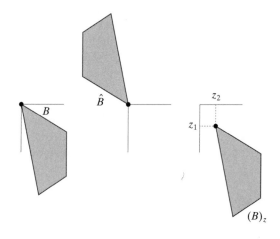

a b c

FIGURE 9.1
(a) A set, (b) its reflection, and (c) its translation by z.

[†]When working with graphics, such as the sets in Fig. 9.1, we use shading to indicate points (pixels) that are members of the set under consideration. When working with binary images, the sets of interest are pixels corresponding to objects. We show these in white, and all other pixels in black. The terms *foreground* and *background* are used often to denote the sets of pixels in an image defined to be objects and non-objects, respectively.

The *translation* of a set B by point $z = (z_1, z_2)$, denoted $(B)_z$, is defined as

$$(B)_z = \{c | c = b + z, \quad \text{for} \quad b \in B\} \tag{9.1-2}$$

If B is the set of pixels representing an object in an image, then $(B)_z$ is the set of points in B whose (x, y) coordinates have been replaced by $(x + z_1, y + z_2)$. Figure 9.1(c) illustrates this concept using the set B from Fig. 9.1(a).

Set reflection and translation are employed extensively in morphology to formulate operations based on so-called *structuring elements* (SEs): small sets or subimages used to probe an image under study for properties of interest. The first row of Fig. 9.2 shows several examples of structuring elements where each shaded square denotes a member of the SE. When it does not matter whether a location in a given structuring element is or is not a member of the SE set, that location is marked with an "×" to denote a "don't care" condition, as defined later in Section 9.5.4. In addition to a definition of which elements are members of the SE, the origin of a structuring element also must be specified. The origins of the various SEs in Fig. 9.2 are indicated by a black dot (although placing the center of an SE at its center of gravity is common, the choice of origin is problem dependent in general). When the SE is symmetric and no dot is shown, the assumption is that the origin is at the center of symmetry.

When working with images, we require that structuring elements be rectangular arrays. This is accomplished by appending the smallest possible number of background elements (shown nonshaded in Fig. 9.2) necessary to form a rectangular array. The first and last SEs in the second row of Fig. 9.2 illustrate the procedure. The other SEs in that row already are in rectangular form.

As an introduction to how structuring elements are used in morphology, consider Fig. 9.3. Figures 9.3(a) and (b) show a simple set and a structuring element. As mentioned in the previous paragraph, a computer implementation requires that set A be converted also to a rectangular array by adding background elements. The background border is made large enough to accommodate the entire structuring element when its origin is on the border of the

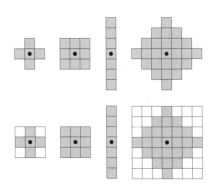

FIGURE 9.2 First row: Examples of structuring elements. Second row: Structuring elements converted to rectangular arrays. The dots denote the centers of the SEs.

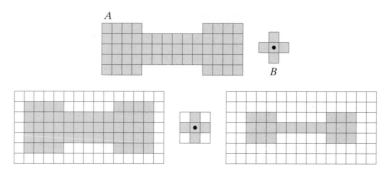

FIGURE 9.3 (a) A set (each shaded square is a member of the set). (b) A structuring element. (c) The set padded with background elements to form a rectangular array and provide a background border. (d) Structuring element as a rectangular array. (e) Set processed by the structuring element.

In future illustrations, we add enough background points to form rectangular arrays, but let the padding be implicit when the meaning is clear in order to simplify the figures.

original set (this is analogous to padding for spatial correlation and convolution, as discussed in Section 3.4.2). In this case, the structuring element is of size 3×3 with the origin in the center, so a one-element border that encompasses the entire set is sufficient, as Fig. 9.3(c) shows. As in Fig. 9.2, the structuring element is filled with the smallest possible number of background elements necessary to make it into a rectangular array [Fig. 9.3(d)].

Suppose that we define an operation on set A using structuring element B, as follows: Create a new set by running B over A so that the origin of B visits every element of A. At each location of the origin of B, if B is completely contained in A, mark that location as a member of the new set (shown shaded); else mark it as not being a member of the new set (shown not shaded). Figure 9.3(e) shows the result of this operation. We see that, when the origin of B is on a border element of A, part of B ceases to be contained in A, thus eliminating the location on which B is centered as a possible member for the new set. The net result is that the boundary of the set is *eroded*, as Fig. 9.3(e) shows. When we use terminology such as "the structuring element is contained in the set," we mean specifically that the elements of A and B fully overlap. In other words, although we showed A and B as arrays containing both shaded and nonshaded elements, only the shaded elements of both sets are considered in determining whether or not B is contained in A. These concepts form the basis of the material in the next section, so it is important that you understand the ideas in Fig. 9.3 fully before proceeding.

9.2 ■ Erosion and Dilation

We begin the discussion of morphology by studying two operations: *erosion* and *dilation*. These operations are fundamental to morphological processing. In fact, many of the morphological algorithms discussed in this chapter are based on these two primitive operations.

9.2.1 Erosion

With A and B as sets in Z^2, the erosion of A by B, denoted $A \ominus B$, is defined as

$$A \ominus B = \{z | (B)_z \subseteq A\} \qquad (9.2\text{-}1)$$

In words, this equation indicates that the erosion of A by B is the set of all points z such that B, translated by z, is contained in A. In the following discussion, set B is assumed to be a structuring element. Equation (9.2-1) is the mathematical formulation of the example in Fig. 9.3(e), discussed at the end of the last section. Because the statement that B has to be contained in A is equivalent to B not sharing any common elements with the background, we can express erosion in the following equivalent form:

$$A \ominus B = \{z | (B)_z \cap A^c = \varnothing\} \qquad (9.2\text{-}2)$$

where, as defined in Section 2.6.4, A^c is the complement of A and $\varnothing$ is the empty set.

Figure 9.4 shows an example of erosion. The elements of A and B are shown shaded and the background is white. The solid boundary in Fig. 9.4(c) is the limit beyond which further displacements of the origin of B would cause the structuring element to cease being completely contained in A. Thus, the locus of points (locations of the origin of B) within (and including) this boundary, constitutes the erosion of A by B. We show the erosion shaded in Fig. 9.4(c). Keep in mind that that erosion is simply the *set* of

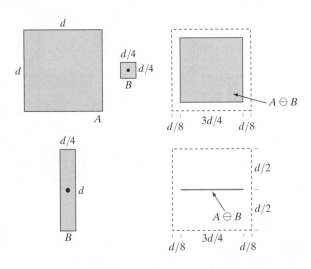

a b c
d   e

FIGURE 9.4 (a) Set A. (b) Square structuring element, B. (c) Erosion of A by B, shown shaded. (d) Elongated structuring element. (e) Erosion of A by B using this element. The dotted border in (c) and (e) is the boundary of set A, shown only for reference.

values of z that satisfy Eq. (9.2-1) or (9.2-2). The boundary of set A is shown dashed in Figs. 9.4(c) and (e) only as a reference; it is not part of the erosion operation. Figure 9.4(d) shows an elongated structuring element, and Fig. 9.4(e) shows the erosion of A by this element. Note that the original set was eroded to a line.

Equations (9.2-1) and (9.2-2) are not the only definitions of erosion (see Problems 9.9 and 9.10 for two additional, equivalent definitions.) However, these equations have the distinct advantage over other formulations in that they are more intuitive when the structuring element B is viewed as a spatial mask (see Section 3.4.1).

EXAMPLE 9.1:
Using erosion to remove image components.

■ Suppose that we wish to remove the lines connecting the center region to the border pads in Fig. 9.5(a). Eroding the image with a square structuring element of size 11×11 whose components are all 1s removed most of the lines, as Fig. 9.5(b) shows. The reason the two vertical lines in the center were thinned but not removed completely is that their width is greater than 11 pixels. Changing the SE size to 15×15 and eroding the original image again did remove all the connecting lines, as Fig. 9.5(c) shows (an alternate approach would have been to erode the image in Fig. 9.5(b) again using the same 11×11 SE). Increasing the size of the structuring element even more would eliminate larger components. For example, the border pads can be removed with a structuring element of size 45×45, as Fig. 9.5(d) shows.

a b
c d

FIGURE 9.5 Using erosion to remove image components. (a) A 486×486 binary image of a wirebond mask. (b)–(d) Image eroded using square structuring elements of sizes 11×11, 15×15, and 45×45, respectively. The elements of the SEs were all 1s.

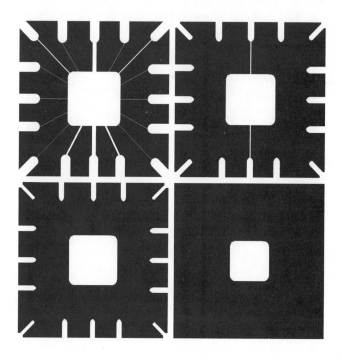

We see from this example that erosion shrinks or thins objects in a binary image. In fact, we can view erosion as a *morphological filtering* operation in which image details smaller than the structuring element are filtered (removed) from the image. In Fig. 9.5, erosion performed the function of a "line filter." We return to the concept of a morphological filter in Sections 9.3 and 9.6.3.   ■

9.2.2 Dilation

With A and B as sets in Z^2, the *dilation* of A by B, denoted $A \oplus B$, is defined as

$$A \oplus B = \left\{z | (\hat{B})_z \cap A \neq \varnothing\right\} \qquad (9.2\text{-}3)$$

This equation is based on reflecting B about its origin, and shifting this reflection by z (see Fig. 9.1). The dilation of A by B then is the set of all displacements, z, such that $\hat{B}$ and A overlap by at least one element. Based on this interpretation, Eq. (9.2-3) can be written equivalently as

$$A \oplus B = \left\{z | [(\hat{B})_z \cap A] \subseteq A\right\} \qquad (9.2\text{-}4)$$

As before, we assume that B is a structuring element and A is the set (image objects) to be dilated.

Equations (9.2-3) and (9.2-4) are not the only definitions of dilation currently in use (see Problems 9.11 and 9.12 for two different, yet equivalent, definitions). However, the preceding definitions have a distinct advantage over other formulations in that they are more intuitive when the structuring element B is viewed as a convolution mask. The basic process of flipping (rotating) B about its origin and then successively displacing it so that it slides over set (image) A is analogous to spatial convolution, as introduced in Section 3.4.2. Keep in mind, however, that dilation is based on set operations and therefore is a nonlinear operation, whereas convolution is a linear operation.

Unlike erosion, which is a shrinking or thinning operation, dilation "grows" or "thickens" objects in a binary image. The specific manner and extent of this thickening is controlled by the shape of the structuring element used. Figure 9.6(a) shows the same set used in Fig. 9.4, and Fig. 9.6(b) shows a structuring element (in this case $\hat{B} = B$ because the SE is symmetric about its origin). The dashed line in Fig. 9.6(c) shows the original set for reference, and the solid line shows the limit beyond which any further displacements of the origin of $\hat{B}$ by z would cause the intersection of $\hat{B}$ and A to be empty. Therefore, all points on and inside this boundary constitute the dilation of A by B. Figure 9.6(d) shows a structuring element designed to achieve more dilation vertically than horizontally, and Fig. 9.6(e) shows the dilation achieved with this element.

FIGURE 9.6
(a) Set A.
(b) Square
structuring ele-
ment (the dot de-
notes the origin).
(c) Dilation of A
by B, shown
shaded.
 (d) Elongated
structuring ele-
ment. (e) Dilation
of A using this
element. The
dotted border in
(c) and (e) is the
boundary of set A,
shown only for
reference

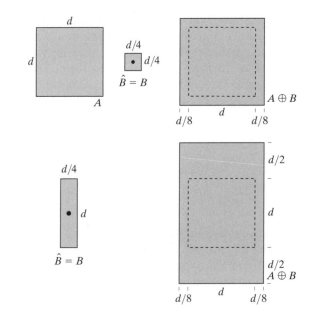

EXAMPLE 9.2:
An illustration of
dilation.

■ One of the simplest applications of dilation is for bridging gaps. Figure 9.7(a) shows the same image with broken characters that we studied in Fig. 4.49 in connection with lowpass filtering. The maximum length of the breaks is known to be two pixels. Figure 9.7(b) shows a structuring element that can be used for repairing the gaps (note that instead of shading, we used 1s to denote the elements of the SE and 0s for the background; this is because the SE is now being treated as a subimage and not as a graphic). Figure 9.7(c) shows the result of dilating the original image with this structuring element. The gaps were bridged. One immediate advantage of the morphological approach over the lowpass filtering method we used to bridge the gaps in Fig. 4.49 is

FIGURE 9.7
(a) Sample text of
poor resolution
with broken
characters (see
magnified view).
(b) Structuring
element.
(c) Dilation of (a)
by (b). Broken
segments were
joined.

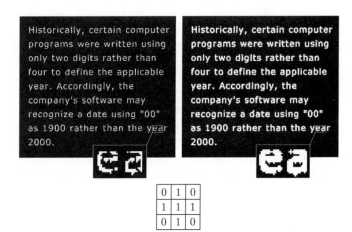

that the morphological method resulted directly in a binary image. Lowpass filtering, on the other hand, started with a binary image and produced a gray-scale image, which would require a pass with a thresholding function to convert it back to binary form.   ■

9.2.3 Duality

Erosion and dilation are duals of each other with respect to set complementation and reflection. That is,

$$(A \ominus B)^c = A^c \oplus \hat{B} \tag{9.2-5}$$

and

$$(A \oplus B)^c = A^c \ominus \hat{B} \tag{9.2-6}$$

Equation (9.2-5) indicates that erosion of A by B is the complement of the dilation of A^c by $\hat{B}$, and vice versa. The duality property is useful particularly when the structuring element is symmetric with respect to its origin (as often is the case), so that $\hat{B} = B$. Then, we can obtain the erosion of an image by B simply by dilating its background (i.e., dilating A^c) with the same structuring element and complementing the result. Similar comments apply to Eq. (9.2-6).

We proceed to prove formally the validity of Eq. (9.2-5) in order to illustrate a typical approach for establishing the validity of morphological expressions. Starting with the definition of erosion, it follows that

$$(A \ominus B)^c = \left\{ z | (B)_z \subseteq A \right\}^c$$

If set $(B)_z$ is contained in A, then $(B)_z \cap A^c = \varnothing$, in which case the preceding expression becomes

$$(A \ominus B)^c = \left\{ z | (B)_z \cap A^c = \varnothing \right\}^c$$

But the *complement* of the set of z's that satisfy $(B)_z \cap A^c = \varnothing$ is the set of z's such that $(B)_z \cap A^c \neq \varnothing$. Therefore,

$$(A \ominus B)^c = \left\{ z | (B)_z \cap A^c \neq \varnothing \right\}$$

$$= A^c \oplus \hat{B}$$

where the last step follows from Eq. (9.2-3). This concludes the proof. A similar line of reasoning can be used to prove Eq. (9.2-6) (see Problem 9.13).

9.3 Opening and Closing

As you have seen, dilation expands the components of an image and erosion shrinks them. In this section we discuss two other important morphological operations: opening and closing. *Opening* generally smoothes the contour of an object, breaks narrow isthmuses, and eliminates thin protrusions. *Closing* also tends to smooth sections of contours but, as opposed to opening, it generally fuses narrow breaks and long thin gulfs, eliminates small holes, and fills gaps in the contour.

The *opening* of set A by structuring element B, denoted $A \circ B$, is defined as

$$A \circ B = (A \ominus B) \oplus B \qquad (9.3\text{-}1)$$

Thus, the opening A by B is the erosion of A by B, followed by a dilation of the result by B.

Similarly, the *closing* of set A by structuring element B, denoted $A \bullet B$, is defined as

$$A \bullet B = (A \oplus B) \ominus B \qquad (9.3\text{-}2)$$

which says that the closing of A by B is simply the dilation of A by B, followed by the erosion of the result by B.

The opening operation has a simple geometric interpretation (Fig. 9.8). Suppose that we view the structuring element B as a (flat) "rolling ball." The *boundary* of $A \circ B$ is then established by the points in B that reach the *farthest* into the boundary of A as B is rolled around the *inside* of this boundary. This geometric *fitting* property of the opening operation leads to a set-theoretic formulation, which states that the opening of A by B is obtained by taking the union of all translates of B that fit into A. That is, opening can be expressed as a fitting process such that

$$A \circ B = \bigcup \{(B)_z | (B)_z \subseteq A\} \qquad (9.3\text{-}3)$$

where $\cup \{\cdot\}$ denotes the union of all the sets inside the braces.

Closing has a similar geometric interpretation, except that now we roll B on the outside of the boundary (Fig. 9.9). As discussed below, opening and closing are duals of each other, so having to roll the ball on the outside is not unexpected. Geometrically, a point w is an element of $A \bullet B$ if and only if $(B)_z \cap A \neq \varnothing$ for any translate of $(B)_z$ that contains w. Figure 9.9 illustrates the basic geometrical properties of closing.

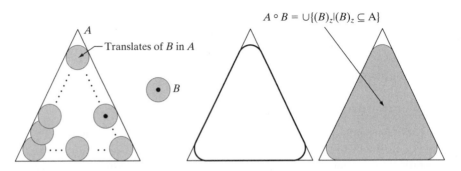

a b c d

FIGURE 9.8 (a) Structuring element B "rolling" along the inner boundary of A (the dot indicates the origin of B). (b) Structuring element. (c) The heavy line is the outer boundary of the opening. (d) Complete opening (shaded). We did not shade A in (a) for clarity.

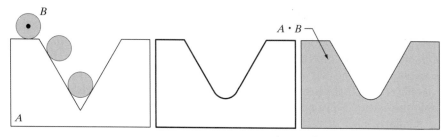

a b c

FIGURE 9.9 (a) Structuring element B "rolling" on the outer boundary of set A. (b) The heavy line is the outer boundary of the closing. (c) Complete closing (shaded). We did not shade A in (a) for clarity.

■ Figure 9.10 further illustrates the opening and closing operations. Figure 9.10(a) shows a set A, and Fig. 9.10(b) shows various positions of a disk structuring element during the erosion process. When completed, this process resulted in the disjoint figure in Fig. 9.10(c). Note the elimination of the bridge between the two main sections. Its width was thin in relation to the diameter of

EXAMPLE 9.3:
A simple illustration of morphological opening and closing.

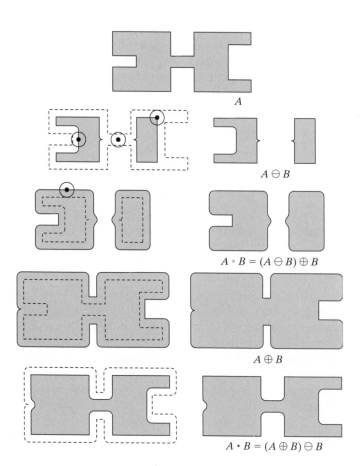

a
b c
d e
f g
h i

FIGURE 9.10
Morphological opening and closing. The structuring element is the small circle shown in various positions in (b). The SE was not shaded here for clarity. The dark dot is the center of the structuring element.

the structuring element; that is, the structuring element could not be completely contained in this part of the set, thus violating the conditions of Eq. (9.2-1). The same was true of the two rightmost members of the object. Protruding elements where the disk did not fit were eliminated. Figure 9.10(d) shows the process of dilating the eroded set, and Fig. 9.10(e) shows the final result of opening. Note that outward pointing corners were rounded, whereas inward pointing corners were not affected.

Similarly, Figs. 9.10(f) through (i) show the results of closing A with the same structuring element. We note that the inward pointing corners were rounded, whereas the outward pointing corners remained unchanged. The leftmost intrusion on the boundary of A was reduced in size significantly, because the disk did not fit there. Note also the smoothing that resulted in parts of the object from both opening and closing the set A with a circular structuring element.   ■

As in the case with dilation and erosion, opening and closing are duals of each other with respect to set complementation and reflection. That is,

$$(A \bullet B)^c = (A^c \circ \hat{B}) \tag{9.3-4}$$

and

$$(A \circ B)^c = (A^c \bullet \hat{B}) \tag{9.3-5}$$

We leave the proof of this result as an exercise (Problem 9.14).

The opening operation satisfies the following properties:

(a) $A \circ B$ is a subset (subimage) of A.
(b) If C is a subset of D, then $C \circ B$ is a subset of $D \circ B$.
(c) $(A \circ B) \circ B = A \circ B$.

Similarly, the closing operation satisfies the following properties:

(a) A is a subset (subimage) of $A \bullet B$.
(b) If C is a subset of D, then $C \bullet B$ is a subset of $D \bullet B$.
(c) $(A \bullet B) \bullet B = A \bullet B$.

Note from condition (c) in both cases that multiple openings or closings of a set have no effect after the operator has been applied once.

EXAMPLE 9.4:
Use of opening and closing for morphological filtering.

■ Morphological operations can be used to construct filters similar in concept to the spatial filters discussed in Chapter 3. The binary image in Fig. 9.11(a) shows a section of a fingerprint corrupted by noise. Here the noise manifests itself as random light elements on a dark background and as dark elements on the light components of the fingerprint. The objective is to eliminate the noise and its effects on the print while distorting it as little as possible. A morphological filter consisting of opening followed by closing can be used to accomplish this objective.

Figure 9.11(b) shows the structuring element used. The rest of Fig. 9.11 shows a step-by-step sequence of the filtering operation. Figure 9.11(c) is the

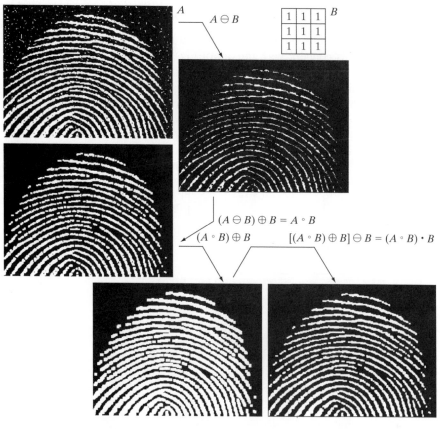

FIGURE 9.11
(a) Noisy image.
(b) Structuring
element.
(c) Eroded image.
(d) Opening of A.
(e) Dilation of the
opening.
(f) Closing of the
opening.
(Original image
courtesy of the
National Institute
of Standards and
Technology.)

result of eroding A with the structuring element. The background noise was completely eliminated in the erosion stage of opening because in this case all noise components are smaller than the structuring element. The size of the noise elements (dark spots) contained within the fingerprint actually increased in size. The reason is that these elements are inner boundaries that increase in size as the object is eroded. This enlargement is countered by performing dilation on Fig. 9.11(c). Figure 9.11(d) shows the result. The noise components contained in the fingerprint were reduced in size or deleted completely.

The two operations just described constitute the opening of A by B. We note in Fig. 9.11(d) that the net effect of opening was to eliminate virtually all noise components in both the background and the fingerprint itself. However, new gaps between the fingerprint ridges were created. To counter this undesirable effect, we perform a dilation on the opening, as shown in Fig. 9.11(e). Most of the breaks were restored, but the ridges were thickened, a condition that can be remedied by erosion. The result, shown in Fig. 9.11(f), constitutes the closing of the opening of Fig. 9.11(d). This final result is remarkably clean of noise specks, but it has the disadvantage that some of the print ridges were not fully repaired, and thus contain breaks. This is not totally unexpected, because no conditions were built into the procedure for maintaining connectivity (we discuss this issue again in Example 9.8 and demonstrate ways to address it in Section 11.1.7).   ■

9.4 The Hit-or-Miss Transformation

The morphological hit-or-miss transform is a basic tool for shape detection. We introduce this concept with the aid of Fig. 9.12, which shows a set A consisting of three shapes (subsets), denoted C, D, and E. The shading in Figs. 9.12(a) through (c) indicates the original sets, whereas the shading in Figs. 9.12(d) and (e) indicates the result of morphological operations. The objective is to find the location of one of the shapes, say, D.

FIGURE 9.12
(a) Set A. (b) A window, W, and the local background of D with respect to W, $(W - D)$.
(c) Complement of A. (d) Erosion of A by D.
(e) Erosion of A^c by $(W - D)$.
(f) Intersection of (d) and (e), showing the location of the origin of D, as desired. The dots indicate the origins of C, D, and E.

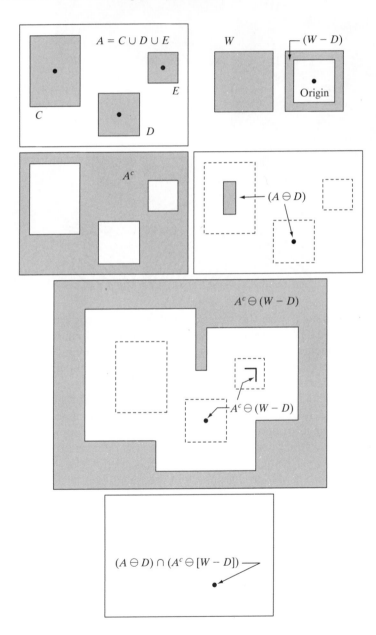

Let the origin of each shape be located at its center of gravity. Let D be enclosed by a small window, W. The *local background* of D with respect to W is defined as the set difference $(W - D)$, as shown in Fig. 9.12(b). Figure 9.12(c) shows the complement of A, which is needed later. Figure 9.12(d) shows the erosion of A by D (the dashed lines are included for reference). Recall that the erosion of A by D is the set of locations of the *origin* of D, such that D is completely contained in A. Interpreted another way, $A \ominus D$ may be viewed geometrically as the set of all locations of the origin of D at which D found a match (hit) in A. Keep in mind that in Fig. 9.12 A consists only of the three *disjoint* sets C, D, and E.

Figure 9.12(e) shows the erosion of the complement of A by the local background set $(W - D)$. The outer shaded region in Fig. 9.12(e) is part of the erosion. We note from Figs. 9.12(d) and (e) that the set of locations for which D *exactly* fits inside A is the *intersection* of the erosion of A by D and the erosion of A^c by $(W - D)$ as shown in Fig. 9.12(f). This intersection is precisely the location sought. In other words, if B denotes the set composed of D and its background, the match (or set of matches) of B in A, denoted $A \circledast B$, is

$$A \circledast B = (A \ominus D) \cap \left[A^c \ominus (W - D) \right] \tag{9.4-1}$$

We can generalize the notation somewhat by letting $B = (B_1, B_2)$, where B_1 is the set formed from elements of B associated with an object and B_2 is the set of elements of B associated with the corresponding background. From the preceding discussion, $B_1 = D$ and $B_2 = (W - D)$. With this notation, Eq. (9.4-1) becomes

$$A \circledast B = (A \ominus B_1) \cap (A^c \ominus B_2) \tag{9.4-2}$$

Thus, set $A \circledast B$ contains all the (origin) points at which, simultaneously, B_1 found a match ("hit") in A and B_2 found a match in A^c. By using the definition of set differences given in Eq. (2.6-19) and the dual relationship between erosion and dilation given in Eq. (9.2-5), we can write Eq. (9.4-2) as

$$A \circledast B = (A \ominus B_1) - (A \oplus \hat{B}_2) \tag{9.4-3}$$

However, Eq. (9.4-2) is considerably more intuitive. We refer to any of the preceding three equations as the *morphological hit-or-miss transform*.

The reason for using a structuring element B_1 associated with objects and an element B_2 associated with the background is based on an assumed definition that two or more objects are distinct only if they form disjoint (disconnected) sets. This is guaranteed by requiring that each object have at least a one-pixel-thick background around it. In some applications, we may be interested in detecting certain patterns (combinations) of 1s and 0s within a set, in which case a background is not required. In such instances, the hit-or-miss transform reduces to simple erosion. As indicated previously, erosion is still a set of matches, but without the additional requirement of a background match for detecting individual objects. This simplified pattern detection scheme is used in some of the algorithms developed in the following section.

9.5 Some Basic Morphological Algorithms

With the preceding discussion as foundation, we are now ready to consider some practical uses of morphology. When dealing with binary images, one of the principal applications of morphology is in extracting image components that are useful in the representation and description of shape. In particular, we consider morphological algorithms for extracting boundaries, connected components, the convex hull, and the skeleton of a region. We also develop several methods (for region filling, thinning, thickening, and pruning) that are used frequently in conjunction with these algorithms as pre- or post-processing steps. We make extensive use in this section of "mini-images," designed to clarify the mechanics of each morphological process as we introduce it. These images are shown graphically with 1s shaded and 0s in white.

9.5.1 Boundary Extraction

The boundary of a set A, denoted by $\beta(A)$, can be obtained by first eroding A by B and then performing the set difference between A and its erosion. That is,

$$\beta(A) = A - (A \ominus B) \tag{9.5-1}$$

where B is a suitable structuring element.

Figure 9.13 illustrates the mechanics of boundary extraction. It shows a simple binary object, a structuring element B, and the result of using Eq. (9.5-1). Although the structuring element in Fig. 9.13(b) is among the most frequently used, it is by no means unique. For example, using a 5×5 structuring element of 1s would result in a boundary between 2 and 3 pixels thick.

From this point on, we do not show border padding explicitly.

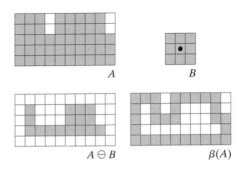

A B

$A \ominus B$ $\beta(A)$

a b
c d

FIGURE 9.13 (a) Set A. (b) Structuring element B. (c) A eroded by B. (d) Boundary, given by the set difference between A and its erosion.

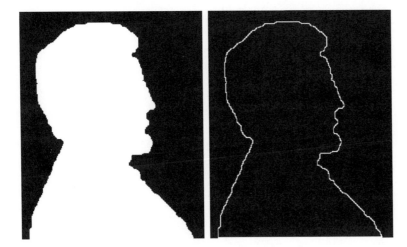

FIGURE 9.14
(a) A simple
binary image, with
1s represented in
white. (b) Result
of using
Eq. (9.5-1) with
the structuring
element in
Fig. 9.13(b).

■ Figure 9.14 further illustrates the use of Eq. (9.5-1) with a 3 × 3 structuring element of 1s. As for all binary images in this chapter, binary 1s are shown in white and 0s in black, so the elements of the structuring element, which are 1s, also are treated as white. Because of the size of the structuring element used, the boundary in Fig. 9.14(b) is one pixel thick. ■

EXAMPLE 9.5:
Boundary
extraction by
morphological
processing.

9.5.2 Hole Filling

A *hole* may be defined as a background region surrounded by a connected border of foreground pixels. In this section, we develop an algorithm based on set dilation, complementation, and intersection for filling holes in an image. Let A denote a set whose elements are 8-connected boundaries, each boundary enclosing a background region (i.e., a hole). Given a point in each hole, the objective is to fill all the holes with 1s.

We begin by forming an array, X_0, of 0s (the same size as the array containing A), except at the locations in X_0 corresponding to the given point in each hole, which we set to 1. Then, the following procedure fills all the holes with 1s:

$$X_k = (X_{k-1} \oplus B) \cap A^c \qquad k = 1, 2, 3, \dots \qquad (9.5\text{-}2)$$

where B is the symmetric structuring element in Fig. 9.15(c). The algorithm terminates at iteration step k if $X_k = X_{k-1}$. The set X_k then contains all the filled holes. The set union of X_k and A contains all the filled holes and their boundaries.

The dilation in Eq. (9.5-2) would fill the entire area if left unchecked. However, the intersection at each step with A^c limits the result to inside the region of interest. This is our first example of how a morphological process can be *conditioned* to meet a desired property. In the current application, it is appropriately called *conditional dilation*. The rest of Fig. 9.15 illustrates further the mechanics of Eq. (9.5-2). Although this example only has one hole, the concept clearly applies to any finite number of holes, assuming that a point inside each hole region is given.

FIGURE 9.15 Hole filling. (a) Set A (shown shaded). (b) Complement of A. (c) Structuring element B. (d) Initial point inside the boundary. (e)–(h) Various steps of Eq. (9.5-2). (i) Final result [union of (a) and (h)].

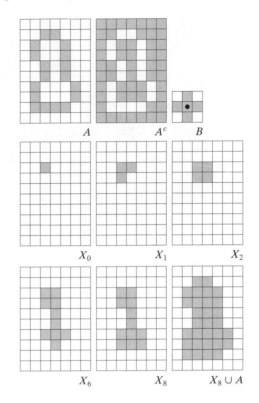

A A^c B

X_0 X_1 X_2

X_6 X_8 $X_8 \cup A$

EXAMPLE 9.6:
Morphological hole filling.

■ Figure 9.16(a) shows an image composed of white circles with black inner spots. An image such as this might result from thresholding into two levels a scene containing polished spheres (e.g., ball bearings). The dark spots inside the spheres could be the result of reflections. The objective is to eliminate the reflections by hole filling. Figure 9.16(a) shows one point selected inside one of the spheres, and Fig. 9.16(b) shows the result of filling that component. Finally,

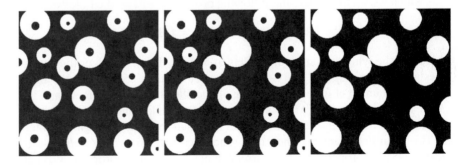

a b c

FIGURE 9.16 (a) Binary image (the white dot inside one of the regions is the starting point for the hole-filling algorithm). (b) Result of filling that region. (c) Result of filling all holes.

Fig. 9.16(c) shows the result of filling all the spheres. Because it must be known whether black points are background points or sphere inner points, fully automating this procedure requires that additional "intelligence" be built into the algorithm. We give a fully automatic approach in Section 9.5.9 based on morphological reconstruction. (See also Problem 9.23.)   ■

9.5.3 Extraction of Connected Components

The concepts of connectivity and connected components were introduced in Section 2.5.2. Extraction of connected components from a binary image is central to many automated image analysis applications. Let A be a set containing one or more connected components, and form an array X_0 (of the same size as the array containing A) whose elements are 0s (background values), except at each location known to correspond to a point in each connected component in A, which we set to 1 (foreground value). The objective is to start with X_0 and find all the connected components. The following iterative procedure accomplishes this objective:

$$X_k = (X_{k-1} \oplus B) \cap A \qquad k = 1, 2, 3, \ldots \qquad (9.5\text{-}3)$$

where B is a suitable structuring element (as in Fig. 9.17). The procedure terminates when $X_k = X_{k-1}$, with X_k containing all the connected components

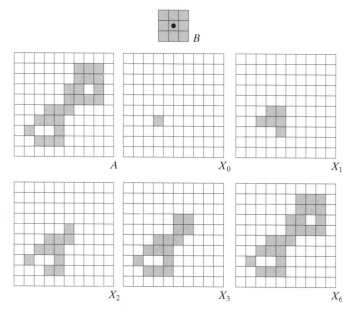

FIGURE 9.17 Extracting connected components. (a) Structuring element. (b) Array containing a set with one connected component. (c) Initial array containing a 1 in the region of the connected component. (d)–(g) Various steps in the iteration of Eq. (9.5-3).

of the input image. Note the similarity in Eqs. (9.5-3) and (9.5-2), the only difference being the use of A as opposed to A^c. This is not surprising, because here we are looking for foreground points, while the objective in Section 9.5.2 was to find background points.

Figure 9.17 illustrates the mechanics of Eq. (9.5-3), with convergence being achieved for $k = 6$. Note that the shape of the structuring element used is based on 8-connectivity between pixels. If we had used the SE in Fig. 9.15, which is based on 4-connectivity, the leftmost element of the connected component toward the bottom of the image would not have been detected because it is 8-connected to the rest of the figure. As in the hole-filling algorithm, Eq. (9.5-3) is applicable to any finite number of connected components contained in A, assuming that a point is known in each.

See Problem 9.24 for an algorithm that does not require that a point in each connected component be known a priori.

EXAMPLE 9.7:
Using connected components to detect foreign objects in packaged food.

■ Connected components are used frequently for automated inspection. Figure 9.18(a) shows an X-ray image of a chicken breast that contains bone fragments. It is of considerable interest to be able to detect such objects in processed food before packaging and/or shipping. In this particular case, the density of the bones is such that their nominal intensity values are different from the background. This makes extraction of the bones from the background

a
b
c d

FIGURE 9.18
(a) X-ray image of chicken filet with bone fragments.
(b) Thresholded image. (c) Image eroded with a 5 × 5 structuring element of 1s.
(d) Number of pixels in the connected components of (c).
(Image courtesy of NTB Elektronische Geraete GmbH, Diepholz, Germany, www.ntbxray.com.)

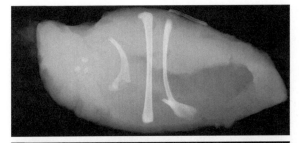

Connected component	No. of pixels in connected comp
01	11
02	9
03	9
04	39
05	133
06	1
07	1
08	743
09	7
10	11
11	11
12	9
13	9
14	674
15	85

a simple matter by using a single threshold (thresholding was introduced in Section 3.1 and is discussed in more detail in Section 10.3). The result is the binary image in Fig. 9.18(b).

The most significant feature in this figure is the fact that the points that remain are clustered into objects (bones), rather than being isolated, irrelevant points. We can make sure that only objects of "significant" size remain by eroding the thresholded image. In this example, we define as significant any object that remains after erosion with a 5×5 structuring element of 1s. The result of erosion is shown in Fig. 9.18(c). The next step is to analyze the size of the objects that remain. We label (identify) these objects by extracting the connected components in the image. The table in Fig. 9.18(d) lists the results of the extraction. There are a total of 15 connected components, with four of them being dominant in size. This is enough to determine that significant undesirable objects are contained in the original image. If needed, further characterization (such as shape) is possible using the techniques discussed in Chapter 11. ■

9.5.4 Convex Hull

A set A is said to be *convex* if the straight line segment joining any two points in A lies entirely within A. The *convex hull H* of an arbitrary set S is the smallest convex set containing S. The set difference $H - S$ is called the *convex deficiency* of S. As discussed in more detail in Sections 11.1.6 and 11.3.2, the convex hull and convex deficiency are useful for object description. Here, we present a simple morphological algorithm for obtaining the convex hull, $C(A)$, of a set A.

Let $B^i, i = 1, 2, 3, 4$, represent the four structuring elements in Fig. 9.19(a). The procedure consists of implementing the equation:

$$X_k^i = (X_{k-1} \circledast B^i) \cup A \quad i = 1, 2, 3, 4 \quad \text{and} \quad k = 1, 2, 3, \ldots \quad (9.5\text{-}4)$$

with $X_0^i = A$. When the procedure converges (i.e., when $X_k^i = X_{k-1}^i$), we let $D^i = X_k^i$. Then the convex hull of A is

$$C(A) = \bigcup_{i=1}^{4} D^i \quad (9.5\text{-}5)$$

In other words, the method consists of iteratively applying the hit-or-miss transform to A with B^1; when no further changes occur, we perform the union with A and call the result D^1. The procedure is repeated with B^2 (applied to A) until no further changes occur, and so on. The union of the four resulting Ds constitutes the convex hull of A. Note that we are using the simplified implementation of the hit-or-miss transform in which no background match is required, as discussed at the end of Section 9.4.

Figure 9.19 illustrates the procedure given in Eqs. (9.5-4) and (9.5-5). Figure 9.19(a) shows the structuring elements used to extract the convex hull. The origin of each element is at its center. The $\times$ entries indicate "don't care" conditions. This means that a structuring element is said to have found a match

FIGURE 9.19
(a) Structuring
elements. (b) Set
A. (c)–(f) Results
of convergence
with the
structuring
elements shown
in (a). (g) Convex
hull. (h) Convex
hull showing the
contribution of
each structuring
element.

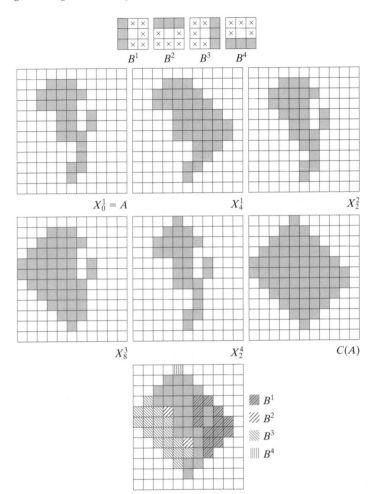

in *A* if the 3 × 3 region of *A* under the structuring element mask at that location matches the pattern of the mask. For a particular mask, a pattern match occurs when the center of the 3 × 3 region in *A* is 0, and the three pixels under the shaded mask elements are 1. The values of the other pixels in the 3 × 3 region do not matter. Also, with respect to the notation in Fig. 9.19(a), B^i is a clockwise rotation of B^{i-1} by 90°.

Figure 9.19(b) shows a set *A* for which the convex hull is sought. Starting with $X_0^1 = A$ resulted in the set in Fig. 9.19(c) after four iterations of Eq. (9.5-4). Then, letting $X_0^2 = A$ and again using Eq. (9.5-4) resulted in the set in Fig. 9.19(d) (convergence was achieved in only two steps in this case). The next two results were obtained in the same way. Finally, forming the union of the sets in Figs. 9.19(c), (d), (e), and (f) resulted in the convex hull shown in Fig. 9.19(g). The contribution of each structuring element is highlighted in the composite set shown in Fig. 9.19(h).

One obvious shortcoming of the procedure just outlined is that the convex hull can grow beyond the minimum dimensions required to guarantee convexity. One simple approach to reduce this effect is to limit growth so that it does not extend past the vertical and horizontal dimensions of the original set of points. Imposing this limitation on the example in Fig. 9.19 resulted in the image shown in Fig. 9.20. Boundaries of greater complexity can be used to limit growth even further in images with more detail. For example, we could use the maximum dimensions of the original set of points along the vertical, horizontal, and diagonal directions. The price paid for refinements such as this is additional complexity and increased computational requirements of the algorithm.

9.5.5 Thinning

The thinning of a set A by a structuring element B, denoted $A \otimes B$, can be defined in terms of the hit-or-miss transform:

$$A \otimes B = A - (A \circledast B)$$

$$= A \cap (A \circledast B)^c \qquad (9.5\text{-}6)$$

As in the previous section, we are interested only in pattern matching with the structuring elements, so no background operation is required in the hit-or-miss transform. A more useful expression for thinning A symmetrically is based on a *sequence* of structuring elements:

$$\{B\} = \{B^1, B^2, B^3, \ldots, B^n\} \qquad (9.5\text{-}7)$$

where B^i is a rotated version of B^{i-1}. Using this concept, we now define thinning by a sequence of structuring elements as

$$A \otimes \{B\} = ((\ldots((A \otimes B^1) \otimes B^2) \ldots) \otimes B^n) \qquad (9.5\text{-}8)$$

The process is to thin A by *one pass* with B^1, then thin the result with one pass of B^2, and so on, until A is thinned with one pass of B^n. The entire process is repeated until no further changes occur. Each individual thinning pass is performed using Eq. (9.5-6).

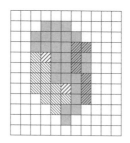

FIGURE 9.20
Result of limiting growth of the convex hull algorithm to the maximum dimensions of the original set of points along the vertical and horizontal directions.

Figure 9.21(a) shows a set of structuring elements commonly used for thinning, and Fig. 9.21(b) shows a set A to be thinned by using the procedure just discussed. Figure 9.21(c) shows the result of thinning after one pass of A with B^1, and Figs. 9.21(d) through (k) show the results of passes with the other structuring elements. Convergence was achieved after the second pass of B^6. Figure 9.21(l) shows the thinned result. Finally, Fig. 9.21(m) shows the thinned set converted to m-connectivity (see Section 2.5.2) to eliminate multiple paths.

9.5.6 Thickening

Thickening is the morphological dual of thinning and is defined by the expression

$$A \odot B = A \cup (A \circledast B) \qquad (9.5-9)$$

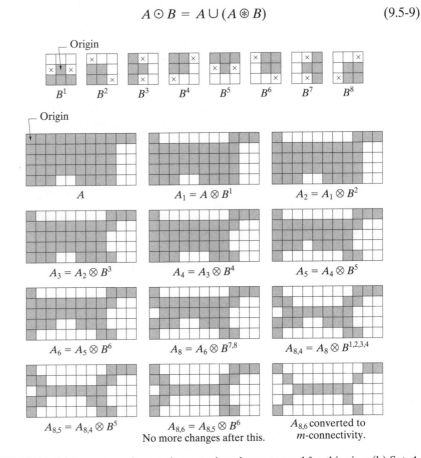

FIGURE 9.21 (a) Sequence of rotated structuring elements used for thinning. (b) Set A. (c) Result of thinning with the first element. (d)–(i) Results of thinning with the next seven elements (there was no change between the seventh and eighth elements). (j) Result of using the first four elements again. (l) Result after convergence. (m) Conversion to m-connectivity.

where B is a structuring element suitable for thickening. As in thinning, thickening can be defined as a sequential operation:

$$A \odot \{B\} = ((\ldots((A \odot B^1) \odot B^2)\ldots) \odot B^n) \qquad (9.5\text{-}10)$$

The structuring elements used for thickening have the same form as those shown in Fig. 9.21(a), but with all 1s and 0s interchanged. However, a separate algorithm for thickening is seldom used in practice. Instead, the usual procedure is to thin the background of the set in question and then complement the result. In other words, to thicken a set A, we form $C = A^c$, thin C, and then form C^c. Figure 9.22 illustrates this procedure.

Depending on the nature of A, this procedure can result in disconnected points, as Fig. 9.22(d) shows. Hence thickening by this method usually is followed by postprocessing to remove disconnected points. Note from Fig. 9.22(c) that the thinned background forms a boundary for the thickening process. This useful feature is not present in the direct implementation of thickening using Eq. (9.5-10), and it is one of the principal reasons for using background thinning to accomplish thickening.

9.5.7 Skeletons

As Fig. 9.23 shows, the notion of a skeleton, $S(A)$, of a set A is intuitively simple. We deduce from this figure that

(a) If z is a point of $S(A)$ and $(D)_z$ is the largest disk centered at z and contained in A, one cannot find a larger disk (not necessarily centered at z) containing $(D)_z$ and included in A. The disk $(D)_z$ is called a *maximum disk*.

(b) The disk $(D)_z$ touches the boundary of A at two or more different places.

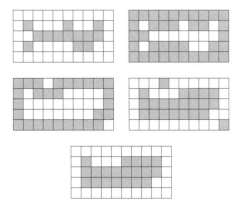

a	b
c	d
	e

FIGURE 9.22 (a) Set A. (b) Complement of A. (c) Result of thinning the complement of A. (d) Thickened set obtained by complementing (c). (e) Final result, with no disconnected points.

a b
c d

FIGURE 9.23
(a) Set A.
(b) Various positions of maximum disks with centers on the skeleton of A.
(c) Another maximum disk on a different segment of the skeleton of A.
(d) Complete skeleton.

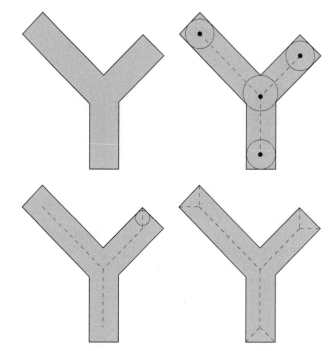

The skeleton of A can be expressed in terms of erosions and openings. That is, it can be shown (Serra [1982]) that

$$S(A) = \bigcup_{k=0}^{K} S_k(A) \tag{9.5-11}$$

with

$$S_k(A) = (A \ominus kB) - (A \ominus kB) \circ B \tag{9.5-12}$$

where B is a structuring element, and $(A \ominus kB)$ indicates k successive erosions of A:

$$(A \ominus kB) = ((\dots((A \ominus B) \ominus B) \ominus \dots) \ominus B) \tag{9.5-13}$$

k times, and K is the last iterative step before A erodes to an empty set. In other words,

$$K = \max\{k | (A \ominus kB) \neq \varnothing\} \tag{9.5-14}$$

The formulation given in Eqs. (9.5-11) and (9.5-12) states that $S(A)$ can be obtained as the union of the *skeleton subsets* $S_k(A)$. Also, it can be shown that A can be *reconstructed* from these subsets by using the equation

$$A = \bigcup_{k=0}^{K} (S_k(A) \oplus kB) \tag{9.5-15}$$

where $(S_k(A) \oplus kB)$ denotes k successive dilations of $S_k(A)$; that is,

$$(S_k(A) \oplus kB) = ((\dots((S_k(A) \oplus B) \oplus B) \oplus \dots) \oplus B) \tag{9.5-16}$$

■ Figure 9.24 illustrates the concepts just discussed. The first column shows the original set (at the top) and two erosions by the structuring element B. Note that one more erosion of A would yield the empty set, so $K = 2$ in this case. The second column shows the opening of the sets in the first column by B. These results are easily explained by the fitting characterization of the opening operation discussed in connection with Fig. 9.8. The third column simply contains the set differences between the first and second columns.

The fourth column contains two partial skeletons and the final result (at the bottom of the column). The final skeleton not only is thicker than it needs to be but, more important, it is not connected. This result is not unexpected, as nothing in the preceding formulation of the morphological skeleton guarantees connectivity. Morphology produces an elegant formulation in terms of erosions and openings of the given set. However, heuristic formulations such as the algorithm developed in Section 11.1.7 are needed if, as is usually the case, the skeleton must be maximally thin, connected, and minimally eroded.

EXAMPLE 9.8:
Computing the skeleton of a simple figure.

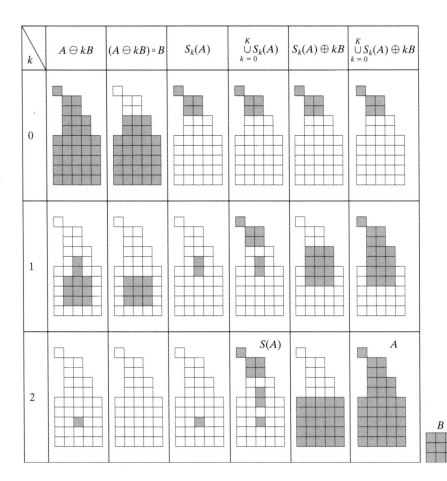

FIGURE 9.24 Implementation of Eqs. (9.5-11) through (9.5-15). The original set is at the top left, and its morphological skeleton is at the bottom of the fourth column. The reconstructed set is at the bottom of the sixth column.

The fifth column shows $S_0(A)$, $S_1(A) \oplus B$, and $(S_2(A) \oplus 2B) = (S_2(A) \oplus B) \oplus B$. Finally, the last column shows reconstruction of set A, which, according to Eq. (9.5-15), is the union of the dilated skeleton subsets shown in the fifth column.   ■

9.5.8 Pruning

Pruning methods are an essential complement to thinning and skeletonizing algorithms because these procedures tend to leave parasitic components that need to be "cleaned up" by postprocessing. We begin the discussion with a pruning problem and then develop a morphological solution based on the material introduced in the preceding sections. Thus, we take this opportunity to illustrate how to go about solving a problem by combining several of the techniques discussed up to this point.

A common approach in the automated recognition of hand-printed characters is to analyze the shape of the skeleton of each character. These skeletons often are characterized by "spurs" (parasitic components). Spurs are caused during erosion by non uniformities in the strokes composing the characters. We develop a morphological technique for handling this problem, starting with the assumption that the length of a parasitic component does not exceed a specified number of pixels.

Figure 9.25(a) shows the skeleton of a hand-printed "a." The parasitic component on the leftmost part of the character is illustrative of what we are interested in removing. The solution is based on suppressing a parasitic branch by successively eliminating its end point. Of course, this also shortens (or eliminates) other branches in the character but, in the absence of other structural information, the assumption in this example is that any branch with three or less pixels is to be eliminated. Thinning of an input set A with a sequence of structuring elements designed to detect only end points achieves the desired result. That is, let

$$X_1 = A \otimes \{B\} \tag{9.5-17}$$

where $\{B\}$ denotes the structuring element sequence shown in Figs. 9.25(b) and (c) [see Eq. (9.5-7) regarding structuring-element sequences]. The sequence of structuring elements consists of two different structures, each of which is rotated 90° for a total of eight elements. The $\times$ in Fig. 9.25(b) signifies a "don't care" condition, in the sense that it does not matter whether the pixel in that location has a value of 0 or 1. Numerous results reported in the literature on morphology are based on the use of a *single* structuring element, similar to the one in Fig. 9.25(b), but having "don't care" conditions along the entire first column. This is incorrect. For example, this element would identify the point located in the eighth row, fourth column of Fig. 9.25(a) as an end point, thus eliminating it and breaking connectivity in the stroke.

Applying Eq. (9.5-17) to A three times yields the set X_1 in Fig. 9.25(d). The next step is to "restore" the character to its original form, but with the parasitic

We may define an *end point* as the center point of a 3×3 region that satisfies any of the arrangements in Figs. 9.25(b) or (c).

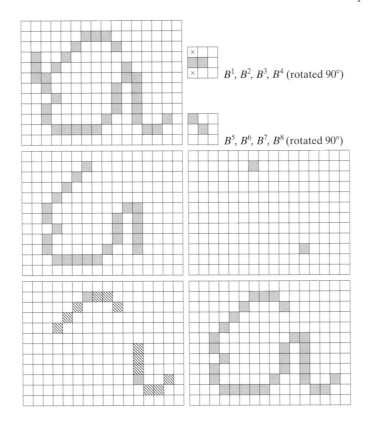

FIGURE 9.25
(a) Original
image. (b) and
(c) Structuring
elements used for
deleting end
points. (d) Result
of three cycles of
thinning. (e) End
points of (d).
(f) Dilation of end
points condi-
tioned on (a).
(g) Pruned image.

B^1, B^2, B^3, B^4 (rotated 90°)

B^5, B^6, B^7, B^8 (rotated 90°)

branches removed. To do so first requires forming a set X_2 containing all end
points in X_1 [Fig. 9.25(e)]:

$$X_2 = \bigcup_{k=1}^{8}(X_1 \circledast B^k) \qquad (9.5\text{-}18)$$

where the B^k are the same end-point detectors shown in Figs. 9.25(b) and (c).
The next step is dilation of the end points three times, using set A as a delimiter:

$$X_3 = (X_2 \oplus H) \cap A \qquad (9.5\text{-}19)$$

Equation (9.5-19) is the
basis for morphological
reconstruction by dila-
tion, as explained in the
next section.

where H is a 3×3 structuring element of 1s and the intersection with A is
applied after each step. As in the case of region filling and extraction of connected components, this type of conditional dilation prevents the creation
of 1-valued elements outside the region of interest, as evidenced by the result shown in Fig. 9.25(f). Finally, the union of X_3 and X_1 yields the desired
result,

$$X_4 = X_1 \cup X_3 \qquad (9.5\text{-}20)$$

in Fig. 9.25(g).

In more complex scenarios, use of Eq. (9.5-19) sometimes picks up the
"tips" of some parasitic branches. This condition can occur when the end

points of these branches are near the skeleton. Although Eq. (9.5-17) may eliminate them, they can be picked up again during dilation because they are valid points in A. Unless entire parasitic elements are picked up again (a rare case if these elements are short with respect to valid strokes), detecting and eliminating them is easy because they are disconnected regions.

A natural thought at this juncture is that there must be easier ways to solve this problem. For example, we could just keep track of all deleted points and simply reconnect the appropriate points to all end points left after application of Eq. (9.5-17). This option is valid, but the advantage of the formulation just presented is that the use of simple morphological constructs solved the entire problem. In practical situations when a set of such tools is available, the advantage is that no new algorithms have to be written. We simply combine the necessary morphological functions into a sequence of operations.

9.5.9 Morphological Reconstruction

The morphological concepts discussed thus far involve an image and a structuring element. In this section, we discuss a powerful morphological transformation called *morphological reconstruction* that involves two images and a structuring element. One image, the *marker*, contains the starting points for the transformation. The other image, the *mask*, constrains the transformation. The structuring element is used to define connectivity.[†]

Geodesic dilation and erosion

Central to morphological reconstruction are the concepts of geodesic dilation and geodesic erosion. Let F denote the marker image and G the mask image. It is assumed in this discussion that both are binary images and that $F \subseteq G$. The *geodesic dilation* of size 1 of the marker image with respect to the mask, denoted by $D_G^{(1)}(F)$, is defined as

$$D_G^{(1)}(F) = (F \oplus B) \cap G \tag{9.5-21}$$

where $\cap$ denotes the set intersection (here $\cap$ may be interpreted as a logical AND because the set intersection and logical AND operations are the same for binary sets). The geodesic dilation of size n of F with respect to G is defined as

$$D_G^{(n)}(F) = D_G^{(1)}\left[D_G^{(n-1)}(F)\right] \tag{9.5-22}$$

with $D_G^{(0)}(F) = F$. In this recursive expression, the set intersection in Eq. (9.5-21) is performed at each step.[‡] Note that the intersection operator guarantees that

---

[†]In much of the literature on morphological reconstruction, the structuring element is tacitly assumed to be isotropic and typically is called an *elementary isotropic structuring element*. In the context of this chapter, an example of such an SE is simply a 3×3 array of 1s with the origin at the center.

[‡]Although it is more intuitive to develop morphological-reconstruction methods using recursive formulations (as we do here), their practical implementation typically is based on more computationally efficient algorithms (see, for example, Vincent [1993] and Soille [2003]). All image-based examples in this section were generated using such algorithms.

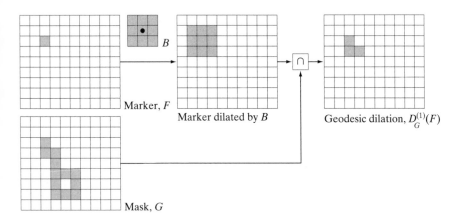

FIGURE 9.26
Illustration of geodesic dilation.

mask G will limit the growth (dilation) of marker F. Figure 9.26 shows a simple example of a geodesic dilation of size 1. The steps in the figure are a direct implementation of Eq. (9.5-21).

Similarly, the *geodesic erosion* of size 1 of marker F with respect to mask G is defined as

$$E_G^{(1)}(F) = (F \ominus B) \cup G \qquad (9.5\text{-}23)$$

where $\cup$ denotes set union (or OR operation). The geodesic erosion of size n of F with respect to G is defined as

$$E_G^{(n)}(F) = E_G^{(1)}\left[E_G^{(n-1)}(F)\right] \qquad (9.5\text{-}24)$$

with $E_G^{(0)}(F) = F$. The set union operation in Eq. (9.5-23) is performed at each iterative step, and guarantees that geodesic erosion of an image remains greater than or equal to its mask image. As expected from the forms in Eqs. (9.5-21) and (9.5-23), geodesic dilation and erosion are *duals* with respect to set complementation (see Problem 9.29). Figure 9.27 shows a simple example of geodesic erosion of size 1. The steps in the figure are a direct implementation of Eq. (9.5-23).

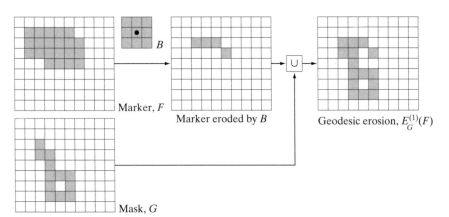

FIGURE 9.27
Illustration of geodesic erosion.

Geodesic dilation and erosion of finite images always converge after a finite number of iterative step because propagation or shrinking of the marker image is constrained by the mask.

Morphological reconstruction by dilation and by erosion

Based on the preceding concepts, *morphological reconstruction by dilation* of a mask image G from a marker image F, denoted $R_G^D(F)$, is defined as the geodesic dilation of F with respect to G, iterated until stability is achieved; that is,

$$R_G^D(F) = D_G^{(k)}(F) \tag{9.5-25}$$

with k such that $D_G^{(k)}(F) = D_G^{(k+1)}(F)$.

Figure 9.28 illustrates reconstruction by dilation. Figure 9.28(a) continues the process begun in Fig. 9.26; that is, the next step in reconstruction after obtaining $D_G^{(1)}(F)$ is to dilate this result and then AND it with the mask G to yield $D_G^{(2)}(F)$, as Fig. 9.28(b) shows. Dilation of $D_G^{(2)}(F)$ and masking with G then yields $D_G^{(3)}(F)$, and so on. This procedure is repeated until stability is reached. If we carried this example one more step, we would find that $D_G^{(5)}(F) = D_G^{(6)}(F)$, so the morphologically reconstructed image by dilation is given by $R_G^D(F) = D_G^{(5)}(F)$, as indicated in Eq. (9.5-25). Note that the reconstructed image in this case is identical to the mask because F contained a single 1-valued pixel (this is analogous to convolution of an image with an impulse, which simply copies the image at the location of the impulse, as explained in Section 3.4.2).

In a similar manner, the *morphological reconstruction by erosion* of a mask image G from a marker image F, denoted $R_G^E(F)$, is defined as the geodesic erosion of F with respect to G, iterated until stability; that is,

$$R_G^E(F) = E_G^{(k)}(F) \tag{9.5-26}$$

with k such that $E_G^{(k)}(F) = E_G^{(k+1)}(F)$. As an exercise, you should generate a figure similar to Fig. 9.28 for morphological reconstruction by erosion.

a b c d
e f g h

FIGURE 9.28
Illustration of morphological reconstruction by dilation. F, G, B and $D_G^{(1)}(F)$ are from Fig. 9.26.

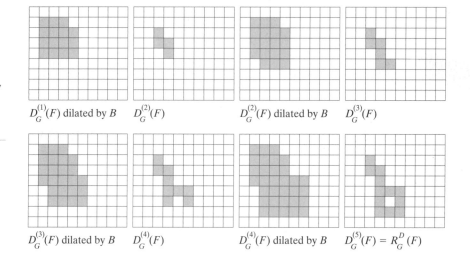

$D_G^{(1)}(F)$ dilated by B $D_G^{(2)}(F)$ $D_G^{(2)}(F)$ dilated by B $D_G^{(3)}(F)$

$D_G^{(3)}(F)$ dilated by B $D_G^{(4)}(F)$ $D_G^{(4)}(F)$ dilated by B $D_G^{(5)}(F) = R_G^D(F)$

Reconstruction by dilation and erosion are duals with respect to set complementation (see Problem 9.30).

Sample applications

Morphological reconstruction has a broad spectrum of practical applications, each determined by the selection of the marker and mask images, by the structuring elements used, and by combinations of the primitive operations defined in the preceding discussion. The following examples illustrate the usefulness of these concepts.

Opening by reconstruction: In a morphological opening, erosion removes small objects and the subsequent dilation attempts to restore the shape of objects that remain. However, the accuracy of this restoration is highly dependent on the similarity of the shapes of the objects and the structuring element used. *Opening by reconstruction* restores *exactly* the shapes of the objects that remain after erosion. The opening by reconstruction of size n of an image F is defined as the reconstruction by dilation of F from the erosion of size n of F; that is,

$$O_R^{(n)}(F) = R_F^D\left[(F \ominus nB)\right] \tag{9.5-27}$$

where $(F \ominus nB)$ indicates n erosions of F by B, as explained in Section 9.5.7. Note that F is used as the mask in this application. A similar expression can be written for closing by reconstruction (see Table 9.1).

Figure 9.29 shows an example of opening by reconstruction. In this illustration, we are interested in extracting from Fig. 9.29(a) the characters that contain long, vertical strokes. Opening by reconstruction requires at least one erosion, so we perform that step first. Figure 9.29(b) shows the erosion

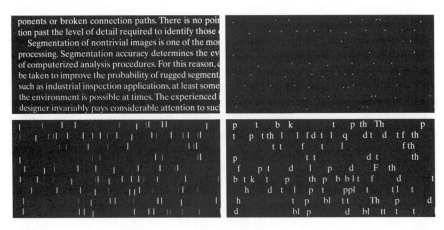

a b
c d

FIGURE 9.29 (a) Text image of size 918×2018 pixels. The approximate average height of the tall characters is 50 pixels. (b) Erosion of (a) with a structuring element of size 51×1 pixels. (c) Opening of (a) with the same structuring element, shown for reference. (d) Result of opening by reconstruction.

of Fig. 9.29(a) with a structuring element of length proportional to the average height of the tall characters (51 pixels) and width of one pixel. For the purpose of comparison, we computed the opening of the image using the same structuring element. Figure 9.29(c) shows the result. Finally, Fig. 9.29(d) is the opening by reconstruction (of size 1) of F [i.e., $O_R^{(1)}(F)$] given in Eq. (9.5-27). This result shows that characters containing long vertical strokes were restored accurately; all other characters were removed.

Filling holes: In Section 9.5.2, we developed an algorithm for filling holes based on knowing a starting point in each hole in the image. Here, we develop a fully automated procedure based on morphological reconstruction. Let $I(x, y)$ denote a binary image and suppose that we form a marker image F that is 0 everywhere, except at the image border, where it is set to $1 - I$; that is,

$$F(x, y) = \begin{cases} 1 - I(x, y) & \text{if } (x, y) \text{ is on the border of } I \\ 0 & \text{otherwise} \end{cases} \tag{9.5-28}$$

Then

$$H = \left[R_{I^c}^D(F) \right]^c \tag{9.5-29}$$

is a binary image equal to I with all holes filled.

Let us consider the individual components of Eq. (9.5-29) to see how this expression in fact leads to all holes in an image being filled. Figure 9.30(a) shows a simple image I containing one hole, and Fig. 9.30(b) shows its complement. Note that because the complement of I sets all foreground (1-valued) pixels to background (0-valued) pixels, and vice versa, this operation in effect builds a "wall" of 0s around the hole. Because I^c is used as an AND mask, all we are doing here is protecting all foreground pixels (including the wall around the hole) from changing during iteration of the procedure. Figure 9.30(c) is array F formed according to Eq. (9.5-28) and Fig. 9.30(d) is F dilated with a 3×3 SE whose elements are all 1s. Note that marker F has a border of 1s (except at locations where I is 1), so the dilation of F of the marker points starts at the border and proceeds inward. Figure 9.30(e) shows the geodesic dilation of F using I^c as the mask. As was just indicated, we see that all locations in this result corresponding to foreground pixels from I are 0, and that this is true now for the hole pixels as well. Another iteration will yield the same result which, when complemented as required by Eq. (9.5-29), gives the result in Fig. 9.30(f). As desired, the hole is now filled and the rest of image I was unchanged. The operation $H \cap I^c$ yields an image containing 1-valued pixels in the locations corresponding to the holes in I, as Fig. 9.30(g) shows.

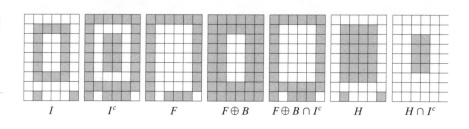

a b c d e f g

FIGURE 9.30
Illustration of hole filling on a simple image.

I I^c F $F \oplus B$ $F \oplus B \cap I^c$ H $H \cap I^c$

ponents or broken connection paths. There is no poir tion past the level of detail required to identify those

Segmentation of nontrivial images is one of the mos processing. Segmentation accuracy determines the ev of computerized analysis procedures. For this reason, c be taken to improve the probability of rugged segment such as industrial inspection applications, at least some the environment is possible at times. The experienced i designer invariably pays considerable attention to sucl

ponents or broken connection paths. There is no poir tion past the level of detail required to identify those

Segmentation of nontrivial images is one of the mos processing. Segmentation accuracy determines the ev of computerized analysis procedures. For this reason, c be taken to improve the probability of rugged segment. such as industrial inspection applications, at least some the environment is possible at times. The experienced i designer invariably pays considerable attention to sucl

ponents or broken connection paths. There is no poir tion past the level of detail required to identify those

Segmentation of nontrivial images is one of the mos processing. Segmentation accuracy determines the ev of computerized analysis procedures. For this reason, c be taken to improve the probability of rugged segment such as industrial inspection applications, at least some the environment is possible at times. The experienced i designer invariably pays considerable attention to sucl

a b
c d

FIGURE 9.31
(a) Text image of size 918×2018 pixels. (b) Complement of (a) for use as a mask image. (c) Marker image. (d) Result of hole-filling using Eq. (9.5-29).

Figure 9.31 shows a more practical example. Figure 9.31(b) shows the complement of the text image in Fig. 9.31(a), and Fig. 9.31(c) is the marker image, F, generated using Eq. (9.5-28). This image has a border of 1s, except at locations corresponding to 1s in the border of the original image. Finally, Fig. 9.31(d) shows the image with all the holes filled.

Border clearing: The extraction of objects from an image for subsequent shape analysis is a fundamental task in automated image processing. An algorithm for removing objects that touch (i.e., are connected to) the border is a useful tool because (1) it can be used to screen images so that only complete objects remain for further processing, or (2) it can be used as a signal that partial objects are present in the field of view. As a final illustration of the concepts introduced in this section, we develop a border-clearing procedure based on morphological reconstruction. In this application, we use the original image as the mask and the following marker image:

$$F(x, y) = \begin{cases} I(x, y) & \text{if } (x, y) \text{ is on the border of } I \\ 0 & \text{otherwise} \end{cases} \qquad (9.5\text{-}30)$$

The border-Eclearing algorithm first computes the morphological reconstruction $R_I^D(F)$ (which simply extracts the objects touching the border) and then computes the difference

$$X = I - R_I^D(F) \qquad (9.5\text{-}31)$$

to obtain an image, X, with no objects touching the border.

a b

FIGURE 9.32
Border clearing. (a) Marker image. (b) Image with no objects touching the border. The original image is Fig. 9.29(a).

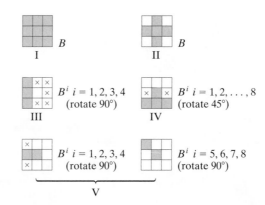

As an example, consider the text image again. Figure 9.32(a) in the previous
page shows the reconstruction $R_I^D(F)$ obtained using a 3×3 structuring element of all 1s (note the objects touching the boundary on the right side), and
Fig. 9.32(b) shows image X, computed using Eq. (9.5-31). If the task at hand
were automated character recognition, having an image in which no characters
touch the border is most useful because the problem of having to recognize
partial characters (a difficult task at best) is avoided.

9.5.10 Summary of Morphological Operations on Binary Images

Table 9.1 summarizes the morphological results developed in the preceding
sections, and Fig. 9.33 summarizes the basic types of structuring elements used
in the various morphological processes discussed thus far. The Roman numer-
als in the third column of Table 9.1 refer to the structuring elements in Fig. 9.33.

TABLE 9.1
Summary of
morphological
operations and
their properties.

Operation	Equation	Comments (The Roman numerals refer to the structuring elements in Fig. 9.33.)
Translation	$(B)_z = \{w \mid w = b + z,$ for $b \in B\}$	Translates the origin of B to point z.
Reflection	$\hat{B} = \{w \mid w = -b, \text{ for } b \in B\}$	Reflects all elements of B about the origin of this set.
Complement	$A^c = \{w \mid w \notin A\}$	Set of points not in A.
Difference	$A - B = \{w \mid w \in A, w \notin B\}$ $= A \cap B^c$	Set of points that belong to A but not to B.
Dilation	$A \oplus B = \{z \mid (\hat{B}_z) \cap A \neq \varnothing\}$	"Expands" the boundary of A. (I)
Erosion	$A \ominus B = \{z \mid (B)_z \subseteq A\}$	"Contracts" the boundary of A. (I)
Opening	$A \circ B = (A \ominus B) \oplus B$	Smoothes contours, breaks narrow isthmuses, and eliminates small islands and sharp peaks. (I)

(Continued)

Operation	Equation	Comments (The Roman numerals refer to the structuring elements in Fig. 9.33.)
Closing	$A \bullet B = (A \oplus B) \ominus B$	Smoothes contours, fuses narrow breaks and long thin gulfs, and eliminates small holes. (I)
Hit-or-miss transform	$A \circledast B = (A \ominus B_1) \cap (A^c \ominus B_2)$ $= (A \ominus B_1) - (A \oplus \hat{B}_2)$	The set of points (coordinates) at which, simultaneously, B_1 found a match ("hit") in A and B_2 found a match in A^c
Boundary extraction	$\beta(A) = A - (A \ominus B)$	Set of points on the boundary of set A. (I)
Hole filling	$X_k = (X_{k-1} \oplus B) \cap A^c$; $k = 1, 2, 3, \ldots$	Fills holes in A; $X_0 =$ array of 0s with a 1 in each hole. (II)
Connected components	$X_k = (X_{k-1} \oplus B) \cap A$; $k = 1, 2, 3, \ldots$	Finds connected components in A; $X_0 =$ array of 0s with a 1 in each connected component. (I)
Convex hull	$X_k^i = (X_{k-1}^i \circledast B^i) \cup A$; $i = 1, 2, 3, 4$; $k = 1, 2, 3, \ldots$; $X_0^i = A$; and $D^i = X_{conv}^i$	Finds the convex hull $C(A)$ of set A, where "conv" indicates convergence in the sense that $X_k^i = X_{k-1}^i$. (III)
Thinning	$A \otimes B = A - (A \circledast B)$ $= A \cap (A \circledast B)^c$ $A \otimes \{B\} =$ $((\ldots((A \otimes B^1) \otimes B^2) \ldots) \otimes B^n)$ $\{B\} = \{B^1, B^2, B^3, \ldots, B^n\}$	Thins set A. The first two equations give the basic definition of thinning. The last equations denote thinning by a sequence of structuring elements. This method is normally used in practice. (IV)
Thickening	$A \odot B = A \cup (A \circledast B)$ $A \odot \{B\} =$ $((\ldots(A \odot B^1) \odot B^2 \ldots) \odot B^n)$	Thickens set A. (See preceding comments on sequences of structuring elements.) Uses IV with 0s and 1s reversed.
Skeletons	$S(A) = \bigcup_{k=0}^{K} S_k(A)$ $S_k(A) = \bigcup_{k=0}^{K} \{(A \ominus kB)$ $- [(A \ominus kB) \circ B]\}$ Reconstruction of A: $A = \bigcup_{k=0}^{K} (S_k(A) \oplus kB)$	Finds the skeleton $S(A)$ of set A. The last equation indicates that A can be reconstructed from its skeleton subsets $S_k(A)$. In all three equations, K is the value of the iterative step after which the set A erodes to the empty set. The notation $(A \ominus kB)$ denotes the kth iteration of successive erosions of A by B. (I)

TABLE 9.1
(Continued)

(Continued)

TABLE 9.1
(Continued)

Operation	Equation	Comments (The Roman numerals refer to the structuring elements in Fig. 9.33.)
Pruning	$X_1 = A \otimes \{B\}$ $X_2 = \bigcup_{k=1}^{8}(X_1 \circledast B^k)$ $X_3 = (X_2 \oplus H) \cap A$ $X_4 = X_1 \cup X_3$	X_4 is the result of pruning set A. The number of times that the first equation is applied to obtain X_1 must be specified. Structuring elements V are used for the first two equations. In the third equation H denotes structuring element I.
Geodesic dilation of size 1	$D_G^{(1)}(F) = (F \oplus B) \cap G$	F and G are called the *marker* and *mask* images, respectively.
Geodesic dilation of size n	$D_G^{(n)}(F) = D_G^{(1)}\left[D_G^{(n-1)}(F)\right];$ $D_G^{(0)}(F) = F$	
Geodesic erosion of size 1	$E_G^{(1)}(F) = (F \ominus B) \cup G$	
Geodesic erosion of size n	$E_G^{(n)}(F) = E_G^{(1)}\left[E_G^{(n-1)}(F)\right];$ $E_G^{(0)}(F) = F$	
Morphological reconstruction by dilation	$R_G^D(F) = D_G^{(k)}(F)$	k is such that $D_G^{(k)}(F) = D_G^{(k+1)}(F)$
Morphological reconstruction by erosion	$R_G^E(F) = E_G^{(k)}(F)$	k is such that $E_G^{(k)}(F) = E_G^{(k+1)}(F)$
Opening by reconstruction	$O_R^{(n)}(F) = R_F^D[(F \ominus nB)]$	$(F \ominus nB)$ indicates n erosions of F by B.
Closing by reconstruction	$C_R^{(n)}(F) = R_F^E[(F \oplus nB)]$	$(F \oplus nB)$ indicates n dilations of F by B.
Hole filling	$H = \left[R_{I^c}^D(F)\right]^c$	H is equal to the input image I, but with all holes filled. See Eq. (9.5-28) for the definition of the marker image F.
Border clearing	$X = I - R_I^D(F)$	X is equal to the input image I, but with all objects that touch (are connected to) the boundary removed. See Eq. (9.5-30) for the definition of the marker image F.

9.6 Gray-Scale Morphology

In this section, we extend to gray-scale images the basic operations of dilation, erosion, opening, and closing. We then use these operations to develop several basic gray-scale morphological algorithms.

Throughout the discussion that follows, we deal with digital functions of the form $f(x, y)$ and $b(x, y)$, where $f(x, y)$ is a gray-scale image and $b(x, y)$ is a structuring element. The assumption is that these functions are discrete in the sense introduced in Section 2.4.2. That is, if Z denotes the set of real integers, then the coordinates (x, y) are integers from the Cartesian product Z^2 and f and b are functions that assign an intensity value (a real number from the set of real numbers, R) to each distinct pair of coordinates (x, y). If the intensity levels are integers also, then Z replaces R.

Structuring elements in gray-scale morphology perform the same basic functions as their binary counterparts: They are used as "probes" to examine a given image for specific properties. Structuring elements in gray-scale morphology belong to one of two categories: *nonflat* and *flat*. Figure 9.34 shows an example of each. Figure 9.34(a) is a hemispherical gray-scale SE shown as an image, and Fig. 9.34(c) is a horizontal intensity profile through its center. Figure 9.34(b) shows a flat structuring element in the shape of a disk and Fig. 9.34(d) is its corresponding intensity profile (the shape of this profile explains the origin of the word "flat"). The elements in Fig. 9.34 are shown as continuous quantities for clarity; their computer implementation is based on digital approximations (e.g., see the rightmost disk SE in Fig. 9.2). Due to a number of difficulties discussed later in this section, gray-scale SEs are used infrequently in practice. Finally, we mention that, as in the binary case, the origin of structuring elements must be clearly identified. Unless mentioned otherwise, all the examples in this section are based on symmetrical, flat structuring elements of unit height whose origins are at the center. The *reflection* of an SE in gray-scale morphology is as defined in Section 9.1, and we denote it in the following discussion by $\hat{b}(x, y) = b(-x - y)$.

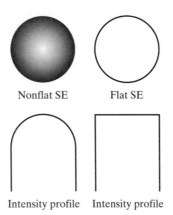

Nonflat SE   Flat SE

Intensity profile   Intensity profile

a b
c d

FIGURE 9.34
Nonflat and flat structuring elements, and corresponding horizontal intensity profiles through their center. All examples in this section are based on flat SEs.

9.6.1 Erosion and Dilation

The *erosion* of f by a *flat* structuring element b at any location (x, y) is defined as the *minimum* value of the image in the region coincident with b when the origin of b is at (x, y). In equation form, the erosion at (x, y) of an image f by a structuring element b is given by

$$\left[f \ominus b\right](x, y) = \min_{(s, t) \in b} \left\{f(x + s, y + t)\right\} \tag{9.6-1}$$

where, in a manner similar to the correlation procedure discussed in Section 3.4.2, x and y are incremented through all values required so that the origin of b visits every pixel in f. That is, to find the erosion of f by b, we place the origin of the structuring element at every pixel location in the image. The erosion at any location is determined by selecting the minimum value of f from all the values of f contained in the region coincident with b. For example, if b is a square structuring element of size 3×3, obtaining the erosion at a point requires finding the minimum of the nine values of f contained in the 3×3 region defined by b when its origin is at that point.

Similarly, the *dilation* of f by a *flat* structuring element b at any location (x, y) is defined as the *maximum* value of the image in the window outlined by $\hat{b}$ when the origin of $\hat{b}$ is at (x, y). That is,

$$\left[f \oplus b\right](x, y) = \max_{(s, t) \in b} \left\{f(x - s, y - t)\right\} \tag{9.6-2}$$

where we used the fact stated earlier that $\hat{b} = b(-x, -y)$. The explanation of this equation is identical to the explanation in the previous paragraph, but using the maximum, rather than the minimum, operation and keeping in mind that the structuring element is reflected about its origin, which we take into account by using $(-s, -t)$ in the argument of the function. This is analogous to spatial convolution, as explained in Section 3.4.2.

EXAMPLE 9.9:
Illustration of gray-scale erosion and dilation.

■ Because gray-scale erosion with a flat SE computes the minimum intensity value of f in every neighborhood of (x, y) coincident with b, we expect in general that an eroded gray-scale image will be darker than the original, that the sizes (with respect to the size of the SE) of bright features will be reduced, and that the sizes of dark features will be increased. Figure 9.35(b) shows the erosion of Fig. 9.35(a) using a disk SE of unit height and a radius of two pixels. The effects just mentioned are clearly visible in the eroded image. For instance, note how the intensities of the small bright dots were reduced, making them barely visible in Fig. 9.35(b), while the dark features grew in thickness. The general background of the eroded image is slightly darker than the background of the original image. Similarly, Fig. 9.35(c) shows the result of dilation with the same SE. The effects are the opposite of those obtained with erosion. The bright features were thickened and the intensities of the dark features were reduced. Note in particular how the thin black connecting wires in the left, middle, and right, bottom of Fig. 9.35(a) are barely visible in Fig. 9.35(c). The sizes of the dark dots were reduced as a result of dilation but, unlike the eroded small white dots in Fig. 9.35(b), they still are easily visible in the dilated

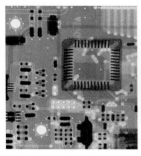  

a b c

FIGURE 9.35 (a) A gray-scale X-ray image of size 448 × 425 pixels. (b) Erosion using a flat disk SE with a radius of two pixels. (c) Dilation using the same SE. (Original image courtesy of Lixi, Inc.)

image. The reason is that the black dots were originally larger than the white dots with respect to the size of the SE. Finally, note that the background of the dilated image is slightly lighter than that of Fig. 9.35(a).   ■

Nonflat SEs have gray-scale values that vary over their domain of definition. The erosion of image f by nonflat structuring element, b_N, is defined as

$$\left[f \ominus b_N\right](x, y) = \min_{(s, t) \in b_N} \left\{f(x + s, y + t) - b_N(s, t)\right\} \qquad (9.6\text{-}3)$$

Here, we actually subtract values from f to determine the erosion at any point. This means that, unlike Eq. (9.6-1), erosion using a nonflat SE is not bounded in general by the values of f, which can present problems in interpreting results. Gray-scale SEs are seldom used in practice because of this, in addition to potential difficulties in selecting meaningful elements for b_N, and the added computational burden when compared with Eq. (9.6-1).

In a similar manner, dilation using a nonflat SE is defined as

$$\left[f \oplus b_N\right](x, y) = \max_{(s, t) \in b_N} \left\{f(x - s, y - t) + b_N(s, t)\right\} \qquad (9.6\text{-}4)$$

The same comments made in the previous paragraph are applicable to dilation with nonflat SEs. When all the elements of b_N are constant (i.e., the SE is flat), Eqs. (9.6-3) and (9.6-4) reduce to Eqs. (9.6-1) and (9.6-2), respectively, within a scalar constant equal to the amplitude of the SE.

As in the binary case, erosion and dilation are duals with respect to function complementation and reflection; that is,

$$(f \ominus b)^c(x, y) = (f^c \oplus \hat{b})(x, y)$$

where $f^c = -f(x, y)$ and $\hat{b} = b(-x, -y)$. The same expression holds for nonflat structuring elements. Except as needed for clarity, we simplify the notation in the following discussion by omitting the arguments of all functions, in which case the preceding equation is written as

$$(f \ominus b)^c = (f^c \oplus \hat{b}) \qquad (9.6\text{-}5)$$

Similarly,

$$(f \oplus b)^c = (f^c \ominus \hat{b}) \tag{9.6-6}$$

Erosion and dilation by themselves are not particularly useful in gray-scale image processing. As with their binary counterparts, these operations become powerful when used in combination to derive higher-level algorithms, as the material in the following sections demonstrates.

Although we deal with flat SEs in the examples in the remainder of this section, the concepts discussed are applicable also to nonflat structuring elements.

9.6.2 Opening and Closing

The expressions for opening and closing gray-scale images have the same form as their binary counterparts. The *opening* of image f by structuring element b, denoted $f \circ b$, is

$$f \circ b = (f \ominus b) \oplus b \tag{9.6-7}$$

As before, opening is simply the erosion of f by b, followed by a dilation of the result with b. Similarly, the *closing* of f by b, denoted $f \cdot b$, is

$$f \bullet b = (f \oplus b) \ominus b \tag{9.6-8}$$

The opening and closing for gray-scale images are duals with respect to complementation and SE reflection:

$$(f \bullet b)^c = f^c \circ \hat{b} \tag{9.6-9}$$

and

$$(f \circ b)^c = f^c \bullet \hat{b} \tag{9.6-10}$$

Because $f^c = -f(x, y)$, Eq. (9.6-9) can be written also as $-(f \bullet b) = (-f \circ \hat{b})$ and similarly for Eq. (9.6-10).

Opening and closing of images have a simple geometric interpretation. Suppose that an image function $f(x, y)$ is viewed as a 3-D surface; that is, its intensity values are interpreted as height values over the xy-plane, as in Fig. 2.18(a). Then the opening of f by b can be interpreted geometrically as pushing the structuring element up from below against the undersurface of f. At each location of the origin of b, the opening is the highest value reached by any part of b as it pushes up against the undersurface of f. The complete opening is then the set of all such values obtained by having the origin of b visit every (x, y) coordinate of f.

Sometimes opening and closing are illustrated by rolling a circle on the under and upper sides of a curve. In 3-D, the circle becomes a sphere and the resulting procedures are called rolling-ball algorithms.

Figure 9.36 illustrates the concept in one dimension. Suppose that the curve in Fig. 9.36(a) is the intensity profile along a single row of an image. Figure 9.36(b) shows a flat structuring element in several positions, pushed up against the bottom of the curve. The solid curve in Fig. 9.36(c) is the complete opening. Because the structuring element is too large to fit completely inside the upward peaks of the curve, the tops of the peaks are clipped by the opening, with the amount removed being proportional to how far the structuring element was able to reach into the peak. In general, openings are used to remove small, bright details, while leaving the overall intensity levels and larger bright features relatively undisturbed.

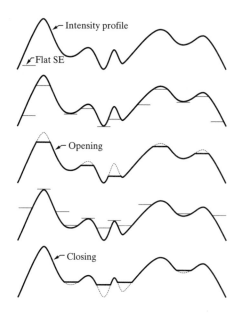

FIGURE 9.36
Opening and clos-
ing in one dimen-
sion. (a) Original
1-D signal. (b) Flat
structuring
element pushed up
underneath the
signal.
(c) Opening.
(d) Flat structuring
element pushed
down along the top
of the signal.
(e) Closing.

Figure 9.36(d) is a graphical illustration of closing. Observe that the structuring element is pushed down on top of the curve while being translated to all
locations. The closing, shown in Fig. 9.36(e), is constructed by finding the lowest points reached by any part of the structuring element as it slides against the
upper side of the curve.

The gray-scale opening operation satisfies the following properties:

(a) $f \circ b \unlhd f$
(b) If $f_1 \unlhd f_2$, then $(f_1 \circ b) \unlhd (f_2 \circ b)$
(c) $(f \circ b) \circ b = f \circ b$

The notation $e \unlhd r$ is used to indicate that the domain of e is a subset of the domain of r, and also that $e(x, y) \leq r(x, y)$ for any (x, y) in the domain of e.

Similarly, the closing operation satisfies the following properties:

(a) $f \unlhd f \bullet b$
(b) If $f_1 \unlhd f_2$, then $(f_1 \bullet b) \unlhd (f_2 \bullet b)$
(c) $(f \bullet b) \bullet b = f \bullet b$

The usefulness of these properties is similar to that of their binary counterparts.

■ Figure 9.37 extends to 2-D the 1-D concepts illustrated in Fig. 9.36. Figure
9.37(a) is the same image we used in Example 9.9, and Fig. 9.37(b) is the opening
obtained using a disk structuring element of unit height and radius of 3 pixels. As
expected, the intensity of all bright features decreased, depending on the sizes of
the features relative to the size of the SE. Comparing this figure with Fig. 9.35(b),
we see that, unlike the result of erosion, opening had negligible effect on the dark
features of the image, and the effect on the background was negligible. Similarly,
Fig. 9.37(c) shows the closing of the image with a disk of radius 5 (the small round

EXAMPLE 9.10:
Illustration of
gray-scale
opening and
closing.

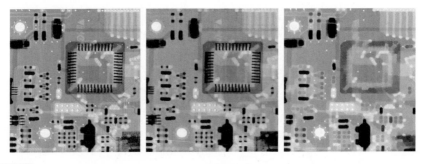

a b c

FIGURE 9.37 (a) A gray-scale X-ray image of size 448×425 pixels. (b) Opening using a disk SE with a radius of 3 pixels. (c) Closing using an SE of radius 5.

black dots are larger than the small white dots, so a larger disk was needed to achieve results comparable to the opening). In this image, the bright details and background were relatively unaffected, but the dark features were attenuated, with the degree of attenuation being dependent on the relative sizes of the features with respect to the SE.   ◼

9.6.3 Some Basic Gray-Scale Morphological Algorithms

Numerous morphological techniques are based on the gray-scale morphological concepts introduced thus far. We illustrate some of these algorithms in the following discussion.

Morphological smoothing

Because opening suppresses bright details smaller than the specified SE, and closing suppresses dark details, they are used often in combination as *morphological filters* for image smoothing and noise removal. Consider Fig. 9.38(a), which shows an image of the Cygnus Loop supernova taken in the X-ray band (see Fig. 1.7 for details about this image). For purposes of the present discussion, suppose that the central light region is the object of interest and that the smaller components are noise. The objective is to remove the noise. Figure 9.38(b) shows the result of opening the original image with a flat disk of radius 2 and then closing the opening with an SE of the same size. Figures 9.38(c) and (d) show the results of the same operation using SEs of radii 3 and 5, respectively. As expected, this sequence shows progressive removal of small components as a function of SE size. In the last result, we see that the object of interest has been extracted. The noise components on the lower side of the image could not be removed completely because of their density.

The results in Fig. 9.38 are based on opening the original image and then closing the opening. A procedure used sometimes is to perform *alternating sequential filtering*, in which the opening–closing sequence starts with the original image, but subsequent steps perform the opening and closing on the results

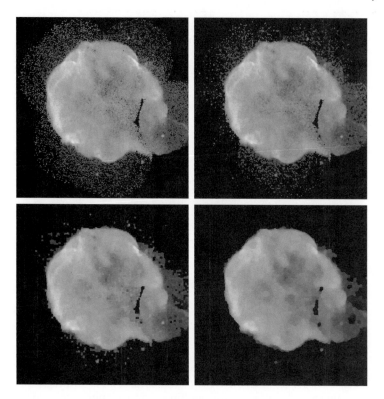

FIGURE 9.38
(a) 566×566
image of the
Cygnus Loop
supernova, taken
in the X-ray band
by NASA's
Hubble Telescope.
(b)–(d) Results of
performing
opening and
closing sequences
on the original
image with disk
structuring
elements of radii,
1, 3, and 5,
respectively.
(Original image
courtesy of
NASA.)

of the previous step. This type of filtering is useful in automated image analysis, in which results at each step are compared against a specified metric. Generally, this approach produces more blurring for the same size SE than the method illustrated in Fig. 9.38.

Morphological gradient

Dilation and erosion can be used in combination with image subtraction to obtain the morphological gradient of an image, denoted by g, where

$$g = (f \oplus b) - (f \ominus b) \tag{9.6-11}$$

See Section 3.6.4 for a definition of the image gradient.

The dilation thickens regions in an image and the erosion shrinks them. Their difference emphasizes the boundaries between regions. Homogenous areas are not affected (as long as the SE is relatively small) so the subtraction operation tends to eliminate them. The net result is an image in which the edges are enhanced and the contribution of the homogeneous areas are suppressed, thus producing a "derivative-like" (gradient) effect.

Figure 9.39 shows an example. Figure 9.39(a) is a head CT scan, and the next two figures are the opening and closing with a 3×3 SE of all 1s. Note the thickening and shrinking just mentioned. Figure 9.39(d) is the morphological gradient obtained using Eq. (9.6-11), in which the boundaries between regions are clearly delineated, as expected of a 2-D derivative image.

FIGURE 9.39
(a) 512×512
image of a head
CT scan.
(b) Dilation.
(c) Erosion.
(d) Morphological
gradient, compu-
ted as the
difference be-
tween (b) and (c).
(Original image
courtesy of Dr.
David R. Pickens,
Vanderbilt
University.)

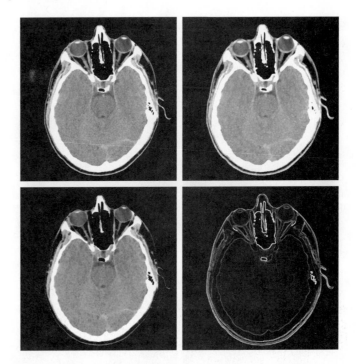

Top-hat and bottom-hat transformations

Combining image subtraction with openings and closings results in so-called
top-hat and bottom-hat transformations. The *top-hat transformation* of a gray-
scale image f is defined as f minus its opening:

$$T_{\text{hat}}(f) = f - (f \circ b) \tag{9.6-12}$$

Similarly, the *bottom-hat transformation* of f is defined as the closing of f
minus f:

$$B_{\text{hat}}(f) = (f \bullet b) - f \tag{9.6-13}$$

One of the principal applications of these transformations is in removing objects from an image by using a structuring element in the opening or closing
operation that does not fit the objects to be removed. The difference operation
then yields an image in which only the removed components remain. The top-
hat transform is used for light objects on a dark background, and the bottom-
hat transform is used for the converse. For this reason, the names *white top-hat*
and *black top-hat*, respectively, are used frequently when referring to these
two transformations.

   An important use of top-hat transformations is in correcting the effects of
nonuniform illumination. As we will see in the next chapter, proper (uniform)
illumination plays a central role in the process of extracting objects from the
background. This process, called *segmentation*, is one of the first steps performed in automated image analysis. A commonly used segmentation approach is to threshold the input image.

To illustrate, consider Fig. 9.40(a), which shows a 600×600 image of grains of rice. This image was obtained under nonuniform lighting, as evidenced by the darker area in the bottom, rightmost part of the image. Figure 9.40(b) shows the result of thresholding using Otsu's method, an optimal thresholding method discussed in Section 10.3.3. The net result of nonuniform illumination was to cause segmentation errors in the dark area (several grains of rice were not extracted from the background), as well as in the top left part of the image, where parts of the background were misclassified. Figure 9.40(c) shows the opening of the image with a disk of radius 40. This SE was large enough so that it would not fit in any of the objects. As a result, the objects were eliminated, leaving only an approximation of the background. The shading pattern is clear in this image. By subtracting this image from the original (i.e., performing a top-hat transformation), the background should become more uniform. This is indeed the case, as Fig. 9.40(d) shows. The background is not perfectly uniform, but the differences between light and dark extremes are less, and this was enough to yield a correct

a b
c d e

FIGURE 9.40 Using the top-hat transformation for *shading correction*. (a) Original image of size 600×600 pixels. (b) Thresholded image. (c) Image opened using a disk SE of radius 40. (d) Top-hat transformation (the image minus its opening). (e) Thresholded top-hat image.

segmentation result in which all rice grains were detected, as Fig. 9.40(e) shows. This image was obtained using Otsu's method, as before.

Granulometry

In terms of image processing, *granulometry* is a field that deals with determining the size distribution of particles in an image. In practice, particles seldom are neatly separated, which makes particle counting by identifying individual particles a difficult task. Morphology can be used to estimate particle size distribution indirectly, without having to identify and measure every particle in the image.

The approach is simple in principle. With particles having regular shapes that are lighter than the background, the method consists of applying openings with SEs of increasing size. The basic idea is that opening operations of a particular size should have the most effect on regions of the input image that contain particles of similar size. For each opening, the *sum* of the pixel values in the opening is computed. This sum, sometimes called the *surface area*, decreases as a function of increasing SE size because, as we noted earlier, openings decrease the intensity of light features. This procedure yields a 1-D array of such numbers, with each element in the array being equal to the sum of the pixels in the opening for the size SE corresponding to that location in the array. To emphasize changes between successive openings, we compute the difference between adjacent elements of the 1-D array. To visualize the results, the differences are plotted. The peaks in the plot are an indication of the predominant size distributions of the particles in the image.

As an example, consider Fig. 9.41(a) which is an image of wood dowel plugs of two dominant sizes. The wood grain in the dowels are likely to introduce variations in the openings, so smoothing is a sensible pre-processing step. Figure 9.41(b) shows the image smoothed using the morphological smoothing

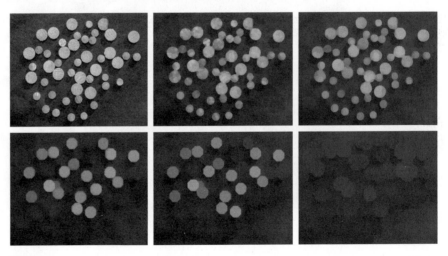

a b c
d e f

FIGURE 9.41 (a) 531×675 image of wood dowels. (b) Smoothed image. (c)–(f) Openings of (b) with disks of radii equal to 10, 20, 25, and 30 pixels, respectively. (Original image courtesy of Dr. Steve Eddins, The MathWorks, Inc.)

filter discussed earlier, with a disk of radius 5. Figures 9.41(c) through (f) show examples of image openings with disks of radii 10, 20, 25, and 30. Note in Fig. 9.41(d) that the intensity contribution due to the small dowels has been almost eliminated. In Fig. 9.41(e) the contribution of the large dowels has been significantly reduced, and in Fig. 9.41(f) even more so. (Observe in Fig. 9.41(e) that the large dowel near the top right of the image is much darker than the others because of its smaller size. This would be useful information if we had been attempting to detect defective dowels.)

Figure 9.42 shows a plot of the difference array. As mentioned previously, we expect significant differences (peaks in the plot) around radii at which the SE is large enough to encompass a set of particles of approximately the same diameter. The result in Fig. 9.42 has two distinct peaks, clearly indicating the presence of two dominant object sizes in the image.

Textural segmentation

Figure 9.43(a) shows a noisy image of dark blobs superimposed on a light background. The image has two textural regions: a region composed on large blobs on the right and a region on the left composed of smaller blobs. The objective is to find a boundary between the two regions based on their textural content (we discuss texture in Section 11.3.3). As noted earlier, the process of subdividing an image into regions is called *segmentation*, which is the topic of Chapter 10.

The objects of interest are darker than the background, and we know that if we close the image with a structuring element larger than the small blobs, these blobs will be removed. The result in Fig. 9.43(b), obtained by closing the input image using a disk with a radius of 30 pixels, shows that indeed this is the case (the radius of the blobs is approximately 25 pixels). So, at this point, we have an image with large, dark blobs on a light background. If we *open* this image with a structuring element that is large relative to the separation between these blobs, the net result should be an image in which the light patches between the blobs are removed, leaving the dark blobs and now equally dark patches between these blobs. Figure 9.43(c) shows the result, obtained using a disk of radius 60.

Performing a morphological gradient on this image with, say, a 3×3 SE of 1s, will give us the boundary between the two regions. Figure 9.43(d) shows the boundary obtained from the morphological gradient operation superimposed

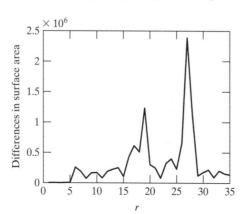

FIGURE 9.42
Differences in surface area as a function of SE disk radius, r. The two peaks are indicative of two dominant particle sizes in the image.

a b
c d

FIGURE 9.43
Textural
segmentation.
(a) A 600 × 600
image consisting
of two types of
blobs. (b) Image
with small blobs
removed by
closing (a).
(c) Image with
light patches
between large
blobs removed by
opening (b).
(d) Original
image with
boundary
between the two
regions in (c)
superimposed.
The boundary was
obtained using a
morphological
gradient
operation.

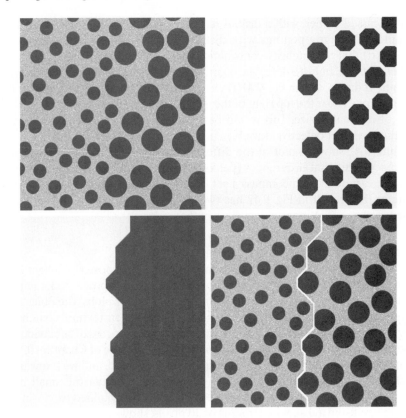

on the original image. All pixels to the right of this boundary are said to belong
to the texture region characterized by large blobs, and conversely for the pixels on the left of the boundary. You will find it instructive to work through this
example in more detail using the graphical analogy for opening and closing illustrated in Fig. 9.36.

9.6.4 Gray-Scale Morphological Reconstruction

Gray-scale morphological reconstruction is defined basically in the same manner introduced in Section 9.5.9 for binary images. Let f and g denote the
marker and *mask* images, respectively. We assume that both are gray-scale images of the same size and that $f \leq g$. The *geodesic dilation* of size 1 of f with
respect to g is defined as

It is understood that
these expressions are
functions of (x, y). We
omit the coordinates to
simplify the notation.

$$D_g^{(1)}(f) = (f \oplus b) \wedge g \qquad (9.6\text{-}14)$$

where $\wedge$ denotes the point-wise minimum operator. This equation indicates
that the geodesic dilation of size 1 is obtained by first computing the dilation
of f by b and then selecting the minimum between the result and g at every
point (x,y). The dilation is given by Eq. (9.6-2) if b is a flat SE or by Eq. (9.6-4)
if it is not. The geodesic dilation of size n of f with respect to g is defined as

$$D_g^{(n)}(f) = D_g^{(1)}[D_g^{(n-1)}(f)] \qquad (9.6\text{-}15)$$

with $D_g^{(0)}(f) = f$.

Similarly, the *geodesic erosion* of size 1 of f with respect to g is defined as

$$E_g^{(1)}(f) = (f \ominus b) \vee g \qquad (9.6\text{-}16)$$

where $\vee$ denotes the point-wise maximum operator. The geodesic erosion of size n is defined as

See Problem 9.33 for a list of dual relationships between expressions in this section.

$$E_g^{(n)}(f) = E_g^{(1)}\big[E_g^{(n-1)}(f)\big] \qquad (9.6\text{-}17)$$

with $E_g^{(0)}(f) = f$.

The *morphological reconstruction by dilation* of a gray-scale mask image, g, by a gray-scale marker image, f, is defined as the geodesic dilation of f with respect to g, iterated until stability is reached; that is,

$$R_g^D(f) = D_g^{(k)}(f) \qquad (9.6\text{-}18)$$

with k *such that* $D_g^{(k)}(f) = D_g^{(k+1)}(f)$. The *morphological reconstruction by erosion* of g by f is similarly defined as

$$R_g^E(f) = E_g^{(k)}(f) \qquad (9.6\text{-}19)$$

with k *such that* $E_g^{(k)}(f) = E_g^{(k+1)}(f)$.

As in the binary case, opening by reconstruction of gray-scale images first erodes the input image and uses it as a marker. The *opening by reconstruction* of size n of an image f is defined as the reconstruction by dilation of f from the erosion of size n of f; that is,

$$O_R^{(n)}(f) = R_f^D\big[(f \ominus nb)\big] \qquad (9.6\text{-}20)$$

where $(f \ominus nb)$ denotes n erosions of f by b, as explained in Section 9.5.7. Recall from the discussion of Eq. (9.5-27) for binary images that the objective of opening by reconstruction is to preserve the shape of the image components that remain after erosion.

Similarly, the *closing by reconstruction* of size n of an image f is defined as the reconstruction by erosion of f from the dilation of size n of f; that is,

$$C_R^{(n)}(f) = R_f^E\big[(f \oplus nb)\big] \qquad (9.6\text{-}21)$$

where $(f \oplus nb)$ denotes n dilations of f by b. Because of duality, the closing by reconstruction of an image can be obtained by complementing the image, obtaining the opening by reconstruction, and complementing the result. Finally, as the following example shows, a useful technique called *top-hat by reconstruction* consists of subtracting from an image its opening by reconstruction.

■ In this example, we illustrate the use of gray-scale reconstruction in several steps to normalize the irregular background of the image in Fig. 9.44(a), leaving only the text on a background of constant intensity. The solution of this problem is a good illustration of the power of morphological concepts. We begin by suppressing the horizontal reflection on the top of the keys. The reflections are wider than any single character in the image, so we should be able to suppress them by performing an opening by reconstruction using a long horizontal line in the erosion operation. This operation will yield the background containing the keys and their reflections. Subtracting this from

EXAMPLE 9.11:
Using morphological reconstruction to flatten a complex background.

a b c
d e f
g h i

FIGURE 9.44 (a) Original image of size 1134×1360 pixels. (b) Opening by reconstruction of (a) using a horizontal line 71 pixels long in the erosion. (c) Opening of (a) using the same line. (d) Top-hat by reconstruction. (e) Top-hat. (f) Opening by reconstruction of (d) using a horizontal line 11 pixels long. (g) Dilation of (f) using a horizontal line 21 pixels long. (h) Minimum of (d) and (g). (i) Final reconstruction result. (Images courtesy of Dr. Steve Eddins, The MathWorks, Inc.)

the original image (i.e., performing a top-hat by reconstruction) will eliminate the horizontal reflections and variations in background from the original image.

Figure 9.44(b) shows the result of opening by reconstruction of the original image using a horizontal line of size 1×71 pixels in the erosion operation. We could have used just an opening to remove the characters, but the resulting background would not have been as uniform, as Fig. 9.44(c) shows (for example, compare the regions between the keys in the two images). Figure 9.44(d) shows the result of subtracting Fig. 9.44(b) from Fig. 9.44(a). As expected, the horizontal reflections and variations in background were suppressed. For comparison, Fig. 9.44(e) shows the result of performing just a top-hat transformation (i.e., subtracting the "standard" opening from the image, as discussed earlier in this section). As expected from the characteristics of the background in Fig. 9.44(c), the background in Fig. 9.44(e) is not nearly as uniform as in Fig. 9.44(d).

The next step is to remove the vertical reflections from the edges of keys, which are quite visible in Fig. 9.44(d). We can do this by performing an opening by reconstruction with a line SE whose width is approximately equal to the reflections (about 11 pixels in this case). Figure 9.44(f) shows the result of performing this operation on Fig. 9.44(d). The vertical reflections were suppressed, but so were thin, vertical strokes that are valid characters (for example, the I in SIN), so we have to find a way to restore the latter. The suppressed characters are very close to the other characters so, if we dilate the remaining characters horizontally, the dilated characters will overlap the area previously occupied by the suppressed characters. Figure 9.44(g), obtained by dilating Fig. 9.44(f) with a line SE of size 1×21, shows that indeed this is case.

All that remains at this point is to restore the suppressed characters. Consider an image formed as the point-wise minimum between the dilated image in Fig. 9.44(g) and the top-hat by reconstruction in Fig. 9.44(d). Figure 9.44(h) shows the minimum image (although this result appears to be close to our objective, note that the I in SIN is still missing). By using this image as a marker and the dilated image as the mask in gray-scale reconstruction [Eq. (9.6-18)] we obtain the final result in Fig. 9.44(i). This image shows that all characters were properly extracted from the original, irregular background, including the background of the keys. The background in Fig. 9.44(i) is uniform throughout. ■

Summary

The morphological concepts and techniques introduced in this chapter constitute a powerful set of tools for extracting features of interest in an image. One of the most appealing aspects of morphological image processing is the extensive set-theoretic foundation from which morphological techniques have evolved. A significant advantage in terms of implementation is the fact that dilation and erosion are primitive operations that are the basis for a broad class of morphological algorithms. As shown in the following chapter, morphology can be used as the basis for developing image segmentation procedures with numerous applications. As discussed in Chapter 11, morphological techniques also play a major role in procedures for image description.

References and Further Reading

The book by Serra [1982] is a fundamental reference on morphological image processing. See also Serra [1988], Giardina and Dougherty [1988], and Haralick and Shapiro [1992]. Additional early references relevant to our discussion include Blum [1967], Lantuéjoul [1980], Maragos [1987], and Haralick et al. [1987]. For an overview of both binary and gray-scale morphology, see Basart and Gonzalez [1992] and Basart et al.

[1992]. This set of references provides ample basic background for the material covered in Sections 9.1 through 9.4. For a good overview of the material in Sections 9.5 and 9.6, see the book by Soille [2003].

Important issues of implementing morphological algorithms such as the ones given in Section 9.5 and 9.6 are exemplified in the papers by Jones and Svalbe [1994], Park and Chin [1995], Sussner and Ritter [1997], Anelli et al. [1998], and Shaked and Bruckstein [1998]. A paper by Vincent [1993] is especially important in terms of practical details for implementing gray-scale morphological algorithms. See also the book by Gonzalez, Woods, and Eddins [2004].

For additional reading on the theory and applications of morphological image processing, see the book by Goutsias and Bloomberg [2000] and a special issue of *Pattern Recognition* [2000]. See also a compilation of references by Rosenfeld [2000]. The books by Marchand-Maillet and Sharaiha [2000] on binary image processing and by Ritter and Wilson [2001] on image algebra also are of interest. Current work in the application of morphological techniques for image processing is exemplified in the papers by Kim [2005] and Evans and Liu [2006].

Problems

Detailed solutions to the problems marked with a star can be found in the book Web site. The site also contains suggested projects based on the material in this chapter.

9.1   Digital images in this book are embedded in square grid arrangements and pixels are allowed to be 4-, 8-, or m-connected. However, other grid arrangements are possible. Specifically, a hexagonal grid arrangement that leads to 6-connectivity, is used sometimes (see the following figure).

  (a) How would you convert an image from a square grid to a hexagonal grid?

  (b) Discuss the shape invariance to rotation of objects represented in a square grid as opposed to a hexagonal grid.

  (c) Is it possible to have ambiguous diagonal configurations in a hexagonal grid, as is the case with 8-connectivity? (See Section 2.5.2.)

9.2 ★ **(a)** Give a morphological algorithm for converting an 8-connected binary boundary to an m-connected boundary (see Section 2.5.2). You may assume that the boundary is fully connected and that it is one pixel thick.

  (b) Does the operation of your algorithm require more than one iteration with each structuring element? Explain your reasoning.

  (c) Is the performance of your algorithm independent of the order in which the structuring elements are applied? If your answer is yes, prove it; otherwise give an example that illustrates the dependence of your procedure on the order of application of the structuring elements.

9.3   Erosion of a set A by structuring element B is a subset of A as long as the origin of B is contained by B. Give an example in which the erosion $A \ominus B$ lies outside, or partially outside, A.

9.4 The following four statements are true. Advance an argument that establishes the reason(s) for their validity. Part (a) is true in general. Parts (b) through (d) are true only for *digital* sets. To show the validity of (b) through (d), draw a discrete, square grid (as shown in Problem 9.1) and give an example for each case using sets composed of points on this grid. (*Hint:* Keep the number of points in each case as small as possible while still establishing the validity of the statements.)

★ **(a)** The erosion of a convex set by a convex structuring element is a convex set.

★ **(b)** The dilation of a convex set by a convex structuring element is not necessarily always convex.

(c) The points in a convex digital set are not always connected.

(d) It is possible to have a set of points in which a line joining every pair of points in the set lies within the set but the set is not convex.

★**9.5** With reference to the image shown, give the structuring element and morphological operation(s) that produced each of the results shown in images (a) through (d). Show the origin of each structuring element clearly. The dashed lines show the boundary of the original set and are included only for reference. Note that in (d) all corners are rounded.

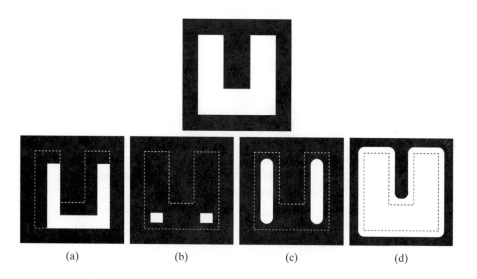

(a)   (b)   (c)   (d)

9.6 Let A denote the set shown shaded in the following figure. Refer to the structuring elements shown (the black dots denote the origin). Sketch the result of the following morphological operations:

(a) $(A \ominus B^4) \oplus B^2$

(b) $(A \ominus B^1) \oplus B^3$

(c) $(A \oplus B^1) \oplus B^3$

(d) $(A \oplus B^3) \ominus B^2$

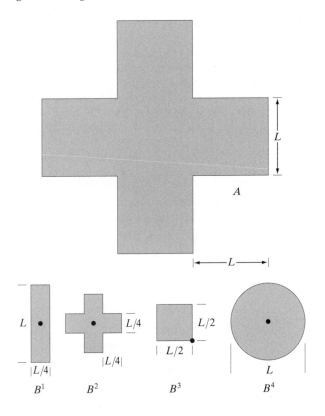

★9.7   **(a)** What is the limiting effect of repeatedly dilating an image? Assume that a trivial (one point) structuring element is not used.

(b) What is the smallest image from which you can start in order for your answer in part (a) to hold?

9.8   **(a)** What is the limiting effect of repeatedly eroding an image? Assume that a trivial (one point) structuring element is not used.

(b) What is the smallest image from which you can start in order for your answer in part (a) to hold?

★9.9   An alternative definition of erosion is

$$A \ominus B = \{w \in Z^2 | w + b \in A, \text{ for every } b \in B\}$$

Show that this definition is equivalent to the definition in Eq. (9.2-1).

9.10   **(a)** Show that the definition of erosion given in Problem 9.9 is equivalent to yet another definition of erosion:

$$A \ominus B = \bigcap_{b \in B}(A)_{-b}$$

(If $-b$ is replaced with b, this expression is called the *Minkowsky subtraction* of two sets.)

(b) Show that the expression in (a) also is equivalent to the definition in Eq. (9.2-1).

★**9.11** An alternative definition of dilation is

$$A \oplus B = \{w \in Z^2 | w = a + b, \text{ for some } a \in A \text{ and } b \in B\}$$

Show that this definition and the definition in Eq. (9.2-3) are equivalent.

9.12 **(a)** Show that the definition of dilation given in Problem 9.11 is equivalent to yet another definition of dilation:

$$A \oplus B = \bigcup_{b \in B} (A)_b$$

(This expression also is called the *Minkowsky addition* of two sets.)

(b) Show that the expression in (a) also is equivalent to the definition in Eq. (9.2-3).

9.13 Prove the validity of the duality expression in Eq. (9.2-6).

★**9.14** Prove the validity of the duality expressions $(A \bullet B)^c = (A^c \circ \hat{B})$ and $(A \circ B)^c = (A^c \bullet \hat{B})$.

9.15 Prove the validity of the following expressions:

★**(a)** $A \circ B$ is a subset (subimage) of A.

(b) If C is a subset of D, then $C \circ B$ is a subset of $D \circ B$.

(c) $(A \circ B) \circ B = A \circ B$.

9.16 Prove the validity of the following expressions (assume that the origin of B is contained in B and that Problems 9.14 and 9.15 are true):

(a) A is a subset (subimage) of $A \bullet B$.

(b) If C is a subset of D, then $C \bullet B$ is a subset of $D \bullet B$.

(c) $(A \bullet B) \bullet B = A \bullet B$.

9.17 Refer to the image and structuring element shown. Sketch what the sets C, D, E, and F would look like in the following sequence of operations: $C = A \ominus B$; $D = C \oplus B$; $E = D \oplus B$; and $F = E \ominus B$. The initial set A consists of all the image components shown in white, with the exception of the structuring element B. Note that this sequence of operations is simply the opening of A by B, followed by the closing of that opening by B. You may assume that B is just large enough to enclose each of the noise components.

★**9.18**   Consider the three binary images shown in the following figure. The image on the left is composed of squares of sizes 1, 3, 5, 7, 9, and 15 pixels on the side. The image in the middle was generated by eroding the image on the left with a square structuring element of 1s, of size 13 × 13 pixels, with the objective of eliminating all the squares, except the largest ones. Finally, the image on the right is the result of dilating the image in the center with the same structuring element, with the objective of restoring the largest squares. You know that erosion followed by dilation is the opening of an image, and you know also that opening generally does not restore objects to their original form. Explain why full reconstruction of the large squares was possible in this case.

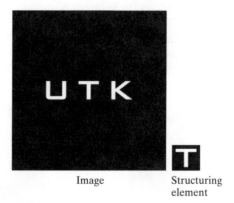

9.19   Sketch the result of applying the hit-or-miss transform to the image and structuring element shown. Indicate clearly the origin and border you selected for the structuring element.

Image   Structuring
                     element

★**9.20**   Three features (lake, bay, and line segment) useful for differentiating thinned objects in an image are shown in the following page. Develop a morphological/logical algorithm for differentiating among these shapes. The input to your algorithm would be one of these three shapes. The output must be the identity of the input. You may assume that the features are 1 pixel thick and that each is fully connected. However, they can appear in any orientation.

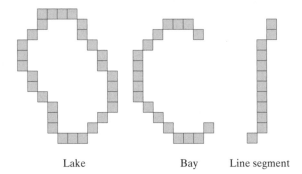

Lake   Bay   Line segment

9.21 Discuss what you would expect the result to be in each of the following cases:

    (a) The starting point of the hole filling algorithm of Section 9.5.2 is a point on the boundary of the object.

    (b) The starting point in the hole filling algorithm is outside of the boundary.

    (c) Sketch what the convex hull of the figure in Problem 9.6 would look like as computed with the algorithm given in Section 9.5.4. Assume that $L = 3$ pixels.

9.22 ★(a) Discuss the effect of using the structuring element in Fig. 9.15(c) for boundary extraction, instead of the one shown in Fig. 9.13(b).

    (b) What would be the effect of using a 3×3 structuring element composed of all 1s in the hole filling algorithm of Eq. (9.5-2), instead of the structuring element shown in Fig. 9.15(c)?

9.23 ★(a) Propose a method (using any of the techniques from Sections 9.1 through 9.5) for automating the example in Fig. 9.16. You may assume that the spheres do not touch each other and that none touch the border of the image.

    (b) Repeat (a), but allowing the spheres to touch in arbitrary ways, including touching the border.

★9.24 The algorithm given in Section 9.5.3 for extracting connected components requires that a point be known in each connected component in order to extract them all. Suppose that you are given a binary image containing an arbitrary (unknown) number of connected components. Propose a completely automated procedure for extracting all connected components. Assume that points belonging to connected components are labeled 1 and background points are labeled 0.

9.25 Give an expression based on reconstruction by dilation capable of extracting all the holes in a binary image.

9.26 With reference to the hole-filling algorithm in Section 9.5.9:

    (a) Explain what would happen if all border points of f are 1.

    (b) If the result in (a) gives the result that you would expect, explain why. If it does not, explain how you would modify the algorithm so that it works as expected.

★**9.27** Explain what would happen in binary erosion and dilation if the structuring element is a single point, valued 1. Give the reason(s) for your answer.

9.28 As explained in Eq. (9.5-27) and Section 9.6.4, opening by reconstruction preserves the shape of the image components that remain after erosion. What does closing by reconstruction do?

★**9.29** Show that geodesic erosion and dilation (Section 9.5.9) are duals with respect to set complementation. That is, show that $E_G^{(n)}(F) = \left[D_{G^c}^{(1)}\left[D_{G^c}^{(n-1)}(F^c)\right]\right]^c$ and, conversely, that $D_G^{(n)}(F) = \left[E_{G^c}^{(1)}\left[E_{G^c}^{(n-1)}(F^c)\right]\right]^c$. Assume that the structuring element is symmetric about its origin.

9.30 Show that reconstruction by dilation and reconstruction by erosion (Section 9.5.9) are duals with respect to set complementation. That is, show that $R_G^D(F) = \left[R_{G^c}^E(F^c)\right]^c$ and, vice versa, that $R_G^E(F) = \left[R_{G^c}^D(F^c)\right]^c$. Assume that the structuring element is symmetric about its origin.

★**9.31** Advance an argument showing that:

(a) $[(F \ominus nB)]^c = (F^c \oplus n\hat{B})$, where $(F \ominus nB)$ indicates n erosions of F by B.

(b) $[(F \oplus nB)]^c = (F^c \ominus n\hat{B})$.

9.32 Show that binary closing by reconstruction is the dual of opening by reconstruction with respect to set complementation: $O_R^{(n)}(F) = \left[C_R^{(n)}(F^c)\right]^c$, and similarly that $C_R^{(n)}(F) = \left[O_R^{(n)}(F^c)\right]^c$. Assume that the structuring element is symmetric with respect to its origin.

9.33 Prove the validity of the following gray-scale morphology expressions. You may assume that b is a flat structuring element. Recall that $f^c(x, y) = -f(x, y)$ and that $\hat{b}(x, y) = b(-x - y)$.

★(a) Duality of erosion and dilation: $(f \ominus b)^c = f^c \oplus \hat{b}$ and $(f \oplus b)^c = f^c \ominus \hat{b}$.

(b) $(f \bullet b)^c = f^c \circ \hat{b}$ and $(f \circ b)^c = f^c \bullet \hat{b}$.

★(c) $D_g^{(n)}(f) = \left[E_{g^c}^{(1)}\left[E_{g^c}^{(n-1)}(f^c)\right]\right]^c$ and $E_g^{(n)}(f) = \left[D_{g^c}^{(1)}\left[D_{g^c}^{(n-1)}(f^c)\right]\right]^c$. Assume a symmetric structuring element.

(d) $R_g^D(f) = \left[R_{g^c}^E(f^c)\right]^c$ and $R_g^E(f) = \left[R_{g^c}^D(f^c)\right]^c$.

(e) $[(f \ominus nb)]^c = (f^c \oplus n\hat{b})$, where $(f \ominus nb)$ indicates n erosions of f by b. Also, $[(f \oplus nb)]^c = (f^c \ominus n\hat{b})$.

(f) $O_R^{(n)}(f) = \left[C_R^{(n)}(f^c)\right]^c$ and $C_R^{(n)}(f) = \left[O_R^{(n)}(f^c)\right]^c$. Assume that the structuring element is symmetric with respect to its origin.

9.34 In Fig. 9.43, a boundary between distinct texture regions was established without difficulty. Consider the image at the top of the facing page, which shows a region of small circles enclosed by a region of larger circles.

(a) Would the method used to generate Fig. 9.43(d) work with this image as well? Explain your reasoning, including any assumptions that you need to make for the method to work.

(b) If your answer was yes, sketch what the boundary will look like.

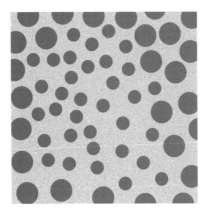

9.35 A gray-scale image, $f(x, y)$, is corrupted by nonoverlapping noise spikes that can be modeled as small, cylindrical artifacts of radii $R_{min} \leq r \leq R_{max}$ and amplitude $A_{min} \leq a \leq A_{max}$.

★ **(a)** Develop a morphological filtering approach for cleaning up the image.

(b) Repeat (a), but now assume that there is overlapping of, at most, four noise spikes.

9.36 A preprocessing step in an application of microscopy is concerned with the issue of isolating individual round particles from similar particles that overlap in groups of two or more particles (see following image). Assuming that all particles are of the same size, propose a morphological algorithm that produces three images consisting respectively of

★ **(a)** Only of particles that have merged with the boundary of the image.

(b) Only overlapping particles.

(c) Only nonoverlapping particles.

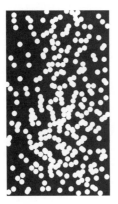

9.37 A high-technology manufacturing plant wins a government contract to manufacture high-precision washers of the form shown in the following figure. The terms of the contract require that the shape of all washers be inspected by an imaging system. In this context, shape inspection refers to deviations from round on the inner and outer edges of the washers. You may assume the following: (1) A "golden" (perfect with respect to the problem) image of an acceptable washer is available; and (2) the imaging and positioning systems ultimately used in the system will have an accuracy high enough to allow you to ignore errors due to digitalization and positioning. You are hired as a consultant to help specify the visual inspection part of the system. Propose a solution based on morphological/logic operations. Your answer should be in the form of a block diagram.

10 Image Segmentation

The whole is equal to the sum of its parts.

Euclid

The whole is greater than the sum of its parts.

Max Wertheimer

Preview

The material in the previous chapter began a transition from image processing methods whose inputs and outputs are images, to methods in which the inputs are images but the outputs are attributes extracted from those images (in the sense defined in Section 1.1). Segmentation is another major step in that direction.

Segmentation subdivides an image into its constituent regions or objects. The level of detail to which the subdivision is carried depends on the problem being solved. That is, segmentation should stop when the objects or regions of interest in an application have been detected. For example, in the automated inspection of electronic assemblies, interest lies in analyzing images of products with the objective of determining the presence or absence of specific anomalies, such as missing components or broken connection paths. There is no point in carrying segmentation past the level of detail required to identify those elements.

Segmentation of nontrivial images is one of the most difficult tasks in image processing. Segmentation accuracy determines the eventual success or failure of computerized analysis procedures. For this reason, considerable care should be taken to improve the probability of accurate segmentation. In some situations, such as in industrial inspection applications, at least some measure of control over the environment typically is possible. The experienced image processing system designer invariably pays considerable attention to such opportunities. In other applications, such as autonomous target acquisition, the system designer has no control over the operating environment, and the usual approach is to focus on selecting the types of sensors most likely to enhance the objects of interest while diminishing the contribution of irrelevant image detail. A good example is the use of infrared imaging by the military to detect objects with strong heat signatures, such as equipment and troops in motion.

Most of the segmentation algorithms in this chapter are based on one of two basic properties of intensity values: discontinuity and similarity. In the first category, the approach is to partition an image based on abrupt changes in intensity, such as edges. The principal approaches in the second category are based on partitioning an image into regions that are similar according to a set of predefined criteria. Thresholding, region growing, and region splitting and merging are examples of methods in this category. In this chapter, we discuss and illustrate a number of these approaches and show that improvements in segmentation performance can be achieved by combining methods from distinct categories, such as techniques in which edge detection is combined with thresholding. We discuss also image segmentation based on morphology. This approach is particularly attractive because it combines several of the positive attributes of segmentation based on the techniques presented in the first part of the chapter. We conclude the chapter with a brief discussion on the use of motion cues for segmentation.

See Sections 6.7 and 10.3.8 for a discussion of segmentation techniques based on more than just gray (intensity) values.

10.1 Fundamentals

Let R represent the entire spatial region occupied by an image. We may view image segmentation as a process that partitions R into n subregions, $R_1, R_2, \ldots, R_n$, such that

(a) $\displaystyle\bigcup_{i=1}^{n} R_i = R.$

See Section 2.5.2 regarding connected sets.

(b) R_i is a connected set, $i = 1, 2, \ldots, n.$

(c) $R_i \cap R_j = \varnothing$ for all i and $j, i \neq j.$

(d) $Q(R_i) = \text{TRUE}$ for $i = 1, 2, \ldots, n.$

(e) $Q(R_i \cup R_j) = \text{FALSE}$ for any adjacent regions R_i and $R_j.$

Here, $Q(R_k)$ is a logical predicate defined over the points in set R_k, and $\varnothing$ is the null set. The symbols $\cup$ and $\cap$ represent set union and intersection, respectively, as defined in Section 2.6.4. Two regions R_i and R_j are said to be *adjacent* if their union forms a connected set, as discussed in Section 2.5.2.

Condition (a) indicates that the segmentation must be complete; that is, every pixel must be in a region. Condition (b) requires that points in a region be connected in some predefined sense (e.g., the points must be 4- or 8-connected, as defined in Section 2.5.2). Condition (c) indicates that the regions must be disjoint. Condition (d) deals with the properties that must be satisfied by the pixels in a segmented region—for example, $Q(R_i) = \text{TRUE}$ if all pixels in R_i have the same intensity level. Finally, condition (e) indicates that two adjacent regions R_i and R_j must be different in the sense of predicate $Q.^\dagger$

---

†In general, Q can be a compound expression such as, for example, $Q(R_i) = \text{TRUE}$ if the average intensity of the pixels in R_i is less than m_i, AND if the standard deviation of their intensity is greater than σ_i, where m_i and σ_i are specified constants.

Thus, we see that the fundamental problem in segmentation is to partition an image into regions that satisfy the preceding conditions. Segmentation algorithms for monochrome images generally are based on one of two basic categories dealing with properties of intensity values: *discontinuity* and *similarity*. In the first category, the assumption is that boundaries of regions are sufficiently different from each other and from the background to allow boundary detection based on local discontinuities in intensity. *Edge-based segmentation* is the principal approach used in this category. *Region-based segmentation* approaches in the second category are based on partitioning an image into regions that are similar according to a set of predefined criteria.

Figure 10.1 illustrates the preceding concepts. Figure 10.1(a) shows an image of a region of constant intensity superimposed on a darker background, also of constant intensity. These two regions comprise the overall image region. Figure 10.1(b) shows the result of computing the boundary of the inner region based on intensity discontinuities. Points on the inside and outside of the boundary are black (zero) because there are no discontinuities in intensity in those regions. To segment the image, we assign one level (say, white) to the pixels on, or interior to, the boundary and another level (say, black) to all points exterior to the boundary. Figure 10.1(c) shows the result of such a procedure. We see that conditions (a) through (c) stated at the beginning of this section are satisfied by

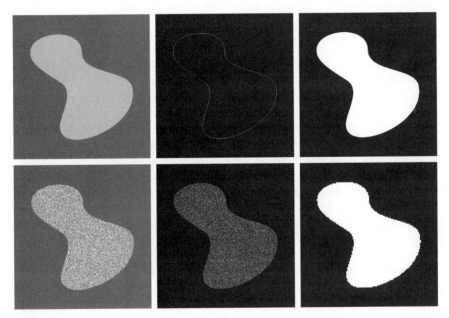

a b c
d e f

FIGURE 10.1 (a) Image containing a region of constant intensity. (b) Image showing the boundary of the inner region, obtained from intensity discontinuities. (c) Result of segmenting the image into two regions. (d) Image containing a textured region. (e) Result of edge computations. Note the large number of small edges that are connected to the original boundary, making it difficult to find a unique boundary using only edge information. (f) Result of segmentation based on region properties.

this result. The predicate of condition (d) is: If a pixel is on, or inside the boundary, label it white; otherwise label it black. We see that this predicate is TRUE for the points labeled black and white in Fig. 10.1(c). Similarly, the two segmented regions (object and background) satisfy condition (e).

The next three images illustrate region-based segmentation. Figure 10.1(d) is similar to Fig. 10.1(a), but the intensities of the inner region form a textured pattern. Figure 10.1(e) shows the result of computing the edges of this image. Clearly, the numerous spurious changes in intensity make it difficult to identify a unique boundary for the original image because many of the nonzero intensity changes are connected to the boundary, so edge-based segmentation is not a suitable approach. We note however, that the outer region is constant, so all we need to solve this simple segmentation problem is a predicate that differentiates between textured and constant regions. The standard deviation of pixel values is a measure that accomplishes this, because it is nonzero in areas of the texture region and zero otherwise. Figure 10.1(f) shows the result of dividing the original image into subregions of size 4×4. Each subregion was then labeled white if the standard deviation of its pixels was positive (i.e., if the predicate was TRUE) and zero otherwise. The result has a "blocky" appearance around the edge of the region because groups of 4×4 squares were labeled with the same intensity. Finally, note that these results also satisfy the five conditions stated at the beginning of this section.

10.2 Point, Line, and Edge Detection

The focus of this section is on segmentation methods that are based on detecting sharp, *local* changes in intensity. The three types of image features in which we are interested are isolated points, lines, and edges. *Edge pixels* are pixels at which the intensity of an image function changes abruptly, and *edges* (or *edge segments*) are sets of connected edge pixels (see Section 2.5.2 regarding connectivity). *Edge detectors* are local image processing methods designed to detect edge pixels. A line may be viewed as an edge segment in which the intensity of the background on either side of the line is either much higher or much lower than the intensity of the line pixels. In fact, as we discuss in the following section and in Section 10.2.4, lines give rise to so-called "roof edges." Similarly, an isolated point may be viewed as a line whose length and width are equal to one pixel.

When we refer to lines, we are referring to thin structures, typically just a few pixels thick. Such lines may correspond, for example, to elements of a digitized architectural drawing or roads in a satellite image.

10.2.1 Background

As we saw in Sections 2.6.3 and 3.5, local averaging smooths an image. Given that averaging is analogous to integration, it should come as no surprise that abrupt, local changes in intensity can be detected using derivatives. For reasons that will become evident shortly, first- and second-order derivatives are particularly well suited for this purpose.

Derivatives of a digital function are defined in terms of differences. There are various ways to approximate these differences but, as explained in Section 3.6.1, we require that any approximation used for a first derivative (1) must be zero in areas of constant intensity; (2) must be nonzero at the onset of an intensity step or ramp; and (3) must be nonzero at points along an intensity ramp. Similarly, we require that an approximation used for a second derivative (1) must be zero in areas of constant intensity; (2) must be nonzero at the onset *and* end of an intensity step or ramp; and (3) must be zero along intensity ramps. Because we are dealing with digital quantities whose values are finite, the maximum possible intensity change is also finite, and the shortest distance over which a change can occur is between adjacent pixels.

We obtain an approximation to the first-order derivative at point x of a one-dimensional function $f(x)$ by expanding the function $f(x + \Delta x)$ into a Taylor series about x, letting $\Delta x = 1$, and keeping only the linear terms (Problem 10.1). The result is the digital difference

$$\frac{\partial f}{\partial x} = f'(x) = f(x + 1) - f(x) \tag{10.2-1}$$

Recall from Section 2.4.2 that increments between image samples are defined as unity for notational clarity, hence the use of $\Delta x = 1$ in the derivation of Eq. (10.2-1).

We used a partial derivative here for consistency in notation when we consider an image function of two variables, $f(x, y)$, at which time we will be dealing with partial derivatives along the two spatial axes. Clearly, $\partial f / \partial x = df / dx$ when f is a function of only one variable.

We obtain an expression for the second derivative by differentiating Eq. (10.2-1) with respect to x:

$$\frac{\partial^2 f}{\partial x^2} = \frac{\partial f'(x)}{\partial x} = f'(x + 1) - f'(x)$$

$$= f(x + 2) - f(x + 1) - f(x + 1) + f(x)$$

$$= f(x + 2) - 2f(x + 1) + f(x)$$

where the second line follows from Eq. (10.2-1). This expansion is about point $x + 1$. Our interest is on the second derivative about point x, so we subtract 1 from the arguments in the preceding expression and obtain the result

$$\frac{\partial^2 f}{\partial x^2} = f''(x) = f(x + 1) + f(x - 1) - 2f(x) \tag{10.2-2}$$

It easily is verified that Eqs. (10.2-1) and (10.2-2) satisfy the conditions stated at the beginning of this section regarding derivatives of the first and second order. To illustrate this, and also to highlight the fundamental similarities and differences between first- and second-order derivatives in the context of image processing, consider Fig. 10.2.

Figure 10.2(a) shows an image that contains various solid objects, a line, and a single noise point. Figure 10.2(b) shows a horizontal intensity profile (scan line) of the image approximately through its center, including the isolated point. Transitions in intensity between the solid objects and the background along the scan line show two types of edges: *ramp edges* (on the left) and *step edges* (on the right). As we discuss later, intensity transitions involving thin objects such as lines often are referred to as *roof edges*. Figure 10.2(c) shows a simplification of the profile, with just enough points to make it possible for us to analyze numerically how the first- and second-order derivatives behave as they encounter a noise point, a line, and the edges of objects. In this simplified diagram the transition in the ramp spans four pixels, the noise point is a single pixel, the line is three pixels thick, and the transition of the intensity step takes place between adjacent pixels. The number of intensity levels was limited to eight for simplicity.

Consider the properties of the first and second derivatives as we traverse the profile from left to right. Initially, we note that the first-order derivative is nonzero at the onset and along the entire intensity ramp, while the second-order derivative is nonzero only at the onset and end of the ramp. Because edges of digital images resemble this type of transition, we conclude that first-order derivatives produce "thick" edges and second-order derivatives much finer ones. Next we encounter the isolated noise point. Here, the magnitude of the response at the point is much stronger for the second- than for the first-order derivative. This is not unexpected, because a second-order derivative is much

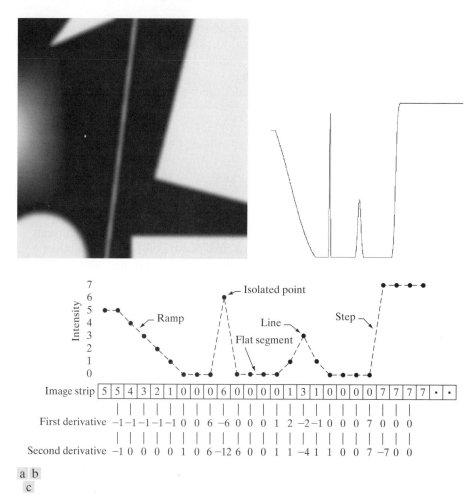

a b
c

FIGURE 10.2 (a) Image. (b) Horizontal intensity profile through the center of the image, including the isolated noise point. (c) Simplified profile (the points are joined by dashes for clarity). The image strip corresponds to the intensity profile, and the numbers in the boxes are the intensity values of the dots shown in the profile. The derivatives were obtained using Eqs. (10.2-1) and (10.2-2).

more aggressive than a first-order derivative in enhancing sharp changes. Thus, we can expect second-order derivatives to enhance fine detail (including noise) much more than first-order derivatives. The line in this example is rather thin, so it too is fine detail, and we see again that the second derivative has a larger magnitude. Finally, note in both the ramp and step edges that the second derivative has opposite signs (negative to positive or positive to negative) as it transitions into and out of an edge. This "double-edge" effect is an important characteristic that, as we show in Section 10.2.6, can be used to locate edges. The sign of the second derivative is used also to determine whether an edge is a transition from light to dark (negative second derivative) or from dark to light (positive second derivative), where the sign is observed as we move *into* the edge.

In summary, we arrive at the following conclusions: (1) First-order derivatives generally produce thicker edges in an image. (2) Second-order derivatives have a stronger response to fine detail, such as thin lines, isolated points, and noise. (3) Second-order derivatives produce a double-edge response at ramp and step transitions in intensity. (4) The sign of the second derivative can be used to determine whether a transition into an edge is from light to dark or dark to light.

The approach of choice for computing first and second derivatives at every pixel location in an image is to use spatial filters. For the 3×3 filter mask in Fig. 10.3, the procedure is to compute the sum of products of the mask coefficients with the intensity values in the region encompassed by the mask. That is, with reference to Eq. (3.4.3), the *response* of the mask at the center point of the region is

$$R = w_1 z_1 + w_2 z_2 + \cdots + w_9 z_9$$

$$= \sum_{k=1}^{9} w_k z_k$$

(10.2-3)

where z_k is the intensity of the pixel whose spatial location corresponds to the location of the kth coefficient in the mask. The details of implementing this operation over all pixels in an image are discussed in detail in Sections 3.4 and 3.6. In other words, computation of derivatives based on spatial masks is spatial filtering of an image with those masks, as explained in those sections.[†]

w_1	w_2	w_3
w_4	w_5	w_6
w_7	w_8	w_9

FIGURE 10.3
A general 3×3 spatial filter mask.

[†]As explained in Section 3.4.3, Eq. (10.2-3) is simplified notation either for spatial correlation, given by Eq. (3.4-1), or spatial convolution, given by Eq. (3.4-2). Therefore, when R is evaluated at all locations in an image, the result is an array. All spatial filtering in this chapter is done using correlation. In some instances, we use the term *convolving a mask with an image* as a matter of convention. However, we use this terminology only when the filter masks are symmetric, in which case correlation and convolution yield the same result.

10.2.2 Detection of Isolated Points

Based on the conclusions reached in the preceding section, we know that point detection should be based on the second derivative. From the discussion in Section 3.6.2, this implies using the Laplacian:

$$\nabla^2 f(x, y) = \frac{\partial^2 f}{\partial x^2} + \frac{\partial^2 f}{\partial y^2} \tag{10.2-4}$$

where the partials are obtained using Eq. (10.2-2):

$$\frac{\partial^2 f(x, y)}{\partial x^2} = f(x + 1, y) + f(x - 1, y) - 2f(x, y) \tag{10.2-5}$$

and

$$\frac{\partial^2 f(x, y)}{\partial y^2} = f(x, y + 1) + f(x, y - 1) - 2f(x, y) \tag{10.2-6}$$

The Laplacian is then

$$\nabla^2 f(x, y) = f(x + 1, y) + f(x - 1, y) + f(x, y + 1)$$
$$+ f(x, y - 1) - 4f(x, y) \tag{10.2-7}$$

As explained in Section 3.6.2, this expression can be implemented using the mask in Fig. 3.37(a). Also, as explained in that section, we can extend Eq. (10.2-7) to include the diagonal terms, and use the mask in Fig. 3.37(b). Using the Laplacian mask in Fig. 10.4(a), which is the same as the mask in Fig. 3.37(b), we say that a point has been detected at the location (x, y) on which the mask is centered if the absolute value of the response of the mask at that point exceeds a specified threshold. Such points are labeled 1 in the output image and all others are labeled 0, thus producing a binary image. In other words, the output is obtained using the following expression:

$$g(x, y) = \begin{cases} 1 & \text{if } |R(x, y)| \geq T \\ 0 & \text{otherwise} \end{cases} \tag{10.2-8}$$

where g is the output image, T is a nonnegative threshold, and R is given by Eq. (10.2-3). This formulation simply measures the weighted differences between a pixel and its 8-neighbors. Intuitively, the idea is that the intensity of an isolated point will be quite different from its surroundings and thus will be easily detectable by this type of mask. The only differences in intensity that are considered of interest are those large enough (as determined by T) to be considered isolated points. Note that, as usual for a derivative mask, the coefficients sum to zero, indicating that the mask response will be zero in areas of constant intensity.

FIGURE 10.4
(a) Point detection (Laplacian) mask.
(b) X-ray image of turbine blade with a porosity. The porosity contains a single black pixel.
(c) Result of convolving the mask with the image. (d) Result of using Eq. (10.2-8) showing a single point (the point was enlarged to make it easier to see). (Original image courtesy of X-TEK Systems, Ltd.)

■ We illustrate segmentation of isolated points in an image with the aid of Fig. 10.4(b), which is an X-ray image of a turbine blade from a jet engine. The blade has a porosity in the upper-right quadrant of the image, and there is a single black pixel embedded within the porosity. Figure 10.4(c) is the result of applying the point detector mask to the X-ray image, and Fig. 10.4(d) shows the result of using Eq. (10.2-8) with T equal to 90% of the highest absolute pixel value of the image in Fig. 10.4(c). The single pixel is clearly visible in this image (the pixel was enlarged manually to enhance its visibility). This type of detection process is rather specialized, because it is based on abrupt intensity changes at single-pixel locations that are surrounded by a homogeneous background in the area of the detector mask. When this condition is not satisfied, other methods discussed in this chapter are more suitable for detecting intensity changes. ■

EXAMPLE 10.1:
Detection of isolated points in an image.

10.2.3 Line Detection

The next level of complexity is line detection. Based on the discussion in Section 10.2.1, we know that for line detection we can expect second derivatives to result in a stronger response and to produce thinner lines than first derivatives. Thus, we can use the Laplacian mask in Fig. 10.4(a) for line detection also, keeping in mind that the double-line effect of the second derivative must be handled properly. The following example illustrates the procedure.

EXAMPLE 10.2:
Using the
Laplacian for line
detection.

■ Figure 10.5(a) shows a 486 × 486 (binary) portion of a wire-bond mask for an electronic circuit, and Fig. 10.5(b) shows its Laplacian image. Because the Laplacian image contains negative values,[†] scaling is necessary for display. As the magnified section shows, mid gray represents zero, darker shades of gray represent negative values, and lighter shades are positive. The double-line effect is clearly visible in the magnified region.

At first, it might appear that the negative values can be handled simply by taking the absolute value of the Laplacian image. However, as Fig. 10.5(c) shows, this approach doubles the thickness of the lines. A more suitable approach is to use only the positive values of the Laplacian (in noisy situations we use the values that exceed a positive threshold to eliminate random variations about zero caused by the noise). As the image in Fig. 10.5(d) shows, this approach results in thinner lines, which are considerably more useful. Note in Figs. 10.5(b) through (d) that when the lines are wide with respect to the size of the Laplacian mask, the lines are separated by a zero "valley."

a b
c d

FIGURE 10.5
(a) Original image.
(b) Laplacian
image; the
magnified section
shows the
positive/negative
double-line effect
characteristic of the
Laplacian.
(c) Absolute value
of the Laplacian.
(d) Positive values
of the Laplacian.

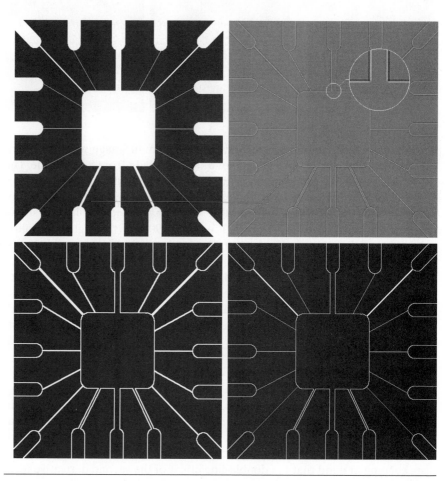

[†]When a mask whose coefficients sum to zero is convolved with an image, the pixels in the resulting image will sum to zero also (Problem 3.16), implying the existence of both positive and negative pixels in the result. Scaling so that all values are nonnegative is required for display purposes.

This is not unexpected. For example, when the 3×3 filter is centered on a line of constant intensity 5 pixels wide, the response will be zero, thus producing the effect just mentioned. When we talk about line detection, the assumption is that lines are thin with respect to the size of the detector. Lines that do not satisfy this assumption are best treated as regions and handled by the edge detection methods discussed later in this section.   ■

The Laplacian detector in Fig. 10.4(a) is isotropic, so its response is independent of direction (with respect to the four directions of the 3×3 Laplacian mask: vertical, horizontal, and two diagonals). Often, interest lies in detecting lines in *specified* directions. Consider the masks in Fig. 10.6. Suppose that an image with a constant background and containing various lines (oriented at $0°$, $\pm 45°$, and $90°$) is filtered with the first mask. The maximum responses would occur at image locations in which a horizontal line passed through the middle row of the mask. This is easily verified by sketching a simple array of 1s with a line of a different intensity (say, 5s) running horizontally through the array. A similar experiment would reveal that the second mask in Fig. 10.6 responds best to lines oriented at $+45°$; the third mask to vertical lines; and the fourth mask to lines in the $-45°$ direction. The preferred direction of each mask is weighted with a larger coefficient (i.e., 2) than other possible directions. The coefficients in each mask sum to zero, indicating a zero response in areas of constant intensity.

Let R_1, R_2, R_3, and R_4 denote the responses of the masks in Fig. 10.6, from left to right, where the Rs are given by Eq. (10.2-3). Suppose that an image is filtered (individually) with the four masks. If, at a given point in the image, $|R_k| > |R_j|$, for all $j \neq k$, that point is said to be more likely associated with a line in the direction of mask k. For example, if at a point in the image, $|R_1| > |R_j|$ for $j = 2, 3, 4$, that particular point is said to be more likely associated with a horizontal line. Alternatively, we may be interested in detecting lines in a specified direction. In this case, we would use the mask associated with that direction and threshold its output, as in Eq. (10.2-8). In other words, if we are interested in detecting all the lines in an image in the direction defined by a given mask, we simply run the mask through the image and threshold the absolute value of the result. The points that are left are the strongest responses which, for lines 1 pixel thick, correspond closest to the direction defined by the mask. The following example illustrates this procedure.

Recall from Section 2.4.2 that the image axis convention has the origin at the top left, with the positive x-axis pointing down and the positive y-axis extending to the right. The angles of the lines discussed in this section are measured with respect to the positive x-axis. For example, a vertical line has an angle of $0°$, and a $+45°$ line extends downward and to the right.

Do not confuse our use of R to designate mask response with the same symbol to denote regions in Section 10.1.

−1	−1	−1
2	2	2
−1	−1	−1

2	−1	−1
−1	2	−1
−1	−1	2

−1	2	−1
−1	2	−1
−1	2	−1

−1	−1	2
−1	2	−1
2	−1	−1

| Horizontal | +45° | Vertical | −45° |

FIGURE 10.6 Line detection masks. Angles are with respect to the axis system in Fig. 2.18(b).

EXAMPLE 10.3:
Detection of lines
in specified
directions.

■ Figure 10.7(a) shows the image used in the previous example. Suppose that we are interested in finding all the lines that are 1 pixel thick and oriented at +45°. For this purpose, we use the second mask in Fig. 10.6. Figure 10.7(b) is the result of filtering the image with that mask. As before, the shades darker than the gray background in Fig. 10.7(b) correspond to negative values. There are two principal segments in the image oriented in the +45° direction, one at the top left and one at the bottom right. Figures 10.7(c) and (d) show zoomed sections of Fig. 10.7(b) corresponding to these two areas. Note how much brighter the straight line segment in Fig. 10.7(d) is than the segment in Fig. 10.7(c). The reason is that the line segment in the bottom right of Fig. 10.7(a) is 1 pixel thick, while the one at the top left is not. The mask is "tuned" to detect 1-pixel-thick lines in the +45° direction, so we expect its response to be stronger when such lines are detected. Figure 10.7(e) shows the positive values of Fig. 10.7(b). Because we are interested in the strongest response, we let T equal the maximum value in Fig. 10.7(e). Figure 10.7(f) shows in white the points whose values satisfied the condition $g \geq T$, where g is the image in Fig. 10.7(e). The isolated points in the figure are points that also had similarly strong responses to the mask. In the original image, these points and their immediate neighbors are oriented in such a way that the mask produced a maximum response at those locations. These isolated points can be detected using the mask in Fig. 10.4(a) and then deleted, or they can be deleted using morphological operators, as discussed in the last chapter.   ■

10.2.4 Edge Models

Edge detection is the approach used most frequently for segmenting images based on abrupt (local) changes in intensity. We begin by introducing several ways to model edges and then discuss a number of approaches for edge detection.

Edge models are classified according to their intensity profiles. A *step edge* involves a transition between two intensity levels occurring ideally over the distance of 1 pixel. Figure 10.8(a) shows a section of a vertical step edge and a horizontal intensity profile through the edge. Step edges occur, for example, in images generated by a computer for use in areas such as solid modeling and animation. These clean, *ideal* edges can occur over the distance of 1 pixel, provided that no additional processing (such as smoothing) is used to make them look "real." Digital step edges are used frequently as edge models in algorithm development. For example, the Canny edge detection algorithm discussed in Section 10.2.6 was derived using a step-edge model.

In practice, digital images have edges that are blurred and noisy, with the degree of blurring determined principally by limitations in the focusing mechanism (e.g., lenses in the case of optical images), and the noise level determined principally by the electronic components of the imaging system. In such situations, edges are more closely modeled as having an intensity *ramp* profile, such as the edge in Fig. 10.8(b). The slope of the ramp is inversely proportional to the degree of blurring in the edge. In this model, we no longer have a thin (1 pixel thick) path. Instead, an edge point now is any point contained in the ramp, and an edge segment would then be a set of such points that are connected.

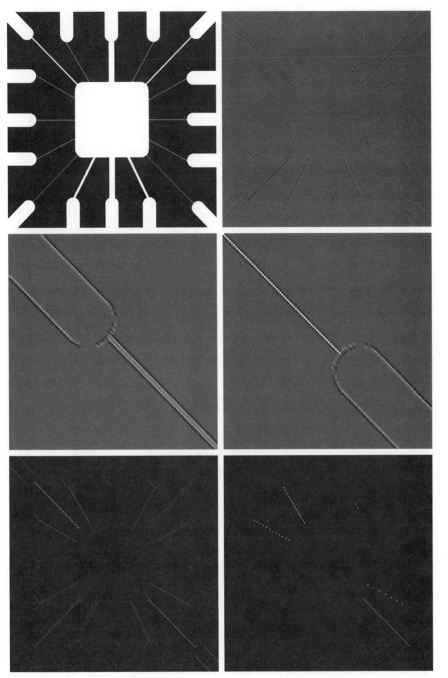

FIGURE 10.7
(a) Image of a wire-bond template.
(b) Result of processing with the +45° line detector mask in Fig. 10.6.
(c) Zoomed view of the top left region of (b).
(d) Zoomed view of the bottom right region of (b). (e) The image in (b) with all negative values set to zero. (f) All points (in white) whose values satisfied the condition $g \geq T$, where g is the image in (e). (The points in (f) were enlarged to make them easier to see.)

A third model of an edge is the so-called *roof edge*, having the characteristics illustrated in Fig. 10.8(c). Roof edges are models of lines through a region, with the base (width) of a roof edge being determined by the thickness and sharpness of the line. In the limit, when its base is 1 pixel wide, a roof edge is

FIGURE 10.8
From left to right,
models (ideal
representations) of
a step, a ramp, and
a roof edge, and
their corresponding
intensity profiles.

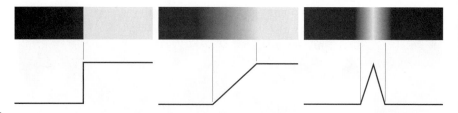

really nothing more than a 1-pixel-thick line running through a region in an image. Roof edges arise, for example, in range imaging, when thin objects (such as pipes) are closer to the sensor than their equidistant background (such as walls). The pipes appear brighter and thus create an image similar to the model in Fig. 10.8(c). As mentioned earlier, other areas in which roof edges appear routinely are in the digitization of line drawings and also in satellite images, where thin features, such as roads, can be modeled by this type of edge.

It is not unusual to find images that contain all three types of edges. Although blurring and noise result in deviations from the ideal shapes, edges in images that are reasonably sharp and have a moderate amount of noise do *resemble* the characteristics of the edge models in Fig. 10.8, as the profiles in Fig. 10.9 illustrate.[†] What the models in Fig. 10.8 allow us to do is write mathematical expressions for edges in the development of image processing algorithms. The performance of these algorithms will depend on the differences between actual edges and the models used in developing the algorithms.

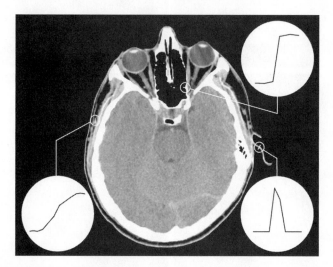

FIGURE 10.9 A 1508 × 1970 image showing (zoomed) actual ramp (bottom, left), step (top, right), and roof edge profiles. The profiles are from dark to light, in the areas indicated by the short line segments shown in the small circles. The ramp and "step" profiles span 9 pixels and 2 pixels, respectively. The base of the roof edge is 3 pixels. (Original image courtesy of Dr. David R. Pickens, Vanderbilt University.)

[†]Ramp edges with a sharp slope of a few pixels often are treated as step edges in order to differentiate them from ramps in the same image whose slopes are more gradual.

Figure 10.10(a) shows the image from which the segment in Fig. 10.8(b) was extracted. Figure 10.10(b) shows a horizontal intensity profile. This figure shows also the first and second derivatives of the intensity profile. As in the discussion in Section 10.2.1, moving from left to right along the intensity profile, we note that the first derivative is positive at the onset of the ramp and at points on the ramp, and it is zero in areas of constant intensity. The second derivative is positive at the beginning of the ramp, negative at the end of the ramp, zero at points on the ramp, and zero at points of constant intensity. The signs of the derivatives just discussed would be reversed for an edge that transitions from light to dark. The intersection between the zero intensity axis and a line extending between the extrema of the second derivative marks a point called the *zero crossing* of the second derivative.

We conclude from these observations that the *magnitude* of the first derivative can be used to detect the presence of an edge at a point in an image. Similarly, the *sign* of the second derivative can be used to determine whether an edge pixel lies on the dark or light side of an edge. We note two additional properties of the second derivative around an edge: (1) it produces two values for every edge in an image (an undesirable feature); and (2) its zero crossings can be used for locating the centers of thick edges, as we show later in this section. Some edge models make use of a smooth transition into and out of the ramp (Problem 10.7). However, the conclusions reached using those models are the same as with an ideal ramp, and working with the latter simplifies theoretical formulations. Finally, although attention thus far has been limited to a 1-D horizontal profile, a similar argument applies to an edge of any orientation in an image. We simply define a profile perpendicular to the edge direction at any desired point and interpret the results in the same manner as for the vertical edge just discussed.

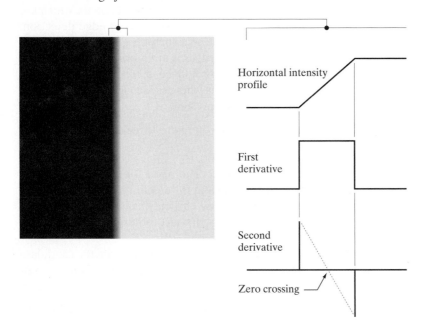

a b

FIGURE 10.10
(a) Two regions of constant intensity separated by an ideal vertical ramp edge.
(b) Detail near the edge, showing a horizontal intensity profile, together with its first and second derivatives.

Horizontal intensity profile

First derivative

Second derivative

Zero crossing

EXAMPLE 10.4:
Behavior of the
first and second
derivatives of a
noisy edge.

■ The edges in Fig. 10.8 are free of noise. The image segments in the first column in Fig. 10.11 show close-ups of four ramp edges that transition from a black region on the left to a white region on the right (keep in mind that the entire transition from black to white is a single edge). The image segment at the top left is free of noise. The other three images in the first column are corrupted by additive Gaussian noise with zero mean and standard deviation of 0.1, 1.0, and 10.0 intensity levels, respectively. The graph below each image is a horizontal intensity profile passing through the center of the image. All images have 8 bits of intensity resolution, with 0 and 255 representing black and white, respectively.

Computation of the derivatives for the entire image segment is discussed in the following section. For now, our interest lies on analyzing just the intensity profiles.

Consider the image at the top of the center column. As discussed in connection with Fig. 10.10(b), the derivative of the scan line on the left is zero in the constant areas. These are the two black bands shown in the derivative image. The derivatives at points on the ramp are constant and equal to the slope of the ramp. These constant values in the derivative image are shown in gray. As we move down the center column, the derivatives become increasingly different from the noiseless case. In fact, it would be difficult to associate the last profile in the center column with the first derivative of a ramp edge. What makes these results interesting is that the noise is almost invisible in the images on the left column. These examples are good illustrations of the sensitivity of derivatives to noise.

As expected, the second derivative is even more sensitive to noise. The second derivative of the noiseless image is shown at the top of the right column. The thin white and black vertical lines are the positive and negative components of the second derivative, as explained in Fig. 10.10. The gray in these images represents zero (as discussed earlier, scaling causes zero to show as gray). The only noisy second derivative image that barely resembles the noiseless case is the one corresponding to noise with a standard deviation of 0.1. The remaining second-derivative images and profiles clearly illustrate that it would be difficult indeed to detect their positive and negative components, which are the truly useful features of the second derivative in terms of edge detection.

The fact that such little visual noise can have such a significant impact on the two key derivatives used for detecting edges is an important issue to keep in mind. In particular, image smoothing should be a serious consideration prior to the use of derivatives in applications where noise with levels similar to those we have just discussed is likely to be present. ■

We conclude this section by noting that there are three fundamental steps performed in edge detection:

1. *Image smoothing for noise reduction.* The need for this step is amply illustrated by the results in the second and third columns of Fig. 10.11.
2. *Detection of edge points.* As mentioned earlier, this is a local operation that extracts from an image all points that are potential candidates to become edge points.
3. *Edge localization.* The objective of this step is to select from the candidate edge points only the points that are true members of the set of points comprising an edge.

The remainder of this section deals with techniques for achieving these objectives.

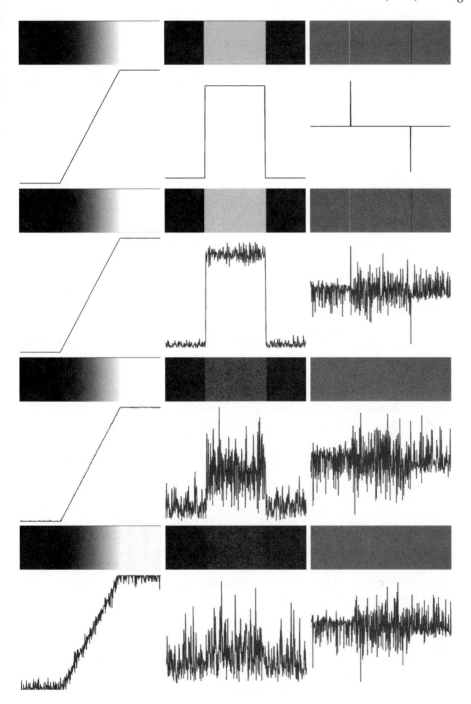

FIGURE 10.11 First column: Images and intensity profiles of a ramp edge corrupted by random Gaussian noise of zero mean and standard deviations of 0.0, 0.1, 1.0, and 10.0 intensity levels, respectively. Second column: First-derivative images and intensity profiles. Third column: Second-derivative images and intensity profiles.

10.2.5 Basic Edge Detection

As illustrated in the previous section, detecting changes in intensity for the purpose of finding edges can be accomplished using first- or second-order derivatives. We discuss first-order derivatives in this section and work with second-order derivatives in Section 10.2.6.

The image gradient and its properties

The tool of choice for finding edge strength *and* direction at location (x, y) of an image, f, is the gradient, denoted by ∇f, and defined as the *vector*

For convenience, we repeat here some equations from Section 3.6.4.

$$\nabla f \equiv \text{grad}(f) \equiv \begin{bmatrix} g_x \\ g_y \end{bmatrix} = \begin{bmatrix} \dfrac{\partial f}{\partial x} \\ \dfrac{\partial f}{\partial y} \end{bmatrix} \tag{10.2-9}$$

This vector has the important geometrical property that it points in the direction of the greatest rate of change of f at location (x, y).

The *magnitude* (*length*) of vector ∇f, denoted as $M(x, y)$, where

$$M(x, y) = \text{mag}(\nabla f) = \sqrt{g_x^2 + g_y^2} \tag{10.2-10}$$

is the *value* of the rate of change in the direction of the gradient vector. Note that g_x, g_y, and $M(x, y)$ are images of the same size as the original, created when x and y are allowed to vary over all pixel locations in f. It is common practice to refer to the latter image as the *gradient image*, or simply as the *gradient* when the meaning is clear. The summation, square, and square root operations are *array operations*, as defined in Section 2.6.1.

The *direction* of the gradient vector is given by the angle

$$\alpha(x, y) = \tan^{-1}\left[\frac{g_y}{g_x}\right] \tag{10.2-11}$$

measured with respect to the x-axis. As in the case of the gradient image, $\alpha(x, y)$ also is an image of the same size as the original created by the array division of image g_y by image g_x. The direction of an edge at an arbitrary point (x, y) is *orthogonal* to the direction, $\alpha(x, y)$, of the gradient vector at the point.

EXAMPLE 10.5:
Properties of the gradient.

■ Figure 10.12(a) shows a zoomed section of an image containing a straight edge segment. Each square shown corresponds to a pixel, and we are interested in obtaining the strength and direction of the edge at the point highlighted with a box. The pixels in gray have value 0 and the pixels in white have value 1. We show following this example that an approach for computing the derivatives in the x- and y-directions using a 3×3 neighborhood centered about a point consists simply of subtracting the pixels in the top row of the neighborhood from the pixels in the bottom row to obtain the partial derivative in the x-direction. Similarly, we subtract the pixels in the left column from the pixels in the right column to obtain the partial derivative in the y-direction. It then

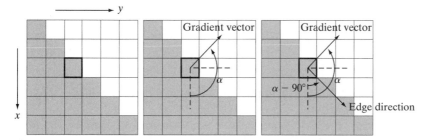

FIGURE 10.12 Using the gradient to determine edge strength and direction at a point. Note that the edge is perpendicular to the direction of the gradient vector at the point where the gradient is computed. Each square in the figure represents one pixel.

follows, using these differences as our estimates of the partials, that $\partial f / \partial x = -2$ and $\partial f / \partial y = 2$ at the point in question. Then,

$$
\nabla f = \begin{bmatrix} g_x \\ g_y \end{bmatrix} = \begin{bmatrix} \dfrac{\partial f}{\partial x} \\ \dfrac{\partial f}{\partial y} \end{bmatrix} = \begin{bmatrix} -2 \\ 2 \end{bmatrix}
$$

Recall from Section 2.4.2 that the origin of the image coordinate system is at the top left, with the positive x- and y-axes extending down and to the right, respectively.

from which we obtain $M(x, y) = 2\sqrt{2}$ at that point. Similarly, the direction of the gradient vector at the same point follows from Eq. (10.2-11): $\alpha(x, y) = \tan^{-1}(g_y/g_x) = -45°$, which is the same as $135°$ measured in the positive direction with respect to the x-axis. Figure 10.12(b) shows the gradient vector and its direction angle.

Figure 10.12(c) illustrates the important fact mentioned earlier that the edge at a point is orthogonal to the gradient vector at that point. So the direction angle of the edge in this example is $\alpha - 90° = 45°$. All edge points in Fig. 10.12(a) have the same gradient, so the entire edge segment is in the same direction. The gradient vector sometimes is called the *edge normal*. When the vector is normalized to unit length by dividing it by its magnitude [Eq. (10.2-10)], the resulting vector is commonly referred to as the *edge unit normal*. ■

Gradient operators

Obtaining the gradient of an image requires computing the partial derivatives $\partial f / \partial x$ and $\partial f / \partial y$ at every pixel location in the image. We are dealing with digital quantities, so a digital approximation of the partial derivatives over a neighborhood about a point is required. From Section 10.2.1 we know that

$$
g_x = \frac{\partial f(x, y)}{\partial x} = f(x + 1, y) - f(x, y) \tag{10.2-12}
$$

and

$$
g_y = \frac{\partial f(x, y)}{\partial y} = f(x, y + 1) - f(x, y) \tag{10.2-13}
$$

a b

FIGURE 10.13
One-dimensional masks used to implement Eqs. (10.2-12) and (10.2-13).

These two equations can be implemented for all pertinent values of x and y by filtering $f(x, y)$ with the 1-D masks in Fig. 10.13.

When diagonal edge direction is of interest, we need a 2-D mask. The *Roberts cross-gradient operators* (Roberts [1965]) are one of the earliest attempts to use 2-D masks with a diagonal preference. Consider the 3×3 region in Fig. 10.14(a). The Roberts operators are based on implementing the diagonal differences

In the remainder of this section we assume implicitly that f is a function of two variables, and omit the variables to simplify the notation.

and

$$g_x = \frac{\partial f}{\partial x} = (z_9 - z_5) \tag{10.2-14}$$

$$g_y = \frac{\partial f}{\partial y} = (z_8 - z_6) \tag{10.2-15}$$

a
b c
d e
f g

FIGURE 10.14
A 3×3 region of an image (the z's are intensity values) and various masks used to compute the gradient at the point labeled z_5.

Filter masks used to compute the derivatives needed for the gradient are often called *gradient operators, difference operators, edge operators,* or *edge detectors*.

z_1	z_2	z_3
z_4	z_5	z_6
z_7	z_8	z_9

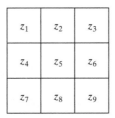

Roberts

−1	−1	−1
0	0	0
1	1	1

−1	0	1
−1	0	1
−1	0	1

Prewitt

−1	−2	−1
0	0	0
1	2	1

−1	0	1
−2	0	2
−1	0	1

Sobel

These derivatives can be implemented by filtering an image with the masks in Figs. 10.14(b) and (c).

Masks of size 2×2 are simple conceptually, but they are not as useful for computing edge direction as masks that are symmetric about the center point, the smallest of which are of size 3×3. These masks take into account the nature of the data on opposite sides of the center point and thus carry more information regarding the direction of an edge. The simplest digital approximations to the partial derivatives using masks of size 3×3 are given by

$$g_x = \frac{\partial f}{\partial x} = (z_7 + z_8 + z_9) - (z_1 + z_2 + z_3) \qquad (10.2\text{-}16)$$

and

$$g_y = \frac{\partial f}{\partial y} = (z_3 + z_6 + z_9) - (z_1 + z_4 + z_7) \qquad (10.2\text{-}17)$$

> Although these equations encompass a larger neighborhood, we are still dealing with *differences* between intensity values, so the conclusions from earlier discussions regarding first-order derivatives still apply.

In these formulations, the difference between the third and first *rows* of the 3×3 region approximates the derivative in the x-direction, and the difference between the third and first columns approximate the derivate in the y-direction. Intuitively, we would expect these approximations to be more accurate than the approximations obtained using the Roberts operators. Equations (10.2-16) and (10.2-17) can be implemented over an entire image by filtering f with the two masks in Figs. 10.14(d) and (e). These masks are called the *Prewitt operators* (Prewitt [1970]).

A slight variation of the preceding two equations uses a weight of 2 in the center coefficient:

$$g_x = \frac{\partial f}{\partial x} = (z_7 + 2z_8 + z_9) - (z_1 + 2z_2 + z_3) \qquad (10.2\text{-}18)$$

and

$$g_y = \frac{\partial f}{\partial y} = (z_3 + 2z_6 + z_9) - (z_1 + 2z_4 + z_7) \qquad (10.2\text{-}19)$$

It can be shown (Problem 10.10) that using a 2 in the center location provides image smoothing. Figures 10.14(f) and (g) show the masks used to implement Eqs. (10.2-18) and (10.2-19). These masks are called the *Sobel operators* (Sobel [1970]).

The Prewitt masks are simpler to implement than the Sobel masks, but, the slight computational difference between them typically is not an issue. The fact that the Sobel masks have better noise-suppression (smoothing) characteristics makes them preferable because, as mentioned in the previous section, noise suppression is an important issue when dealing with derivatives. Note that the coefficients of all the masks in Fig. 10.14 sum to zero, thus giving a response of zero in areas of constant intensity, as expected of a derivative operator.

a b
c d

FIGURE 10.15
Prewitt and Sobel masks for detecting diagonal edges.

0	1	1	−1	−1	0
−1	0	1	−1	0	1
−1	−1	0	0	1	1

Prewitt

0	1	2	−2	−1	0
−1	0	1	−1	0	1
−2	−1	0	0	1	2

Sobel

The masks just discussed are used to obtain the gradient components g_x and g_y at every pixel location in an image. These two partial derivatives are then used to estimate edge strength and direction. Computing the magnitude of the gradient requires that g_x and g_y be combined in the manner shown in Eq. (10.2-10). However, this implementation is not always desirable because of the computational burden required by squares and square roots. An approach used frequently is to approximate the magnitude of the gradient by absolute values:

$$M(x, y) \approx |g_x| + |g_y| \tag{10.2-20}$$

This equation is more attractive computationally, and it still preserves relative changes in intensity levels. The price paid for this advantage is that the resulting filters will not be isotropic (invariant to rotation) in general. However, this is not an issue when masks such as the Prewitt and Sobel masks are used to compute g_x and g_y, because these masks give isotropic results only for vertical and horizontal edges. Results would be isotropic only for edges in those two directions, regardless of which of the two equations is used. In addition, Eqs. (10.2-10) and (10.2-20) give identical results for vertical and horizontal edges when the Sobel or Prewitt masks are used (Problem 10.8).

It is possible to modify the 3×3 masks in Fig. 10.14 so that they have their strongest responses along the diagonal directions. Figure 10.15 shows the two additional Prewitt and Sobel masks needed for detecting edges in the diagonal directions.

EXAMPLE 10.6:
Illustration of the 2-D gradient magnitude and angle.

■ Figure 10.16 illustrates the absolute value response of the two components of the gradient, $|g_x|$ and $|g_y|$, as well as the gradient image formed from the sum of these two components. The directionality of the horizontal and vertical components of the gradient is evident in Figs. 10.16(b) and (c). Note, for example, how strong the roof tile, horizontal brick joints, and horizontal segments of the windows are in Fig. 10.16(b) compared to other edges. By contrast,

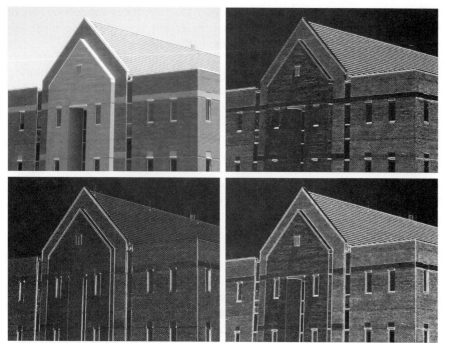

FIGURE 10.16
(a) Original image
of size
834 × 1114 pixels,
with intensity
values scaled to
the range [0, 1].
(b) $|g_x|$, the
component of the
gradient in the
x-direction,
obtained using
the Sobel mask in
Fig. 10.14(f) to
filter the image.
(c) $|g_y|$, obtained
using the mask in
Fig. 10.14(g).
(d) The gradient
image, $|g_x| + |g_y|$.

Fig. 10.16(c) favors features such as the vertical components of the façade and windows. It is common terminology to use the term *edge map* when referring to an image whose principal features are edges, such as gradient magnitude images. The intensities of the image in Fig. 10.16(a) were scaled to the range [0, 1]. We use values in this range to simplify parameter selection in the various methods for edge detection discussed in this section.

Figure 10.17 shows the gradient angle image computed using Eq. (10.2-11). In general, angle images are not as useful as gradient magnitude images for edge detection, but they do complement the information extracted from an image using the magnitude of the gradient. For instance, the constant intensity areas in Fig. 10.16(a), such as the front edge of the sloping roof and top horizontal bands of the front wall, are constant in Fig. 10.17, indicating that the gradient vector direction at all the pixel locations in those regions is the same.

FIGURE 10.17
Gradient angle
image computed
using
Eq. (10.2-11).
Areas of constant
intensity in this
image indicate
that the direction
of the gradient
vector is the same
at all the pixel
locations in those
regions.

As we show in Section 10.2.6, angle information plays a key supporting role in the implementation of the Canny edge detection algorithm, the most advanced edge detection method we discuss in this chapter.   ■

The original image in Fig. 10.16(a) is of reasonably high resolution (834 × 1114 pixels), and at the distance the image was acquired, the contribution made to image detail by the wall bricks is significant. This level of fine detail often is undesirable in edge detection because it tends to act as noise, which is enhanced by derivative computations and thus complicates detection of the principal edges in an image. One way to reduce fine detail is to smooth the image. Figure 10.18 shows the same sequence of images as in Fig. 10.16, but with the original image smoothed first using a 5 × 5 averaging filter (see Section 3.5 regarding smoothing filters). The response of each mask now shows almost no contribution due to the bricks, with the results being dominated mostly by the principal edges.

The maximum edge strength (magnitude) of a smoothed image decreases inversely as a function of the size of the smoothing mask (Problem 10.13).

It is evident in Figs. 10.16 and 10.18 that the horizontal and vertical Sobel masks do not differentiate between edges oriented in the ±45° directions. If it is important to emphasize edges along the diagonal directions, then one of the masks in Fig. 10.15 should be used. Figures 10.19(a) and (b) show the absolute responses of the 45° and −45° Sobel masks, respectively. The stronger diagonal response of these masks is evident in these figures. Both diagonal masks have similar response to horizontal and vertical edges but, as expected, their response in these directions is weaker than the response of the horizontal and vertical masks, as discussed earlier.

a b
c d

FIGURE 10.18
Same sequence as in Fig. 10.16, but with the original image smoothed using a 5 × 5 averaging filter prior to edge detection.

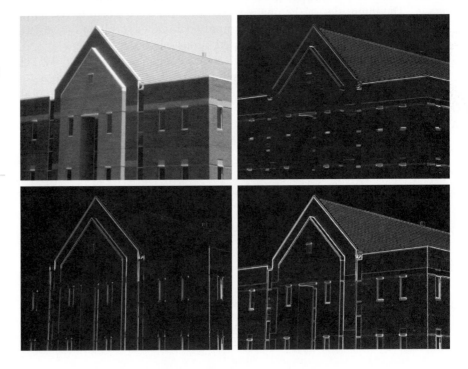

a b

FIGURE 10.19
Diagonal edge
detection.
(a) Result of
using the mask in
Fig. 10.15(c).
(b) Result of
using the mask in
Fig. 10.15(d). The
input image in
both cases was
Fig. 10.18(a).

Combining the gradient with thresholding

The results in Fig. 10.18 show that edge detection can be made more selective by smoothing the image prior to computing the gradient. Another approach aimed at achieving the same basic objective is to threshold the gradient image. For example, Fig. 10.20(a) shows the gradient image from Fig. 10.16(d) thresholded, in the sense that pixels with values greater than or equal to 33% of the maximum value of the gradient image are shown in white, while pixels below the threshold value are shown in black. Comparing this image with Fig. 10.18(d), we see that there are fewer edges in the thresholded image, and that the edges in this image are much sharper (see, for example, the edges in the roof tile). On the other hand, numerous edges, such as the 45° line defining the far edge of the roof, are broken in the thresholded image.

When interest lies both in highlighting the principal edges and on maintaining as much connectivity as possible, it is common practice to use both smoothing and thresholding. Figure 10.20(b) shows the result of thresholding Fig. 10.18(d), which is the gradient of the smoothed image. This result shows a

The threshold used to generate Fig. 10.20(a) was selected so that most of the small edges caused by the bricks were eliminated. Recall that this was the original objective for smoothing the image in Fig. 10.16 prior to computing the gradient.

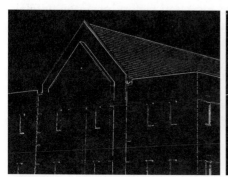
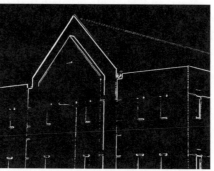

a b

FIGURE 10.20 (a) Thresholded version of the image in Fig. 10.16(d), with the threshold selected as 33% of the highest value in the image; this threshold was just high enough to eliminate most of the brick edges in the gradient image. (b) Thresholded version of the image in Fig. 10.18(d), obtained using a threshold equal to 33% of the highest value in that image.

reduced number of broken edges; for instance, compare the 45° edges in Figs. 10.20(a) and (b). Of course, edges whose intensity values were severely attenuated due to blurring (e.g., the edges in the tile roof) are likely to be totally eliminated by thresholding. We return to the problem of broken edges in Section 10.2.7.

10.2.6 More Advanced Techniques for Edge Detection

The edge-detection methods discussed in the previous section are based simply on filtering an image with one or more masks, with no provisions being made for edge characteristics and noise content. In this section, we discuss more advanced techniques that make an attempt to improve on simple edge-detection methods by taking into account factors such as image noise and the nature of edges themselves.

The Marr-Hildreth edge detector

One of the earliest successful attempts at incorporating more sophisticated analysis into the edge-finding process is attributed to Marr and Hildreth [1980]. Edge-detection methods in use at the time were based on using small operators (such as the Sobel masks), as discussed in the previous section. Marr and Hildreth argued (1) that intensity changes are not independent of image scale and so their detection requires the use of operators of different sizes; and (2) that a sudden intensity change will give rise to a peak or trough in the first derivative or, equivalently, to a zero crossing in the second derivative (as we saw in Fig. 10.10).

These ideas suggest that an operator used for edge detection should have two salient features. First and foremost, it should be a differential operator capable of computing a digital approximation of the first or second derivative at every point in the image. Second, it should be capable of being "tuned" to act at any desired scale, so that large operators can be used to detect blurry edges and small operators to detect sharply focused fine detail.

Marr and Hildreth argued that the most satisfactory operator fulfilling these conditions is the filter $\nabla^2 G$ where, as defined in Section 3.6.2, ∇^2 is the Laplacian operator, $(\partial^2/\partial x^2 + \partial^2/\partial y^2)$, and G is the 2-D Gaussian function

> To convince yourself that edge detection is not independent of scale, consider, for example, the roof edge in Fig. 10.8(c). If the scale of the image is reduced, the edge will appear thinner.

> It is customary for Eq. (10.2-21) to differ from the definition of a 2-D Gaussian PDF by the constant term $1/2\pi\sigma^2$. If an exact expression is desired in a given application, then the multiplying constant can be appended to the final result in Eq. (10.2-23).

$$G(x, y) = e^{-\frac{x^2+y^2}{2\sigma^2}} \qquad (10.2\text{-}21)$$

with standard deviation σ (sometimes σ is called the *space constant*). To find an expression for $\nabla^2 G$ we perform the following differentiations:

$$\nabla^2 G(x, y) = \frac{\partial^2 G(x, y)}{\partial x^2} + \frac{\partial^2 G(x, y)}{\partial y^2}$$

$$= \frac{\partial}{\partial x}\left[\frac{-x}{\sigma^2}e^{-\frac{x^2+y^2}{2\sigma^2}}\right] + \frac{\partial}{\partial y}\left[\frac{-y}{\sigma^2}e^{-\frac{x^2+y^2}{2\sigma^2}}\right] \qquad (10.2\text{-}22)$$

$$= \left[\frac{x^2}{\sigma^4} - \frac{1}{\sigma^2}\right]e^{-\frac{x^2+y^2}{2\sigma^2}} + \left[\frac{y^2}{\sigma^4} - \frac{1}{\sigma^2}\right]e^{-\frac{x^2+y^2}{2\sigma^2}}$$

Collecting terms gives the final expression:

$$\nabla^2 G(x, y) = \left[\frac{x^2 + y^2 - 2\sigma^2}{\sigma^4} \right] e^{-\frac{x^2+y^2}{2\sigma^2}} \qquad (10.2\text{-}23)$$

This expression is called the *Laplacian of a Gaussian* (LoG).

Figures 10.21(a) through (c) show a 3-D plot, image, and cross section of the *negative* of the LoG function (note that the zero crossings of the LoG occur at $x^2 + y^2 = 2\sigma^2$, which defines a circle of radius $\sqrt{2}\sigma$ centered on the origin). Because of the shape illustrated in Fig. 10.21(a), the LoG function sometimes is called the *Mexican hat* operator. Figure 10.21(d) shows a 5×5 mask that approximates the shape in Fig. 10.21(a) (in practice we would use the *negative* of this mask). This approximation is not unique. Its purpose is to capture the essential *shape* of the LoG function; in terms of Fig. 10.21(a), this means a positive, central term surrounded by an adjacent, negative region whose values increase as a function of distance from the origin, and a zero outer region. The coefficients must sum to zero so that the response of the mask is zero in areas of constant intensity.

Masks of arbitrary size can be generated by sampling Eq. (10.2-23) and scaling the coefficients so that they sum to zero. A more effective approach for generating a LoG filter is to sample Eq. (10.2-21) to the desired $n \times n$ size and

Note the similarity between the cross section in Fig. 10.21(c) and the highpass filter in Fig. 4.37(d). Thus, we can expect the LoG to behave as a highpass filter.

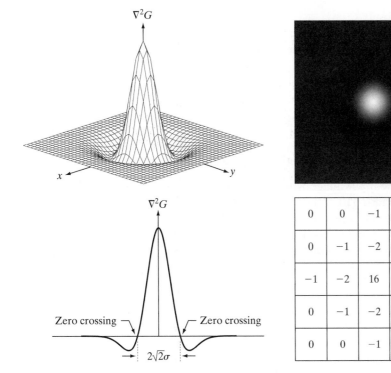

a b
c d

FIGURE 10.21
(a) Three-dimensional plot of the *negative* of the LoG. (b) Negative of the LoG displayed as an image. (c) Cross section of (a) showing zero crossings. (d) 5×5 mask approximation to the shape in (a). The negative of this mask would be used in practice.

0	0	−1	0	0
0	−1	−2	−1	0
−1	−2	16	−2	−1
0	−1	−2	−1	0
0	0	−1	0	0

then convolve[†] the resulting array with a Laplacian mask, such as the mask in Fig. 10.4(a). Because convolving an image array with a mask whose coefficients sum to zero yields a result whose elements also sum to zero (see Problems 3.16 and 10.14), this approach automatically satisfies the requirement that the sum of the LoG filter coefficients be zero. We discuss the issue of selecting the size of LoG filter later in this section.

There are two fundamental ideas behind the selection of the operator $\nabla^2 G$. First, the Gaussian part of the operator blurs the image, thus reducing the intensity of structures (including noise) at scales much smaller than σ. Unlike averaging of the form discussed in Section 3.5 and used in Fig. 10.18, the Gaussian function is smooth in both the spatial and frequency domains (see Section 4.8.3), and is thus less likely to introduce artifacts (e.g., ringing) not present in the original image. The other idea concerns ∇^2, the second derivative part of the filter. Although first derivatives can be used for detecting abrupt changes in intensity, they are directional operators. The Laplacian, on the other hand, has the important advantage of being isotropic (invariant to rotation), which not only corresponds to characteristics of the human visual system (Marr [1982]) but also responds equally to changes in intensity in any mask direction, thus avoiding having to use multiple masks to calculate the strongest response at any point in the image.

The Marr-Hildreth algorithm consists of convolving the LoG filter with an input image, $f(x, y)$,

$$g(x, y) = [\nabla^2 G(x, y)] \star f(x, y) \tag{10.2-24}$$

and then finding the zero crossings of $g(x, y)$ to determine the locations of edges in $f(x, y)$. Because these are linear processes, Eq. (10.2-24) can be written also as

This expression is implemented in the spatial domain using Eq. (3.4-2). It can be implemented also in the frequency domain using Eq. (4.7-1).

$$g(x, y) = \nabla^2 [G(x, y) \star f(x, y)] \tag{10.2-25}$$

indicating that we can smooth the image first with a Gaussian filter and then compute the Laplacian of the result. These two equations give identical results.

The Marr-Hildreth edge-detection algorithm may be summarized as follows:

1. Filter the input image with an $n \times n$ Gaussian lowpass filter obtained by sampling Eq. (10.2-21).
2. Compute the Laplacian of the image resulting from Step 1 using, for example, the 3×3 mask in Fig. 10.4(a). [Steps 1 and 2 implement Eq. (10.2-25).]
3. Find the zero crossings of the image from Step 2.

To specify the size of the Gaussian filter, recall that about 99.7% of the volume under a 2-D Gaussian surface lies between $\pm 3\sigma$ about the mean. Thus, as a rule

---

[†]The LoG is a symmetric filter, so spatial filtering using correlation or convolution yields the same result. We use the convolution terminology here to indicate linear filtering for consistency with the literature on this topic. Also, this gives you exposure to terminology that you will encounter in other contexts. It is important that you keep in mind the comments made at the end of Section 3.4.2 regarding this topic.

of thumb, the size of an $n \times n$ LoG discrete filter should be such that n is the smallest odd integer greater than or equal to 6σ. Choosing a filter mask smaller than this will tend to "truncate" the LoG function, with the degree of truncation being inversely proportional to the size of the mask; using a larger mask would make little difference in the result.

One approach for finding the zero crossings at any pixel, p, of the filtered image, $g(x, y)$, is based on using a 3×3 neighborhood centered at p. A zero crossing at p implies that the *signs* of at least two of its opposing neighboring pixels must differ. There are four cases to test: left/right, up/down, and the two diagonals. If the values of $g(x, y)$ are being compared against a threshold (a common approach), then not only must the signs of opposing neighbors be different, but the absolute value of their numerical difference must also exceed the threshold before we can call p a zero-crossing pixel. We illustrate this approach in Example 10.7 below.

Zero crossings are the key feature of the Marr-Hildreth edge-detection method. The approach discussed in the previous paragraph is attractive because of its simplicity of implementation and because it generally gives good results. If the accuracy of the zero-crossing locations found using this method is inadequate in a particular application, then a technique proposed by Huertas and Medioni [1986] for finding zero crossings with subpixel accuracy can be employed.

Attempting to find the zero crossings by finding the coordinates (x, y), such that $g(x, y) = 0$ is impractical because of noise and/or computational inaccuracies.

EXAMPLE 10.7:
Illustration of the Marr-Hildreth edge-detection method.

■ Figure 10.22(a) shows the original building image used earlier and Fig. 10.22(b) is the result of Steps 1 and 2 of the Marr-Hildreth algorithm, using $\sigma = 4$ (approximately 0.5% of the short dimension of the image) and $n = 25$ (the smallest odd integer greater than or equal to 6σ, as discussed earlier). As in Fig. 10.5, the gray tones in this image are due to scaling. Figure 10.22(c) shows the zero crossings obtained using the 3×3 neighborhood approach discussed above with a threshold of zero. Note that all the edges form closed loops. This so-called "spaghetti" effect is a serious drawback of this method when a threshold value of zero is used (Problem 10.15). We avoid closed-loop edges by using a positive threshold.

Figure 10.22(d) shows the result of using a threshold approximately equal to 4% of the maximum value of the LoG image. Note that the majority of the principal edges were readily detected and "irrelevant" features, such as the edges due to the bricks and the tile roof, were filtered out. As we show in the next section, this type of performance is virtually impossible to obtain using the gradient-based edge-detection techniques discussed in the previous section. Another important consequence of using zero crossings for edge detection is that the resulting edges are 1 pixel thick. This property simplifies subsequent stages of processing, such as edge linking.   ■

A procedure used sometimes to take into account the fact mentioned earlier that intensity changes are scale dependent is to filter an image with various values of σ. The resulting zero-crossings edge maps are then combined by keeping only the edges that are common to all maps. This approach can yield

a b
c d

FIGURE 10.22
(a) Original image
of size 834×1114
pixels, with
intensity values
scaled to the range
$[0, 1]$. (b) Results
of Steps 1 and 2 of
the Marr-Hildreth
algorithm using
$\sigma = 4$ and $n = 25$.
(c) Zero crossings
of (b) using a
threshold of 0
(note the closed-
loop edges).
(d) Zero crossings
found using a
threshold equal to
4% of the
maximum value of
the image in (b).
Note the thin
edges.

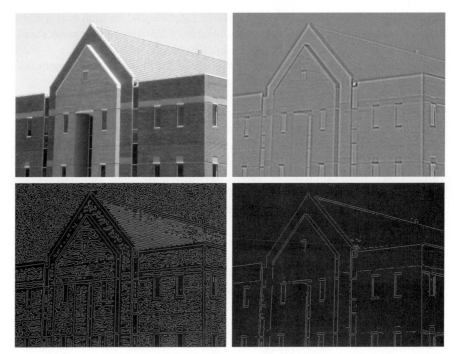

useful information, but, due to its complexity, it is used in practice mostly as a
design tool for selecting an appropriate value of σ to use with a single filter.

Marr and Hildreth [1980] noted that it is possible to approximate the LoG
filter in Eq. (10.2-23) by a difference of Gaussians (DoG):

The difference of
Gaussians is a highpass
filter, as discussed in
Section 4.7.4.

$$\text{DoG}(x, y) = \frac{1}{2\pi\sigma_1^2} e^{-\frac{x^2+y^2}{2\sigma_1^2}} - \frac{1}{2\pi\sigma_2^2} e^{-\frac{x^2+y^2}{2\sigma_2^2}} \tag{10.2-26}$$

with $\sigma_1 > \sigma_2$. Experimental results suggest that certain "channels" in the
human vision system are selective with respect to orientation and frequency,
and can be modeled using Eq. (10.2-26) with a ratio of standard deviations of
1.75:1. Marr and Hildreth suggested that using the ratio 1.6:1 preserves the
basic characteristics of these observations and also provides a closer "engineering" approximation to the LoG function. To make meaningful comparisons between the LoG and DoG, the value of σ for the LoG must be selected
as in the following equation so that the LoG and DoG have the same zero
crossings (Problem 10.17):

$$\sigma^2 = \frac{\sigma_1^2 \sigma_2^2}{\sigma_1^2 - \sigma_2^2} \ln\left[\frac{\sigma_1^2}{\sigma_2^2}\right] \tag{10.2-27}$$

Although the zero crossings of the LoG and DoG will be the same when this
value of σ is used, their amplitude scales will be different. We can make them
compatible by scaling both functions so that they have the same value at the
origin.

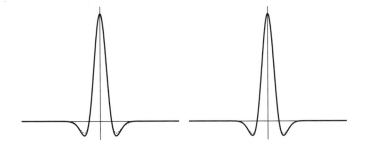

FIGURE 10.23
(a) Negatives of the LoG (solid) and DoG (dotted) profiles using a standard deviation ratio of 1.75:1.
(b) Profiles obtained using a ratio of 1.6:1.

The profiles in Figs. 10.23(a) and (b) were generated with standard deviation ratios of 1:1.75 and 1:1.6, respectively (by convention, the curves shown are inverted, as in Fig. 10.21). The LoG profiles are shown as solid lines while the DoG profiles are dotted. The curves shown are intensity profiles through the center of LoG and DoG arrays generated by sampling Eq. (10.2-23) (with the constant in $1/2\pi\sigma^2$ in front) and Eq. (10.2-26), respectively. The amplitude of all curves at the origin were normalized to 1. As Fig. 10.23(b) shows, the ratio 1:1.6 yielded a closer approximation between the LoG and DoG functions.

Both the LoG and the DoG filtering operations can be implemented with 1-D convolutions instead of using 2-D convolutions directly (Problem 10.19). For an image of size $M \times N$ and a filter of size $n \times n$, doing so reduces the number of multiplications and additions for each convolution from being proportional to n^2MN for 2-D convolutions to being proportional to nMN for 1-D convolutions. This implementation difference is significant. For example, if $n = 25$, a 1-D implementation will require on the order of 12 times fewer multiplication and addition operations than using 2-D convolution.

The Canny edge detector

Although the algorithm is more complex, the performance of the Canny edge detector (Canny [1986]) discussed in this section is superior in general to the edge detectors discussed thus far. Canny's approach is based on three basic objectives:

1. *Low error rate.* All edges should be found, and there should be no spurious responses. That is, the edges detected must be as close as possible to the true edges.
2. *Edge points should be well localized.* The edges located must be as close as possible to the true edges. That is, the distance between a point marked as an edge by the detector and the center of the true edge should be minimum.
3. *Single edge point response.* The detector should return only one point for each true edge point. That is, the number of local maxima around the true edge should be minimum. This means that the detector should not identify multiple edge pixels where only a single edge point exists.

The essence of Canny's work was in expressing the preceding three criteria mathematically and then attempting to find optimal solutions to these formulations. In general, it is difficult (or impossible) to find a closed-form solution

Recall that *white noise* is
noise having a frequency
spectrum that is continu-
ous and uniform over a
specified frequency band.
White Gaussian noise is
white noise in which the
distribution of amplitude
values is Gaussian.
Gaussian white noise is a
good approximation of
many real-world situa-
tions and generates
mathematically tractable
models. It has the useful
property that its values
are statistically
independent.

that satisfies all the preceding objectives. However, using numerical optimization with 1-D step edges corrupted by additive white Gaussian noise led to the
conclusion that a good approximation[†] to the optimal step edge detector is the
first derivative of a Gaussian:

$$\frac{d}{dx}e^{-\frac{x^2}{2\sigma^2}} = \frac{-x}{\sigma^2}e^{-\frac{x^2}{2\sigma^2}} \tag{10.2-28}$$

Generalizing this result to 2-D involves recognizing that the 1-D approach *still
applies* in the direction of the edge normal (see Fig. 10.12). Because the direction of the normal is unknown beforehand, this would require applying the
1-D edge detector in all possible directions. This task can be approximated by
first smoothing the image with a *circular* 2-D Gaussian function, computing
the gradient of the result, and then using the gradient magnitude and direction
to estimate edge strength and direction at every point.

Let $f(x, y)$ denote the input image and $G(x, y)$ denote the Gaussian function:

$$G(x, y) = e^{-\frac{x^2+y^2}{2\sigma^2}} \tag{10.2-29}$$

We form a smoothed image, $f_s(x, y)$, by convolving G and f:

$$f_s(x, y) = G(x, y)\star f(x, y) \tag{10.2-30}$$

This operation is followed by computing the gradient magnitude and direction
(angle), as discussed in Section 10.2.5:

$$M(x, y) = \sqrt{g_x^2 + g_y^2} \tag{10.2-31}$$

and

$$\alpha(x, y) = \tan^{-1}\left[\frac{g_y}{g_x}\right] \tag{10.2-32}$$

with $g_x = \partial f_s/\partial x$ and $g_y = \partial f_s/\partial y$. Any of the filter mask pairs in Fig. 10.14 can
be used to obtain g_x and g_y. Equation (10.2-30) is implemented using an $n \times n$
Gaussian mask whose size is discussed below. Keep in mind that $M(x, y)$ and
$\alpha(x, y)$ are arrays of the same size as the image from which they are computed.

Because it is generated using the gradient, $M(x, y)$ typically contains wide
ridges around local maxima (recall the discussion in Section 10.2.1 regarding
edges obtained using the gradient). The next step is to thin those ridges. One
approach is to use *nonmaxima suppression.* This can be done in several ways,
but the essence of the approach is to specify a number of discrete orientations

---

[†]Canny [1986] showed that using a Gaussian approximation proved only about 20% worse than using the
optimized numerical solution. A difference of this magnitude generally is imperceptible in most applications.

of the edge normal (gradient vector). For example, in a 3 × 3 region we can define four orientations[†] for an edge passing through the center point of the region: horizontal, vertical, +45° and −45°. Figure 10.24(a) shows the situation for the two possible orientations of a horizontal edge. Because we have to quantize all possible edge directions into four, we have to define a range of directions over which we consider an edge to be horizontal. We determine edge direction from the direction of the edge normal, which we obtain directly from the image data using Eq. (10.2-32). As Fig. 10.24(b) shows, if the edge normal is in the range of directions from −22.5° to 22.5° or from −157.5° to 157.5°, we call the edge a horizontal edge. Figure 10.24(c) shows the angle ranges corresponding to the four directions under consideration.

Let d_1, d_2, d_3, and d_4 denote the four basic edge directions just discussed for a 3 × 3 region: horizontal, −45°, vertical, and +45°, respectively. We can formulate the following nonmaxima suppression scheme for a 3 × 3 region centered at *every* point (x, y) in $\alpha(x, y)$:

1. Find the direction d_k that is closest to $\alpha(x, y)$.
2. If the value of $M(x, y)$ is less than at least one of its two neighbors along d_k, let $g_N(x, y) = 0$ (suppression); otherwise, let $g_N(x, y) = M(x, y)$

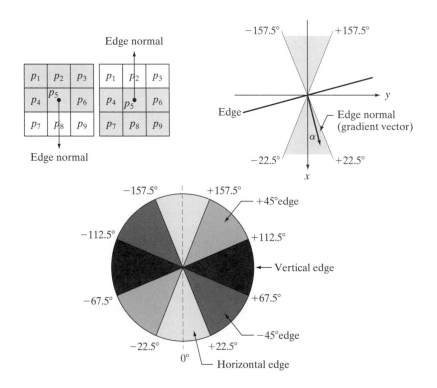

a b
c

FIGURE 10.24
(a) Two possible orientations of a horizontal edge (in gray) in a 3 × 3 neighborhood. (b) Range of values (in gray) of α, the direction angle of the *edge normal*, for a horizontal edge. (c) The angle ranges of the edge normals for the four types of edge directions in a 3 × 3 neighborhood. Each edge direction has two ranges, shown in corresponding shades of gray.

[†]Keep in mind that every edge has two possible orientations. For example, an edge whose normal is oriented at 0° and an edge whose normal is oriented at 180° are the same *horizontal* edge.

where $g_N(x, y)$ is the nonmaxima-suppressed image. For example, with reference to Fig. 10.24(a), letting (x, y) be at p_5 and assuming a horizontal edge through p_5, the pixels in which we would be interested in Step 2 are p_2 and p_8. Image $g_N(x, y)$ contains only the thinned edges; it is equal to $M(x, y)$ with the nonmaxima edge points suppressed.

The final operation is to threshold $g_N(x, y)$ to reduce false edge points. In Section 10.2.5 we did this using a single threshold, in which all values below the threshold were set to 0. If we set the threshold too low, there will still be some false edges (called *false positives*). If the threshold is set too high, then actual valid edge points will be eliminated (*false negatives*). Canny's algorithm attempts to improve on this situation by using *hysteresis thresholding* which, as we discuss in Section 10.3.6, uses two thresholds: a low threshold, T_L, and a high threshold, T_H. Canny suggested that the ratio of the high to low threshold should be two or three to one.

We can visualize the thresholding operation as creating two additional images

$$g_{NH}(x, y) = g_N(x, y) \geq T_H \qquad (10.2\text{-}33)$$

and

$$g_{NL}(x, y) = g_N(x, y) \geq T_L \qquad (10.2\text{-}34)$$

where, initially, both $g_{NH}(x, y)$ and $g_{NL}(x, y)$ are set to 0. After thresholding, $g_{NH}(x, y)$ will have fewer nonzero pixels than $g_{NL}(x, y)$ in general, but all the nonzero pixels in $g_{NH}(x, y)$ will be contained in $g_{NL}(x, y)$ because the latter image is formed with a lower threshold. We eliminate from $g_{NL}(x, y)$ all the nonzero pixels from $g_{NH}(x, y)$ by letting

$$g_{NL}(x, y) = g_{NL}(x, y) - g_{NH}(x, y) \qquad (10.2\text{-}35)$$

The nonzero pixels in $g_{NH}(x, y)$ and $g_{NL}(x, y)$ may be viewed as being "strong" and "weak" edge pixels, respectively.

After the thresholding operations, all strong pixels in $g_{NH}(x, y)$ are assumed to be valid edge pixels and are so marked immediately. Depending on the value of T_H, the edges in $g_{NH}(x, y)$ typically have gaps. Longer edges are formed using the following procedure:

(a) Locate the next unvisited edge pixel, p, in $g_{NH}(x, y)$.
(b) Mark as valid edge pixels all the weak pixels in $g_{NL}(x, y)$ that are connected to p using, say, 8-connectivity.
(c) If all nonzero pixels in $g_{NH}(x, y)$ have been visited go to Step d. Else, return to Step a.
(d) Set to zero all pixels in $g_{NL}(x, y)$ that were not marked as valid edge pixels.

At the end of this procedure, the final image output by the Canny algorithm is formed by appending to $g_{NH}(x, y)$ all the nonzero pixels from $g_{NL}(x, y)$.

We used two additional images, $g_{NH}(x, y)$ and $g_{NL}(x, y)$, to simplify the discussion. In practice, hysteresis thresholding can be implemented directly during nonmaxima suppression, and thresholding can be implemented directly on $g_N(x, y)$ by forming a list of strong pixels and the weak pixels connected to them.

Summarizing, the Canny edge detection algorithm consists of the following basic steps:

1. Smooth the input image with a Gaussian filter.
2. Compute the gradient magnitude and angle images.
3. Apply nonmaxima suppression to the gradient magnitude image.
4. Use double thresholding and connectivity analysis to detect and link edges.

Although the edges after nonmaxima suppression are thinner than raw gradient edges, edges thicker than 1 pixel can still remain. To obtain edges 1 pixel thick, it is typical to follow Step 4 with one pass of an edge-thinning algorithm (see Section 9.5.5).

As mentioned earlier, smoothing is accomplished by convolving the input image with a Gaussian mask whose size, $n \times n$, must be specified. We can use the approach discussed in the previous section in connection with the Marr-Hildreth algorithm to determine a value of n. That is, a filter mask generated by sampling Eq. (10.2-29) so that n is the smallest odd integer greater than or equal to 6σ provides essentially the "full" smoothing capability of the Gaussian filter. If practical considerations require a smaller filter mask, then the tradeoff is less smoothing for smaller values of n.

Some final comments on implementation: As noted earlier in the discussion of the Marr-Hildreth edge detector, the 2-D Gaussian function in Eq. (10.2-29) is separable into a product of two 1-D Gaussians. Thus, Step 1 of the Canny algorithm can be formulated as 1-D convolutions that operate on the rows (columns) of the image one at a time and then work on the columns (rows) of the result. Furthermore, if we use the approximations in Eqs. (10.2-12) and (10.2-13), we can also implement the gradient computations required for Step 2 as 1-D convolutions (Problem 10.20).

■ Figure 10.25(a) shows the familiar building image. For comparison, Figs. 10.25(b) and (c) show, respectively, the results obtained earlier in Fig. 10.20(b) using the thresholded gradient and Fig. 10.22(d) using the Marr-Hildreth detector. Recall that the parameters used in generating those two images were selected to detect the principal edges while attempting to reduce "irrelevant" features, such as the edges due to the bricks and the tile roof.

Figure 10.25(d) shows the result obtained with the Canny algorithm using the parameters $T_L = 0.04$, $T_H = 0.10$ (2.5 times the value of the low threshold), $\sigma = 4$ and a mask of size 25×25, which corresponds to the smallest odd integer greater than 6σ. These parameters were chosen interactively to achieve the objectives stated in the previous paragraph for the gradient and Marr-Hildreth images. Comparing the Canny image with the other two images, we

EXAMPLE 10.8:
Illustration of the Canny edge-detection method.

a b
c d

FIGURE 10.25
(a) Original image of size 834 × 1114 pixels, with intensity values scaled to the range [0, 1].
(b) Thresholded gradient of smoothed image.
(c) Image obtained using the Marr-Hildreth algorithm.
(d) Image obtained using the Canny algorithm. Note the significant improvement of the Canny image compared to the other two.

The threshold values given here should be considered only in relative terms. Implementation of most algorithms involves various scaling steps, such as scaling the range of values of the input image to the range [0, 1]. Different scaling schemes obviously would require different values of thresholds from those used in this example.

see significant improvements in detail of the principal edges and, at the same time, more rejection of irrelevant features in the Canny result. Note, for example, that both edges of the concrete band lining the bricks in the upper section of the image were detected by the Canny algorithm, whereas the thresholded gradient lost both of these edges and the Marr-Hildreth image contains only the upper one. In terms of filtering out irrelevant detail, the Canny image does not contain a single edge due to the roof tiles; this is not true in the other two images. The quality of the lines with regard to continuity, thinness, and straightness is also superior in the Canny image. Results such as these have made the Canny algorithm a tool of choice for edge detection.   ■

EXAMPLE 10.9:
Another illustration of the three principal edge detection methods discussed in this section.

■ As another comparison of the three principal edge-detection methods discussed in this section, consider Fig. 10.26(a) which shows a 512 × 512 head CT image. Our objective in this example is to extract the edges of the outer contour of the brain (the gray region in the image), the contour of the spinal region (shown directly behind the nose, toward the front of the brain), and the outer contour of the head. We wish to generate the thinnest, continuous contours possible, while eliminating edge details related to the gray content in the eyes and brain areas.

Figure 10.26(b) shows a thresholded gradient image that was first smoothed with a 5 × 5 averaging filter. The threshold required to achieve the result shown was 15% of the maximum value of the gradient image. Figure 10.26(c) shows the result obtained with the Marr-Hildreth edge-detection algorithm with a threshold of 0.002, $\sigma = 3$, and a mask of size 19 × 19 pixels. Figure 10.26(d) was obtained using the Canny algorithm with $T_L = 0.05$, $T_H = 0.15$ (3 times the

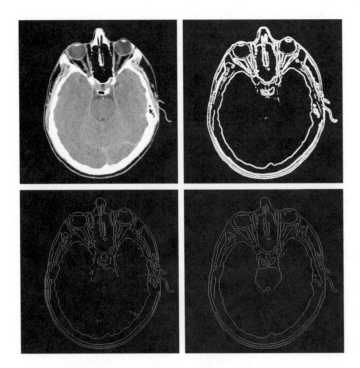

a b
c d

FIGURE 10.26
(a) Original head
CT image of size
512 × 512 pixels,
with intensity
values scaled to
the range [0, 1].
(b) Thresholded
gradient of
smoothed image.
(c) Image
obtained using
the Marr-Hildreth
algorithm.
(d) Image
obtained using
the Canny
algorithm.
(Original image
courtesy of Dr.
David R. Pickens,
Vanderbilt
University.)

value of the low threshold), $\sigma = 2$, and a mask of size 13×13, which, as in the Marr-Hildreth case, corresponds to the smallest odd integer greater than 6σ.

The results in Fig. 10.26 correspond closely to the results and conclusions in the previous example in terms of edge quality and the ability to eliminate irrelevant detail. Note also that the Canny algorithm was the only procedure capable of yielding a totally unbroken edge for the posterior boundary of the brain. It was also the only procedure capable of finding the best contours while eliminating all the edges associated with the gray matter in the original image. ■

As might be expected, the price paid for the improved performance of the Canny algorithm is a more complex implementation than the two approaches discussed earlier, requiring also considerably more execution time. In some applications, such as real-time industrial image processing, cost and speed requirements usually dictate the use of simpler techniques, principally the thresholded gradient approach. When edge quality is the driving force, then the Marr-Hildreth and Canny algorithms, especially the latter, offer superior alternatives.

10.2.7 Edge Linking and Boundary Detection

Ideally, edge detection should yield sets of pixels lying only on edges. In practice, these pixels seldom characterize edges completely because of noise, breaks in the edges due to nonuniform illumination, and other effects that introduce spurious discontinuities in intensity values. Therefore, edge detection typically is followed by linking algorithms designed to assemble edge pixels into meaningful edges and/or region boundaries. In this section, we discuss three fundamental approaches to edge linking that are representative of techniques used in practice.

The first requires knowledge about edge points in a local region (e.g., a 3 × 3 neighborhood); the second requires that points on the boundary of a region be known; and the third is a global approach that works with an entire edge image.

Local processing

One of the simplest approaches for linking edge points is to analyze the characteristics of pixels in a small neighborhood about every point (x, y) that has been declared an edge point by one of the techniques discussed in the previous section. All points that are similar according to predefined criteria are linked, forming an edge of pixels that share common properties according to the specified criteria.

The two principal properties used for establishing similarity of edge pixels in this kind of analysis are (1) the strength (magnitude) and (2) the direction of the gradient vector. The first property is based on Eq. (10.2-10). Let S_{xy} denote the set of coordinates of a neighborhood centered at point (x, y) in an image. An edge pixel with coordinates (s, t) in S_{xy} is similar in *magnitude* to the pixel at (x, y) if

$$|M(s, t) - M(x, y)| \leq E \tag{10.2-36}$$

where E is a positive threshold.

The direction angle of the gradient vector is given by Eq. (10.2-11). An edge pixel with coordinates (s, t) in S_{xy} has an *angle* similar to the pixel at (x, y) if

$$|\alpha(s, t) - \alpha(x, y)| \leq A \tag{10.2-37}$$

where A is a positive angle threshold. As noted in Section 10.2.5, the direction of the edge at (x, y) is *perpendicular* to the direction of the gradient vector at that point.

A pixel with coordinates (s, t) in S_{xy} is linked to the pixel at (x, y) if both magnitude and direction criteria are satisfied. This process is repeated at every location in the image. A record must be kept of linked points as the center of the neighborhood is moved from pixel to pixel. A simple bookkeeping procedure is to assign a different intensity value to each set of linked edge pixels.

The preceding formulation is computationally expensive because all neighbors of every point have to be examined. A simplification particularly well suited for real time applications consists of the following steps:

1. Compute the gradient magnitude and angle arrays, $M(x, y)$ and $\alpha(x, y)$, of the input image, $f(x, y)$.
2. Form a binary image, g, whose value at any pair of coordinates (x, y) is given by:

$$g(x, y) = \begin{cases} 1 & \text{if } M(x, y) > T_M \text{ AND } \alpha(x, y) = A \pm T_A \\ 0 & \text{otherwise} \end{cases}$$

where T_M is a threshold, A is a specified angle direction, and $\pm T_A$ defines a "band" of acceptable directions about A.

3. Scan the rows of g and fill (set to 1) all gaps (sets of 0s) in each row that do not exceed a specified length, K. Note that, by definition, a gap is bounded at both ends by one or more 1s. The rows are processed individually, with no memory between them.
4. To detect gaps in any other direction, θ, rotate g by this angle and apply the horizontal scanning procedure in Step 3. Rotate the result back by $-\theta$.

When interest lies in horizontal and vertical edge linking, Step 4 becomes a simple procedure in which g is rotated ninety degrees, the rows are scanned, and the result is rotated back. This is the application found most frequently in practice and, as the following example shows, this approach can yield good results. In general, image rotation is an expensive computational process so, when linking in numerous angle directions is required, it is more practical to combine Steps 3 and 4 into a single, radial scanning procedure.

■ Figure 10.27(a) shows an image of the rear of a vehicle. The objective of this example is to illustrate the use of the preceding algorithm for finding rectangles whose sizes makes them suitable candidates for license plates. The formation of these rectangles can be accomplished by detecting strong horizontal and vertical edges. Figure 10.27(b) shows the gradient magnitude image, $M(x, y)$, and Figs. 10.27(c) and (d) show the result of Steps (3) and (4) of the algorithm obtained by letting T_M equal to 30% of the maximum gradient value,

EXAMPLE 10.10: Edge linking using local processing.

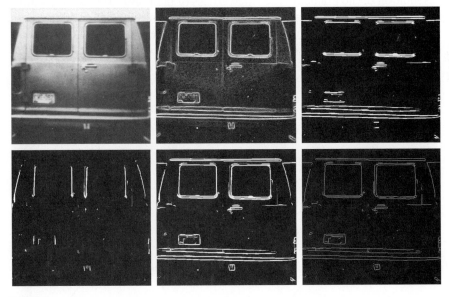

a b c
d e f

FIGURE 10.27 (a) A 534 × 566 image of the rear of a vehicle. (b) Gradient magnitude image. (c) Horizontally connected edge pixels. (d) Vertically connected edge pixels. (e) The logical OR of the two preceding images. (f) Final result obtained using morphological thinning. (Original image courtesy of Perceptics Corporation.)

$A = 90°$, $T_A = 45°$, and filling in all gaps of 25 or fewer pixels (approximately 5% of the image width). Use of a large range of allowable angle directions was required to detect the rounded corners of the license plate enclosure, as well as the rear windows of the vehicle. Figure 10.27(e) is the result of forming the logical OR of the two preceding images, and Fig. 10.27(f) was obtained by thinning 10.27(e) with the thinning procedure discussed in Section 9.5.5. As Fig. 10.16(f) shows, the rectangle corresponding to the license plate was clearly detected in the image. It would be a simple matter to isolate the license plate from all the rectangles in the image using the fact that the width-to-height ratio of license plates in the U.S. has a distinctive 2:1 proportion. ■

Regional processing

Often, the location of regions of interest in an image are known or can be determined. This implies that knowledge is available regarding the regional membership of pixels in the corresponding edge image. In such situations, we can use techniques for linking pixels on a regional basis, with the desired result being an approximation to the boundary of the region. One approach to this type of processing is functional approximation, where we fit a 2-D curve to the known points. Typically, interest lies in fast-executing techniques that yield an approximation to essential features of the boundary, such as extreme points and concavities. Polygonal approximations are particularly attractive because they can capture the essential shape features of a region while keeping the representation of the boundary (i.e., the vertices of the polygon) relatively simple. In this section, we develop and illustrate an algorithm suitable for this purpose.

Before stating the algorithm, we discuss the mechanics of the procedure using a simple example. Figure 10.28 shows a set of points representing an open curve in which the end points have been labeled A and B. These two

FIGURE 10.28
Illustration of the iterative polygonal fit algorithm.

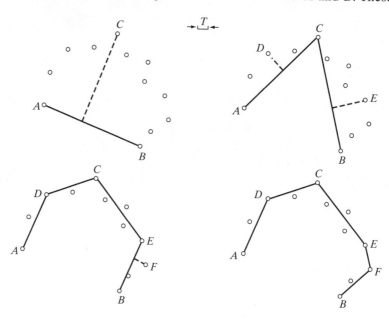

points are by definition vertices of the polygon. We begin by computing the parameters of a straight line passing through A and B. Then, we compute the perpendicular distance from all other points in the curve to this line and select the point that yielded the maximum distance (ties are resolved arbitrarily). If this distance exceeds a specified threshold, T, the corresponding point, labeled C, is declared a vertex, as Fig. 10.28(a) shows. Lines from A to C and from C to B are then established, and distances from all points between A and C to line AC are obtained. The point corresponding to the maximum distance is declared a vertex, D, if the distance exceeds T; otherwise no new vertices are declared for that segment. A similar procedure is applied to the points between C and B. Figure 10.28(b) shows the result and Fig. 10.28(c) shows the next step. This iterative procedure is continued until no points satisfy the threshold test. Figure 10.28(d) shows the final result which, as you can see, is a reasonable approximation to the shape of a curve fitting the given points.

Two important requirements are implicit in the procedure just explained. First, two starting points must be specified; second, all the points must be ordered (e.g., in a clockwise or counterclockwise direction). When an arbitrary set of points in 2-D does not form a connected path (as is typically the case in edge images) it is not always obvious whether the points belong to a boundary segment (open curve) or a boundary (closed curve). Given that the points are ordered, we can infer whether we are dealing with an open or closed curve by analyzing the distances between points. A large distance between two consecutive points in the ordered sequence relative to the distance between other points as we traverse the sequence of points is a good indication that the curve is open. The end points are then used to start the procedure. If the separation between points tends to be uniform, then we are most likely dealing with a closed curve. In this case, we have several options for selecting the two starting points. One approach is to choose the rightmost and leftmost points in the set. Another is to find the extreme points of the curve (we discuss a way to do this in Section 11.2.1). An algorithm for finding a polygonal fit to open and closed curves may be stated as follows:

1. Let P be a sequence of ordered, distinct, 1-valued points of a binary image. Specify two starting points, A and B. These are the two starting vertices of the polygon.

   See Section 11.1.1 for an algorithm that creates ordered point sequences.

2. Specify a threshold, T, and two empty stacks, OPEN and CLOSED.

3. If the points in P correspond to a closed curve, put A into OPEN and put B into OPEN *and* into CLOSED. If the points correspond to an open curve, put A into OPEN and B into CLOSED.

   The use of OPEN and CLOSED for the stack names is not *related to open and closed curves. The stack names indicate simply a stack to store final (CLOSED) vertices or vertices that are in transition (OPEN).*

4. Compute the parameters of the line passing from the last vertex in CLOSED to the last vertex in OPEN.

5. Compute the distances from the line in Step 4 to all the points in P whose sequence places them between the vertices from Step 4. Select the point, V_{max}, with the maximum distance, D_{max} (ties are resolved arbitrarily).

6. If $D_{max} > T$, place V_{max} at the end of the OPEN stack as a new vertex. Go to Step 4.

7. Else, remove the last vertex from OPEN and insert it as the last vertex of CLOSED.

8. If OPEN is not empty, go to Step 4.

9. Else, exit. The vertices in CLOSED are the vertices of the polygonal fit to the points in P.

The mechanics of the algorithm are illustrated in the following two examples.

EXAMPLE 10.11:
Edge linking
using a polygonal
approximation.

■ Consider the set of points, P, in Fig. 10.29(a). Assume that these points belong to a closed curve, that they are ordered in a clockwise direction (note that some of the points are not adjacent), and that A and B are selected to be the leftmost and rightmost points in P, respectively. These are the starting vertices, as Table 10.1 shows. Select the first point in the sequence to be the leftmost point, A. Figure 10.29(b) shows the only point (labeled C) in the upper curve segment between A and B that satisfied Step 6 of the algorithm, so it is designated as a new vertex and added to the vertices in the OPEN stack. The second row in Table 10.1 shows C being detected, and the third row shows it being added as the last vertex in OPEN. The threshold, T, in Fig. 10.29(b) is approximately equal to 1.5 subdivisions in the figure grid.

Note in Fig. 10.29(b) that there is a point below line AB that also satisfies Step 6. However, because the points are ordered, only one subset of the points between these two vertices is detected at one time. The other point in the lower segment will be detected later, as Fig. 10.29(e) shows. The key is always to follow the points in the order in which they are given.

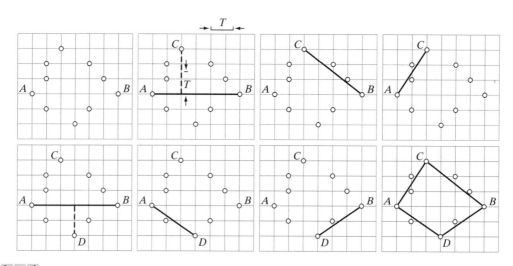

a b c d
e f g h

FIGURE 10.29 (a) A set of points in a clockwise path (the points labeled A and B were chosen as the starting vertices). (b) The distance from point C to the line passing through A and B is the largest of all the points between A and B and also passed the threshold test, so C is a new vertex. (d)–(g) Various stages of the algorithm. (h) The final vertices, shown connected with straight lines to form a polygon. Table 10.1 shows step-by-step details.

CLOSED	OPEN	Curve segment processed	Vertex generated
B	*B, A*	–	*A, B*
B	*B, A*	(*BA*)	*C*
B	*B, A, C*	(*BC*)	–
B, C	*B, A*	(*CA*)	–
B, C, A	*B*	(*AB*)	*D*
B, C, A	*B, D*	(*AD*)	–
B, C, A, D	*B*	(*DB*)	–
B, C, A, D, B	Empty	–	–

TABLE 10.1
Step-by-step details of the mechanics in Example 10.11.

Table 10.1 shows the individual steps leading to the solution in Fig. 10.29(h). Four vertices were detected, and the figure shows them connected with straight line segments to form a polygon approximating the given boundary points. Note in the table that the vertices detected, *B*, *C*, *A*, *D*, *B* are in the counterclockwise direction, even though the points were followed in a clockwise direction to generate the vertices. Had the input been an open curve, the vertices would have been in a clockwise order. The reason for the discrepancy is the way in which the OPEN and CLOSED stacks are initialized. The difference in which stack CLOSED is formed for open and closed curves also leads to the first and last vertices in a closed curve being repeated. This is consistent with how one would differentiate between open and closed polygons given only the vertices. ■

■ Figure 10.30 shows a more practical example of polygonal fitting. The input image in Fig. 10.30(a) is a 550 × 566 X-ray image of a human tooth with intensities scaled to the interval [0, 1]. The objective of this example is to extract the boundary of the tooth, a process useful in areas such as matching against a database for forensics purposes. Figure 10.30(b) is a gradient image obtained using the Sobel masks and thresholded with $T = 0.1$ (10% of the maximum intensity). As expected for an X-ray image, the noise content is high, so the first step is noise reduction. Because the image is binary, morphological techniques are well suited for this purpose. Figure 10.30(c) shows the result of *majority filtering*, which sets a pixel to 1 if five or more pixels in its 3 × 3 neighborhood are 1 and sets the pixel to 0 otherwise. Although the noise was reduced, some noise points are still clearly visible. Figure 10.30(d) shows the result of morphological shrinking, which further reduced the noise to isolated points. These were eliminated [Fig. 10.30(e)] by morphological filtering in the manner described in Example 9.4. At this point, the image consists of thick boundaries, which can be thinned by obtaining the morphological skeleton, as Fig. 10.30(f) shows. Finally, Fig. 10.30(g) shows the last step in preprocessing using spur reduction, as discussed in Section 9.5.8.

Next, we fit the points in Fig. 10.30(g) with a polygon. Figures 10.30(h)–(j) show the result of using the polygon fitting algorithm with thresholds equal to 0.5%, 1%, and 2% of the image width ($T = 3, 6,$ and 12). The first two results are good approximations to the boundary, but the third is marginal. Excessive jaggedness in all three cases clearly indicates that boundary smoothing is

EXAMPLE 10.12:
Polygonal fitting of an image boundary.

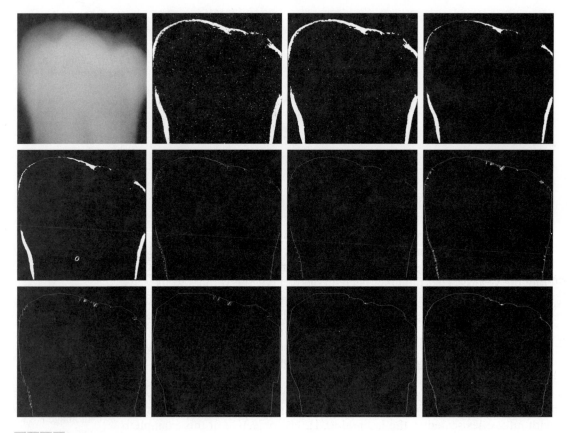

a b c d
e f g h
i j k l

FIGURE 10.30 (a) A 550 × 566 X-ray image of a human tooth. (b) Gradient image. (c) Result of majority filtering. (d) Result of morphological shrinking. (e) Result of morphological cleaning. (f) Skeleton. (g) Spur reduction. (h)–(j) Polygonal fit using thresholds of approximately 0.5%, 1%, and 2% of image width ($T = 3$, 6, and 12). (k) Boundary in (j) smoothed with a 1-D averaging filter of size 1 × 31 (approximately 5% of image width). (l) Boundary in (h) smoothed with the same filter.

required. Figures 10.30(k) and (l) show the result of convolving a 1-D averaging mask with the boundaries in (j) and (h), respectively. The mask used was a 1 × 31 array of 1s, corresponding approximately to 5% of the image width. As expected, the result in Fig. 10.30(k) again is marginal in terms of preserving important shape features (e.g., the right side is severely distorted). On the other hand, the result in Fig. 10.30(l) shows significant boundary smoothing and reasonable preservation of shape features. For example, the roundness of the left-upper cusp and the details of the right-upper cusp were preserved with reasonable fidelity.   ■

The results in the preceding example are typical of what can be achieved with the polygon fitting algorithm discussed in this section. The advantage of this algorithm is that it is simple to implement and yields results that generally are quite acceptable. In Section 11.1.3, we discuss a more sophisticated procedure capable of yielding closer fits by computing minimum-perimeter polygons.

Global processing using the Hough transform

The methods discussed in the previous two sections are applicable in situations where knowledge about pixels belonging to individual objects is at least partially available. For example, in regional processing, it makes sense to link a given set of pixels only if we know that they are part of the boundary of a meaningful region. Often, we have to work with unstructured environments in which all we have is an edge image and no knowledge about where objects of interest might be. In such situations, all pixels are candidates for linking and thus have to be accepted or eliminated based on predefined *global* properties. In this section, we develop an approach based on whether sets of pixels lie on curves of a specified shape. Once detected, these curves form the edges or region boundaries of interest.

Given n points in an image, suppose that we want to find subsets of these points that lie on straight lines. One possible solution is to find first all lines determined by every pair of points and then find all subsets of points that are close to particular lines. This approach involves finding $n(n - 1)/2 \sim n^2$ lines and then performing $(n)(n(n - 1))/2 \sim n^3$ comparisons of every point to all lines. This is a computationally prohibitive task in all but the most trivial applications.

Hough [1962] proposed an alternative approach, commonly referred to as the *Hough transform*. Consider a point (x_i, y_i) in the xy-plane and the general equation of a straight line in slope-intercept form, $y_i = ax_i + b$. Infinitely many lines pass through (x_i, y_i), but they all satisfy the equation $y_i = ax_i + b$ for varying values of a and b. However, writing this equation as $b = -x_i a + y_i$ and considering the ab-plane (also called *parameter space*) yields the equation of a *single* line for a fixed pair (x_i, y_i). Furthermore, a second point (x_j, y_j) also has a line in parameter space associated with it, and, unless they are parallel, this line intersects the line associated with (x_i, y_i) at some point (a', b'), where a' is the slope and b' the intercept of the line containing *both* (x_i, y_i) and (x_j, y_j) in the xy-plane. In fact, *all* the points on this line have lines in parameter space that intersect at (a', b'). Figure 10.31 illustrates these concepts.

In principle, the parameter-space lines corresponding to all points (x_k, y_k) in the xy-plane could be plotted, and the principal lines in that plane could be found by identifying points in parameter space where large numbers of parameter-space lines intersect. A practical difficulty with this approach, however, is that a

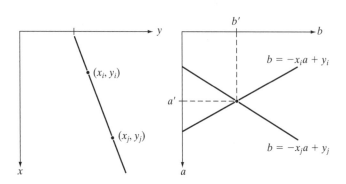

a b

FIGURE 10.31
(a) xy-plane.
(b) Parameter space.

(the slope of a line) approaches infinity as the line approaches the vertical direction. One way around this difficulty is to use the normal representation of a line:

$$x \cos \theta + y \sin \theta = \rho \qquad (10.2\text{-}38)$$

Figure 10.32(a) illustrates the geometrical interpretation of the parameters ρ and θ. A horizontal line has $\theta = 0°$, with ρ being equal to the positive x-intercept. Similarly, a vertical line has $\theta = 90°$, with ρ being equal to the positive y-intercept, or $\theta = -90°$, with ρ being equal to the negative y-intercept. Each sinusoidal curve in Figure 10.32(b) represents the family of lines that pass through a particular point (x_k, y_k) in the xy-plane. The intersection point (ρ', θ') in Fig. 10.32(b) corresponds to the line that passes through both (x_i, y_i) and (x_j, y_j) in Fig. 10.32(a).

The computational attractiveness of the Hough transform arises from subdividing the $\rho\theta$ parameter space into so-called *accumulator cells*, as Fig. 10.32(c) illustrates, where (ρ_{min}, ρ_{max}) and $(\theta_{min}, \theta_{max})$ are the expected ranges of the parameter values: $-90° \le \theta \le 90°$ and $-D \le \rho \le D$, where D is the maximum distance between opposite corners in an image. The cell at coordinates (i, j), with accumulator value $A(i, j)$, corresponds to the square associated with parameter-space coordinates (ρ_i, θ_j). Initially, these cells are set to zero. Then, for every non-background point (x_k, y_k) in the xy-plane, we let θ equal each of the allowed subdivision values on the θ-axis and solve for the corresponding ρ using the equation $\rho = x_k \cos \theta + y_k \sin \theta$. The resulting ρ values are then rounded off to the nearest allowed cell value along the ρ axis. If a choice of θ_p results in solution ρ_q, then we let $A(p, q) = A(p, q) + 1$. At the end of this procedure, a value of P in $A(i, j)$ means that P points in the xy-plane lie on the line $x \cos \theta_j + y \sin \theta_j = \rho_i$. The number of subdivisions in the $\rho\theta$-plane determines the accuracy of the colinearity of these points. It can be shown (Problem 10.24) that the number of computations in the method just discussed is linear with respect to n, the number of non-background points in the xy-plane.

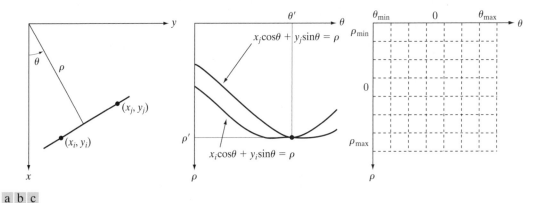

a b c

FIGURE 10.32 (a) (ρ, θ) parameterization of line in the xy-plane. (b) Sinusoidal curves in the $\rho\theta$-plane; the point of intersection (ρ', θ') corresponds to the line passing through points (x_i, y_i) and (x_j, y_j) in the xy-plane. (c) Division of the $\rho\theta$-plane into accumulator cells.

■ Figure 10.33 illustrates the Hough transform based on Eq. (10.2-38). Figure 10.33(a) shows an image of size 101×101 pixels with five labeled points, and Fig. 10.33(b) shows each of these points mapped onto the $\rho\theta$-plane using subdivisions of one unit for the ρ and θ axes. The range of θ values is $\pm90°$, and the range of the ρ axis is $\pm\sqrt{2}D$, where D is the distance between corners in the image. As Fig. 10.33(c) shows, each curve has a different sinusoidal shape. The horizontal line resulting from the mapping of point 1 is a special case of a sinusoid with zero amplitude.

The points labeled A (not to be confused with accumulator values) and B in Fig. 10.33(b) show the colinearity detection property of the Hough transform.

EXAMPLE 10.13:
An illustration of basic Hough transform properties.

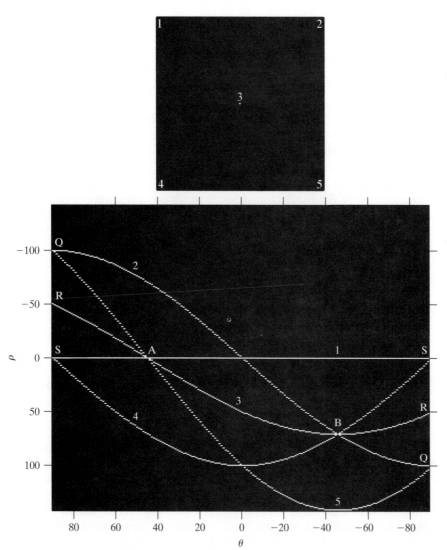

a
b

FIGURE 10.33
(a) Image of size 101×101 pixels, containing five points.
(b) Corresponding parameter space. (The points in (a) were enlarged to make them easier to see.)

Point A denotes the intersection of the curves corresponding to points 1, 3, and 5 in the xy image plane. The location of point A indicates that these three points lie on a straight line passing through the origin ($\rho = 0$) and oriented at 45° [see Fig. 10.32(a)]. Similarly, the curves intersecting at point B in the parameter space indicate that points 2, 3, and 4 lie on a straight line oriented at −45°, and whose distance from the origin is $\rho = 71$ (one-half the diagonal distance from the origin of the image to the opposite corner, rounded to the nearest integer value). Finally, the points labeled Q, R, and S in Fig. 10.33(b) illustrate the fact that the Hough transform exhibits a reflective adjacency relationship at the right and left edges of the parameter space. This property is the result of the manner in which θ and ρ change sign at the ±90° boundaries.   ■

Although the focus thus far has been on straight lines, the Hough transform is applicable to any function of the form $g(\mathbf{v}, \mathbf{c}) = 0$, where $\mathbf{v}$ is a vector of coordinates and $\mathbf{c}$ is a vector of coefficients. For example, points lying on the circle

$$(x - c_1)^2 + (y - c_2)^2 = c_3^2 \tag{10.2-39}$$

can be detected by using the basic approach just discussed. The difference is the presence of three parameters (c_1, c_2, and c_3), which results in a 3-D parameter space with cube-like cells and accumulators of the form $A(i, j, k)$. The procedure is to increment c_1 and c_2, solve for the c_3 that satisfies Eq. (10.2-39), and update the accumulator cell associated with the triplet (c_1, c_2, c_3). Clearly, the complexity of the Hough transform depends on the number of coordinates and coefficients in a given functional representation. Further generalizations of the Hough transform to detect curves with no simple analytic representations are possible, as is the application of the transform to gray-scale images. Several references dealing with these extensions are included at the end of this chapter.

We return now to the edge-linking problem. An approach based on the Hough transform is as follows:

1. Obtain a *binary* edge image using any of the techniques discussed earlier in this section.
2. Specify subdivisions in the $\rho\theta$-plane.
3. Examine the counts of the accumulator cells for high pixel concentrations.
4. Examine the relationship (principally for continuity) between pixels in a chosen cell.

Continuity in this case usually is based on computing the distance between disconnected pixels corresponding to a given accumulator cell. A gap in a line associated with a given cell is bridged if the length of the gap is less than a specified threshold. Note that the mere fact of being able to group lines based on direction is a *global* concept applicable over the entire image, requiring only that we examine pixels associated with specific accumulator cells. This is a significant advantage over the methods discussed in the previous two sections. The following example illustrates these concepts.

■ Figure 10.34(a) shows an aerial image of an airport. The objective of this example is to use the Hough transform to extract the two edges of the principal runway. A solution to such a problem might be of interest, for instance, in applications involving autonomous navigation of air vehicles.

EXAMPLE 10.14:
Using the Hough transform for edge linking.

The first step is to obtain an edge image. Figure 10.34(b) shows the edge image obtained using Canny's algorithm with the same parameters and procedure used in Example 10.9. For the purpose of computing the Hough transform, similar results can be obtained using any of the edge-detection techniques discussed in Sections 10.2.5 or 10.2.6. Figure 10.34(c) shows the Hough parameter space obtained using 1° increments for θ and 1 pixel increments for ρ.

The runway of interest is oriented approximately 1° off the north direction, so we select the cells corresponding to $\pm90°$ and containing the highest count because the runways are the longest lines oriented in these directions. The small white boxes on the edges of Fig. 10.34(c) highlight these cells. As mentioned earlier in connection with Fig. 10.33(b), the Hough transform exhibits adjacency at the edges. Another way of interpreting this property is that a line oriented at $+90°$ and a line oriented at $-90°$ are equivalent (i.e., they are both vertical). Figure 10.34(d) shows the lines corresponding to the two accumulator cells just discussed, and Fig. 10.34(e) shows the lines superimposed on the

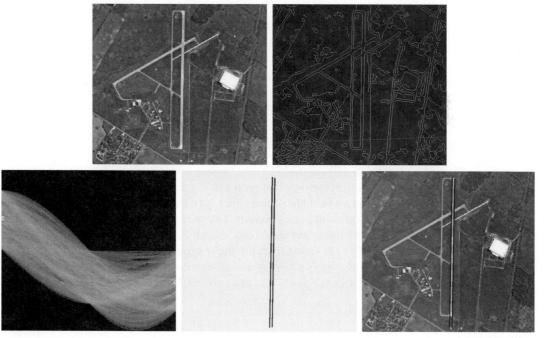

a b
c d e

FIGURE 10.34 (a) A 502 × 564 aerial image of an airport. (b) Edge image obtained using Canny's algorithm. (c) Hough parameter space (the boxes highlight the points associated with long vertical lines). (d) Lines in the image plane corresponding to the points highlighted by the boxes. (e) Lines superimposed on the original image.

original image. The lines were obtained by joining all gaps not exceeding 20% of the image height (approximately 100 pixels). These lines clearly correspond to the edges of the runway of interest.

Note that the only key knowledge needed to solve this problem was the orientation of the runway and the observer's position relative to it. In other words, a vehicle navigating autonomously would know that if the runway of interest faces north, and the vehicle's direction of travel also is north, the runway should appear vertically in the image. Other relative orientations are handled in a similar manner. The orientations of runways throughout the world are available in flight charts, and direction of travel is easily obtainable using GPS (Global Positioning System) information. This information also could be used to compute the distance between the vehicle and the runway, thus allowing estimates of parameters such as expected length of lines relative to image size, as we did in this example.   ■

10.3 Thresholding

Because of its intuitive properties, simplicity of implementation, and computational speed, image thresholding enjoys a central position in applications of image segmentation. Thresholding was introduced in Section 3.1.1, and we have used it in various discussions since then. In this section, we discuss thresholding in a more formal way and develop techniques that are considerably more general than what has been presented thus far.

10.3.1 Foundation

In the previous section, regions were identified by first finding edge segments and then attempting to link the segments into boundaries. In this section, we discuss techniques for partitioning images directly into regions based on intensity values and/or properties of these values.

The basics of intensity thresholding

Suppose that the intensity histogram in Fig. 10.35(a) corresponds to an image, $f(x, y)$, composed of light objects on a dark background, in such a way that object and background pixels have intensity values grouped into two dominant modes. One obvious way to extract the objects from the background is to select a threshold, T, that separates these modes. Then, any point (x, y) in the image at which $f(x, y) > T$ is called an *object point*; otherwise, the point is called a *background point*. In other words, the segmented image, $g(x, y)$, is given by

Although we follow convention in using 0 intensity for the background and 1 for object pixels, any two distinct values can be used in Eq. (10.3-1).

$$g(x, y) = \begin{cases} 1 & \text{if } f(x, y) > T \\ 0 & \text{if } f(x, y) \leq T \end{cases} \tag{10.3-1}$$

When T is a constant applicable over an entire image, the process given in this equation is referred to as *global thresholding*. When the value of T changes over an image, we use the term *variable thresholding*. The term *local* or *regional thresholding* is used sometimes to denote variable thresholding in

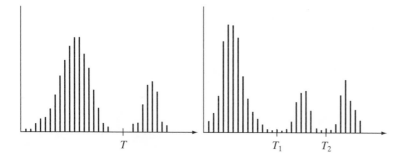

FIGURE 10.35
Intensity
histograms that
can be partitioned
(a) by a single
threshold, and
(b) by dual
thresholds.

which the value of T at any point (x, y) in an image depends on properties of a neighborhood of (x, y) (for example, the average intensity of the pixels in the neighborhood). If T depends on the spatial coordinates (x, y) themselves, then variable thresholding is often referred to as *dynamic* or *adaptive* thresholding. Use of these terms is not universal, and one is likely to see them used interchangeably in the literature on image processing.

Figure 10.35(b) shows a more difficult thresholding problem involving a histogram with three dominant modes corresponding, for example, to two types of light objects on a dark background. Here, *multiple thresholding* classifies a point (x, y) as belonging to the background if $f(x, y) \leq T_1$, to one object class if $T_1 < f(x, y) \leq T_2$, and to the other object class if $f(x, y) > T_2$. That is, the segmented image is given by

$$g(x, y) = \begin{cases} a & \text{if } f(x, y) > T_2 \\ b & \text{if } T_1 < f(x, y) \leq T_2 \\ c & \text{if } f(x, y) \leq T_1 \end{cases} \qquad (10.3\text{-}2)$$

where a, b, and c are any three distinct intensity values. We discuss dual thresholding in Section 10.3.6. Segmentation problems requiring more than two thresholds are difficult (often impossible) to solve, and better results usually are obtained using other methods, such as variable thresholding, as discussed in Section 10.3.7, or region growing, as discussed in Section 10.4.

Based on the preceding discussion, we may infer intuitively that the success of intensity thresholding is directly related to the width and depth of the valley(s) separating the histogram modes. In turn, the key factors affecting the properties of the valley(s) are: (1) the separation between peaks (the further apart the peaks are, the better the chances of separating the modes); (2) the noise content in the image (the modes broaden as noise increases); (3) the relative sizes of objects and background; (4) the uniformity of the illumination source; and (5) the uniformity of the reflectance properties of the image.

The role of noise in image thresholding

As an illustration of how noise affects the histogram of an image, consider Fig. 10.36(a). This simple synthetic image is free of noise, so its histogram consists of two "spike" modes, as Fig. 10.36(d) shows. Segmenting this image into two regions is a trivial task involving a threshold placed anywhere between the two

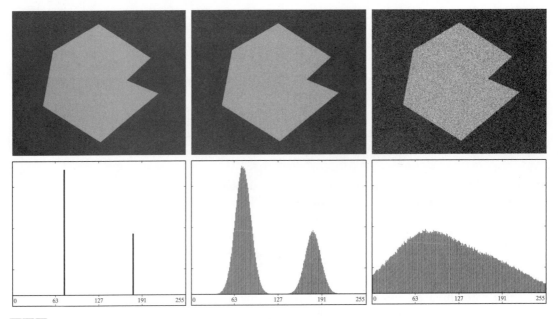

a b c
d e f

FIGURE 10.36 (a) Noiseless 8-bit image. (b) Image with additive Gaussian noise of mean 0 and standard deviation of 10 intensity levels. (c) Image with additive Gaussian noise of mean 0 and standard deviation of 50 intensity levels. (d)–(f) Corresponding histograms.

modes. Figure 10.36(b) shows the original image corrupted by Gaussian noise of zero mean and a standard deviation of 10 intensity levels. Although the corresponding histogram modes are now broader [Fig. 10.36(e)], their separation is large enough so that the depth of the valley between them is sufficient to make the modes easy to separate. A threshold placed midway between the two peaks would do a nice job of segmenting the image. Figure 10.36(c) shows the result of corrupting the image with Gaussian noise of zero mean and a standard deviation of 50 intensity levels. As the histogram in Fig. 10.36(f) shows, the situation is much more serious now, as there is no way to differentiate between the two modes. Without additional processing (such as the methods discussed in Sections 10.3.4 and 10.3.5) we have little hope of finding a suitable threshold for segmenting this image.

The role of illumination and reflectance

Figure 10.37 illustrates the effect that illumination can have on the histogram of an image. Figure 10.37(a) is the noisy image from Fig. 10.36(b), and Fig. 10.37(d) shows its histogram. As before, this image is easily segmentable with a single threshold. We can illustrate the effects of nonuniform illumination by multiplying the image in Fig. 10.37(a) by a variable intensity function, such as the intensity ramp in Fig. 10.37(b), whose histogram is shown in Fig. 10.37(e). Figure 10.37(c) shows the product of the image and this shading pattern. As Fig. 10.37(f) shows, the deep valley between peaks was corrupted to the point

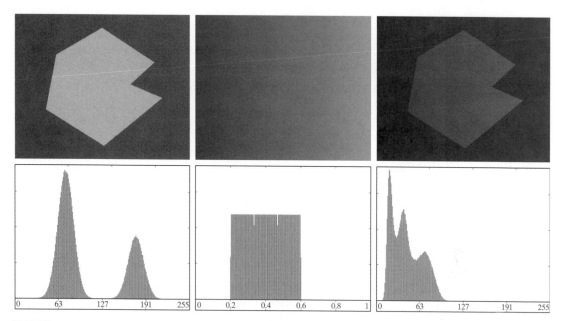

a b c
d e f

FIGURE 10.37 (a) Noisy image. (b) Intensity ramp in the range [0.2, 0.6]. (c) Product of (a) and (b). (d)–(f) Corresponding histograms.

where separation of the modes without additional processing (see Sections 10.3.4 and 10.3.5) is no longer possible. Similar results would be obtained if the illumination was perfectly uniform, but the reflectance of the image was not, due, for example, to natural reflectivity variations in the surface of objects and/or background.

The key point in the preceding paragraph is that illumination and reflectance play a central role in the success of image segmentation using thresholding or other segmentation techniques. Therefore, controlling these factors when it is possible to do so should be the first step considered in the solution of a segmentation problem. There are three basic approaches to the problem when control over these factors is not possible. One is to correct the shading pattern directly. For example, nonuniform (but fixed) illumination can be corrected by multiplying the image by the inverse of the pattern, which can be obtained by imaging a flat surface of constant intensity. The second approach is to attempt to correct the global shading pattern via processing using, for example, the top-hat transformation introduced in Section 9.6.3. The third approach is to "work around" nonuniformities using variable thresholding, as discussed in Section 10.3.7.

In theory, the histogram of a ramp image is uniform. In practice, achieving perfect uniformity depends on the size of the image and number of intensity bits. For example, a 256×256, 256-level ramp image has a uniform histogram, but a 256×257 ramp image with the same number of intensities does not.

10.3.2 Basic Global Thresholding

As noted in the previous section, when the intensity distributions of objects and background pixels are sufficiently distinct, it is possible to use a single (*global*) threshold applicable over the entire image. In most applications, there is usually enough variability between images that, even if global thresholding is a suitable approach, an algorithm capable of estimating automatically the threshold value for each image is required. The following iterative algorithm can be used for this purpose:

1. Select an initial estimate for the global threshold, T.
2. Segment the image using T in Eq. (10.3-1). This will produce two groups of pixels: G_1 consisting of all pixels with intensity values $> T$, and G_2 consisting of pixels with values $\leq T$.
3. Compute the average (mean) intensity values m_1 and m_2 for the pixels in G_1 and G_2, respectively.
4. Compute a new threshold value:

$$T = \frac{1}{2}(m_1 + m_2)$$

5. Repeat Steps 2 through 4 until the difference between values of T in successive iterations is smaller than a predefined parameter ΔT.

This simple algorithm works well in situations where there is a reasonably clear valley between the modes of the histogram related to objects and background. Parameter ΔT is used to control the number of iterations in situations where speed is an important issue. In general, the larger ΔT is, the fewer iterations the algorithm will perform. The initial threshold must be chosen greater than the minimum and less than maximum intensity level in the image (Problem 10.28). The average intensity of the image is a good initial choice for T.

EXAMPLE 10.15:
Global
thresholding.

■ Figure 10.38 shows an example of segmentation based on a threshold estimated using the preceding algorithm. Figure 10.38(a) is the original image, and Fig. 10.38(b) is the image histogram, showing a distinct valley. Application of the preceding iterative algorithm resulted in the threshold $T = 125.4$ after three iterations, starting with $T = m$ (the average image intensity), and using $\Delta T = 0$. Figure 10.38(c) shows the result obtained using $T = 125$ to segment the original image. As expected from the clear separation of modes in the histogram, the segmentation between object and background was quite effective.   ■

The preceding algorithm was stated in terms of successively thresholding the input image and calculating the means at each step because it is more intuitive to introduce it in this manner. However, it is possible to develop a more efficient procedure by expressing all computations in the terms of the image histogram, which has to be computed only once (Problem 10.26).

10.3.3 Optimum Global Thresholding Using Otsu's Method

Thresholding may be viewed as a statistical-decision theory problem whose objective is to minimize the average error incurred in assigning pixels to two or more groups (also called *classes*). This problem is known to have an elegant closed-form solution known as the *Bayes decision rule* (see Section 12.2.2). The solution is based on only two parameters: the probability density function (PDF) of the intensity levels of each class and the probability that each class occurs in a given application. Unfortunately, estimating PDFs is not a trivial

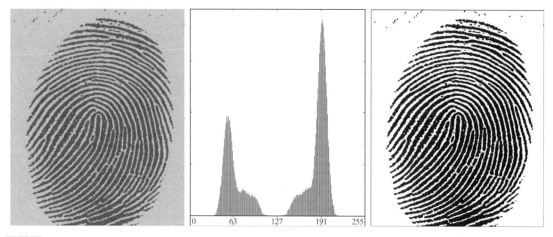

a b c

FIGURE 10.38 (a) Noisy fingerprint. (b) Histogram. (c) Segmented result using a global threshold (the border was added for clarity). (Original courtesy of the National Institute of Standards and Technology.)

matter, so the problem usually is simplified by making workable assumptions about the form of the PDFs, such as assuming that they are Gaussian functions. Even with simplifications, the process of implementing solutions using these assumptions can be complex and not always well-suited for practical applications.

The approach discussed in this section, called *Otsu's method* (Otsu [1979]), is an attractive alternative. The method is optimum in the sense that it maximizes the *between-class variance*, a well-known measure used in statistical discriminant analysis. The basic idea is that well-thresholded classes should be distinct with respect to the intensity values of their pixels and, conversely, that a threshold giving the best separation between classes in terms of their intensity values would be the best (optimum) threshold. In addition to its optimality, Otsu's method has the important property that it is based entirely on computations performed on the histogram of an image, an easily obtainable 1-D array.

Let $\{0, 1, 2, \ldots, L - 1\}$ denote the L distinct intensity levels in a digital image of size $M \times N$ pixels, and let n_i denote the number of pixels with intensity i. The total number, MN, of pixels in the image is $MN = n_0 + n_1 + n_2 + \cdots + n_{L-1}$. The normalized histogram (see Section 3.3) has components $p_i = n_i/MN$, from which it follows that

$$\sum_{i=0}^{L-1} p_i = 1, \qquad p_i \geq 0 \tag{10.3-3}$$

Now, suppose that we select a threshold $T(k) = k, 0 < k < L - 1$, and use it to threshold the input image into two classes, C_1 and C_2, where C_1 consists of all the pixels in the image with intensity values in the range $[0, k]$ and C_2 consists of the pixels with values in the range $[k + 1, L - 1]$. Using this threshold, the probability, $P_1(k)$, that a pixel is assigned to (i.e., thresholded into) class C_1 is given by the cumulative sum

$$P_1(k) = \sum_{i=0}^{k} p_i \tag{10.3-4}$$

Viewed another way, this is the probability of class C_1 occurring. For example, if we set $k = 0$, the probability of class C_1 having any pixels assigned to it is zero. Similarly, the probability of class C_2 occurring is

$$P_2(k) = \sum_{i=k+1}^{L-1} p_i = 1 - P_1(k) \tag{10.3-5}$$

From Eq. (3.3-18), the mean intensity value of the pixels assigned to class C_1 is

$$
\begin{aligned}
m_1(k) &= \sum_{i=0}^{k} iP(i/C_1) \\
&= \sum_{i=0}^{k} iP(C_1/i)P(i)/P(C_1) \\
&= \frac{1}{P_1(k)} \sum_{i=0}^{k} ip_i
\end{aligned}
\tag{10.3-6}
$$

where $P_1(k)$ is given in Eq. (10.3-4). The term $P(i/C_1)$ in the first line of Eq. (10.3-6) is the probability of value i, given that i comes from class C_1. The second line in the equation follows from Bayes' formula:

$$P(A/B) = P(B/A)P(A)/P(B)$$

The third line follows from the fact that $P(C_1/i)$, the probability of C_1 given i, is 1 because we are dealing only with values of i from class C_1. Also, $P(i)$ is the probability of the ith value, which is simply the ith component of the histogram, p_i. Finally, $P(C_1)$ is the probability of class C_1, which we know from Eq. (10.3-4) is equal to $P_1(k)$.

Similarly, the mean intensity value of the pixels assigned to class C_2 is

$$
\begin{aligned}
m_2(k) &= \sum_{i=k+1}^{L-1} iP(i/C_2) \\
&= \frac{1}{P_2(k)} \sum_{i=k+1}^{L-1} ip_i
\end{aligned}
\tag{10.3-7}
$$

The cumulative mean (average intensity) up to level k is given by

$$m(k) = \sum_{i=0}^{k} ip_i \tag{10.3-8}$$

and the average intensity of the entire image (i.e., the *global* mean) is given by

$$m_G = \sum_{i=0}^{L-1} ip_i \tag{10.3-9}$$

The validity of the following two equations can be verified by direct substitution of the preceding results:

$$P_1 m_1 + P_2 m_2 = m_G \qquad (10.3\text{-}10)$$

and

$$P_1 + P_2 = 1 \qquad (10.3\text{-}11)$$

where we have omitted the ks temporarily in favor of notational clarity.

In order to evaluate the "goodness" of the threshold at level k we use the normalized, dimensionless metric

$$\eta = \frac{\sigma_B^2}{\sigma_G^2} \qquad (10.3\text{-}12)$$

where σ_G^2 is the *global variance* [i.e., the intensity variance of all the pixels in the image, as given in Eq. (3.3-19)],

$$\sigma_G^2 = \sum_{i=0}^{L-1} (i - m_G)^2 p_i \qquad (10.3\text{-}13)$$

and σ_B^2 is the *between-class variance*, defined as

$$\sigma_B^2 = P_1 (m_1 - m_G)^2 + P_2 (m_2 - m_G)^2 \qquad (10.3\text{-}14)$$

This expression can be written also as

$$\sigma_B^2 = P_1 P_2 (m_1 - m_2)^2$$
$$= \frac{(m_G P_1 - m)^2}{P_1 (1 - P_1)} \qquad (10.3\text{-}15)$$

The second step in Eq. (10.3-15) makes sense only if P_1 is greater than 0 and less than 1, which, in view of Eq. (10.3-11), implies that P_2 must satisfy the same condition.

where m_G and m are as stated earlier. The first line of this equation follows from Eqs. (10.3-14), (10.3-10), and (10.3-11). The second line follows from Eqs. (10.3-5) through (10.3-9). This form is slightly more efficient computationally because the global mean, m_G, is computed only once, so only two parameters, m and P_1, need to be computed for any value of k.

We see from the first line in Eq. (10.3-15) that the farther the two means m_1 and m_2 are from each other the larger σ_B^2 will be, indicating that the between-class variance is a measure of *separability* between classes. Because σ_G^2 is a constant, it follows that η also is a measure of separability, and maximizing this metric is equivalent to maximizing σ_B^2. The objective, then, is to determine the threshold value, k, that maximizes the between-class variance, as stated at the beginning of this section. Note that Eq. (10.3-12) assumes implicitly that $\sigma_G^2 > 0$. This variance can be zero only when all the intensity levels in the image are the same, which implies the existence of only one class of pixels. This in turn means that $\eta = 0$ for a constant image since the separability of a single class from itself is zero.

Reintroducing k, we have the final results:

$$\eta(k) = \frac{\sigma_B^2(k)}{\sigma_G^2} \tag{10.3-16}$$

and

$$\sigma_B^2(k) = \frac{\left[m_G P_1(k) - m(k)\right]^2}{P_1(k)\left[1 - P_1(k)\right]} \tag{10.3-17}$$

Then, the optimum threshold is the value, k^*, that maximizes $\sigma_B^2(k)$:

$$\sigma_B^2(k^*) = \max_{0 \le k \le L-1} \sigma_B^2(k) \tag{10.3-18}$$

In other words, to find k^* we simply evaluate Eq. (10.3-18) for all *integer* values of k (such that the condition $0 < P_1(k) < 1$ holds) and select that value of k that yielded the maximum $\sigma_B^2(k)$. If the maximum exists for more than one value of k, it is customary to average the various values of k for which $\sigma_B^2(k)$ is maximum. It can be shown (Problem 10.33) that a maximum always exists, subject to the condition that $0 < P_1(k) < 1$. Evaluating Eqs. (10.3-17) and (10.3-18) for all values of k is a relatively inexpensive computational procedure, because the maximum number of integer values that k can have is L.

Once k^* has been obtained, the input image $f(x, y)$ is segmented as before:

$$g(x, y) = \begin{cases} 1 & \text{if } f(x, y) > k^* \\ 0 & \text{if } f(x, y) \le k^* \end{cases} \tag{10.3-19}$$

for $x = 0, 1, 2, \ldots, M - 1$ and $y = 0, 1, 2, \ldots, N - 1$. Note that all the quantities needed to evaluate Eq. (10.3-17) are obtained using only the histogram of $f(x, y)$. In addition to the optimum threshold, other information regarding the segmented image can be extracted from the histogram. For example, $P_1(k^*)$ and $P_2(k^*)$, the class probabilities evaluated at the optimum threshold, indicate the portions of the areas occupied by the classes (groups of pixels) in the thresholded image. Similarly, the means $m_1(k^*)$ and $m_2(k^*)$ are estimates of the average intensity of the classes in the original image.

The normalized metric η, evaluated at the optimum threshold value, $\eta(k^*)$, can be used to obtain a quantitative estimate of the separability of classes, which in turn gives an idea of the ease of thresholding a given image. This measure has values in the range

Although our interest is in the value of η at the optimum threshold, k^*, this inequality holds in general for any value of k in the range $[0, L - 1]$.

$$0 \le \eta(k^*) \le 1 \tag{10.3-20}$$

The lower bound is attainable only by images with a single, constant intensity level, as mentioned earlier. The upper bound is attainable only by 2-valued images with intensities equal to 0 and $L - 1$ (Problem 10.34).

Otsu's algorithm may be summarized as follows:

1. Compute the normalized histogram of the input image. Denote the components of the histogram by p_i, $i = 0, 1, 2, \ldots, L - 1$.
2. Compute the cumulative sums, $P_1(k)$, for $k = 0, 1, 2, \ldots, L - 1$, using Eq. (10.3-4).
3. Compute the cumulative means, $m(k)$, for $k = 0, 1, 2, \ldots, L - 1$, using Eq. (10.3-8).
4. Compute the global intensity mean, m_G, using (10.3-9).
5. Compute the between-class variance, $\sigma_B^2(k)$, for $k = 0, 1, 2, \ldots, L - 1$, using Eq. (10.3-17).
6. Obtain the Otsu threshold, k^*, as the value of k for which $\sigma_B^2(k)$ is maximum. If the maximum is not unique, obtain k^* by averaging the values of k corresponding to the various maxima detected.
7. Obtain the separability measure, η^*, by evaluating Eq. (10.3-16) at $k = k^*$.

The following example illustrates the preceding concepts.

■ Figure 10.39(a) shows an optical microscope image of polymersome cells, and Fig. 10.39(b) shows its histogram. The objective of this example is to segment the molecules from the background. Figure 10.39(c) is the result of using the basic global thresholding algorithm developed in the previous section. Because the histogram has no distinct valleys and the intensity difference between the background and objects is small, the algorithm failed to achieve the desired segmentation. Figure 10.39(d) shows the result obtained using Otsu's method. This result obviously is superior to Fig. 10.39(c). The threshold value computed by the basic algorithm was 169, while the threshold computed by Otsu's method was 181, which is closer to the lighter areas in the image defining the cells. The separability measure η was 0.467.

As a point of interest, applying Otsu's method to the fingerprint image in Example 10.15 yielded a threshold of 125 and a separability measure of 0.944. The threshold is identical to the value (rounded to the nearest integer) obtained with the basic algorithm. This is not unexpected, given the nature of the histogram. In fact, the separability measure is high due primarily to the relatively large separation between modes and the deep valley between them. ■

EXAMPLE 10.16:
Optimum global thresholding using Otsu's method.

Polymersomes are cells artificially engineered using polymers. Polymersomes are invisible to the human immune system and can be used, for example, to deliver medication to targeted regions of the body.

10.3.4 Using Image Smoothing to Improve Global Thresholding

As noted in Fig. 10.36, noise can turn a simple thresholding problem into an unsolvable one. When noise cannot be reduced at the source, and thresholding is the segmentation method of choice, a technique that often enhances performance is to smooth the image prior to thresholding. We illustrate the approach with an example.

Figure 10.40(a) is the image from Fig. 10.36(c), Fig. 10.40(b) shows its histogram, and Fig. 10.40(c) is the image thresholded using Otsu's method. Every black point in the white region and every white point in the black region is a

a b
c d

FIGURE 10.39
(a) Original
image.
(b) Histogram
(high peaks were
clipped to
highlight details in
the lower values).
(c) Segmentation
result using the
basic global
algorithm from
Section 10.3.2.
(d) Result
obtained using
Otsu's method.
(Original image
courtesy of
Professor Daniel
A. Hammer, the
University of
Pennsylvania.)

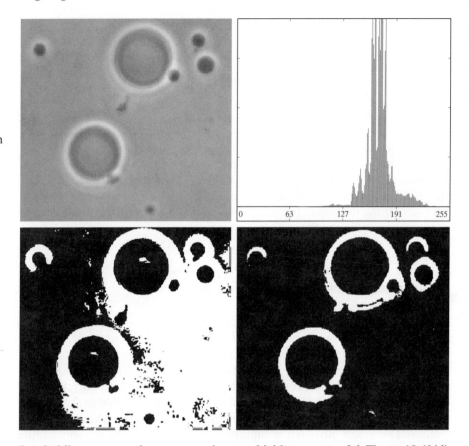

thresholding error, so the segmentation was highly unsuccessful. Figure 10.40(d) shows the result of smoothing the noisy image with an averaging mask of size 5 × 5 (the image is of size 651 × 814 pixels), and Fig. 10.40(e) is its histogram. The improvement in the shape of the histogram due to smoothing is evident, and we would expect thresholding of the smoothed image to be nearly perfect. As Fig. 10.40(f) shows, this indeed was the case. The slight distortion of the boundary between object and background in the segmented, smoothed image was caused by the blurring of the boundary. In fact, the more aggressively we smooth an image, the more boundary errors we should anticipate in the segmented result.

Next we consider the effect of reducing the size of the region in Fig. 10.40(a) with respect to the background. Figure 10.41(a) shows the result. The noise in this image is additive Gaussian noise with zero mean and a standard deviation of 10 intensity levels (as opposed to 50 in the previous example). As Fig. 10.41(b) shows, the histogram has no clear valley, so we would expect segmentation to fail, a fact that is confirmed by the result in Fig. 10.41(c). Figure 10.41(d) shows the image smoothed with an averaging mask of size 5 × 5, and Fig. 10.40(e) is the corresponding histogram. As expected, the net effect was to reduce the spread of the histogram, but the distribution still is unimodal. As Fig. 10.40(f) shows, segmentation failed again. The reason for the failure can be traced to the fact that the region is so small that its contribution to the histogram is insignificant compared to the intensity spread caused by noise. In

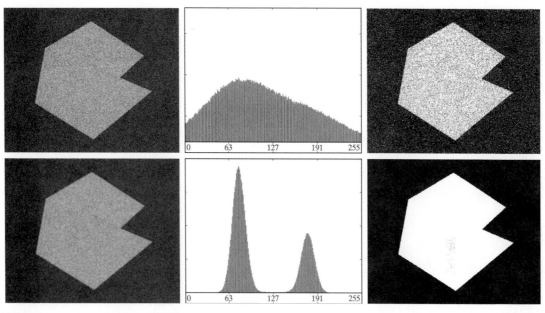

a b c
d e f

FIGURE 10.40 (a) Noisy image from Fig. 10.36 and (b) its histogram. (c) Result obtained using Otsu's method. (d) Noisy image smoothed using a 5 × 5 averaging mask and (e) its histogram. (f) Result of thresholding using Otsu's method.

situations such as this, the approach discussed in the following section is more likely to succeed.

10.3.5 Using Edges to Improve Global Thresholding

Based on the discussion in the previous four sections, we conclude that the chances of selecting a "good" threshold are enhanced considerably if the histogram peaks are tall, narrow, symmetric, and separated by deep valleys. One approach for improving the shape of histograms is to consider only those pixels that lie on or near the edges between objects and the background. An immediate and obvious improvement is that histograms would be less dependent on the relative sizes of objects and the background. For instance, the histogram of an image composed of a small object on a large background area (or vice versa) would be dominated by a large peak because of the high concentration of one type of pixels. We saw in the previous section that this can lead to failure in thresholding.

If only the pixels on or near the edges between objects and background were used, the resulting histogram would have peaks of approximately the same height. In addition, the probability that any of those pixels lies on an object would be approximately equal to the probability that it lies on the background, thus improving the symmetry of the histogram modes. Finally, as indicated in the following paragraph, using pixels that satisfy some simple measures based on gradient and Laplacian operators has a tendency to deepen the valley between histogram peaks.

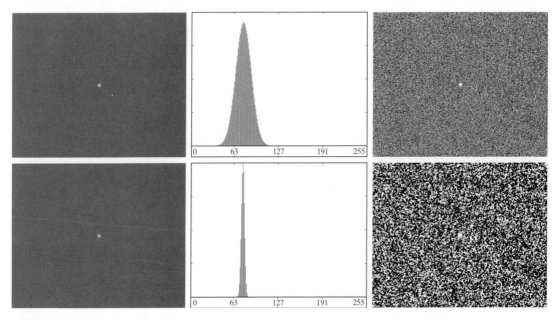

a b c
d e f

FIGURE 10.41 (a) Noisy image and (b) its histogram. (c) Result obtained using Otsu's method. (d) Noisy image smoothed using a 5×5 averaging mask and (e) its histogram. (f) Result of thresholding using Otsu's method. Thresholding failed in both cases.

The approach just discussed assumes that the edges between objects and background are known. This information clearly is not available during segmentation, as finding a division between objects and background is precisely what segmentation is all about. However, with reference to the discussion in Section 10.2, an indication of whether a pixel is on an edge may be obtained by computing its gradient or Laplacian. For example, the average value of the Laplacian is 0 at the transition of an edge (see Fig. 10.10), so the valleys of histograms formed from the pixels selected by a Laplacian criterion can be expected to be sparsely populated. This property tends to produce the desirable deep valleys discussed above. In practice, comparable results typically are obtained using either the gradient or Laplacian images, with the latter being favored because it is computationally more attractive and is also an isotropic edge detector.

The preceding discussion is summarized in the following algorithm, where $f(x, y)$ is the input image:

It is possible to modify this algorithm so that both the magnitude of the gradient and the absolute value of the Laplacian images are used. In this case, we would specify a threshold for each image and form the logical OR of the two results to obtain the marker image. This approach is useful when more control is desired over the points deemed to be valid edge points.

1. Compute an edge image as either the magnitude of the gradient, or absolute value of the Laplacian, of $f(x, y)$ using any of the methods discussed in Section 10.2.
2. Specify a threshold value, T.
3. Threshold the image from Step 1 using the threshold from Step 2 to produce a binary image, $g_T(x, y)$. This image is used as a *mask image* in the following step to select pixels from $f(x, y)$ corresponding to "strong" edge pixels.

4. Compute a histogram using only the pixels in $f(x, y)$ that correspond to the locations of the 1-valued pixels in $g_T(x, y)$.
5. Use the histogram from Step 4 to segment $f(x, y)$ globally using, for example, Otsu's method.

If T is set to any value less than the minimum value of the edge image then, according to Eq. (10.3-1), $g_T(x, y)$ will consist of all 1s, implying that all pixels of $f(x, y)$ will be used to compute the image histogram. In this case, the preceding algorithm becomes global thresholding in which the histogram of the original image is used. It is customary to specify the value of T corresponding to a percentile, which typically is set high (e.g., in the high 90s) so that few pixels in the gradient/Laplacian image will be used in the computation. The following examples illustrate the concepts just discussed. The first example uses the gradient and the second uses the Laplacian. Similar results can be obtained in both examples using either approach. The important issue is to generate a suitable derivative image.

The nth percentile is the smallest number that is greater than $n\%$ of the numbers in a given set. For example, if you received a 95 in a test and this score was greater than 85% of all the students taking the test, then you would be in the 85th percentile with respect to the test scores.

■ Figures 10.42(a) and (b) show the image and histogram from Fig. 10.41. You saw that this image could not be segmented by smoothing followed by thresholding. The objective of this example is to solve the problem using edge information. Figure 10.42(c) is the gradient magnitude image thresholded at the

EXAMPLE 10.17:
Using edge information based on the gradient to improve global thresholding.

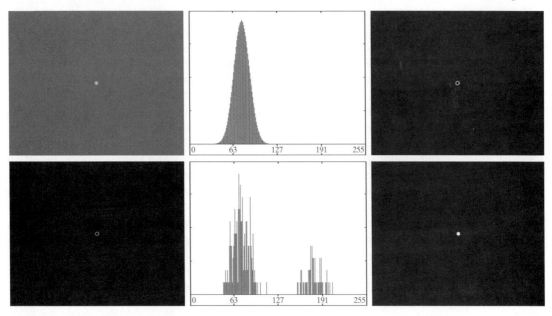

a b c
d e f

FIGURE10.42 (a) Noisy image from Fig. 10.41(a) and (b) its histogram. (c) Gradient magnitude image thresholded at the 99.7 percentile. (d) Image formed as the product of (a) and (c). (e) Histogram of the nonzero pixels in the image in (d). (f) Result of segmenting image (a) with the Otsu threshold based on the histogram in (e). The threshold was 134, which is approximately midway between the peaks in this histogram.

99.7 percentile. Figure 10.42(d) is the image formed by multiplying this (mask) image by the input image. Figure 10.42(e) is the histogram of the nonzero elements in Fig. 10.42(d). Note that this histogram has the important features discussed earlier; that is, it has reasonably symmetrical modes separated by a deep valley. Thus, while the histogram of the original noisy image offered no hope for successful thresholding, the histogram in Fig. 10.42(e) indicates that thresholding of the small object from the background is indeed possible. The result in Fig. 10.42(f) shows that indeed this is the case. This image was obtained by using Otsu's method to obtain a threshold based on the histogram in Fig. 10.42(e) and then applying this threshold globally to the noisy image in Fig. 10.42(a). The result is nearly perfect. ■

EXAMPLE 10.18:
Using edge information based on the Laplacian to improve global thresholding.

■ In this example we consider a more complex thresholding problem. Figure 10.43(a) shows an 8-bit image of yeast cells in which we wish to use global thresholding to obtain the regions corresponding to the bright spots. As a starting point, Fig. 10.43(b) shows the image histogram, and Fig. 10.43(c) is the result obtained using Otsu's method directly on the image, using the histogram shown. We see that Otsu's method failed to achieve the original objective of detecting the bright spots, and, although the method was able to isolate some of the cell regions themselves, several of the segmented regions on the right are not disjoint. The threshold computed by the Otsu method was 42 and the separability measure was 0.636.

Figure 10.43(d) shows the image $g_T(x, y)$ obtained by computing the absolute value of the Laplacian image and then thresholding it with T set to 115 on an intensity scale in the range $[0, 255]$. This value of T corresponds approximately to the 99.5 percentile of the values in the absolute Laplacian image, so thresholding at this level should result in a sparse set of pixels, as Fig. 10.43(d) shows. Note in this image how the points cluster near the edges of the bright spots, as expected from the preceding discussion. Figure 10.43(e) is the histogram of the nonzero pixels in the product of (a) and (d). Finally, Fig. 10.43(f) shows the result of globally segmenting the original image using Otsu's method based on the histogram in Fig. 10.43(e). This result agrees with the locations of the bright spots in the image. The threshold computed by the Otsu method was 115 and the separability measure was 0.762, both of which are higher than the values obtained by using the original histogram.

By varying the percentile at which the threshold is set we can even improve on the segmentation of the cell regions. For example, Fig. 10.44 shows the result obtained using the same procedure as in the previous paragraph, but with the threshold set at 55, which is approximately 5% of the maximum value of the absolute Laplacian image. This value is at the 53.9 percentile of the values in that image. This result clearly is superior to the result in Fig. 10.43(c) obtained using Otsu's method with the histogram of the original image. ■

10.3.6 Multiple Thresholds

Thus far, we have focused attention on image segmentation using a single global threshold. The thresholding method introduced in Section 10.3.3 can be extended to an arbitrary number of thresholds, because the separability measure

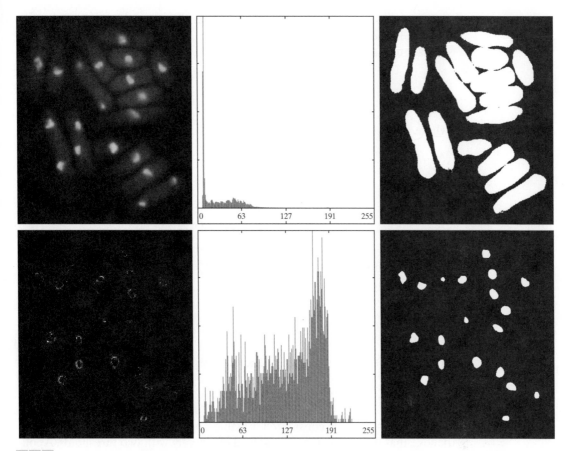

a b c
d e f

FIGURE 10.43 (a) Image of yeast cells. (b) Histogram of (a). (c) Segmentation of (a) with Otsu's method using the histogram in (b). (d) Thresholded absolute Laplacian. (e) Histogram of the nonzero pixels in the product of (a) and (d). (f) Original image thresholded using Otsu's method based on the histogram in (e). (Original image courtesy of Professor Susan L. Forsburg, University of Southern California.)

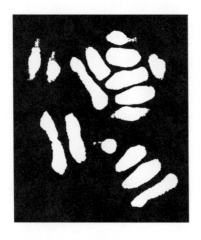

FIGURE 10.44
Image in
Fig. 10.43(a)
segmented using
the same
procedure as
explained in
Figs. 10.43(d)–(f),
but using a lower
value to threshold
the absolute
Laplacian image.

on which it is based also extends to an arbitrary number of classes (Fukunaga [1972]). In the case of K classes, $C_1, C_2, \ldots C_K$, the between-class variance generalizes to the expression

$$\sigma_B^2 = \sum_{k=1}^{K} P_k (m_k - m_G)^2 \qquad (10.3\text{-}21)$$

where

$$P_k = \sum_{i \in C_k} p_i \qquad (10.3\text{-}22)$$

$$m_k = \frac{1}{P_k} \sum_{i \in C_k} i p_i \qquad (10.3\text{-}23)$$

and m_G is the global mean given in Eq. (10.3-9). The K classes are separated by $K - 1$ thresholds whose values, $k_1^*, k_2^*, \ldots, k_{K-1}^*$, are the values that maximize Eq. (10.3-21):

$$\sigma_B^2(k_1^*, k_2^*, \ldots, k_{K-1}^*) = \max_{0 < k_1 < k_2 < \ldots k_{n-1} < L-1} \sigma_B^2(k_1, k_2, \ldots k_{K-1}) \qquad (10.3\text{-}24)$$

Although this result is perfectly general, it begins to lose meaning as the number of classes increases, because we are dealing with only one variable (intensity). In fact, the between-class variance usually is cast in terms of multiple variables expressed as vectors (Fukunaga [1972]). In practice, using multiple global thresholding is considered a viable approach when there is reason to believe that the problem can be solved effectively with two thresholds. Applications that require more than two thresholds generally are solved using more than just intensity values. Instead, the approach is to use additional descriptors (e.g., color) and the application is cast as a pattern recognition problem, as explained in Section 10.3.8.

For three classes consisting of three intensity intervals (which are separated by two thresholds) the between-class variance is given by:

Thresholding with two thresholds sometimes is referred to as hysteresis thresholding.

$$\sigma_B^2 = P_1 (m_1 - m_G)^2 + P_2 (m_2 - m_G)^2 + P_3 (m_3 - m_G)^2 \qquad (10.3\text{-}25)$$

where

$$P_1 = \sum_{i=0}^{k_1} p_i$$

$$P_2 = \sum_{i=k_1+1}^{k_2} p_i \qquad (10.3\text{-}26)$$

$$P_3 = \sum_{i=k_2+1}^{L-1} p_i$$

and

$$m_1 = \frac{1}{P_1} \sum_{i=0}^{k_1} i p_i$$

$$m_2 = \frac{1}{P_2} \sum_{i=k_1+1}^{k_2} i p_i \qquad (10.3\text{-}27)$$

$$m_3 = \frac{1}{P_3} \sum_{i=k_2+1}^{L-1} i p_i$$

As in Eqs. (10.3-10) and (10.3-11), the following relationships hold:

$$P_1 m_1 + P_2 m_2 + P_3 m_3 = m_G \qquad (10.3\text{-}28)$$

and

$$P_1 + P_2 + P_3 = 1 \qquad (10.3\text{-}29)$$

We see that the P and m terms and, therefore σ_B^2, are functions of k_1 and k_2. The two optimum threshold values, k_1^* and k_2^*, are the values that maximize $\sigma_B^2(k_1, k_2)$. In other words, as in the single-threshold case discussed in Section 10.3.3, we find the optimum thresholds by finding

$$\sigma_B^2(k_1^*, k_2^*) = \max_{0 < k_1 < k_2 < L-1} \sigma_B^2(k_1, k_2) \qquad (10.3\text{-}30)$$

The procedure starts by selecting the first value of k_1 (that value is 1 because looking for a threshold at 0 intensity makes no sense; also, keep in mind that the increment values are integers because we are dealing with intensities). Next, k_2 is incremented through all its values greater than k_1 and less than $L - 1$ (i.e., $k_2 = k_1 + 1, \ldots, L - 2$). Then k_1 is incremented to its next value and k_2 is incremented again through all its values greater than k_1. This procedure is repeated until $k_1 = L - 3$. The result of this process is a 2-D array, $\sigma_B^2(k_1, k_2)$, and the last step is to look for the maximum value in this array. The values of k_1 and k_2 corresponding to that maximum are the optimum thresholds, k_1^* and k_2^*. If there are several maxima, the corresponding values of k_1 and k_2 are averaged to obtain the final thresholds. The thresholded image is then given by

$$g(x, y) = \begin{cases} a & \text{if } f(x, y) \le k_1^* \\ b & \text{if } k_1^* < f(x, y) \le k_2^* \\ c & \text{if } f(x, y) > k_2^* \end{cases} \qquad (10.3\text{-}31)$$

where a, b, and c are any three valid intensity values.

Finally, we note that the separability measure defined in Section 10.3.3 for one threshold extends directly to multiple thresholds:

$$\eta(k_1^*, k_2^*) = \frac{\sigma_B^2(k_1^*, k_2^*)}{\sigma_G^2} \qquad (10.3\text{-}32)$$

where σ_G^2 is the total image variance from Eq. (10.3-13).

EXAMPLE 10.19:
Multiple global
thresholding.

■ Figure 10.45(a) shows an image of an iceberg. The objective of this example is to segment the image into three regions: the dark background, the illuminated area of the iceberg, and the area in shadows. It is evident from the image histogram in Fig. 10.45(b) that two thresholds are required to solve this problem. The procedure discussed above resulted in the thresholds $k_1^* = 80$ and $k_2^* = 177$, which we note from Fig. 10.45(b) are near the centers of the two histogram valleys. Figure 10.45(c) is the segmentation that resulted using these two thresholds in Eq. (10.3-31). The separability measure was 0.954. The principal reason this example worked out so well can be traced to the histogram having three distinct modes separated by reasonably wide, deep valleys. ■

10.3.7 Variable Thresholding

As discussed in Section 10.3.1, factors such as noise and nonuniform illumination play a major role in the performance of a thresholding algorithm. We showed in Sections 10.3.4 and 10.3.5 that image smoothing and using edge information can help significantly. However, it frequently is the case that this type of preprocessing is either impractical or simply ineffective in improving the situation to the point where the problem is solvable by any of the methods discussed thus far. In such situations, the next level of thresholding complexity involves variable thresholding. In this section, we discuss various techniques for choosing variable thresholds.

Image partitioning

One of the simplest approaches to variable thresholding is to subdivide an image into nonoverlapping rectangles. This approach is used to compensate for non-uniformities in illumination and/or reflectance. The rectangles are chosen small enough so that the illumination of each is approximately uniform. We illustrate this approach with an example.

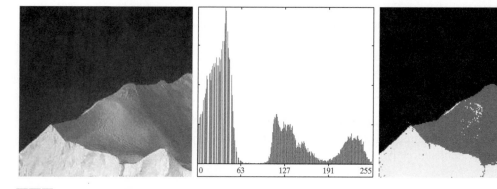

a b c

FIGURE 10.45 (a) Image of iceberg. (b) Histogram. (c) Image segmented into three regions using dual Otsu thresholds. (Original image courtesy of NOAA.)

■ Figure 10.46(a) shows the image from Fig. 10.37(c), and Fig. 10.46(b) shows its histogram. When discussing Fig. 10.37(c) we concluded that this image could not be segmented with a global threshold, a fact confirmed by Figs. 10.46(c) and (d), which show the results of segmenting the image using the iterative scheme discussed in Section 10.3.2 and Otsu's method, respectively. Both methods produced comparable results, in which numerous segmentation errors are visible.

Figure 10.46(e) shows the original image subdivided into six rectangular regions, and Fig. 10.46(f) is the result of applying Otsu's global method to each subimage. Although some errors in segmentation are visible, image subdivision produced a reasonable result on an image that is quite difficult to segment. The reason for the improvement is explained easily by analyzing the histogram of each subimage. As Fig. 10.47 shows, each subimage is characterized by a bimodal histogram with a deep valley between the modes, a fact that we know will lead to effective global thresholding.

Image subdivision generally works well when the objects of interest and the background occupy regions of reasonably comparable size, as in Fig. 10.46. When this is not the case, the method typically fails because of the likelihood of subdivisions containing only object or background pixels. Although this situation can be addressed by using additional techniques to determine when a subdivision contains both types of pixels, the logic required to address different

EXAMPLE 10.20:
Variable thresholding via image partitioning.

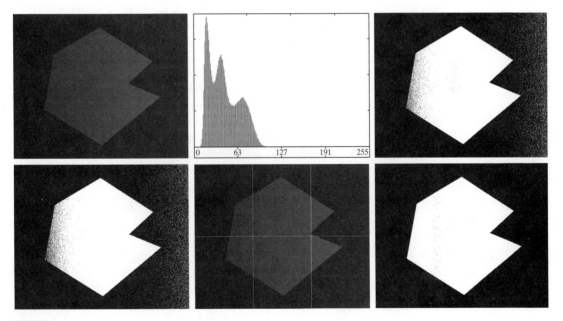

a b c
d e f

FIGURE 10.46 (a) Noisy, shaded image and (b) its histogram. (c) Segmentation of (a) using the iterative global algorithm from Section 10.3.2. (d) Result obtained using Otsu's method. (e) Image subdivided into six subimages. (f) Result of applying Otsu's method to each subimage individually.

FIGURE 10.47
Histograms of the
six subimages in
Fig. 10.46(e).

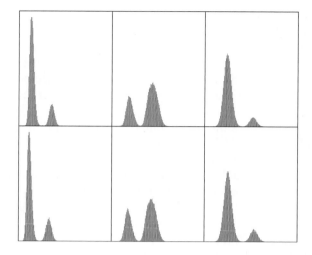

scenarios can get complicated. In such situations, methods such as those discussed in the remainder of this section typically are preferable.   ▥

Variable thresholding based on local image properties

A more general approach than the image subdivision method discussed in the previous section is to compute a threshold at every point, (x, y), in the image based on one or more specified properties computed in a neighborhood of (x, y). Although this may seem like a laborious process, modern algorithms and hardware allow for fast neighborhood processing, especially for common functions such as logical and arithmetic operations.

We illustrate the basic approach to local thresholding using the standard deviation and mean of the pixels in a neighborhood of every point in an image. These two quantities are quite useful for determining local thresholds because they are descriptors of local contrast and average intensity. Let σ_{xy} and m_{xy} denote the standard deviation and mean value of the set of pixels contained in a neighborhood, S_{xy}, centered at coordinates (x, y) in an image (see Section 3.3.4 regarding computation of the local mean and standard deviation). The following are common forms of variable, local thresholds:

$$T_{xy} = a\sigma_{xy} + bm_{xy} \qquad (10.3\text{-}33)$$

where a and b are nonnegative constants, and

$$T_{xy} = a\sigma_{xy} + bm_G \qquad (10.3\text{-}34)$$

where m_G is the global image mean. The segmented image is computed as

$$g(x, y) = \begin{cases} 1 & \text{if } f(x, y) > T_{xy} \\ 0 & \text{if } f(x, y) \le T_{xy} \end{cases} \qquad (10.3\text{-}35)$$

where $f(x, y)$ is the input image. This equation is evaluated for all pixel locations in the image, and a different threshold is computed at each location (x, y) using the pixels in the neighborhood S_{xy}.

Significant power (with a modest increase in computation) can be added to local thresholding by using predicates based on the parameters computed in the neighborhoods of (x, y):

$$g(x, y) = \begin{cases} 1 & \text{if } Q(\text{local parameters}) \text{ is true} \\ 0 & \text{if } Q(\text{local parameters}) \text{ is false} \end{cases} \qquad (10.3\text{-}36)$$

where Q is a *predicate* based on parameters computed using the pixels in neighborhood S_{xy}. For example, consider the following predicate, $Q(\sigma_{xy}, m_{xy})$, based on the local mean and standard deviation:

$$Q(\sigma_{xy}, m_{xy}) = \begin{cases} \text{true} & \text{if } f(x, y) > a\sigma_{xy} \text{ AND } f(x, y) > bm_{xy} \\ \text{false} & \text{otherwise} \end{cases} \qquad (10.3\text{-}37)$$

Note that Eq. (10.3-35) is a special case of Eq. (10.3-36), obtained by letting Q be true if $f(x, y) > T_{xy}$ and false otherwise. In this case, the predicate is based simply on the intensity at a point.

■ Figure 10.48(a) shows the yeast image from Example 10.18. This image has three predominant intensity levels, so it is reasonable to assume that perhaps dual thresholding could be a good segmentation approach. Figure 10.48(b) is the result of using the dual thresholding method explained in Section 10.3.6. As the figure shows, it was possible to isolate the bright areas from the background, but the mid-gray regions on the right side of the image were not segmented properly (recall that we encountered a similar problem with Fig. 10.43(c) in Example 10.18). To illustrate the use of local thresholding, we computed the local standard deviation σ_{xy} for all (x, y) in the input image using a neighborhood of size 3×3. Figure 10.48(c) shows the result. Note how the faint outer lines correctly delineate the boundaries of the cells. Next, we formed a predicate of the form shown in Eq. (10.3-37) but using the global mean instead of m_{xy}. Choosing the global mean generally gives better results when the background is nearly constant and all the object intensities are above or below the background intensity. The values $a = 30$ and $b = 1.5$ were used in completing the specification of the predicate (these values were determined experimentally, as is usually the case in applications such as this). The image was then segmented using Eq. (10.3-36). As Fig. 10.48(d) shows, the result agrees quite closely with the two types of intensity regions prevalent in the input image. Note in particular that all the outer regions were segmented properly and that most of the inner, brighter regions were isolated correctly. ■

EXAMPLE 10.21:
Variable thresholding based on local image properties.

Using moving averages

A special case of the local thresholding method just discussed is based on computing a moving average along scan lines of an image. This implementation is quite useful in document processing, where speed is a fundamental requirement. The scanning typically is carried out line by line in a zigzag pattern to

a b
c d

FIGURE 10.48
(a) Image from
Fig. 10.43.
(b) Image
segmented using
the dual
thresholding
approach
discussed in
Section 10.3.6.
(c) Image of local
standard
deviations.
(d) Result
obtained using
local thresholding.

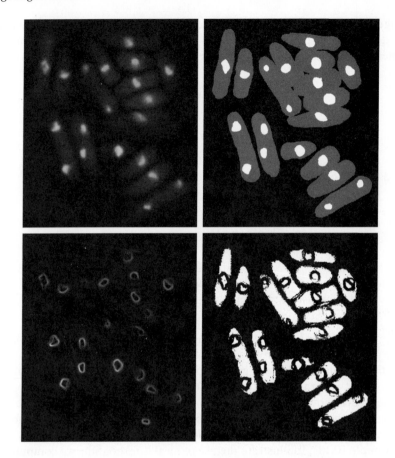

reduce illumination bias. Let z_{k+1} denote the intensity of the point encountered in the scanning sequence at step $k + 1$. The moving average (mean intensity) at this new point is given by

The first expression is valid for $k \geq n - 1$. When k is less than $n - 1$, averages are formed with the available points. Similarly, the second expression is valid for $k \geq n + 1$.

$$m(k + 1) = \frac{1}{n} \sum_{i=k+2-n}^{k+1} z_i$$

$$= m(k) + \frac{1}{n}(z_{k+1} - z_{k-n}) \qquad (10.3\text{-}38)$$

where n denotes the number of points used in computing the average and $m(1) = z_1/n$. This initial value is not strictly correct because the average of a single point is the value of the point itself. However, we use $m(1) = z_1/n$ so that no special computations are required when Eq. (10.3-38) first starts up. Another way of viewing it is that this is the value we would obtain if the border of the image were padded with $n - 1$ zeros. The algorithm is initialized only once, not at every row. Because a moving average is computed for every point in the image, segmentation is implemented using Eq. (10.3-35) with $T_{xy} = bm_{xy}$ where b is constant and m_{xy} is the moving average from Eq. (10.3-38) at point (x, y) in the input image.

■ Figure 10.49(a) shows an image of handwritten text shaded by a spot intensity pattern. This form of intensity shading is typical of images obtained with a photographic flash. Figure 10.49(b) is the result of segmentation using the Otsu global thresholding method. It is not unexpected that global thresholding could not overcome the intensity variation. Figure 10.49(c) shows successful segmentation with local thresholding using moving averages. A rule of thumb is to let n equal 5 times the average stroke width. In this case, the average width was 4 pixels, so we let $n = 20$ in Eq. (10.3-38) and used $b = 0.5$.

EXAMPLE 10.22:
Document
thresholding using
moving averages.

As another illustration of the effectiveness of this segmentation approach we used the same parameters as in the previous paragraph to segment the image in Fig. 10.50(a), which is corrupted by a sinusoidal intensity variation typical of the variation that may occur when the power supply in a document scanner is not grounded properly. As Figs. 10.50(b) and (c) show, the segmentation results are comparable to those in Fig. 10.49.

It is of interest to note that successful segmentation results were obtained in both cases using the same values for n and b, which shows the relative ruggedness of the approach. In general, thresholding based on moving averages works well when the objects of interest are small (or thin) with respect to the image size, a condition satisfied by images of typed or handwritten text.   ■

10.3.8 Multivariable Thresholding

Thus far, we have been concerned with thresholding based on a single variable: gray-scale intensity. In some cases, a sensor can make available more than one variable to characterize each pixel in an image, and thus allow *multivariable thresholding*. A notable example is color imaging, where red (R), green (G), and blue (B) components are used to form a composite color image (see Chapter 6). In this case, each "pixel" is characterized by three values, and can be represented as a 3-D vector, $\mathbf{z} = (z_1, z_2, z_3)^T$, whose components are the RGB colors at a point. These 3-D points often are referred to as *voxels*, to denote *volumetric* elements, as opposed to *image* elements.

a b c

FIGURE 10.49 (a) Text image corrupted by spot shading. (b) Result of global thresholding using Otsu's method. (c) Result of local thresholding using moving averages.

As discussed in some detail in Section 6.7, multivariable thresholding may be viewed as a distance computation. Suppose that we want to extract from a color image all regions having a specified color range: say, reddish hues. Let **a** denote the average reddish color in which we are interested. One way to segment a color image based on this parameter is to compute a distance measure, $D(\mathbf{z}, \mathbf{a})$, between an arbitrary color point, **z**, and the average color, **a**. Then, we segment the input image as follows:

$$
g = \begin{cases} 1 & \text{if } D(\mathbf{z}, \mathbf{a}) < T \\ 0 & \text{otherwise} \end{cases} \tag{10.3-39}
$$

where T is a threshold, and it is understood that the distance computation is performed at all coordinates in the input image to generate the corresponding segmented values in g. Note that the inequalities in this equation are the opposite of the inequalities we used in Eq. (10.3-1) for thresholding a single variable. The reason is that the equation $D(\mathbf{z}, \mathbf{a}) = T$ defines a volume (see Fig. 6.43) and it is more intuitive to think of segmented pixel values as being contained within the volume and background pixel values as being on the surface or outside the volume. Equation (10.3-39) reduces to Eq. (10.3-1) by letting $D(\mathbf{z}, \mathbf{a}) = -f(x, y)$.

Observe that the condition $f(x, y) > T$ basically says that the Euclidean distance between the value of f and the origin of the real line exceeds the value of T. Thus, thresholding is based on the computation of a distance measure, and the form of Eq. (10.3-39) depends on the measure used. If, in general, **z** in an n-dimensional vector, we know from Section 2.6.6 that the n-dimensional *Euclidean distance* is defined as

$$
D(\mathbf{z}, \mathbf{a}) = \|\mathbf{z} - \mathbf{a}\|
$$

$$
= \left[(\mathbf{z} - \mathbf{a})^T (\mathbf{z} - \mathbf{a}) \right]^{\frac{1}{2}} \tag{10.3-40}
$$

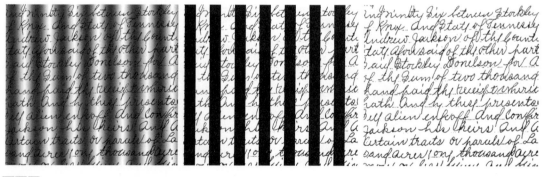

a b c

FIGURE 10.50 (a) Text image corrupted by sinusoidal shading. (b) Result of global thresholding using Otsu's method. (c) Result of local thresholding using moving averages.

The equation $D(\mathbf{z}, \mathbf{a}) = T$ describes a sphere (called a *hypersphere*) in n-dimensional Euclidean space (Fig. 6.43 shows a 3-D example). A more powerful distance measure is the so-called *Mahalanobis distance*, defined as

$$D(\mathbf{z}, \mathbf{a}) = \left[(\mathbf{z} - \mathbf{a})^T \mathbf{C}^{-1} (\mathbf{z} - \mathbf{a}) \right]^{\frac{1}{2}} \qquad (10.3\text{-}41)$$

where $\mathbf{C}$ is the covariance matrix of the $\mathbf{z}$s, as discussed Section 12.2.2. $D(\mathbf{z}, \mathbf{a}) = T$ describes an n-dimensional hyperellipse (Fig. 6.43 shows a 3-D example). This expression reduces to Eq. (10.3-40) when $\mathbf{C} = \mathbf{I}$, the identity matrix.

We gave a detailed example in Section 6.7 regarding the use of these expressions. We also discuss in Section 12.2 the problem of segmenting regions out of an image using pattern recognition techniques based on decision functions, which may be viewed as a multiclass, multivariable thresholding problem.

10.4 Region-Based Segmentation

As discussed in Section 10.1, the objective of segmentation is to partition an image into regions. In Section 10.2, we approached this problem by attempting to find boundaries between regions based on discontinuities in intensity levels, whereas in Section 10.3, segmentation was accomplished via thresholds based on the distribution of pixel properties, such as intensity values or color. In this section, we discuss segmentation techniques that are based on finding the regions directly.

You should review the terminology introduced in Section 10.1 before proceeding.

10.4.1 Region Growing

As its name implies, *region growing* is a procedure that groups pixels or subregions into larger regions based on predefined criteria for growth. The basic approach is to start with a set of "seed" points and from these grow regions by appending to each seed those neighboring pixels that have predefined properties similar to the seed (such as specific ranges of intensity or color).

Selecting a set of one or more starting points often can be based on the nature of the problem, as shown later in Example 10.23. When a priori information is not available, the procedure is to compute at every pixel the same set of properties that ultimately will be used to assign pixels to regions during the growing process. If the result of these computations shows clusters of values, the pixels whose properties place them near the centroid of these clusters can be used as seeds.

The selection of similarity criteria depends not only on the problem under consideration, but also on the type of image data available. For example, the analysis of land-use satellite imagery depends heavily on the use of color. This problem would be significantly more difficult, or even impossible, to solve without the inherent information available in color images. When the images are monochrome, region analysis must be carried out with a set of descriptors based on intensity levels and spatial properties (such as moments or texture). We discuss descriptors useful for region characterization in Chapter 11.

Descriptors alone can yield misleading results if connectivity properties are not used in the region-growing process. For example, visualize a random arrangement of pixels with only three distinct intensity values. Grouping pixels with the same intensity level to form a "region" without paying attention to connectivity would yield a segmentation result that is meaningless in the context of this discussion.

Another problem in region growing is the formulation of a stopping rule. Region growth should stop when no more pixels satisfy the criteria for inclusion in that region. Criteria such as intensity values, texture, and color are local in nature and do not take into account the "history" of region growth. Additional criteria that increase the power of a region-growing algorithm utilize the concept of size, likeness between a candidate pixel and the pixels grown so far (such as a comparison of the intensity of a candidate and the average intensity of the grown region), and the shape of the region being grown. The use of these types of descriptors is based on the assumption that a model of expected results is at least partially available.

Let: $f(x, y)$ denote an input image array; $S(x, y)$ denote a *seed* array containing 1s at the locations of seed points and 0s elsewhere; and Q denote a predicate to be applied at each location (x, y). Arrays f and S are assumed to be of the same size. A basic region-growing algorithm based on 8-connectivity may be stated as follows.

See Sections 2.5.2 and 9.5.3 regarding connected components, and Section 9.2.1 regarding erosion.

1. Find all connected components in $S(x, y)$ and erode each connected component to one pixel; label all such pixels found as 1. All other pixels in S are labeled 0.

2. Form an image f_Q such that, at a pair of coordinates (x, y), let $f_Q(x, y) = 1$ if the input image satisfies the given predicate, Q, at those coordinates; otherwise, let $f_Q(x, y) = 0$.

3. Let g be an image formed by appending to each seed point in S all the 1-valued points in f_Q that are 8-connected to that seed point.

4. Label each connected component in g with a different region label (e.g., $1, 2, 3, \ldots$). This is the segmented image obtained by region growing.

We illustrate the mechanics of this algorithm by an example.

EXAMPLE 10.23:
Segmentation by region growing.

■ Figure 10.51(a) shows an 8-bit X-ray image of a weld (the horizontal dark region) containing several cracks and porosities (the bright regions running horizontally through the center of the image). We illustrate the use of region growing by segmenting the defective weld regions. These regions could be used in applications such as weld inspection, for inclusion in a database of historical studies, or for controlling an automated welding system.

The first order of business is to determine the seed points. From the physics of the problem, we know that cracks and porosities will attenuate X-rays considerably less than solid welds, so we expect the regions containing these types of defects to be significantly brighter than other parts of the X-ray image. We can extract the seed points by thresholding the original image, using a threshold set at a high percentile. Figure 10.51(b) shows the histogram of the image

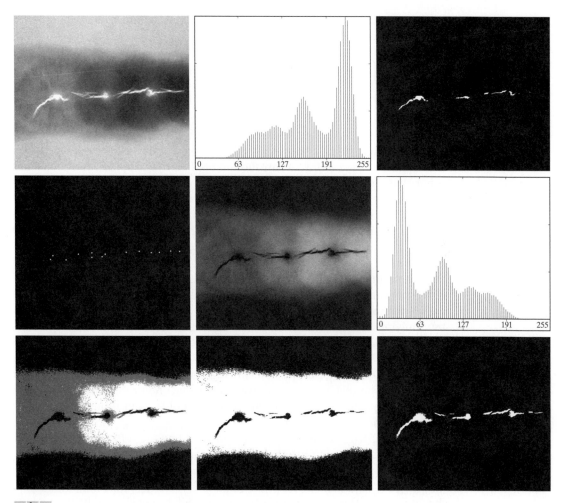

a b c
d e f
g h i

FIGURE 10.51 (a) X-ray image of a defective weld. (b) Histogram. (c) Initial seed image. (d) Final seed image (the points were enlarged for clarity). (e) Absolute value of the difference between (a) and (c). (f) Histogram of (e). (g) Difference image thresholded using dual thresholds. (h) Difference image thresholded with the smallest of the dual thresholds. (i) Segmentation result obtained by region growing. (Original image courtesy of X-TEK Systems, Ltd.)

and Fig. 10.51(c) shows the thresholded result obtained with a threshold equal to the 99.9 percentile of intensity values in the image, which in this case was 254 (see Section 10.3.5 regarding percentiles). Figure 10.51(d) shows the result of morphologically eroding each connected component in Fig. 10.51(c) to a single point.

Next, we have to specify a predicate. In this example, we are interested in appending to each seed all the pixels that (a) are 8-connected to that seed and

(b) are "similar" to it. Using intensity differences as a measure of similarity, our predicate applied at each location (x, y) is

$$Q = \begin{cases} \text{TRUE} & \text{if the absolute difference of the intensities} \\ & \text{between the seed and the pixel at } (x, y) \text{ is } \leq T \\ \text{FALSE} & \text{otherwise} \end{cases}$$

where T is a specified threshold. Although this predicate is based on intensity differences and uses a single threshold, we could specify more complex schemes in which a different threshold is applied to each pixel, and properties other than differences are used. In this case, the preceding predicate is sufficient to solve the problem, as the rest of this example shows.

From the previous paragraph, we know that the smallest seed value is 255 because the image was thresholded with a threshold of 254. Figure 10.51(e) shows the absolute value of the difference between the images in Figs. 10.51(a) and (c). The image in Fig. 10.51(e) contains all the differences needed to compute the predicate at each location (x, y). Figure 10.51(f) shows the corresponding histogram. We need a threshold to use in the predicate to establish similarity. The histogram has three principal modes, so we can start by applying to the difference image the dual thresholding technique discussed in Section 10.3.6. The resulting two thresholds in this case were $T_1 = 68$ and $T_2 = 126$, which we see correspond closely to the valleys of the histogram. (As a brief digression, we segmented the image using these two thresholds. The result in Fig. 10.51(g) shows that the problem of segmenting the defects cannot be solved using dual thresholds, even though the thresholds are in the main valleys.)

Figure 10.51(h) shows the result of thresholding the difference image with only T_1. The black points are the pixels for which the predicate was TRUE; the others failed the predicate. The important result here is that the points in the good regions of the weld failed the predicate, so they will not be included in the final result. The points in the outer region will be considered by the region-growing algorithm as candidates. However, Step 3 will reject the outer points, because they are not 8-connected to the seeds. In fact, as Fig. 10.51(i) shows, this step resulted in the correct segmentation, indicating that the use of connectivity was a fundamental requirement in this case. Finally, note that in Step 4 we used the same value for all the regions found by the algorithm. In this case, it was visually preferable to do so.   ■

10.4.2 Region Splitting and Merging

The procedure discussed in the last section grows regions from a set of seed points. An alternative is to subdivide an image initially into a set of arbitrary, disjoint regions and then merge and/or split the regions in an attempt to satisfy the conditions of segmentation stated in Section 10.1. The basics of splitting and merging are discussed next.

Let R represent the entire image region and select a predicate Q. One approach for segmenting R is to subdivide it successively into smaller and smaller quadrant regions so that, for any region R_i, $Q(R_i) = $ TRUE. We start with the entire region. If $Q(R) = $ FALSE, we divide the image into quadrants. If Q is FALSE for any quadrant, we subdivide that quadrant into subquadrants, and so on. This particular splitting technique has a convenient representation in the form of so-called *quadtrees*, that is, trees in which each node has exactly four descendants, as Fig. 10.52 shows (the images corresponding to the nodes of a quadtree sometimes are called *quadregions* or *quadimages*). Note that the root of the tree corresponds to the entire image and that each node corresponds to the subdivision of a node into four descendant nodes. In this case, only R_4 was subdivided further.

If only splitting is used, the final partition normally contains adjacent regions with identical properties. This drawback can be remedied by allowing merging as well as splitting. Satisfying the constraints of segmentation outlined in Section 10.1 requires merging only adjacent regions whose combined pixels satisfy the predicate Q. That is, two adjacent regions R_j and R_k are merged only if $Q(R_j \cup R_k) = $ TRUE.

See Section 2.5.2 regarding region adjacency.

The preceding discussion can be summarized by the following procedure in which, at any step, we

1. Split into four disjoint quadrants any region R_i for which $Q(R_i) = $ FALSE.
2. When no further splitting is possible, merge any adjacent regions R_j and R_k for which $Q(R_j \cup R_k) = $ TRUE.
3. Stop when no further merging is possible.

It is customary to specify a minimum quadregion size beyond which no further splitting is carried out.

Numerous variations of the preceding basic theme are possible. For example, a significant simplification results if in Step 2 we allow merging of any two adjacent regions R_i and R_j if each one satisfies the predicate individually. This results in a much simpler (and faster) algorithm, because testing of the predicate is limited to individual quadregions. As the following example shows, this simplification is still capable of yielding good segmentation results.

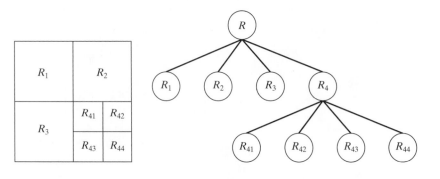

a b

FIGURE 10.52
(a) Partitioned image.
(b) Corresponding quadtree. R represents the entire image region.

EXAMPLE 10.24:
Segmentation by
region splitting
and merging.

■ Figure 10.53(a) shows a 566 × 566 X-ray band image of the Cygnus Loop. The objective of this example is to segment out of the image the "ring" of less dense matter surrounding the dense center. The region of interest has some obvious characteristics that should help in its segmentation. First, we note that the data in this region has a random nature, indicating that its standard deviation should be greater than the standard deviation of the background (which is near 0) and of the large central region, which is fairly smooth. Similarly, the mean value (average intensity) of a region containing data from the outer ring should be greater than the mean of the darker background and less than the mean of the large, lighter central region. Thus, we should be able to segment the region of interest using the following predicate:

$$Q = \begin{cases} \text{TRUE} & \text{if } \sigma > a \quad \text{AND} \quad 0 < m < b \\ \text{FALSE} & \text{otherwise} \end{cases}$$

where m and σ are the mean and standard deviation of the pixels in a quadregion, and a and b are constants.

Analysis of several regions in the outer area of interest revealed that the mean intensity of pixels in those regions did not exceed 125 and the standard deviation was always greater than 10. Figures 10.53(b) through (d) show the results obtained using these values for a and b, and varying the minimum size allowed for the quadregions from 32 to 8. The pixels in a quadregion whose

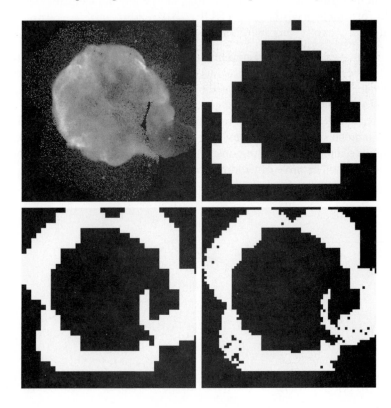

a b
c d

FIGURE 10.53
(a) Image of the
Cygnus Loop
supernova, taken
in the X-ray band
by NASA's
Hubble Telescope.
(b)–(d) Results of
limiting the
smallest allowed
quadregion to
sizes of
32 × 32, 16 × 16,
and 8 × 8 pixels,
respectively.
(Original image
courtesy of
NASA.)

pixels satisfied the predicate were set to white; all others in that region were set to black. The best result in terms of capturing the shape of the outer region was obtained using quadregions of size 16×16. The black squares in Fig. 10.53(d) are quadregions of size 8×8 whose pixels did not satisfied the predicate. Using smaller quadregions would result in increasing numbers of such black regions. Using regions larger than the one illustrated here results in a more "block-like" segmentation. Note that in all cases the segmented regions (white pixels) completely separate the inner, smoother region from the background. Thus, the segmentation effectively partitioned the image into three distinct areas that correspond to the three principal features in the image: background, dense, and sparse regions. Using any of the white regions in Fig. 10.53 as a mask would make it a relatively simple task to extract these regions from the original image (Problem 10.40). As in Example 10.23, these results could not have been obtained using edge- or threshold-based segmentation.   ■

As used in the preceding example, properties based on the mean and standard deviation of pixel intensities in a region attempt to quantify the *texture* of the region (see Section 11.3.3 for a discussion on texture). The concept of *texture segmentation* is based on using measures of texture in the predicates. In other words, we can perform texture segmentation by any of the methods discussed in this section simply by specifying predicates based on texture content.

10.5 Segmentation Using Morphological Watersheds

Thus far, we have discussed segmentation based on three principal concepts: (a) edge detection, (b) thresholding, and (c) region growing. Each of these approaches was found to have advantages (for example, speed in the case of global thresholding) and disadvantages (for example, the need for post-processing, such as edge linking, in edge-based segmentation). In this section we discuss an approach based on the concept of so-called *morphological watersheds*. As will become evident in the following discussion, segmentation by watersheds embodies many of the concepts of the other three approaches and, as such, often produces more stable segmentation results, including connected segmentation boundaries. This approach also provides a simple framework for incorporating knowledge-based constraints (see Fig. 1.23) in the segmentation process.

10.5.1 Background

The concept of watersheds is based on visualizing an image in three dimensions: two spatial coordinates versus intensity, as in Fig. 2.18(a). In such a "topographic" interpretation, we consider three types of points: (a) points belonging to a regional minimum; (b) points at which a drop of water, if placed at the location of any of those points, would fall with certainty to a single minimum; and (c) points at which water would be equally likely to fall to more than one such minimum. For a particular regional minimum, the set of points satisfying condition (b) is called the *catchment basin* or *watershed* of that minimum. The points satisfying condition (c) form crest lines on the topographic surface and are termed *divide lines* or *watershed lines*.

The principal objective of segmentation algorithms based on these concepts is to find the watershed lines. The basic idea is simple, as the following analogy illustrates. Suppose that a hole is punched in each regional minimum and that the entire topography is flooded from below by letting water rise through the holes at a uniform rate. When the rising water in distinct catchment basins is about to merge, a dam is built to prevent the merging. The flooding will eventually reach a stage when only the tops of the dams are visible above the water line. These dam boundaries correspond to the divide lines of the watersheds. Therefore, they are the (connected) boundaries extracted by a watershed segmentation algorithm.

These ideas can be explained further with the aid of Fig. 10.54. Figure 10.54(a) shows a gray-scale image and Fig. 10.54(b) is a topographic view, in which the height of the "mountains" is proportional to intensity values in the input image. For ease of interpretation, the backsides of structures are shaded. This is not to be confused with intensity values; only the general topography of the three-dimensional representation is of interest. In order to prevent the rising water from spilling out through the edges of the image, we imagine the

a b
c d

FIGURE 10.54
(a) Original image.
(b) Topographic
view. (c)–(d) Two
stages of flooding.

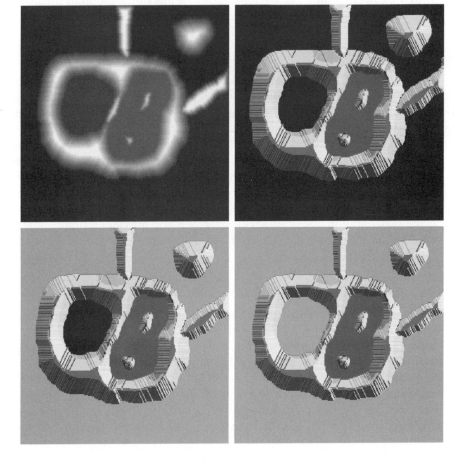

perimeter of the entire topography (image) being enclosed by dams of height greater than the highest possible mountain, whose value is determined by the highest possible intensity value in the input image.

Suppose that a hole is punched in each regional minimum [shown as dark areas in Fig. 10.54(b)] and that the entire topography is flooded from below by letting water rise through the holes at a uniform rate. Figure 10.54(c) shows the first stage of flooding, where the "water," shown in light gray, has covered only areas that correspond to the very dark background in the image. In Figs. 10.54(d) and (e) we see that the water now has risen into the first and second catchment basins, respectively. As the water continues to rise, it will eventually overflow from one catchment basin into another. The first indication of this is shown in 10.54(f). Here, water from the left basin actually overflowed into the basin on the right and a short "dam" (consisting of single pixels) was built to prevent water from merging at that level of flooding (the details of dam building are discussed in the following section). The effect is more pronounced as water continues to rise, as shown in Fig. 10.54(g). This figure shows a longer dam between the two catchment basins and another dam in the top part of the right basin. The latter dam was built to prevent merging of water from that basin with water from areas corresponding to the background. This process is continued until the maximum

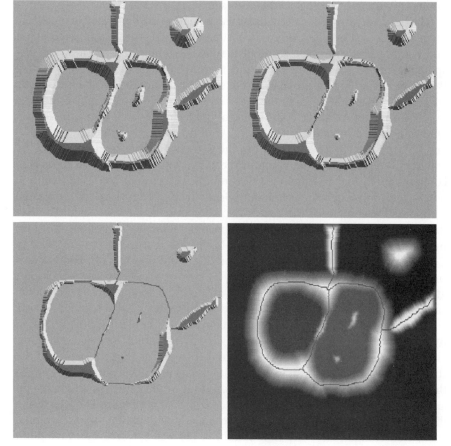

e f
g h

FIGURE 10.54
(Continued)
(e) Result of further flooding.
(f) Beginning of merging of water from two catchment basins (a short dam was built between them). (g) Longer dams. (h) Final watershed (segmentation) lines.
(Courtesy of Dr. S. Beucher, CMM/Ecole des Mines de Paris.)

level of flooding (corresponding to the highest intensity value in the image) is reached. The final dams correspond to the watershed lines, which are the desired segmentation result. The result for this example is shown in Fig. 10.54(h) as dark, 1-pixel-thick paths superimposed on the original image. Note the important property that the watershed lines form connected paths, thus giving continuous boundaries between regions.

One of the principal applications of watershed segmentation is in the extraction of nearly uniform (bloblike) objects from the background. Regions characterized by small variations in intensity have small gradient values. Thus, in practice, we often see watershed segmentation applied to the gradient of an image, rather than to the image itself. In this formulation, the regional minima of catchment basins correlate nicely with the small value of the gradient corresponding to the objects of interest.

10.5.2 Dam Construction

Before proceeding, let us consider how to construct the dams or watershed lines required by watershed segmentation algorithms. Dam construction is based on binary images, which are members of 2-D integer space Z^2 (see Section 2.4.2). The simplest way to construct dams separating sets of binary points is to use morphological dilation (see Section 9.2.2).

The basics of how to construct dams using dilation are illustrated in Fig. 10.55. Figure 10.55(a) shows portions of two catchment basins at flooding step $n - 1$ and Fig. 10.55(b) shows the result at the next flooding step, n. The water has spilled from one basin to the other and, therefore, a dam must be built to keep this from happening. In order to be consistent with notation to be introduced shortly, let M_1 and M_2 denote the sets of coordinates of points in two regional minima. Then let the set of coordinates of points in the *catchment basin* associated with these two minima at stage $n - 1$ of flooding be denoted by $C_{n-1}(M_1)$ and $C_{n-1}(M_2)$, respectively. These are the two gray regions in Fig. 10.55(a).

Let $C[n - 1]$ denote the union of these two sets. There are two connected components in Fig. 10.55(a) (see Section 2.5.2 regarding connected components) and only one connected component in Fig. 10.55(b). This connected component encompasses the earlier two components, shown dashed. The fact that two connected components have become a *single* component indicates that water between the two catchment basins has merged at flooding step n. Let this connected component be denoted q. Note that the two components from step $n - 1$ can be extracted from q by performing the simple AND operation $q \cap C[n - 1]$. We note also that all points belonging to an individual catchment basin form a single connected component.

Suppose that each of the connected components in Fig. 10.55(a) is dilated by the structuring element shown in Fig. 10.55(c), subject to two conditions: (1) The dilation has to be constrained to q (this means that the center of the structuring element can be located only at points in q during dilation), and (2) the dilation cannot be performed on points that would cause the sets being dilated to merge (become a single connected component). Figure 10.55(d) shows that a first dilation pass (in light gray) expanded the boundary of each original connected component. Note that condition (1) was satisfied by every point

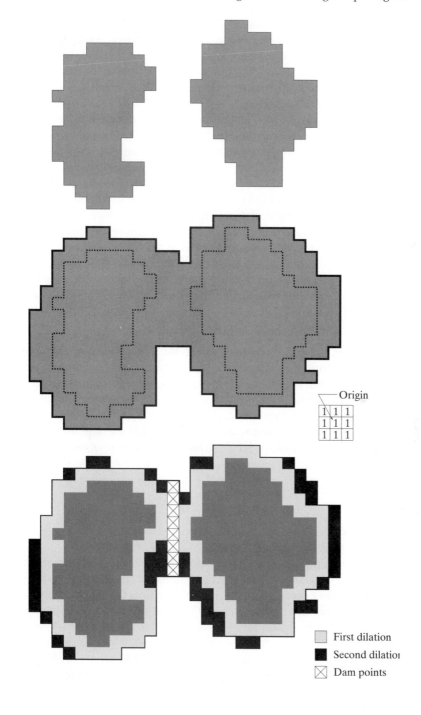

Origin

1	1	1
1	1	1
1	1	1

First dilation

Second dilation

Dam points

a
b
d c

FIGURE 10.55 (a) Two partially flooded catchment basins at stage $n - 1$ of flooding.
(b) Flooding at stage n, showing that water has spilled between basins. (c) Structuring
element used for dilation. (d) Result of dilation and dam construction.

during dilation, and condition (2) did not apply to any point during the dilation process; thus the boundary of each region was expanded uniformly.

In the second dilation (shown in black), several points failed condition (1) while meeting condition (2), resulting in the broken perimeter shown in the figure. It also is evident that the only points in q that satisfy the two conditions under consideration describe the 1-pixel-thick connected path shown crosshatched in Fig. 10.55(d). This path constitutes the desired separating dam at stage n of flooding. Construction of the dam at this level of flooding is completed by setting all the points in the path just determined to a value greater than the maximum intensity value of the image. The height of all dams is generally set at 1 plus the maximum allowed value in the image. This will prevent water from crossing over the part of the completed dam as the level of flooding is increased. It is important to note that dams built by this procedure, which are the desired segmentation boundaries, are connected components. In other words, this method eliminates the problems of broken segmentation lines.

Although the procedure just described is based on a simple example, the method used for more complex situations is exactly the same, including the use of the 3×3 symmetric structuring element shown in Fig. 10.55(c).

10.5.3 Watershed Segmentation Algorithm

Let $M_1, M_2, \ldots, M_R$ be sets denoting the *coordinates* of the points in the regional minima of an image $g(x, y)$. As indicated at the end of Section 10.5.1, this typically will be a gradient image. Let $C(M_i)$ be a set denoting the coordinates of the points in the catchment basin associated with regional minimum M_i (recall that the points in any catchment basin form a connected component). The notation min and max will be used to denote the minimum and maximum values of $g(x, y)$. Finally, let $T[n]$ represent the set of coordinates (s, t) for which $g(s, t) < n$. That is,

$$T[n] = \{(s, t) \mid g(s, t) < n\} \tag{10.5-1}$$

Geometrically, $T[n]$ is the set of coordinates of points in $g(x, y)$ lying below the plane $g(x, y) = n$.

The topography will be flooded in *integer* flood increments, from $n = \text{min} + 1$ to $n = \text{max} + 1$. At any step n of the flooding process, the algorithm needs to know the number of points below the flood depth. Conceptually, suppose that the coordinates in $T[n]$ that are below the plane $g(x, y) = n$ are "marked" black, and all other coordinates are marked white. Then when we look "down" on the xy-plane at any increment n of flooding, we will see a binary image in which black points correspond to points in the function that are below the plane $g(x, y) = n$. This interpretation is quite useful in helping clarify the following discussion.

Let $C_n(M_i)$ denote the set of coordinates of points in the catchment basin associated with minimum M_i that are flooded at stage n. With reference to the discussion in the previous paragraph, $C_n(M_i)$ may be viewed as a binary image given by

$$C_n(M_i) = C(M_i) \cap T[n] \tag{10.5-2}$$

In other words, $C_n(M_i) = 1$ at location (x, y) if $(x, y) \in C(M_i)$ AND $(x, y) \in T[n]$; otherwise $C_n(M_i) = 0$. The geometrical interpretation of this result is straightforward. We are simply using the AND operator to isolate at stage n of flooding the portion of the binary image in $T[n]$ that is associated with regional minimum M_i.

Next, we let $C[n]$ denote the union of the flooded catchment basins at stage n:

$$C[n] = \bigcup_{i=1}^{R} C_n(M_i) \tag{10.5-3}$$

Then $C[\text{max} + 1]$ is the union of all catchment basins:

$$C[\text{max} + 1] = \bigcup_{i=1}^{R} C(M_i) \tag{10.5-4}$$

It can be shown (Problem 10.41) that the elements in both $C_n(M_i)$ and $T[n]$ are never replaced during execution of the algorithm, and that the number of elements in these two sets either increases or remains the same as n increases. Thus, it follows that $C[n - 1]$ is a subset of $C[n]$. According to Eqs. (10.5-2) and (10.5-3), $C[n]$ is a subset of $T[n]$, so it follows that $C[n - 1]$ is a subset of $T[n]$. From this we have the important result that each connected component of $C[n - 1]$ is contained in exactly one connected component of $T[n]$.

The algorithm for finding the watershed lines is initialized with $C[\text{min} + 1] = T[\text{min} + 1]$. The algorithm then proceeds recursively, computing $C[n]$ from $C[n - 1]$. A procedure for obtaining $C[n]$ from $C[n - 1]$ is as follows. Let Q denote the set of connected components in $T[n]$. Then, for each connected component $q \in Q[n]$, there are three possibilities:

1. $q \cap C[n - 1]$ is empty.
2. $q \cap C[n - 1]$ contains one connected component of $C[n - 1]$.
3. $q \cap C[n - 1]$ contains more than one connected component of $C[n - 1]$.

Construction of $C[n]$ from $C[n - 1]$ depends on which of these three conditions holds. Condition 1 occurs when a new minimum is encountered, in which case connected component q is incorporated into $C[n - 1]$ to form $C[n]$. Condition 2 occurs when q lies within the catchment basin of some regional minimum, in which case q is incorporated into $C[n - 1]$ to form $C[n]$. Condition 3 occurs when all, or part, of a ridge separating two or more catchment basins is encountered. Further flooding would cause the water level in these catchment basins to merge. Thus a dam (or dams if more than two catchment basins are involved) must be built within q to prevent overflow between the catchment basins. As explained in the previous section, a one-pixel-thick dam can be constructed when needed by dilating $q \cap C[n - 1]$ with a 3×3 structuring element of 1s, and constraining the dilation to q.

Algorithm efficiency is improved by using only values of n that correspond to existing intensity values in $g(x, y)$; we can determine these values, as well as the values of min and max, from the histogram of $g(x, y)$.

a b
c d

FIGURE 10.56
(a) Image of blobs.
(b) Image gradient.
(c) Watershed lines.
(d) Watershed lines
superimposed on
original image.
(Courtesy of Dr.
S. Beucher,
CMM/Ecole des
Mines de Paris.)

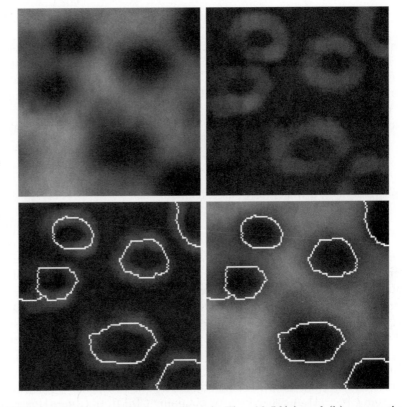

EXAMPLE 10.25:
Illustration of the
watershed
segmentation
algorithm.

■ Consider the image and its gradient in Figs. 10.56(a) and (b), respectively. Application of the watershed algorithm just described yielded the watershed lines (white paths) of the gradient image in Fig. 10.56(c). These segmentation boundaries are shown superimposed on the original image in Fig. 10.56(d). As noted at the beginning of this section, the segmentation boundaries have the important property of being connected paths. ■

10.5.4 The Use of Markers

Direct application of the watershed segmentation algorithm in the form discussed in the previous section generally leads to *oversegmentation* due to noise and other local irregularities of the gradient. As Fig. 10.57 shows, oversegmentation can be serious enough to render the result of the algorithm virtually useless. In this case, this means a large number of segmented regions. A practical solution to this problem is to limit the number of allowable regions by incorporating a preprocessing stage designed to bring additional knowledge into the segmentation procedure.

An approach used to control oversegmentation is based on the concept of markers. A *marker* is a connected component belonging to an image. We have *internal* markers, associated with objects of interest, and *external* markers, associated with the background. A procedure for marker selection typically will consist of two principal steps: (1) preprocessing; and (2) definition of a set of criteria that markers must satisfy. To illustrate, consider Fig. 10.57(a) again.

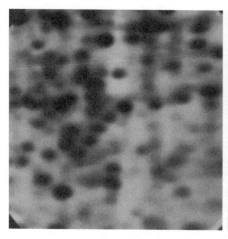
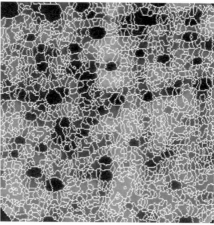

a b
FIGURE 10.57
(a) Electrophoresis image. (b) Result of applying the watershed segmentation algorithm to the gradient image. Oversegmentation is evident. (Courtesy of Dr. S. Beucher, CMM/Ecole des Mines de Paris.)

Part of the problem that led to the oversegmented result in Fig. 10.57(b) is the large number of potential minima. Because of their size, many of these minima are irrelevant detail. As has been pointed out several times in earlier discussions, an effective method for minimizing the effect of small spatial detail is to filter the image with a smoothing filter. This is an appropriate preprocessing scheme in this particular case.

Suppose that we define an *internal marker* as (1) a region that is surrounded by points of higher "altitude"; (2) such that the points in the region form a connected component; and (3) in which all the points in the connected component have the same intensity value. After the image was smoothed, the internal markers resulting from this definition are shown as light gray, bloblike regions in Fig. 10.58(a). Next, the watershed algorithm was applied to the

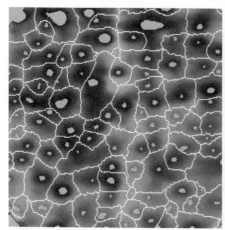
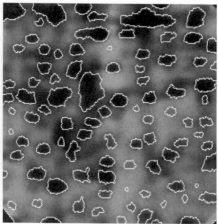

a b

FIGURE 10.58 (a) Image showing internal markers (light gray regions) and external markers (watershed lines). (b) Result of segmentation. Note the improvement over Fig. 10.47(b). (Courtesy of Dr. S. Beucher, CMM/Ecole des Mines de Paris.)

smoothed image, under the restriction that these internal markers be the *only* allowed regional minima. Figure 10.58(a) shows the resulting watershed lines. These watershed lines are defined as the *external markers*. Note that the points along the watershed line pass along the highest points between neighboring markers.

The external markers in Fig. 10.58(a) effectively partition the image into regions, with each region containing a single internal marker and part of the background. The problem is thus reduced to partitioning each of these regions into two: a single object and its background. We can bring to bear on this simplified problem many of the segmentation techniques discussed earlier in this chapter. Another approach is simply to apply the watershed segmentation algorithm to each individual region. In other words, we simply take the gradient of the smoothed image [as in Fig. 10.56(b)] and then restrict the algorithm to operate on a single watershed that contains the marker in that particular region. The result obtained using this approach is shown in 10.58(b). The improvement over the image in 10.57(b) is evident.

Marker selection can range from simple procedures based on intensity values and connectivity, as was just illustrated, to more complex descriptions involving size, shape, location, relative distances, texture content, and so on (see Chapter 11 regarding descriptors). The point is that using markers brings a priori knowledge to bear on the segmentation problem. The reader is reminded that humans often aid segmentation and higher-level tasks in everyday vision by using a priori knowledge, one of the most familiar being the use of context. Thus, the fact that segmentation by watersheds offers a framework that can make effective use of this type of knowledge is a significant advantage of this method.

10.6 The Use of Motion in Segmentation

Motion is a powerful cue used by humans and many other animals to extract objects or regions of interest from a background of irrelevant detail. In imaging applications, motion arises from a relative displacement between the sensing system and the scene being viewed, such as in robotic applications, autonomous navigation, and dynamic scene analysis. In the following sections we consider the use of motion in segmentation both spatially and in the frequency domain.

10.6.1 Spatial Techniques

Basic approach

One of the simplest approaches for detecting changes between two image frames $f(x, y, t_i)$ and $f(x, y, t_j)$ taken at times t_i and t_j, respectively, is to compare the two images pixel by pixel. One procedure for doing this is to form a difference image. Suppose that we have a reference image containing only stationary components. Comparing this image against a subsequent image of the same scene, but including a moving object, results in the difference of the two images canceling the stationary elements, leaving only nonzero entries that correspond to the nonstationary image components.

A difference image between two images taken at times t_i and t_j may be defined as

$$d_{ij}(x, y) = \begin{cases} 1 & \text{if } |f(x, y, t_i) - f(x, y, t_j)| > T \\ 0 & \text{otherwise} \end{cases} \quad (10.6\text{-}1)$$

where T is a specified threshold. Note that $d_{ij}(x, y)$ has a value of 1 at spatial coordinates (x, y) only if the intensity difference between the two images is *appreciably different* at those coordinates, as determined by the specified threshold T. It is assumed that all images are of the same size. Finally, we note that the values of the coordinates (x, y) in Eq. (10.6-1) span the dimensions of these images, so that the difference image $d_{ij}(x, y)$ is of the same size as the images in the sequence.

In dynamic image processing, all pixels in $d_{ij}(x, y)$ with value 1 are considered the result of object motion. This approach is applicable only if the two images are registered spatially and if the illumination is relatively constant within the bounds established by T. In practice, 1-valued entries in $d_{ij}(x, y)$ may arise as a result of noise. Typically, these entries are isolated points in the difference image, and a simple approach to their removal is to form 4- or 8-connected regions of 1s in $d_{ij}(x, y)$ and then ignore any region that has less than a predetermined number of elements. Although it may result in ignoring small and/or slow-moving objects, this approach improves the chances that the remaining entries in the difference image actually are the result of motion.

Accumulative differences

Consider a sequence of image frames $f(x, y, t_1), f(x, y, t_2), \ldots, f(x, y, t_n)$ and let $f(x, y, t_1)$ be the *reference image*. An *accumulative difference image* (ADI) is formed by comparing this reference image with every subsequent image in the sequence. A counter for each pixel location in the accumulative image is incremented every time a difference occurs at that pixel location between the reference and an image in the sequence. Thus when the kth frame is being compared with the reference, the entry in a given pixel of the accumulative image gives the number of times the intensity at that position was different [as determined by T in Eq. (10.6-1)] from the corresponding pixel value in the reference image.

Consider the following three types of accumulative difference images: *absolute*, *positive*, and *negative* ADIs. Assuming that the intensity values of the moving objects are larger than the background, these three types of ADIs are defined as follows. Let $R(x, y)$ denote the reference image and, to simplify the notation, let k denote t_k, so that $f(x, y, k) = f(x, y, t_k)$. We assume that $R(x, y) = f(x, y, 1)$. Then, for any $k > 1$, and keeping in mind that the values of the ADIs are *counts*, we define the following for all relevant values of (x, y):

$$A_k(x, y) = \begin{cases} A_{k-1}(x, y) + 1 & \text{if } |R(x, y) - f(x, y, k)| > T \\ A_{k-1}(x, y) & \text{otherwise} \end{cases} \quad (10.6\text{-}2)$$

$$P_k(x, y) = \begin{cases} P_{k-1}(x, y) + 1 & \text{if } \big[R(x, y) - f(x, y, k)\big] > T \\ P_{k-1}(x, y) & \text{otherwise} \end{cases} \qquad (10.6\text{-}3)$$

and

$$N_k(x, y) = \begin{cases} N_{k-1}(x, y) + 1 & \text{if } \big[R(x, y) - f(x, y, k)\big] < -T \\ N_{k-1}(x, y) & \text{otherwise} \end{cases} \qquad (10.6\text{-}4)$$

where $A_k(x, t)$, $P_k(x, y)$, and $N_k(x, y)$ are the absolute, positive, and negative ADIs, respectively, after the kth image in the sequence is encountered.

It is understood that these ADIs start out with all zero values (counts). Note also that the ADIs are of the same size as the images in the sequence. Finally, we note that the order of the inequalities and signs of the thresholds in Eqs. (10.6-3) and (10.6-4) are reversed if the intensity values of the background pixels are greater than the values of the moving objects.

EXAMPLE 10.26:
Computation of the absolute, positive, and negative accumulative difference images.

■ Figure 10.59 shows the three ADIs displayed as intensity images for a rectangular object of dimension 75×50 pixels that is moving in a southeasterly direction at a speed of $5\sqrt{2}$ pixels per frame. The images are of size 256×256 pixels. We note the following: (1) The nonzero area of the positive ADI is equal to the size of the moving object. (2) The location of the positive ADI corresponds to the location of the moving object in the reference frame. (3) The number of counts in the positive ADI stops increasing when the moving object is displaced completely with respect to the same object in the reference frame. (4) The absolute ADI contains the regions of the positive and negative ADI. (5) The direction and speed of the moving object can be determined from the entries in the absolute and negative ADIs.   ■

Establishing a reference image

A key to the success of the techniques discussed in the preceding two sections is having a reference image against which subsequent comparisons can be

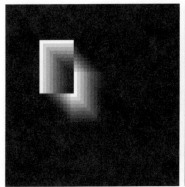

a b c

FIGURE 10.59 ADIs of a rectangular object moving in a southeasterly direction. (a) Absolute ADI. (b) Positive ADI. (c) Negative ADI.

made. The difference between two images in a dynamic imaging problem has the tendency to cancel all stationary components, leaving only image elements that correspond to noise and to the moving objects.

In practice, obtaining a reference image with only stationary elements is not always possible, and building a reference from a set of images containing one or more moving objects becomes necessary. This applies particularly to situations describing busy scenes or in cases where frequent updating is required. One procedure for generating a reference image is as follows. Consider the first image in a sequence to be the reference image. When a nonstationary component has moved completely out of its position in the reference frame, the corresponding background in the present frame can be duplicated in the location originally occupied by the object in the reference frame. When all moving objects have moved completely out of their original positions, a reference image containing only stationary components will have been created. Object displacement can be established by monitoring the changes in the positive ADI, as indicated in the preceding section.

■ Figures 10.60(a) and (b) show two image frames of a traffic intersection. The first image is considered the reference, and the second depicts the same scene some time later. The objective is to remove the principal moving objects in the reference image in order to create a static image. Although there are other smaller moving objects, the principal moving feature is the automobile at the intersection moving from left to right. For illustrative purposes we focus on this object. By monitoring the changes in the positive ADI, it is possible to determine the initial position of a moving object, as explained previously. Once the area occupied by this object is identified, the object can be removed from the image by subtraction. By looking at the frame in the sequence at which the positive ADI stopped changing, we can copy from this image the area previously occupied by the moving object in the initial frame. This area then is pasted onto the image from which the object was cut out, thus restoring the background of that area. If this is done for all moving objects, the result is a reference image with only static components against which we can compare subsequent frames for motion detection. The result of removing the east-bound moving vehicle in this case is shown in Fig. 10.60(c). ■

EXAMPLE 10.27:
Building a reference image.

a b c

FIGURE 10.60 Building a static reference image. (a) and (b) Two frames in a sequence. (c) Eastbound automobile subtracted from (a) and the background restored from the corresponding area in (b). (Jain and Jain.)

10.6.2 Frequency Domain Techniques

In this section we consider the problem of determining motion via a Fourier transform formulation. Consider a sequence $f(x, y, t)$, $t = 0, 1, \ldots, K - 1$, of K digital image frames of size $M \times N$ generated by a stationary camera. We begin the development by assuming that all frames have a homogeneous background of zero intensity. The exception is a single, 1-pixel object of unit intensity that is moving with constant velocity. Suppose that for frame one ($t = 0$), the object is at location (x', y') and that the image plane is *projected* onto the x-axis; that is, the pixel intensities are summed across the columns in the image. This operation yields a 1-D array with M entries that are zero, except at x', which is the x-coordinate of the single-point object. If we now multiply all the components of the 1-D array by the quantity $\exp[j2\pi a_1 x \Delta t]$ for $x = 0, 1, 2, \ldots,$ $M - 1$ and sum the results, we obtain the single term $\exp[j2\pi a_1 x' \Delta t]$. In this notation, a_1 is a positive integer, and Δt is the time interval between frames.

Suppose that in frame two ($t = 1$) the object has moved to coordinates $(x' + 1, y')$; that is, it has moved 1 pixel parallel to the x-axis. Then repeating the projection procedure discussed in the previous paragraph yields the sum $\exp[j2\pi a_1(x' + 1) \Delta t]$. If the object continues to move 1 pixel location per frame, then, at any integer instant of time, t, the result is $\exp[j2\pi a_1(x' + t) \Delta t]$, which, using Euler's formula, may be expressed as

$$e^{j2\pi a_1(x'+t) \Delta t} = \cos[2\pi a_1(x' + t) \Delta t] + j \sin[2\pi a_1(x' + t) \Delta t] \quad (10.6\text{-}5)$$

for $t = 0, 1, \ldots, K - 1$. In other words, this procedure yields a complex sinusoid with frequency a_1. If the object were moving V_1 pixels (in the x-direction) between frames, the sinusoid would have frequency $V_1 a_1$. Because t varies between 0 and $K - 1$ in integer increments, restricting a_1 to integer values causes the discrete Fourier transform of the complex sinusoid to have two peaks—one located at frequency $V_1 a_1$ and the other at $K - V_1 a_1$. This latter peak is the result of symmetry in the discrete Fourier transform, as discussed in Section 4.6.4, and may be ignored. Thus a peak search in the Fourier spectrum yields $V_1 a_1$. Division of this quantity by a_1 yields V_1, which is the velocity component in the x-direction, as the frame rate is assumed to be known. A similar argument would yield V_2, the component of velocity in the y-direction.

A sequence of frames in which no motion takes place produces identical exponential terms, whose Fourier transform would consist of a single peak at a frequency of 0 (a single dc term). Therefore, because the operations discussed so far are linear, the general case involving one or more moving objects in an arbitrary static background would have a Fourier transform with a peak at dc corresponding to static image components and peaks at locations proportional to the velocities of the objects.

These concepts may be summarized as follows. For a sequence of K digital images of size $M \times N$, the sum of the weighted projections onto the x axis at any integer instant of time is

$$g_x(t, a_1) = \sum_{x=0}^{M-1} \sum_{y=0}^{N-1} f(x, y, t) e^{j2\pi a_1 x \Delta t} \quad t = 0, 1, \ldots, K - 1 \quad (10.6\text{-}6)$$

Similarly, the sum of the projections onto the y-axis is

$$g_y(t, a_2) = \sum_{y=0}^{N-1} \sum_{x=0}^{M-1} f(x, y, t)e^{j2\pi a_2 y \Delta t} \quad t = 0, 1, \ldots, K-1 \tag{10.6-7}$$

where, as noted already, a_1 and a_2 are positive integers.

The 1-D Fourier transforms of Eqs. (10.6-6) and (10.6-7), respectively, are

$$G_x(u_1, a_1) = \sum_{t=0}^{K-1} g_x(t, a_1)e^{-j2\pi u_1 t/K} \quad u_1 = 0, 1, \ldots, K-1 \tag{10.6-8}$$

and

$$G_y(u_2, a_2) = \sum_{t=0}^{K-1} g_y(t, a_2)e^{-j2\pi u_2 t/K} \quad u_2 = 0, 1, \ldots, K-1 \tag{10.6-9}$$

In practice, computation of these transforms is carried out using an FFT algorithm, as discussed in Section 4.11.

The frequency-velocity relationship is

$$u_1 = a_1 V_1 \tag{10.6-10}$$

and

$$u_2 = a_2 V_2 \tag{10.6-11}$$

In this formulation the unit of velocity is in pixels per total frame time. For example, $V_1 = 10$ is interpreted as a motion of 10 pixels in K frames. For frames that are taken uniformly, the actual physical speed depends on the frame rate and the distance between pixels. Thus if $V_1 = 10$, $K = 30$, the frame rate is two images per second, and the distance between pixels is 0.5 m, then the actual physical speed in the x-direction is

$$V_1 = (10 \text{ pixels})(0.5 \text{ m/pixel})(2 \text{ frames/s})/(30 \text{ frames})$$

$$= 1/3 \text{ m/s}$$

The sign of the x-component of the velocity is obtained by computing

$$S_{1x} = \left. \frac{d^2 \text{Re}[g_x(t, a_1)]}{dt^2} \right|_{t=n} \tag{10.6-12}$$

and

$$S_{2x} = \left. \frac{d^2 \text{Im}[g_x(t, a_1)]}{dt^2} \right|_{t=n} \tag{10.6-13}$$

Because g_x is sinusoidal, it can be shown (Problem 10.47) that S_{1x} and S_{2x} will have the same sign at an arbitrary point in time, n, if the velocity component V_1 is positive. Conversely, opposite signs in S_{1x} and S_{2x} indicate a negative component. If either S_{1x} or S_{2x} is zero, we consider the next closest point in time, $t = n \pm \Delta t$. Similar comments apply to computing the sign of V_2.

FIGURE 10.61
LANDSAT frame.
(Cowart, Snyder,
and Ruedger.)

EXAMPLE 10.28:
Detection of a
small moving
object via the
frequency
domain.

■ Figures 10.61 through 10.64 illustrate the effectiveness of the approach just derived. Figure 10.61 shows one of a 32-frame sequence of LANDSAT images generated by adding white noise to a reference image. The sequence contains a superimposed target moving at 0.5 pixel per frame in the x-direction and 1 pixel per frame in the y-direction. The target, shown circled in Fig. 10.62, has a Gaussian intensity distribution spread over a small (9-pixel) area and is not easily discernible by eye. Figures 10.63 and 10.64 show the results of computing Eqs. (10.6-8) and (10.6-9) with $a_1 = 6$ and $a_2 = 4$, respectively. The peak at $u_1 = 3$ in Fig. 10.63 yields $V_1 = 0.5$ from Eq. (10.6-10). Similarly, the peak at $u_2 = 4$ in Fig. 10.64 yields $V_2 = 1.0$ from Eq. (10.6-11). ■

Guidelines for the selection of a_1 and a_2 can be explained with the aid of Figs. 10.63 and 10.64. For instance, suppose that we had used $a_2 = 15$ instead of $a_2 = 4$. In that case the peaks in Fig. 10.64 would now be at $u_2 = 15$ and 17 because $V_2 = 1.0$, which would be a seriously aliased result. As discussed in Section 4.5.4, aliasing is caused by undersampling (too few frames in the present discussion, as the range of u is determined by K). Because $u = aV$, one possibility is to select

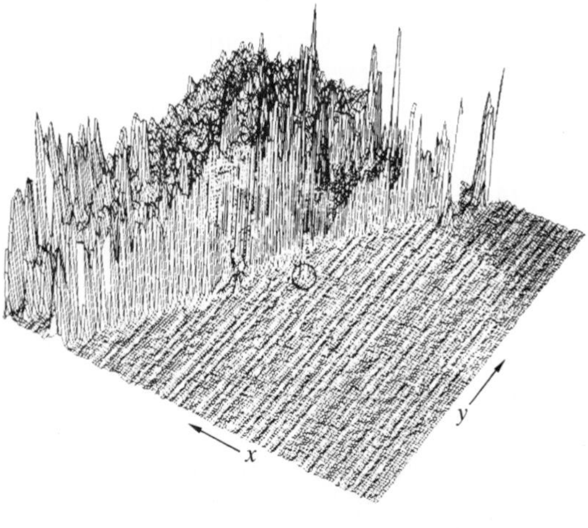

FIGURE 10.62
Intensity plot of
the image in Fig.
10.61, with the
target circled.
(Rajala, Riddle,
and Snyder.)

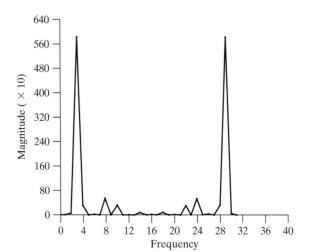

FIGURE 10.63
Spectrum of Eq.
(10.6-8) showing a
peak at $u_1 = 3$.
(Rajala, Riddle,
and Snyder.)

a as the integer closest to $a = u_{max}/V_{max}$, where u_{max} is the aliasing frequency limitation established by K and V_{max} is the maximum expected object velocity.

Summary

Image segmentation is an essential preliminary step in most automatic pictorial pattern recognition and scene analysis applications. As indicated by the range of examples presented in the previous sections, the choice of one segmentation technique over another is dictated mostly by the peculiar characteristics of the problem being considered. The methods discussed in this chapter, although far from exhaustive, are representative of techniques commonly used in practice. The following references can be used as the basis for further study of this topic.

References and Further Reading

Because of its central role in autonomous image processing, segmentation is a topic covered in most books dealing with image processing, image analysis, and computer vision. The following books provide complementary and/or supplementary reading for our coverage of this topic: Umbaugh [2005]; Davies [2005]; Gonzalez, Woods, and Eddins [2004]; Shapiro and Stockman [2001]; Sonka et al. [1999]; and Petrou and Bosdogianni [1999].

Work dealing with the use of masks to detect intensity discontinuities (Section 10.2) has a long history. Numerous masks have been proposed over the years: Roberts [1965], Prewitt [1970], Kirsh [1971], Robinson [1976], Frei and Chen [1977], and Canny [1986]. A review article by Fram and Deutsch [1975] contains numerous masks and an evaluation of

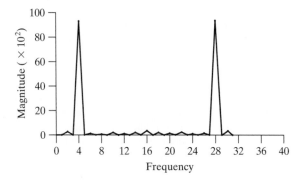

FIGURE 10.64
Spectrum of Eq.
(10.6-9) showing a
peak at $u_2 = 4$.
(Rajala, Riddle,
and Snyder.)

their performance. The issue of mask performance, especially for edge detection, still is an area of considerable interest, as exemplified by Qian and Huang [1996], Wang et al. [1996], Heath et al. [1997, 1998], and Ando [2000]. Edge detection on color images has been increasing in popularity for a number of multisensing applications. See, for example, Salinas, Abidi, and Gonzalez [1996]; Zugaj and Lattuati [1998]; Mirmehdi and Petrou [2000]; and Plataniotis and Venetsanopoulos [2000]. The interplay between image characteristics and mask performance also is a topic of current interest, as exemplified by Ziou [2001]. Our presentation of the zero-crossing properties of the Laplacian is based on a paper by Marr and Hildredth [1980] and on the book by Marr [1982]. See also a paper by Clark [1989] on authenticating edges produced by zero-crossing algorithms. (Corrections of parts of the Clark paper are given by Piech [1990].) As mentioned in Section 10.2, zero crossing via the Laplacian of a Gaussian is an important approach whose relative performance is still an active topic of research (Gunn [1998, 1999]). As the name implies, the Canny edge detector discussed in Section 10.2.6 is due to Canny [1986]. For an example of work on this topic twenty years later, see Zhang and Rockett [2006].

The Hough transform (Hough [1962]) is a practical method for global pixel linking and curve detection. Numerous generalizations to the basic transform discussed in this chapter have been proposed over the years. For example, Lo and Tsai [1995] discuss an approach for detecting thick lines, Guil et al. [1995, 1997] deal with fast implementations of the Hough transform and detection of primitive curves, Daul at al. [1998] discuss further generalizations for detecting elliptical arcs, and Shapiro [1996] deals with implementation of the Hough transform for gray-scale images.

As mentioned at the beginning of Section 10.3, thresholding techniques enjoy a significant degree of popularity because they are simple to implement. It is not surprising that there is a considerable body of work reported in the literature on this topic. A good appreciation of the extent of this literature can be gained from the review papers by Sahoo et al. [1988] and by Lee et al. [1990]. In addition to the techniques discussed in this chapter, other approaches used to deal with the effects of illumination and reflectance (Section 10.3.1) are illustrated by the work of Perez and Gonzalez [1987], Parker [1991], Murase and Nayar [1994], Bischsel [1998], Drew et al. [1999], and Toro and Funt [2007]. For additional reading on the material in Section 10.3.2, see Jain et al. [1995].

Early work on optimal global thresholding (Section 10.3.3) is exemplified in the classic paper by Chow and Kaneko [1972] (we discuss this method in Section 12.2.2 in the more general context of object recognition). Although it is optimal in theory, applications of this method in intensity thresholding are limited because of the need to estimate probability density functions. The optimum approach we developed in Section 10.3.3, due to Otsu [1979], has gained much more acceptance because it combines excellent performance with simplicity of implementation, requiring only estimation of image histograms. The basic idea of using preprocessing (Sections 10.3.4 and 10.3.5) dates back to an early paper by White and Rohrer [1983]), which combined thresholding, the gradient, and the Laplacian in the solution of a difficult segmentation problem. It is interesting to compare the fundamental similarities in terms of image segmentation capability between the methods discussed in the preceding three articles and work on thresholding done almost twenty years later by Cheriet et al. [1998], Sauvola and Pietikainen [2000]), Liang et al. [2000], and Chan et al. [2000]. For additional reading on multiple thresholding (Section 10.3.6), see Yin and Chen [1997], Liao et al. [2001], and Zahara et al. [2005]. For additional reading on variable thresholding (Section 10.3.7), see Parker [1997]. See also Delon et al. [2007].

See Fu and Mui [1981] for an early survey on the topic of region-oriented segmentation. The work of Haddon and Boyce [1990] and of Pavlidis and Liow [1990] are among the earliest efforts to integrate region and boundary information for the purpose of segmentation. A newer region-growing approach proposed by Hojjatoleslami and Kittler [1998] also is of interest. For current basic coverage of region-oriented segmentation concepts, see Shapiro and Stockman [2001] and Sonka et al. [1999].

Segmentation by watersheds was shown in Section 10.5 to be a powerful concept. Early references dealing with segmentation by watersheds are Serra [1988], Beucher [1990], and Beucher and Meyer [1992]. The paper by Baccar et al. [1996] discusses segmentation based on data fusion and morphological watersheds. Progress ten years later is evident in a special issue of *Pattern Recognition* [2000], devoted entirely to this topic. As indicated in our discussion in Section 10.5, one of the key issues with watersheds is the problem of over segmentation. The papers by Najmanand and Schmitt [1996], Haris et al. [1998], and Bleau and Leon [2000] are illustrative of approaches for dealing with this problem. Bieniek and Moga [2000] discuss a watershed segmentation algorithm based on connected components.

The material in Section 10.6.1 is from Jain, R. [1981]. See also Jain, Kasturi, and Schunck [1995]. The material in Section 10.6.2 is from Rajala, Riddle, and Snyder [1983]. See also the papers by Shariat and Price [1990] and by Cumani et al. [1991]. The books by Sonka et al. [1999], Shapiro and Stockman [2001], Snyder and Qi [2004], and Davies [2005] provide additional reading on motion estimation. See also Alexiadis and Sergiadis [2007].

Problems

★**10.1**   Prove the validity of Eq. (10.2-1). (*Hint:* Use a Taylor series expansion and keep only the linear terms.)

★**10.2**   A binary image contains straight lines oriented horizontally, vertically, at $45°$, and at $-45°$. Give a set of 3×3 masks that can be used to detect 1-pixel breaks in these lines. Assume that the intensities of the lines and background are 1 and 0, respectively.

10.3   Propose a technique for detecting gaps of length ranging between 1 and K pixels in line segments of a binary image. Assume that the lines are 1 pixel thick. Base your technique on 8-neighbor connectivity analysis, rather than attempting to construct masks for detecting the gaps.

10.4   Refer to Fig. 10.7 in answering the following questions.

    ★**(a)**   Some of the lines joining the pads and center element in Fig. 10.7(e) are single lines, while others are double lines. Explain why.

    (b)   Propose a method for eliminating the components in Fig. 10.7(f) that are not part of the line oriented at $-45°$.

10.5   Refer to the edge models in Fig. 10.8.

    ★**(a)**   Suppose that we compute the gradient magnitude of each of these models using the Prewitt operators in Fig. 10.14. Sketch what a horizontal profile through the center of each gradient image would look like.

    (b)   Sketch a horizontal profile for each corresponding angle image.

    (*Note:* Answer this question without generating the gradient and angle images. Simply provide sketches of the profiles that show what you would *expect* the profiles of the magnitude and angle images to look like.)

10.6   Consider a horizontal intensity profile through the middle of a binary image that contains a step edge running vertically through the center of the image. Draw what the profile would look like after the image has been blurred by an averaging mask of size $n \times n$, with coefficients equal to $1/n^2$. For simplicity, assume that the image was scaled so that its intensity levels are 0 on the left of the edge and 1 on its right. Also, assume that the size of the mask is much smaller than the image, so that image border effects are not a concern near the center of the horizontal intensity profile.

★**10.7**   Suppose that we had used the edge models shown in the next page, instead of the ramp model in Fig. 10.10. Sketch the gradient and Laplacian of each profile.

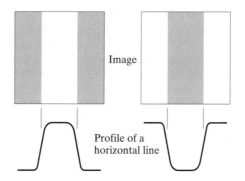

10.8   Refer to Fig. 10.14 in answering the following questions.

    (a) Assume that the Sobel masks are used to obtain g_x and g_y. Show that in this case the magnitude of the gradient computed using Eqs. (10.2-10) and (10.2-20) give identical results.

    (b) Show that this is true also for the Prewitt masks.

★10.9   Show that the Sobel and Prewitt masks in Figs. 10.14 and 10.15 give isotropic results only for horizontal and vertical edges and for edges oriented at $\pm 45°$, respectively.

10.10   The results obtained by a single pass through an image of some 2-D masks can be achieved also by two passes using 1-D masks. For example, the same result of using a 3×3 smoothing mask with coefficients $1/9$ can be obtained by a pass of the mask [1 1 1] through an image. The result of this pass is then followed by a pass of the mask

$$\begin{bmatrix} 1 \\ 1 \\ 1 \end{bmatrix}$$

The final result is then scaled by $1/9$. Show that the response of Sobel masks (Fig. 10.14) can be implemented similarly by one pass of the *differencing* mask $[-1\ 0\ 1]$ (or its vertical counterpart) followed by the *smoothing* mask [1 2 1] (or its vertical counterpart).

10.11   The so-called compass gradient operators of size 3×3 are designed to measure gradients of edges oriented in eight directions: E, NE, N, NW, W, SW, S, and SE.

    ★**(a)** Give the form of these eight operators using coefficients valued 0, 1, or -1.

    (b) Specify the gradient vector direction of each mask, keeping in mind that the gradient direction is orthogonal to the edge direction.

10.12   The rectangle in the binary image in the next page is of size $m \times n$ pixels.

    (a) What would the magnitude of the gradient of this image look like based on using the approximation given in Eq. (10.2-20)? Assume that g_x and g_y are obtained using the Sobel operators. Show all relevant different pixel values in the gradient image.

    (b) Sketch the histogram of edge *directions* computed using Eq. (10.2-11). Be precise in labeling the height of each component of the histogram.

    (c) What would the Laplacian of this image look like based on using the approximation in Eq. (10.2-7)? Show all relevant different pixel values in the Laplacian image.

10.13 Suppose that an image $f(x, y)$ is convolved with a mask of size $n \times n$ (with co-efficients $1/n^2$) to produce a smoothed image $\bar{f}(x, y)$.

★**(a)** Derive an expression for *edge strength* (edge magnitude) of the smoothed image as a function of mask size. Assume for simplicity that n is odd and that edges are obtained using the partial derivatives

$$\partial\bar{f}/\partial x = \bar{f}(x + 1, y) - \bar{f}(x, y) \quad \text{and} \quad \partial\bar{f}/\partial y = \bar{f}(x, y + 1) - \bar{f}(x, y).$$

(b) Show that the ratio of the maximum edge strength of the smoothed image to the maximum edge strength of the original image is $1/n$. In other words, edge strength is inversely proportional to the size of the smoothing mask.

10.14 With reference to Eq. (10.2-23):

★**(a)** Show that the average value of the Laplacian of a Gaussian operator, $\nabla^2 G(x, y)$, is zero.

(b) Show that the average value of any image convolved with this operator also is zero. (*Hint*: Consider solving this problem in the frequency domain, using the convolution theorem and the fact that the average value of a function is proportional to its Fourier transform evaluated at the origin.)

(c) Would (b) be true in general if we (1) used the mask in Fig. 10.4(a) to compute the Laplacian of a Gaussian lowpass filter using a Laplacian mask of size 3×3, and (2) convolved this result with any image? Explain. (*Hint*: Refer to Problem 3.16.)

10.15 Refer to Fig. 10.22(c).

(a) Explain why the edges form closed contours.

★**(b)** Does the zero-crossing method for finding edge location always result in closed contours? Explain.

10.16 One often finds in the literature a derivation of the Laplacian of a Gaussian (LoG) that starts with the expression

$$G(r) = e^{-r^2/2\sigma^2}$$

where $r^2 = x^2 + y^2$. The LoG is then found by taking the second partial derivative: $\nabla^2 G(r) = \partial^2 G/\partial r^2$. Finally, $x^2 + y^2$ is substituted for r^2 to get the (incorrect) result

$$\nabla^2 G(x, y) = \left[(x^2 + y^2 - \sigma^2)/\sigma^4\right]\exp\left[-(x^2 + y^2)/2\sigma^2\right]$$

Derive this result and explain the reason for the difference between this expression and Eq. (10.2-23).

10.17 **(a)** Derive Eq. (10.2-27).

(b) Let $k = \sigma_1/\sigma_2$ denote the standard deviation ratio discussed in connection with the DoG function. Express Eq. (10.2-27) in terms of k and σ_2.

10.18 In the following, assume that G and f are discrete arrays of size $n \times n$ and $M \times N$, respectively.

★**(a)** Show that the 2-D convolution of the Gaussian function $G(x, y)$ in Eq. (10.2-21) with an image $f(x, y)$ can be expressed as a 1-D convolution along the rows (columns) of $f(x, y)$ followed by a 1-D convolution along the columns (rows) of the result. (See Section 3.4.2 regarding discrete convolution.)

(b) Derive an expression for the computational advantage of using the 1-D convolution approach in (a) as opposed to implementing the 2-D convolution directly. Assume that $G(x, y)$ is sampled to produce an array of size $n \times n$ and that $f(x, y)$ is of size $M \times N$. The computational advantage is the ratio of the number of multiplications required for 2-D convolution to the number required for 1-D convolution.

★**10.19** **(a)** Show that Steps 1 and 2 of the Marr-Hildreth algorithm can be implemented using four, 1-D convolutions. (*Hints*: Refer to Problem 10.18(a) and express the Laplacian operator as the sum of two partial derivatives, given by Eqs. (10.2-5) and (10.2-6), and implement each derivative using a 1-D mask, as in Problem 10.10.)

(b) Derive an expression for the computational advantage of using the 1-D convolution approach in (a) as opposed to implementing the 2-D convolution directly. Assume that $G(x, y)$ is sampled to produce an array of size $n \times n$ and that $f(x, y)$ is of size $M \times N$. The computational advantage is the ratio of the number of multiplications required for 2-D convolution to the number required for 1-D convolution (see Problem 10.18).

10.20 **(a)** Formulate Step 1 and the gradient magnitude image computation in Step 2 of the Canny algorithm using 1-D instead of 2-D convolutions.

(b) What is the computational advantage of using the 1-D convolution approach as opposed to implementing a 2-D convolution. Assume that the 2-D Gaussian filter in Step 1 is sampled into an array of size $n \times n$ and the input image is of size $M \times N$. Express the computational advantage as the ratio of the number of multiplications required by each method.

10.21 Refer to the three vertical edge models and corresponding profiles in Fig. 10.8.

★**(a)** Suppose that we compute the gradient magnitude of each of the three edge models using the Sobel masks. Sketch the horizontal intensity profiles of the three gradient images.

★**(b)** Sketch the horizontal intensity profiles of the three Laplacian images, assuming that the Laplacian is computed using the 3×3 mask in Fig. 10.4(a).

★**(c)** Repeat for an image generated using only the first two steps of the Marr-Hildreth edge detector.

(d) Repeat for the first two steps of the Canny edge detector. You may ignore the angle images.

(e) Sketch the horizontal profile of the angle images for the Canny edge detector.

(*Note*: Answer this question without generating the images. Simply provide sketches of the profiles that show what you would *expect* the profiles of the images to look like.)

10.22 Refer to the Hough transform discussed in Section 10.2.7.

(a) Develop a general procedure for obtaining the normal representation of a line from its slope-intercept form, $y = ax + b$.

★**(b)** Find the normal representation of the line $y = -2x + 1$.

★**10.23** Refer to the Hough transform discussed in Section 10.2.7.

(a) Explain why the Hough mapping of point 1 in Fig. 10.33(a) is a straight line in Fig. 10.33(b).

(b) Is this the only point that would produce that result? Explain.

(c) Explain the reflective adjacency relationship illustrated by, for example, the curve labeled Q in Fig. 10.33(b).

10.24 Show that the number of operations required to implement the accumulator-cell approach discussed in Section 10.2.7 is linear in n, the number of non-background points in the image plane (i.e., the xy-plane).

10.25 An important area of application for image segmentation techniques is in processing images resulting from so-called bubble chamber events. These images arise from experiments in high-energy physics in which a beam of particles of known properties is directed onto a target of known nuclei. A typical event consists of incoming tracks, any one of which, in the event of a collision, branches out into secondary tracks of particles emanating from the point of collision. Propose a segmentation approach for detecting all tracks that contain at least 100 pixels and are angled at any of the following six directions off the horizontal: $\pm 25°$, $\pm 50°$, and $\pm 75°$. The allowed estimation error in any of these six directions is $\pm 5°$. For a track to be valid it must be at least 100 pixels long and not have more than three gaps, any of which cannot exceed 10 pixels. You may assume that the images have been preprocessed so that they are binary and that all tracks are 1 pixel wide, except at the point of collision from which they emanate. Your procedure should be able to differentiate between tracks that have the same direction but different origins. (*Hint*: Base your solution on the Hough transform.)

★**10.26** Restate the basic global thresholding algorithm in Section 10.3.2 so that it uses the histogram of an image instead of the image itself.

★**10.27** Prove that the basic global thresholding algorithm in Section 10.3.2 converges in a finite number of steps. (*Hint*: Use the histogram formulation from Problem 10.26.)

10.28 Give an explanation why the initial threshold in the basic global thresholding algorithm in Section 10.3.2 must be between the minimum and maximum values in the image. (*Hint*: Construct an example that shows the algorithm failing for a threshold value selected outside this range.)

★**10.29** Is the threshold obtained with the basic global thresholding algorithm in Section 10.3.2 independent of the starting point? If your answer is yes, prove it. If your answer is no, give an example.

10.30 You may assume in both of the following cases that the threshold value during iteration is bounded in the open interval $(0, L - 1)$.

★**(a)** Prove that if the histogram of an image is uniform over all possible intensity levels, the basic global thresholding algorithm in Section 10.3.2 converges to the average intensity of the image, $(L - 1)/2$.

(b) Prove that if the histogram of an image is bimodal, with identical modes that are symmetric about their means, then the basic global algorithm will converge to the point halfway between the means of the modes.

10.31 Refer to the thresholding algorithm in Section 10.3.2. Assume that in a given problem the histogram is bimodal with modes that are Gaussian curves of the form $A_1 \exp[-(z - m_1)^2/2\sigma_1^2]$ and $A_2 \exp[-(z - m_2)^2/2\sigma_2^2]$. Assume that $m_1 > m_2$ and that the initial T is between the max and min image intensities. Give conditions (in terms of the parameters of these curves) for the following to be true when the algorithm converges:

(a) The threshold is equal to $(m_1 + m_2)/2$.

(b) The threshold is to the left of m_2.

(c) The threshold is in the interval $(m_1 + m_2)/2 < T < m_1$.

If it is not possible for any of these conditions to exist, so state, and give a reason.

★10.32 (a) Show how the first line in Eq. (10.3-15) follows from Eqs. (10.3-14), (10.3-10), and (10.3-11).

(b) Show how the second line in Eq. (10.3-15) follows from the first.

10.33 Show that a maximum value for Eq. (10.3-18) always exists for k in the range $0 \le k \le L - 1$.

10.34 With reference to Eq. (10.3-20), advance an argument that establishes that $0 \le \eta(k) \le 1$, for k in the range $0 \le k \le L - 1$, where the minimum is achievable only by images with constant intensity, and the maximum occurs only for 2-valued images with values 0 and $L - 1$.

★10.35 (a) Suppose that the intensities of an image $f(x, y)$ are in the range $[0, 1]$ and that a threshold, T, successfully segmented the image into objects and background. Show that the threshold $T' = 1 - T$ will successfully segment the negative of $f(x, y)$ into the same regions. The term *negative* is used here in the sense defined in Section 3.2.1.

(b) The intensity transformation function in (a) that maps an image into its negative is a linear function with negative slope. State the conditions that an arbitrary intensity transformation function must satisfy for the segmentability of the original image with respect to a threshold, T, to be preserved. What would be the value of the threshold after the intensity transformation?

10.36 The objects and background in the image shown have a mean intensity of 170 and 60, respectively, on a $[0, 255]$ scale. The image is corrupted by Gaussian noise with 0 mean and a standard deviation of 10 intensity levels. Propose a thresholding method capable of yielding a correct segmentation rate of 90% or higher. (Recall that 99.7% of the area of a Gaussian curve lies in a $\pm 3\sigma$ interval about the mean, where σ is the standard deviation.)

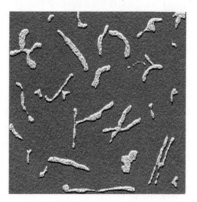

10.37 Refer to the intensity ramp image in Fig. 10.37(b) and the moving-average algorithm discussed in Section 10.3.7. Assume that the image is of size 500×700 pixels and that its minimum and maximum values are 0 and 1, where 0s are contained only in the first column.

★(a) What would be the result of segmenting this image with the moving-average algorithm using $b = 0$ and an arbitrary value for n. Explain what the image would look like.

(b) Now reverse the direction of the ramp so that its leftmost value is 1 and the rightmost value is 0 and repeat (a).

(c) Repeat (a) but with $n = 2$ and $b = 1$.

(d) Repeat (a) but with $n = 100$ and $b = 1$.

10.38 Propose a region-growing algorithm to segment the image in Problem 10.36.

★**10.39** Segment the image shown by using the split and merge procedure discussed in Section 10.4.2. Let $Q(R_i) = \text{TRUE}$ if all pixels in R_i have the same intensity. Show the quadtree corresponding to your segmentation.

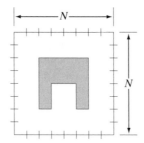

10.40 Consider the region of 1s resulting from the segmentation of the sparse regions in the image of the Cygnus Loop in Example 10.24. Propose a technique for using this region as a mask to isolate the three main components of the image: (1) background, (2) dense inner region, and (3) sparse outer region.

10.41 Refer to the discussion in Section 10.5.3.

★**(a)** Show that the elements of $C_n(M_i)$ and $T[n]$ are never replaced during execution of the watershed segmentation algorithm.

(b) Show that the number of elements in sets $C_n(M_i)$ and $T[n]$ either increases or remains the same as n increases.

10.42 The boundaries illustrated in Section 10.5, obtained using the watershed segmentation algorithm, form closed loops (for example, see Figs. 10.56 and 10.58). Advance an argument that establishes whether or not closed boundaries always result from application of this algorithm.

★**10.43** Give a step-by-step implementation of the dam-building procedure for the one-dimensional intensity cross section shown. Show a drawing of the cross section at each step, showing "water" levels and dams constructed.

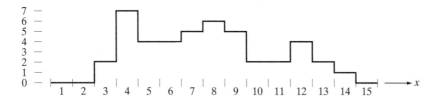

10.44 What would the negative ADI image in Fig. 10.59(c) look like if we tested against T (instead of testing against $-T$) in Eq. (10.6-4)?

10.45 Are the following statements true or false? Explain the reason for your answer in each.

★**(a)** The nonzero entries in the absolute ADI continue to grow in dimension, provided that the object is moving.

(b) The nonzero entries in the positive ADI always occupy the same area, regardless of the motion undergone by the object.

(c) The nonzero entries in the negative ADI continue to grow in dimension, provided that the object is moving.

10.46 Suppose that in Example 10.28 motion along the x-axis is set to zero. The object now moves only along the y-axis at 1 pixel per frame for 32 frames and then (instantaneously) reverses direction and moves in exactly the opposite direction for another 32 frames. What would Figs. 10.63 and 10.64 look like under these conditions?

★**10.47** Advance an argument that demonstrates that when the signs of S_{1x} and S_{2x} in Eqs. (10.6-12) and (10.6-13) are the same, the velocity component V_1 is positive.

10.48 An automated pharmaceutical plant uses image processing in measuring the shapes of medication tablets for the purpose of quality control. The segmentation stage of the system is based on Otsu's method. The speed of the inspection lines is so high that a very high rate flash illumination is required to "stop" motion. When new, the illumination lamps project a uniform pattern of light. However, as the lamps age, the illumination pattern deteriorates as a function of time and spatial coordinates according to the equation

$$i(x, y) = A(t) - t^2 e^{-[(x-M/2)^2+(y-N/2)^2]}$$

where $(M/2, N/2)$ is the center of the viewing area and t is time measured in increments of months. The lamps are experimental and the behavior of $A(t)$ is not fully understood by the manufacturer. All that is known is that, during the life of the lamps, $A(t)$ is always greater than the negative component in the preceding equation because illumination cannot be negative. It has been observed that Otsu's algorithm works well when the lamps are new, and their pattern of illumination is nearly constant over the entire image. However, segmentation performance deteriorates with time. Being experimental, the lamps are exceptionally expensive, so you are employed as a consultant to help solve the problem computationally and thus extend the useful life of the lamps. You are given flexibility to install any special markers or other visual cues near the edges of the viewing area of the imaging cameras. Propose a solution in sufficient detail that the engineering plant manager can understand your approach. (*Hint:* Review the image model discussed in Section 2.3.4 and consider using a small target of known reflectivity.)

10.49 The speed of a bullet in flight is to be estimated by using high-speed imaging techniques. The method of choice involves the use of a TV camera and flash that exposes the scene for K s. The bullet is 2.5 cm long, 1 cm wide, and its range of speed is 750 ± 250 m/s. The camera optics produce an image in which the bullet occupies 10% of the horizontal resolution of a 256×256 digital image.

★**(a)** Determine the maximum value of K that will guarantee that the blur from motion does not exceed 1 pixel.

(b) Determine the minimum number of frames per second that would have to be acquired in order to guarantee that at least two complete images of the bullet are obtained during its path through the field of view of the camera.

(c) Propose a segmentation procedure for automatically extracting the bullet from a sequence of frames.

(d) Propose a method for automatically determining the speed of the bullet.

11 Representation and Description

Well, but reflect; have we not several times acknowledged that names rightly given are the likenesses and images of the things which they name?

Socrates

Preview

After an image has been segmented into regions by methods such as those discussed in Chapter 10, the resulting aggregate of segmented pixels usually is represented and described in a form suitable for further computer processing. Basically, representing a region involves two choices: (1) We can represent the region in terms of its external characteristics (its boundary), or (2) we can represent it in terms of its internal characteristics (the pixels comprising the region). Choosing a representation scheme, however, is only part of the task of making the data useful to a computer. The next task is to *describe* the region based on the chosen representation. For example, a region may be *represented* by its boundary, and the boundary *described* by features such as its length, the orientation of the straight line joining its extreme points, and the number of concavities in the boundary.

An external representation is chosen when the primary focus is on shape characteristics. An internal representation is selected when the primary focus is on regional properties, such as color and texture. Sometimes it may be necessary to use both types of representation. In either case, the features selected as descriptors should be as insensitive as possible to variations in size, translation, and rotation. For the most part, the descriptors discussed in this chapter satisfy one or more of these properties.

11.1 Representation

The segmentation techniques discussed in Chapter 10 yield raw data in the form of pixels along a boundary or pixels contained in a region. It is standard practice to use schemes that compact the segmented data into representations that facilitate the computation of descriptors. In this section, we discuss various representation approaches.

11.1.1 Boundary (Border) Following

You will find it helpful to review Sections 2.5.2 and 9.5.3 before proceeding.

Several of the algorithms discussed in this chapter require that the points in the boundary of a region be ordered in a clockwise (or counterclockwise) direction. Consequently, we begin our discussion by introducing a boundary-following algorithm whose output is an *ordered* sequence of points. We assume (1) that we are working with binary images in which object and background points are labeled 1 and 0, respectively, and (2) that images are padded with a border of 0s to eliminate the possibility of an object merging with the image border. For convenience, we limit the discussion to single regions. The approach is extended to multiple, disjoint regions by processing the regions individually.

Given a binary region R or its boundary, an algorithm for following the border of R, or the given boundary, consists of the following steps:

1. Let the starting point, b_0, be the *uppermost, leftmost* point[†] in the image that is labeled 1. Denote by c_0 the *west* neighbor of b_0 [see Fig. 11.1(b)]. Clearly, c_0 always is a background point. Examine the 8-neighbors of b_0, starting at c_0 and proceeding in a clockwise direction. Let b_1 denote the *first* neighbor encountered whose value is 1, and let c_1 be the (background) point immediately preceding b_1 in the sequence. Store the locations of b_0 and b_1 for use in Step 5.
2. Let $b = b_1$ and $c = c_1$ [see Fig. 11.1(c)].
3. Let the 8-neighbors of b, starting at c and proceeding in a clockwise direction, be denoted by $n_1, n_2, \ldots, n_8$. Find the first n_k labeled 1.

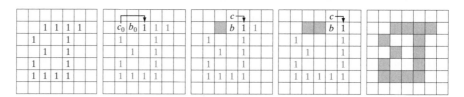

a b c d e

FIGURE 11.1 Illustration of the first few steps in the boundary-following algorithm. The point to be processed next is labeled in black, the points yet to be processed are gray, and the points found by the algorithm are labeled as gray squares.

[†]As you will see later in this chapter, the uppermost, leftmost point in a boundary has the important property that a polygonal approximation to the boundary has a convex vertex at that location. Also, the left and north neighbors of the point are guaranteed to be background points. These properties make it a good "standard" point at which to start boundary-following algorithms.

4. Let $b = n_k$ and $c = n_{k-1}$.
5. Repeat Steps 3 and 4 until $b = b_0$ *and* the next boundary point found is b_1. The sequence of b points found when the algorithm stops constitutes the set of ordered boundary points.

Note that c in Step 4 always is a background point because n_k is the first 1-valued point found in the clockwise scan. This algorithm sometimes is referred to as the *Moore boundary tracking algorithm* after Moore [1968]. The stopping rule in Step 5 of the algorithm frequently is found stated incorrectly in the literature as stopping the first time that b_0 is encountered again. As you will see shortly, this can lead to erroneous results.

Figure 11.1 shows the first few steps of the boundary-following algorithm just discussed. It easily is verified that continuing with this procedure will yield the correct boundary shown in Fig. 11.1(e), whose points are a clockwise-ordered sequence.

To examine the need for the stopping rule as stated in Step 5 of the algorithm, consider the boundary in Fig. 11.2. The segment on the upper side of the boundary could arise, for example, from incomplete spur removal (see Section 9.5.8 regarding spurs). Starting at the topmost leftmost point results in the steps shown. We see in Fig. 11.2(c) that the algorithm has returned to the starting point. If the procedure were stopped because we have reached the starting point again, it is evident that the rest of the boundary would not be found. Using the stopping rule in Step 5 allows the algorithm to continue, and it is a simple matter to show that the entire boundary in Fig. 11.2 would be found.

The boundary-following algorithm works equally well if a region, rather than its boundary (as in the preceding illustrations), is given. That is, the procedure extracts the *outer boundary* of a binary region. If the objective is to find the boundaries of holes in a region (these are called the *inner boundaries* of the region), a simple approach is to extract the holes (see Section 9.5.9) and treat them as 1-valued regions on a background of 0s. Applying the boundary-following algorithm to these regions will yield the inner boundaries of the original region.

We could have stated the algorithm just as easily based on following a boundary in the counterclockwise direction. In fact, you will encounter algorithms formulated on the assumption that boundary points are ordered in that

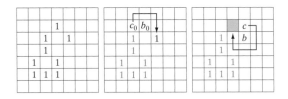

a b c

FIGURE 11.2 Illustration of an erroneous result when the stopping rule is such that boundary-following stops when the starting point, b_0, is encountered again.

direction. We use both directions interchangeably (but consistently) in the following sections to help you build familiarity with both approaches.

11.1.2 Chain Codes

Chain codes are used to represent a boundary by a connected sequence of straight-line segments of specified length and direction. Typically, this representation is based on 4- or 8-connectivity of the segments. The *direction* of each segment is coded by using a numbering scheme, as in Fig. 11.3. A boundary code formed as a sequence of such directional numbers is referred to as a *Freeman chain code*.

Digital images usually are acquired and processed in a grid format with equal spacing in the *x*- and *y*-directions, so a chain code can be generated by following a boundary in, say, a clockwise direction and assigning a direction to the segments connecting every pair of pixels. This method generally is unacceptable for two principal reasons: (1) The resulting chain tends to be quite long and (2) any small disturbances along the boundary due to noise or imperfect segmentation cause changes in the code that may not be related to the principal shape features of the boundary.

An approach frequently used to circumvent these problems is to resample the boundary by selecting a larger grid spacing, as Fig. 11.4(a) shows. Then, as the boundary is traversed, a boundary point is assigned to each node of the large grid, depending on the proximity of the original boundary to that node, as in Fig. 11.4(b). The resampled boundary obtained in this way then can be represented by a 4- or 8-code. Figure 11.4(c) shows the coarser boundary points represented by an 8-directional chain code. It is a simple matter to convert from an 8-code to a 4-code, and vice versa (see Problems 2.12 and 2.13). The starting point in Fig. 11.4(c) is (arbitrarily) at the topmost, leftmost point of the boundary, which gives the chain code 0766 . . . 12. As might be expected, the accuracy of the resulting code representation depends on the spacing of the sampling grid.

The chain code of a boundary depends on the starting point. However, the code can be normalized with respect to the starting point by a straightforward procedure: We simply treat the chain code as a circular sequence of direction numbers and redefine the starting point so that the resulting sequence of numbers forms an integer of minimum magnitude. We can normalize also for rotation (in angles that are integer multiples of the directions in Fig. 11.3) by using the *first difference* of the chain code instead of the code

a b

FIGURE 11.3
Direction
numbers for
(a) 4-directional
chain code, and
(b) 8-directional
chain code.

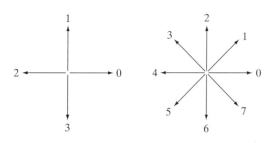

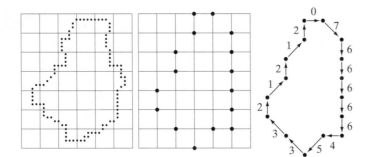

FIGURE 11.4
(a) Digital boundary with resampling grid superimposed.
(b) Result of resampling.
(c) 8-directional chain-coded boundary.

itself. This difference is obtained by counting the number of direction changes (in a counterclockwise direction in Fig. 11.3) that separate two adjacent elements of the code. For instance, the first difference of the 4-direction chain code 10103322 is 3133030. If we treat the code as a circular sequence to normalize with respect to the starting point, then the first element of the difference is computed by using the transition between the last and first components of the chain. Here, the result is 33133030. Size normalization can be achieved by altering the size of the resampling grid.

These normalizations are exact only if the boundaries themselves are invariant to rotation (again, in angles that are integer multiples of the directions in Fig. 11.3) and scale change, which seldom is the case in practice. For instance, the same object digitized in two different orientations will have different boundary shapes in general, with the degree of dissimilarity being proportional to image resolution. This effect can be reduced by selecting chain elements that are long in proportion to the distance between pixels in the digitized image and/or by orienting the resampling grid along the principal axes of the object to be coded, as discussed in Section 11.2.2, or along its eigen axes, as discussed in Section 11.4.

■ Figure 11.5(a) shows a 570 × 570, 8-bit gray-scale image of a circular stroke embedded in small specular fragments. The objective of this example is to obtain the Freeman chain code, the integer of minimum magnitude, and the first difference of the outer boundary of the largest object in Fig. 11.5(a). Because the object of interest is embedded in small fragments, extracting its boundary would result is a noisy curve that would not be descriptive of the general shape of the object. Smoothing is a routine process when working with noisy boundaries. Figure 11.5(b) shows the original image smoothed with an averaging mask of size 9 × 9, and Fig. 11.5(c) is the result of thresholding this image with a global threshold obtained using Otsu's method. Note that the number of regions has been reduced to two (one of which is a dot), significantly simplifying the problem.

Figure 11.5(d) is the outer boundary of the largest region in Fig. 11.5(c). Obtaining the chain code of this boundary directly would result in a long sequence with small variations that are not representative of the shape of the

EXAMPLE 11.1:
Freeman chain code and some of its variations.

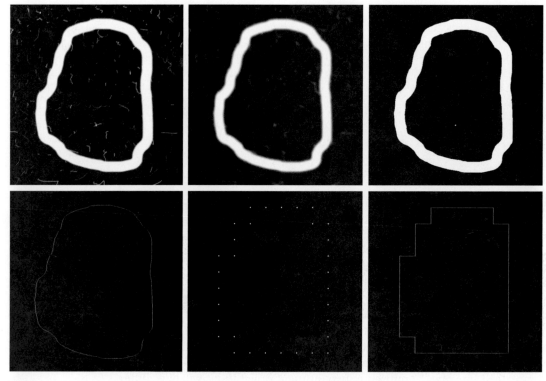

FIGURE 11.5 (a) Noisy image. (b) Image smoothed with a 9×9 averaging mask. (c) Smoothed image, thresholded using Otsu's method. (d) Longest outer boundary of (c). (e) Subsampled boundary (the points are shown enlarged for clarity). (f) Connected points from (e).

boundary. As mentioned earlier in this section, it is customary to resample a boundary before obtaining its chain code in order to reduce variability. Figure 11.5(e) is the result of resampling the boundary in a grid with nodes 50 pixels apart (approximately 10% of the image width) and Fig. 11.5(f) is the result of joining the resulting vertices by straight lines. This simpler approximation retained the principal features of the original boundary.

The 8-directional Freeman chain code of the simplified boundary is

0 0 0 0 6 0 6 6 6 6 6 6 6 6 4 4 4 4 4 4 2 4 2 2 2 2 2 0 2 2 0 2

The starting point of the boundary is at coordinates $(2, 5)$ in the subsampled grid. This is the uppermost leftmost point in Fig. 11.5(f). The integer of minimum magnitude of the code happens in this case to be the same as the chain code:

0 0 0 0 6 0 6 6 6 6 6 6 6 6 4 4 4 4 4 4 2 4 2 2 2 2 2 0 2 2 0 2

The first difference of either code is

0 0 0 6 2 6 0 0 0 0 0 0 0 6 0 0 0 0 0 6 2 6 0 0 0 0 6 2 0 6 2 6

Using any of these codes to represent the boundary results in a significant reduction in the amount of data needed to store the boundary. In addition, working with code numbers offers a unified way to analyze the shape of a boundary, as we discuss in Section 11.2. Finally, keep in mind that the subsampled boundary can be recovered from any of the preceding codes.   ■

11.1.3 Polygonal Approximations Using Minimum-Perimeter Polygons

A digital boundary can be approximated with arbitrary accuracy by a polygon. For a closed boundary, the approximation becomes exact when the number of segments of the polygon is equal to the number of points in the boundary so that each pair of adjacent points defines a segment of the polygon. The goal of a polygonal approximation is to capture the essence of the shape in a given boundary using the fewest possible number of segments. This problem is not trivial in general and can turn into a time-consuming iterative search. However, approximation techniques of modest complexity are well suited for image processing tasks. Among these, one of the most powerful is representing a boundary by a *minimum-perimeter polygon* (MPP), as defined in the following discussion.

Foundation

An intuitively appealing approach for generating an algorithm to compute MPPs is to enclose a boundary [Fig. 11.6(a)] by a set of concatenated cells, as in Fig. 11.6(b). Think of the boundary as a rubber band. As it is allowed to shrink, the rubber band will be constrained by the inner and outer walls of

a b c

FIGURE 11.6 (a) An object boundary (black curve). (b) Boundary enclosed by cells (in gray). (c) Minimum-perimeter polygon obtained by allowing the boundary to shrink. The vertices of the polygon are created by the corners of the inner and outer walls of the gray region.

the bounding region defined by the cells. Ultimately, this shrinking produces the shape of a polygon of minimum perimeter (with respect to this geometrical arrangement) that circumscribes the region enclosed by the cell strip, as Fig. 11.6(c) shows. Note in this figure that all the vertices of the MPP coincide with corners of either the inner or the outer wall.

The size of the cells determines the accuracy of the polygonal approximation. In the limit, if the size of each (square) cell corresponds to a pixel in the boundary, the error in each cell between the boundary and the MPP approximation at most would be $\sqrt{2}d$, where d is the minimum possible distance between pixels (i.e., the distance between pixels established by the resolution of the original sampled boundary). This error can be reduced in half by forcing each cell in the polygonal approximation to be centered on its corresponding pixel in the original boundary. The objective is to use the largest possible cell size acceptable in a given application, thus producing MPPs with the fewest number of vertices. Our objective in this section is to formulate a procedure for finding these MPP vertices.

The cellular approach just described reduces the shape of the object enclosed by the original boundary to the area circumscribed by the gray wall in Fig. 11.6(b). Figure 11.7(a) shows this shape in dark gray. We see that its boundary consists of 4-connected straight line segments. Suppose that we traverse this boundary in a *counterclockwise* direction. Every turn encountered in the traversal will be either a *convex* or a *concave* vertex, with the angle of a vertex being an *interior* angle of the 4-connected boundary. Convex and concave vertices are

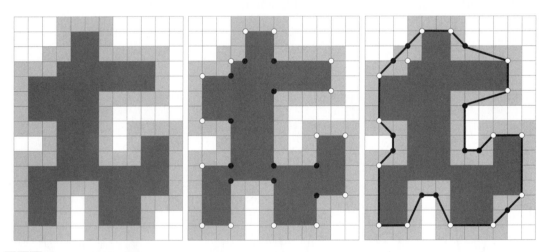

a b c

FIGURE 11.7 (a) Region (dark gray) resulting from enclosing the original boundary by cells (see Fig. 11.6). (b) Convex (white dots) and concave (black dots) vertices obtained by following the boundary of the dark gray region in the counterclockwise direction. (c) Concave vertices (black dots) displaced to their diagonal mirror locations in the outer wall of the bounding region; the convex vertices are not changed. The MPP (black boundary) is superimposed for reference.

shown, respectively, as white and black dots in Fig. 11.7(b). Note that these vertices are the vertices of the inner wall of the light-gray bounding region in Fig. 11.7(b), and that every concave (black) vertex in the dark gray region has a corresponding "mirror" vertex in the light gray wall, located diagonally opposite the vertex. Figure 11.7(c) shows the mirrors of all the concave vertices, with the MPP from Fig. 11.6(c) superimposed for reference. We see that the vertices of the MPP coincide either with convex vertices in the inner wall (white dots) or with the mirrors of the concave vertices (black dots) in the outer wall. A little thought will reveal that only convex vertices of the inner wall and concave vertices of the outer wall can be vertices of the MPP. Thus, our algorithm needs to focus attention on only these vertices.

MPP algorithm

The set of cells enclosing a digital boundary, described in the previous paragraphs, is called a *cellular complex*. We assume that the boundaries under consideration are not self intersecting, which leads to *simply connected* cellular complexes. Based on these assumptions, and letting *white* (*W*) and *black* (*B*) denote *convex* and *mirrored concave* vertices, respectively, we state the following observations:

1. The MPP bounded by a simply connected cellular complex is not self-intersecting.
2. Every *convex* vertex of the MPP is a *W* vertex, but not every *W* vertex of a boundary is a vertex of the MPP.
3. Every *mirrored concave* vertex of the MPP is a *B* vertex, but not every *B* vertex of a boundary is a vertex of the MPP.
4. All *B* vertices are on or outside the MPP, and all *W* vertices are on or inside the MPP.
5. The uppermost, leftmost vertex in a sequence of vertices contained in a cellular complex is always a *W* vertex of the MPP.

These assertions can be proved formally (Sklansky et al. [1972], Sloboda et al. [1998], and Klette and Rosenfeld [2004]). However, their correctness is evident for our purposes (Fig. 11.7), so we do not dwell on the proofs here. Unlike the angles of the vertices of the dark gray region in Fig. 11.7, the angles sustained by the vertices of the MPP are not necessarily multiples of 90°.

In the discussion that follows, we will need to calculate the orientation of triplets of points. Consider the triplet of points, (a, b, c), and let the coordinates of these points be $a = (x_1, y_1)$, $b = (x_2, y_2)$, and $c = (x_3, y_3)$. If we arrange these points as the rows of the matrix

$$\mathbf{A} = \begin{bmatrix} x_1 & y_1 & 1 \\ x_2 & y_2 & 1 \\ x_3 & y_3 & 1 \end{bmatrix} \tag{11.1-1}$$

then it follows from elementary matrix analysis that

$$\det(\mathbf{A}) = \begin{cases} > 0 & \text{if } (a, b, c) \text{ is a counterclockwise sequence} \\ = 0 & \text{if the points are collinear} \\ < 0 & \text{if } (a, b, c) \text{ is a clockwise sequence} \end{cases} \quad (11.1\text{-}2)$$

where $\det(\mathbf{A})$ is the determinant of $\mathbf{A}$. In terms of this equation, movement in a counterclockwise or clockwise direction is with respect to a right-handed coordinate system (see the footnote in Section 2.4.2). For example, using this image coordinate system (Fig. 2.18), in which the origin is at the top left, the positive x-axis extends vertically downward, and the positive y-axis extends horizontally to the right, the sequence $a = (3, 4), b = (2, 3),$ and $c = (3, 2)$ is in the counterclockwise direction and would give $\det(\mathbf{A}) > 0$ when substituted into Eq. (11.1-2). It is notationally convenient when describing the algorithm to define

$$\mathrm{sgn}(a, b, c) \equiv \det(\mathbf{A}) \quad (11.1\text{-}3)$$

so that $\mathrm{sgn}(a, b, c) > 0$ for a counterclockwise sequence, $\mathrm{sgn}(a, b, c) < 0$ for a clockwise sequence, and $\mathrm{sgn}(a, b, c) = 0$ when the points are collinear. Geometrically, $\mathrm{sgn}(a, b, c) > 0$ indicates that point c lies on the positive side of pair (a, b) (i.e., c lies on the positive side of the line passing through points a and b). If $\mathrm{sgn}(a, b, c) < 0$, point c lies on the negative side of that line. Equations (11.1-2) and (11.1-3) give the same result if the sequence (c, a, b) or (b, c, a) is used because the direction of travel in the sequence is the same as for (a, b, c). However, the geometrical interpretation is different. For example, $\mathrm{sgn}(c, a, b) > 0$ indicates that point b lies on the positive side of the line through points c and a.

To prepare the data for the MPP algorithm, we form a list whose rows are the coordinates of each vertex and an additional element denoting whether the vertex is W or B. It is important that the concave vertices be mirrored, as in Fig. 11.7(c), that the vertices be in sequential order,[†] and that the first vertex be the uppermost leftmost vertex, which we know from property 5 is a W vertex of the MPP. Let V_0 denote this vertex. We assume that the vertices are arranged in the counterclockwise direction. The algorithm for finding MPPs uses two "crawler" points: a white crawler (W_C) and a black (B_C) crawler. W_C crawls along convex (W) vertices, and B_C crawls along mirrored concave (B) vertices. These two crawler points, the last MPP vertex found, and the vertex being examined are all that is necessary to implement the procedure.

The algorithm starts by setting $W_C = B_C = V_0$ (recall that V_0 is an MPP-vertex). Then, at any step in the algorithm, let V_L denote the last MPP vertex found, and let V_k denote the current vertex being examined. One of three conditions can exist between V_L, V_k, and the two crawler points:

(a) V_k lies to the positive side of the line through pair (V_L, W_C); that is, $\mathrm{sgn}(V_L, W_C, V_k) > 0$.
(b) V_k lies on the negative side of the line though pair (V_L, W_C) or is collinear with it; that is $\mathrm{sgn}(V_L, W_C, V_k) \leq 0$. At the same time, V_k lies to the positive

Assuming the coordinate system defined in Fig. 2.18(b), when traversing the boundary of a polygon in a counterclockwise direction, all points to the right of the direction of travel are *outside* the polygon. All points to the left of the direction of travel are *inside* the polygon.

[†]Vertices of a boundary can be ordered by tracking the boundary using, for example, the algorithm described in Section 11.1.1.

side of the line through (V_L, B_C) or is collinear with it; that is, $\text{sgn}(V_L, B_C, V_k) \geq 0$.

(c) V_k lies on the negative side of the line though pair (V_L, B_C); that is, $\text{sgn}(V_L, B_C, V_k) < 0$.

If condition (a) holds, the next MPP vertex is W_C, and we let $V_L = W_C$; then we reinitialize the algorithm by setting $W_C = B_C = V_L$, and continue with the next vertex after V_L.

If condition (b) holds, V_k becomes a *candidate* MPP vertex. In this case, we set $W_C = V_k$ if V_k is convex (i.e., it is a W vertex); otherwise we set $B_C = V_k$. We then continue with the next vertex in the list.

If condition (c) holds, the next MPP vertex is B_C and we let $V_L = B_C$; then we reinitialize the algorithm by setting $W_C = B_C = V_L$ and continue with the next vertex after V_L.

The algorithm terminates when it reaches the first vertex again, and thus has processed all the vertices in the polygon. The V_L vertices found by the algorithm are the vertices of the MPP. It has been proved that this algorithm finds all the MPP vertices of a polygon enclosed by a simply connected cellular complex (Sloboda et al. [1998]; Klette and Rosenfeld [2004]).

■ A manual example will help clarify the preceding concepts. Consider the vertices in Fig. 11.7(c). In our image coordinate system, the top left point of the grid is at coordinates $(0, 0)$. Assuming that the grid divisions are unity, the first few rows of the (counterclockwise) vertex list are:

EXAMPLE 11.2:
Illustration of the MPP algorithm.

$$
\begin{array}{lll}
V_0 & (1, 4) & W \\
V_1 & (2, 3) & B \\
V_2 & (3, 3) & W \\
V_3 & (3, 2) & B \\
V_4 & (4, 1) & W \\
V_5 & (7, 1) & W \\
V_6 & (8, 2) & B \\
V_7 & (9, 2) & B \\
\end{array}
$$

The first element of the list is always our first MPP, so we start by letting $W_C = B_C = V_0 = V_L = (1, 4)$. The next vertex is $V_1 = (2, 3)$. Evaluating the sgn function gives $\text{sgn}(V_L, W_C, V_1) = 0$ and $\text{sgn}(V_L, B_C, V_1) = 0$, so condition (b) holds. We let $B_C = V_1 = (2, 3)$ because V_1 is a B (concave) vertex. W_C remains unchanged. At this stage, crawler W_C is at $(1, 4)$, crawler B_C is at $(2, 3)$ and V_L is still at $(1, 4)$ because no new MPP-vertex was found.

Next, we look at $V_2 = (3, 3)$. The values of the sgn function are: $\text{sgn}(V_L, W_C, V_2) = 0$, and $\text{sgn}(V_L, B_C, V_2) = 1$, so condition (b) of the algorithm holds again. Because V_2 is a W (convex) vertex, we let $W_C = V_2 = (3, 3)$. At this stage, the crawlers are at $W_C = (3, 3)$ and $B_C = (2, 3)$; V_L remains unchanged.

The next vertex is $V_3 = (3, 2)$. The values of the sgn function are $sgn(V_L, W_C, V_3) = -2$ and $sgn(V_L, B_C, V_3) = 0$, so condition (b) holds again. Because V_3 is a B vertex, we update the black crawler, $B_C = (3, 2)$. Crawler W_C remains unchanged, as does V_L.

The next vertex is $V_4 = (4, 1)$ and we have $sgn(V_L, W_C, V_4) = -3$ and $sgn(V_L, B_C, V_4) = 0$ so condition (b) holds yet again. Because V_4 is a white vertex, we update the white crawler, $W_C = (4, 1)$. Black crawler B_C remains at $(3, 2)$, and V_L is still back at $(1, 4)$.

The next vertex is $V_5 = (7, 1)$ and $sgn(V_L, W_C, V_5) = 9$, so condition (a) holds, and we set $V_L = W_C = (4, 1)$. Because a new MPP vertex was found, we reinitialize the algorithm by setting $W_C = B_C = V_L$ and start again with the next vertex being the vertex after the newly found V_L. The next vertex is V_5, so we visit it again.

With $V_5 = (7, 1)$ and the new values of $V_L, W_C,$ and $B_C,$ we obtain $sgn(V_L, W_C, V_5) = 0$ and $sgn(V_L, B_C, V_5) = 0$, so condition (b) holds. Therefore, we let $W_C = V_5 = (7, 1)$ because V_5 is a W vertex.

The next vertex is $V_6 = (8, 2)$ and $sgn(V_L, W_C, V_6) = 3$, so condition (a) holds. Thus, we let $V_L = W_C = (7, 1)$ and reinitialize the algorithm by setting $W_C = B_C = V_L$.

Because of the reinitialization at $(7, 1)$, the next vertex considered is again $V_6 = (8, 2)$. Continuing as above with this and the remaining vertices yields the MPP vertices in Fig. 11.7(c). As mentioned earlier, the mirrored B vertices at $(2, 3), (3, 2)$ and on the lower-right side at $(13, 10)$, while being on the boundary of the MPP, are collinear and therefore are *not* considered vertices of the MPP. Appropriately, the algorithm did not detect them as such.   ■

EXAMPLE 11.3:
Applying the
MPP algorithm.

■ Figure 11.8(a) is a 566 × 566 binary image of a maple leaf and Fig. 11.8(b) is its 8-connected boundary. The sequence in Figs. 11.8(c) through (i) shows MMP representations of this boundary using square cellular complex cells of sizes 2, 3, 4, 6, 8, 16, and 32, respectively (the vertices in each figure were connected with straight lines to form a closed boundary). The leaf has two major features: a stem and three main lobes. The stem begins to be lost for cell sizes greater than 4 × 4, as Fig. 11.8(f) shows. The three main lobes are preserved reasonably well, even for a cell size of 16 × 16, as Fig. 11.8(h) shows. However, we see in Fig. 11.8(i) that by the time the cell size is increased to 32 × 32 this distinctive feature has been nearly lost.

The number of points in the original boundary [Fig. 11.8(b)] is 1900. The numbers of vertices in Figs. 11.8(c) through (i) are 206, 160, 127, 92, 66, 32, and 13, respectively. Figure 11.8(e), which has 127 vertices, retained all the major features of the original boundary while achieving a data reduction of over 90%. So here we see a significant advantage of MMPs for representing a boundary. Another important advantange is that MPPs perform boundary smoothing. As explained in the previous section, this is a usual requirement when representing a boundary by a chain code.   ■

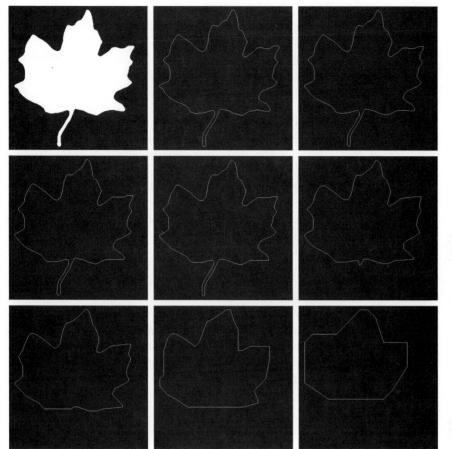

a b c
d e f
g h i

FIGURE 11.8
(a) 566×566 binary image.
(b) 8-connected boundary.
(c) through (i), MMPs obtained using square cells of sizes 2, 3, 4, 6, 8, 16, and 32, respectively (the vertices were joined by straight lines for display). The number of boundary points in (b) is 1900. The numbers of vertices in (c) through (i) are 206, 160, 127, 92, 66, 32, and 13, respectively.

11.1.4 Other Polygonal Approximation Approaches

At times, approaches that are conceptually simpler than the MPP algorithm discussed in the previous section can be used for polygonal approximations. In this section, we discuss two such approaches.

Merging techniques

Merging techniques based on average error or other criteria have been applied to the problem of polygonal approximation. One approach is to merge points along a boundary until the least square error line fit of the points merged so far exceeds a preset threshold. When this condition occurs, the parameters of the line are stored, the error is set to 0, and the procedure is repeated, merging new points along the boundary until the error again exceeds the threshold. At the end of the procedure the intersections of adjacent line segments form the vertices of the polygon. One of the principal difficulties with this method is that vertices in the resulting approximation do not always correspond to inflections (such as corners) in the original boundary, because a new line is not started until the error threshold is exceeded. If, for instance, a long straight line were being tracked and it turned a corner, a number (depending on the threshold) of points past the corner would be absorbed before the threshold was exceeded. However, splitting (discussed next) along with merging can be used to alleviate this difficulty.

Splitting techniques

One approach to boundary segment *splitting* is to subdivide a segment successively into two parts until a specified criterion is satisfied. For instance, a requirement might be that the maximum perpendicular distance from a boundary segment to the line joining its two end points not exceed a preset threshold. If it does, the point having the greatest distance from the line becomes a vertex, thus subdividing the initial segment into two subsegments. This approach has the advantage of seeking prominent inflection points. For a closed boundary, the best starting points usually are the two farthest points in the boundary. For example, Fig. 11.9(a) shows an object boundary, and Fig. 11.9(b) shows a subdivision of this boundary about its farthest points. The point marked c is the farthest point (in terms of perpendicular distance) from the top boundary segment to line ab. Similarly, point d is the farthest point in the bottom segment. Figure 11.9(c) shows the result of using the splitting procedure with a threshold equal to 0.25 times the length of line ab. As no point in the new boundary segments has a perpendicular distance (to its corresponding straight-line segment) that exceeds this threshold, the procedure terminates with the polygon in Fig. 11.9(d).

11.1.5 Signatures

A signature is a 1-D functional representation of a boundary and may be generated in various ways. One of the simplest is to plot the distance from the centroid to the boundary as a function of angle, as illustrated in Fig. 11.10. Regardless of how a signature is generated, however, the basic idea is to reduce the boundary representation to a 1-D function that presumably is easier to describe than the original 2-D boundary.

a b
c d

FIGURE 11.9
(a) Original boundary.
(b) Boundary divided into segments based on extreme points. (c) Joining of vertices.
(d) Resulting polygon.

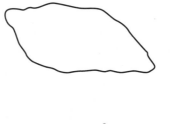
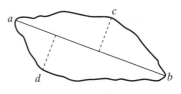
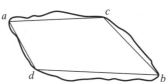
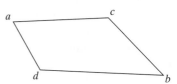

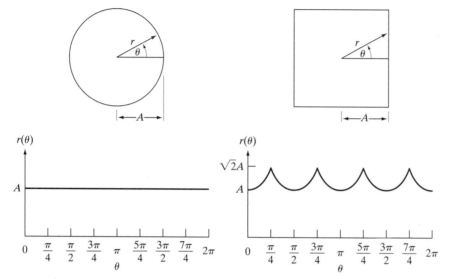

FIGURE 11.10
Distance-versus-angle signatures. In (a) $r(\theta)$ is constant. In (b), the signature consists of repetitions of the pattern $r(\theta) = A \sec\theta$ for $0 \le \theta \le \pi/4$ and $r(\theta) = A \csc\theta$ for $\pi/4 < \theta \le \pi/2$.

Signatures generated by the approach just described are invariant to translation, but they do depend on rotation and scaling. Normalization with respect to rotation can be achieved by finding a way to select the same starting point to generate the signature, regardless of the shape's orientation. One way to do so is to select the starting point as the point farthest from the centroid, assuming that this point is unique for each shape of interest. Another way is to select the point on the eigen axis (see Section 11.4) that is farthest from the centroid. This method requires more computation but is more rugged because the direction of the eigen axis is determined by using all contour points. Yet another way is to obtain the chain code of the boundary and then use the approach discussed in Section 11.1.2, assuming that the coding is coarse enough so that rotation does not affect its circularity.

Based on the assumptions of uniformity in scaling with respect to both axes, and that sampling is taken at equal intervals of θ, changes in size of a shape result in changes in the amplitude values of the corresponding signature. One way to normalize for this is to scale all functions so that they always span the same range of values, e.g., $[0, 1]$. The main advantage of this method is simplicity, but it has the potentially serious disadvantage that scaling of the entire function depends on only two values: the minimum and maximum. If the shapes are noisy, this dependence can be a source of significant error from object to object. A more rugged (but also more computationally intensive) approach is to divide each sample by the variance of the signature, assuming that the variance is not zero—as in the case of Fig. 11.10(a)—or so small that it creates computational difficulties. Use of the variance yields a variable scaling factor that is inversely proportional to changes in size and works much as automatic gain control does. Whatever the method used, keep in mind that the basic idea is to remove dependency on size while preserving the fundamental shape of the waveforms.

Distance versus angle is not the only way to generate a signature. For example, another way is to traverse the boundary and, corresponding to each point on the boundary, plot the angle between a line tangent to the boundary at that point and a reference line. The resulting signature, although quite different from the $r(\theta)$ curves in Fig. 11.10, would carry information about basic shape characteristics. For instance, horizontal segments in the curve would correspond to straight lines along the boundary, because the tangent angle would be constant there. A variation of this approach is to use the so-called *slope density function* as a signature. This function is a histogram of tangent-angle values. Because a histogram is a measure of concentration of values, the slope density function responds strongly to sections of the boundary with constant tangent angles (straight or nearly straight segments) and has deep valleys in sections producing rapidly varying angles (corners or other sharp inflections).

EXAMPLE 11.4:
Signatures of two simple objects.

■ Figures 11.11(a) and (b) show two binary objects and Figs. 11.11(c) and (d) are their boundaries. The corresponding $r(\theta)$ signatures in Figs. 11.11(e) and (f) range from 0° to 360° in increments of 1°. The number of prominent peaks in the signatures is sufficient to differentiate between the shapes of the two objects.   ■

11.1.6 Boundary Segments

Decomposing a boundary into segments is often useful. Decomposition reduces the boundary's complexity and thus simplifies the description process. This approach is particularly attractive when the boundary contains one or more significant concavities that carry shape information. In this case, use of the convex hull of the region enclosed by the boundary is a powerful tool for robust decomposition of the boundary.

As defined in Section 9.5.4, the *convex hull H* of an arbitrary set S is the smallest convex set containing S. The set difference $H - S$ is called the *convex deficiency D* of the set S. To see how these concepts might be used to partition a boundary into meaningful segments, consider Fig. 11.12(a), which shows an object (set S) and its convex deficiency (shaded regions). The region boundary can be partitioned by following the contour of S and marking the points at which a transition is made into or out of a component of the convex deficiency. Figure 11.12(b) shows the result in this case. Note that, in principle, this scheme is independent of region size and orientation.

In practice, digital boundaries tend to be irregular because of digitization, noise, and variations in segmentation. These effects usually result in convex deficiencies that have small, meaningless components scattered randomly throughout the boundary. Rather than attempt to sort out these irregularities by postprocessing, a common approach is to smooth a boundary prior to partitioning. There are a number of ways to do so. One way is to traverse the boundary and replace the coordinates of each pixel by the average coordinates of k of its neighbors along the boundary. This approach works for small irregularities, but it is time-consuming and difficult to control. Large values of k can result in excessive smoothing, whereas small values of k might not be sufficient in some

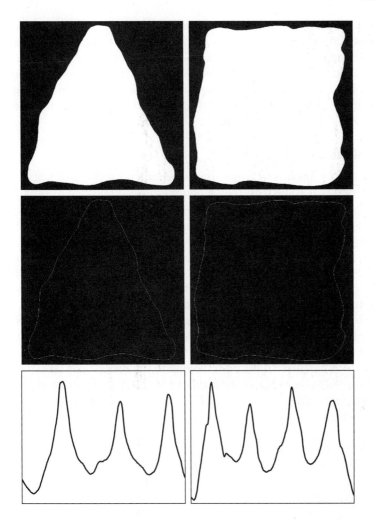

FIGURE 11.11
Two binary regions, their external boundaries, and their corresponding $r(\theta)$ signatures. The horizontal axes in (e) and (f) correspond to angles from $0°$ to $360°$, in increments of $1°$.

segments of the boundary. A more rugged technique is to use a polygonal approximation prior to finding the convex deficiency of a region. Most digital boundaries of interest are simple polygons (recall from Section 11.1.3 that these are polygons without self-intersection). Graham and Yao [1983] give an algorithm for finding the convex hull of such polygons.

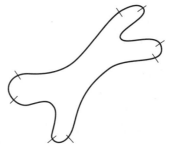

FIGURE 11.12
(a) A region, S, and its convex deficiency (shaded).
(b) Partitioned boundary.

The concepts of a convex hull and its deficiency are equally useful for describing an entire region, as well as just its boundary. For example, description of a region might be based on its area and the area of its convex deficiency, the number of components in the convex deficiency, the relative location of these components, and so on. Recall that a morphological algorithm for finding the convex hull was developed in Section 9.5.4. References cited at the end of this chapter contain other formulations.

11.1.7 Skeletons

An important approach to representing the structural shape of a plane region is to reduce it to a graph. This reduction may be accomplished by obtaining the *skeleton* of the region via a thinning (also called *skeletonizing*) algorithm. Thinning procedures play a central role in a broad range of problems in image processing, ranging from automated inspection of printed circuit boards to counting of asbestos fibers in air filters. We already discussed in Section 9.5.7 the basics of skeletonizing using morphology. However, as noted in that section, the procedure discussed there made no provisions for keeping the skeleton connected. The algorithm developed here corrects that problem.

The skeleton of a region may be defined via the medial axis transformation (MAT) proposed by Blum [1967]. The MAT of a region R with border B is as follows. For each point p in R, we find its closest neighbor in B. If p has more than one such neighbor, it is said to belong to the *medial axis* (skeleton) of R. The concept of "closest" (and the resulting MAT) depend on the definition of a distance (see Section 2.5.3). Figure 11.13 shows some examples using the Euclidean distance. The same results would be obtained with the maximum disk of Section 9.5.7.

The MAT of a region has an intuitive definition based on the so-called "prairie fire concept." Consider an image region as a prairie of uniform, dry grass, and suppose that a fire is lit along its border. All fire fronts will advance into the region at the same speed. The MAT of the region is the set of points reached by more than one fire front at the same time.

Although the MAT of a region yields an intuitively pleasing skeleton, direct implementation of this definition is expensive computationally. Implementation potentially involves calculating the distance from every interior

a b c

FIGURE 11.13
Medial axes (dashed) of three simple regions.

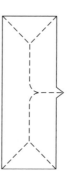
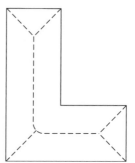

p_9	p_2	p_3
p_8	p_1	p_4
p_7	p_6	p_5

FIGURE 11.14
Neighborhood arrangement used by the thinning algorithm.

point to every point on the boundary of a region. Numerous algorithms have been proposed for improving computational efficiency while at the same time attempting to produce a medial axis representation of a region. Typically, these are thinning algorithms that iteratively delete boundary points of a region subject to the constraints that deletion of these points (1) does not remove end points, (2) does not break connectivity, and (3) does not cause excessive erosion of the region.

In this section we present an algorithm for thinning binary regions. Region points are assumed to have value 1 and background points to have value 0. The method consists of successive passes of two basic steps applied to the border points of the given region, where, based on the definition given in Section 2.5.2, a *border point* is any pixel with value 1 and having at least one neighbor valued 0. With reference to the 8-neighborhood notation in Fig. 11.14, Step 1 *flags* a contour point p_1 for *deletion* if the following conditions are satisfied:

(a) $2 \leq N(p_1) \leq 6$
(b) $T(p_1) = 1$
(c) $p_2 \cdot p_4 \cdot p_6 = 0$
(d) $p_4 \cdot p_6 \cdot p_8 = 0$ (11.1-4)

where $N(p_1)$ is the number of nonzero neighbors of p_1; that is,

$$N(p_1) = p_2 + p_3 + \cdots + p_8 + p_9 \qquad (11.1\text{-}5)$$

where p_i is either 0 or 1, and $T(p_1)$ is the number of 0–1 transitions in the ordered sequence $p_2, p_3, \ldots, p_8, p_9, p_2$. For example, $N(p_1) = 4$ and $T(p_1) = 3$ in Fig. 11.15.

In Step 2, conditions (a) and (b) remain the same, but conditions (c) and (d) are changed to

(c′) $p_2 \cdot p_4 \cdot p_8 = 0$
(d′) $p_2 \cdot p_6 \cdot p_8 = 0$ (11.1-6)

0	0	1
1	p_1	0
1	0	1

FIGURE 11.15
Illustration of conditions (a) and (b) in Eq. (11.1-4). In this case $N(p_1) = 4$ and $T(p_1) = 3$.

Step 1 is applied to every border pixel in the binary region under consideration. If one or more of conditions (a)–(d) are violated, the value of the point in question is not changed. If all conditions are satisfied, the point is flagged for deletion. However, the point is not deleted until all border points have been processed. This delay prevents changing the structure of the data during execution of the algorithm. After Step 1 has been applied to all border points, those that were flagged are deleted (changed to 0). Then Step 2 is applied to the resulting data in exactly the same manner as Step 1.

Thus, one iteration of the thinning algorithm consists of (1) applying Step 1 to flag border points for deletion; (2) deleting the flagged points; (3) applying Step 2 to flag the remaining border points for deletion; and (4) deleting the flagged points. This basic procedure is applied iteratively until no further points are deleted, at which time the algorithm terminates, yielding the skeleton of the region.

Condition (a) is violated when contour point p_1 has only one or seven 8-neighbors valued 1. Having only one such neighbor implies that p_1 is the end point of a skeleton stroke and obviously should not be deleted. Deleting p_1 if it had seven such neighbors would cause erosion into the region. Condition (b) is violated when it is applied to points on a stroke 1 pixel thick. Hence this condition prevents breaking segments of a skeleton during the thinning operation. Conditions (c) and (d) are satisfied simultaneously by the minimum set of values: $(p_4 = 0 \ or \ p_6 = 0)$ or $(p_2 = 0 \ and \ p_8 = 0)$. Thus with reference to the neighborhood arrangement in Fig. 11.14, a point that satisfies these conditions, as well as conditions (a) and (b), is an east or south boundary point or a northwest corner point in the boundary. In either case, p_1 is not part of the skeleton and should be removed. Similarly, conditions (c′) and (d′) are satisfied simultaneously by the following minimum set of values: $(p_2 = 0 \ or \ p_8 = 0)$ or $(p_4 = 0 \ and \ p_6 = 0)$. These correspond to north or west boundary points, or a southeast corner point. Note that northeast corner points have $p_2 = 0$ and $p_4 = 0$ and thus satisfy conditions (c) and (d), as well as (c′) and (d′). The same is true for southwest corner points, which have $p_6 = 0$ and $p_8 = 0$.

FIGURE 11.16
Human leg bone and skeleton of the region shown superimposed.

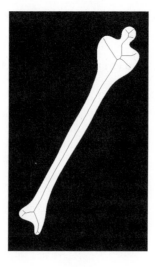

▪ Figure 11.16 shows a segmented image of a human leg bone and, superimposed, the skeleton of the region. For the most part, the skeleton looks intuitively correct. There is a double branch on the right side of the "shoulder" of the bone that at first glance one would expect to be a single branch, as on the corresponding left side. Note, however, that the right shoulder is somewhat broader (in the long direction) than the left shoulder. That is what caused the branch to be created by the algorithm. This type of unpredictable behavior is not unusual in skeletonizing algorithms.   ▪

EXAMPLE 11.5:
The skeleton of a region.

11.2 Boundary Descriptors

In this section, we consider several approaches to describing the boundary of a region, and in Section 11.3 we focus on regional descriptors. Parts of Sections 11.4 and 11.5 are applicable to both boundaries and regions.

11.2.1 Some Simple Descriptors

The *length* of a boundary is one of its simplest descriptors. The number of pixels along a boundary gives a rough approximation of its length. For a chain-coded curve with unit spacing in both directions, the number of vertical and horizontal components plus $\sqrt{2}$ times the number of diagonal components gives its exact length.

The *diameter* of a boundary B is defined as

$$\text{Diam}(B) = \max_{i,j}\left[D(p_i, p_j)\right] \tag{11.2-1}$$

where D is a distance measure (see Section 2.5.3) and p_i and p_j are points on the boundary. The value of the diameter and the orientation of a line segment connecting the two extreme points that comprise the diameter (this line is called the *major axis* of the boundary) are useful descriptors of a boundary. The *minor axis* of a boundary is defined as the line perpendicular to the major axis, and of such length that a box passing through the outer four points of intersection of the boundary with the two axes completely encloses the boundary.[†] The box just described is called the *basic rectangle*, and the ratio of the major to the minor axis is called the *eccentricity* of the boundary. This also is a useful descriptor.

Curvature is defined as the rate of change of slope. In general, obtaining reliable measures of curvature at a point in a digital boundary is difficult because these boundaries tend to be locally "ragged." However, using the difference between the slopes of adjacent boundary segments (which have been represented as straight lines) as a descriptor of curvature at the point of intersection of the segments sometimes proves useful. For example, the vertices of boundaries such as those shown in Fig. 11.6(c) lend themselves well to curvature descriptions. As the boundary is traversed in the clockwise direction, a vertex point p is said to be part of a *convex* segment if the change in slope at p is nonnegative;

[†]Do not confuse this definition of major and minor axes with the eigen axes defined in Section 11.4.

otherwise, p is said to belong to a segment that is *concave*. The description of curvature at a point can be refined further by using ranges in the change of slope. For instance, p could be part of a nearly straight segment if the change is less than $10°$ or a *corner* point if the change exceeds $90°$. These descriptors must be used with care because their interpretation depends on the length of the individual segments relative to the overall length of the boundary.

11.2.2 Shape Numbers

As explained in Section 11.1.2, the first difference of a chain-coded boundary depends on the starting point. The *shape number* of such a boundary, based on the 4-directional code of Fig. 11.3(a), is defined as the first difference of smallest magnitude. The *order n* of a shape number is defined as the number of digits in its representation. Moreover, n is even for a closed boundary, and its value limits the number of possible different shapes. Figure 11.17 shows all the shapes of order 4, 6, and 8, along with their chain-code representations, first differences, and corresponding shape numbers. Note that the first difference is computed by treating the chain code as a circular sequence, as discussed in Section 11.1.2. Although the first difference of a chain code is independent of rotation, in general the coded boundary depends on the orientation of the grid. One way to normalize the grid orientation is by aligning the chain-code grid with the sides of the basic rectangle defined in the previous section.

In practice, for a desired shape order, we find the rectangle of order n whose eccentricity (defined in the previous section) best approximates that of the basic rectangle and use this new rectangle to establish the grid size. For

FIGURE 11.17
All shapes of order 4, 6, and 8. The directions are from Fig. 11.3(a), and the dot indicates the starting point.

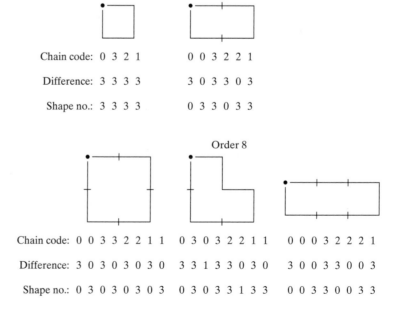

example, if $n = 12$, all the rectangles of order 12 (that is, those whose perimeter length is 12) are $2 \times 4, 3 \times 3$, and 1×5. If the eccentricity of the 2×4 rectangle best matches the eccentricity of the basic rectangle for a given boundary, we establish a 2×4 grid centered on the basic rectangle and use the procedure outlined in Section 11.1.2 to obtain the chain code. The shape number follows from the first difference of this code. Although the order of the resulting shape number usually equals n because of the way the grid spacing was selected, boundaries with depressions comparable to this spacing sometimes yield shape numbers of order greater than n. In this case, we specify a rectangle of order lower than n and repeat the procedure until the resulting shape number is of order n.

■ Suppose that $n = 18$ is specified for the boundary in Fig. 11.18(a). To obtain a shape number of this order requires following the steps just discussed. The first step is to find the basic rectangle, as shown in Fig. 11.18(b). The closest rectangle of order 18 is a 3×6 rectangle, requiring subdivision of the basic rectangle as shown in Fig. 11.18(c), where the chain-code directions are aligned with the resulting grid. The final step is to obtain the chain code and use its first difference to compute the shape number, as shown in Fig. 11.18(d). ■

EXAMPLE 11.6:
Computing shape numbers.

a	b
c	d

FIGURE 11.18
Steps in the generation of a shape number.

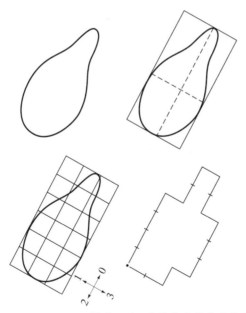

Chain code: 0 0 0 0 3 0 0 3 2 2 3 2 2 2 1 2 1 1

Difference: 3 0 0 0 3 1 0 3 3 0 1 3 0 0 3 1 3 0

Shape no.: 0 0 0 3 1 0 3 3 0 1 3 0 0 3 1 3 0 3

11.2.3 Fourier Descriptors

Figure 11.19 shows a K-point digital boundary in the xy-plane. Starting at an arbitrary point (x_0, y_0), coordinate pairs $(x_0, y_0), (x_1, y_1), (x_2, y_2), \ldots,$ (x_{K-1}, y_{K-1}) are encountered in traversing the boundary, say, in the counterclockwise direction. These coordinates can be expressed in the form $x(k) = x_k$ and $y(k) = y_k$. With this notation, the boundary itself can be represented as the sequence of coordinates $s(k) = [x(k), y(k)]$, for $k = 0, 1, 2, \ldots, K - 1$. Moreover, each coordinate pair can be treated as a complex number so that

$$s(k) = x(k) + jy(k) \tag{11.2-2}$$

for $k = 0, 1, 2, \ldots, K - 1$. That is, the x-axis is treated as the real axis and the y-axis as the imaginary axis of a sequence of complex numbers. Although the interpretation of the sequence was recast, the nature of the boundary itself was not changed. Of course, this representation has one great advantage: It reduces a 2-D to a 1-D problem.

From Eq. (4.4-6), the discrete Fourier transform (DFT) of $s(k)$ is

$$a(u) = \sum_{k=0}^{K-1} s(k)e^{-j2\pi uk/K} \tag{11.2-3}$$

for $u = 0, 1, 2, \ldots, K - 1$. The complex coefficients $a(u)$ are called the *Fourier descriptors* of the boundary. The inverse Fourier transform of these coefficients restores $s(k)$. That is, from Eq. (4.4-7),

$$s(k) = \frac{1}{K} \sum_{u=0}^{K-1} a(u)e^{j2\pi uk/K} \tag{11.2-4}$$

for $k = 0, 1, 2, \ldots, K - 1$. Suppose, however, that instead of all the Fourier coefficients, only the first P coefficients are used. This is equivalent to setting $a(u) = 0$ for $u > P - 1$ in Eq. (11.2-4). The result is the following *approximation* to $s(k)$:

$$\hat{s}(k) = \frac{1}{P} \sum_{u=0}^{P-1} a(u)e^{j2\pi uk/P} \tag{11.2-5}$$

FIGURE 11.19
A digital boundary and its representation as a complex sequence. The points (x_0, y_0) and (x_1, y_1) shown are (arbitrarily) the first two points in the sequence.

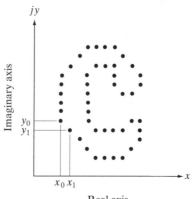

for $k = 0, 1, 2, \ldots, K - 1$. Although only P terms are used to obtain each component of $\hat{s}(k)$, k still ranges from 0 to $K - 1$. That is, the *same* number of points exists in the approximate boundary, but not as many terms are used in the reconstruction of each point. Recall from discussions of the Fourier transform in Chapter 4 that high-frequency components account for fine detail, and low-frequency components determine global shape. Thus, the smaller P becomes, the more detail that is lost on the boundary, as the following example demonstrates.

■ Figure 11.20(a) shows the boundary of a human chromosome, consisting of 2868 points. The corresponding 2868 Fourier descriptors were obtained for this boundary using Eq. (11.2-3). The objective of this example is to examine the effects of reconstructing the boundary based on decreasing the number of Fourier descriptors. Figure 11.20(b) shows the boundary reconstructed using one-half of the 2868 descriptors. It is interesting to note that there is no perceptible difference between this boundary and the original. Figures 11.20(c) through (h) show the boundaries reconstructed with the number of Fourier

EXAMPLE 11.7:
Using Fourier descriptors.

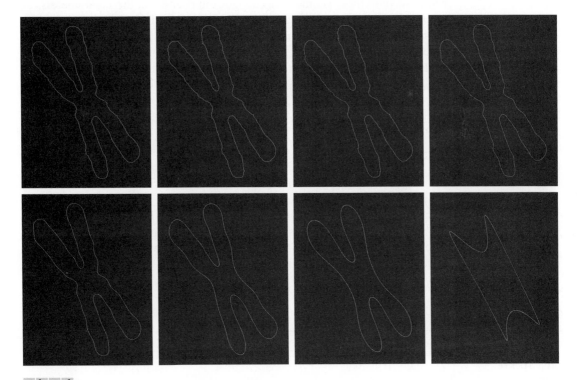

a b c d
e f g h

FIGURE 11.20 (a) Boundary of human chromosome (2868 points). (b)–(h) Boundaries reconstructed using 1434, 286, 144, 72, 36, 18, and 8 Fourier descriptors, respectively. These numbers are approximately 50%, 10%, 5%, 2.5%, 1.25%, 0.63%, and 0.28% of 2868, respectively.

descriptors being 10%, 5%, 2.5%, 1.25%, 0.63% and 0.28% of 2868, respectively. These percentages are equal approximately to 286, 144, 72, 36, 18, and 8 descriptors, respectively, where the numbers were rounded to the nearest even integer. The important point here is that 18 descriptors, a mere six-tenths of one percent of the original 2868 descriptors, were sufficient to retain the principal shape features of the original boundary: four long protrusions and two deep bays. Figure 11.20(h), obtained with 8 descriptors, is an unacceptable result because the principal features are lost. Further reductions to 4 and 2 descriptors would result in an ellipse and a circle, respectively (Problem 11.13).   ■

As the preceding example demonstrates, a few Fourier descriptors can be used to capture the gross essence of a boundary. This property is valuable, because these coefficients carry shape information. Thus they can be used as the basis for differentiating between distinct boundary shapes, as we discuss in Chapter 12.

We have stated several times that descriptors should be as insensitive as possible to translation, rotation, and scale changes. In cases where results depend on the order in which points are processed, an additional constraint is that descriptors should be insensitive to the starting point. Fourier descriptors are not directly insensitive to these geometrical changes, but changes in these parameters can be related to simple transformations on the descriptors. For example, consider rotation, and recall from basic mathematical analysis that rotation of a point by an angle θ about the origin of the complex plane is accomplished by multiplying the point by $e^{j\theta}$. Doing so to every point of $s(k)$ rotates the entire sequence about the origin. The rotated sequence is $s(k)e^{j\theta}$, whose Fourier descriptors are

$$a_r(u) = \sum_{k=0}^{K-1} s(k)e^{j\theta}e^{-j2\pi uk/K}$$

$$= a(u)e^{j\theta} \tag{11.2-6}$$

for $u = 0, 1, 2, \ldots, K - 1$. Thus rotation simply affects all coefficients equally by a multiplicative *constant* term $e^{j\theta}$.

Table 11.1 summarizes the Fourier descriptors for a boundary sequence $s(k)$ that undergoes rotation, translation, scaling, and changes in starting point. The symbol Δ_{xy} is defined as $\Delta_{xy} = \Delta x + j\Delta y$, so the notation $s_t(k) = s(k) + \Delta_{xy}$ indicates redefining (translating) the sequence as

$$s_t(k) = [x(k) + \Delta x] + j[y(k) + \Delta y] \tag{11.2-7}$$

TABLE 11.1
Some basic properties of Fourier descriptors.

Transformation	Boundary	Fourier Descriptor
Identity	$s(k)$	$a(u)$
Rotation	$s_r(k) = s(k)e^{j\theta}$	$a_r(u) = a(u)e^{j\theta}$
Translation	$s_t(k) = s(k) + \Delta_{xy}$	$a_t(u) = a(u) + \Delta_{xy}\delta(u)$
Scaling	$s_s(k) = \alpha s(k)$	$a_s(u) = \alpha a(u)$
Starting point	$s_p(k) = s(k - k_0)$	$a_p(u) = a(u)e^{-j2\pi k_0 u/K}$

In other words, translation consists of adding a constant displacement to all coordinates in the boundary. Note that translation has no effect on the descriptors, except for $u = 0$, which has the impulse $\delta(u)$.[†] Finally, the expression
$s_p(k) = s(k - k_0)$ means redefining the sequence as

$$s_p = x(k - k_0) + jy(k - k_0) \tag{11.2-8}$$

which merely changes the starting point of the sequence to $k = k_0$ from
$k = 0$. The last entry in Table 11.1 shows that a change in starting point affects
all descriptors in a different (but known) way, in the sense that the term multiplying $a(u)$ depends on u.

11.2.4 Statistical Moments

The shape of boundary segments (and of signature waveforms) can be described
quantitatively by using statistical moments, such as the mean, variance, and higher-
order moments. To see how this can be accomplished, consider Fig. 11.21(a), which
shows the segment of a boundary, and Fig. 11.21(b), which shows the segment
represented as a 1-D function $g(r)$ of an arbitrary variable r. This function is obtained by connecting the two end points of the segment and rotating the line
segment until it is horizontal. The coordinates of the points are rotated by the
same angle.

Consult the book Web site
for a brief review of probability theory.

Let us treat the amplitude of g as a discrete random variable v and form
an amplitude histogram $p(v_i)$, $i = 0, 1, 2, \ldots, A - 1$, where A is the number
of discrete amplitude increments in which we divide the amplitude scale.
Then, keeping in mind that $p(v_i)$ is an estimate of the probability of value v_i
occurring, it follows from Eq. (3.3-17) that the nth moment of v about its
mean is

$$\mu_n(v) = \sum_{i=0}^{A-1} (v_i - m)^n p(v_i) \tag{11.2-9}$$

where

$$m = \sum_{i=0}^{A-1} v_i p(v_i) \tag{11.2-10}$$

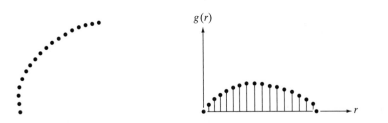

$g(r)$

a b

FIGURE 11.21
(a) Boundary
segment.
(b) Representation
as a 1-D function.

[†]Recall from Chapter 4 that the Fourier transform of a constant is an impulse located at the origin.
Recall also that the impulse is zero everywhere else.

The quantity m is recognized as the mean or average value of v and μ_2 as its variance. Generally, only the first few moments are required to differentiate between signatures of clearly distinct shapes.

An alternative approach is to normalize $g(r)$ to unit area and treat it as a histogram. In other words, $g(r_i)$ is now treated as the probability of value r_i occurring. In this case, r is treated as the random variable and the moments are

$$\mu_n(r) = \sum_{i=0}^{K-1} (r_i - m)^n g(r_i) \tag{11.2-11}$$

where

$$m = \sum_{i=0}^{K-1} r_i g(r_i) \tag{11.2-12}$$

In this notation, K is the number of points on the boundary, and $\mu_n(r)$ is directly related to the shape of $g(r)$. For example, the second moment $\mu_2(r)$ measures the spread of the curve about the mean value of r and the third moment $\mu_3(r)$ measures its symmetry with reference to the mean.

Basically, what we have accomplished is to reduce the description task to that of describing 1-D functions. Although moments are by far the most popular method, they are not the only descriptors used for this purpose. For instance, another method involves computing the 1-D discrete Fourier transform, obtaining its spectrum, and using the first q components of the spectrum to describe $g(r)$. The advantage of moments over other techniques is that implementation of moments is straightforward and they also carry a "physical" interpretation of boundary shape. The insensitivity of this approach to rotation is clear from Fig. 11.21. Size normalization, if desired, can be achieved by scaling the range of values of g and r.

11.3 Regional Descriptors

In this section we consider various approaches for describing image regions. Keep in mind that it is common practice to use both boundary and regional descriptors combined.

11.3.1 Some Simple Descriptors

The *area* of a region is defined as the number of pixels in the region. The *perimeter* of a region is the length of its boundary. Although area and perimeter are sometimes used as descriptors, they apply primarily to situations in which the size of the regions of interest is invariant. A more frequent use of these two descriptors is in measuring *compactness* of a region, defined as (perimeter)^2/area. A slightly different (within a scalar multiplier) descriptor of compactness is the *circularity ratio*, defined as the ratio of the area of a region to the area of a circle (the most compact shape) having the *same* perimeter. The area of a circle with perimeter length P is $P^2/4\pi$. Therefore, the circularity ratio, R_c, is given by the expression

$$R_c = \frac{4\pi A}{P^2} \tag{11.3-1}$$

where A is the area of the region in question and P is the length of its perimeter. The value of this measure is 1 for a circular region and $\pi/4$ for a square. Compactness is a dimensionless measure and thus is insensitive to uniform scale changes; it is insensitive also to orientation, ignoring, of course, computational errors that may be introduced in resizing and rotating a digital region.

Other simple measures used as region descriptors include the mean and median of the intensity levels, the minimum and maximum intensity values, and the number of pixels with values above and below the mean.

■ Even a simple region descriptor such as normalized area can be quite useful in extracting information from images. For instance, Fig. 11.22 shows a satellite infrared image of the Americas. As discussed in Section 1.3.4, images such as these provide a global inventory of human settlements. The sensor used to collect these images has the capability to detect visible and near infrared emissions, such as lights, fires, and flares. The table alongside the images shows (by region from top to bottom) the ratio of the area occupied by white (the lights) to the total light area in all four regions. A simple measurement like this can give, for example, a relative estimate by region of electrical energy consumed. The data can be refined by normalizing it with respect to land mass per region, with respect to population numbers, and so on.   ■

EXAMPLE 11.8:
Using area computations to extract information from images.

11.3.2 Topological Descriptors

Topological properties are useful for global descriptions of regions in the image plane. Simply defined, *topology* is the study of properties of a figure that are unaffected by any deformation, as long as there is no tearing or joining of the figure (sometimes these are called *rubber-sheet* distortions). For example, Fig. 11.23 shows a region with two holes. Thus if a topological descriptor is defined by the number of holes in the region, this property obviously will not be affected by a stretching or rotation transformation. In general, however, the number of holes will change if the region is torn or folded. Note that, as stretching affects distance, topological properties do not depend on the notion of distance or any properties implicitly based on the concept of a distance measure.

Another topological property useful for region description is the number of connected components. A *connected component* of a region was defined in Section 2.5.2. Figure 11.24 shows a region with three connected components. (See Section 9.5.3 regarding an algorithm for computing connected components.)

The number of holes H and connected components C in a figure can be used to define the *Euler number E*:

$$E = C - H \tag{11.3-2}$$

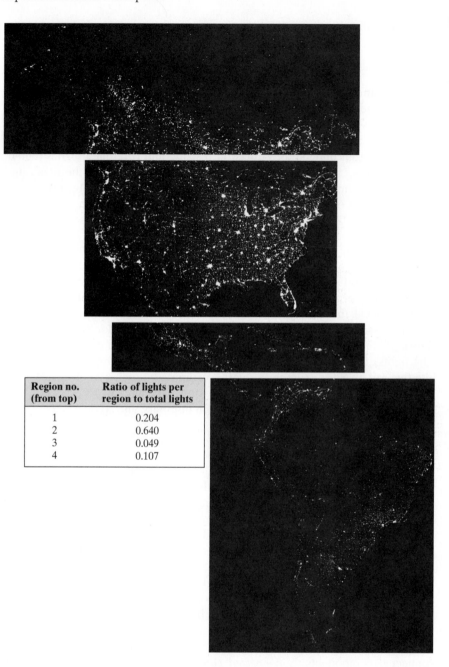

Region no. (from top)	Ratio of lights per region to total lights
1	0.204
2	0.640
3	0.049
4	0.107

FIGURE 11.22 Infrared images of the Americas at night. (Courtesy of NOAA.)

FIGURE 11.23
A region with
two holes.

The Euler number is also a topological property. The regions shown in Fig. 11.25, for example, have Euler numbers equal to 0 and -1, respectively, because the "A" has one connected component and one hole and the "B" one connected component but two holes.

Regions represented by straight-line segments (referred to as *polygonal networks*) have a particularly simple interpretation in terms of the Euler number. Figure 11.26 shows a polygonal network. Classifying interior regions of such a network into faces and holes is often important. Denoting the number of vertices by V, the number of edges by Q, and the number of faces by F gives the following relationship, called the *Euler formula*:

$$V - Q + F = C - H$$

which, in view of Eq. (11.3-2), is equal to the Euler number:

$$V - Q + F = C - H$$
$$= E \qquad\qquad (11.3\text{-}3)$$

The network in Fig. 11.26 has 7 vertices, 11 edges, 2 faces, 1 connected region, and 3 holes; thus the Euler number is -2:

$$7 - 11 + 2 = 1 - 3 = -2$$

Topological descriptors provide an additional feature that is often useful in characterizing regions in a scene.

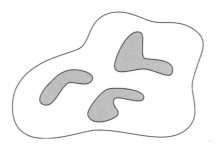

FIGURE 11.24
A region with
three connected
components.

FIGURE 11.25
Regions with
Euler numbers
equal to 0 and −1,
respectively.

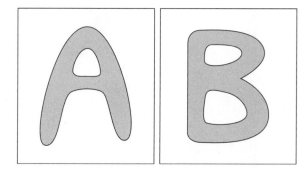

FIGURE 11.26 A
region containing
a polygonal
network.

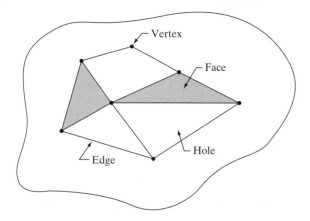

EXAMPLE 11.9:
Use of connected
components for
extracting the
largest features in
a segmented
image.

■ Figure 11.27(a) shows a 512 × 512, 8-bit image of Washington, D.C. taken by a NASA LANDSAT satellite. This particular image is in the near infrared band (see Fig. 1.10 for details). Suppose that we want to segment the river using only this image (as opposed to using several multispectral images, which would simplify the task). Since the river is a rather dark, uniform region of the image, thresholding is an obvious thing to try. The result of thresholding the image with the highest possible threshold value before the river became a disconnected region is shown in Fig. 11.27(b). The threshold was selected manually to illustrate the point that it would be impossible in this case to segment the river by itself without other regions of the image also appearing in the thresholded result. The objective of this example is to illustrate how connected components can be used to "finish" the segmentation.

The image in Fig. 11.27(b) has 1591 connected components (obtained using 8-connectivity) and its Euler number is 1552, from which we deduce that the number of holes is 39. Figure 11.27(c) shows the connected component with the largest number of elements (8479). This is the desired result, which we already know cannot be segmented by itself from the image using a threshold. Note how clean this result is. If we wanted to perform measurements, like the length of each branch of the river, we could use the skeleton of the connected component [Fig. 11.27(d)] to do so. In other words, the

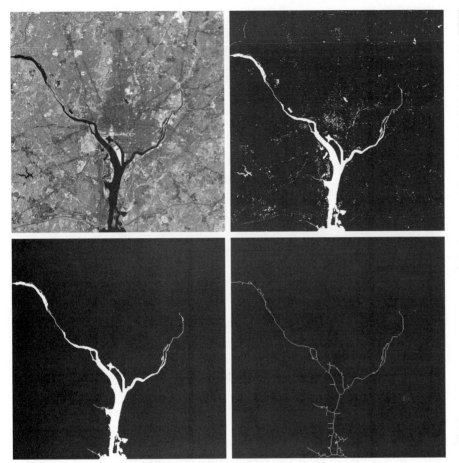

FIGURE 11.27
(a) Infrared
image of the
Washington,
D.C. area.
(b) Thresholded
image. (c) The
largest connected
component of
(b). Skeleton
of (c).

length of each branch in the skeleton would be a reasonably close approximation to the length of the river branch it represents.   ■

11.3.3 Texture

An important approach to region description is to quantify its *texture* content. Although no formal definition of texture exists, intuitively this descriptor provides measures of properties such as smoothness, coarseness, and regularity (Fig. 11.28 shows some examples). The three principal approaches used in image processing to describe the texture of a region are statistical, structural, and spectral. Statistical approaches yield characterizations of textures as smooth, coarse, grainy, and so on. Structural techniques deal with the arrangement of image primitives, such as the description of texture based on regularly spaced parallel lines. Spectral techniques are based on properties of the Fourier spectrum and are used primarily to detect global periodicity in an image by identifying high-energy, narrow peaks in the spectrum.

FIGURE 11.28
The white squares mark, from left to right, smooth, coarse, and regular textures. These are optical microscope images of a superconductor, human cholesterol, and a microprocessor. (Courtesy of Dr. Michael W. Davidson, Florida State University.)

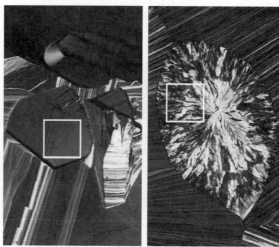

Statistical approaches

One of the simplest approaches for describing texture is to use statistical moments of the intensity histogram of an image or region. Let z be a random variable denoting intensity and let $p(z_i), i = 0, 1, 2, \ldots, L - 1$, be the corresponding histogram, where L is the number of distinct intensity levels. From Eq. (3.3-17), the nth moment of z about the mean is

$$\mu_n(z) = \sum_{i=0}^{L-1} (z_i - m)^n p(z_i) \tag{11.3-4}$$

where m is the mean value of z (the average intensity):

$$m = \sum_{i=0}^{L-1} z_i p(z_i) \tag{11.3-5}$$

Note from Eq. (11.3-4) that $\mu_0 = 1$ and $\mu_1 = 0$. The second moment [the *variance* $\sigma^2(z) = \mu_2(z)$] is of particular importance in texture description. It is a measure of intensity contrast that can be used to establish descriptors of relative smoothness. For example, the measure

$$R(z) = 1 - \frac{1}{1 + \sigma^2(z)} \tag{11.3-6}$$

is 0 for areas of constant intensity (the variance is zero there) and approaches 1 for large values of $\sigma^2(z)$. Because variance values tend to be large for gray-scale images with values, for example, in the range 0 to 255, it is a good idea to normalize the variance to the interval $[0, 1]$ for use in Eq. (11.3-6). This is done simply by dividing $\sigma^2(z)$ by $(L - 1)^2$ in Eq. (11.3-6). The standard deviation, $\sigma(z)$, also is used frequently as a measure of texture because values of the standard deviation tend to be more intuitive to many people.

The third moment,

$$\mu_3(z) = \sum_{i=0}^{L-1} (z_i - m)^3 p(z_i) \tag{11.3-7}$$

is a measure of the skewness of the histogram while the fourth moment is a measure of its relative flatness. The fifth and higher moments are not so easily related to histogram shape, but they do provide further quantitative discrimination of texture content. Some useful additional texture measures based on histograms include a measure of "uniformity," given by

$$U(z) = \sum_{i=0}^{L-1} p^2(z_i) \tag{11.3-8}$$

and an *average entropy* measure, which, you will recall from basic information theory, is defined as

$$e(z) = -\sum_{i=0}^{L-1} p(z_i) \log_2 p(z_i) \tag{11.3-9}$$

Because the ps have values in the range $[0, 1]$ and their sum equals 1, measure U is maximum for an image in which all intensity levels are equal (maximally uniform), and decreases from there. Entropy is a measure of variability and is 0 for a constant image.

■ Table 11.2 summarizes the values of the preceding measures for the three types of textures highlighted in Fig. 11.28. The mean just tells us the average intensity of each region and is useful only as a rough idea of intensity, not really texture. The standard deviation is much more informative; the numbers clearly show that the first texture has significantly less variability in intensity levels (it is smoother) than the other two textures. The coarse texture shows up clearly in this measure. As expected, the same comments hold for R, because it measures essentially the same thing as the standard deviation. The third moment generally is useful for determining the degree of symmetry of histograms and whether they are skewed to the left (negative value) or the right (positive value). This gives a rough idea of whether the intensity levels are biased toward the dark or light side of the mean. In terms of texture, the information derived from the third moment is useful only when variations between measurements are large. Looking at the measure of uniformity, we again conclude

EXAMPLE 11.10:
Texture measures based on histograms.

Texture	Mean	Standard deviation	R (normalized)	Third moment	Uniformity	Entropy
Smooth	82.64	11.79	0.002	−0.105	0.026	5.434
Coarse	143.56	74.63	0.079	−0.151	0.005	7.783
Regular	99.72	33.73	0.017	0.750	0.013	6.674

TABLE 11.2
Texture measures for the subimages shown in Fig. 11.28.

that the first subimage is smoother (more uniform than the rest) and that the most random (lowest uniformity) corresponds to the coarse texture. This is not surprising. Finally, the entropy values are in the opposite order and thus lead us to the same conclusions as the uniformity measure did. The first subimage has the lowest variation in intensity levels and the coarse image the most. The regular texture is in between the two extremes with respect to both these measures. ■

Measures of texture computed using only histograms carry no information regarding the relative position of pixels with respect to each other. This is important when describing texture, and one way to incorporate this type of information into the texture-analysis process is to consider not only the distribution of intensities, but also the *relative positions* of pixels in an image.

Let Q be an operator that defines the position of two pixels relative to each other, and consider an image, f, with L possible intensity levels. Let $\mathbf{G}$ be a matrix whose element g_{ij} is the number of times that pixel pairs with intensities z_i and z_j occur in f in the position specified by Q, where $1 \leq i, j \leq L$. A matrix formed in this manner is referred to as a *gray-level* (or *intensity*) *co-occurrence matrix*. When the meaning is clear, $\mathbf{G}$ is referred to simply as a *co-occurrence matrix*.

Figure 11.29 shows an example of how to construct a co-occurrence matrix using $L = 8$ and a position operator Q defined as "one pixel immediately to the right" (i.e., the neighbor of a pixel is defined as the pixel immediately to its right). The array on the left is a small image under consideration and the array on the right is matrix $\mathbf{G}$. We see that element $(1, 1)$ of $\mathbf{G}$ is 1, because there is only one occurrence in f of a pixel valued 1 having a pixel valued 1 immediately to its right. Similarly, element $(6, 2)$ of $\mathbf{G}$ is 3, because there are three occurrences in f of a pixel with a value of 6 having a pixel valued 2 immediately to its right. The other elements of $\mathbf{G}$ are computed in this manner. If we had defined Q as, say, "one pixel to the right and one pixel above," then

Note that we are using the intensity range $[1, L]$ instead of our usual $[0, L - 1]$. This is done so that intensity values will correspond with "traditional" matrix indexing (i.e., intensity value 1 corresponds to the first row and column indices of $\mathbf{G}$).

FIGURE 11.29
How to generate a co-occurrence matrix.

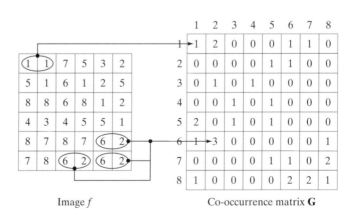

Image f Co-occurrence matrix $\mathbf{G}$

position $(1, 1)$ in **G** would have been 0, because there are no instances in f of a 1 with another 1 in the position specified by Q. On the other hand, positions $(1, 3), (1, 5)$, and $(1, 7)$ in **G** would all be 1s because intensity value 1 occurs in f with neighbors valued 3, 5, and 7 in the position specified by Q, one time each. As an exercise, you should compute all the elements of **G** using this definition of Q.

The number of possible intensity levels in the image determines the size of matrix **G**. For an 8-bit image (256 possible levels) **G** will be of size 256×256. This is not a problem when working with one matrix, but as Example 11.11 shows, co-occurrence matrices sometimes are used in sequences. In order to reduce computation load, an approach used frequently is to quantize the intensities into a few bands in order to keep the size of matrix **G** manageable. For example, in the case of 256 intensities we can do this by letting the first 32 intensity levels equal to 1, the next 32 equal to 2, and so on. This will result in a co-occurrence matrix of size 8×8.

The total number, n, of pixel pairs that satisfy Q is equal to the sum of the elements of **G** ($n = 30$ in the preceding example). Then, the quantity

$$p_{ij} = g_{ij}/n$$

is an estimate of the probability that a pair of points satisfying Q will have values (z_i, z_j). These probabilities are in the range $[0, 1]$ and their sum is 1:

$$\sum_{i=1}^{K} \sum_{j=1}^{K} p_{ij} = 1$$

where K is the row (or column) dimension of square matrix **G**.

Because **G** depends on Q, the presence of intensity texture patterns can be detected by choosing an appropriate position operator and analyzing the elements of **G**. A set of descriptors useful for characterizing the contents of **G** are listed in Table 11.3. The quantities used in the correlation descriptor (second row in the table) are defined as follows:

$$m_r = \sum_{i=1}^{K} i \sum_{j=1}^{K} p_{ij}$$

$$m_c = \sum_{j=1}^{K} j \sum_{i=1}^{K} p_{ij}$$

and

$$\sigma_r^2 = \sum_{i=1}^{K} (i - m_r)^2 \sum_{j=1}^{K} p_{ij}$$

$$\sigma_c^2 = \sum_{j=1}^{K} (j - m_c)^2 \sum_{i=1}^{K} p_{ij}$$

TABLE 11.3
Descriptors used for characterizing co-occurrence matrices of size $K \times K$. The term p_{ij} is the ijth term of **G** divided by the sum of the elements of **G**.

Descriptor	Explanation	Formula
Maximum probability	Measures the strongest response of **G**. The range of values is $[0, 1]$.	$\max\limits_{i,j}(p_{ij})$
Correlation	A measure of how correlated a pixel is to its neighbor over the entire image. Range of values is 1 to -1, corresponding to perfect positive and perfect negative correlations. This measure is not defined if either standard deviation is zero.	$\sum\limits_{i=1}^{K}\sum\limits_{j=1}^{K}\dfrac{(i - m_r)(j - m_c)p_{ij}}{\sigma_r\sigma_c}$ $\sigma_r \neq 0; \sigma_c \neq 0$
Contrast	A measure of intensity contrast between a pixel and its neighbor over the entire image. The range of values is 0 (when **G** is constant) to $(K - 1)^2$.	$\sum\limits_{i=1}^{K}\sum\limits_{j=1}^{K}(i - j)^2 p_{ij}$
Uniformity (also called Energy)	A measure of uniformity in the range $[0, 1]$. Uniformity is 1 for a constant image.	$\sum\limits_{i=1}^{K}\sum\limits_{j=1}^{K}p_{ij}^2$
Homogeneity	Measures the spatial closeness of the distribution of elements in **G** to the diagonal. The range of values is $[0, 1]$, with the maximum being achieved when **G** is a diagonal matrix.	$\sum\limits_{i=1}^{K}\sum\limits_{i=1}^{K}\dfrac{p_{ij}}{1 + |i - j|}$
Entropy	Measures the randomness of the elements of **G**. The entropy is 0 when all p_{ij}'s are 0 and is maximum when all p_{ij}'s are equal. The maximum value is $2 \log_2 K$. (See Eq. (11.3-9) regarding entropy).	$-\sum\limits_{i=1}^{K}\sum\limits_{i=1}^{K}p_{ij} \log_2 p_{ij}$

If we let

$$P(i) = \sum_{j=1}^{K} p_{ij}$$

and

$$P(j) = \sum_{i=1}^{K} p_{ij}$$

then the preceding equations can be written as

$$m_r = \sum_{i=1}^{K} iP(i)$$

$$m_c = \sum_{j=1}^{K} jP(j)$$

$$\sigma_r^2 = \sum_{i=1}^{K} (i - m_r)^2 P(i)$$

$$\sigma_c^2 = \sum_{j=1}^{K} (j - m_c)^2 P(j)$$

With reference to Eqs. (11.3-4), (11.3-5), and to their explanation, we see that m_r is in the form of a mean computed along rows of the normalized **G** and m_c is a mean computed along the columns. Similarly, σ_r and σ_c are in the form of standard deviations (square roots of the variances) computed along rows and columns respectively. Each of these terms is a scalar, independently of the size of **G**.

Keep in mind when studying Table 11.3 that "neighbors" are with respect to the way in which Q is defined (i.e., neighbors do not necessarily have to be adjacent), and also that the p_{ij}'s are nothing more than normalized counts of the number of times that pixels having that intensities z_i and z_j occur in f relative to the position specified in Q. Thus, all we are doing here is trying to find patterns (texture) in those counts.

■ Figures 11.30(a) through (c) show images consisting of random, horizontally periodic (sine), and mixed pixel patterns, respectively. This example has two objectives: (1) to show values of the descriptors in Table 11.3 for the three co-occurrence matrices, $\mathbf{G}_1$, $\mathbf{G}_2$, and $\mathbf{G}_3$ corresponding (from top to bottom) to these images, and (2) to illustrate how sequences of co-occurrence matrices can be used to detect texture patterns in an image.

Figure 11.31 shows co-occurrence matrices $\mathbf{G}_1$, $\mathbf{G}_2$, and $\mathbf{G}_3$ displayed as images. These matrices were obtained using $L = 256$ and the position operator "one pixel immediately to the right." The value at coordinates (i, j) in these

EXAMPLE 11.11:
Using descriptors to characterize co-occurrence matrices.

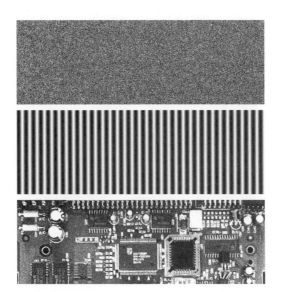

a
b
c

FIGURE 11.30
Images whose pixels have (a) random, (b) periodic, and (c) mixed texture patterns. Each image is of size 263×800 pixels.

a b c

FIGURE 11.31
256×256 co-occurrence matrices, $\mathbf{G}_1$, $\mathbf{G}_2$, and $\mathbf{G}_3$, corresponding from left to right to the images in Fig. 11.30.

images is the number of times that pixels pairs with intensities z_i and z_j occur in f in the position specified by Q, so it is not surprising that Fig. 11.31(a) is a random image, given the nature of the image from which it was obtained.

Figure 11.31(b) is more interesting. The first obvious feature is the symmetry about the main diagonal. Due to the symmetry of the sine wave, the number of counts for a pair (z_i, z_j) is the same as for the pair (z_j, z_i), which produces a symmetric co-occurrence matrix. The non-zero elements of $\mathbf{G}_2$ are sparse because value differences between horizontally adjacent pixels in a horizontal sine wave are relatively small. It helps to remember in interpreting these concepts that a digitized sine wave is a staircase, with the height and width of each step depending on frequency and the number of amplitude levels used in representing the function.

The structure of co-occurrence matrix $\mathbf{G}_3$ in Fig. 11.31(c) is more complex. High count values are grouped along the main diagonal also, but their distribution is more dense than for $\mathbf{G}_2$, a property that is indicative of an image with a rich variation in intensity values, but few large jumps in intensity between adjacent pixels. Examining Fig. 11.30(c), we see that there are large areas characterized by low variability in intensities. The high transitions in intensity occur at object boundaries, but these counts are low with respect to the moderate intensity transitions over large areas, so they are obscured by the ability of an image display to show high and low values simultaneously, as we discussed in Chapter 3.

The preceding observations are qualitative. To quantify the "content" of co-occurrence matrices we need descriptors such as those in Table 11.3. Table 11.4 shows values of these descriptors computed for the three co-occurrence matrices

TABLE 11.4
Descriptors evaluated using the co-occurrence matrices displayed in Fig. 11.31.

Normalized Co-occurrence Matrix	Descriptor					
	Max Probability	Correlation	Contrast	Uniformity	Homogeneity	Entropy
$\mathbf{G}_1/n_1$	0.00006	−0.0005	10838	0.00002	0.0366	15.75
$\mathbf{G}_2/n_2$	0.01500	0.9650	570	0.01230	0.0824	6.43
$\mathbf{G}_3/n_3$	0.06860	0.8798	1356	0.00480	0.2048	13.58

in Fig. 11.31. Note that to use these descriptors the co-occurrence matrices must be normalized by dividing them by the sum of their elements, as discussed earlier. The entries in Table 11.4 agree with what one would expect from looking at the images in Fig. 11.30 and their corresponding co-occurrence matrices in Fig. 11.31. For example, consider the Maximum Probability column in Table 11.4. The highest probability corresponds to the third co-occurrence matrix, which tells us that this matrix has the highest number of counts (largest number of pixel pairs occurring in the image relative to the positions in Q) than the other two matrices. This agrees with our earlier analysis of $\mathbf{G}_3$. The second column indicates that the highest correlation corresponds to $\mathbf{G}_2$, which in turn tells us that the intensities in the second image are highly correlated. The repetitiveness of the sinusoidal pattern over and over again in Fig. 11.30(b) reveals why this is so. Note that the correlation for $\mathbf{G}_1$ is essentially zero, indicating virtually no correlation between adjacent pixels, a characteristic of random images, such as the image in Fig. 11.30(a).

The contrast descriptor is highest for $\mathbf{G}_1$ and lowest for $\mathbf{G}_2$. Thus, we see that the less random an image is, the lower its contrast tends to be. We can see the reason by studying the matrices displayed in Fig. 11.31. The $(i - j)^2$ terms are differences of integers for $1 \leq i, j \leq L$ so they are the same for any $\mathbf{G}$. Therefore, the probabilities in the elements of the normalized co-occurrence matrices are the factors that determine the value of contrast. Although $\mathbf{G}_1$ has the lowest maximum probability, the other two matrices have many more zero or near zero probabilities (the dark areas in Fig. 11.31). Keeping in mind that the sum of the values of $\mathbf{G}/n$ is 1, it is easy to see why the contrast descriptor tends to increase as a function of randomness.

The remaining three descriptors are explained in a similar manner. Uniformity increases as a function of the values of the probabilities squared. Thus the less randomness there is in an image, the higher the uniformity descriptor will be, as the fifth column in Table 11.4 shows. Homogeneity measures the concentration of values of $\mathbf{G}$ with respect to the main diagonal. The values of the denominator term $(1 + |i - j|)$ are the same for all three co-occurrence matrices, and they decrease as i and j become closer in value (i.e., closer to the main diagonal). Thus, the matrix with the highest values of probabilities (numerator terms) near the main diagonal will have the highest value of homogeneity. As we discussed earlier, such a matrix will correspond to images with a "rich" gray-level content and areas of slowly varying intensity values. The entries in the sixth column of Table 11.4 are consistent with this interpretation.

The entries in the last column of the table are measures of randomness in co-occurrence matrices, which in turn translate into measures of randomness in the corresponding images. As expected, $\mathbf{G}_1$ had the highest value because the image from which it was derived was totally random. The other two entries are self-explanatory. Note that the entropy measure for $\mathbf{G}_1$ is near the theoretical maximum of 16 ($2 \log_2 256 = 16$). The image in Fig. 11.30(a) is composed of uniform noise, so each intensity level has approximately an equal probability of occurrence, which is the condition stated in Table 11.3 for maximum entropy.

Thus far, we have dealt with single images and their co-occurrence matrices. Suppose that we want to "discover" (without looking at the images) if there are any sections in these images that contain repetitive components (i.e., periodic textures). One way to accomplish this goal is to examine the correlation descriptor for sequences of co-occurrence matrices, derived from these images by increasing the distance between neighbors. As mentioned earlier, it is customary when working with sequences of co-occurrence matrices to quantize the number of intensities in order to reduce matrix size and corresponding computational load. The following results were obtained using $L = 8$.

Figure 11.32 shows plots of the correlation descriptors as a function of horizontal "offset" (i.e., horizontal distance between neighbors) from 1 (for adjacent pixels) to 50. Figure 11.32(a) shows that all correlation values are near 0, indicating that no such patterns were found in the random image. The shape of the correlation in Fig. 11.32(b) is a clear indication that the input image is sinusoidal in the horizontal direction. Note that the correlation function starts at a high value and then decreases as the distance between neighbors increases, and then repeats itself.

Figure 11.32(c) shows that the correlation descriptor associated with the circuit board image decreases initially, but has a strong peak for an offset distance of 16 pixels. Analysis of the image in Fig. 11.30(c) shows that the upper solder joints form a repetitive pattern approximately 16 pixels apart (see Fig. 11.33). The next major peak is at 32, caused by the same pattern, but the amplitude of the peak is lower because the number of repetitions at this distance is less than at 16 pixels. A similar observation explains the even smaller peak at an offset of 48 pixels.   ■

There are other repetitive patterns in the image, but they were obscured by the coarse quantization of 256 intensity levels into 8.

Structural approaches

As mentioned at the beginning of this section, a second category of texture description is based on structural concepts. Suppose that we have a rule of the form $S \rightarrow aS$, which indicates that the symbol S may be rewritten as aS (for example, three applications of this rule would yield the string $aaaS$). If a

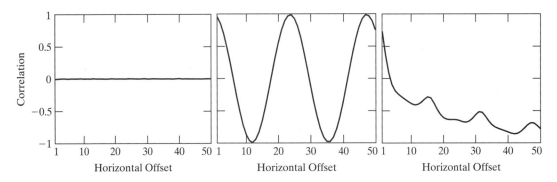

a b c

FIGURE 11.32 Values of the correlation descriptor as a function of offset (distance between "adjacent" pixels) corresponding to the (a) noisy, (b) sinusoidal, and (c) circuit board images in Fig. 11.30.

16 pixels

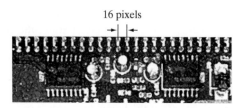

FIGURE 11.33
A zoomed section of the circuit board image showing periodicity of components.

represents a circle [Fig. 11.34(a)] and the meaning of "circles to the right" is assigned to a string of the form *aaa*..., the rule $S \rightarrow aS$ allows generation of the texture pattern shown in Fig. 11.34(b).

Suppose next that we add some new rules to this scheme: $S \rightarrow bA$, $A \rightarrow cA$, $A \rightarrow c$, $A \rightarrow bS$, $S \rightarrow a$, where the presence of a *b* means "circle down" and the presence of a *c* means "circle to the left." We can now generate a string of the form *aaabccbaa* that corresponds to a 3 × 3 matrix of circles. Larger texture patterns, such as the one in Fig. 11.34(c), can be generated easily in the same way. (Note, however, that these rules can also generate structures that are not rectangular.)

The basic idea in the foregoing discussion is that a simple "texture primitive" can be used to form more complex texture patterns by means of some rules that limit the number of possible arrangements of the primitive(s). These concepts lie at the heart of relational descriptions, a topic that we treat in more detail in Section 11.5.

Spectral approaches

As discussed in Section 5.4, the Fourier spectrum is ideally suited for describing the directionality of periodic or almost periodic 2-D patterns in an image. These global texture patterns are easily distinguishable as concentrations of high-energy bursts in the spectrum. Here, we consider three features of the Fourier spectrum that are useful for texture description: (1) Prominent peaks in the spectrum give

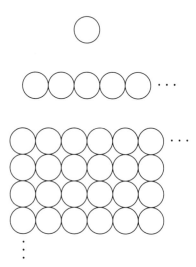

a
b
c

FIGURE 11.34
(a) Texture primitive.
(b) Pattern generated by the rule $S \rightarrow aS$.
(c) 2-D texture pattern generated by this and other rules.

the principal direction of the texture patterns. (2) The location of the peaks in the frequency plane gives the fundamental spatial period of the patterns. (3) Eliminating any periodic components via filtering leaves nonperiodic image elements, which can then be described by statistical techniques. Recall that the spectrum is symmetric about the origin, so only half of the frequency plane needs to be considered. Thus for the purpose of analysis, every periodic pattern is associated with only one peak in the spectrum, rather than two.

Detection and interpretation of the spectrum features just mentioned often are simplified by expressing the spectrum in polar coordinates to yield a function $S(r, \theta)$, where S is the spectrum function and r and θ are the variables in this coordinate system. For each direction θ, $S(r, \theta)$ may be considered a 1-D function $S_\theta(r)$. Similarly, for each frequency r, $S_r(\theta)$ is a 1-D function. Analyzing $S_\theta(r)$ for a fixed value of θ yields the behavior of the spectrum (such as the presence of peaks) along a radial direction from the origin, whereas analyzing $S_r(\theta)$ for a fixed value of r yields the behavior along a circle centered on the origin.

A more global description is obtained by integrating (summing for discrete variables) these functions:

$$S(r) = \sum_{\theta=0}^{\pi} S_\theta(r) \qquad (11.3\text{-}10)$$

and

$$S(\theta) = \sum_{r=1}^{R_0} S_r(\theta) \qquad (11.3\text{-}11)$$

where R_0 is the radius of a circle centered at the origin.

The results of Eqs. (11.3-10) and (11.3-11) constitute a pair of values $[S(r), S(\theta)]$ for *each* pair of coordinates (r, θ). By varying these coordinates, we can generate two 1-D functions, $S(r)$ and $S(\theta)$, that constitute a spectral-energy description of texture for an entire image or region under consideration. Furthermore, descriptors of these functions themselves can be computed in order to characterize their behavior quantitatively. Descriptors typically used for this purpose are the location of the highest value, the mean and variance of both the amplitude and axial variations, and the distance between the mean and the highest value of the function.

EXAMPLE 11.12:
Spectral texture.

■ Figure 11.35(a) shows an image containing randomly distributed matches and Fig. 11.35(b) shows an image in which these objects are arranged periodically. Figures 11.35(c) and (d) show the corresponding Fourier spectra. The periodic bursts of energy extending quadrilaterally in two dimensions in both Fourier spectra are due to the periodic texture of the coarse background material on which the matches rest. The other dominant components in the spectra in Fig. 11.35(c) are caused by the random orientation of the object edges in Fig. 11.35(a). On the other hand, the main energy in Fig. 11.35(d) not associated with the background is along the horizontal axis, corresponding to the strong vertical edges in Fig. 11.35(b).

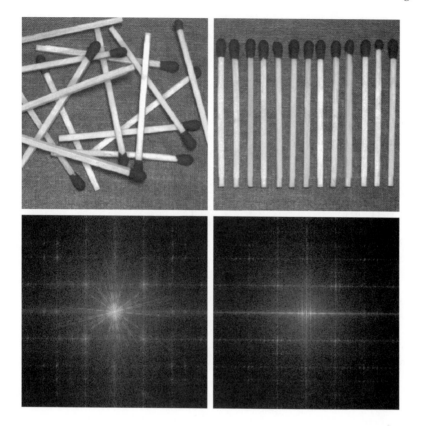

FIGURE 11.35
(a) and (b) Images
of random and
ordered objects.
(c) and (d) Corres-
ponding Fourier
spectra. All images
are of size
600×600 pixels.

Figures 11.36(a) and (b) are plots of $S(r)$ and $S(\theta)$ for the random matches
and similarly in (c) and (d) for the ordered matches. The plot of $S(r)$ for the
random matches shows no strong periodic components (i.e., there are no dominant peaks in the spectrum besides the peak at the origin, which is the dc component). Conversely, the plot of $S(r)$ for the ordered matches shows a strong
peak near $r = 15$ and a smaller one near $r = 25$, corresponding to the periodic horizontal repetition of the light (matches) and dark (background) regions in Fig. 11.35(b). Similarly, the random nature of the energy bursts in
Fig. 11.35(c) is quite apparent in the plot of $S(\theta)$ in Fig. 11.36(b). By contrast,
the plot in Fig. 11.36(d) shows strong energy components in the region near
the origin and at 90° and 180°. This is consistent with the energy distribution
of the spectrum in Fig. 11.35(d). ■

11.3.4 Moment Invariants

The 2-D *moment* of order $(p + q)$ of a digital image $f(x, y)$ of size $M \times N$ is
defined as

$$m_{pq} = \sum_{x=0}^{M-1} \sum_{y=0}^{N-1} x^p y^q f(x, y) \qquad (11.3\text{-}12)$$

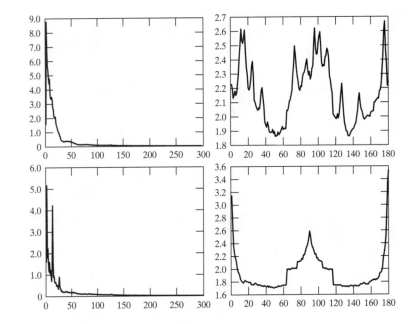

a b
c d

FIGURE 11.36
Plots of (a) $S(r)$
and (b) $S(\theta)$ for
Fig. 11.35(a).
(c) and (d) are
plots of $S(r)$ and
$S(\theta)$ for Fig.
11.35(b). All
vertical axes are
$\times 10^5$.

where $p = 0, 1, 2, \ldots$ and $q = 0, 1, 2, \ldots$ are integers. The corresponding *central moment* of order $(p + q)$ is defined as

$$\mu_{pq} = \sum_{x=0}^{M-1} \sum_{y=0}^{N-1} (x - \bar{x})^p (y - \bar{y})^q f(x, y) \tag{11.3-13}$$

for $p = 0, 1, 2, \ldots$ and $q = 0, 1, 2, \ldots$, where

$$\bar{x} = \frac{m_{10}}{m_{00}} \quad \text{and} \quad \bar{y} = \frac{m_{01}}{m_{00}} \tag{11.3-14}$$

The *normalized central moments*, denoted η_{pq}, are defined as

$$\eta_{pq} = \frac{\mu_{pq}}{\mu_{00}^{\gamma}} \tag{11.3-15}$$

where

$$\gamma = \frac{p + q}{2} + 1 \tag{11.3-16}$$

for $p + q = 2, 3, \ldots$.

A set of seven *invariant moments* can be derived from the second and third moments.[†]

[†]Derivation of these results involves concepts that are beyond the scope of this discussion. The book by Bell [1965] and the paper by Hu [1962] contain detailed discussions of these concepts. For generating moment invariants of order higher than 7, see Flusser [2000]. Moment invariants can be generalized to n dimensions (Mamistvalov [1998]).

$$\phi_1 = \eta_{20} + \eta_{02} \tag{11.3-17}$$

$$\phi_2 = (\eta_{20} - \eta_{02})^2 + 4\eta_{11}^2 \tag{11.3-18}$$

$$\phi_3 = (\eta_{30} - 3\eta_{12})^2 + (3\eta_{21} - \eta_{03})^2 \tag{11.3-19}$$

$$\phi_4 = (\eta_{30} + \eta_{12})^2 + (\eta_{21} + \eta_{03})^2 \tag{11.3-20}$$

$$\phi_5 = (\eta_{30} - 3\eta_{12})(\eta_{30} + \eta_{12})[(\eta_{30} + \eta_{12})^2$$

$$- 3(\eta_{21} + \eta_{03})^2] + (3\eta_{21} - \eta_{03})(\eta_{21} + \eta_{03}) \tag{11.3-21}$$

$$[3(\eta_{30} + \eta_{12})^2 - (\eta_{21} + \eta_{03})^2]$$

$$\phi_6 = (\eta_{20} - \eta_{02})[(\eta_{30} + \eta_{12})^2 - (\eta_{21} + \eta_{03})^2] \tag{11.3-22}$$

$$+ 4\eta_{11}(\eta_{30} + \eta_{12})(\eta_{21} + \eta_{03})$$

$$\phi_7 = (3\eta_{21} - \eta_{03})(\eta_{30} + \eta_{12})[(\eta_{30} + \eta_{12})^2$$

$$- 3(\eta_{21} + \eta_{03})^2] + (3\eta_{12} - \eta_{30})(\eta_{21} + \eta_{03}) \tag{11.3-23}$$

$$[3(\eta_{30} + \eta_{12})^2 - (\eta_{21} + \eta_{03})^2]$$

This set of moments is invariant to translation, scale change, mirroring (within a minus sign) and rotation.

■ The objective of this example is to compute and compare the preceding moment invariants using the image in Fig. 11.37(a). The black (0) border was added to make all images in this example be of the same size; the zeros do not affect computation of the moment invariants. Figures 11.37(b) through (f) show the original image translated, scaled by 0.5 in both spatial dimensions, mirrored, rotated by 45° and rotated by 90°, respectively. Table 11.5 summarizes the values of the seven moment invariants for these six images. To reduce dynamic range and thus simplify interpretation, the values shown are $\text{sgn}(\phi_i) \log_{10}(|\phi_i|)$. The absolute value is needed because many of the values are fractional and/or negative; the sgn function preserves the sign (interest here is on the *invariance* and relative signs of the moments, not on their actual values). The two key points in Table 11.5 are (1) the closeness of the values of the moments, independent of translation, scale change, mirroring and rotation; and (2) the fact that the *sign* of ϕ_7 is different for the mirrored image (a property used in practice to detect whether an image has been mirrored). ■

EXAMPLE 11.13:
Moment
invariants.

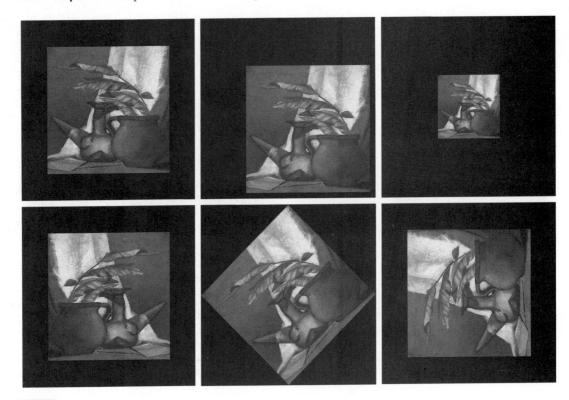

FIGURE 11.37 (a) Original image. (b)–(f) Images translated, scaled by one-half, mirrored, rotated by 45° and rotated by 90°, respectively.

TABLE 11.5
Moment invariants for the images in Fig. 11.37.

Moment Invariant	Original Image	Translated	Half Size	Mirrored	Rotated 45°	Rotated 90°
ϕ_1	2.8662	2.8662	2.8664	2.8662	2.8661	2.8662
ϕ_2	7.1265	7.1265	7.1257	7.1265	7.1266	7.1265
ϕ_3	10.4109	10.4109	10.4047	10.4109	10.4115	10.4109
ϕ_4	10.3742	10.3742	10.3719	10.3742	10.3742	10.3742
ϕ_5	21.3674	21.3674	21.3924	21.3674	21.3663	21.3674
ϕ_6	13.9417	13.9417	13.9383	13.9417	13.9417	13.9417
ϕ_7	−20.7809	−20.7809	−20.7724	20.7809	−20.7813	−20.7809

11.4 Use of Principal Components for Description

Consult the book Web site for a brief review of vectors and matrices.

The material discussed in this section is applicable to boundaries and regions. In addition, it can be used as the basis for describing sets of images that are registered spatially, but whose corresponding pixel values are different (e.g., the three component images of an RGB image). Suppose that we are given the three component images of such a color image. The three images can be treated as a unit by expressing each group of three corresponding pixels as a vector. For example, let x_1, x_2, and x_3, respectively, be the values of a pixel in each of the three RGB component images. These three elements can be expressed in the form of a 3-D *column* vector, $\mathbf{x}$, where

$$\mathbf{x} = \begin{bmatrix} x_1 \\ x_2 \\ x_3 \end{bmatrix}$$

This vector represents *one* common pixel in all three images. If the images are of size $M \times N$, there will be a total of $K = MN$ 3-D vectors after all the pixels are represented in this manner. If we have n registered images, the vectors will be n-dimensional:

$$\mathbf{x} = \begin{bmatrix} x_1 \\ x_2 \\ \vdots \\ x_n \end{bmatrix} \tag{11.4-1}$$

Throughout this section, the assumption is that all vectors are column vectors (i.e., matrices of order $n \times 1$). We can write them on a line of text simply by expressing them as $\mathbf{x} = (x_1, x_2, \ldots, x_n)^T$, where "$T$" indicates transpose.

We can treat the vectors as random quantities, just like we did when constructing an intensity histogram. The only difference is that, instead of talking about quantities like the mean and variance of the random variables, we now talk about *mean vectors* and *covariance matrices* of the random vectors. The mean vector of the population is defined as

$$\mathbf{m_x} = E\{\mathbf{x}\} \tag{11.4-2}$$

where $E\{\cdot\}$ is the expected value of the argument, and the subscript denotes that $\mathbf{m}$ is associated with the population of $\mathbf{x}$ vectors. Recall that the expected value of a vector or matrix is obtained by taking the expected value of each element.

The *covariance matrix* of the vector population is defined as

$$\mathbf{C_x} = E\{(\mathbf{x} - \mathbf{m_x})(\mathbf{x} - \mathbf{m_x})^T\} \tag{11.4-3}$$

Because $\mathbf{x}$ is n dimensional, $\mathbf{C_x}$ and $(\mathbf{x} - \mathbf{m_x})(\mathbf{x} - \mathbf{m_x})^T$ are matrices of order $n \times n$. Element c_{ii} of $\mathbf{C_x}$ is the variance of x_i, the ith component of the $\mathbf{x}$ vectors in the population, and element c_{ij} of $\mathbf{C_x}$ is the covariance[†] between elements x_i and x_j of these vectors. The matrix $\mathbf{C_x}$ is real and symmetric. If elements x_i and x_j are uncorrelated, their covariance is zero and, therefore, $c_{ij} = c_{ji} = 0$. All these definitions reduce to their familiar one-dimensional counterparts when $n = 1$.

[†]Recall that the variance of a random variable x with mean m can be defined as $E\{(x - m)^2\}$. The covariance of two random variables x_i and x_j is defined as $E\{(x_i - m_i)(x_j - m_j)\}$.

For K vector samples from a random population, the mean vector can be approximated from the samples by using the familiar averaging expression

$$\mathbf{m_x} = \frac{1}{K} \sum_{k=1}^{K} \mathbf{x}_k \tag{11.4-4}$$

Similarly, by expanding the product $(\mathbf{x} - \mathbf{m_x})(\mathbf{x} - \mathbf{m_x})^T$ and using Eqs. (11.4-2) and (11.4-4) we would find that the covariance matrix can be approximated from the samples as follows:

$$\mathbf{C_x} = \frac{1}{K} \sum_{k=1}^{K} \mathbf{x}_k \mathbf{x}_k^T - \mathbf{m_x}\mathbf{m_x}^T \tag{11.4-5}$$

EXAMPLE 11.14:
Computation of the mean vector and covariance matrix.

■ To illustrate the mechanics of Eqs. (11.4-4) and (11.4-5), consider the four vectors $\mathbf{x}_1 = (0, 0, 0)^T$, $\mathbf{x}_2 = (1, 0, 0)^T$, $\mathbf{x}_3 = (1, 1, 0)^T$, and $\mathbf{x}_4 = (1, 0, 1)^T$. Applying Eq. (11.4-4) yields the following mean vector:

$$\mathbf{m_x} = \frac{1}{4} \begin{bmatrix} 3 \\ 1 \\ 1 \end{bmatrix}$$

Similarly, using Eq. (11.4-5) yields the following covariance matrix:

$$\mathbf{C_x} = \frac{1}{16} \begin{bmatrix} 3 & 1 & 1 \\ 1 & 3 & -1 \\ 1 & -1 & 3 \end{bmatrix}$$

All the elements along the main diagonal are equal, which indicates that the three components of the vectors in the population have the same variance. Also, elements x_1 and x_2, as well as x_1 and x_3, are positively correlated; elements x_2 and x_3 are negatively correlated. ■

Because $\mathbf{C_x}$ is real and symmetric, finding a set of n orthonormal eigenvectors always is possible (Noble and Daniel [1988]). Let $\mathbf{e}_i$ and λ_i, $i = 1, 2, \ldots, n$, be the eigenvectors and corresponding eigenvalues of $\mathbf{C_x}$,[†] arranged (for convenience) in descending order so that $\lambda_j \geq \lambda_{j+1}$ for $j = 1, 2, \ldots, n - 1$. Let $\mathbf{A}$ be a matrix whose rows are formed from the eigenvectors of $\mathbf{C_x}$, ordered so that the first row of $\mathbf{A}$ is the eigenvector corresponding to the largest eigenvalue, and the last row is the eigenvector corresponding to the smallest eigenvalue.

Suppose that we use $\mathbf{A}$ as a transformation matrix to map the $\mathbf{x}$s into vectors denoted by $\mathbf{y}$s, as follows:

$$\mathbf{y} = \mathbf{A}(\mathbf{x} - \mathbf{m_x}) \tag{11.4-6}$$

[†]By definition, the eigenvectors and eigenvalues of an $n \times n$ matrix, $\mathbf{C}$, satisfy the relation $\mathbf{Ce}_i = \lambda_i \mathbf{e}_i$, for $i = 1, 2, \ldots, n$.

This expression is called the *Hotelling transform*, which, as will be shown shortly, has some interesting and useful properties.

The Hotelling transform is the same as the discrete *Karhunen-Loève transform* (Karhunen [1947]), so the two names are used interchangeably in the literature.

It is not difficult to show that the mean of the **y** vectors resulting from this transformation is zero; that is,

$$\mathbf{m_y} = E\{\mathbf{y}\} = \mathbf{0} \tag{11.4-7}$$

It follows from basic matrix theory that the covariance matrix of the **y**s is given in terms of **A** and $\mathbf{C_x}$ by the expression

$$\mathbf{C_y} = \mathbf{A}\mathbf{C_x}\mathbf{A}^T \tag{11.4-8}$$

Furthermore, because of the way **A** was formed, $\mathbf{C_y}$ is a diagonal matrix whose elements along the main diagonal are the eigenvalues of $\mathbf{C_x}$; that is,

$$\mathbf{C_y} = \begin{bmatrix} \lambda_1 & & & 0 \\ & \lambda_2 & & \\ & & \ddots & \\ 0 & & & \lambda_n \end{bmatrix} \tag{11.4-9}$$

The off-diagonal elements of this covariance matrix are 0, so the elements of the **y** vectors are uncorrelated. Keep in mind that the λ_j's are the eigenvalues of $\mathbf{C_x}$ and that the elements along the main diagonal of a diagonal matrix are its eigenvalues (Noble and Daniel [1988]). Thus $\mathbf{C_x}$ and $\mathbf{C_y}$ have the same eigenvalues.

Another important property of the Hotelling transform deals with the reconstruction of **x** from **y**. Because the rows of **A** are orthonormal vectors, it follows that $\mathbf{A}^{-1} = \mathbf{A}^T$, and any vector **x** can be recovered from its corresponding **y** by using the expression

$$\mathbf{x} = \mathbf{A}^T\mathbf{y} + \mathbf{m_x} \tag{11.4-10}$$

Suppose, however, that instead of using all the eigenvectors of $\mathbf{C_x}$ we form matrix $\mathbf{A}_k$ from the k eigenvectors corresponding to the k largest eigenvalues, yielding a transformation matrix of order $k \times n$. The **y** vectors would then be k dimensional, and the reconstruction given in Eq. (11.4-10) would no longer be exact (this is somewhat analogous to the procedure we used in Section 11.2.3 to describe a boundary with a few Fourier coefficients).

The vector reconstructed by using $\mathbf{A}_k$ is

$$\hat{\mathbf{x}} = \mathbf{A}_k^T\mathbf{y} + \mathbf{m_x} \tag{11.4-11}$$

It can be shown that the mean square error between **x** and $\hat{\mathbf{x}}$ is given by the expression

$$e_{ms} = \sum_{j=1}^{n}\lambda_j - \sum_{j=1}^{k}\lambda_j$$

$$= \sum_{j=k+1}^{n}\lambda_j \tag{11.4-12}$$

The first line of Eq. (11.4-12) indicates that the error is zero if $k = n$ (that is, if all the eigenvectors are used in the transformation). Because the λ_j's decrease monotonically, Eq. (11.4-12) also shows that the error can be minimized by selecting the k eigenvectors associated with the largest eigenvalues. Thus the Hotelling transform is optimal in the sense that it minimizes the mean square error between the vectors $\mathbf{x}$ and their approximations $\hat{\mathbf{x}}$. Due to this idea of using the eigenvectors corresponding to the largest eigenvalues, the Hotelling transform also is known as the *principal components* transform.

EXAMPLE 11.15:
Using principal components for image description.

■ Figure 11.38 shows six multispectral satellite images corresponding to six spectral bands: visible blue (450–520 nm), visible green (520–600 nm), visible red (630–690 nm), near infrared (760–900 nm), middle infrared (1550–1750 nm), and thermal infrared (10,400–12500 nm). The objective of this example is to illustrate how to use principal components for image description.

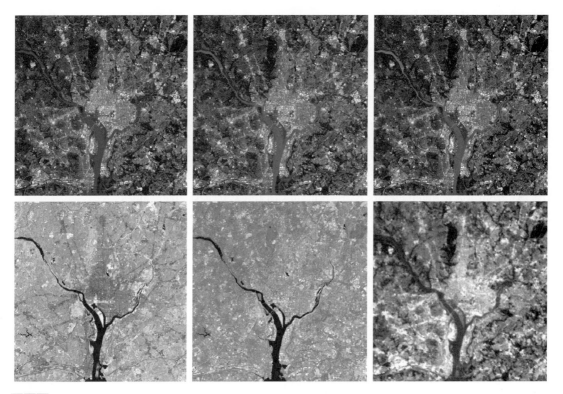

a b c
d e f

FIGURE 11.38 Multispectral images in the (a) visible blue, (b) visible green, (c) visible red, (d) near infrared, (e) middle infrared, and (f) thermal infrared bands. (Images courtesy of NASA.)

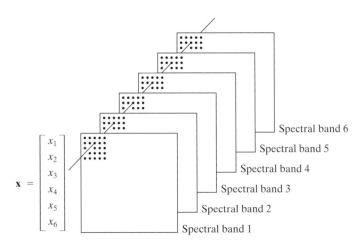

FIGURE 11.39
Formation of a
vector from
corresponding
pixels in six
images.

Spectral band 6

Spectral band 5

Spectral band 4

Spectral band 3

Spectral band 2

Spectral band 1

$$\mathbf{x} = \begin{bmatrix} x_1 \\ x_2 \\ x_3 \\ x_4 \\ x_5 \\ x_6 \end{bmatrix}$$

Organizing the images as in Fig. 11.39 leads to the formation of a six-element vector $\mathbf{x} = (x_1, x_2, \ldots x_6)^T$ from each set of corresponding pixels in the images, as discussed at the beginning of this section. The images in this example are of size 564×564 pixels, so the population consisted of $(564)^2 = 318{,}096$ vectors from which the mean vector, covariance matrix, and corresponding eigenvalues and eigenvectors were computed. The eigenvectors were then used as the rows of matrix $\mathbf{A}$, and a set of $\mathbf{y}$ vectors were obtained using Eq. (11.4-6). Similarly, we used Eq. (11.4-8) to obtain $\mathbf{C_y}$. Table 11.6 shows the eigenvalues of this matrix. Note the dominance of the first two eigenvalues.

A set of principal component images was generated using the $\mathbf{y}$ vectors mentioned in the previous paragraph (images are constructed from vectors by applying Fig. 11.39 in reverse). Figure 11.40 shows the results. Figure 11.40(a) was formed from the first component of the 318,096 $\mathbf{y}$ vectors, Fig. 11.40(b) from the second component of these vectors, and so on, so these images are of the same size as the original images in Fig. 11.38. The most obvious feature in the principal component images is that a significant portion of the contrast detail is contained in the first two images, and it decreases rapidly from there. The reason can be explained by looking at the eigenvalues. As Table 11.6 shows, the first two eigenvalues are much larger than the others. Because the eigenvalues are the variances of the elements of the $\mathbf{y}$ vectors and variance is a measure of intensity contrast, it is not unexpected that the images formed from the vector components corresponding to the largest eigenvalues would exhibit the highest contrast. In fact, the first two images in Fig. 11.40 account

λ_1	λ_2	λ_3	λ_4	λ_5	λ_6
10344	2966	1401	203	94	31

TABLE 11.6
Eigenvalues of
the covariance
matrices obtained
from the images
in Fig. 11.38.

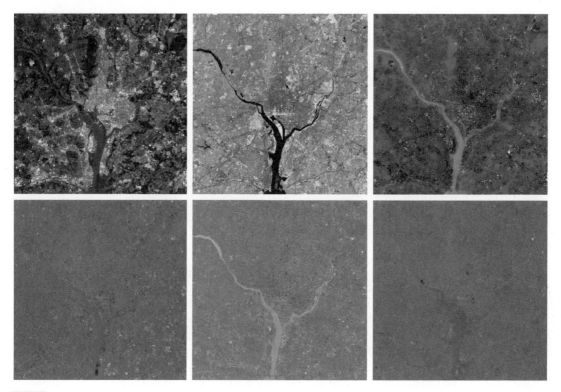

a b c
d e f

FIGURE 11.40 The six principal component images obtained from vectors computed using Eq. (11.4-6). Vectors are converted to images by applying Fig. 11.39 in reverse.

for about 89% of the total variance. The other four images have low contrast detail because they account for only the remaining 11%.

According to Eqs. (11.4-11) and (11.4-12), if we used all the eigenvectors in matrix **A** we could reconstruct the original images (vectors) from the principal component images (vectors) with zero error between the original and reconstructed images. That is, the original and reconstructed images would be identical. If the objective were to store and/or transmit the principal component images and the transformation matrix for later reconstruction of the original images, it would make no sense to store and/or transmit all the principal component images because nothing would be gained. Suppose, however, that we keep and/or transmit only the two principal component images (they have most of the contrast detail). Then there would be a significant savings in storage and/or transmission (matrix **A** would be of size 2 × 6, so its impact would be negligible).

Figure 11.41 shows the results of reconstructing the six multispectral images from the two principal component images corresponding the largest eigenvalues. The first five images are quite close in appearance to the originals in Fig. 11.38,

When referring to images, we use the term "vectors" interchangeably because there is a one-to-one correspondence between the two in the present context.

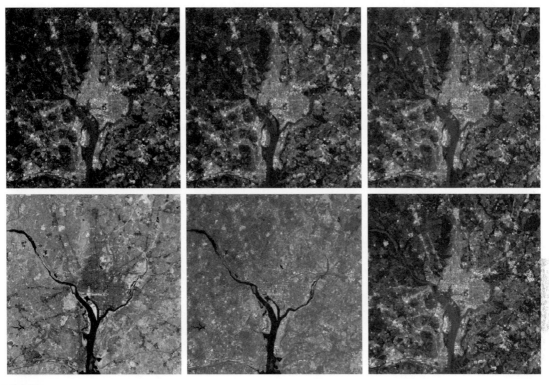

a b c
d e f

FIGURE 11.41 Multispectral images reconstructed using only the two principal component images corresponding to the two principal component images with the largest eigenvalues (variance). Compare these images with the originals in Fig. 11.38.

but this is not true for the sixth image. The reason is that the original sixth image is actually blurry, but the two principal component images used in the reconstruction are sharp, therefore, the blurry "detail" is lost. Figure 11.42 shows the differences between the original and reconstructed images. The images in Fig. 11.42 were enhanced to highlight the differences between them. If they were shown without enhancement, the first five images would appear almost all black. As expected, the sixth difference image shows the most variability. ■

■ As mentioned earlier in this chapter, representation and description should be as independent as possible with respect to size, translation, and rotation. Principal components provide a convenient way to normalize boundaries and/or regions for variations in these three parameters. Consider the object in Fig. 11.43, and assume that its size, location, and orientation (rotation) are arbitrary. The points in the region (or its boundary) may be treated as two dimensional vectors, $\mathbf{x} = (x_1, x_2)^T$, where x_1 and x_2 are the coordinate values of any point along the x_1- and x_2-axis, respectively. All the points in the region or

EXAMPLE 11.16:
Using principal components for normalizing with respect to variations in size, translation, and rotation.

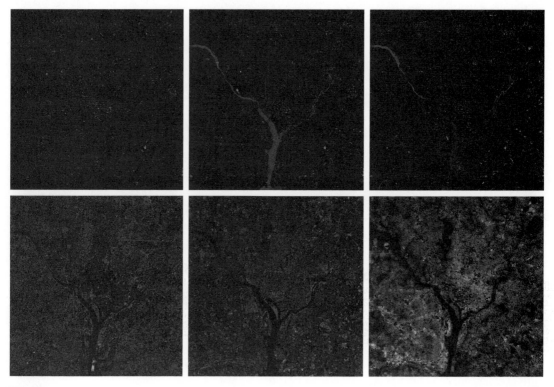

a b c
d e f

FIGURE 11.42 Differences between the original and reconstructed images. All difference images were enhanced by scaling them to the full $[0, 255]$ range to facilitate visual analysis.

boundary constitute a 2-D vector population which can be used to compute the covariance matrix $\mathbf{C_x}$ and mean vector $\mathbf{m_x}$, as before. One eigenvector of $\mathbf{C_x}$ points in the direction of maximum variance (data spread) of the population, while the second eigenvector is perpendicular to the first, as Fig. 11.43(b) shows. In terms of the present discussion, the principal components transform in Eq. (11.4-6) accomplishes two things: (1) It establishes the center of the transformed coordinates system at the center of gravity (mean) of the population because $\mathbf{m_x}$ is subtracted from each $\mathbf{x}$; and (2) the $\mathbf{y}$ coordinates (vectors) it generates are rotated versions of the $\mathbf{x}$'s, so that the data align with the eigenvectors. If we define a (y_1, y_2) axis system so that y_1 is along the first eigenvector and y_2 along the second, then the geometry that results is as illustrated in Fig. 11.43(c). That is, the dominant data directions are aligned with the axis system. The same result will be obtained regardless of the size, translation, or rotation of the object, provided that all points in the region or boundary undergo the same changes. If we wished to size-normalize the transformed data, we would divide the coordinates by the corresponding eigenvalues.

Observe in Fig. 11.43(c) that the points in the y-axes system can have both positive and negative values. To convert all coordinates to positive values, we

The y-axis system could be in a direction 180° opposite to the direction shown in Fig. 11.43(c), depending on the orientation of the original object. For example, if the nose of the airplane in Fig. 11.43(a) had been pointing in the opposite direction, the resulting eigenvectors would point to the left and down.

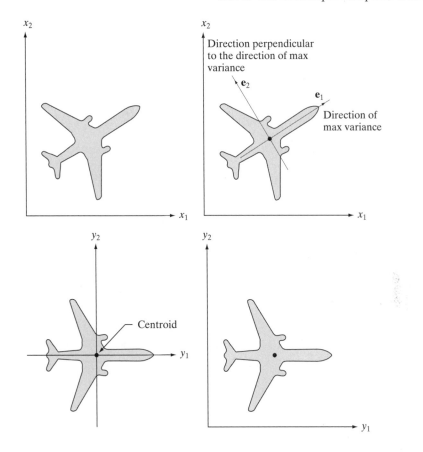

FIGURE 11.43
(a) An object.
(b) Object
showing
eigenvectors of its
covariance matrix.
(c) Transformed
object, obtained
using Eq. (11.4-6).
(d) Object
translated so that
all its coordinate
values are greater
than 0.

simply subtract the vector $(y_{1min}, y_{2min})^T$ from all the **y** vectors. To displace the resulting points so that they are all greater than 0, as in Fig. 11.43(d), we add to them a vector $(a, b)^T$ where a and b are greater than 0.

Although the preceding discussion is straightforward in principle, the mechanics are a frequent source of confusion. Thus, we conclude this example with a simple manual illustration. Figure 11.44(a) shows four points with coordinates $(1, 1), (2, 4), (4, 2),$ and $(5, 5)$. The mean vector, covariance matrix, and normalized (unit length) eigenvectors of this population are

$$\mathbf{m_x} = \begin{bmatrix} 3 \\ 3 \end{bmatrix}$$

$$\mathbf{C_x} = \begin{bmatrix} 3.333 & 2.00 \\ 2.00 & 3.333 \end{bmatrix}$$

and

$$\mathbf{e_1} = \begin{bmatrix} 0.707 \\ 0.707 \end{bmatrix}, \quad \mathbf{e_2} = \begin{bmatrix} -0.707 \\ 0.707 \end{bmatrix}$$

a b
c d

FIGURE 11.44
A manual
example.
(a) Original
points.
(b) Eigenvectors
of the covariance
matrix of the
points in (a).
(c) Transformed
points obtained
using Eq. (11.4-6).
(d) Points from
(c), rounded and
translated so that
all coordinate
values are
integers greater
than 0. The
dashed lines are
included to
facilitate viewing.
They are not part
of the data.

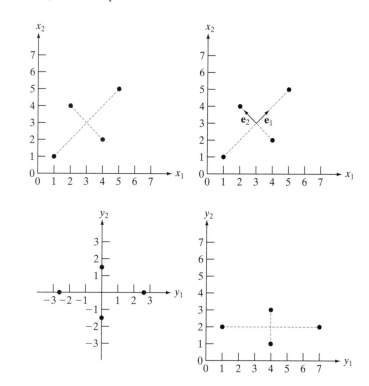

The corresponding eigenvalues are $\lambda_1 = 5.333$ and $\lambda_2 = 1.333$. Figure 11.44(b) shows the eigenvectors superimposed on the data. From Eq. (11.4-6), the transformed points (the **y**s) are $(-2.828, 0)$, $(0, 1.414)$, $(0, -1.414)$, and $(2.828, 0)$. These points are plotted in Fig. 11.44(c). Note that they are aligned with the y-axes and that they have fractional values. When working with images, values generally are integers, making it necessary to round all fractions to their nearest integer value. Figure 11.44(d) shows the points rounded to the nearest integer values and their location shifted so that all coordinate values are integers greater than 0, as in the original figure.   ■

11.5 Relational Descriptors

We introduced in Section 11.3.3 the concept of rewriting rules for describing texture. In this section, we expand that concept in the context of relational descriptors. These apply equally well to boundaries or regions, and their main purpose is to capture in the form of rewriting rules basic repetitive patterns in a boundary or region.

Consider the simple staircase structure shown in Fig. 11.45(a). Assume that this structure has been segmented out of an image and that we want to describe it in some formal way. By defining the two *primitive elements* a and b shown, we may code Fig. 11.45(a) in the form shown in Fig. 11.45(b). The most obvious property of the coded structure is the repetitiveness of the elements

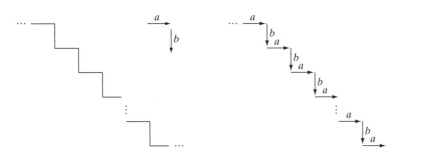

FIGURE 11.45
(a) A simple staircase structure.
(b) Coded structure.

a and *b*. Therefore, a simple description approach is to formulate a recursive relationship involving these primitive elements. One possibility is to use the *rewriting rules*:

(1) $S \rightarrow aA$,
(2) $A \rightarrow bS$, and
(3) $A \rightarrow b$,

where S and A are variables and the elements a and b are constants corresponding to the primitives just defined. Rule 1 indicates that S, called the *starting symbol*, can be replaced by primitive a and variable A. This variable, in turn, can be replaced by b and S or by b alone. Replacing A with bS, leads back to the first rule and the procedure can be repeated. Replacing A with b terminates the procedure, because no variables remain in the expression. Figure 11.46 illustrates some sample derivations of these rules, where the numbers below the structures represent the order in which rules 1, 2, and 3 were applied. The relationship between a and b is preserved, because these rules force an a always to be followed by a b. Notably, these three simple rewriting rules can be used to generate (or describe) infinitely many "similar" structures.

Because strings are 1-D structures, their application to image description requires establishing an appropriate method for reducing 2-D positional relations to 1-D form. Most applications of strings to image description are based on the idea of extracting connected line segments from the objects of interest. One approach is to follow the contour of an object and code the result with segments of specified direction and/or length. Figure 11.47 illustrates this procedure.

Another, somewhat more general, approach is to describe sections of an image (such as small homogeneous regions) by directed line segments, which

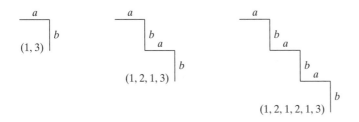

FIGURE 11.46
Sample derivations for the rules
$S \rightarrow aA$, $A \rightarrow bS$, and $A \rightarrow b$.

FIGURE 11.47
Coding a region
boundary with
directed line
segments.

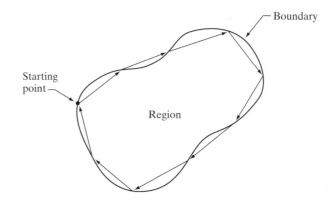

can be joined in other ways besides head-to-tail connections. Figure 11.48(a)
illustrates this approach, and Fig. 11.48(b) shows some typical operations
that can be defined on abstracted primitives. Figure 11.48(c) shows a set of
specific primitives consisting of line segments defined in four directions, and

a b
c
d

FIGURE 11.48
(a) Abstracted
primitives.
(b) Operations
among primitives.
(c) A set of
specific primitives.
(d) Steps in
building a
structure.

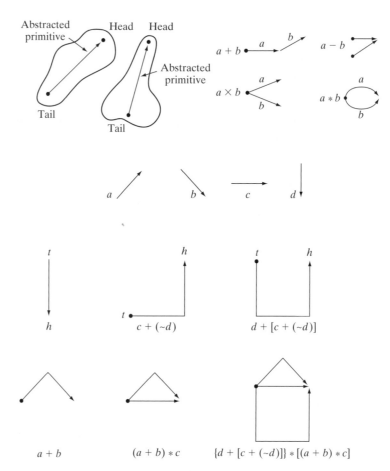

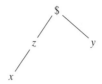

FIGURE 11.49 A simple tree with root $ and frontier xy.

Fig. 11.48(d) shows a step-by-step generation of a specific shape, where ($\sim d$) indicates the primitive d with its direction reversed. Note that each composite structure has a single head and a single tail. The result of interest is the last string, which describes the complete structure.

String descriptions are best suited for applications in which connectivity of primitives can be expressed in a head-to-tail or other continuous manner. Sometimes regions that are similar in terms of texture or other descriptor may not be contiguous, and techniques are required for describing such situations. One of the most useful approaches for doing so is to use tree descriptors.

A *tree T* is a finite set of one or more nodes for which

(a) there is a unique node $ designated the *root*, and
(b) the remaining nodes are partitioned into m disjointed sets $T_1, \ldots, T_m$, each of which in turn is a tree called a *subtree* of T.

The *tree frontier* is the set of nodes at the bottom of the tree (the *leaves*), taken in order from left to right. For example, the tree shown in Fig. 11.49 has root $ and frontier xy.

Generally, two types of information in a tree are important: (1) information about a node stored as a set of words describing the node, and (2) information relating a node to its neighbors, stored as a set of pointers to those neighbors. As used in image description, the first type of information identifies an image substructure (e.g., region or boundary segment), whereas the second type defines the physical relationship of that substructure to other substructures. For example, Fig. 11.50(a) can be represented by a tree by using the relationship "inside of." Thus, if the root of the tree is denoted $, Fig. 11.50(a) shows that the first level of complexity involves a and c inside $, which produces two branches emanating from the root, as shown in Fig. 11.50(b). The next level involves b inside a, and d and e inside c. Finally, f inside e completes the tree.

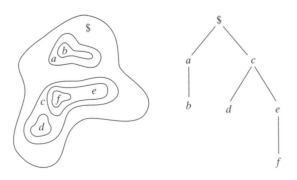

FIGURE 11.50
(a) A simple composite region.
(b) Tree representation obtained by using the relationship "inside of."

Summary

The representation and description of objects or regions that have been segmented out of an image are early steps in the operation of most automated processes involving images. These descriptions, for example, constitute the input to the object recognition methods developed in the following chapter. As indicated by the range of description techniques covered in this chapter, the choice of one method over another is determined by the problem under consideration. The objective is to choose descriptors that "capture" essential differences between objects, or classes of objects, while maintaining as much independence as possible to changes in location, size, and orientation.

References and Further Reading

The boundary-following algorithm in Section 11.1.1 was first proposed by Moore [1968]. The chain-code representation discussed in Section 11.1.2 was first proposed by Freeman [1961, 1974]. For current work using chain codes, see Bribiesca [1999], who also has extended chain codes to 3-D (Bribiesca [2000]). For a detailed discussion and algorithm to compute minimum-perimeter polygons (Section 11.1.3), see Klette and Rosenfeld [2004]. See also Sloboda et al. [1998] and Coeurjolly and Klette [2004]. Additional topics of interest for the material in Section 11.1.4 include invariant polygonal fitting (Voss and Suesse [1997]), methods for evaluating the performance of polygonal approximation algorithms (Rosin [1997]), generic implementations (Huang and Sun [1999]), and computational speed (Davis [1999]).

References for the discussion of signatures (Section 11.1.5) are Ballard and Brown [1982] and Gupta and Srinath [1988]. See Preparata and Shamos [1985] regarding fundamental formulations for finding the convex hull and convex deficiency (Section 11.1.6). See also the paper by Liu-Yu and Antipolis [1993]. Katzir et al. [1994] discuss the detection of partially occluded curves. Zimmer et al. [1997] discuss an improved algorithm for computing the convex hull, and Latecki and Lakämper [1999] discuss a convexity rule for shape decomposition.

The skeletonizing algorithm discussed in Section 11.1.7 is based on Zhang and Suen [1984]. Some useful additional comments on the properties and implementation of this algorithm are included in a paper by Lu and Wang [1986]. A paper by Jang and Chin [1990] provides an interesting tie between the discussion in Section 11.1.7 and the morphological concept of thinning introduced in Section 9.5.5. For thinning approaches in the presence of noise, see Shi and Wong [1994] and Chen and Yu [1996]. Shaked and Bruckstein [1998] discuss a pruning algorithm useful for removing spurs from a skeleton. Fast computation of the medial axis transform is discussed by Sahni and Jenq [1992] and by Ferreira and Ubéda [1999]. The survey paper by Loncaric [1998] is of interest regarding many of the approaches discussed in Section 11.1.

Freeman and Shapira [1975] give an algorithm for finding the basic rectangle of a closed, chain-coded curve (Section 11.2.1). The discussion on shape numbers in Section 11.2.2 is based on the work of Bribiesca and Guzman [1980] and Bribiesca [1981]. For additional reading on Fourier descriptors (Section 11.2.3), see the early papers by Zahn and Roskies [1972] and by Persoon and Fu [1977]. See also Aguado et al. [1998] and Sonka et al. [1999]. Reddy and Chatterji [1996] discuss an interesting approach using the FFT to achieve invariance to translation, rotation, and scale change. The material in Section 11.2.4 is based on elementary probability theory (see, for example, Peebles [1993] and Popoulis [1991]).

For additional reading on Section 11.3.2, see Rosenfeld and Kak [1982] and Ballard and Brown [1982]. For an excellent introduction to texture (Section 11.3.3), see Haralick and Shapiro [1992]. For an early survey on texture, see Wechsler [1980]. The papers by Murino et al. [1998] and Garcia [1999], and the discussion by Shapiro and Stockman [2001], are representative of current work in this field.

The moment-invariant approach discussed in Section 11.3.4 is from Hu [1962]. Also see Bell [1965]. To get an idea of the range of applications of moment invariants, see Hall [1979] regarding image matching and Cheung and Teoh [1999] regarding the use of moments for describing symmetry. Moment invariants were generalized to n dimensions by Mamistvalov [1998]. For generating moments of arbitrary order, see Flusser [2000].

Hotelling [1933] was the first to derive and publish the approach that transforms discrete variables into uncorrelated coefficients. He referred to this technique as *the method of principal components*. His paper gives considerable insight into the method and is worth reading. Hotelling's transformation was rediscovered by Kramer and Mathews [1956] and by Huang and Schultheiss [1963]. Principal components are still a basic tool for image description used in numerous applications, as exemplified by Swets and Weng [1996] and by Duda, Heart, and Stork [2001]. References for the material in Section 11.5 are Gonzalez and Thomason [1978] and Fu [1982]. See also Sonka et al. [1999]. For additional reading on the topics of this chapter with a focus on implementation, see Nixon and Aguado [2002] and Gonzalez, Woods, and Eddins [2004].

Problems

11.1 ★(a) Show that redefining the starting point of a chain code so that the resulting sequence of numbers forms an integer of minimum magnitude makes the code independent of the initial starting point on the boundary.

(b) Find the normalized starting point of the code 11076765543322.

11.2 (a) Show that the first difference of a chain code normalizes it to rotation, as explained in Section 11.1.2.

(b) Compute the first difference of the code 0101030303323232212111.

11.3 ★(a) Show that the rubber-band polygonal approximation approach discussed in Section 11.1.3 yields a polygon with minimum perimeter.

(b) Show that if each cell corresponds to a pixel on the boundary, the maximum possible error in that cell is $\sqrt{2}d$, where d is the minimum possible horizontal or vertical distance between adjacent pixels (i.e., the distance between lines in the sampling grid used to produce the digital image).

11.4 Explain how the MPP algorithm in Section 11.1.3 behaves under the following conditions:

★(a) 1-pixel wide, 1-pixel deep indentations.

★(b) 1-pixel wide, 2-or-more pixel deep indentations.

(c) 1-pixel wide, 1-pixel long protrusions.

(d) 1-pixel wide, n-pixel long protrusions.

11.5 ★(a) Discuss the effect on the resulting polygon if the error threshold is set to zero in the merging method discussed in Section 11.1.4.

(b) What would be the effect on the splitting method?

11.6 ★(a) Plot the signature of a square boundary using the tangent angle method discussed in Section 11.1.5.

(b) Repeat for the slope density function.

Assume that the square is aligned with the x- and y-axes, and let the x-axis be the reference line. Start at the corner closest to the origin.

11.7 Find an expression for the signature of each of the following boundaries, and plot the signatures.

★(a) An equilateral triangle

(b) A rectangle

(c) An ellipse

11.8 Draw the medial axis of

★**(a)** A circle
★**(b)** A square
(c) A rectangle
(d) An equilateral triangle

11.9 For each of the figures shown,

★**(a)** Discuss the action taken at point p by Step 1 of the skeletonizing algorithm presented in Section 11.1.7.
(b) Repeat for Step 2 of the algorithm. Assume that $p = 1$ in all cases.

1	1	0		0	0	0		0	1	0		1	1	0
1	p	0		1	p	0		1	p	1		0	p	1
1	1	0		0	0	0		0	1	0		0	0	0

11.10 With reference to the skeletonizing algorithm in Section 11.1.7, what would the figure shown look like after

★**(a)** One pass of Step 1 of the algorithm?
(b) One pass of Step 2 (on the result of Step 1, not the original image)?

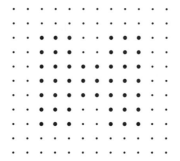

11.11 ★**(a)** What is the order of the shape number for the figure shown?
(b) Obtain the shape number.

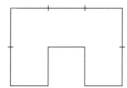

11.12 The procedure discussed in Section 11.2.3 for using Fourier descriptors consists of expressing the coordinates of a contour as complex numbers, taking the DFT of these numbers, and keeping only a few components of the DFT as descriptors of the boundary shape. The inverse DFT is then an approximation to the original contour. What class of contour shapes would have a DFT consisting of real numbers and how would the axis system in Fig. 11.19 have to be set up to obtain these real numbers?

11.13 Show that if you use only two Fourier descriptors ($u = 0$ and $u = 1$) to reconstruct a boundary with Eq. (11.2-5), the result will always be a circle. (*Hint:* Use the parametric representation of a circle in the complex plane and express the equation of a circle in polar coordinates.)

★**11.14** Give the smallest number of statistical moment descriptors needed to differentiate between the signatures of the figures shown in Fig. 11.10.

11.15 Give two boundary shapes that have the same mean and third statistical moment descriptors, but different second moments.

★**11.16** Propose a set of descriptors capable of differentiating between the shapes of the characters 0, 1, 8, 9, and X. (*Hint:* Use topological descriptors in conjunction with the convex hull.)

11.17 Consider a binary image of size 200×200 pixels, with a vertical black band extending from columns 1 to 99 and a vertical white band extending from columns 100 to 200.

    (a) Obtain the co-occurrence matrix of this image using the position operator "one pixel to the right."

    ★**(b)** Normalize this matrix so that its elements become probability estimates, as explained in Section 11.3.1.

    (c) Use your matrix from (b) to compute the six descriptors in Table 11.3.

11.18 Consider a checkerboard image composed of alternating black and white squares, each of size $m \times m$. Give a position operator that would yield a diagonal co-occurrence matrix.

11.19 Obtain the gray-level co-occurrence matrix of a 5×5 image composed of a checkerboard of alternating 1s and 0s if

    ★**(a)** the position operator Q is defined as "one pixel to the right," and

    (b) the position operator Q is defined as "two pixels to the right."

    Assume that the top left pixel has value 0.

11.20 Prove the validity of Eqs. (11.4-7), (11.4-8), and (11.4-9).

★**11.21** It was mentioned in Example 11.13 that a credible job could be done of reconstructing approximations to the six original images by using only the two principal-component images associated with the largest eigenvalues. What would be the mean square error incurred in doing so? Express your answer as a percentage of the maximum possible error.

11.22 For a set of images of size 64×64, assume that the covariance matrix given in Eq. (11.4-9) turns out to be the identity matrix. What would be the mean square error between the original images and images reconstructed using Eq. (11.4-11) with only half of the original eigenvectors?

★**11.23** Under what conditions would you expect the major axes of a boundary, defined in Section 11.2.1, to be equal to the eigen axes of that boundary?

11.24 Give a spatial relationship and corresponding tree representation for a checkerboard pattern of black and white squares. Assume that the top left element is black and that the root of the tree corresponds to that element. Your tree can have no more than two branches emanating from each node.

★**11.25** You are contracted to design an image processing system for detecting imperfections on the inside of certain solid plastic wafers. The wafers are examined using an X-ray imaging system, which yields 8-bit images of size 512×512. In the absence of imperfections, the images appear "bland," having a mean intensity of 100 and variance of 400. The imperfections appear as bloblike regions in which about 70% of the pixels have excursions in intensity of 50 intensity levels or less about a mean of 100. A wafer is considered defective if such a region occupies an area exceeding 20×20 pixels in size. Propose a system based on texture analysis.

11.26 A company that bottles a variety of industrial chemicals has heard of your success solving imaging problems and hires you to design an approach for detecting when bottles are not full. The bottles appear as shown in the following figure as they move along a conveyor line past an automatic filling and capping station. A bottle is considered imperfectly filled when the level of the liquid is below the midway point between the bottom of the neck and the shoulder of the bottle. The shoulder is defined as the region of the bottle where the sides and slanted portion of the bottle intersect. The bottles are moving, but the company has an imaging system equipped with a illumination flash front end that effectively stops motion, so you will be given images that look very close to the sample shown here. Based on the material you have learned up to this point, propose a solution for detecting bottles that are not filled properly. State clearly all assumptions that you make and that are likely to impact the solution you propose.

11.27 Having heard about your success with the bottling problem, you are contacted by a fluids company that wishes to automate bubble-counting in certain processes for quality control. The company has solved the imaging problem and can obtain 8-bit images of size 700×700 pixels, such as the one shown. Each image represents an area of 7 cm^2. The company wishes to do two things with each image: (1) Determine the ratio of the area occupied by bubbles to the total area of the image, and (2) count the number of distinct bubbles. Based on the material you have learned up to this point, propose a solution to this problem. In your solution, make sure to state the physical dimensions of the smallest bubble your solution can detect. State clearly all assumptions that you make and that are likely to impact the solution you propose.

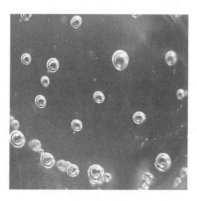

12 *Object Recognition*

> One of the most interesting aspects of the world is that it can be considered to be made up of patterns.
> A pattern is essentially an arrangement. It is characterized by the order of the elements of which it is made, rather than by the intrinsic nature of these elements.
>
> *Norbert Wiener*

Preview

We conclude our coverage of digital image processing with an introduction to techniques for object recognition. As noted in Section 1.1, we have defined the scope covered by our treatment of digital image processing to include recognition of *individual* image regions, which in this chapter we call *objects* or *patterns*.

The approaches to pattern recognition developed in this chapter are divided into two principal areas: decision-theoretic and structural. The first category deals with patterns described using quantitative descriptors, such as length, area, and texture. The second category deals with patterns best described by qualitative descriptors, such as the relational descriptors discussed in Section 11.5.

Central to the theme of recognition is the concept of "learning" from sample patterns. Learning techniques for both decision-theoretic and structural approaches are developed and illustrated in the material that follows.

12.1 Patterns and Pattern Classes

A *pattern* is an *arrangement of descriptors*, such as those discussed in Chapter 11. The name *feature* is used often in the pattern recognition literature to denote a descriptor. A *pattern class* is a family of patterns that share some common properties. Pattern classes are denoted $\omega_1, \omega_2, \ldots, \omega_W$, where W is the number of classes. Pattern recognition by machine involves

and

$$d_2(\mathbf{x}) = \mathbf{x}^T\mathbf{m}_2 - \frac{1}{2}\mathbf{m}_2^T\mathbf{m}_2$$
$$= 1.5x_1 + 0.3x_2 - 1.17$$

From Eq. (12.2-6), the equation of the boundary is

$$d_{12}(\mathbf{x}) = d_1(\mathbf{x}) - d_2(\mathbf{x})$$
$$= 2.8x_1 + 1.0x_2 - 8.9 = 0$$

Figure 12.6 shows a plot of this boundary (note that the axes are not to the same scale). Substitution of any pattern vector from class ω_1 would yield $d_{12}(\mathbf{x}) > 0$. Conversely, any pattern from class ω_2 would yield $d_{12}(\mathbf{x}) < 0$. In other words, given an unknown pattern belonging to one of these two classes, the sign of $d_{12}(\mathbf{x})$ would be sufficient to determine the pattern's class membership. ■

In practice, the minimum distance classifier works well when the distance between means is large compared to the spread or randomness of each class with respect to its mean. In Section 12.2.2 we show that the minimum distance classifier yields optimum performance (in terms of minimizing the average loss of misclassification) when the distribution of each class about its mean is in the form of a spherical "hypercloud" in n-dimensional pattern space.

The simultaneous occurrence of large mean separations and relatively small class spread occur seldomly in practice unless the system designer controls the nature of the input. An excellent example is provided by systems designed to read stylized character fonts, such as the familiar American Banker's Association E-13B font character set. As Fig. 12.7 shows, this particular font set consists of 14 characters that were purposely designed on a 9×7 grid in order to facilitate their reading. The characters usually are printed in ink that contains finely ground magnetic material. Prior to being read, the ink is subjected to a magnetic field, which accentuates each character to simplify detection. In other words, the segmentation problem is solved by artificially highlighting the key characteristics of each character.

The characters typically are scanned in a horizontal direction with a single-slit reading head that is narrower but taller than the characters. As the head moves across a character, it produces a 1-D electrical signal (a signature) that is conditioned to be proportional to the rate of increase or decrease of the character area under the head. For example, consider the waveform associated with the number 0 in Fig. 12.7. As the reading head moves from left to right, the area seen by the head begins to increase, producing a positive derivative (a positive rate of change). As the head begins to leave the left leg of the 0, the area under the head begins to decrease, producing a negative derivative. When the head is in the middle zone of the character, the area remains nearly constant, producing a zero derivative. This pattern repeats itself as the head enters the right leg of the character. The design of the font ensures that the waveform of each character is distinct from that of all others. It also

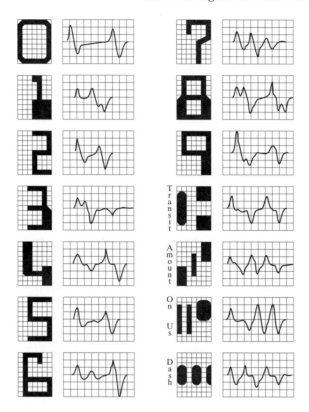

FIGURE 12.7
American
Bankers
Association
E-13B font
character set and
corresponding
waveforms.

ensures that the peaks and zeros of each waveform occur approximately on the vertical lines of the background grid on which these waveforms are displayed, as shown in Fig. 12.7. The E-13B font has the property that sampling the waveforms only at these points yields enough information for their proper classification. The use of magnetized ink aids in providing clean waveforms, thus minimizing scatter.

Designing a minimum distance classifier for this application is straightforward. We simply store the sample values of each waveform and let each set of samples be represented as a prototype vector $\mathbf{m}_j, j = 1, 2, \ldots, 14$. When an unknown character is to be classified, the approach is to scan it in the manner just described, express the grid samples of the waveform as a vector, $\mathbf{x}$, and identify its class by selecting the class of the prototype vector that yields the highest value in Eq. (12.2-5). High classification speeds can be achieved with analog circuits composed of resistor banks (see Problem 12.4).

Matching by correlation

We introduced the basic idea of spatial correlation in Section 3.4.2 and used it extensively for spatial filtering in that section. We also mentioned the correlation theorem briefly in Section 4.6.7 and Table 4.3. From Eq. (3.4-1), we know that correlation of a mask $w(x, y)$ of size $m \times n$, with an image $f(x, y)$ may be expressed in the form

To be formal, we should refer to correlation as *crosscorrelation* when the functions are different and as *autocorrelation* when they are same. However, it is customary to use the generic term *correlation* when it is clear whether the two functions in a given application are equal or different.

$$c(x, y) = \sum_s \sum_t w(s, t) f(x + s, y + t) \qquad (12.2\text{-}7a)$$

where the limits of summation are taken over the region shared by w and f. This equation is evaluated for all values of the displacement variables x and y so that all elements of w visit every pixel of f, where f is assumed to be larger than w. Just as spatial convolution is related to the Fourier transform of the functions via the convolution theorem, spatial correlation is related to the transforms of the functions via the correlation theorem:

$$f(x, y) \star w(x, y) \Leftrightarrow F^*(u, v) W(u, v) \qquad (12.2\text{-}7b)$$

where "$\star$" indicates spatial convolution and F^* is the complex conjugate of F. The other half of the correlation theorem stated in Table 4.3 is of no interest in the present discussion. Equation (12.2-7b) is a Fourier transform pair whose interpretation is identical to the discussion of Eq. (4.6-24), except that we use the complex conjugate of one of the functions. The inverse Fourier transform of Eq. (12.2-7b) yields a two-dimensional circular correlation analogous to Eq. (4.6-23), and the padding issues discussed in Section 4.6.6 regarding convolution are applicable also to correlation.

We do not dwell on either of the preceding equations because they are both sensitive to scale changes in f and w. Instead, we use the following *normalized correlation coefficient*

$$\gamma(x, y) = \frac{\sum_s \sum_t \left[w(s, t) - \overline{w} \right] \sum_s \sum_t \left[f(x + s, y + t) - \overline{f}(x + s, y + t) \right]}{\left\{ \sum_s \sum_t \left[w(s, t) - \overline{w} \right]^2 \sum_s \sum_t \left[f(x + s, y + t) - \overline{f}(x + s, y + t) \right]^2 \right\}^{\frac{1}{2}}}$$

$$(12.2\text{-}8)$$

You will find it helpful to review Section 3.4.2 regarding the mechanics of spatial correlation.

where the limits of summation are taken over the region shared by w and f, $\overline{w}$ is the average value of the mask (computed only once), and $\overline{f}(x + s, y + t)$ is the average value of f in the region coincident with w. Often, w is referred to as a *template* and correlation is referred to as *template matching*. It can be shown (Problem 12.7) that $\gamma(x, y)$ has values in the range $[-1, 1]$ and is thus normalized to changes in the amplitudes of w and f. The maximum value of $\gamma(x, y)$ occurs when the normalized w and the corresponding normalized region in f are identical. This indicates *maximum correlation* (i.e., the best possible match). The minimum occurs with the two normalized functions exhibit the least similarity in the sense of Eq. (12.2-8). The correlation coefficient cannot be computed using the Fourier transform because of the nonlinear terms in the equation (division and squares).

Figure 12.8 illustrates the mechanics of the procedure just described. The border around f is the padding necessary to provide for the situation when the center of w is on the border of f, as explained in Section 3.4.2. (In template matching, values of correlation when the center of the template is past the border of the image generally are of no interest, so the padding is limited to half the mask width.) As usual, we limit attention to templates of odd size for notational convenience.

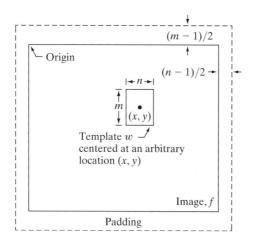

FIGURE 12.8
The mechanics of
template
matching.

Figure 12.8 shows a template of size $m \times n$ whose center is at an arbitrary location (x, y). The correlation at this point is obtained by applying Eq. (12.2-8). Then the center of the template is incremented to an adjacent location and the procedure is repeated. The complete correlation coefficient $\gamma(x, y)$ is obtained by moving the center of the template (i.e., by incrementing x and y) so that the *center* of w visits every pixel in f. At the end of the procedure, we look for the maximum in $\gamma(x, y)$ to find where the best match occurred. It is possible to have multiple locations in $\gamma(x, y)$ with the same maximum value, indicating several matches between w and f.

■ Figure 12.9(a) shows a 913 × 913 satellite image of Hurricane Andrew, in which the eye of the storm is clearly visible. As an example of correlation we wish to find the location of the best match in (a) of the template in Fig. 12.9(b), which is a small (31 × 31) subimage of the eye of the storm. Figure 12.9(c) shows the result of computing the correlation coefficient in Eq. (12.2-8). The original size of this image was 943 × 943 pixels due to padding (see Fig. 12.8), but we cropped it to the size of the original image for display purposes. Intensity in this image is proportional to correlation value, and all negative correlations were clipped at 0 (black) to simplify the visual analysis of the image. The brightest point of the correlation image is clearly visible near the eye of the storm. Figure 12.9(d) shows as a white dot the location of the maximum correlation (in this case there was a unique match whose maximum value was 1), which we see corresponds closely with the location of the eye in Fig. 12.9(a). ■

EXAMPLE 12.2:
Matching by
correlation.

The preceding discussion shows that it is possible to normalize correlation for changes in intensity values of the functions being processed. Normalizing for size and rotation is a more complicated problem. Normalizing for size involves spatial scaling, which, as explained in Sections 2.6.5 and 4.5.4, is image resampling. In order for resampling to make sense, the size to which an image should be rescaled must be known. In some situations, this can become a difficult issue unless spatial cues are available. For example, in a remote sensing application, if the viewing geometry of the imaging sensors

a b
c d

FIGURE12.9
(a) Satellite image
of Hurricane
Andrew, taken on
August 24, 1992.
(b) Template of
the eye of the
storm. (c) Corre-
lation coefficient
shown as an
image (note the
brightest point).
(d) Location of
the best match.
This point is a
single pixel, but
its size was
enlarged to make
it easier to see.
(Original image
courtesy of
NOAA.)

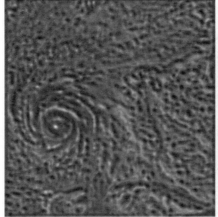

is known (which typically is the case), then knowing the altitude of the sensor with respect to the area being imaged may be sufficient to be able to
normalize image size, assuming a fixed viewing angle. Normalizing for rotation similarly requires that the angle to which images should be rotated be
known. This again requires spatial cues. In the remote sensing example just
given, the direction of flight may be sufficient to be able to rotate the sensed
images into a standard orientation. In unconstrained situations, normalizing
for size and orientation can become a truly challenging task, requiring the
automated detection of images features (as discussed in Chapter 11) that
can be used as spatial cues.

12.2.2 Optimum Statistical Classifiers

In this section we develop a probabilistic approach to recognition. As is true in
most fields that deal with measuring and interpreting physical events, probability considerations become important in pattern recognition because of the
randomness under which pattern classes normally are generated. As shown in
the following discussion, it is possible to derive a classification approach that is

optimal in the sense that, on average, its use yields the lowest probability of committing classification errors (see Problem 12.10).

Foundation

The probability that a particular pattern $\mathbf{x}$ comes from class ω_i is denoted $p(\omega_i/\mathbf{x})$. If the pattern classifier decides that $\mathbf{x}$ came from ω_j when it actually came from ω_i, it incurs a loss, denoted L_{ij}. As pattern $\mathbf{x}$ may belong to any one of W classes under consideration, the average loss incurred in assigning $\mathbf{x}$ to class ω_j is

$$r_j(\mathbf{x}) = \sum_{k=1}^{W} L_{kj} p(\omega_k/\mathbf{x}) \tag{12.2-9}$$

This equation often is called the *conditional average risk* or *loss* in decision-theory terminology.

From basic probability theory, we know that $p(A/B) = [p(A)p(B/A)]/p(B)$. Using this expression, we write Eq. (12.2-9) in the form

Consult the book Web site for a brief review of probability theory.

$$r_j(\mathbf{x}) = \frac{1}{p(\mathbf{x})} \sum_{k=1}^{W} L_{kj} p(\mathbf{x}/\omega_k) P(\omega_k) \tag{12.2-10}$$

where $p(\mathbf{x}/\omega_k)$ is the probability density function of the patterns from class ω_k and $P(\omega_k)$ is the probability of occurrence of class ω_k (sometimes these probabilities are referred to as *a priori*, or simply *prior*, *probabilities*). Because $1/p(\mathbf{x})$ is positive and common to all the $r_j(\mathbf{x})$, $j = 1, 2, \ldots, W$, it can be dropped from Eq. (12.2-10) without affecting the relative order of these functions from the smallest to the largest value. The expression for the average loss then reduces to

$$r_j(\mathbf{x}) = \sum_{k=1}^{W} L_{kj} p(\mathbf{x}/\omega_k) P(\omega_k) \tag{12.2-11}$$

The classifier has W possible classes to choose from for any given unknown pattern. If it computes $r_1(\mathbf{x}), r_2(\mathbf{x}), \ldots, r_W(\mathbf{x})$ for each pattern $\mathbf{x}$ and assigns the pattern to the class with the smallest loss, the total average loss with respect to all decisions will be minimum. The classifier that minimizes the total average loss is called the *Bayes classifier*. Thus the Bayes classifier assigns an unknown pattern $\mathbf{x}$ to class ω_i if $r_i(\mathbf{x}) < r_j(\mathbf{x})$ for $j = 1, 2, \ldots, W$; $j \neq i$. In other words, $\mathbf{x}$ is assigned to class ω_i if

$$\sum_{k=1}^{W} L_{ki} p(\mathbf{x}/\omega_k) P(\omega_k) < \sum_{q=1}^{W} L_{qj} p(\mathbf{x}/\omega_q) P(\omega_q) \tag{12.2-12}$$

for all j; $j \neq i$. The "loss" for a correct decision generally is assigned a value of zero, and the loss for any incorrect decision usually is assigned the same nonzero value (say, 1). Under these conditions, the loss function becomes

$$L_{ij} = 1 - \delta_{ij} \tag{12.2-13}$$

where $\delta_{ij} = 1$ if $i = j$ and $\delta_{ij} = 0$ if $i \neq j$. Equation (12.2-13) indicates a loss of unity for incorrect decisions and a loss of zero for correct decisions. Substituting Eq. (12.2-13) into Eq. (12.2-11) yields

$$r_j(\mathbf{x}) = \sum_{k=1}^{W}(1 - \delta_{kj})p(\mathbf{x}/\omega_k)P(\omega_k)$$

$$= p(\mathbf{x}) - p(\mathbf{x}/\omega_j)P(\omega_j) \qquad (12.2\text{-}14)$$

The Bayes classifier then assigns a pattern $\mathbf{x}$ to class ω_i if, for all $j \neq i$,

$$p(\mathbf{x}) - p(\mathbf{x}/\omega_i)P(\omega_i) < p(\mathbf{x}) - p(\mathbf{x}/\omega_j)P(\omega_j) \qquad (12.2\text{-}15)$$

or, equivalently, if

$$p(\mathbf{x}/\omega_i)P(\omega_i) > p(\mathbf{x}/\omega_j)P(\omega_j) \qquad j = 1, 2, \ldots, W; j \neq i \quad (12.2\text{-}16)$$

With reference to the discussion leading to Eq. (12.2-1), we see that the Bayes classifier for a 0-1 loss function is nothing more than computation of decision functions of the form

$$d_j(\mathbf{x}) = p(\mathbf{x}/\omega_j)P(\omega_j) \qquad j = 1, 2, \ldots, W \qquad (12.2\text{-}17)$$

where a pattern vector $\mathbf{x}$ is assigned to the class whose decision function yields the largest numerical value.

The decision functions given in Eq. (12.2-17) are optimal in the sense that they minimize the average loss in misclassification. For this optimality to hold, however, the probability density functions of the patterns in each class, as well as the probability of occurrence of each class, must be known. The latter requirement usually is not a problem. For instance, if all classes are equally likely to occur, then $P(\omega_j) = 1/W$. Even if this condition is not true, these probabilities generally can be inferred from knowledge of the problem. Estimation of the probability density functions $p(\mathbf{x}/\omega_j)$ is another matter. If the pattern vectors, $\mathbf{x}$, are n-dimensional, then $p(\mathbf{x}/\omega_j)$ is a function of n variables, which, if its form is not known, requires methods from multivariate probability theory for its estimation. These methods are difficult to apply in practice, especially if the number of representative patterns from each class is not large or if the underlying form of the probability density functions is not well behaved. For these reasons, use of the Bayes classifier generally is based on the assumption of an analytic expression for the various density functions and then an estimation of the necessary parameters from sample patterns from each class. By far the most prevalent form assumed for $p(\mathbf{x}/\omega_j)$ is the Gaussian probability density function. The closer this assumption is to reality, the closer the Bayes classifier approaches the minimum average loss in classification.

Bayes classifier for Gaussian pattern classes

To begin, let us consider a 1-D problem ($n = 1$) involving two pattern classes ($W = 2$) governed by Gaussian densities, with means m_1 and m_2 and standard deviations σ_1 and σ_2, respectively. From Eq. (12.2-17) the Bayes decision functions have the form

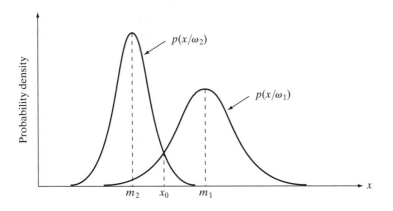

FIGURE 12.10
Probability
density functions
for two 1-D
pattern classes.
The point x_0
shown is the
decision boundary
if the two classes
are equally likely
to occur.

$$d_j(x) = p(x/\omega_j)P(\omega_j)$$

$$= \frac{1}{\sqrt{2\pi}\sigma_j}e^{-\frac{(x-m_j)^2}{2\sigma_j^2}}P(\omega_j) \quad j = 1, 2 \tag{12.2-18}$$

where the patterns are now scalars, denoted by x. Figure 12.10 shows a plot of the probability density functions for the two classes. The boundary between the two classes is a single point, denoted x_0 such that $d_1(x_0) = d_2(x_0)$. If the two classes are equally likely to occur, then $P(\omega_1) = P(\omega_2) = 1/2$, and the decision boundary is the value of x_0 for which $p(x_0/\omega_1) = p(x_0/\omega_2)$. This point is the intersection of the two probability density functions, as shown in Fig. 12.10. Any pattern (point) to the right of x_0 is classified as belonging to class ω_1. Similarly, any pattern to the left of x_0 is classified as belonging to class ω_2. When the classes are not equally likely to occur, x_0 moves to the left if class ω_1 is more likely to occur or, conversely, to the right if class ω_2 is more likely to occur. This result is to be expected, because the classifier is trying to minimize the loss of misclassification. For instance, in the extreme case, if class ω_2 never occurs, the classifier would never make a mistake by always assigning all patterns to class ω_1 (that is, x_0 would move to negative infinity).

In the n-dimensional case, the Gaussian density of the vectors in the jth pattern class has the form

$$p(\mathbf{x}/\omega_j) = \frac{1}{(2\pi)^{n/2}|\mathbf{C}_j|^{1/2}}e^{-\frac{1}{2}(\mathbf{x}-\mathbf{m}_j)^T\mathbf{C}_j^{-1}(\mathbf{x}-\mathbf{m}_j)} \tag{12.2-19}$$

where each density is specified completely by its mean vector $\mathbf{m}_j$ and covariance matrix $\mathbf{C}_j$, which are defined as

$$\mathbf{m}_j = E_j\{\mathbf{x}\} \tag{12.2-20}$$

and

$$\mathbf{C}_j = E_j\{(\mathbf{x} - \mathbf{m}_j)(\mathbf{x} - \mathbf{m}_j)^T\} \tag{12.2-21}$$

where $E_j\{\cdot\}$ denotes the expected value of the argument over the patterns of class ω_j. In Eq. (12.2-19), n is the dimensionality of the pattern vectors, and

See the remarks at the end of this section regarding the fact that the Bayes classifier for one variable is an optimum *thresholding* function, as mentioned in Section 10.3.3.

$|\mathbf{C}_j|$ is the determinant of the matrix $\mathbf{C}_j$. Approximating the expected value E_j by the average value of the quantities in question yields an estimate of the mean vector and covariance matrix:

$$\mathbf{m}_j = \frac{1}{N_j} \sum_{\mathbf{x} \in \omega_j} \mathbf{x} \tag{12.2-22}$$

and

$$\mathbf{C}_j = \frac{1}{N_j} \sum_{\mathbf{x} \in \omega_j} \mathbf{x}\mathbf{x}^T - \mathbf{m}_j \mathbf{m}_j^T \tag{12.2-23}$$

where N_j is the number of pattern vectors from class ω_j, and the summation is taken over these vectors. Later in this section we give an example of how to use these two expressions.

Consult the book Web site for a brief review of vectors and matrices.

The covariance matrix is symmetric and positive semidefinite. As explained in Section 11.4, the diagonal element c_{kk} is the variance of the kth element of the pattern vectors. The off-diagonal element c_{jk} is the covariance of x_j and x_k. The multivariate Gaussian density function reduces to the product of the univariate Gaussian density of each element of $\mathbf{x}$ when the off-diagonal elements of the covariance matrix are zero. This happens when the vector elements x_j and x_k are uncorrelated.

According to Eq. (12.2-17), the Bayes decision function for class ω_j is $d_j(\mathbf{x}) = p(\mathbf{x}/\omega_j)P(\omega_j)$. However, because of the exponential form of the Gaussian density, working with the natural logarithm of this decision function is more convenient. In other words, we can use the form

$$d_j(\mathbf{x}) = \ln\left[p(\mathbf{x}/\omega_j)P(\omega_j) \right]$$
$$= \ln p(\mathbf{x}/\omega_j) + \ln P(\omega_j) \tag{12.2-24}$$

This expression is equivalent to Eq. (12.2-17) in terms of classification performance because the logarithm is a monotonically increasing function. In other words, the numerical *order* of the decision functions in Eqs. (12.2-17) and (12.2-24) is the same. Substituting Eq. (12.2-19) into Eq. (12.2-24) yields

$$d_j(\mathbf{x}) = \ln P(\omega_j) - \frac{n}{2}\ln 2\pi - \frac{1}{2}\ln|\mathbf{C}_j| - \frac{1}{2}\left[(\mathbf{x} - \mathbf{m}_j)^T\mathbf{C}_j^{-1}(\mathbf{x} - \mathbf{m}_j)\right] \tag{12.2-25}$$

The term $(n/2)\ln 2\pi$ is the same for all classes, so it can be eliminated from Eq. (12.2-25), which then becomes

$$d_j(\mathbf{x}) = \ln P(\omega_j) - \frac{1}{2}\ln|\mathbf{C}_j| - \frac{1}{2}\left[(\mathbf{x} - \mathbf{m}_j)^T\mathbf{C}_j^{-1}(\mathbf{x} - \mathbf{m}_j)\right] \tag{12.2-26}$$

for $j = 1, 2, \ldots, W$. Equation (12.2-26) represents the Bayes decision functions for Gaussian pattern classes under the condition of a 0-1 loss function.

The decision functions in Eq. (12.2-26) are hyperquadrics (quadratic functions in n-dimensional space), because no terms higher than the second degree in the components of $\mathbf{x}$ appear in the equation. Clearly, then, the best that a

Bayes classifier for Gaussian patterns can do is to place a general second-order decision surface between each pair of pattern classes. If the pattern populations are truly Gaussian, however, no other surface would yield a lesser average loss in classification.

If all covariance matrices are equal, then $\mathbf{C}_j = \mathbf{C}$, for $j = 1, 2, \ldots, W$. By expanding Eq. (12.2-26) and dropping all terms independent of j, we obtain

$$d_j(\mathbf{x}) = \ln P(\omega_j) + \mathbf{x}^T \mathbf{C}^{-1} \mathbf{m}_j - \frac{1}{2} \mathbf{m}_j^T \mathbf{C}^{-1} \mathbf{m}_j \qquad (12.2\text{-}27)$$

which are linear decision functions (*hyperplanes*) for $j = 1, 2, \ldots, W$.

If, in addition, $\mathbf{C} = \mathbf{I}$, where $\mathbf{I}$ is the identity matrix, and also $P(\omega_j) = 1/W$, for $j = 1, 2, \ldots, W$, then

$$d_j(\mathbf{x}) = \mathbf{x}^T \mathbf{m}_j - \frac{1}{2} \mathbf{m}_j^T \mathbf{m}_j \qquad j = 1, 2, \ldots, W \qquad (12.2\text{-}28)$$

These are the decision functions for a minimum distance classifier, as given in Eq. (12.2-5). Thus the minimum distance classifier is optimum in the Bayes sense if (1) the pattern classes are Gaussian, (2) all covariance matrices are equal to the identity matrix, and (3) all classes are equally likely to occur. Gaussian pattern classes satisfying these conditions are spherical clouds of identical shape in n dimensions (called *hyperspheres*). The minimum distance classifier establishes a hyperplane between every pair of classes, with the property that the hyperplane is the perpendicular bisector of the line segment joining the center of the pair of hyperspheres. In two dimensions, the classes constitute circular regions, and the boundaries become lines that bisect the line segment joining the center of every pair of such circles.

■ Figure 12.11 shows a simple arrangement of two pattern classes in three dimensions. We use these patterns to illustrate the mechanics of implementing the Bayes classifier, assuming that the patterns of each class are samples from a Gaussian distribution.

EXAMPLE 12.3: A Bayes classifier for three-dimensional patterns.

Applying Eq. (12.2-22) to the patterns of Fig. 12.11 yields

$$\mathbf{m}_1 = \frac{1}{4} \begin{bmatrix} 3 \\ 1 \\ 1 \end{bmatrix} \quad \text{and} \quad \mathbf{m}_2 = \frac{1}{4} \begin{bmatrix} 1 \\ 3 \\ 3 \end{bmatrix}$$

Similarly, applying Eq. (12.2-23) to the two pattern classes in turn yields two covariance matrices, which in this case are equal:

$$\mathbf{C}_1 = \mathbf{C}_2 = \frac{1}{16} \begin{bmatrix} 3 & 1 & 1 \\ 1 & 3 & -1 \\ 1 & -1 & 3 \end{bmatrix}$$

FIGURE 12.11
Two simple
pattern classes
and their Bayes
decision boundary
(shown shaded).

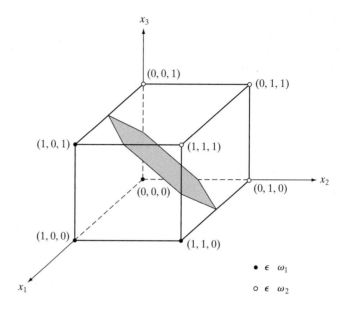

Because the covariance matrices are equal the Bayes decision functions are given by Eq. (12.2-27). If we assume that $P(\omega_1) = P(\omega_2) = 1/2$, then Eq. (12.2-28) applies, giving

$$d_j(\mathbf{x}) = \mathbf{x}^T\mathbf{C}^{-1}\mathbf{m}_j - \frac{1}{2}\mathbf{m}_j^T\mathbf{C}^{-1}\mathbf{m}_j$$

in which

$$\mathbf{C}^{-1} = \begin{bmatrix} 8 & -4 & -4 \\ -4 & 8 & 4 \\ -4 & 4 & 8 \end{bmatrix}$$

Carrying out the vector-matrix expansion for $d_j(\mathbf{x})$ provides the decision functions:

$$d_1(\mathbf{x}) = 4x_1 - 1.5 \quad \text{and} \quad d_2(\mathbf{x}) = -4x_1 + 8x_2 + 8x_3 - 5.5$$

The decision surface separating the two classes then is

$$d_1(\mathbf{x}) - d_2(\mathbf{x}) = 8x_1 - 8x_2 - 8x_3 + 4 = 0$$

Figure 12.11 shows a section of this surface, where we note that the classes were separated effectively.   ■

One of the most successful applications of the Bayes classifier approach is in the classification of remotely sensed imagery generated by multispectral scanners aboard aircraft, satellites, or space stations. The voluminous image data generated by these platforms make automatic image classification and analysis a task of considerable interest in remote sensing. The applications of remote sensing are varied and include land use, crop inventory, crop disease detection, forestry, air and water quality monitoring, geological studies, weather prediction, and a score of other applications having environmental significance. The following example shows a typical application.

■ As discussed in Sections 1.3.4 and 11.4, a multispectral scanner responds to selected bands of the electromagnetic energy spectrum; for example, 0.45–0.52, 0.52–0.60, 0.63–0.69, and 0.76–0.90 microns. These ranges are in the visible blue, visible green, visible red, and near infrared bands, respectively. A region on the ground scanned in this manner produces four digital images of the region, one for each band. If the images are registered spatially, a condition generally met in practice, they can be visualized as being stacked one behind the other, as Fig. 12.12 shows. Thus, just as we did in Section 11.4, every point on the ground can be represented by a 4-element pattern vector of the form $\mathbf{x} = (x_1, x_2, x_3, x_4)^T$, where x_1 is a shade of blue, x_2 a shade of green, and so on. If the images are of size 512×512 pixels, each stack of four multispectral images can be represented by 266,144 four-dimensional pattern vectors. As noted previously, the Bayes classifier for Gaussian patterns requires estimates of the mean vector and covariance matrix for each class. In remote sensing applications, these estimates are obtained by collecting multispectral data whose class is known from each region of interest. The resulting vectors are then used to estimate the required mean vectors and covariance matrices, as in Example 12.3.

Figures 12.13(a) through (d) show four 512×512 multispectral images of the Washington, D.C. area taken in the bands mentioned in the previous paragraph. We are interested in classifying the pixels in the region encompassed by the images into one of three pattern classes: water, urban development, or vegetation. The masks in Fig. 12.13(e) were superimposed on the images to extract

EXAMPLE 12.4:
Classification of multispectral data using a Bayes classifier.

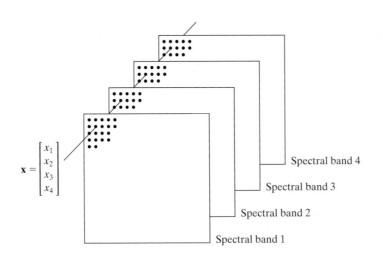

FIGURE 12.12
Formation of a pattern vector from registered pixels of four digital images generated by a multispectral scanner.

$$\mathbf{x} = \begin{bmatrix} x_1 \\ x_2 \\ x_3 \\ x_4 \end{bmatrix}$$

Spectral band 4

Spectral band 3

Spectral band 2

Spectral band 1

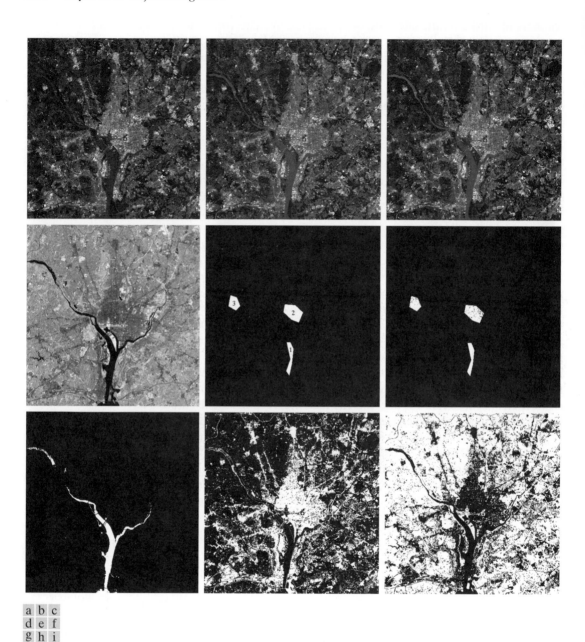

a b c
d e f
g h i

FIGURE 12.13 Bayes classification of multispectral data. (a)–(d) Images in the visible blue, visible green, visible red, and near infrared wavelengths. (e) Mask showing sample regions of water (1), urban development (2), and vegetation (3). (f) Results of classification; the black dots denote points classified incorrectly. The other (white) points were classified correctly. (g) All image pixels classified as water (in white). (h) All image pixels classified as urban development (in white). (i) All image pixels classified as vegetations (in white).

TABLE 12.1
Bayes classification of multispectral image data.

	Training Patterns					Independent Patterns				
	No. of	**Classified into Class**			**%**		**No. of**	**Classified into Class**			**%**
Class	**Samples**	**1**	**2**	**3**	**Correct**	**Class**	**Samples**	**1**	**2**	**3**	**Correct**
1	484	482	2	0	99.6	1	483	478	3	2	98.9
2	933	0	885	48	94.9	2	932	0	880	52	94.4
3	483	0	19	464	96.1	3	482	0	16	466	96.7

samples representative of these three classes. Half of the samples were used for training (i.e., for estimating the mean vectors and covariance matrices), and the other half were used for independent testing to assess preliminary classifier performance. The a priori probabilities, $P(\omega_i)$, seldom are known in unconstrained multispectral data classification, so we assume here that they are equal: $P(\omega_i) = 1/3, i = 1, 2, 3$.

Table 12.1 summarizes the recognition results obtained with the training and independent data sets. The percentage of training and independent pattern vectors recognized correctly was about the same with both data sets, indicating stability in the parameter estimates. The largest error in both cases was with patterns from the urban area. This is not unexpected, as vegetation is present there also (note that no patterns in the vegetation or urban areas were misclassified as water). Figure 12.13(f) shows as black dots the patterns that were misclassified and as white dots the patterns that were classified correctly. No black dots are readily visible in region 1, because the 7 misclassified points are very close to the boundary of the white region.

Figures 12.13(g) through (i) are much more interesting. Here, we used the mean vectors and covariance matrices obtained from the training data to classify *all* image pixels into one of the three categories. Figure 12.13(g) shows in white all pixels that were classified as water. Pixels not classified as water are shown in black. We see that the Bayes classifier did an excellent job of determining which parts of the image were water. Figure 12.13(h) shows in white all pixels classified as urban development; observe how well the system performed in recognizing urban features, such as the bridges and highways. Figure 12.13(i) shows the pixels classified as vegetation. The center area in Fig. 12.13(h) shows a high concentration of white pixels in the downtown area, with the density decreasing as a function of distance from the center of the image. Figure 12.13(i) shows the opposite effect, indicating the least vegetation toward the center of the image, when urban development is at its maximum. ■

We mentioned at the beginning of Section 10.3.3 that thresholding may be viewed as a Bayes classification problem, which optimally assigns patterns to two or more classes. In fact, as the previous problem shows, pixel-by-pixel classification is really a segmentation problem that partitions an image into two or more possible types of regions. If only one single variable (e.g., intensity) is used, then Eq. (12.2-17) becomes an optimum function that similarly partitions an image based on the intensity of its pixels, as we did in Section 10.3. Keep in mind that optimality requires that the PDF and a priori probability of each class be known. As we have mentioned previously, estimating these densities is not a trivial task. If assumptions have to be made (e.g., as in assuming Gaussian densities), then the degree of optimality achieved in segmentation is proportional to how close the assumptions are to reality.

12.2.3 Neural Networks

The approaches discussed in the preceding two sections are based on the use of sample patterns to estimate statistical parameters of each pattern class. The minimum distance classifier is specified completely by the mean vector of each class. Similarly, the Bayes classifier for Gaussian populations is specified completely by the mean vector and covariance matrix of each class. The patterns (of *known* class membership) used to estimate these parameters usually are called *training patterns*, and a set of such patterns from each class is called a *training set*. The process by which a training set is used to obtain decision functions is called *learning* or *training*.

In the two approaches just discussed, training is a simple matter. The training patterns of each class are used to compute the parameters of the decision function corresponding to that class. After the parameters in question have been estimated, the structure of the classifier is fixed, and its eventual performance will depend on how well the actual pattern populations satisfy the underlying statistical assumptions made in the derivation of the classification method being used.

The statistical properties of the pattern classes in a problem often are unknown or cannot be estimated (recall our brief discussion in the preceding section regarding the difficulty of working with multivariate statistics). In practice, such decision-theoretic problems are best handled by methods that yield the required decision functions directly via training. Then, making assumptions regarding the underlying probability density functions or other probabilistic information about the pattern classes under consideration is unnecessary. In this section we discuss various approaches that meet this criterion.

Background

The essence of the material that follows is the use of a multitude of elemental nonlinear computing elements (called *neurons*) organized as networks reminiscent of the way in which neurons are believed to be interconnected in the brain. The resulting models are referred to by various names, including *neural networks, neurocomputers, parallel distributed processing* (PDP) *models, neuromorphic systems, layered self-adaptive networks*, and *connectionist models*. Here, we use the name *neural networks*, or *neural nets* for short. We use these networks as vehicles for adaptively developing the coefficients of decision functions via successive presentations of training sets of patterns.

Interest in neural networks dates back to the early 1940s, as exemplified by the work of McCulloch and Pitts [1943]. They proposed neuron models in the form of binary threshold devices and stochastic algorithms involving sudden 0-1 and 1-0 changes of states in neurons as the bases for modeling neural systems. Subsequent work by Hebb [1949] was based on mathematical models that attempted to capture the concept of learning by reinforcement or association.

During the mid-1950s and early 1960s, a class of so-called *learning machines* originated by Rosenblatt [1959, 1962] caused significant excitement among researchers and practitioners of pattern recognition theory. The reason for the great interest in these machines, called *perceptrons*, was the development of mathematical proofs showing that perceptrons, when trained with linearly separable training sets (i.e., training sets separable by a hyperplane), would converge to a solution in a finite number of iterative steps. The solution took the form of coefficients of hyperplanes capable of correctly separating the classes represented by patterns of the training set.

Unfortunately, the expectations following discovery of what appeared to be a well-founded theoretic model of learning soon met with disappointment. The basic perceptron and some of its generalizations at the time were simply inadequate for most pattern recognition tasks of practical significance. Subsequent attempts to extend the power of perceptron-like machines by considering multiple layers of these devices, although conceptually appealing, lacked effective training algorithms such as those that had created interest in the perceptron itself. The state of the field of learning machines in the mid-1960s was summarized by Nilsson [1965]. A few years later, Minsky and Papert [1969] presented a discouraging analysis of the limitation of perceptron-like machines. This view was held as late as the mid-1980s, as evidenced by comments by Simon [1986]. In this work, originally published in French in 1984, Simon dismisses the perceptron under the heading "Birth and Death of a Myth."

More recent results by Rumelhart, Hinton, and Williams [1986] dealing with the development of new training algorithms for multilayer perceptrons have changed matters considerably. Their basic method, often called the *generalized delta rule for learning by backpropagation*, provides an effective training method for multilayer machines. Although this training algorithm cannot be shown to converge to a solution in the sense of the analogous proof for the single-layer perceptron, the generalized delta rule has been used successfully in numerous problems of practical interest. This success has established multilayer perceptron-like machines as one of the principal models of neural networks currently in use.

Perceptron for two pattern classes

In its most basic form, the perceptron learns a linear decision function that dichotomizes two linearly separable training sets. Figure 12.14(a) shows schematically the perceptron model for two pattern classes. The response of this basic device is based on a weighted sum of its inputs; that is,

$$d(\mathbf{x}) = \sum_{i=1}^{n} w_i x_i + w_{n+1} \tag{12.2-29}$$

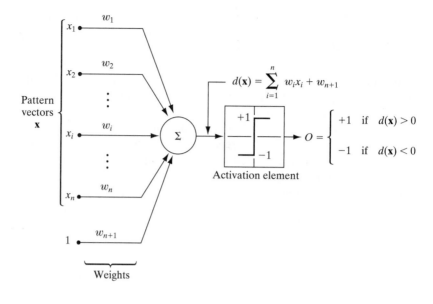

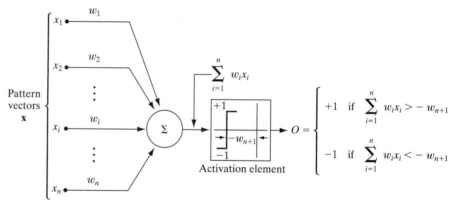

a
b
FIGURE 12.14 Two equivalent representations of the perceptron model for two pattern classes.

which is a linear decision function with respect to the components of the pattern vectors. The coefficients w_i, $i = 1, 2, \ldots, n, n + 1$, called *weights*, modify the inputs before they are summed and fed into the threshold element. In this sense, weights are analogous to synapses in the human neural system. The function that maps the output of the summing junction into the final output of the device sometimes is called the *activation function*.

When $d(\mathbf{x}) > 0$, the threshold element causes the output of the perceptron to be +1, indicating that the pattern $\mathbf{x}$ was recognized as belonging to class ω_1. The reverse is true when $d(\mathbf{x}) < 0$. This mode of operation agrees with the comments made earlier in connection with Eq. (12.2-2) regarding the use of a single decision function for two pattern classes. When $d(\mathbf{x}) = 0$, $\mathbf{x}$ lies on the decision surface separating the two pattern classes, giving an indeterminate condition. The decision boundary implemented by the perceptron is obtained by setting Eq. (12.2-29) equal to zero:

$$d(\mathbf{x}) = \sum_{i=1}^{n} w_i x_i + w_{n+1} = 0 \tag{12.2-30}$$

or

$$w_1 x_1 + w_2 x_2 + \cdots + w_n x_n + w_{n+1} = 0 \tag{12.2-31}$$

which is the equation of a hyperplane in n-dimensional pattern space. Geometrically, the first n coefficients establish the orientation of the hyperplane, whereas the last coefficient, w_{n+1}, is proportional to the perpendicular distance from the origin to the hyperplane. Thus if $w_{n+1} = 0$, the hyperplane goes through the origin of the pattern space. Similarly, if $w_j = 0$, the hyperplane is parallel to the x_j-axis.

The output of the threshold element in Fig. 12.14(a) depends on the sign of $d(\mathbf{x})$. Instead of testing the entire function to determine whether it is positive or negative, we could test the summation part of Eq. (12.2-29) against the term w_{n+1}, in which case the output of the system would be

$$O = \begin{cases} +1 & \text{if } \sum_{i=1}^{n} w_i x_i > -w_{n+1} \\ -1 & \text{if } \sum_{i=1}^{n} w_i x_i < -w_{n+1} \end{cases} \tag{12.2-32}$$

This implementation is equivalent to Fig. 12.14(a) and is shown in Fig. 12.14(b), the only differences being that the threshold function is displaced by an amount $-w_{n+1}$ and that the constant unit input is no longer present. We return to the equivalence of these two formulations later in this section when we discuss implementation of multilayer neural networks.

Another formulation used frequently is to augment the pattern vectors by appending an additional $(n + 1)$st element, which is always equal to 1, regardless of class membership. That is, an augmented pattern vector $\mathbf{y}$ is created from a pattern vector $\mathbf{x}$ by letting $y_i = x_i, i = 1, 2, \ldots, n$, and appending the additional element $y_{n+1} = 1$. Equation (12.2-29) then becomes

$$d(\mathbf{y}) = \sum_{i=1}^{n+1} w_i y_i \tag{12.2-33}$$

$$= \mathbf{w}^T \mathbf{y}$$

where $\mathbf{y} = (y_1, y_2, \ldots, y_n, 1)^T$ is now an *augmented pattern vector*, and $\mathbf{w} = (w_1, w_2, \ldots, w_n, w_{n+1})^T$ is called the *weight vector*. This expression is usually more convenient in terms of notation. Regardless of the formulation used, however, the key problem is to find $\mathbf{w}$ by using a given training set of pattern vectors from each of two classes.

Training algorithms

The algorithms developed in the following discussion are representative of the numerous approaches proposed over the years for training perceptrons.

Linearly separable classes: A simple, iterative algorithm for obtaining a solution weight vector for two linearly separable training sets follows. For two training sets of augmented pattern vectors belonging to pattern classes ω_1 and ω_2, respectively, let $\mathbf{w}(1)$ represent the initial weight vector, which may be chosen arbitrarily. Then, at the kth iterative step, if $\mathbf{y}(k) \in \omega_1$ and $\mathbf{w}^T(k)\mathbf{y}(k) \le 0$, replace $\mathbf{w}(k)$ by

$$\mathbf{w}(k + 1) = \mathbf{w}(k) + c\mathbf{y}(k) \tag{12.2-34}$$

where c is a positive correction increment. Conversely, if $\mathbf{y}(k) \in \omega_2$ and $\mathbf{w}^T(k)\mathbf{y}(k) \ge 0$, replace $\mathbf{w}(k)$ with

$$\mathbf{w}(k + 1) = \mathbf{w}(k) - c\mathbf{y}(k) \tag{12.2-35}$$

Otherwise, leave $\mathbf{w}(k)$ unchanged:

$$\mathbf{w}(k + 1) = \mathbf{w}(k) \tag{12.2-36}$$

This algorithm makes a change in $\mathbf{w}$ only if the pattern being considered at the kth step in the training sequence is misclassified. The correction increment c is assumed to be positive and, for now, to be constant. This algorithm sometimes is referred to as the *fixed increment correction rule.*

Convergence of the algorithm occurs when the entire training set for both classes is cycled through the machine without any errors. The fixed increment correction rule converges in a finite number of steps if the two training sets of patterns are linearly separable. A proof of this result, sometimes called the *perceptron training theorem,* can be found in the books by Duda, Hart, and Stork [2001]; Tou and Gonzalez [1974]; and Nilsson [1965].

EXAMPLE 12.5:
Illustration of the perceptron algorithm.

■ Consider the two training sets shown in Fig. 12.15(a), each consisting of two patterns. The training algorithm will be successful because the two training sets are linearly separable. Before the algorithm is applied the patterns are augmented, yielding the training set $\{(0, 0, 1)^T, (0, 1, 1)^T\}$ for class ω_1 and $\{(1, 0, 1)^T, (1, 1, 1)^T\}$ for class ω_2. Letting $c = 1$, $\mathbf{w}(1) = \mathbf{0}$, and presenting the patterns in order results in the following sequence of steps:

$$\mathbf{w}^T(1)\mathbf{y}(1) = [0, 0, 0]\begin{bmatrix} 0 \\ 0 \\ 1 \end{bmatrix} = 0 \qquad \mathbf{w}(2) = \mathbf{w}(1) + \mathbf{y}(1) = \begin{bmatrix} 0 \\ 0 \\ 1 \end{bmatrix}$$

$$\mathbf{w}^T(2)\mathbf{y}(2) = [0, 0, 1]\begin{bmatrix} 0 \\ 1 \\ 1 \end{bmatrix} = 1 \qquad \mathbf{w}(3) = \mathbf{w}(2) = \begin{bmatrix} 0 \\ 0 \\ 1 \end{bmatrix}$$

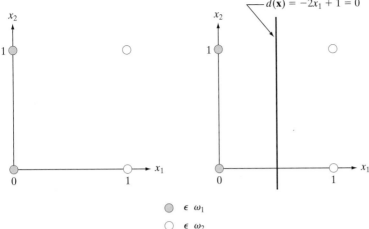

FIGURE 12.15
(a) Patterns belonging to two classes.
(b) Decision boundary determined by training.

$$\mathbf{w}^T(3)\mathbf{y}(3) = [0, 0, 1]\begin{bmatrix} 1 \\ 0 \\ 1 \end{bmatrix} = 1 \qquad \mathbf{w}(4) = \mathbf{w}(3) - \mathbf{y}(3) = \begin{bmatrix} -1 \\ 0 \\ 0 \end{bmatrix}$$

$$\mathbf{w}^T(4)\mathbf{y}(4) = [-1, 0, 0]\begin{bmatrix} 1 \\ 1 \\ 1 \end{bmatrix} = -1 \qquad \mathbf{w}(5) = \mathbf{w}(4) = \begin{bmatrix} -1 \\ 0 \\ 0 \end{bmatrix}$$

where corrections in the weight vector were made in the first and third steps because of misclassifications, as indicated in Eqs. (12.2-34) and (12.2-35). Because a solution has been obtained only when the algorithm yields a complete error-free iteration through all training patterns, the training set must be presented again. The machine learning process is continued by letting $\mathbf{y}(5) = \mathbf{y}(1)$, $\mathbf{y}(6) = \mathbf{y}(2)$, $\mathbf{y}(7) = \mathbf{y}(3)$, and $\mathbf{y}(8) = \mathbf{y}(4)$, and proceeding in the same manner. Convergence is achieved at $k = 14$, yielding the solution weight vector $\mathbf{w}(14) = (-2, 0, 1)^T$. The corresponding decision function is $d(\mathbf{y}) = -2y_1 + 1$. Going back to the original pattern space by letting $x_i = y_i$ yields $d(\mathbf{x}) = -2x_1 + 1$, which, when set equal to zero, becomes the equation of the decision boundary shown in Fig. 12.15(b). ■

Nonseparable classes: In practice, linearly separable pattern classes are the (rare) exception, rather than the rule. Consequently, a significant amount of research effort during the 1960s and 1970s went into development of techniques designed to handle nonseparable pattern classes. With recent advances in the training of neural networks, many of the methods dealing with nonseparable behavior have become merely items of historical interest. One of the early methods, however, is directly relevant to this discussion: the original delta rule. Known as the *Widrow-Hoff*, or *least-mean-square* (LMS) *delta rule* for training perceptrons, the method minimizes the error between the actual and desired response at any training step.

Consider the criterion function

$$J(\mathbf{w}) = \frac{1}{2}(r - \mathbf{w}^T\mathbf{y})^2 \tag{12.2-37}$$

where r is the desired response (that is, $r = +1$ if the augmented training pattern vector $\mathbf{y}$ belongs to class ω_1, and $r = -1$ if $\mathbf{y}$ belongs to class ω_2). The task is to adjust $\mathbf{w}$ incrementally in the direction of the negative gradient of $J(\mathbf{w})$ in order to seek the minimum of this function, which occurs when $r = \mathbf{w}^T\mathbf{y}$; that is, the minimum corresponds to correct classification. If $\mathbf{w}(k)$ represents the weight vector at the kth iterative step, a general gradient descent algorithm may be written as

$$\mathbf{w}(k + 1) = \mathbf{w}(k) - \alpha\left[\frac{\partial J(\mathbf{w})}{\partial \mathbf{w}}\right]_{\mathbf{w}=\mathbf{w}(k)} \tag{12.2-38}$$

where $\mathbf{w}(k + 1)$ is the new value of $\mathbf{w}$, and $\alpha > 0$ gives the magnitude of the correction. From Eq. (12.2-37),

$$\frac{\partial J(\mathbf{w})}{\partial \mathbf{w}} = -(r - \mathbf{w}^T\mathbf{y})\mathbf{y} \tag{12.2-39}$$

Substituting this result into Eq. (12.2-38) yields

$$\mathbf{w}(k + 1) = \mathbf{w}(k) + \alpha\left[r(k) - \mathbf{w}^T(k)\mathbf{y}(k)\right]\mathbf{y}(k) \tag{12.2-40}$$

with the starting weight vector, $\mathbf{w}(1)$, being arbitrary.

By defining the change (delta) in weight vector as

$$\Delta\mathbf{w} = \mathbf{w}(k + 1) - \mathbf{w}(k) \tag{12.2-41}$$

we can write Eq. (12.2-40) in the form of a *delta correction algorithm:*

$$\Delta\mathbf{w} = \alpha e(k)\mathbf{y}(k) \tag{12.2-42}$$

where

$$e(k) = r(k) - \mathbf{w}^T(k)\mathbf{y}(k) \tag{12.2-43}$$

is the error committed with weight vector $\mathbf{w}(k)$ when pattern $\mathbf{y}(k)$ is presented.

Equation (12.2-43) gives the error with weight vector $\mathbf{w}(k)$. If we change it to $\mathbf{w}(k + 1)$, but leave the pattern the same, the error becomes

$$e(k) = r(k) - \mathbf{w}^T(k + 1)\mathbf{y}(k) \tag{12.2-44}$$

The change in error then is

$$\begin{aligned}
\Delta e(k) &= \left[r(k) - \mathbf{w}^T(k + 1)\mathbf{y}(k)\right] - \left[r(k) - \mathbf{w}^T(k)\mathbf{y}(k)\right] \\
&= -\left[\mathbf{w}^T(k + 1) - \mathbf{w}^T(k)\right]\mathbf{y}(k) \tag{12.2-45} \\
&= -\Delta\mathbf{w}^T\mathbf{y}(k)
\end{aligned}$$

But $\Delta\mathbf{w} = \alpha e(k)\mathbf{y}(k)$, so

$$\Delta e = -\alpha e(k)\mathbf{y}^T(k)\mathbf{y}(k)$$
$$= -\alpha e(k)\|\mathbf{y}(k)\|^2 \qquad (12.2\text{-}46)$$

Hence changing the weights reduces the error by a factor $\alpha\|\mathbf{y}(k)\|^2$. The next input pattern starts the new adaptation cycle, reducing the next error by a factor $\alpha\|\mathbf{y}(k+1)\|^2$, and so on.

The choice of α controls stability and speed of convergence (Widrow and Stearns [1985]). Stability requires that $0 < \alpha < 2$. A practical range for α is $0.1 < \alpha < 1.0$. Although the proof is not shown here, the algorithm of Eq. (12.2-40) or Eqs. (12.2-42) and (12.2-43) converges to a solution that minimizes the mean square error over the patterns of the training set. When the pattern classes are separable, the solution given by the algorithm just discussed may or may not produce a separating hyperplane. That is, a mean-square-error solution does not imply a solution in the sense of the perceptron training theorem. This uncertainty is the price of using an algorithm that converges under both the separable and nonseparable cases in this particular formulation.

The two perceptron training algorithms discussed thus far can be extended to more than two classes and to nonlinear decision functions. Based on the historical comments made earlier, exploring multiclass training algorithms here has little merit. Instead, we address multiclass training in the context of neural networks.

Multilayer feedforward neural networks

In this section we focus on decision functions of multiclass pattern recognition problems, independent of whether or not the classes are separable, and involving architectures that consist of layers of perceptron computing elements.

Basic architecture: Figure 12.16 shows the architecture of the neural network model under consideration. It consists of layers of structurally identical computing nodes (neurons) arranged so that the output of every neuron in one layer feeds into the input of every neuron in the next layer. The number of neurons in the first layer, called layer A, is N_A. Often, $N_A = n$, the dimensionality of the input pattern vectors. The number of neurons in the output layer, called layer Q, is denoted N_Q. The number N_Q equals W, the number of pattern classes that the neural network has been trained to recognize. The network recognizes a pattern vector $\mathbf{x}$ as belonging to class ω_i if the ith output of the network is "high" while all other outputs are "low," as explained in the following discussion.

As the blowup in Fig. 12.16 shows, each neuron has the same form as the perceptron model discussed earlier (see Fig. 12.14), with the exception that the hard-limiting activation function has been replaced by a soft-limiting "sigmoid" function. Differentiability along all paths of the neural network is required in the development of the training rule. The following sigmoid activation function has the necessary differentiability:

$$h_j(I_j) = \frac{1}{1 + e^{-(I_j+\theta_j)/\theta_o}} \qquad (12.2\text{-}47)$$

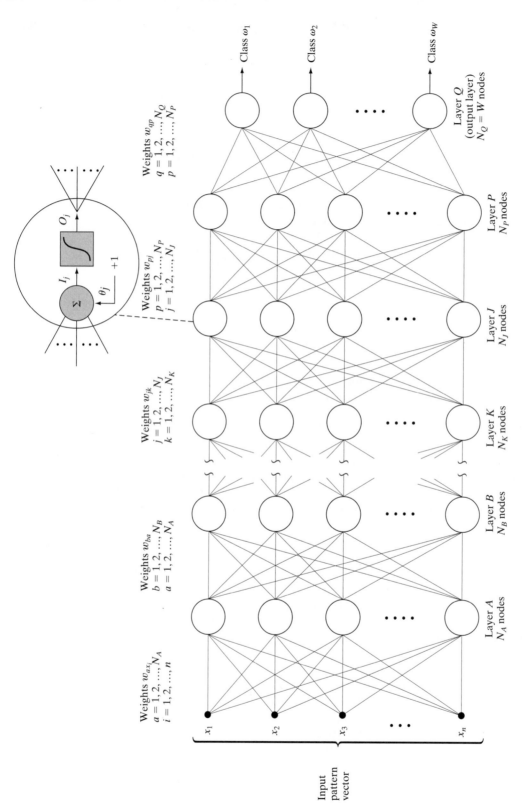

FIGURE 12.16 Multilayer feedforward neural network model. The blowup shows the basic structure of each neuron element throughout the network. The offset, θ_j, is treated as just another weight.

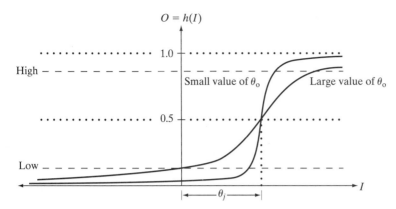

FIGURE 12.17
The sigmoidal
activation
function of
Eq. (12.2-47).

where $I_j, j = 1, 2, \ldots, N_J$, is the input to the activation element of each node in layer J of the network, θ_j is an offset, and θ_o controls the shape of the sigmoid function.

Equation (12.2-47) is plotted in Fig. 12.17, along with the limits for the "high" and "low" responses out of each node. Thus when this particular function is used, the system outputs a high reading for any value of I_j greater than θ_j. Similarly, the system outputs a low reading for any value of I_j less than θ_j. As Fig. 12.17 shows, the sigmoid activation function always is positive, and it can reach its limiting values of 0 and 1 only if the input to the activation element is infinitely negative or positive, respectively. For this reason, values near 0 and 1 (say, 0.05 and 0.95) define low and high values at the output of the neurons in Fig. 12.16. In principle, different types of activation functions could be used for different layers or even for different nodes in the same layer of a neural network. In practice, the usual approach is to use the same form of activation function throughout the network.

With reference to Fig. 12.14(a), the offset θ_j shown in Fig. 12.17 is analogous to the weight coefficient w_{n+1} in the earlier discussion of the perceptron. Implementation of this displaced threshold function can be done in the form of Fig. 12.14(a) by absorbing the offset θ_j as an additional coefficient that modifies a constant unity input to all nodes in the network. In order to follow the notation predominantly found in the literature, we do not show a separate constant input of +1 into all nodes of Fig. 12.16. Instead, this input and its modifying weight θ_j are integral parts of the network nodes. As noted in the blowup in Fig. 12.16, there is one such coefficient for each of the N_J nodes in layer J.

In Fig. 12.16, the input to a node in any layer is the weighted sum of the outputs from the previous layer. Letting layer K denote the layer preceding layer J (no alphabetical order is implied in Fig. 12.16) gives the input to the activation element of each node in layer J, denoted I_j:

$$I_j = \sum_{k=1}^{N_K} w_{jk} O_k \qquad (12.2\text{-}48)$$

for $j = 1, 2, \ldots, N_J$, where N_J is the number of nodes in layer J, N_K is the number of nodes in layer K, and w_{jk} are the weights modifying the outputs O_k of the nodes in layer K before they are fed into the nodes in layer J. The outputs of layer K are

$$O_k = h_k(I_k) \tag{12.2-49}$$

for $k = 1, 2, \ldots, N_K$.

A clear understanding of the subscript notation used in Eq. (12.2-48) is important, because we use it throughout the remainder of this section. First, note that $I_j, j = 1, 2, \ldots, N_J$, represents the input to the *activation element* of the jth node in layer J. Thus I_1 represents the input to the activation element of the first (topmost) node in layer J, I_2 represents the input to the activation element of the second node in layer J, and so on. There are N_K inputs to every node in layer J, but *each* individual input can be weighted differently. Thus the N_K inputs to the first node in layer J are weighted by coefficients $w_{1k}, k = 1, 2, \ldots, N_K$; the inputs to the second node are weighted by coefficients $w_{2k}, k = 1, 2, \ldots, N_K$; and so on. Hence a total of $N_J \times N_K$ coefficients are necessary to specify the weighting of the outputs of layer K as they are fed into layer J. An additional N_J offset coefficients, θ_j, are needed to specify completely the nodes in layer J.

Substitution of Eq. (12.2-48) into (12.2-47) yields

$$h_j(I_j) = \frac{1}{1 + e^{-\left(\sum\limits_{k=1}^{N_k} w_{jk}o_k + \theta_j\right)/\theta_o}} \tag{12.2-50}$$

which is the form of activation function used in the remainder of this section.

During training, adapting the neurons in the output layer is a simple matter because the desired output of each node is known. The main problem in training a multilayer network lies in adjusting the weights in the so-called *hidden layers*. That is, in those other than the output layer.

Training by back propagation: We begin by concentrating on the output layer. The total squared error between the desired responses, r_q, and the corresponding actual responses, O_q, of nodes in (output) layer Q, is

$$E_Q = \frac{1}{2} \sum_{q=1}^{N_Q} (r_q - O_q)^2 \tag{12.2-51}$$

where N_Q is the number of nodes in output layer Q and the $\frac{1}{2}$ is used for convenience in notation for taking the derivative later.

The objective is to develop a training rule, similar to the delta rule, that allows adjustment of the weights in each of the layers in a way that seeks a minimum to an error function of the form shown in Eq. (12.2-51). As before, adjusting the weights in proportion to the partial derivative of the error with respect to the weights achieves this result. In other words,

$$\Delta w_{qp} = -\alpha \frac{\partial E_Q}{\partial w_{qp}} \tag{12.2-52}$$

where layer P precedes layer Q, Δw_{qp} is as defined in Eq. (12.2-42), and α is a positive correction increment.

The error E_Q is a function of the outputs, O_q, which in turn are functions of the inputs I_q. Using the chain rule, we evaluate the partial derivative of E_Q as follows:

$$\frac{\partial E_Q}{\partial w_{qp}} = \frac{\partial E_Q}{\partial I_q} \frac{\partial I_q}{\partial w_{qp}} \tag{12.2-53}$$

From Eq. (12.2-48),

$$\frac{\partial I_q}{\partial w_{qp}} = \frac{\partial}{\partial w_{qp}} \sum_{p=1}^{N_P} w_{qp} O_p = O_p \tag{12.2-54}$$

Substituting Eqs. (12.2-53) and (12.2-54) into Eq. (12.2-52) yields

$$\Delta w_{qp} = -\alpha \frac{\partial E_Q}{\partial I_q} O_p$$

$$= \alpha \delta_q O_p \tag{12.2-55}$$

where

$$\delta_q = -\frac{\partial E_Q}{\partial I_q} \tag{12.2-56}$$

In order to compute $\partial E_Q / \partial I_q$, we use the chain rule to express the partial derivative in terms of the rate of change of E_Q with respect to O_q and the rate of change of O_q with respect to I_q. That is,

$$\delta_q = -\frac{\partial E_Q}{\partial I_q} = -\frac{\partial E_Q}{\partial O_q} \frac{\partial O_q}{\partial I_q} \tag{12.2-57}$$

From Eq. (12.2-51),

$$\frac{\partial E_Q}{\partial O_q} = -(r_q - O_q) \tag{12.2-58}$$

and, from Eq. (12.2-49),

$$\frac{\partial O_Q}{\partial I_q} = \frac{\partial}{\partial I_q} h_q(I_q) = h'_q(I_q) \tag{12.2-59}$$

Substituting Eqs. (12.2-58) and (12.2-59) into Eq. (12.2-57) gives

$$\delta_q = (r_q - O_q)h_q'(I_q) \tag{12.2-60}$$

which is proportional to the error quantity $(r_q - O_q)$. Substitution of Eqs. (12.2-56) through (12.2-58) into Eq. (12.2-55) finally yields

$$\Delta w_{qp} = \alpha(r_q - O_q)h_q'(I_q)O_p$$
$$= \alpha\delta_q O_p \tag{12.2-61}$$

After the function $h_q(I_q)$ has been specified, all the terms in Eq. (12.2-61) are known or can be observed in the network. In other words, upon presentation of any training pattern to the input of the network, we know what the desired response, r_q, of each output node should be. The value O_q of each output node can be observed as can I_q, the input to the activation elements of layer Q, and O_p, the output of the nodes in layer P. Thus we know how to adjust the weights that modify the links between the last and next-to-last layers in the network.

Continuing to work our way back from the output layer, let us now analyze what happens at layer P. Proceeding in the same manner as above yields

$$\Delta w_{pj} = \alpha(r_p - O_p)h_p'(I_p)O_j$$
$$= \alpha\delta_p O_j \tag{12.2-62}$$

where the error term is

$$\delta_p = (r_p - O_p)h_p'(I_p) \tag{12.2-63}$$

With the exception of r_p, all the terms in Eqs. (12.2-62) and (12.2-63) either are known or can be observed in the network. The term r_p makes no sense in an internal layer because we do not know what the response of an internal node in terms of pattern membership should be. We may specify what we want the response r to be only at the outputs of the network where final pattern classification takes place. If we knew that information at internal nodes, there would be no need for further layers. Thus we have to find a way to restate δ_p in terms of quantities that are known or can be observed in the network.

Going back to Eq. (12.2-57), we write the error term for layer P as

$$\delta_p = -\frac{\partial E_p}{\partial I_p} = -\frac{\partial E_p}{\partial O_p}\frac{\partial O_p}{\partial I_p} \tag{12.2-64}$$

The term $\partial O_p / \partial I_p$ presents no difficulties. As before, it is

$$\frac{\partial O_p}{\partial I_p} = \frac{\partial h_p(I_p)}{\partial I_p} = h_p'(I_p) \tag{12.2-65}$$

which is known once h_p is specified because I_p can be observed. The term that produced r_p was the derivative $\partial E_p / \partial O_p$, so this term must be expressed in a way that does not contain r_p. Using the chain rule, we write the derivative as

$$-\frac{\partial E_p}{\partial O_p} = -\sum_{q=1}^{N_Q}\frac{\partial E_p}{\partial I_q}\frac{\partial I_q}{\partial O_p} = \sum_{q=1}^{N_Q}\left(-\frac{\partial E_p}{\partial I_q}\right)\frac{\partial}{\partial O_p}\sum_{p=1}^{N_P}w_{qp}O_p$$

$$= \sum_{q=1}^{N_Q} \left(-\frac{\partial E_P}{\partial I_q} \right) w_{qp}$$

$$= \sum_{q=1}^{N_Q} \delta_q w_{qp} \qquad (12.2\text{-}66)$$

where the last step follows from Eq. (12.2-56). Substituting Eqs. (12.2-65) and (12.2-66) into Eq. (12.2-64) yields the desired expression for δ_p:

$$\delta_p = h'_p(I_p) \sum_{q=1}^{N_Q} \delta_q w_{qp} \qquad (12.2\text{-}67)$$

The parameter δ_p can be computed now because all its terms are known. Thus Eqs. (12.2-62) and (12.2-67) establish completely the training rule for layer P. The importance of Eq. (12.2-67) is that it computes δ_p from the quantities δ_q and w_{qp}, which are terms that were computed in the layer immediately following layer P. After the error term and weights have been computed for layer P, these quantities may be used similarly to compute the error and weights for the layer immediately preceding layer P. In other words, we have found a way to propagate the error back into the network, starting with the error at the output layer.

We may summarize and generalize the training procedure as follows. For any layers K and J, where layer K immediately precedes layer J, compute the weights w_{jk}, which modify the connections between these two layers, by using

$$\Delta w_{jk} = \alpha \delta_j O_k \qquad (12.2\text{-}68)$$

If layer J is the output layer, δ_j is

$$\delta_j = (r_j - O_j) h'_j(I_j) \qquad (12.2\text{-}69)$$

If layer J is an internal layer and layer P is the next layer (to the right), then δ_j is given by

$$\delta_j = h'_j(I_j) \sum_{p=1}^{N_P} \delta_p w_{jp} \qquad (12.2\text{-}70)$$

for $j = 1, 2, \ldots, N_j$. Using the activation function in Eq. (12.2-50) with $\theta_o = 1$ yields

$$h'_j(I_j) = O_j(1 - O_j) \qquad (12.2\text{-}71)$$

in which case Eqs. (12.2-69) and (12.2-70) assume the following, particularly attractive forms:

$$\delta_j = (r_j - O_j) O_j(1 - O_j) \qquad (12.2\text{-}72)$$

for the output layer, and

$$\delta_j = O_j(1 - O_j) \sum_{p=1}^{N_P} \delta_p w_{jp} \qquad (12.2\text{-}73)$$

for internal layers. In both Eqs. (12.2-72) and (12.2-73), $j = 1, 2, \ldots, N_J$.

Equations (12.2-68) through (12.2-70) constitute the generalized delta rule for training the multilayer feedforward neural network of Fig. 12.16. The process starts with an arbitrary (but not all equal) set of weights throughout the network. Then application of the generalized delta rule at any iterative step involves two basic phases. In the first phase, a training vector is presented to the network and is allowed to propagate through the layers to compute the output O_j for each node. The outputs O_q of the nodes in the output layer are then compared against their desired responses, r_p, to generate the error terms δ_q. The second phase involves a backward pass through the network during which the appropriate error signal is passed to each node and the corresponding weight changes are made. This procedure also applies to the bias weights θ_j. As discussed earlier in some detail, these are treated simply as additional weights that modify a unit input into the summing junction of every node in the network.

Common practice is to track the network error, as well as errors associated with individual patterns. In a successful training session, the network error decreases with the number of iterations and the procedure converges to a stable set of weights that exhibit only small fluctuations with additional training. The approach followed to establish whether a pattern has been classified correctly during training is to determine whether the response of the node in the output layer associated with the pattern class from which the pattern was obtained is high, while all the other nodes have outputs that are low, as defined earlier.

After the system has been trained, it classifies patterns using the parameters established during the training phase. In normal operation, all feedback paths are disconnected. Then any input pattern is allowed to propagate through the various layers, and the pattern is classified as belonging to the class of the output node that was high, while all the others were low. If more than one output is labeled high, or if none of the outputs is so labeled, the choice is one of declaring a misclassification or simply assigning the pattern to the class of the output node with the highest numerical value.

EXAMPLE 12.6:
Shape classification using a neural network.

■ We illustrate now how a neural network of the form shown in Fig. 12.16 was trained to recognize the four shapes shown in Fig. 12.18(a), as well as noisy versions of these shapes, samples of which are shown in Fig. 12.18(b).

Pattern vectors were generated by computing the normalized signatures of the shapes (see Section 11.1.3) and then obtaining 48 uniformly spaced samples of each signature. The resulting 48-dimensional vectors were the inputs to the three-layer feedforward neural network shown in Fig. 12.19. The number of neuron nodes in the first layer was chosen to be 48, corresponding to the dimensionality of the input pattern vectors. The four neurons in the third (output) layer correspond to the number of pattern classes, and the number of neurons in the middle layer was heuristically specified as 26 (the average of the number of neurons in the input and output layers). There are no known rules for specifying the number of nodes in the internal layers of a neural network, so this number generally is based either on prior experience or simply chosen arbitrarily and then refined by testing. In the output layer, the four nodes from

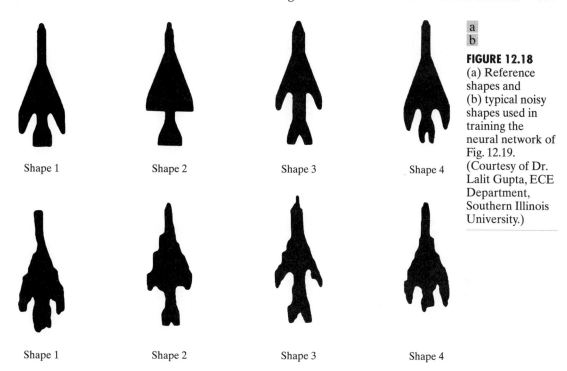

FIGURE 12.18
(a) Reference shapes and (b) typical noisy shapes used in training the neural network of Fig. 12.19. (Courtesy of Dr. Lalit Gupta, ECE Department, Southern Illinois University.)

top to bottom in this case represent the classes $\omega_j, j = 1, 2, 3, 4$, respectively. After the network structure has been set, activation functions have to be selected for each unit and layer. All activation functions were selected to satisfy Eq. (12.2-50) with $\theta_o = 1$ so that, according to our earlier discussion, Eqs. (12.2-72) and (12.2-73) apply.

The training process was divided in two parts. In the first part, the weights were initialized to small random values with zero mean, and the network was then trained with pattern vectors corresponding to noise-free samples like the shapes shown in Fig. 12.18(a). The output nodes were monitored during training. The network was said to have learned the shapes from all four classes when, for any training pattern from class ω_i, the elements of the output layer yielded $O_i \geq 0.95$ and $O_q \leq 0.05$, for $q = 1, 2, \ldots, N_Q; q \neq i$. In other words, for any pattern of class ω_i, the output unit corresponding to that class had to be high (≥ 0.95) while, simultaneously, the output of all other nodes had to be low (≤ 0.05).

The second part of training was carried out with noisy samples, generated as follows. Each contour pixel in a noise-free shape was assigned a probability V of retaining its original coordinate in the image plane and a probability $R = 1 - V$ of being randomly assigned to the coordinates of one of its eight neighboring pixels. The degree of noise was increased by decreasing V (that is, increasing R). Two sets of noisy data were generated. The first consisted of 100 noisy patterns of each class generated by varying R between 0.1 and 0.6, giving a total of 400 patterns. This set, called the *test set*, was used to establish system performance after training.

FIGURE 12.19
Three-layer
neural network
used to recognize
the shapes in Fig.
12.18.
(Courtesy of Dr.
Lalit Gupta, ECE
Department,
Southern Illinois
University.)

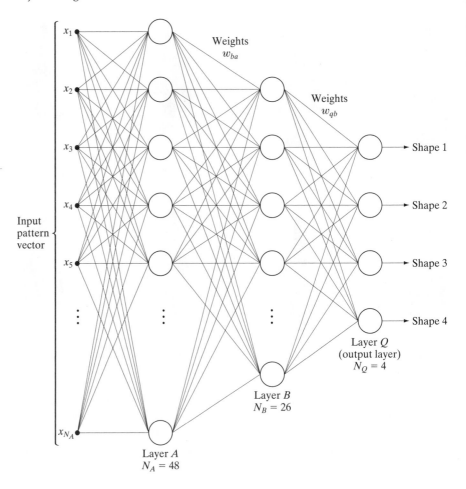

Several noisy sets were generated for training the system with noisy data. The first set consisted of 10 samples for each class, generated by using $R_t = 0$, where R_t denotes a value of R used to generate training data. Starting with the weight vectors obtained in the first (noise-free) part of training, the system was allowed to go through a learning sequence with the new data set. Because $R_t = 0$ implies no noise, this retraining was an extension of the earlier, noise-free training. Using the resulting weights learned in this manner, the network was subjected to the test data set yielding the results shown by the curve labeled $R_t = 0$ in Fig. 12.20. The number of misclassified patterns divided by the total number of patterns tested gives the probability of misclassification, which is a measure commonly used to establish neural network performance.

Next, starting with the weight vectors learned by using the data generated with $R_t = 0$, the system was retrained with a noisy data set generated with $R_t = 0.1$. The recognition performance was then established by running the test samples through the system again with the new weight vectors. Note the significant improvement in performance. Figure 12.20 shows the results obtained by continuing

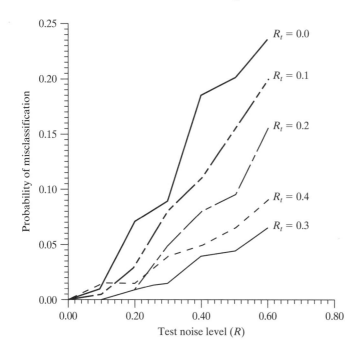

FIGURE 12.20
Performance of
the neural
network as a
function of noise
level. (Courtesy
of Dr. Lalit
Gupta, ECE
Department,
Southern Illinois
University.)

this retraining and retesting procedure for $R_t = 0.2, 0.3$, and 0.4. As expected if the system is learning properly, the probability of misclassifying patterns from the test set decreased as the value of R_t increased because the system was being trained with noisier data for higher values of R_t. The one exception in Fig. 12.20 is the result for $R_t = 0.4$. The reason is the small number of samples used to train the system. That is, the network was not able to adapt itself sufficiently to the larger variations in shape at higher noise levels with the number of samples used. This hypothesis is verified by the results in Fig. 12.21, which show a lower probability of misclassification as the number of training samples was increased. Figure 12.21 also shows as a reference the curve for $R_t = 0.3$ from Fig. 12.20.

The preceding results show that a three-layer neural network was capable of learning to recognize shapes corrupted by noise after a modest level of training. Even when trained with noise-free data ($R_t = 0$ in Fig. 12.20), the system was able to achieve a correct recognition level of close to 77% when tested with data highly corrupted by noise ($R = 0.6$ in Fig. 12.20). The recognition rate on the same data increased to about 99% when the system was trained with noisier data ($R_r = 0.3$ and 0.4). It is important to note that the system was trained by increasing its classification power via systematic, small incremental additions of noise. When the nature of the noise is known, this method is ideal for improving the convergence and stability properties of a neural network during learning. ■

Complexity of decision surfaces: We have already established that a single-layer perceptron implements a hyperplane decision surface. A natural question at this point is: What is the nature of the decision surfaces implemented by

FIGURE 12.21
Improvement in
performance for
$R_t = 0.4$ by
increasing the
number of
training patterns
(the curve for
$R_t = 0.3$ is shown
for reference).
(Courtesy of Dr.
Lalit Gupta, ECE
Department,
Southern Illinois
University.)

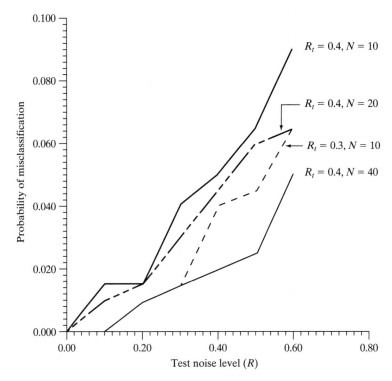

a multilayer network, such as the model in Fig. 12.16? It is demonstrated in the following discussion that a three-layer network is capable of implementing arbitrarily complex decision surfaces composed of intersecting hyperplanes.

As a starting point, consider the two-input, two-layer network shown in Fig. 12.22(a). With two inputs, the patterns are two dimensional, and therefore, each node in the first layer of the network implements a line in 2-D space. We denote by 1 and 0, respectively, the high and low outputs of these two nodes. We assume that a 1 output indicates that the corresponding input vector to a node in the first layer lies on the positive side of the line. Then the possible combinations of outputs feeding the single node in the second layer

a b c

FIGURE 12.22
(a) A two-input,
two-layer,
feedforward
neural network.
(b) and (c)
Examples of
decision
boundaries that
can be
implemented with
this network.

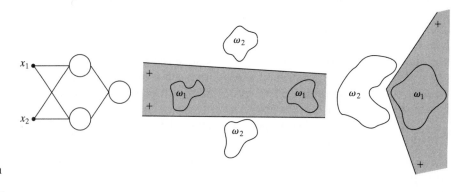

are $(1, 1)$, $(1, 0)$, $(0, 1)$, and $(0, 0)$. If we define two regions, one for class ω_1 lying on the positive side of both lines and the other for class ω_2 lying anywhere else, the output node can classify any input pattern as belonging to one of these two regions simply by performing a logical AND operation. In other words, the output node responds with a 1, indicating class ω_1, only when both outputs from the first layer are 1. The AND operation can be performed by a neural node of the form discussed earlier if θ_j is set to a value in the half-open interval $(1, 2]$. Thus if we assume 0 and 1 responses out of the first layer, the response of the output node will be high, indicating class ω_1, only when the sum performed by the neural node on the two outputs from the first layer is greater than 1. Figures 12.22(b) and (c) show how the network of Fig. 12.22(a) can successfully dichotomize two pattern classes that could not be separated by a single linear surface.

If the number of nodes in the first layer were increased to three, the network of Fig. 12.22(a) would implement a decision boundary consisting of the intersection of three lines. The requirement that class ω_1 lie on the positive side of all three lines would yield a convex region bounded by the three lines. In fact, an arbitrary open or closed convex region can be constructed simply by increasing the number of nodes in the first layer of a two-layer neural network.

The next logical step is to increase the number of layers to three. In this case the nodes of the first layer implement lines, as before. The nodes of the second layer then perform AND operations in order to form regions from the various lines. The nodes in the third layer assign class membership to the various regions. For instance, suppose that class ω_1 consists of two distinct regions, each of which is bounded by a different set of lines. Then two of the nodes in the second layer are for regions corresponding to the same pattern class. One of the output nodes needs to be able to signal the presence of that class when either of the two nodes in the second layer goes high. Assuming that high and low conditions in the second layer are denoted 1 and 0, respectively, this capability is obtained by making the output nodes of the network perform the logical OR operation. In terms of neural nodes of the form discussed earlier, we do so by setting θ_j to a value in the half-open interval $[0, 1)$. Then, whenever at least one of the nodes in the second layer associated with that output node goes high (outputs a 1), the corresponding node in the output layer will go high, indicating that the pattern being processed belongs to the class associated with that node.

Figure 12.23 summarizes the preceding comments. Note in the third row that the complexity of decision regions implemented by a three-layer network is, in principle, arbitrary. In practice, a serious difficulty usually arises in structuring the second layer to respond correctly to the various combinations associated with particular classes. The reason is that lines do not just stop at their intersection with other lines, and, as a result, patterns of the same class may occur on both sides of lines in the pattern space. In practical terms, the second layer may have difficulty figuring out which lines should be included in the AND operation for a given pattern class—or it may even be impossible. The reference to the exclusive-OR problem in the third column of Fig. 12.23 deals with the fact that, if the input patterns were binary, only four different patterns could be constructed in two

FIGURE 12.23
Types of decision
regions that can
be formed by
single- and
multilayer feed-
forward networks
with one and two
layers of hidden
units and two
inputs.
(Lippman.)

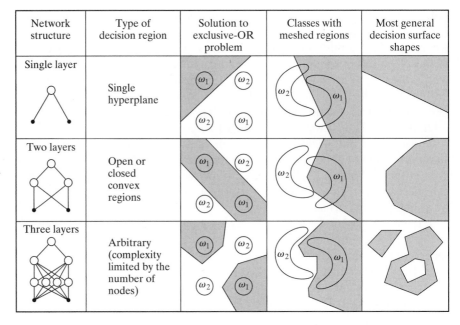

Network structure	Type of decision region	Solution to exclusive-OR problem	Classes with meshed regions	Most general decision surface shapes
Single layer	Single hyperplane	ω_1 ω_2 / ω_2 ω_1	ω_2 ω_1	
Two layers	Open or closed convex regions	ω_1 ω_2 / ω_2 ω_1	ω_2 ω_1	
Three layers	Arbitrary (complexity limited by the number of nodes)	ω_1 ω_2 / ω_2 ω_1	ω_2 ω_1	

dimensions. If the patterns are arranged so that class ω_1 consists of patterns $\{(0, 1), (1, 0)\}$ and class ω_2 consists of the patterns $\{(0, 0), (1, 1)\}$, class membership of the patterns in these two classes is given by the exclusive-OR (XOR) logical function, which is 1 only when one or the other of the two variables is 1, and it is 0 otherwise. Thus an XOR value of 1 indicates patterns of class ω_1, and an XOR value of 0 indicates patterns of class ω_2.

The preceding discussion is generalized to n dimensions in a straightforward way: Instead of lines, we deal with hyperplanes. A single-layer network implements a single hyperplane. A two-layer network implements arbitrarily convex regions consisting of intersections of hyperplanes. A three-layer network implements decision surfaces of arbitrary complexity. The number of nodes used in each layer determines the complexity of the last two cases. The number of classes in the first case is limited to two. In the other two cases, the number of classes is arbitrary, because the number of output nodes can be selected to fit the problem at hand.

Considering the preceding comments, it is logical to ask: Why would anyone be interested in studying neural networks having more than three layers? After all, a three-layer network can implement decision surfaces of arbitrary complexity. The answer lies in the method used to train a network to utilize only three layers. The training rule for the network in Fig. 12.16 minimizes an error measure but says nothing about how to associate groups of hyperplanes with specific nodes in the second layer of a three-layer network of the type discussed earlier. In fact, the problem of how to perform trade-off analyses between the number of layers and the number of nodes in each layer remains unresolved. In practice, the trade-off is generally resolved by trial and error or by previous experience with a given problem domain.

12.3 Structural Methods

The techniques discussed in Section 12.2 deal with patterns quantitatively and largely ignore any structural relationships inherent in a pattern's shape. The structural methods discussed in this section, however, seek to achieve pattern recognition by capitalizing precisely on these types of relationships. In this section, we introduce two basic approaches for the recognition of boundary shapes based on string representations. Strings are the most practical approach in structural pattern recognition.

12.3.1 Matching Shape Numbers

A procedure analogous to the minimum distance concept introduced in Section 12.2.1 for pattern vectors can be formulated for the comparison of region boundaries that are described in terms of shape numbers. With reference to the discussion in Section 11.2.2, the *degree of similarity*, k, between two region boundaries (shapes) is defined as the largest order for which their shape numbers still coincide. For example, let a and b denote shape numbers of closed boundaries represented by 4-directional chain codes. These two shapes have a degree of similarity k if

$$s_j(a) = s_j(b) \qquad \text{for } j = 4, 6, 8, \ldots, k$$
$$s_j(a) \neq s_j(b) \qquad \text{for } j = k + 2, k + 4, \ldots \qquad (12.3\text{-}1)$$

where s indicates shape number and the subscript indicates order. The *distance* between two shapes a and b is defined as the inverse of their degree of similarity:

$$D(a, b) = \frac{1}{k} \qquad (12.3\text{-}2)$$

This distance satisfies the following properties:

$$D(a, b) \geq 0$$
$$D(a, b) = 0 \quad \text{iff } a = b \qquad (12.3\text{-}3)$$
$$D(a, c) \leq \max\left[D(a, b), D(b, c) \right]$$

Either k or D may be used to compare two shapes. If the degree of similarity is used, the larger k is, the more similar the shapes are (note that k is infinite for identical shapes). The reverse is true when the distance measure is used.

■ Suppose that we have a shape f and want to find its closest match in a set of five other shapes (a, b, c, d, and e), as shown in Fig. 12.24(a). This problem is analogous to having five prototype shapes and trying to find the best match to a given unknown shape. The search may be visualized with the aid of the similarity tree shown in Fig. 12.24(b). The root of the tree corresponds to the lowest possible degree of similarity, which, for this example, is 4. Suppose that the shapes are identical up to degree 8, with the exception of shape a, whose degree of similarity with respect to all other shapes is 6. Proceeding down the

EXAMPLE 12.7:
Using shape numbers to compare shapes.

FIGURE 12.24
(a) Shapes.
(b) Hypothetical
similarity tree.
(c) Similarity
matrix.
(Bribiesca and
Guzman.)

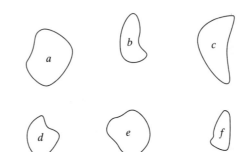

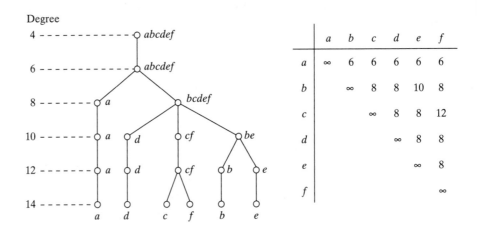

tree, we find that shape d has degree of similarity 8 with respect to all others, and so on. Shapes f and c match uniquely, having a higher degree of similarity than any other two shapes. At the other extreme, if a had been an unknown shape, all we could have said using this method is that a was similar to the other five shapes with degree of similarity 6. The same information can be summarized in the form of a *similarity matrix*, as shown in Fig. 12.24(c).   ■

12.3.2 String Matching

Suppose that two region boundaries, a and b, are coded into strings (see Section 11.5) denoted $a_1 a_2 \ldots a_n$ and $b_1 b_2 \ldots b_m$, respectively. Let α represent the number of matches between the two strings, where a match occurs in the kth position if $a_k = b_k$. The number of symbols that do not match is

$$\beta = \max(|a|, |b|) - \alpha \qquad (12.3\text{-}4)$$

where $|\text{arg}|$ is the length (number of symbols) in the string representation of the argument. It can be shown that $\beta = 0$ if and only if a and b are identical (see Problem 12.21).

A simple measure of similarity between a and b is the ratio

$$R = \frac{\alpha}{\beta} = \frac{\alpha}{\max(|a|, |b|) - \alpha} \qquad (12.3\text{-}5)$$

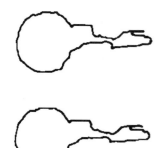

FIGURE 12.25
(a) and (b)
Sample
boundaries of two
different object
classes; (c) and
(d) their
corresponding
polygonal
approximations;
(e)–(g) tabula-
tions of R.
(Sze and Yang.)

R	1.a	1.b	1.c	1.d	1.e	1.f
1.a	∞					
1.b	16.0	∞				
1.c	9.6	26.3	∞			
1.d	5.1	8.1	10.3	∞		
1.e	4.7	7.2	10.3	14.2	∞	
1.f	4.7	7.2	10.3	8.4	23.7	∞

R	2.a	2.b	2.c	2.d	2.e	2.f
2.a	∞					
2.b	33.5	∞				
2.c	4.8	5.8	∞			
2.d	3.6	4.2	19.3	∞		
2.e	2.8	3.3	9.2	18.3	∞	
2.f	2.6	3.0	7.7	13.5	27.0	∞

R	1.a	1.b	1.c	1.d	1.e	1.f
2.a	1.24	1.50	1.32	1.47	1.55	1.48
2.b	1.18	1.43	1.32	1.47	1.55	1.48
2.c	1.02	1.18	1.19	1.32	1.39	1.48
2.d	1.02	1.18	1.19	1.32	1.29	1.40
2.e	0.93	1.07	1.08	1.19	1.24	1.25
2.f	0.89	1.02	1.02	1.24	1.22	1.18

Hence R is infinite for a perfect match and 0 when none of the corresponding
symbols in a and b match ($\alpha = 0$ in this case). Because matching is done symbol by symbol, the starting point on each boundary is important in terms of reducing the amount of computation. Any method that normalizes to, or near, the
same starting point is helpful, so long as it provides a computational advantage
over brute-force matching, which consists of starting at arbitrary points on each
string and then shifting one of the strings (with wraparound) and computing
Eq. (12.3-5) for each shift. The largest value of R gives the best match.

■ Figures 12.25(a) and (b) show sample boundaries from each of two object
classes, which were approximated by a polygonal fit (see Section 11.1.3). Figures
12.25(c) and (d) show the polygonal approximations corresponding to the
boundaries shown in Figs. 12.25(a) and (b), respectively. Strings were formed
from the polygons by computing the interior angle, θ, between segments as each
polygon was traversed clockwise. Angles were coded into one of eight possible
symbols, corresponding to 45° increments; that is, α_1: $0° < \theta \leq 45°$; α_2: $45° < \theta
\leq 90°$; $\ldots$; α_8: $315° < \theta \leq 360°$.

EXAMPLE 12.8:
Illustration of
string matching.

Figure 12.25(e) shows the results of computing the measure R for six samples of object 1 against themselves. The entries correspond to R values and, for example, the notation 1.c refers to the third string from object class 1. Figure 12.25(f) shows the results of comparing the strings of the second object class against themselves. Finally, Fig. 12.25(g) shows a tabulation of R values obtained by comparing strings of one class against the other. Note that, here, all R values are considerably smaller than any entry in the two preceding tabulations, indicating that the R measure achieved a high degree of discrimination between the two classes of objects. For example, if the class membership of string 1.a had been unknown, the *smallest* value of R resulting from comparing this string against sample (prototype) strings of class 1 would have been 4.7 [Fig. 12.25(e)]. By contrast, the *largest* value in comparing it against strings of class 2 would have been 1.24 [Fig. 12.25(g)]. This result would have led to the conclusion that string 1.a is a member of object class 1. This approach to classification is analogous to the minimum distance classifier introduced in Section 12.2.1.   ■

Summary

Starting with Chapter 9, our treatment of digital image processing began a transition from processes whose outputs are images to processes whose outputs are attributes about images, in the sense defined in Section 1.1. Although the material in the present chapter is introductory in nature, the topics covered are fundamental to understanding the state of the art in object recognition. As mentioned at the beginning of this chapter, recognition of individual objects is a logical place to conclude this book. To go past this point, we need concepts that are beyond the scope we set for our journey back in Section 1.4. Specifically, the next logical step would be the development of image analysis methods whose proper development requires concepts from machine intelligence.

As mentioned in Sections 1.1 and 1.4, machine intelligence and some areas that depend on it, such as scene analysis and computer vision, still are in their relatively early stages of practical development. Solutions of image analysis problems today are characterized by heuristic approaches. While these approaches are indeed varied, most of them share a significant base of techniques that are precisely the methods covered in this book.

Having concluded study of the material in the preceding twelve chapters, you are now in the position of being able to understand the principal areas spanning the field of digital image processing, both from a theoretical and practical point of view. Care was taken throughout all discussions to lay a solid foundation upon which further study of this and related fields could be based. Given the task-specific nature of many imaging problems, a clear understanding of basic principles enhances significantly the chances for their successful solution.

References and Further Reading

Background material for Sections 12.1 through 12.2.2 are the books by Theodoridis and Koutroumbas [2006], by Duda, Hart, and Stork [2001], and by Tou and Gonzalez [1974]. The survey article by Jain et al. [2000] also is of interest. The book by Principe et al. [1999] presents a good overview of neural networks. A special issue of *IEEE Trans. Image Processing* [1998] is worth comparing with a similar special issue ten years earlier (*IEEE Computer* [1988]). The material presented in Section 12.2.3 is introductory. In fact, the neural network model used in that discussion is one of numerous models proposed over the years. However, the model we discussed is representative and also is used quite extensively in image processing. The example dealing with the recognition of distorted shapes is adapted from Gupta et al. [1990, 1994]. The paper by Gori and Scarselli [1998] discusses the classification power of multilayer neural networks. An approach reported by Ueda [2000] based on using linear combinations of neural networks to achieve minimum classification error is good additional reading in this context.

For additional reading on the material in Section 12.3.1, see Bribiesca and Guzman [1980]. On string matching, see Sze and Yang [1981], Oommen and Loke [1997], and Gdalyahu and Weinshall [1999]. Additional references on structural pattern recognition are Gonzalez and Thomason [1978], Fu [1982], Bunke and Sanfeliu [1990], Tanaka [1995], Vailaya et al. [1998], Aizaka and Nakamura [1999], and Jonk et al. [1999]. See also the book by Huang [2002].

Problems

Detailed solutions to the problems marked with a star can be found in the book Web site. The site also contains suggested projects based on the material in this chapter.

12.1 **(a)** Compute the decision functions of a minimum distance classifier for the patterns shown in Fig. 12.1. You may obtain the required mean vectors by (careful) inspection.

    (b) Sketch the decision surfaces implemented by the decision functions in (a).

★**12.2** Show that Eqs. (12.2-4) and (12.2-5) perform the same function in terms of pattern classification.

12.3 Show that the surface given by Eq. (12.2-6) is the perpendicular bisector of the line joining the n-dimensional points $\mathbf{m}_i$ and $\mathbf{m}_j$.

★**12.4** Show how the minimum distance classifier discussed in connection with Fig. 12.7 could be implemented by using W resistor banks (W is the number of classes), a summing junction at each bank (for summing currents), and a maximum selector capable of selecting the maximum of W inputs, where the inputs are currents.

12.5 Show that the correlation coefficient of Eq. (12.2-8) has values in the range $[-1, 1]$. (*Hint:* Express $\gamma(x, y)$ in vector form.)

★**12.6** An experiment produces binary images of blobs that are nearly elliptical in shape (see the following figure). The blobs are of three sizes, with the average values of the principal axes of the ellipses being $(1.3, 0.7)$, $(1.0, 0.5)$, and $(0.75, 0.25)$. The dimensions of these axes vary $\pm 10\%$ about their average values. Develop an image processing system capable of rejecting incomplete or overlapping ellipses and then classifying the remaining single ellipses into one of the three size classes given. Show your solution in block diagram form, giving specific details regarding the operation of each block. Solve the classification problem using a minimum distance classifier, indicating clearly how you would go about obtaining training samples and how you would use these samples to train the classifier.

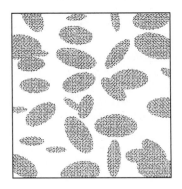

12.7 The following pattern classes have Gaussian probability density functions: ω_1: $\{(0, 0)^T, (2, 0)^T, (2, 2)^T, (0, 2)^T\}$ and ω_2: $\{(4, 4)^T, (6, 4)^T, (6, 6)^T, (4, 6)^T\}$.

   (a) Assume that $P(\omega_1) = P(\omega_2) = \frac{1}{2}$ and obtain the equation of the Bayes decision boundary between these two classes.

   (b) Sketch the boundary.

★**12.8** Repeat Problem 12.7, but use the following pattern classes: ω_1: $\{(-1, 0)^T, (0, -1)^T, (1, 0)^T, (0, 1)^T\}$ and ω_2: $\{(-2, 0)^T, (0, -2)^T, (2, 0)^T, (0, 2)^T\}$. Observe that these classes are not linearly separable.

12.9 Repeat Problem 12.6, but use a Bayes classifier (assume Gaussian densities). Indicate clearly how you would go about obtaining training samples and how you would use these samples to train the classifier.

★**12.10** The Bayes decision functions $d_j(\mathbf{x}) = p(\mathbf{x}/\omega_j)P(\omega_j), j = 1, 2, \ldots, W$, were derived using a 0-1 loss function. Prove that these decision functions minimize the probability of error. (*Hint:* The probability of error $p(e)$ is $1 - p(c)$, where $p(c)$ is the probability of being correct. For a pattern vector $\mathbf{x}$ belonging to class ω_i, $p(c/\mathbf{x}) = p(\omega_i/\mathbf{x})$. Find $p(c)$ and show that $p(c)$ is maximum [$p(e)$ is minimum] when $p(\mathbf{x}/\omega_i)P(\omega_i)$ is maximum.)

12.11 **(a)** Apply the perceptron algorithm to the following pattern classes: ω_1: $\{(0, 0, 0)^T, (1, 0, 0)^T, (1, 0, 1)^T, (1, 1, 0)^T\}$ and ω_2: $\{(0, 0, 1)^T, (0, 1, 1)^T, (0, 1, 0)^T, (1, 1, 1)^T\}$. Let $c = 1$, and $\mathbf{w}(1) = (-1, -2, -2, 0)^T$.

   (b) Sketch the decision surface obtained in (a). Show the pattern classes and indicate the positive side of the surface.

★**12.12** The perceptron algorithm given in Eqs. (12.2-34) through (12.2-36) can be expressed in a more concise form by multiplying the patterns of class ω_2 by -1, in which case the correction steps in the algorithm become $\mathbf{w}(k + 1) = \mathbf{w}(k)$, if $\mathbf{w}^T(k)\mathbf{y}(k) > 0$, and $\mathbf{w}(k + 1) = \mathbf{w}(k) + c\mathbf{y}(k)$ otherwise. This is one of several perceptron algorithm formulations that can be derived by starting from the general gradient descent equation

$$\mathbf{w}(k + 1) = \mathbf{w}(k) - c\left[\frac{\partial J(\mathbf{w}, \mathbf{y})}{\partial \mathbf{w}}\right]_{\mathbf{w}=\mathbf{w}(k)}$$

where $c > 0$, $J(\mathbf{w}, \mathbf{y})$ is a criterion function, and the partial derivative is evaluated at $\mathbf{w} = \mathbf{w}(k)$. Show that the perceptron algorithm formulation is obtainable from this general gradient descent procedure by using the criterion function

$$J(\mathbf{w}, \mathbf{y}) = \frac{1}{2}(|\mathbf{w}^T\mathbf{y}| - \mathbf{w}^T\mathbf{y}),$$ where $|arg|$ is the absolute value of the argument.

(*Note:* The partial derivative of $\mathbf{w}^T\mathbf{y}$ with respect to $\mathbf{w}$ equals $\mathbf{y}$.)

12.13 Prove that the perceptron training algorithm given in Eqs. (12.2-34) through (12.2-36) converges in a finite number of steps if the training pattern sets are linearly separable. [*Hint:* Multiply the patterns of class ω_2 by -1 and consider a nonnegative threshold, T, so that the perceptron training algorithm (with $c = 1$) is expressed as $\mathbf{w}(k + 1) = \mathbf{w}(k)$, if $\mathbf{w}^T(k)\mathbf{y}(k) > T$, and $\mathbf{w}(k + 1) = \mathbf{w}(k) + \mathbf{y}(k)$ otherwise. You may need to use the Cauchy-Schwartz inequality: $\|\mathbf{a}\|^2\|\mathbf{b}\|^2 \geq (\mathbf{a}^T\mathbf{b})^2$.]

★**12.14** Specify the structure and weights of a neural network capable of performing *exactly* the same function as a minimum distance classifier for two pattern classes in n-dimensional space.

12.15 Specify the structure and weights of a neural network capable of performing *exactly* the same function as a Bayes classifier for two pattern classes in n-dimensional space. The classes are Gaussian with different means but equal covariance matrices.

★**12.16** **(a)** Under what conditions are the neural networks in Problems 12.14 and 12.15 identical?

   (b) Would the generalized delta rule for multilayer feedforward neural networks developed in Section 12.2.3 yield the particular neural network in (a) if trained with a sufficiently large number of samples?

12.17 Two pattern classes in two dimensions are distributed in such a way that the patterns of class ω_1 lie randomly along a circle of radius r_1. Similarly, the patterns of class ω_2 lie randomly along a circle of radius r_2, where $r_2 = 2r_1$. Specify the structure of a neural network with the minimum number of layers and nodes needed to classify properly the patterns of these two classes.

★**12.18** Repeat Problem 12.6, but use a neural network. Indicate clearly how you would go about obtaining training samples and how you would use these samples to train the classifier. Select the simplest possible neural network that, in your opinion, is capable of solving the problem.

12.19 Show that the expression $h_j'(I_j) = O_j(1 - O_j)$ given in Eq. (12.2-71), where $h_j'(I_j) = \partial h_j(I_j)/\partial I_j$, follows from Eq. (12.2-50) with $\theta_o = 1$.

★**12.20** Show that the distance measure $D(A, B)$ of Eq. (12.3-2) satisfies the properties given in Eq. (12.3-3).

12.21 Show that $\beta = \max(|a|, |b|) - \alpha$ in Eq. (12.3-4) is 0 if and only if a and b are identical strings.

12.22 A certain factory mass produces small American flags for sporting events. The quality assurance team has observed that, during periods of peak production, some printing machines have a tendency to drop (randomly) between one and three stars and one or two entire stripes. Aside from these errors, the flags are perfect in every other way. Although the flags containing errors represent a small percentage of total production, the plant manager decides to solve the problem. After much investigation, he concludes that automatic inspection using image processing techniques is the most economical way to handle the problem. The basic specifications are as follows: The flags are approximately 7.5 cm by 12.5 cm in size. They move lengthwise down the production line (individually, but with a $\pm15°$ variation in orientation) at approximately 50 cm/s, with a separation between flags of approximately 5 cm. In all cases, "approximately" means $\pm5\%$. The plant manager hires you to design an image processing system for each production line. You are told that cost and simplicity are important parameters in determining the viability of your approach. Design a complete system based on the model of Fig. 1.23. Document your solution (including assumptions and specifications) in a brief (but clear) written report addressed to the plant manager.

A Coding Tables for Image Compression

Preview

This appendix contains code tables for use in CCITT and JPEG compression. Tables A.1 and A.2 are modified Huffman code tables for CCITT Group 3 and 4 compression. Tables A.3 through A.5 are for the coding of JPEG DCT coefficients. For more on the use of these tables, refer to Sections 8.2.5 and 8.2.8 of Chapter 8.

Run Length	White Code Word	Black Code Word	Run Length	White Code Word	Black Code Word
0	00110101	0000110111	32	00011011	000001101010
1	000111	010	33	00010010	000001101011
2	0111	11	34	00010011	000011010010
3	1000	10	35	00010100	000011010011
4	1011	011	36	00010101	000011010100
5	1100	0011	37	00010110	000011010101
6	1110	0010	38	00010111	000011010110
7	1111	00011	39	00101000	000011010111
8	10011	000101	40	00101001	000001101100
9	10100	000100	41	00101010	000001101101
10	00111	0000100	42	00101011	000011011010
11	01000	0000101	43	00101100	000011011011
12	001000	0000111	44	00101101	000001010100
13	000011	00000100	45	00000100	000001010101
14	110100	00000111	46	00000101	000001010110
15	110101	000011000	47	00001010	000001010111
16	101010	0000010111	48	00001011	000001100100
17	101011	0000011000	49	01010010	000001100101
18	0100111	0000001000	50	01010011	000001010010
19	0001100	00001100111	51	01010100	000001010011
20	0001000	00001101000	52	01010101	000000100100
21	0010111	00001101100	53	00100100	000000110111
22	0000011	00000110111	54	00100101	000000111000
23	0000100	00000101000	55	01011000	000000100111
24	0101000	00000010111	56	01011001	000000101000
25	0101011	00000011000	57	01011010	000001011000
26	0010011	000011001010	58	01011011	000001011001
27	0100100	000011001011	59	01001010	000000101011
28	0011000	000011001100	60	01001011	000000101100
29	00000010	000011001101	61	00110010	000001011010
30	00000011	000001101000	62	00110011	000001100110
31	00011010	000001101001	63	00110100	000001100111

TABLE A.1
CCITT terminating codes.

TABLE A.2
CCITT makeup codes.

Run Length	White Code Word	Black Code Word	Run Length	White Code Word	Black Code Word
64	11011	0000001111	960	011010100	0000001110011
128	10010	000011001000	1024	011010101	0000001110100
192	010111	000011001001	1088	011010110	0000001110101
256	0110111	000001011011	1152	011010111	0000001110110
320	00110110	000000110011	1216	011011000	0000001110111
384	00110111	000000110100	1280	011011001	0000001010010
448	01100100	000000110101	1344	011011010	0000001010011
512	01100101	0000001101100	1408	011011011	0000001010100
576	01101000	0000001101101	1472	010011000	0000001010101
640	01100111	0000001001010	1536	010011001	0000001011010
704	011001100	0000001001011	1600	010011010	0000001011011
768	011001101	0000001001100	1664	011000	0000001100100
832	011010010	0000001001101	1728	010011011	0000001100101
896	011010011	0000001110010			

	Code Word			Code Word
1792	00000001000		2240	000000010110
1856	00000001100		2304	000000010111
1920	00000001101		2368	000000011100
1984	000000010010		2432	000000011101
2048	000000010011		2496	000000011110
2112	000000010100		2560	000000011111
2176	000000010101			

TABLE A.3   JPEG coefficient coding categories.

Range	DC Difference Category	AC Category
0	0	N/A
$-1, 1$	1	1
$-3, -2, 2, 3$	2	2
$-7, \ldots, -4, 4, \ldots, 7$	3	3
$-15, \ldots, -8, 8, \ldots, 15$	4	4
$-31, \ldots, -16, 16, \ldots, 31$	5	5
$-63, \ldots, -32, 32, \ldots, 63$	6	6
$-127, \ldots, -64, 64, \ldots, 127$	7	7
$-255, \ldots, -128, 128, \ldots, 255$	8	8
$-511, \ldots, -256, 256, \ldots, 511$	9	9
$-1023, \ldots, -512, 512, \ldots, 1023$	A	A
$-2047, \ldots, -1024, 1024, \ldots, 2047$	B	B
$-4095, \ldots, -2048, 2048, \ldots, 4095$	C	C
$-8191, \ldots, -4096, 4096, \ldots, 8191$	D	D
$-16383, \ldots, -8192, 8192, \ldots, 16383$	E	E
$-32767, \ldots, -16384, 16384, \ldots, 32767$	F	N/A

Category	Base Code	Length	Category	Base Code	Length
0	010	3	6	1110	10
1	011	4	7	11110	12
2	100	5	8	111110	14
3	00	5	9	1111110	16
4	101	7	A	11111110	18
5	110	8	B	111111110	20

TABLE A.4 JPEG default DC code (luminance).

Run/Category	Base Code	Length	Run/Category	Base Code	Length
0/0	**1010 (= EOB)**	**4**			
0/1	00	3	8/1	11111010	9
0/2	01	4	8/2	111111111000000	17
0/3	100	6	8/3	1111111110110111	19
0/4	1011	8	8/4	1111111110111000	20
0/5	11010	10	8/5	1111111110111001	21
0/6	111000	12	8/6	1111111110111010	22
0/7	1111000	14	8/7	1111111110111011	23
0/8	1111110110	18	8/8	1111111110111100	24
0/9	1111111110000010	25	8/9	1111111110111101	25
0/A	1111111110000011	26	8/A	1111111110111110	26
1/1	1100	5	9/1	111111000	10
1/2	111001	8	9/2	1111111110111111	18
1/3	1111001	10	9/3	1111111111000000	19
1/4	111110110	13	9/4	1111111111000001	20
1/5	11111110110	16	9/5	1111111111000010	21
1/6	1111111110000100	22	9/6	1111111111000011	22
1/7	1111111110000101	23	9/7	1111111111000100	23
1/8	1111111110000110	24	9/8	1111111111000101	24
1/9	1111111110000111	25	9/9	1111111111000110	25
1/A	1111111110001000	26	9/A	1111111111000111	26
2/1	11011	6	A/1	111111001	10
2/2	11111000	10	A/2	1111111111001000	18
2/3	1111110111	13	A/3	1111111111001001	19
2/4	1111111110001001	20	A/4	1111111111001010	20
2/5	1111111110001010	21	A/5	1111111111001011	21
2/6	1111111110001011	22	A/6	1111111111001100	22
2/7	1111111110001100	23	A/7	1111111111001101	23
2/8	1111111110001101	24	A/8	1111111111001110	24
2/9	1111111110001110	25	A/9	1111111111001111	25
2/A	1111111110001111	26	A/A	1111111111010000	26
3/1	111010	7	B/1	111111010	10
3/2	111110111	11	B/2	1111111111010001	18
3/3	11111110111	14	B/3	1111111111010010	19
3/4	1111111110010000	20	B/4	1111111111010011	20
3/5	1111111110010001	21	B/5	1111111111010100	21
3/6	1111111110010010	22	B/6	1111111111010101	22
3/7	1111111110010011	23	B/7	1111111111010110	23

TABLE A.5 JPEG default AC code (luminance).

(Continued)

TABLE A.5
(Continued)

Run/ Category	Base Code	Length	Run/ Category	Base Code	Length
3/8	1111111110010100	24	B/8	1111111111010111	24
3/9	1111111110010101	25	B/9	1111111111011000	25
3/A	1111111110010110	26	B/A	1111111111011001	26
4/1	111011	7	C/1	1111111010	11
4/2	1111111000	12	C/2	1111111111011010	18
4/3	1111111110010111	19	C/3	1111111111011011	19
4/4	1111111110011000	20	C/4	1111111111011100	20
4/5	1111111110011001	21	C/5	1111111111011101	21
4/6	1111111110011010	22	C/6	1111111111011110	22
4/7	1111111110011011	23	C/7	1111111111011111	23
4/8	1111111110011100	24	C/8	1111111111100000	24
4/9	1111111110011101	25	C/9	1111111111100001	25
4/A	1111111110011110	26	C/A	1111111111100010	26
5/1	1111010	8	D/1	11111111010	12
5/2	1111111001	12	D/2	1111111111100011	18
5/3	1111111110011111	19	D/3	1111111111100100	19
5/4	1111111110100000	20	D/4	1111111111100101	20
5/5	1111111110100001	21	D/5	1111111111100110	21
5/6	1111111110100010	22	D/6	1111111111100111	22
5/7	1111111110100011	23	D/7	1111111111101000	23
5/8	1111111110100100	24	D/8	1111111111101001	24
5/9	1111111110100101	25	D/9	1111111111101010	25
5/A	1111111110100110	26	D/A	1111111111101011	26
6/1	1111011	8	E/1	111111110110	13
6/2	11111111000	13	E/2	1111111111101100	18
6/3	1111111110100111	19	E/3	1111111111101101	19
6/4	1111111110101000	20	E/4	1111111111101110	20
6/5	1111111110101001	21	E/5	1111111111101111	21
6/6	1111111110101010	22	E/6	1111111111110000	22
6/7	1111111110101011	23	E/7	1111111111110001	23
6/8	1111111110101100	24	E/8	1111111111110010	24
6/9	1111111110101101	25	E/9	1111111111110011	25
6/A	1111111110101110	26	E/A	1111111111110100	26
7/1	11111001	9	**F/0**	**111111110111**	**12**
7/2	11111111001	13	F/1	1111111111110101	17
7/3	1111111110101111	19	F/2	1111111111110110	18
7/4	1111111110110000	20	F/3	1111111111110111	19
7/5	1111111110110001	21	F/4	1111111111111000	20
7/6	1111111110110010	22	F/5	1111111111111001	21
7/7	1111111110110011	23	F/6	1111111111111010	22
7/8	1111111110110100	24	F/7	1111111111111011	23
7/9	1111111110110101	25	F/8	1111111111111100	24
7/A	1111111110110110	26	F/9	1111111111111101	25
			F/A	1111111111111110	26

Bibliography

Abidi, M. A. and Gonzalez, R. C. (eds.) [1992]. *Data Fusion in Robotics and Machine Intelligence*, Academic Press, New York.

Abidi, M. A., Eason, R. O., and Gonzalez, R. C. [1991]. "Autonomous Robotics Inspection and Manipulation Using Multisensor Feedback," *IEEE Computer*, vol. 24, no. 4, pp. 17–31.

Abramson, N. [1963]. *Information Theory and Coding*, McGraw-Hill, New York.

Adiv, G. [1985]. "Determining Three-Dimensional Motion and Structure from Optical Flow Generated by Several Moving Objects," *IEEE Trans. Pattern Anal. Mach. Intell.*, vol. PAMI-7, no. 4, pp. 384–401.

Aggarwal, J. K. and Badler, N. I. (eds.) [1980]. "Motion and Time-Varying Imagery," *IEEE Trans. Pattern Anal. Mach. Intell.*, Special Issue, vol. PAMI-2, no. 6, pp. 493–588.

Aguado, A. S., Nixon, M. S., and Montiel, M. M. [1998], "Parameterizing Arbitrary Shapes via Fourier Descriptors for Evidence-Gathering Extraction," *Computer Vision and Image Understanding*, vol. 69, no. 2, pp. 202–221.

Ahmed, N., Natarajan, T., and Rao, K. R. [1974]. "Discrete Cosine Transforms," *IEEE Trans. Comp.*, vol. C-23, pp. 90–93.

Ahmed, N. and Rao, K. R. [1975]. *Orthogonal Transforms for Digital Signal Processing*, Springer-Verlag, New York.

Aizaka, K. and Nakamura, A. [1999]. "Parsing of Two-Dimensional Images Represented by Quadtree Adjoining Grammars," *Pattern Recog.*, vol. 32, no. 2, pp. 277–294.

Alexiadis, D. S. and Sergiadis, G. D. [2007]. "Estimation of Multiple Accelerated Motions Using Chirp-Fourier Transforms and Clustering," *IEEE Trans. Image Proc.*, vol. 16, no. 1, pp. 142–152.

Alliney, S. [1993]. "Digital Analysis of Rotated Images," *IEEE Trans. Pattern Anal. Machine Intell.*, vol. 15, no. 5, pp. 499–504.

Ando, S. [2000]. "Consistent Gradient Operators," *IEEE Trans. Pattern Anal. Machine Intell.*, vol. 22, no. 3, pp. 252–265.

Andrews, H. C. [1970]. *Computer Techniques in Image Processing*, Academic Press, New York.

Andrews, H. C. and Hunt, B. R. [1977]. *Digital Image Restoration*, Prentice Hall, Englewood Cliffs, N.J.

Anelli, G., Broggi, A., and Destri, G. [1998]. "Decomposition of Arbitrarily-Shaped Morphological Structuring Elements Using Genetic Algorithms," *IEEE Trans. Pattern Anal. Machine Intell.*, vol. 20, no. 2, pp. 217–224.

Ang, P. H., Ruetz, P. A., and Auld, D. [1991]. "Video Compression Makes Big Gains," *IEEE Spectrum*, vol. 28, no. 10, pp. 16–19.

Antonini, M., Barlaud, M., Mathieu, P., and Daubechies, I. [1992]. "Image Coding Using Wavelet Transform," *IEEE Trans. Image Processing*, vol. 1, no. 2, pp. 205–220.

Ascher, R. N. and Nagy, G. [1974]. "A Means for Achieving a High Degree of Compaction on Scan-Digitized Printed Text," *IEEE Transactions on Comp.*, C-23:1174–1179.

Atchison, D. A. and Smith, G. [2000]. *Optics of the Human Eye*, Butterworth-Heinemann, Boston, Mass.

Baccar, M., Gee, L. A., Abidi, M. A., and Gonzalez, R. C. [1996]. "Segmentation of Range Images Via Data Fusion and Morphological Watersheds," *Pattern Recog.*, vol. 29, no. 10, pp. 1671–1685.

Bajcsy, R. and Lieberman, L. [1976]. "Texture Gradient as a Depth Cue," *Comput. Graph. Image Proc.*, vol. 5, no. 1, pp. 52–67.

Bakir, T. and Reeves, J. S. [2000]. "A Filter Design Method for Minimizing Ringing in a Region of Interest in MR Spectroscopic Images," *IEEE Trans. Medical Imaging*, vol. 19, no. 6, pp. 585–600.

Ballard, D. H. [1981]. "Generalizing the Hough Transform to Detect Arbitrary Shapes," *Pattern Recognition*, vol. 13, no. 2, pp. 111–122.

Ballard, D. H. and Brown, C. M. [1982]. *Computer Vision*, Prentice Hall, Englewood Cliffs, N.J.

Banham, M. R., Galatsanos, H. L., Gonzalez, H. L., and Katsaggelos, A. K. [1994]. "Multichannel Restoration of Single Channel Images Using a Wavelet-Based Subband Decomposition," *IEEE Trans. Image Processing*, vol. 3, no. 6, pp. 821–833.

Banham, M. R. and Katsaggelos, A. K. [1996]. "Spatially Adaptive Wavelet-Based Multiscale Image Restoration," *IEEE Trans. Image Processing*, vol. 5, no. 5, pp. 619–634.

Basart, J. P. and Gonzalez, R. C. [1992]. "Binary Morphology," in *Advances in Image Analysis*, Y. Mahdavieh and R. C. Gonzalez (eds.), SPIE Press, Bellingham, Wash., pp. 277–305.

Basart, J. P., Chacklackal, M. S., and Gonzalez, R. C. [1992]. "Introduction to Gray-Scale Morphology," in *Advances in Image Analysis*, Y. Mahdavieh and R. C. Gonzalez (eds.), SPIE Press, Bellingham, Wash., pp. 306–354.

Bates, R. H. T. and McDonnell, M. J. [1986]. *Image Restoration and Reconstruction*, Oxford University Press, New York.

Battle, G. [1987]. "A Block Spin Construction of Ondelettes. Part I: Lemarié Functions," *Commun. Math. Phys.*, vol. 110, pp. 601–615.

Battle, G. [1988]. "A Block Spin Construction of Ondelettes. Part II: the QFT Connection," *Commun. Math. Phys.*, vol. 114, pp. 93–102.

Baumert, L. D., Golomb, S. W., and Hall, M., Jr. [1962]. "Discovery of a Hadamard Matrix of Order 92," *Bull. Am. Math. Soc.*, vol. 68, pp. 237–238.

Baxes, G. A. [1994]. *Digital Image Processing: Principles and Applications*, John Wiley & Sons, New York.

Baylon, D. M. and Lim, J. S. [1990]. "Transform/Subband Analysis and Synthesis of Signals," *Tech. Report*, MIT Research Laboratory of Electronics, Cambridge, Mass.

Bell, E. T. [1965]. *Men of Mathematics*, Simon & Schuster, New York.

Bengtsson, A. and Eklundh, J. O. [1991]. "Shape Representation by Multiscale Contour Approximation," *IEEE Trans. Pattern Anal. Machine Intell.*, vol. 13, no. 1, pp. 85–93.

Benson, K. B. [1985]. *Television Engineering Handbook*, McGraw-Hill, New York.

Berger, T. [1971]. *Rate Distortion Theory*, Prentice Hall, Englewood Cliffs, N.J.

Beucher, S. [1990]. Doctoral Thesis, Centre de Morphologie Mathématique, École des Mines de Paris, France. (The core of this material is contained in the following paper.)

Beucher, S. and Meyer, F. [1992]. "The Morphological Approach of Segmentation: The Watershed Transformation," in *Mathematical Morphology in Image Processing*, E. Dougherty (ed.), Marcel Dekker, New York.

Bezdek, J. C., et al. [2005]. *Fuzzy Models and Algorithms for Pattern Recognition and Image Processing*, Springer, New York.

Bhaskaran, V. and Konstantinos, K. [1997]. *Image and Video Compression Standards: Algorithms and Architectures*, Kluwer, Boston, Mass.

Bhatt, B., Birks, D., Hermreck, D. [1997]. "Digital Television: Making It Work," *IEEE Spectrum*, vol. 34, no. 10, pp. 19–28.

Biberman, L. M. [1973]. "Image Quality," In *Perception of Displayed Information*, Biberman, L. M. (ed.), Plenum Press, New York.

Bichsel, M. [1998]. "Analyzing a Scene's Picture Set under Varying Lighting," *Computer Vision and Image Understanding*, vol. 71, no. 3, pp. 271–280.

Bieniek, A. and Moga, A. [2000]. "An Efficient Watershed Algorithm Based on Connected Components," *Pattern Recog.*, vol. 33, no. 6, pp. 907–916.

Bisignani, W. T., Richards, G. P., and Whelan, J. W. [1966]. "The Improved Grey Scale and Coarse-Fine PCM Systems: Two New Digital TV Bandwidth Reduction Techniques," *Proc. IEEE*, vol. 54, no. 3, pp. 376–390.

Blahut, R. E. [1987]. *Principles and Practice of Information Theory*, Addison-Wesley, Reading, Mass.

Bleau, A. and Leon, L. J. [2000]. "Watershed-Based Segmentation and Region Merging," *Computer Vision and Image Understanding*, vol. 77, no. 3, pp. 317–370.

Blouke, M. M., Sampat, N., and Canosa, J. [2001]. *Sensors and Camera Systems for Scientific, Industrial, and Digital Photography Applications-II*, SPIE Press, Bellingham, Wash.

Blum, H. [1967]. "A Transformation for Extracting New Descriptors of Shape," In *Models for the Perception of Speech and Visual Form*, Wathen-Dunn, W. (ed.), MIT Press, Cambridge, Mass.

Blume, H. and Fand, A. [1989]. "Reversible and Irreversible Image Data Compression Using the S-Transform and Lempel-Ziv Coding," *Proc. SPIE Medical Imaging III: Image Capture and Display*, vol. 1091, pp. 2–18.

Boie, R. A. and Cox, I. J. [1992]. "An Analysis of Camera Noise," *IEEE Trans. Pattern Anal. Machine Intell.*, vol. 14, no. 6, pp. 671–674.

Born, M. and Wolf, E. [1999]. *Principles of Optics: Electromagnetic Theory of Propagation, Interference and Diffraction of Light*, 7th ed., Cambridge University Press, Cambridge, UK.

Boulgouris, N. V., Tzovaras, D., and Strintzis, M. G. [2001]. "Lossless Image Compression Based on Optimal Prediction, Adaptive Lifting, and Conditional Arithmetic Coding," *IEEE Trans. Image Processing*, vol. 10, no. 1, pp. 1–14.

Bouman, C. and Liu, B. [1991]. "Multiple Resolution Segmentation of Textured Images," *IEEE Trans. Pattern. Anal. Machine Intell.*, vol. 13, no. 2, pp. 99–113.

Boyd, J. E. and Meloche, J. [1998]. "Binary Restoration of Thin Objects in Multidimensional Imagery," *IEEE Trans. Pattern Anal. Machine Intell.*, vol. 20, no. 6, pp. 647–651.

Bracewell, R. N. [1995]. *Two-Dimensional Imaging*, Prentice Hall, Upper Saddle River, N.J.

Bracewell, R. N. [2000]. *The Fourier Transform and its Applications*, 3rd ed. McGraw-Hill, New York.

Brechet, L., Lucas, M., Doncarli, C., and Farnia, D. [2007]. "Compression of Biomedical Signals with Mother Wavelet Optimization and Best-Basis Wavelet Packet Selection," *IEEE Trans. on Biomedical Engineering*, in press.

Bribiesca, E. [1981]. "Arithmetic Operations Among Shapes Using Shape Numbers," *Pattern Recog.*, vol. 13, no. 2, pp. 123–138.

Bribiesca, E. [1999]. "A New Chain Code," *Pattern Recog.*, vol. 32, no. 2, pp. 235–251.

Bribiesca, E. [2000]. "A Chain Code for Representing 3–D Curves," *Pattern Recog.*, vol. 33, no. 5, pp. 755–765.

Bribiesca, E. and Guzman, A. [1980]. "How to Describe Pure Form and How to Measure Differences in Shape Using Shape Numbers," *Pattern Recog.*, vol. 12, no. 2, pp. 101–112.

Brigham, E. O. [1988]. *The Fast Fourier Transform and its Applications*, Prentice Hall, Upper Saddle River, N.J.

Brinkman, B. H., Manduca, A., and Robb, R. A. [1998]. "Optimized Homomorphic Unsharp Masking for MR Grayscale Inhomogeneity Correction," *IEEE Trans. Medical Imaging*, vol. 17, no. 2, pp. 161–171.

Brummer, M. E. [1991]. "Hough Transform Detection of the Longitudinal Fissure in Tomographic Head Images," *IEEE Trans. Biomed. Images*, vol. 10, no. 1, pp. 74–83.

Brzakovic, D., Patton, R., and Wang, R. [1991]. "Rule-Based Multi-Template Edge Detection," *Comput. Vision, Graphics, Image Proc: Graphical Models and Image Proc.*, vol. 53, no. 3, pp. 258–268.

Bunke, H. and Sanfeliu, A. (eds.) [1990]. *Syntactic and Structural Pattern Recognition: Theory and Applications*, World Scientific, Teaneck, N.J.

Burrus, C. S., Gopinath, R. A., and Guo, H. [1998]. *Introduction to Wavelets and Wavelet Transforms*, Prentice Hall, Upper Saddle River, N J., pp. 250–251.

Burt, P. J. and Adelson, E. H. [1983]. "The Laplacian Pyramid as a Compact Image Code," *IEEE Trans. Commun.*, vol. COM-31, no. 4, pp. 532–540.

Cameron, J. P. [2005]. *Sets, Logic, and Categories*, Springer, New York.

Campbell, J. D. [1969]. "Edge Structure and the Representation of Pictures," Ph.D. dissertation, Dept. of Elec. Eng., University of Missouri, Columbia.

Candy, J. C., Franke, M. A., Haskell, B. G., and Mounts, F. W. [1971]. "Transmitting Television as Clusters of Frame-to-Frame Differences," *Bell Sys. Tech. J.*, vol. 50, pp. 1889–1919.

Cannon, T. M. [1974]. "Digital Image Deblurring by Non-Linear Homomorphic Filtering," Ph.D. thesis, University of Utah.

Canny, J. [1986]. "A Computational Approach for Edge Detection," *IEEE Trans. Pattern Anal. Machine Intell.*, vol. 8, no. 6, pp. 679–698.

Carey, W. K., Chuang, D. B., and Hamami, S. S. [1999]. "Regularity-Preserving Image Interpolation," *IEEE Trans. Image Processing*, vol. 8, no. 9, pp. 1293–1299.

Caselles, V., Lisani, J.-L., Morel, J.-M., and Sapiro, G. [1999]. "Shape Preserving Local Histogram Modification," *IEEE Trans. Image Processing*, vol. 8, no. 2, pp. 220–230.

Castleman, K. R. [1996]. *Digital Image Processing*, 2nd ed. Prentice Hall, Upper Saddle River, N.J.

Centeno, J. A. S. and Haertel, V. [1997]. "An Adaptive Image Enhancement Algorithm," *Pattern Recog.*, vol. 30, no. 7 pp. 1183–1189.

Chan, R. C., Karl, W. C., and Lees, R. S. [2000]. "A New Model-Based Technique for Enhanced Small-Vessell Measurements in X-Ray Cine-Angiograms," *IEEE Trans. Medical Imaging*, vol. 19, no. 3, pp. 243–255.

Chandler, D. and Hemami, S. [2005]. "Dynamic Contrast-Based Quantization for Lossy Wavelet Image Compression," *IEEE Trans. Image Proc.*, vol. 14, no. 4, pp. 397–410.

Chang, S. G., Yu, B., and Vetterli, M. [2000]. "Spatially Adaptive Wavelet Thresholding with Context Modeling for Image Denoising," *IEEE Trans. Image Processing*, vol. 9, no. 9, pp. 1522–1531.

Chang, S. K. [1989]. *Principles of Pictorial Information Systems Design*, Prentice Hall, Englewood Cliffs, N.J.

Chang, T. and Kuo, C.-C. J. [1993]. "Texture Analysis and Classification with Tree-Structures Wavelet Transforms," *IEEE Trans. Image Processing*, vol. 2, no. 4, pp. 429–441.

Champeney, D. C. [1987]. *A Handbook of Fourier Theorems*, Cambridge University Press, New York.

Chaudhuri, B. B. [1983]. "A Note on Fast Algorithms for Spatial Domain Techniques in Image Processing," *IEEE Trans. Syst. Man Cyb.*, vol. SMC-13, no. 6, pp. 1166–1169.

Chen, M. C. and Wilson, A. N. [2000]. "Motion-Vector Optimization of Control Grid Interpolation and Overlapped Block Motion Compensation Using Iterated Dynamic Programming," *IEEE Trans. Image Processing.*, vol. 9, no. 7, pp. 1145–1157.

Chen, Y.-S. and Yu, Y.-T. [1996]. "Thinning Approach for Noisy Digital Patterns," *Pattern Recog.*, vol. 29, no. 11, pp. 1847–1862.

Cheng, H. D. and Huijuan Xu, H. [2000]. "A Novel Fuzzy Logic Approach to Contrast Enhancement," *Pattern Recog.*, vol. 33, no. 5, pp. 809–819.

Cheriet, M., Said, J. N., and Suen, C. Y. [1998]. "A Recursive Thresholding Technique for Image Segmentation," *IEEE Trans. Image Processing*, vol. 7, no. 6, pp. 918–921.

Cheung, J., Ferris, D., and Kurz, L. [1997]. "On Classification of Multispectral Infrared Image Data," *IEEE Trans. Image Processing*, vol. 6, no. 10, pp. 1456–1464.

Cheung, K. K. T. and Teoh, E. K. [1999]. "Symmetry Detection by Generalized Complex (GC) Moments: A Closed-Form Solution," *IEEE Trans. Pattern Anal. Machine Intell.*, vol. 21, no. 5, pp. 466–476.

Chow, C. K. and Kaneko, T. [1972]. "Automatic Boundary Detection of the Left Ventricle from Cineangiograms," *Comp., and Biomed. Res.*, vol. 5, pp. 388–410.

Chu, C.-C. and Aggarwal, J. K. [1993]. "The Integration of Image Segmentation Maps Using Regions and Edge Information," *IEEE Trans. Pattern Anal. Machine Intell.*, vol. 15, no. 12, pp. 1241–1252.

CIE [1978]. *Uniform Color Spaces—Color Difference Equations—Psychometric Color Terms*, Commission Internationale de L'Eclairage, Publication No. 15, Supplement No. 2, Paris.

Clark, J. J. [1989]. "Authenticating Edges Produced by Zero-Crossing Algorithms," *IEEE Trans. Pattern Anal. Machine Intell.*, vol. 12, no. 8, pp. 830–831.

Clarke, R. J. [1985]. *Transform Coding of Images*, Academic Press, New York.

Cochran, W. T., Cooley, J. W., et al. [1967]. "What Is the Fast Fourier Transform?" *IEEE Trans. Audio and Electroacoustics*, vol. AU-15, no. 2, pp. 45–55.

Coeurjolly, D. and Klette, R. [2004]. "A Comparative Evaluation of Length Estimators of Digital Curves," *IEEE Trans. Pattern. Analysis Machine Int.*, vol. 26, no. XX, pp. 252–258.

Cohen, A. and Daubechies, I. [1992]. *A Stability Criterion for Biorthogonal Wavelet Bases and Their Related Subband Coding Schemes*, Technical Report, AT&T Bell Laboratories.

Cohen, A., Daubechies, I., and Feauveau, J.-C. [1992]. "Biorthogonal Bases of Compactly Supported Wavelets," *Commun. Pure and Appl. Math.*, vol. 45, pp. 485–560.

Coifman, R. R. and Wickerhauser, M. V. [1992]. "Entropy-Based Algorithms for Best Basis Selection," *IEEE Tran. Information Theory*, vol. 38, no. 2, pp. 713–718.

Coltuc, D., Bolon, P., and Chassery, J. M. [2006]. "Exact Histogram Specification," *IEEE Trans. Image Processing*, vol. 15, no. 5, pp. 1143–1152.

Cooley, J. W., Lewis, P. A. W., and Welch, P. D. [1967a]. "Historical Notes on the Fast Fourier Transform," *IEEE Trans. Audio and Electroacoustics*, vol. AU-15, no. 2, pp. 76–79.

Cooley, J. W., Lewis, P. A. W., and Welch, P. D. [1967b]. "Application of the Fast Fourier Transform to Computation of Fourier Integrals," *IEEE Trans. Audio and Electroacoustics*, vol. AU-15, no. 2, pp. 79–84.

Cooley, J. W., Lewis, P. A. W., and Welch, P. D. [1969]. "The Fast Fourier Transform and Its Applications," *IEEE Trans. Educ.*, vol. E-12, no. 1. pp. 27–34.

Cooley, J. W. and Tukey, J. W. [1965]. "An Algorithm for the Machine Calculation of Complex Fourier Series," *Math. of Comput.*, vol. 19, pp. 297–301.

Cornsweet, T. N. [1970]. *Visual Perception*, Academic Press, New York.

Cortelazzo, G. M., Lucchese, L., and Monti, C. M. [1999]. "Frequency Domain Analysis of General Planar Rigid Motion with Finite Duration," *J. Opt. Soc. Amer.-A. Optics, Image Science, and Vision*, vol. 16, no. 6, pp. 1238–1253.

Cowart, A. E., Snyder, W. E., and Ruedger, W. H. [1983]. "The Detection of Unresolved Targets Using the Hough Transform," *Comput. Vision Graph Image Proc.*, vol. 21, pp. 222–238.

Cox, I., Kilian, J., Leighton, F., and Shamoon, T. [1997]. "Secure Spread Spectrum Watermarking for Multimedia," *IEEE Trans. Image Proc.*, vol. 6, no. 12, pp. 1673–1687.

Cox, I., Miller, M., and Bloom, J. [2001]. *Digital Watermarking*, Morgan Kaufmann (Elsevier), New York.

Creath, K. and Wyant, J. C. [1992]. "Moire and Fringe Patterns," in *Optical Shop Testing*, 2nd ed., (D. Malacara, ed.), John Wiley & Sons, New York, pp. 653–685.

Croisier, A., Esteban, D., and Galand, C. [1976]. "Perfect Channel Splitting by Use of Interpolation/Decimation/Tree Decomposition Techniques," *Int. Conf. On Inform. Sciences and Systems*, Patras, Greece, pp. 443–446.

Cumani, A., Guiducci, A., and Grattoni, P. [1991]. "Image Description of Dynamic Scenes," *Pattern Recog.*, vol. 24, no. 7, pp. 661–674.

Cutrona, L. J. and Hall, W. D. [1968]. "Some Considerations in Post-Facto Blur Removal," In *Evaluation of Motion-Degraded Images*, NASA Publ. SP-193, pp. 139–148.

Danielson, G. C. and Lanczos, C. [1942]. "Some Improvements in Practical Fourier Analysis and Their Application to X-Ray Scattering from Liquids," *J. Franklin Institute*, vol. 233, pp. 365–380, 435–452.

Daubechies, I. [1988]. "Orthonormal Bases of Compactly Supported Wavelets," *Commun. On Pure and Appl. Math.*, vol. 41, pp. 909–996.

Daubechies, I. [1990]. "The Wavelet Transform, Time-Frequency Localization and Signal Analysis," *IEEE Transactions on Information Theory*, vol. 36, no. 5, pp. 961–1005.

Daubechies, I. [1992]. *Ten Lectures on Wavelets*, Society for Industrial and Applied Mathematics, Philadelphia, Pa.

Daubechies, I. [1993]. "Orthonormal Bases of Compactly Supported Wavelets II, Variations on a Theme," *SIAM J. Mathematical Analysis*, vol. 24, no. 2, pp. 499–519.

Daubechies, I. [1996]. "Where Do We Go from Here? — A Personal Point of View," *Proc. IEEE*, vol. 84, no. 4, pp. 510–513.

Daul, C., Graebling, P., and Hirsch, E. [1998]. "From the Hough Transform to a New Approach for the Detection of and Approximation of Elliptical Arcs," *Computer Vision and Image Understanding*, vol. 72, no. 3, pp. 215–236.

Davies, E. R. [2005]. *Machine Vision: Theory, Algorithms, Practicalities*, Morgan Kaufmann, San Francisco.

Davis, L. S. [1982]. "Hierarchical Generalized Hough Transforms and Line-Segment Based Generalized Hough Transforms," *Pattern Recog.*, vol. 15, no. 4, pp. 277–285.

Davis, T. J. [1999]. "Fast Decomposition of Digital Curves into Polygons Using the Haar Transform," *IEEE Trans. Pattern Anal. Machine Intell.*, vol. 21, no. 8, pp. 786–790.

Davisson, L. D. [1972]. "Rate-Distortion Theory and Application," *Proc. IEEE*, vol. 60, pp. 800–808.

Delaney, A. H. and Bresler, Y. [1995]. "Multiresolution Tomographic Reconstruction Using Wavelets," *IEEE Trans. Image Processing*, vol. 4, no. 6, pp. 799–813.

Delon, J., Desolneux, A., Lisani, J. L., and Petro, A. B. [2007]. "A Nonparametric Approach for Histogram Segmentation," *IEEE Trans. Image Proc.*, vol. 16, no. 1, pp. 253–261.

Delp, E. J. and Mitchell, O. R. [1979]. "Image Truncation using Block Truncation Coding," *IEEE Trans. Comm.*, vol. COM-27, pp. 1335–1342.

Di Zenzo, S. [1986]. "A Note on the Gradient of a Multi-Image," *Computer Vision, Graphics, and Image Processing*, vol. 33, pp. 116–125.

Dijkstra, E. [1959]. "Note on Two Problems in Connection with Graphs," *Numerische Mathematik*, vol. 1, pp. 269–271.

Djeziri, S., Nouboud, F., and Plamondon, R. [1998]. "Extraction of Signatures from Check Background Based on a Filiformity Criterion," *IEEE Trans. Image Processing*, vol. 7, no. 102, pp. 1425–1438.

Dougherty, E. R. [1992]. *An Introduction to Morphological Image Processing*, SPIE Press, Bellingham, Wash.

Dougherty, E. R. (ed.) [2000]. *Random Processes for Image and Signal Processing*, IEEE Press, New York.

Dougherty. E. R. and Lotufo, R. A. [2003]. *Hands-on Morphological Image Processing*, SPIE Press, Bellingham, WA.

Drew, M. S., Wei, J., and Li, Z.-N. [1999]. "Illumination Invariant Image Retrieval and Video Segmentation," *Pattern Recog.*, vol. 32, no. 8, pp. 1369–1388.

Duda, R. O. and Hart, P. E. [1972]. "Use of the Hough Transformation to Detect Lines and Curves in Pictures," *Comm. ACM*, vol. 15, no. 1, pp. 11–15.

Duda, R. O, Hart, P. E., and Stork, D. G. [2001]. *Pattern Classification*, John Wiley & Sons, New York.

Dugelay, J., Roche, S., Rey, C., and Doerr, G. [2006]. "Still-Image Watermarking Robust to Local Geometric Distortions," *IEEE Trans. Image Proc.*, vol. 15, no. 9, pp. 2831–2842.

Edelman, S. [1999]. *Representation and Recognition in Vision*, The MIT Press, Cambridge, Mass.

Elias, P. [1952]. "Fourier Treatment of Optical Processes," *J. Opt. Soc. Am.*, vol. 42, no. 2, pp. 127–134.

Elliott, D. F. and Rao, K. R. [1983]. *Fast Transforms: Algorithms and Applications*, Academic Press, New York.

Eng, H.-L. and Ma, K.-K. [2001]. "Noise Adaptive Soft-Switching Median Filter," *IEEE Trans. Image Processing*, vol. 10, no. 2, pp. 242–251.

Eng, H.-L. and Ma, K.-K. [2006]. "A Switching Median Filter With Boundary Discriminitative Noise Detection for Extremely Corrupted Images," *IEEE Trans. Image Proc.*, vol. 15, no. 6, pp. 1506–1516.

Equitz, W. H. [1989]. "A New Vector Quantization Clustering Algorithm," *IEEE Trans. Acous. Speech Signal Processing*, vol. ASSP-37, no. 10, pp. 1568–1575.

Etienne, E. K. and Nachtegael, M. (eds.) [2000]. *Fuzzy Techniques in Image Processing*, Springer-Verlag, New York.

Evans, A. N. and Liu, X. U. [2006]. "A Morphological Gradient Approach to Color Edge Detection," *IEEE Trans. Image Proc.*, vol. 15, no. 6, pp. 1454–1463.

Falconer, D. G. [1970]. "Image Enhancement and Film Grain Noise." *Opt. Acta*, vol. 17, pp. 693–705.

Fairchild, M. D. [1998]. *Color Appearance Models*, Addison-Wesley, Reading, Mass.

Federal Bureau of Investigation [1993]. *WSQ Gray-Scale Fingerprint Image Compression Specification*, IAFIS-IC-0110v2, Washington, D. C.

Felsen, L. B. and Marcuvitz, N. [1994]. *Radiation and Scattering of Waves*, IEEE Press, New York.

Ferreira, A. and Ubéda, S. [1999]. "Computing the Medial Axis Transform in Parallel with Eight Scan Operators," *IEEE Trans. Pattern Anal. Machine Intell.*, vol. 21, no. 3, pp. 277–282.

Fischler, M. A. [1980]. "Fast Algorithms for Two Maximal Distance Problems with Applications to Image Analysis," *Pattern Recog.*, vol. 12, pp. 35–40.

Fisher, R. A. [1936]. "The Use of Multiple Measurements in Taxonomic Problems," *Ann. Eugenics*, vol. 7, Part 2, pp. 179–188. (Also in *Contributions to Mathematical Statistics*, John Wiley & Sons, New York, 1950.)

Flusser, J. [2000]. "On the Independence of Rotation Moment Invariants," *Pattern Recog.*, vol. 33, pp. 1405–1410.

Forsyth, D. F. and Ponce, J. [2002]. *Computer Vision—A Modern Approach*, Prentice Hall, Upper Saddle River, NJ.

Fortner, B. and Meyer, T. E. [1997]. *Number by Colors*, Springer-Verlag, New York.

Fox, E. A. [1991]. "Advances in Interactive Digital Multimedia Systems," *Computer*, vol. 24, no. 10, pp. 9–21.

Fram, J. R. and Deutsch, E. S. [1975]. "On the Quantitative Evaluation of Edge Detection Schemes and Their Comparison with Human Performance," *IEEE Trans. Computers*, vol. C-24, no. 6, pp. 616–628.

Freeman, A. (translator) [1878]. J. Fourier, *The Analytical Theory of Heat*, Cambridge: University Press, London.

Freeman, C. [1987]. *Imaging Sensors and Displays*, ISBN 0–89252–800–1, SPIE Press, Bellingham, Wash.

Freeman, H. [1961]. "On the Encoding of Arbitrary Geometric Configurations," *IEEE Trans. Elec. Computers*, vol. EC-10, pp. 260–268.

Freeman, H. [1974]. "Computer Processing of Line Drawings," *Comput. Surveys*, vol. 6, pp. 57–97.

Freeman, H. and Shapira, R. [1975]. "Determining the Minimum-Area Encasing Rectangle for an Arbitrary Closed Curve," *Comm. ACM*, vol. 18, no. 7, pp. 409–413.

Freeman, J. A. and Skapura, D. M. [1991]. *Neural Networks: Algorithms, Applications, and Programming Techniques*, Addison-Wesley, Reading, Mass.

Frei, W. and Chen, C. C. [1977]. "Fast Boundary Detection: A Generalization and a New Algorithm," *IEEE Trans. Computers*, vol. C-26, no. 10, pp. 988–998.

Frendendall, G. L. and Behrend, W. L. [1960]. "Picture Quality—Procedures for Evaluating Subjective Effects of Interference," *Proc. IRE*, vol. 48, pp. 1030–1034.

Fu, K. S. [1982]. *Syntactic Pattern Recognition and Applications*, Prentice Hall, Englewood Cliffs, N.J.

Fu, K. S. and Bhargava, B. K. [1973]. "Tree Systems for Syntactic Pattern Recognition," *IEEE Trans. Comput.*, vol. C-22, no. 12, pp. 1087–1099.

Fu, K. S., Gonzalez, R. C., and Lee, C. S. G. [1987]. *Robotics: Control, Sensing, Vision, and Intelligence*, McGraw-Hill, New York.

Fu, K. S. and Mui, J. K. [1981]. "A Survey of Image Segmentation," *Pattern Recog.*, vol. 13, no. 1, pp. 3–16.

Fukunaga, K. [1972]. *Introduction to Statistical Pattern Recognition*, Academic Press, New York.

Furht, B., Greenberg, J., and Westwater, R. [1997]. *Motion Estimation Algorithms for Video Compression*, Kluwer Academic Publishers, Boston.

Gallager, R. and Voorhis, D. V. [1975]. "Optimal Source Codes for Geometrically Distributed Integer Alphabets," *IEEE Trans. Inform. Theory*, vol. IT-21, pp. 228–230.

Gao, X., Sattar, F., and Vekateswarlu, R. [2007]. "Multiscale Corner Detection of Gray Level Images Based on Log-Gabor Wavelet Transform," *IEEE Trans. Circuits and Systems for Video Technology*, in press.

Garcia, P. [1999]. "The Use of Boolean Model for Texture Analysis of Grey Images," *Computer Vision and Image Understanding*, vol. 74, no. 3, pp. 227–235.

Gdalyahu, Y. and Weinshall, D. [1999]. "Flexible Syntactic Matching of Curves and Its Application to Automated Hierarchical Classification of Silhouettes," *IEEE Trans. Pattern Anal. Machine Intell.*, vol. 21, no. 12, pp. 1312–1328.

Gegenfurtner, K. R. and Sharpe, L. T. (eds.) [1999]. *Color Vision: From Genes to Perception*, Cambridge University Press, New York.

Geladi, P. and Grahn, H. [1996]. *Multivariate Image Analysis*, John Wiley & Sons, New York.

Geman, D. and Reynolds, G. [1992]. "Constrained Restoration and the Recovery of Discontinuities," *IEEE Trans. Pattern Anal. Machine Intell.*, vol. 14, no. 3, pp. 367–383.

Gentleman, W. M. [1968]. "Matrix Multiplication and Fast Fourier Transformations," *Bell System Tech. J.*, vol. 47, pp. 1099–1103.

Gentleman, W. M. and Sande, G. [1966]. "Fast Fourier Transform for Fun and Profit," *Fall Joint Computer Conf.*, vol. 29, pp. 563–578, Spartan, Washington, D. C.

Gharavi, H. and Tabatabai, A. [1988]. "Sub-Band Coding of Monochrome and Color Images," *IEEE Trans. Circuits Sys.*, vol. 35, no. 2, pp. 207–214.

Giannakis, G. B. and Heath, R. W., Jr. [2000]. "Blind Identification of Multichannel FIR Blurs and Perfect Image Restoration," *IEEE Trans. Image Processing*, vol. 9, no. 11, pp. 1877–1896.

Giardina, C. R. and Dougherty, E. R. [1988]. *Morphological Methods in Image and Signal Processing*, Prentice Hall, Upper Saddle River, N.J.

Golomb, S. W. [1966]. "Run-Length Encodings," *IEEE Trans. Inform. Theory*, vol. IT-12, pp. 399–401.

Gonzalez, R. C. [1985]. "Computer Vision," in *Yearbook of Science and Technology*, McGraw-Hill, New York, pp. 128–132.

Gonzalez, R. C. [1985]. "Industrial Computer Vision," in *Advances in Information Systems Science*, Tou, J. T. (ed.), Plenum, New York, pp. 345–385.

Gonzalez, R. C. [1986]. "Image Enhancement and Restoration," in *Handbook of Pattern Recognition and Image Processing*, Young, T. Y., and Fu, K. S. (eds.), Academic Press, New York, pp. 191–213.

Gonzalez, R. C., Edwards, J. J., and Thomason, M. G. [1976]. "An Algorithm for the Inference of Tree Grammars," *Int. J. Comput. Info. Sci.*, vol. 5, no. 2, pp. 145–163.

Gonzalez, R. C. and Fittes, B. A. [1977]. "Gray-Level Transformations for Interactive Image Enhancement," *Mechanism and Machine Theory*, vol. 12, pp. 111–122.

Gonzalez, R. C. and Safabakhsh, R. [1982]. "Computer Vision Techniques for Industrial Applications," *Computer*, vol. 15, no. 12, pp. 17–32.

Gonzalez, R. C. and Thomason, M. G. [1978]. *Syntactic Pattern Recognition: An Introduction*, Addison-Wesley, Reading, Mass.

Gonzalez, R. C. and Woods, R. E. [1992]. *Digital Image Processing*, Addison-Wesley, Reading, Mass.

Gonzalez, R. C. and Woods, R. E. [2002]. *Digital Image Processing*, 2nd ed., Prentice Hall, Upper Saddle River, NJ.

Gonzalez, R. C., Woods, R. E., and Eddins, S. L. [2004]. *Digital Image Processing Using MATLAB*, Prentice Hall, Upper Saddle River, NJ.

Good, I. J. [1958]. "The Interaction Algorithm and Practical Fourier Analysis," *J. R. Stat. Soc. (Lond.)*, vol. B20, pp. 361–367; *Addendum*, vol. 22, 1960, pp. 372–375.

Goodson, K. J. and Lewis, P. H. [1990]. "A Knowledge-Based Line Recognition System," *Pattern Recog. Letters*, vol. 11, no. 4, pp. 295–304.

Gordon, I. E. [1997]. *Theories of Visual Perception*, 2nd ed., John Wiley & Sons, New York.

Gori, M. and Scarselli, F. [1998]. "Are Multilayer Perceptrons Adequate for Pattern Recognition and Verification?" *IEEE Trans. Pattern Anal. Machine Intell.*, vol. 20, no. 11, pp. 1121–1132.

Goutsias, J., Vincent, L., and Bloomberg, D. S. (eds) [2000]. *Mathematical Morphology and Its Applications to Image and Signal Processing*, Kluwer Academic Publishers, Boston, Mass.

Graham, R. E. [1958]. "Predictive Quantizing of Television Signals," *IRE Wescon Conv. Rec.*, vol. 2, pt. 2, pp. 147–157.

Graham, R. L. and Yao, F. F. [1983]. "Finding the Convex Hull of a Simple Polygon," *J. Algorithms*, vol. 4, pp. 324–331.

Gray, R. M. [1984]. "Vector Quantization," *IEEE Trans. Acous. Speech Signal Processing*, vol. ASSP-1, no. 2, pp. 4–29.

Gröchenig, K. and Madych, W. R. [1992]. "Multiresolution Analysis, Haar Bases and Self-Similar Tilings of R^n," *IEEE Trans. Information Theory*, vol. 38, no. 2, pp. 556–568.

Grossman, A. and Morlet, J. [1984]. "Decomposition of Hardy Functions into Square Integrable Wavelets of Constant Shape," *SIAM J. Appl. Math.* vol. 15, pp. 723–736.

Guil, N., Villalba, J. and Zapata, E. L. [1995]. "A Fast Hough Transform for Segment Detection," *IEEE Trans. Image Processing*, vol. 4, no. 11, pp. 1541–1548.

Guil, N. and Zapata, E. L. [1997]. "Lower Order Circle and Ellipse Hough Transform," *Pattern Recog.*, vol. 30, no. 10, pp. 1729–1744.

Gunn, S. R. [1998]. "Edge Detection Error in the Discrete Laplacian of a Gaussian," *Proc. 1998 Int'l Conference on Image Processing*, vol. II, pp. 515–519.

Gunn, S. R. [1999]. "On the Discrete Representation of the Laplacian of a Gaussian," *Pattern Recog.*, vol. 32, no. 8, pp. 1463–1472.

Gupta, L., Mohammad, R. S., and Tammana, R. [1990]. "A Neural Network Approach to Robust Shape Classification," *Pattern Recog.*, vol. 23, no. 6, pp. 563–568.

Gupta, L. and Srinath, M. D. [1988]. "Invariant Planar Shape Recognition Using Dynamic Alignment," *Pattern Recog.*, vol. 21, pp. 235–239.

Gupta, L., Wang., J., Charles, A., and Kisatsky, P. [1994]. "Three-Layer Perceptron Based Classifiers for the Partial Shape Classification Problem," *Pattern Recog.*, vol. 27, no. 1, pp. 91–97.

Haar, A. [1910]. "Zur Theorie der Orthogonalen Funktionensysteme," *Math. Annal.*, vol. 69, pp. 331–371.

Habibi, A. [1971]. "Comparison of Nth Order DPCM Encoder with Linear Transformations and Block Quantization Techniques," *IEEE Trans. Comm. Tech.*, vol. COM-19, no. 6, pp. 948–956.

Habibi, A. [1974]. "Hybrid Coding of Pictorial Data," *IEEE Trans. Comm.*, vol. COM-22, no. 5, pp. 614–624.

Haddon, J. F. and Boyce, J. F. [1990]. "Image Segmentation by Unifying Region and Boundary Information," *IEEE Trans. Pattern Anal. Machine Intell.*, vol. 12, no. 10, pp. 929–948.

Hall, E. L. [1979]. *Computer Image Processing and Recognition*, Academic Press, New York.

Hamming, R. W. [1950]. "Error Detecting and Error Correcting Codes," *Bell Sys. Tech. J.*, vol. 29, pp. 147–160.

Hannah, I., Patel, D., and Davies, R. [1995]. "The Use of Variance and Entropy Thresholding Methods for Image Segmentation," *Pattern Recog.*, vol. 28, no. 8, pp. 1135–1143.

Haralick, R. M. and Lee, J. S. J. [1990]. "Context Dependent Edge Detection and Evaluation," *Pattern Recog.*, vol. 23, no. 1–2, pp. 1–20.

Haralick, R. M. and Shapiro, L. G. [1985]. "Survey: Image Segmentation," *Comput. Vision, Graphics, Image Proc.*, vol. 29, pp. 100–132.

Haralick, R. M. and Shapiro, L. G. [1992]. *Computer and Robot Vision*, vols. 1 & 2, Addison-Wesley, Reading, Mass.

Haralick, R. M., Sternberg, S. R., and Zhuang, X. [1987]. "Image Analysis Using Mathematical Morphology," *IEEE Trans. Pattern Anal. Machine Intell.*, vol. PAMI-9, no. 4, pp. 532–550.

Haralick, R. M., Shanmugan, R., and Dinstein, I. [1973]. "Textural Features for Image Classification," *IEEE Trans Syst. Man Cyb.*, vol. SMC-3, no. 6, pp. 610–621.

Harikumar, G. and Bresler, Y. [1999]. "Perfect Blind Restoration of Images Blurred by Multiple Filters: Theory and Efficient Algorithms," *IEEE Trans. Image Processing*, vol. 8, no. 2, pp. 202–219.

Harmuth, H. F. [1970]. *Transmission of Information by Orthogonal Signals*, Springer-Verlag, New York.

Haris, K., Efstratiadis, S. N, Maglaveras, N., and Katsaggelos, A. K. [1998]. "Hybrid Image Segmentation Using Watersheds and Fast Region Merging," *IEEE Trans. Image Processing*, vol. 7, no. 12, pp. 1684–1699.

Hart, P. E., Nilsson, N. J., and Raphael, B. [1968]. "A Formal Basis for the Heuristic Determination of Minimum-Cost Paths," *IEEE Trans. Syst. Man Cyb*, vol. SMC-4, pp. 100–107.

Hartenstein, H., Ruhl, M., and Saupe, D. [2000]. "Region-Based Fractal Image Compression," *IEEE Trans. Image Processing*, vol. 9, no. 7, pp. 1171–1184.

Haskell, B. G. and Netravali, A. N. [1997]. *Digital Pictures: Representation, Compression, and Standards*, Perseus Publishing, New York.

Haykin, S. [1996]. *Adaptive Filter Theory*, Prentice Hall, Upper Saddle River, N.J.

Healy, D. J. and Mitchell, O. R. [1981]. "Digital Video Bandwidth Compression Using Block Truncation Coding," *IEEE Trans. Comm.*, vol. COM-29, no. 12, pp. 1809–1817.

Heath, M. D., Sarkar, S., Sanocki, T., and Bowyer, K. W. [1997]. "A Robust Visual Method for Assessing the Relative Performance of Edge-Detection Algorithms," *IEEE Trans. Pattern Anal. Machine Intell.*, vol. 19, no. 12, pp. 1338–1359.

Heath, M., Sarkar, S., Sanoki, T., and Bowyer, K. [1998]. "Comparison of Edge Detectors: A Methodology and Initial Study," *Computer Vision and Image Understanding*, vol. 69, no. 1, pp. 38–54.

Hebb, D. O. [1949]. *The Organization of Behavior: A Neuropsychological Theory*, John Wiley & Sons, New York.

Heijmans, H. J. A. M. and Goutsias, J. [2000]. "Nonlinear Multiresolution Signal Decomposition Schemes—Part II: Morphological Wavelets," *IEEE Trans. Image Processing*, vol. 9, no. 11, pp. 1897–1913.

Highnam, R. and Brady, M. [1997]. "Model-Based Image Enhancement of Far Infrared Images," *IEEE Trans. Pattern Anal. Machine Intell.*, vol. 19, no. 4, pp. 410–415.

Hojjatoleslami, S. A. and Kittler, J. [1998]. "Region Growing: A New Approach," *IEEE Trans. Image Processing*, vol. 7, no. 7, pp. 1079–1084.

Hong, Pi, Hung, Li, and Hua, Li [2006]. "A Novel Fractal Image Watermarking," *IEEE Trans. Multimedia*, vol. 8, no. 3, pp. 488–499.

Hoover, R. B. and Doty, F. [1996]. *Hard X-Ray/Gamma-Ray and Neutron Optics, Sensors, and Applications*, SPIE Press, Bellingham, Wash.

Horn, B. K. P. [1986]. *Robot Vision*, McGraw-Hill, New York.

Hotelling, H. [1933]. "Analysis of a Complex of Statistical Variables into Principal Components," *J. Educ. Psychol.*, vol. 24, pp. 417–441, 498–520.

Hough, P. V. C. [1962]. "Methods and Means for Recognizing Complex Patterns," U. S. Patent 3,069,654.

Hsu, C. C. and Huang, J. S. [1990]. "Partitioned Hough Transform for Ellipsoid Detection," *Pattern Recog.*, vol. 23, no. 3–4, pp. 275–282.

Hu. J. and Yan, H. [1997]. "Polygonal Approximation of Digital Curves Based on the Principles of Perceptual Organization," *Pattern Recog.*, vol. 30, no. 5, pp. 701–718.

Hu, M. K. [1962]. "Visual Pattern Recognition by Moment Invariants," *IRE Trans. Info. Theory*, vol. IT-8, pp. 179–187.

Hu, Y., Kwong, S., and Huang, J. [2006]. "An Algorithm for Removable Visible Watermarking," *IEEE Trans. Circuits and Systems for Video Technology*, vol. 16, no. 1, pp. 129–133.

Huang, K.-Y. [2002]. *Syntactic Pattern Recognition for Seismic Oil Exploration*, World Scientific, Hackensack, NJ.

Huang, S.-C. and Sun, Y.-N. [1999]. "Polygonal Approximation Using Generic Algorithms," *Pattern Recog.*, vol. 32, no. 8, pp. 1409–1420.

Huang, T. S. [1965]. "PCM Picture Transmission," *IEEE Spectrum*, vol. 2, no. 12, pp. 57–63.

Huang, T. S. [1966]. "Digital Picture Coding," *Proc. Natl. Electron. Conf.*, pp. 793–797.

Huang, T. S., ed. [1975]. *Picture Processing and Digital Filtering*, Springer-Verlag, New York.

Huang, T. S. [1981]. *Image Sequence Analysis*, Springer-Verlag, New York.

Huang, T. S. and Hussian, A. B. S. [1972]. "Facsimile Coding by Skipping White," *IEEE Trans. Comm.*, vol. COM-23, no. 12, pp. 1452–1466.

Huang, T. S. and Tretiak, O. J. (eds.). [1972]. *Picture Bandwidth Compression*, Gordon and Breech, New York.

Huang, T. S., Yang, G. T., and Tang, G. Y. [1979]. "A Fast Two-Dimensional Median Filtering Algorithm," *IEEE Trans. Acoust., Speech, Sig. Proc.*, vol. ASSP-27, pp. 13–18.

Huang, Y. and Schultheiss, P. M. [1963]. "Block Quantization of Correlated Gaussian Random Variables," *IEEE Trans. Commun. Syst.*, vol. CS-11, pp. 289–296.

Hubbard, B. B. [1998]. *The World According to Wavelets—The Story of a Mathematical Technique in the Making*, 2nd ed, A. K. Peters, Ltd., Wellesley, Mass.

Hubel, D. H. [1988]. *Eye, Brain, and Vision*, Scientific Amer. Library, W. H. Freeman, New York.

Huertas, A. and Medione, G. [1986]. "Detection of Intensity Changes with Subpixel Accuracy using Laplacian-Gaussian Masks," *IEEE Trans. Pattern. Anal. Machine Intell.*, vol. PAMI-8, no. 5, pp. 651–664.

Huffman, D. A. [1952]. "A Method for the Construction of Minimum Redundancy Codes," *Proc. IRE*, vol. 40, no. 10, pp. 1098–1101.

Hufnagel, R. E. and Stanley, N. R. [1964]. "Modulation Transfer Function Associated with Image Transmission Through Turbulent Media," *J. Opt. Soc. Amer.*, vol. 54, pp. 52–61.

Hummel, R. A. [1974]. "Histogram Modification Techniques," Technical Report TR-329. F-44620–72C-0062, Computer Science Center, University of Maryland, College Park, Md.

Hunt, B. R. [1971]. "A Matrix Theory Proof of the Discrete Convolution Theorem," *IEEE Trans. Audio and Electroacoust.*, vol. AU-19, no. 4, pp. 285–288.

Hunt, B. R. [1973]. "The Application of Constrained Least Squares Estimation to Image Restoration by Digital Computer," *IEEE Trans. Comput.*, vol. C-22, no. 9, pp. 805–812.

Hunter, R. and Robinson, A. H. [1980]. "International Digital Facsimile Coding Standards," *Proc. IEEE*, vol. 68, no. 7, pp. 854–867.

Hurn, M. and Jennison, C. [1996]. "An Extension of Geman and Reynolds' Approach to Constrained Restoration and the Recovery of Discontinuities," *IEEE Trans. Pattern Anal. Machine Intell.*, vol. 18, no. 6, pp. 657–662.

Hwang, H. and Haddad, R. A. [1995]. "Adaptive Median Filters: New Algorithms and Results," *IEEE Trans. Image Processing*, vol. 4, no. 4, pp. 499–502.

IEEE Computer [1974]. Special issue on digital image processing. vol. 7, no. 5.

IEEE Computer [1988]. Special issue on artificial neural systems. vol. 21, no. 3.

IEEE Trans. Circuits and Syst. [1975]. Special issue on digital filtering and image processing, vol. CAS-2, pp. 161–304.

IEEE Trans. Computers [1972]. Special issue on two-dimensional signal processing, vol. C-21, no. 7.

IEEE Trans. Comm. [1981]. Special issue on picture communication systems, vol. COM-29, no. 12.

IEEE Trans. on Image Processing [1994]. Special issue on image sequence compression, vol. 3, no. 5.

IEEE Trans. on Image Processing [1996]. Special issue on vector quantization, vol. 5, no. 2.

IEEE Trans. Image Processing [1996]. Special issue on nonlinear image processing, vol. 5, no. 6.

IEEE Trans. Image Processing [1997]. Special issue on automatic target detection, vol. 6, no. 1.

IEEE Trans. Image Processing [1997]. Special issue on color imaging, vol. 6, no. 7.

IEEE Trans. Image Processing [1998]. Special issue on applications of neural networks to image processing, vol. 7, no. 8.

IEEE Trans. Information Theory [1992]. Special issue on wavelet transforms and mulitresolution signal analysis, vol. 11, no. 2, Part II.

IEEE Trans. Pattern Analysis and Machine Intelligence [1989]. Special issue on multi-resolution processing, vol. 11, no. 7.

IEEE Trans. Signal Processing [1993]. Special issue on wavelets and signal processing, vol. 41, no. 12.

IES Lighting Handbook, 9th ed. [2000]. Illuminating Engineering Society Press, New York.

ISO/IEC [1999]. *ISO/IEC 14495-1:1999: Information technology—Lossless and near-lossless compression of continuous-tone still images: Baseline.*

ISO/IEC JTC 1/SC 29/WG 1 [2000]. *ISO/IEC FCD 15444-1: Information technology—JPEG 2000 image coding system: Core coding system.*

Jähne, B. [1997]. *Digital Image Processing: Concepts, Algorithms, and Scientific Applications*, Springer-Verlag, New York.

Jähne, B. [2002]. *Digital Image Processing*, 5th ed., Springer, New York.

Jain, A. K. [1981]. "Image Data Compression: A Review," *Proc. IEEE*, vol. 69, pp. 349–389.

Jain, A. K. [1989]. *Fundamentals of Digital Image Processing*, Prentice Hall, Englewood Cliffs, N.J.

Jain, A. K., Duin, R. P. W., and Mao, J. [2000]. "Statistical Pattern Recognition: A Review," *IEEE Trans. Pattern Anal. Machine Intell.*, vol. 22, no. 1, pp. 4–37.

Jain, J. R. and Jain, A. K. [1981]. "Displacement Measurement and Its Application in Interframe Image Coding," *IEEE Trans. Comm.*, vol. COM-29, pp. 1799–1808.

Jain, R. [1981]. "Dynamic Scene Analysis Using Pixel-Based Processes," *Computer*, vol. 14, no. 8, pp. 12–18.

Jain, R., Kasturi, R., and Schunk, B. [1995]. *Computer Vision*, McGraw-Hill, New York.

Jang, B. K. and Chin, R. T. [1990]. "Analysis of Thinning Algorithms Using Mathematical Morphology," *IEEE Trans. Pattern Anal. Machine Intell.*, vol. 12, no. 6, pp. 541–551.

Jayant, N. S. (ed.) [1976]. *Waveform Quantization and Coding*, IEEE Press, New York.

Jones, R. and Svalbe, I. [1994]. "Algorithms for the Decomposition of Gray-Scale Morphological Operations," *IEEE Trans. Pattern Anal. Machine Intell.*, vol. 16, no. 6, pp. 581–588.

Jonk, A., van den Boomgaard, S., and Smeulders, A. [1999]. "Grammatical Inference of Dashed Lines," *Computer Vision and Image Understanding*, vol. 74, no. 3, pp. 212–226.

Kahaner, D. K. [1970]. "Matrix Description of the Fast Fourier Transform," *IEEE Trans. Audio Electroacoustics*, vol. AU-18, no. 4, pp. 442–450.

Kak, A. C. and Slaney, M. [2001]. *Principles of Computerized Tomographic Imaging*, Society for Industrial and Applied Mathematics, Philadelphia, Pa.

Kamstra, L. and Heijmans, H.J.A.M. [2005]. "Reversible Data Embedding Into Images Using Wavelet Techniques and Sorting," *IEEE Trans. Image Processing*, vol. 14, no. 12, pp. 2082–2090.

Karhunen, K. [1947]. "Über Lineare Methoden in der Wahrscheinlichkeitsrechnung," *Ann. Acad. Sci. Fennicae*, Ser. A137. (Translated by I. Selin in "On Linear Methods in Probability Theory." T-131, 1960, The RAND Corp., Santa Monica, Calif.)

Kasson, J. and Plouffe, W. [1992]. "An Analysis of Selected Computer Interchange Color Spaces," *ACM Trans. on Graphics*, vol. 11, no. 4, pp. 373–405.

Katzir, N., Lindenbaum, M., and Porat, M. [1994]. "Curve Segmentation Under Partial Occlusion," *IEEE Trans. Pattern Anal. Machine Intell.*, vol. 16, no. 5, pp. 513–519.

Kerre, E. E. and Nachtegael, M., eds. [2000]. *Fuzzy Techniques in Image Processing*, Springer-Verlag, New York.

Khanna, T. [1990]. *Foundations of Neural Networks*, Addison-Wesley, Reading, Mass.

Kim, C. [2005]. "Segmenting a Low-Depth-of-Filed Image Using Morphological Filters and Region Merging," *IEEE Trans. Image Proc.*, vol. 14, no. 10, pp. 1503–1511.

Kim, J. K., Park, J. M., Song, K. S., and Park, H. W. [1997]. "Adaptive Mammographic Image Enhancement Using First Derivative and Local Statistics," *IEEE Trans. Medical Imaging*, vol. 16, no. 5, pp. 495–502.

Kimme, C., Ballard, D. H., and Sklansky, J. [1975]. "Finding Circles by an Array of Accumulators," *Comm. ACM*, vol. 18, no. 2, pp. 120–122.

Kirsch, R. [1971]. "Computer Determination of the Constituent Structure of Biological Images," *Comput. Biomed. Res.*, vol. 4, pp. 315–328.

Kiver, M. S. [1965]. *Color Television Fundamentals*, McGraw-Hill, New York.

Klette, R. and Rosenfeld, A. [2004]. *Digital Geometry—Geometric Methods for Digital Picture Analysis*, Morgan Kaufmann, San Francisco.

Klinger, A. [1976]. "Experiments in Picture Representation Using Regular Decomposition," *Comput. Graphics Image Proc.*, vol. 5, pp. 68–105.

Knowlton, K. [1980]. "Progressive Transmission of Gray-Scale and Binary Pictures by Simple, Efficient, and Lossless Encoding Schemes," *Proc. IEEE*, vol. 68, no. 7, pp. 885–896.

Kohler, R. J. and Howell, H. K. [1963]. "Photographic Image Enhancement by Superposition of Multiple Images," *Photogr. Sci. Eng.*, vol. 7, no. 4, pp. 241–245.

Kokaram, A. [1998]. *Motion Picture Restoration*, Springer-Verlag, New York.

Kokare, M., Biswas, P., and Chatterji, B. [2005]. "Texture Image Retrieval Using New Rotated Complex Wavelet Filters," *IEEE Trans. Systems, Man, Cybernetics, Part B*, vol. 35, no. 6, pp. 1168–1178.

Kramer, H. P. and Mathews, M. V. [1956]. "A Linear Coding for Transmitting a Set of Correlated Signals," *IRE Trans. Info. Theory*, vol. IT-2, pp. 41–46.

Langdon, G. C. and Rissanen, J. J. [1981]. "Compression of Black-White Images with Arithmetic Coding," *IEEE Trans. Comm.*, vol. COM-29, no. 6, pp. 858–867.

Lantuéjoul, C. [1980]. "Skeletonization in Quantitative Metallography," in *Issues of Digital Image Processing*, Haralick, R. M., and Simon, J. C. (eds.), Sijthoff and Noordhoff, Groningen, The Netherlands.

Latecki, L. J. and Lakämper, R. [1999]. "Convexity Rule for Shape Decomposition Based on Discrete Contour Evolution," *Computer Vision and Image Understanding*, vol. 73, no. 3, pp. 441–454.

Le Gall, D. and Tabatabai, A. [1988]. "Sub-Band Coding of Digital Images Using Symmetric Short Kernel Filters and Arithmetic Coding Techniques," *IEEE International Conference on Acoustics, Speech, and Signal Processing*, New York, pp. 761–765.

Ledley, R. S. [1964]. "High-Speed Automatic Analysis of Biomedical Pictures," *Science*, vol. 146, no. 3461, pp. 216–223.

Lee, J.-S., Sun, Y.-N., and Chen, C.-H. [1995]. "Multiscale Corner Detection by Using Wavelet Transforms," *IEEE Trans. Image Processing*, vol. 4, no. 1, pp. 100–104.

Lee, S. U., Chung, S. Y., and Park, R. H. [1990]. "A Comparative Performance Study of Several Global Thresholding Techniques for Segmentation," *Comput. Vision, Graphics, Image Proc.*, vol. 52, no. 2, pp. 171–190.

Lehmann, T. M., Gönner, C., and Spitzer, K. [1999]. "Survey: Interpolation Methods in Medical Image Processing," *IEEE Trans. Medical Imaging*, vol. 18, no. 11, pp. 1049–1076.

Lema, M. D. and Mitchell, O. R. [1984]. "Absolute Moment Block Truncation Coding and Its Application to Color Images," *IEEE Trans. Comm.*, vol. COM-32, no. 10. pp. 1148–1157.

Levine, M. D. [1985]. *Vision in Man and Machine*, McGraw-Hill, New York.

Liang, K.-C. and Kuo, C.-C. J. [1991]. "Waveguide: A Joint Wavelet-Based Image Representation and Description System," *IEEE Trans. Image Processing*, vol. 8, no. 11, pp. 1619–1629.

Liang, Q., Wendelhag, J. W., and Gustavsson, T. [2000]. "A Multiscale Dynamic Programming Procedure for Boundary Detection in Ultrasonic Artery Images," *IEEE Trans. Medical Imaging*, vol. 19, no. 2, pp. 127–142.

Liao, P., Chen, T., and Chung, P. [2001]. "A Fast Algorithm for Multilevel Thresholding," *J. Inform. Sc. and Eng.*, vol. 17, pp. 713–727.

Lillesand, T. M. and Kiefer, R. W. [1999]. *Remote Sensing and Image Interpretation*, John Wiley & Sons, New York.

Lim, J. S. [1990]. *Two-Dimensional Signal and Image Processing*, Prentice Hall, Upper Saddle River, N.J.

Limb, J. O. and Rubinstein, C. B. [1978]. "On the Design of Quantizers for DPCM Coders: A Functional Relationship Between Visibility, Probability, and Masking," IEEE Trans. Comm., vol. COM-26, pp. 573–578.

Lindblad, T. and Kinser, J. M. [1998]. *Image Processing Using Pulse-Coupled Neural Networks*, Springer-Verlag, New York.

Linde, Y., Buzo, A., and Gray, R. M. [1980]. "An Algorithm for Vector Quantizer Design," *IEEE Trans. Comm.*, vol. COM-28, no. 1, pp. 84–95.

Lippmann, R. P. [1987]. "An Introduction to Computing with Neural Nets," *IEEE ASSP Magazine*, vol. 4, pp. 4–22.

Liu, J. and Yang, Y.-H. [1994]. "Multiresolution Color Image Segmentation," *IEEE Trans Pattern Anal. Machine Intell.*, vol. 16, no. 7, pp. 689–700.

Liu-Yu, S. and Antipolis, M. [1993]. "Description of Object Shapes by Apparent Boundary and Convex Hull," *Pattern Recog.*, vol. 26, no. 1, pp. 95–107.

Lo, R.-C. and Tsai, W.-H. [1995]. "Gray-Scale Hough Transform for Thick Line Detection in Gray-Scale Images," *Pattern Recog.*, vol. 28, no. 5, pp. 647–661.

Loncaric, S. [1998]. "A Survey of Shape Analysis Techniques," *Pattern Recog.*, vol. 31, no. 8, pp. 983–1010.

Lu, H. E. and Wang, P. S. P. [1986]. "A Comment on 'A Fast Parallel Algorithm for Thinning Digital Patterns,'" *Comm. ACM*, vol. 29, no. 3, pp. 239–242.

Lu, N. [1997]. *Fractal Imaging*, Academic Press, New York.

Lu, W.-S. and Antoniou, A. [1992]. "Two-Dimensional Digital Filters," Marcel Dekker, New York.

MacAdam, D. L. [1942]. "Visual Sensitivities to Color Differences in Daylight," *J. Opt. Soc. Am.*, vol. 32, pp. 247–274.

MacAdam, D. P. [1970]. "Digital Image Restoration by Constrained Deconvolution," *J. Opt. Soc. Am.*, vol. 60, pp. 1617–1627.

Maki, A., Nordlund, P., and Eklundh, J.-O. [2000]. "Attentional Scene Segmentation: Integrating Depth and Motion," *Computer Vision and Image Understanding*, vol. 78, no. 3, pp. 351–373.

Malacara, D. [2001]. *Color Vision and Colorimetry: Theory and Applications*, SPIE Press, Bellingham, Wash.

Mallat, S. [1987]. "A Compact Multiresolution Representation: The Wavelet Model," *Proc. IEEE Computer Society Workshop on Computer Vision*, IEEE Computer Society Press, Washington, D. C., pp. 2–7.

Mallat, S. [1989a]. "A Theory for Multiresolution Signal Decomposition: The Wavelet Representation," *IEEE Trans. Pattern Anal. Mach. Intell.*, vol. PAMI-11, pp. 674–693.

Mallat, S. [1989b]. "Multiresolution Approximation and Wavelet Orthonormal Bases of L^2," *Trans. American Mathematical Society*, vol. 315, pp. 69–87.

Mallat, S. [1989c]. "Multifrequency Channel Decomposition of Images and Wavelet Models," *IEEE Trans. Acoustics, Speech, and Signal Processing*, vol. 37, pp. 2091–2110.

Mallat, S. [1998]. *A Wavelet Tour of Signal Processing*, Academic Press, Boston, Mass.

Mallat, S. [1999]. *A Wavelet Tour of Signal Processing*, 2nd ed., Academic Press, San Diego, Calif.

Mallot, A. H. [2000]. *Computational Vision*, The MIT Press, Cambridge, Mass.

Mamistvalov, A. [1998]. "n-Dimensional Moment Invariants and Conceptual Mathematical Theory of Recognition [of] n-Dimensional Solids," *IEEE Trans. Pattern Anal. Machine Intell.*, vol. 20, no. 8, pp. 819–831.

Manjunath, B., Salembier, P., and Sikora, T. [2001]. *Introduction to MPEG-7*, John Wiley & Sons, West Sussex, UK.

Maragos, P. [1987]. "Tutorial on Advances in Morphological Image Processing and Analysis," Optical Engineering, vol. 26, no. 7, pp. 623–632.

Marchand-Maillet, S. and Sharaiha, Y. M. [2000]. *Binary Digital Image Processing: A Discrete Approach*, Academic Press, New York.

Maren, A. J., Harston, C. T., and Pap, R. M. [1990]. *Handbook of Neural Computing Applications*, Academic Press, New York.

Marr, D. [1982]. *Vision*, Freeman, San Francisco.

Marr, D. and Hildreth, E. [1980]. "Theory of Edge Detection," *Proc. R. Soc. Lond.*, vol. B207, pp. 187–217.

Martelli, A. [1972]. "Edge Detection Using Heuristic Search Methods," *Comput. Graphics Image Proc.*, vol. 1, pp. 169–182.

Martelli, A. [1976]. "An Application of Heuristic Search Methods to Edge and Contour Detection," *Comm. ACM*, vol. 19, no. 2, pp. 73–83.

Martin, M. B. and Bell, A. E. [2001]. "New Image Compression Techniques Using Multiwavelets and Multiwavelet Packets," *IEEE Trans. on Image Proc.*, vol. 10, no. 4, pp. 500–510.

Mather, P. M. [1999]. *Computer Processing of Remotely Sensed Images: An Introduction*, John Wiley & Sons, New York.

Max, J. [1960]. "Quantizing for Minimum Distortion," *IRE Trans. Info. Theory*, vol. IT-6, pp. 7–12.

McClelland, J. L. and Rumelhart, D. E. (eds.) [1986]. *Parallel Distributed Processing: Explorations in the Microstructures of Cognition*, vol. 2: *Psychological and Biological Models*, The MIT Press, Cambridge, Mass.

McCulloch, W. S. and Pitts, W. H. [1943]. "A Logical Calculus of the Ideas Imminent in Nervous Activity," *Bulletin of Mathematical Biophysics*, vol. 5, pp. 115–133.

McFarlane, M. D. [1972]. "Digital Pictures Fifty Years Ago," *Proc. IEEE*, vol. 60, no. 7, pp. 768–770.

McGlamery, B. L. [1967]. "Restoration of Turbulence-Degraded Images," *J. Opt. Soc. Am.*, vol. 57, no. 3, pp. 293–297.

Meijering, H. W., Zuiderveld, K. J., and Viergever, M. A. [1999]. "Image Registration for Digital Subtraction Angiography," *Int. J. Comput. Vision*, vol. 31, pp. 227–246.

Meijering, E. H. W., Niessen, W. J., and Viergever, M. A. [1999]. "Retrospective Motion Correction in Digital Subtraction Angiography: A Review," *IEEE Trans. Medical Imaging*, vol. 18, no. 1, pp. 2–21.

Meijering, E. H. W., et al. [2001]. "Reduction of Patient Motion Artifacts in Digital Subtraction Angiography: Evaluation of a Fast and Fully Automatic Technique," *Radiology*, vol. 219, pp. 288–293.

Memon, N., Neuhoff, D. L., and Shende, S. [2000]. "An Analysis of Some Common Scanning Techniques for Lossless Image Coding," *IEEE Trans. Image Processing*, vol. 9, no. 11, pp. 1837–1848.

Mesarović, V. Z. [2000]. "Iterative Linear Minimum Mean-Square-Error Image Restoration from Partially Known Blur," *J. Opt. Soc. Amer.-A. Optics, Image Science, and Vision*, vol. 17, no. 4, pp. 711–723.

Meyer, Y. [1987]. "L'analyses par Ondelettes," *Pour la Science*.

Meyer, Y. [1990]. *Ondelettes et ope'rateurs*, Hermann, Paris.

Meyer, Y. (ed.) [1992a]. *Wavelets and Applications: Proceedings of the International Conference, Marseille, France*, Mason, Paris, and Springer-Verlag, Berlin.

Meyer, Y. (translated by D. H. Salinger) [1992b]. *Wavelets and Operators*, Cambridge University Press, Cambridge, UK.

Meyer, Y. (translated by R. D. Ryan) [1993]. *Wavelets: Algorithms and Applications*, Society for Industrial and Applied Mathematics, Philadelphia.

Meyer, F. G., Averbuch, A. Z., and Strömberg, J.-O. [2000]. "Fast Adaptive Wavelet Packet Image Compression," *IEEE Trans. Image Processing*, vol. 9, no. 7, pp. 792–800.

Meyer, F. and Beucher, S. [1990]. "Morphological Segmentation," *J. Visual Comm., and Image Representation*, vol. 1, no. 1, pp. 21–46.

Meyer, H., Rosdolsky, H. G., and Huang, T. S. [1973]. "Optimum Run Length Codes," *IEEE Trans. Comm.*, vol. COM-22, no. 6, pp. 826–835.

Minsky, M. and Papert, S. [1969]. *Perceptrons: An Introduction to Computational Geometry*, MIT Press, Cambridge, Mass.

Mirmehdi, M. and Petrou, M. [2000]. "Segmentation of Color Textures," *IEEE Trans. Pattern Anal. Machine Intell.*, vol. 22, no. 2, pp. 142–159.

Misiti, M., Misiti, Y., Oppenheim, G., and Poggi, J.-M. [1996]. *Wavelet Toolbox User's Guide*, The MathWorks, Inc., Natick, Mass.

Mitchell, D. P. and Netravali, A. N. [1988]. "Reconstruction Filters in Computer Graphics," *Comp. Graphics*, vol. 22, no. 4, pp. 221–228.

Mitchell, J., Pennebaker, W., Fogg, C., and LeGall, D. [1997]. *MPEG Video Compression Standard*, Chapman & Hall, New York.

Mitiche, A. [1994]. *Computational Analysis of Visual Motion*, Perseus Publishing, New York.

Mitra, S., Murthy, C., and Kundu, M. [1998]. "Technique for Fractal Image Compression Using Genetic Algorithm," *IEEE Transactions on Image Processing*, vol. 7, no. 4, pp. 586–593.

Mitra, S. K. and Sicuranza, G. L. (eds.) [2000]. *Nonlinear Image Processing*, Academic Press, New York.

Mohanty, S., et al. [1999]. "A Dual Watermarking Technique for Images," *Proc. 7th ACM International Multimedia Conference*, ACM-MM'99, Part 2, pp. 49–51.

Moore, G. A. [1968]. "Automatic Scanning and Computer Processes for the Quantitative Analysis of Micrographs and Equivalent Subjects," in *Pictorial Pattern Recognition*, (G. C. Cheng et al., eds), pp. 275–326, Thomson, Washington, D.C.

Mukherjee, D. and Mitra, S. [2003]. "Vector SPIHT for Embedded Wavelet Video and Image Coding," *IEEE Trans. Circuits and Systems for Video Technology*, vol. 13, no. 3, pp. 231–246.

Murase, H. and Nayar, S. K. [1994]. "Illumination Planning for Object Recognition Using Parametric Eigen Spaces," *IEEE Trans. Pattern Anal. Machine Intell.*, vol. 16, no. 12, pp. 1219–1227.

Murino, V., Ottonello, C., and Pagnan, S. [1998]. "Noisy Texture Classfication: A Higher-Order Statistical Approach," *Pattern Recog.*, vol. 31, no. 4, pp. 383–393.

Nagao, M. and Matsuyama, T. [1980]. *A Structural Analysis of Complex Aerial Photographs*, Plenum Press, New York.

Najman, L. and Schmitt, M. [1996]. "Geodesic Saliency of Watershed Contours and Hierarchical Segmentation," *IEEE Trans. Pattern Anal. Machine Intell.*, vol. 18, no. 12, pp. 1163–1173.

Narendra, P. M. and Fitch, R. C. [1981]. "Real-Time Adaptive Contrast Enhancement," *IEEE Trans. Pattern Anal. Mach. Intell.*, vol. PAMI-3, no. 6, pp. 655–661.

Netravali, A. N. [1977]. "On Quantizers for DPCM Coding of Picture Signals," *IEEE Trans. Info. Theory*, vol. IT-23, no. 3, pp. 360–370.

Netravali, A. N. and Limb, J. O. [1980]. "Picture Coding: A Review," *Proc. IEEE*, vol. 68, no. 3, pp. 366–406.

Nevatia, R. [1982]. *Machine Perception*, Prentice Hall, Upper Saddle River, N.J.

Ngan, K. N., Meier, T., and Chai, D. [1999]. *Advanced Video Coding: Principles and Techniques*, Elsevier, Boston.

Nie, Y. and Barner, K. E. [2006]. "The Fuzzy Transformation and Its Applications in Image Processing," IEEE Trans. Image Proc., vol. 15, no. 4, pp. 910–927.

Nilsson, N. J. [1965]. *Learning Machines: Foundations of Trainable Pattern-Classifying Systems*, McGraw-Hill, New York.

Nilsson, N. J. [1971]. *Problem Solving Methods in Artificial Intelligence*, McGraw-Hill, New York.

Nilsson, N. J. [1980]. *Principles of Artificial Intelligence*, Tioga, Palo Alto, Calif.

Nixon, M. and Aguado, A. [2002]. *Feature Extraction and Image Processing*, Newnes, Boston, MA.

Noble, B. and Daniel, J. W. [1988]. *Applied Linear Algebra*, 3rd ed., Prentice Hall, Upper Saddle River, N.J.

O'Connor, Y. Z. and Fessler, J. A. [2006]. "Fourier-Based Forward and Back-Projections in Iterative Fan-Beam Tomographic Image Reconstruction," *IEEE Trans. Med. Imag.*, vol. 25, no. 5, pp. 582–589.

Odegard, J. E., Gopinath, R. A., and Burrus, C. S. [1992]. "Optimal Wavelets for Signal Decomposition and the Existence of Scale-Limited Signals," *Proceedings of IEEE Int. Conf. On Signal Proc.*, ICASSP-92, San Francisco, CA, vol. IV, 597–600.

Olkkonen, J. and Olkkonen, H. [2007]. Discrete Lattice Wavelet Transform, *IEEE Trans. Circuits and Systems II: Express Briefs*, vol. 54, no. 1, pp. 71–75.

Olson, C. F. [1999]. "Constrained Hough Transforms for Curve Detection," *Computer Vision and Image Understanding*, vol. 73, no. 3, pp. 329–345.

O'Neil, J. B. [1971]. "Entropy Coding in Speech and Television Differential PCM Systems," *IEEE Trans. Info. Theory*, vol. IT-17, pp. 758–761.

Oommen, R. J. and Loke, R. K. S. [1997]. "Pattern Recognition of Strings with Substitutions, Insertions, Deletions, and Generalized Transpositions," *Pattern Recog.*, vol. 30, no. 5, pp. 789–800.

Oppenheim, A. V. and Schafer, R. W. [1975]. *Digital Signal Processing*, Prentice Hall, Englewood Cliffs, N.J.

Oppenheim, A. V., Schafer, R. W., and Stockham, T. G., Jr. [1968]. "Nonlinear Filtering of Multiplied and Convolved Signals," *Proc. IEEE*, vol. 56, no. 8, pp. 1264–1291.

Oster, G. and Nishijima, Y. [1963]. "Moiré Patterns," *Scientific American*, vol. 208, no. 5, pp. 54–63.

Otsu, N. [1979]. "A Threshold Selection Method from Gray-Level Histograms," *IEEE Trans. Systems, Man, and Cybernetics*, vol. 9, no. 1, pp. 62–66.

Oyster, C. W. [1999]. *The Human Eye: Structure and Function*, Sinauer Associates, Sunderland, Mass.

Paez, M. D. and Glisson, T. H. [1972]. "Minimum Mean-Square-Error Quantization in Speech PCM and DPCM Systems," *IEEE Trans. Comm.*, vol. COM-20, pp. 225–230.

Pao, Y. H. [1989]. *Adaptive Pattern Recognition and Neural Networks*, Addison-Wesley, Reading, Mass.

Papamarkos, N. and Atsalakis, A. [2000]. "Gray-Level Reduction Using Local Spatial Features," *Computer Vision and Image Understanding*, vol. 78, no. 3, pp. 336–350.

Papoulis, A. [1991]. *Probability, Random Variables, and Stochastic Processes*, 3rd ed., McGraw-Hill, New York.

Parhi, K. and Nishitani, T. [1999]. "Digital Signal Processing in Multimedia Systems," Chapter 18: *A Review of Watermarking Principles and Practices*, M. Miller, et al., pp. 461–485, Marcel Dekker Inc., New York.

Park, H. and Chin, R. T. [1995]. "Decomposition of Arbitrarily-Shaped Morphological Structuring," *IEEE Trans. Pattern Anal. Machine Intell.*, vol. 17, no. 1, pp. 2–15.

Parker, J. R. [1991]. "Gray Level Thresholding in Badly Illuminated Images," *IEEE Trans. Pattern Anal. Machine Intell.*, vol. 13, no. 8, pp. 813–819.

Parker, J. R. [1997]. *Algorithms for Image Processing and Computer Vision*, John Wiley & Sons, New York.

Patrascu, V. [2004]. "Fuzzy Enhancement Method Using Logarithmic Model," *IEEE-Fuzz'04*, vol. 3, pp. 1431–1436.

Pattern Recognition [2000]. Special issue on mathematical morphology and nonlinear image processing, vol. 33, no. 6, pp. 875–1117.

Pavlidis, T. [1977]. *Structural Pattern Recognition*, Springer-Verlag, New York.

Pavlidis, T. [1982]. *Algorithms for Graphics and Image Processing*, Computer Science Press, Rockville, Md.

Pavlidis, T. and Liow, Y. T. [1990]. "Integrating Region Growing and Edge Detection," *IEEE Trans. Pattern Anal. Mach. Intell.*, vol. 12, no. 3, pp. 225–233.

Peebles, P. Z. [1993]. *Probability, Random Variables, and Random Signal Principles*, 3rd ed., McGraw-Hill, New York.

Pennebaker, W. B. and Mitchell, J. L. [1992]. *JPEG: Still Image Data Compression Standard*, Van Nostrand Reinhold, New York.

Pennebaker, W. B., Mitchell, J. L., Langdon, G. G., Jr., and Arps, R. B. [1988]. "An Overview of the Basic Principles of the Q-coder Adaptive Binary Arithmetic Coder," *IBM J. Res. Dev.*, vol. 32, no. 6, pp. 717–726.

Perez, A. and Gonzalez, R. C. [1987]. "An Iterative Thresholding Algorithm for Image Segmentation," *IEEE Trans. Pattern Anal. Machine Intell.*, vol. PAMI-9, no. 6, pp. 742–751.

Perona, P. and Malik, J. [1990]. "Scale-Space and Edge Detection Using Anisotropic Diffusion," *IEEE Trans. Pattern Anal. Machine Intell.*, vol. 12, no. 7, pp. 629–639.

Persoon, E. and Fu, K. S. [1977]. "Shape Discrimination Using Fourier Descriptors," *IEEE Trans. Systems Man Cyb.*, vol. SMC-7, no. 2, pp. 170–179.

Petrou, M. and Bosdogianni, P. [1999]. *Image Processing: The Fundamentals*, John Wiley & Sons, UK.

Petrou, M. and Kittler, J. [1991]. "Optimal Edge Detector for Ramp Edges," *IEEE Trans. Pattern Anal. Machine Intell.*, vol. 13, no. 5, pp. 483–491.

Piech, M. A. [1990]. "Decomposing the Laplacian," *IEEE Trans. Pattern Anal. Machine Intell.*, vol. 12, no. 8, pp. 830–831.

Pitas, I. and Vanetsanopoulos, A. N. [1990]. *Nonlinear Digital Filters: Principles and Applications*, Kluwer Academic Publishers, Boston, Mass.

Plataniotis, K. N. and Venetsanopoulos, A. N. [2000]. *Color Image Processing and Applications*, Springer-Verlag, New York.

Pokorny, C. K. and Gerald, C. F. [1989]. *Computer Graphics: The Principles Behind the Art and Science*, Franklin, Beedle & Associates, Irvine, Calif.

Porco, C. C., West R. A., et al. [2004]. "Cassini Imaging Science: Instrument Characteristics and Anticipated Scientific Investigations at Saturn," *Space Science Reviews*, vol. 115, pp. 363–497.

Poynton, C. A. [1996]. *A Technical Introduction to Digital Video*, John Wiley & Sons, New York.

Prasad, L. and Iyengar, S. S. [1997]. *Wavelet Analysis with Applications to Image Processing*, CRC Press, Boca Raton, Fla.

Pratt, W. K. [2001]. *Digital Image Processing*, 3rd ed., John Wiley & Sons, New York.

Preparata, F. P. and Shamos, M. I. [1985]. *Computational Geometry: An Introduction*, Springer-Verlag, New York.

Preston, K. [1983]. "Cellular Logic Computers for Pattern Recognition," *Computer*, vol. 16, no. 1, pp. 36–47.

Prewitt, J. M. S. [1970]. "Object Enhancement and Extraction," in *Picture Processing and Psychopictorics*, Lipkin, B. S., and Rosenfeld, A. (eds.), Academic Press, New York.

Prince, J. L. and Links, J. M. [2006]. *Medical Imaging Signals and Systems*, Prentice Hall, Upper Saddle River, NJ.

Principe, J. C., Euliano, N. R., and Lefebre, W. C. [1999]. *Neural and Adaptive Systems: Fundamentals through Simulations*, John Wiley & Sons, New York.

Pritchard, D. H. [1977]. "U. S. Color Television Fundamentals—A Review," *IEEE Trans. Consumer Electronics*, vol. CE-23, no. 4, pp. 467–478.

Proc. IEEE [1967]. Special issue on redundancy reduction, vol. 55, no. 3.

Proc. IEEE [1972]. Special issue on digital picture processing, vol. 60, no. 7.

Proc. IEEE [1980]. Special issue on the encoding of graphics, vol. 68, no. 7.

Proc. IEEE [1985]. Special issue on visual communication systems, vol. 73, no. 2.

Qian, R. J. and Huang, T. S. [1996]. "Optimal Edge Detection in Two-Dimensional Images," *IEEE Trans. Image Processing*, vol. 5, no. 7, pp. 1215–1220.

Rabbani, M. and Jones, P. W. [1991]. *Digital Image Compression Techniques*, SPIE Press, Bellingham, Wash.

Rajala, S. A., Riddle, A. N., and Snyder, W. E. [1983]. "Application of One-Dimensional Fourier Transform for Tracking Moving Objects in Noisy Environments," *Comp., Vision, Image Proc.*, vol. 21, pp. 280–293.

Ramachandran, G. N. and Lakshminarayanan, A. V. [1971]. "Three Dimensional Reconstructions from Radiographs and Electron Micrographs: Application of Convolution Instead of Fourier Transforms," Proc. Nat. Acad. Sci., vol. 68, pp. 2236–2240.

Rane, S. and Sapiro, G. [2001]. "Evaluation of JPEG-LS, the New Lossless and Controlled-Lossy Still Image Compression Standard, for Compression of High-Resolution Elevation Data," *IEEE Trans. Geoscience and Remote Sensing*, vol. 39, no. 10, pp. 2298–2306.

Rangayyan, R. M. [2005]. *Biomedical Image Analysis*, CRC Press, Boca Raton, FL.

Reddy, B. S. and Chatterji, B. N. [1996]. "An FFT-Based Technique for Translation, Rotation, and Scale Invariant Image Registration," *IEEE Trans. Image Processing*, vol. 5, no. 8, pp. 1266–1271.

Regan, D. D. [2000]. *Human Perception of Objects: Early Visual Processing of Spatial Form Defined by Luminance, Color, Texture, Motion, and Binocular Disparity*, Sinauer Associates, Sunderland, Mass.

Rice, R. F. [1979]. "Some Practical Universal Noiseless Coding Techniques," *Tech. Rep. JPL-79-22*, Jet Propulsion Lab., Pasadena, CA.

Ritter, G. X. and Wilson, J. N. [2001]. *Handbook of Computer Vision Algorithms in Image Algebra*, CRC Press, Boca Raton, Fla.

Roberts, L. G. [1965]. "Machine Perception of Three-Dimensional Solids," in *Optical and Electro-Optical Information Processing*, Tippet, J. T. (ed.), MIT Press, Cambridge, Mass.

Robertson, A. R. [1977]. "The CIE 1976 Color Difference Formulae," *Color Res. Appl.*, vol. 2, pp. 7–11.

Robinson, G. S. [1976]. "Detection and Coding of Edges Using Directional Masks," University of Southern California, Image Processing Institute, Report no. 660.

Robinson, J. A. [1965]. "A Machine-Oriented Logic Based on the Resolution Principle," *J. ACM*, vol. 12, no. 1, pp. 23–41.

Robinson, J. [2006]. "Adaptive Prediction Trees for Image Compression," *IEEE Trans. Image Proc.*, vol. 15, no. 8, pp. 2131–2145.

Rock, I. [1984]. *Perception*, W. H. Freeman, New York.

Roese, J. A., Pratt, W. K., and Robinson, G. S. [1977]. "Interframe Cosine Transform Image Coding," *IEEE Trans. Comm.*, vol. COM-25, pp. 1329–1339.

Rosenblatt, F. [1959]. "Two Theorems of Statistical Separability in the Perceptron," In *Mechanisation of Thought Processes: Proc. of Symposium No. 10*, held at the National Physical Laboratory, November 1958, H. M. Stationery Office, London, vol. 1, pp. 421–456.

Rosenblatt, F. [1962]. *Principles of Neurodynamics: Perceptrons and the Theory of Brain Mechanisms*, Spartan, Washington, D. C.

Rosenfeld, A. (ed.) [1984]. *Multiresolution Image Processing and Analysis*, Springer-Verlag, New York.

Rosenfeld, A. [1999]. "Image Analysis and Computer Vision: 1998," *Computer Vision and Image Understanding*, vol. 74, no. 1, pp. 36–95.

Rosenfeld, A. [2000]. "Image Analysis and Computer Vision: 1999," *Computer Vision and Image Understanding*, vol. 78, no. 2, pp. 222–302.

Rosenfeld, A. and Kak, A. C. [1982]. *Digital Picture Processing*, vols. 1 and 2, 2nd ed., Academic Press, New York.

Rosin, P. L. [1997]. "Techniques for Assessing Polygonal Approximations of Curves," *IEEE Trans. Pattern Anal. Machine Intell.*, vol. 19, no. 6, pp. 659–666.

Rudnick, P. [1966]. "Note on the Calculation of Fourier Series," *Math. Comput.*, vol. 20, pp. 429–430.

Rumelhart, D. E., Hinton, G. E., and Williams, R. J. [1986]. "Learning Internal Representations by Error Propagation," In *Parallel Distributed Processing: Explorations in the Microstructures of Cognition, Vol. 1: Foundations*, Rumelhart, D. E., et al. (eds.), MIT Press, Cambridge, Mass., pp. 318–362.

Rumelhart, D. E. and McClelland, J. L. (eds.) [1986]. *Parallel Distributed Processing: Explorations in the Microstructures of Cognition, Vol. 1: Foundations*, MIT Press, Cambridge, Mass.

Runge, C. [1903]. *Zeit. für Math., and Physik*, vol. 48, p. 433.

Runge, C. [1905]. *Zeit. für Math., and Physik*, vol. 53, p. 117.

Runge, C. and König, H. [1924]. "Die Grundlehren der Mathematischen Wissenschaften," *Vorlesungen über Numerisches Rechnen*, vol. 11, Julius Springer, Berlin.

Russ, J. C. [1999]. *The Image Processing Handbook*, 3rd ed., CRC Press, Boca Raton, Fla.

Russo F. and Ramponi, G. [1994]. "Edge Extraction by FIRE Operators," *Fuzz-IEEE '94*, vol. 1, pp. 249–243, IEEE Press, New York.

Sahni, S. and Jenq, J.-F. [1992]. "Serial and Parallel Algorithms for the Medial Axis Transform," *IEEE Trans. Pattern Anal. Machine Intell.*, vol. 14, no. 12, pp. 1218–1224.

Sahoo, S. S. P. K., Wong, A. K. C., and Chen, Y. C. [1988]. "Survey of Thresholding Techniques," *Computer Vision, Graphics and Image Processing*, vol. 41, pp. 233–260.

Saito, N. and Cunningham, M. A. [1990]. "Generalized E-Filter and its Application to Edge Detection," *IEEE Trans. Pattern Anal. Machine Intell.*, vol. 12, no. 8, pp. 814–817.

Sakrison, D. J. and Algazi, V. R. [1971]. "Comparison of Line-by-Line and Two-Dimensional Encoding of Random Images," *IEEE Trans. Info. Theory*, vol. IT-17, no. 4, pp. 386–398.

Salari, E. and Siy, P. [1984]. "The Ridge-Seeking Method for Obtaining the Skeleton of Digital Images," *IEEE Trans. Syst. Man Cyb.*, vol. SMC-14, no. 3, pp. 524–528.

Salinas, R. A., Abidi, M. A., and Gonzalez, R. C. [1996]. "Data Fusion: Color Edge Detection and Surface Reconstruction Through Regularization," *IEEE Trans. Industrial Electronics*, vol. 43, no. 3, pp. 355–363, 1996.

Sato, Y. [1992]. "Piecewise Linear Approximation of Plane Curves by Perimeter Optimization," *Pattern Recog.*, vol. 25, no. 12, pp. 1535–1543.

Sauvola, J. and Pietikainen, M. [2000]. "Adaptive Document Image Binarization," *Pattern Recog.*, vol. 33, no. 2, pp. 225–236.

Schalkoff, R. J. [1989]. *Digital Image Processing and Computer Vision*, John Wiley & Sons, New York.

Schonfeld, D. and Goutsias, J. [1991]. "Optimal Morphological Pattern Restoration from Noisy Binary Images," *IEEE Trans. Pattern Anal. Machine Intell.*, vol. 13, no. 1, pp. 14–29.

Schowengerdt, R. A. [1983]. *Techniques for Image Processing and Classification in Remote Sensing*, Academic Press, New York.

Schreiber, W. F. [1956]. "The Measurement of Third Order Probability Distributions of Television Signals," *IRE Trans. Info. Theory*, vol. IT-2, pp. 94–105.

Schreiber, W. F. [1967]. "Picture Coding," *Proc. IEEE* (Special issue on redundancy reduction), vol. 55, pp. 320–330.

Schreiber, W. F. and Knapp, C. F. [1958]. "TV Bandwidth Reduction by Digital Coding," *Proc. IRE National Convention*, pt. 4, pp. 88–99.

Schwartz, J. W. and Barker, R. C. [1966]. "Bit-Plane Encoding: A Technique for Source Encoding," *IEEE Trans. Aerosp. Elec. Systems*, vol. AES-2, no. 4, pp. 385–392.

Selesnick, I., Baraniuk, R., and Kingsbury, N. [2005]. "The Dual-Tree Complex Wavelet Transform," *IEEE Signal Processing Magazine*, vol. 22, no. 6, pp. 123–151.

Serra, J. [1982]. *Image Analysis and Mathematical Morphology*, Academic Press, New York.

Serra, J. (ed.) [1988]. *Image Analysis and Mathematical Morphology*, vol. 2, Academic Press, New York.

Sezan, M. I., Rabbani, M., and Jones, P. W. [1989]. "Progressive Transmission of Images Using a Prediction/Residual Encoding Approach," *Opt. Eng.*, vol. 28, no. 5, pp. 556–564.

Shack, R. V. [1964]. "The Influence of Image Motion and Shutter Operation on the Photographic Transfer Function," *Appl. Opt.*, vol. 3, pp. 1171–1181.

Shafarenko, L., Petrou, M., and Kittler, J. [1998]. "Histogram-Based Segmentation in a Perceptually Uniform Color Space," *IEEE Trans. Image Processing*, vol. 7, no. 9, pp. 1354–1358.

Shaked, D. and Bruckstein, A. M. [1998]. "Pruning Medial Axes," *Computer Vision and Image Understanding*, vol. 69, no. 2, pp. 156–169.

Shannon, C. E. [1948]. "A Mathematical Theory of Communication," *The Bell Sys. Tech. J.*, vol. XXVII, no. 3, pp. 379–423.

Shapiro, L. G. and Stockman, G. C. [2001]. *Computer Vision*, Prentice Hall, Upper Saddle River, N.J.

Shapiro, V. A. [1996]. "On the Hough Transform of Multi-Level Pictures," *Pattern Recog.*, vol. 29, no. 4, pp. 589–602.

Shariat, H. and Price, K. E. [1990]. "Motion Estimation with More Than Two Frames," *IEEE Trans. Pattern Anal. Machine Intell.*, vol. 12, no. 5, pp. 417–434.

Shepp, L. A. and Logan, B. F. [1974]. "The Fourier Reconstruction of a Head Section," *IEEE Trans. Nucl. Sci.*, vol. NS-21, pp. 21–43.

Sheppard, J. J., Jr., Stratton, R. H., and Gazley, C., Jr. [1969]. "Pseudocolor as a Means of Image Enhancement," *Am. J. Optom. Arch. Am. Acad. Optom.*, vol. 46, pp. 735–754.

Shi, F. Y. and Wong, W.-T. [1994]. "Fully Parallel Thinning with Tolerance to Boundary Noise," *Pattern Recog.*, vol. 27, no. 12, pp. 1677–1695.

Shih, F. Y. C. and Mitchell, O. R. [1989]. "Threshold Decomposition of Gray-Scale Morphology into Binary Morphology," *IEEE Trans. Pattern Anal. Machine Intell.*, vol. 11, no. 1, pp. 31–42.

Shirley, P. [2002]. *Fundamentals of Computer Graphics*, A. K. Peters, Natick, MA.

Sid-Ahmed, M. A. [1995]. *Image Processing: Theory, Algorithms, and Architectures*, McGraw-Hill, New York.

Sikora, T. [1997]. "MPEG Digital Video-Coding Standards," *IEEE Signal Processing*, vol. 14, no. 5, pp. 82–99.

Simon, J. C. [1986]. *Patterns and Operators: The Foundations of Data Representations*, McGraw-Hill, New York.

Sklansky, J., Chazin, R. L., and Hansen, B. J. [1972]. "Minimum-Perimeter Polygons of Digitized Silhouettes," *IEEE Trans. Comput.*, vol. C-21, no. 3, pp. 260–268.

Sloboda, F., Zatko, B., and Stoer, J. [1998]. "On Approximation of Planar One-Dimensional Continua," in *Advances in Digital and Computational Geometry*, R. Klette, A. Rosenfeld, and F. Sloboda (eds.), Springer, Singapore, pp. 113–160.

Smirnov, A. [1999]. *Processing of Multidimensional Signals*, Springer-Verlag, New York.

Smith, A. R. [1978]. "Color Gamut Transform Pairs," *Proc. SIGGRAPH '78*, published as *Computer Graphics*, vol. 12, no. 3, pp. 12–19.

Smith, J. O., III [2003]. *Mathematics of the Discrete Fourier Transform*, W3K Publishing, CCRMA, Stanford, CA. (Also available online at http://ccrma.stanford.edu/~jos/mdft).

Smith, M. J. T. and Barnwell, T. P. III [1984]. "A Procedure for Building Exact Reconstruction Filter Banks for Subband Coders," *Proc. IEEE Int. Conf. Acoust., Speech, and Signal Proc.*, San Diego, Calif.

Smith, M. J. T. and Barnwell, T. P. III [1986]. "Exact Reconstruction Techniques for Tree-Structured Subband Coders," *IEEE Trans. On Acoust., Speech, and Signal Proc.*, vol. 34, no. 3, pp. 434–441.

Snyder, W. E. and Qi, Hairong [2004]. *Machine Vision*, Cambridge University Press, New York.

Sobel, I. E. [1970]. "Camera Models and Machine Perception," Ph.D. dissertation, Stanford University, Palo Alto, Calif.

Sonka, M., Hlavac, V., and Boyle, R. [1999]. *Image Processing, Analysis, and Machine Vision*, 2nd ed., PWS Publishing, New York.

Snyder, W. E. and Qi, Hairong [2004]. *Machine Vision*, Cambridge University Press, New York.

Soille, P. [2003]. *Morphological Image Analysis: Principles and Applications*, 2nd ed., Springer-Verlag, New York.

Solari, S. [1997]. *Digital Video and Audio Compression*, McGraw-Hill, New York.

Stark, H. (ed.) [1987]. *Image Recovery: Theory and Application*, Academic Press, New York.

Stark, J. A. [2000]. "Adaptive Image Contrast Enhancement Using Generalizations of Histogram Equalization," *IEEE Trans. Image Processing*, vol. 9, no. 5, pp. 889–896.

Stockham, T. G., Jr. [1972]. "Image Processing in the Context of a Visual Model," *Proc. IEEE*, vol. 60, no. 7, pp. 828–842.

Storer, J. A. and Reif, J. H., eds. [1991]. *Proceedings of DDC '91*, IEEE Computer Society Press, Los Alamitos, Calif.

Strang, G. and Nguyen, T. [1996]. *Wavelets and Filter Banks*, Wellesley-Cambridge Press, Wellesley, Mass.

Stumpff, K. [1939]. *Tafeln und Aufgaben zur Harmonischen Analyse und Periodogrammrechnung*, Julius Springer, Berlin.

Sussner, P. and Ritter, G. X. [1997]. "Decomposition of Gray-Scale Morphological Templates Using the Rank Method," *IEEE Trans. Pattern Anal. Machine Intell.*, vol. 19, no. 6, pp. 649–658.

Swets, D. L. and Weng, J. [1996]. "Using Discriminant Eigenfeatures for Image Retrieval," *IEEE Trans. Pattern Anal. Machine Intell.*, vol. 18, no. 8, pp. 1831–1836.

Symes, P. D. [2001]. *Video Compression Demystified*, McGraw-Hill, New York.

Sze, T. W. and Yang, Y. H. [1981]. "A Simple Contour Matching Algorithm," *IEEE Trans. Pattern Anal. Mach. Intell.*, vol. PAMI-3, no. 6, pp. 676–678.

Tanaka, E. [1995]. "Theoretical Aspects of Syntactic Pattern Recognition," *Pattern Recog.*, vol. 28, no. 7 pp. 1053–1061.

Tanimoto, S. L. [1979]. "Image Transmission with Gross Information First," *Comput. Graphics Image Proc.*, vol. 9, pp. 72–76.

Tasto, M. and Wintz, P. A. [1971]. "Image Coding by Adaptive Block Quantization," *IEEE Trans. Comm. Tech.*, vol. COM-19, pp. 957–972.

Tasto, M. and Wintz, P. A. [1972]. "A Bound on the Rate-Distortion Function and Application to Images," *IEEE Trans. Info. Theory*, vol. IT-18, pp. 150–159.

Teh, C. H. and Chin, R. T. [1989]. "On the Detection of Dominant Points on Digital Curves," *IEEE Trans. Pattern Anal. Machine Intell.*, vol. 11, no. 8, pp. 859–872.

Theoridis, S. and Konstantinos, K. [2006]. *Pattern Recognition*, 3rd ed., Academic Press, New York.

Thévenaz, P. and Unser, M [2000]. "Optimization of Mutual Information for Multiresolution Image Registration," *IEEE Trans. Image Processing*, vol. 9, no. 12, pp. 2083–2099.

Thomas, L. H. [1963]. "Using a Computer to Solve Problems in Physics," *Application of Digital Computers*, Ginn, Boston.

Thomason, M. G. and Gonzalez, R. C. [1975]. "Syntactic Recognition of Imperfectly Specified Patterns," *IEEE Trans. Comput.*, vol. C-24, no. 1, pp. 93–96.

Thompson, W. B. (ed.) [1989]. Special issue on visual motion, *IEEE Trans. Pattern Anal. Machine Intell.*, vol. 11, no. 5, pp. 449–541.

Thompson, W. B. and Barnard, S. T. [1981]. "Lower-Level Estimation and Interpretation of Visual Motion," *Computer*, vol. 14, no. 8, pp. 20–28.

Thorell, L. G. and Smith, W. J. [1990]. *Using Computer Color Effectively*, Prentice Hall, Upper Saddle River, N.J.

Tian, J. and Wells, R. O., Jr. [1995]. *Vanishing Moments and Wavelet Approximation*, Technical Report CML TR-9501, Computational Mathematics Lab., Rice University, Houston, Texas.

Tizhoosh, H. R. [2000]. "Fuzzy Image Enhancement: An Overview," in *Fuzzy Techniques in Image Processing*, E. Kerre and M. Nachtegael, eds., Springer-Verlag, New York.

Tomita, F., Shirai, Y., and Tsuji, S. [1982]. "Description of Texture by a Structural Analysis," *IEEE Trans. Pattern Anal. Mach. Intell.*, vol. PAMI-4, no. 2, pp. 183–191.

Topiwala, P. N. (ed.) [1998]. *Wavelet Image and Video Compression*, Kluwer Academic Publishers, Boston, Mass.

Toro, J. and Funt, B. [2007]. "A Multilinear Constraint on Dichromatic Planes for Illumination Estimation," *IEEE Trans. Image Proc.*, vol. 16, no. 1, pp. 92–97.

Tou, J. T. and Gonzalez, R. C. [1974]. *Pattern Recognition Principles*, Addison-Wesley, Reading, Mass.

Tourlakis, G. J. [2003]. *Lectures in Logic and Set Theory*, Cambridge University Press, Cambridge, UK.

Toussaint, G. T. [1982]. "Computational Geometric Problems in Pattern Recognition," In *Pattern Recognition Theory and Applications*, Kittler, J., Fu, K. S., and Pau, L. F. (eds.), Reidel, New York, pp. 73–91.

Tsai, J.-C., Hsieh, C.-H, and Hsu, T.-C. [2000]. "A New Dynamic Finite-State Vector Quantization Algorithm for Image Compression," *IEEE Trans. Image Processing*, vol. 9, no. 11, pp. 1825–1836.

Tsujii, O., Freedman, M. T., and Mun, K. S. [1998]. "Anatomic Region-Based Dynamic Range Compression for Chest Radiographs Using Warping Transformation of Correlated Distribution," *IEEE Trans. Medical Imaging*, vol. 17, no. 3, pp. 407–418.

Udpikar, V. R. and Raina, J. P. [1987]. "BTC Image Coding Using Vector Quantization," *IEEE Trans. Comm.*, vol. COM-35, no. 3, pp. 352–356.

Ueda, N. [2000]. "Optimal Linear Combination of Neural Networks for Improving Classification Performance," *IEEE Trans. Pattern Anal. Machine Intell.*, vol. 22, no. 2, pp. 207–215.

Ullman, S. [1981]. "Analysis of Visual Motion by Biological and Computer Systems," *IEEE Computer*, vol. 14, no. 8, pp. 57–69.

Umbaugh, S. E. [2005]. *Computer Imaging: Digital Image Analysis and Processing*, CRC Press, Boca Raton, FL.

Umeyama, S. [1988]. "An Eigendecomposition Approach to Weighted Graph Matching Problems," *IEEE Trans. Pattern Anal. Machine Intell.*, vol. 10, no. 5, pp. 695–703.

Unser, M. [1995]. "Texture Classification and Segmentation Using Wavelet Frames," *IEEE Trans. on Image Processing*, vol. 4, no. 11, pp. 1549–1560.

Unser, M., Aldroubi, A., and Eden, M. [1993]. "A Family of Polynomial Spline Wavelet Transforms," *Signal Proc.*, vol. 30, no. 2, pp. 141–162.

Unser, M., Aldroubi, A., and Eden, M. [1993]. "B-Spline Signal Processing, Parts I and II," *IEEE Trans. Signal Proc.*, vol. 41, no. 2, pp. 821–848.

Unser, M., Aldroubi, A., and Eden, M. [1995]. "Enlargement or Reduction of Digital Images with Minimum Loss of Information," *IEEE Trans. Image Processing*, vol. 4, no. 5, pp. 247–257.

Vaidyanathan, P. P. and Hoang, P.-Q. [1988]. "Lattice Structures for Optimal Design and Robust Implementaion of Two-Channel Perfect Reconstruction Filter Banks," *IEEE Trans. Acoust., Speech, and Signal Proc.*, vol. 36, no. 1, pp. 81–94.

Vailaya, A., Jain, A. and Zhang, H. J. [1998]. "On Image Classification: City Images vs. Landscapes," *Pattern Recog.*, vol. 31, no. 12, pp. 1921–1935.

Vetterli, M. [1986]. "Filter Banks Allowing Perfect Reconstruction," *Signal Proc.*, vol. 10, no. 3, pp. 219–244.

Vetterli, M. and Kovacevic, J. [1995]. *Wavelets and Suband Coding*, Prentice Hall, Englewood Cliffs, N.J.

Vincent, L. [1993]. "Morphological Grayscale Reconstruction in Image Analysis: Applications and Efficient Algorithms," *IEEE Trans. Image Proc.*, vol. 2. no. 2, pp. 176–201.

Voss, K. and Suesse, H. [1997]. "Invariant Fitting of Planar Objects by Primitives," *IEEE Trans. Pattern Anal. Machine Intell.*, vol. 19, no. 1, pp. 80–84.

Vuylsteke, P. and Kittler, J. [1990]. "Edge-Labeling Using Dictionary-Based Relaxation," *IEEE Trans. Pattern Anal. Machine Intell.*, vol. 12, no. 2, pp. 165–181.

Walsh, J. W. T. [1958]. *Photometry*, Dover, New York.

Wang, D., Zhang, L., Vincent, A., and Speranza, F. [2006]. "Curved Wavelet Transform for Image Coding," *IEEE Trans. Image Proc.*, vol. 15, no. 8, pp. 2413–2421.

Wang, G., Zhang, J., and Pan, G.-W. [1995]. "Solution of Inverse Problems in Image Processing by Wavelet Expansion," *IEEE Trans. Image Processing*, vol. 4, no. 5, pp. 579–593.

Wang, Y.-P., Lee, S. L., and Toraichi, K. [1999]. "Multiscale Curvature-Based Shape Representation Using β-Spline Wavelets," *IEEE Trans. Image Processing*, vol. 8, no. 11, pp. 1586–1592.

Wang, Z., Rao, K. R., and Ben-Arie, J. [1996]. "Optimal Ramp Edge Detection Using Expansion Matching," *IEEE Trans. Pattern Anal. Machine Intell.*, vol. 18, no. 11, pp. 1092–1097.

Watt, A. [1993]. *3D Computer Graphics*, 2nd ed., Addison-Wesley, Reading, Mass.

Wechsler [1980]. "Texture Analysis—A Survey," *Signal Proc.*, vol. 2, pp. 271–280.

Wei, D., Tian, J., Wells, R. O., Jr., and Burrus, C. S. [1998]. "A New Class of Biorthogonal Wavelet Systems for Image Transform Coding," *IEEE Trans. Image Processing*, vol. 7, no. 7, pp. 1000–1013.

Weinberger, M. J., Seroussi, G., and Sapiro, G. [2000]. "The LOCO-I Lossless Image Compression Algorithm: Principles and Standardization into JPEG-LS," *IEEE Trans. Image Processing*, vol. 9, no. 8, pp. 1309–1324.

Westenberg, M. A. and Roerdink, J. B. T. M. [2000]. "Frequency Domain Volume Rendering by the Wavelet X-Ray Transform," *IEEE Trans. Image Processing*, vol. 9, no. 7, pp. 1249–1261.

Weszka, J. S. [1978]. "A Survey of Threshold Selection Techniques," *Comput. Graphics Image Proc.*, vol. 7, pp. 259–265.

White, J. M. and Rohrer, G. D. [1983]. "Image Thresholding for Optical Character Recognition and Other Applications Requiring Character Image Extraction," *IBM J. Res. Devel.*, vol. 27, no. 4, pp. 400–411.

Widrow, B. [1962]. "Generalization and Information Storage in Networks of 'Adaline' Neurons," In *Self-Organizing Systems 1962*, Yovitz, M. C., et al. (eds.), Spartan, Washington, D. C., pp. 435–461.

Widrow, B. and Hoff, M. E. [1960]. "Adaptive Switching Circuits," *1960 IRE WESCON Convention Record, Part 4*, pp. 96–104.

Widrow, B. and Stearns, S. D. [1985]. *Adaptive Signal Processing*, Prentice Hall, Englewood Cliffs, N.J.

Wiener, N. [1942]. *Extrapolation, Interpolation, and Smoothing of Stationary Time Series*, the MIT Press, Cambridge, Mass.

Wilburn, J. B. [1998]. "Developments in Generalized Ranked-Order Filters," *J. Opt. Soc. Amer.-A. Optics, Image Science, and Vision*, vol. 15, no. 5, pp. 1084–1099.

Windyga, P. S. [2001]. "Fast Impulsive Noise Removal," *IEEE Trans. Image Processing*, vol. 10, no. 1, pp. 173–179.

Wintz, P. A. [1972]. "Transform Picture Coding," *Proc. IEEE*, vol. 60, no. 7, pp. 809–820.

Witten, I. H., Neal, R. M., and Cleary, J. G. [1987]. "Arithmetic Coding for Data Compression," *Comm. ACM*, vol. 30, no. 6, pp. 520–540.

Wolberg, G. [1990]. *Digital Image Warping*, IEEE Computer Society Press, Los Alamitos, CA.

Wolff, R. S. and Yaeger, L. [1993]. *Visualization of Natural Phenomena*, Springer-Verlag, New York.

Won, C. S. and Gray, R. M. [2004]. *Stochastic Image Processing*, Kluwer Academic/Plenum Publishers, New York.

Woods, J. W. and O'Neil, S. D. [1986]. "Subband Coding of Images," *IEEE Trans. Acous. Speech Signal Proc.*, vol. ASSP-35, no. 5, pp. 1278–1288.

Woods, R. E. and Gonzalez, R. C. [1981]. "Real-Time Digital Image Enhancement," *Proc. IEEE*, vol. 69, no. 5, pp. 643–654.

Xu, Y., Weaver, J. B., Healy, D. M., Jr., and Lu, J. [1994]. "Wavelet Transform Domain Filters: A Spatially Selective Noise Filtration Technique," *IEEE Trans. Image Processing*, vol. 3, no. 6, pp. 747–758.

Xu, R., Pattanaik, S., and Hughes, C. [2005]. "High-Dynamic-Range Still-Image Encoding in JPEG 2000," *IEEE Computer Graphics and Applications*, vol 25, no. 6, pp. 57–64.

Yachida, M. [1983]. "Determining Velocity Maps by Spatio-Temporal Neighborhoods from Image Sequences," *Comput. Vis. Graph. Image Proc.*, vol. 21, no. 2, pp. 262–279.

Yamazaki, Y., Wakahara, Y., and Teramura, H. [1976]. "Digital Facsimile Equipment 'Quick-FAX' Using a New Redundancy Reduction Technique," *NTC '76*, pp. 6.2-1–6.2-5.

Yan, Y. and Cosman, P. [2003]. "Fast and Memory Efficient Text Image Compression with JBIG2," *IEER Trans. Image Proc.*, vol. 12, no. 8, pp. 944–956.

Yang, X. and Ramchandran, K. [2000]. "Scalable Wavelet Video Coding Using Aliasing-Reduced Hierarchical Motion Compensation," *IEEE Trans. Image Processing*, vol. 9, no. 5, pp. 778–791.

Yates, F. [1937]. "The Design and Analysis of Factorial Experiments," Commonwealth Agricultural Bureaux, Farnam Royal, Burks, England.

Yin, P. Y., Yin, L. H., and Chen, L. H. [1997]. "A Fast Iterative Scheme for Multilevel Thresholding Methods," *Signal Processing*, vol. 60, pp. 305–313.

Yitzhaky, Y., Lantzman, A., and Kopeika, N. S. [1998]. "Direct Method for Restoration of Motion Blurred Images," *J. Opt. Soc. Amer.-A. Optics, Image Science, and Vision*, vol. 15, no. 6, pp. 1512–1519.

You, J. and Bhattacharya, P. [2000]. "A Wavelet-Based Coarse-to-Fine Image Matching Scheme in a Parallel Virtual Machine Environment," *IEEE Trans. Image Processing*, vol. 9, no. 9, pp. 1547–1559.

Yu, D. and Yan, H. [2001]. "Reconstruction of Broken Handwritten Digits Based on Structural Morphology," *Pattern Recog.*, vol. 34, no. 2, pp. 235–254.

Yu, S. S. and Tsai, W. H. [1990]. "A New Thinning Algorithm for Gray-Scale Images," *Pattern Recog.*, vol. 23, no. 10, pp. 1067–1076.

Yuan, M. and Li, J. [1987]. "A Production System for LSI Chip Anatomizing," *Pattern Recog. Letters*, vol. 5, no. 3, pp. 227–232.

Zadeh, L. A. [1965]. "Fuzzy Sets," *Inform. and Control*, vol. 8, pp. 338–353.

Zadeh, L. A. [1973]. "Outline of New Approach to the Analysis of Complex Systems and Decision Processes," *IEEE Trans. Systems, Man, Cyb.*, vol. SMC-3, no. 1, pp. 28–44.

Zadeh, L. A. [1976]. "A Fuzzy-Algorithmic Approach to the Definition of Complex or Imprecise Concepts," *Int. J. Man-Machine Studies*, vol. 8, pp. 249–291.

Zahara, E., Shu-Kai, S., and Du-Ming, T. [2005]. "Optimal Multi-Thresholding Using a Hybrid Optimization Approach," *Pattern Recognition Letters*, vol. 26, no. 8, pp. 1082–1095.

Zahn, C. T. and Roskies, R. Z. [1972]. "Fourier Descriptors for Plane Closed Curves," *IEEE Trans. Comput.*, vol. C-21, no. 3, pp. 269–281.

Zhang, T. Y. and Suen, C. Y. [1984]. "A Fast Parallel Algorithm for Thinning Digital Patterns," *Comm. ACM*, vol. 27, no. 3, pp. 236–239.

Zhang, Y. and Rockett, P. I. [2006]. "The Bayesian Operating Point of the Canny Edge Detector," *IEEE Trans. Image Proc.*, vol. 15, no. 11, pp. 3409–3416.

Zhu, H., Chan F. H. Y., and Lam, F. K. [1999]. "Image Contrast Enhancement by Constrained Local Histogram Equalization," *Computer Vision and Image Understanding*, vol. 73, no. 2, pp. 281–290.

Zhu, P. and Chirlian, P. M. [1995]. "On Critical Point Detection of Digital Shapes," *IEEE Trans. Pattern Anal. Machine Intell.*, vol. 17, no. 8, pp. 737–748.

Zimmer, Y., Tepper, R., and Akselrod, S. [1997]. "An Improved Method to Compute the Convex Hull of a Shape in a Binary Image," *Pattern Recog.*, vol. 30, no. 3, pp. 397–402.

Ziou, D. [2001]. "The Influence of Edge Direction on the Estimation of Edge Contrast and Orientation," *Pattern Recog.*, vol. 34, no. 4, pp. 855–863.

Ziv, J. and Lempel, A. [1977]. "A Universal Algorithm for Sequential Data Compression," *IEEE Trans. Info. Theory*, vol. IT-23, no. 3, pp. 337–343.

Ziv, J. and Lempel, A. [1978]. "Compression of Individual Sequences Via Variable-Rate Coding," *IEEE Trans. Info. Theory*, vol. IT-24, no. 5, pp. 530–536.

Zucker, S. W. [1976]. "Region Growing: Childhood and Adolescence," *Comput. Graphics Image Proc.*, vol. 5, pp. 382–399.

Zugaj, D. and Lattuati, V. [1998]. "A New Approach of Color Images Segmentation Based on Fusing Region and Edge Segmentation Outputs," *Pattern Recog.*, vol. 31, no. 2, pp. 105–113.

Index